MULTIVARIABLE
Calculus
Concepts and Contexts ◆ 3E

Calculus and the
Architecture of Curves

Walt Disney Concert Hall, CA / Frank O. Gehry & Associates

The cover photograph shows the Walt Disney Concert Hall in Los Angeles, designed and built 1992–2003 by Frank Gehry and Associates. It is a daring building, a layered composition of curved surfaces in the form of billowing sails with brushed stainless steel cladding.

The highly complex structures that Frank Gehry designs would be impossible to build without the computer. The CATIA software that his architects and engineers use to produce the computer models is based on principles of calculus—fitting curves by matching tangent lines, making sure the curvature isn't too large, and controlling parametric surfaces. "Consequently," says Gehry, "we have a lot of freedom. I can play with shapes."

The process starts with Gehry's initial sketches, which are translated into a succession of physical models. (Hundreds of different physical models were constructed during the design of the building, first with basic wooden blocks and then evolving into more sculptural forms.) Then an engineer uses a digitizer to record the coordinates of a series of points on a physical model. The digitized points are fed into a computer and the CATIA software is used to link these points with smooth curves. (It joins curves so that their tangent lines coincide.) The architect has considerable freedom in creating these curves, guided by displays of the curve, its derivative, and its curvature. Then the

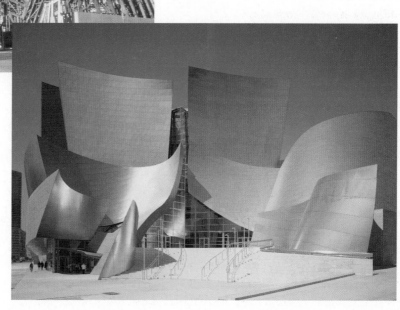

curves are connected to each other by a parametric surface, and again the architect can do so in many possible ways with the guidance of displays of the geometric characteristics of the surface.

The CATIA model is then used to produce another physical model, which, in turn, suggests modifications and leads to additional computer and physical models.

The CATIA program was developed in France by Dassault Systèmes, originally for designing airplanes, and was subsequently employed in the automotive industry. Frank Gehry, because of his complex sculptural shapes, is the first to use it in architecture. It helps him answer his question, "How wiggly can you get and still make a building?"

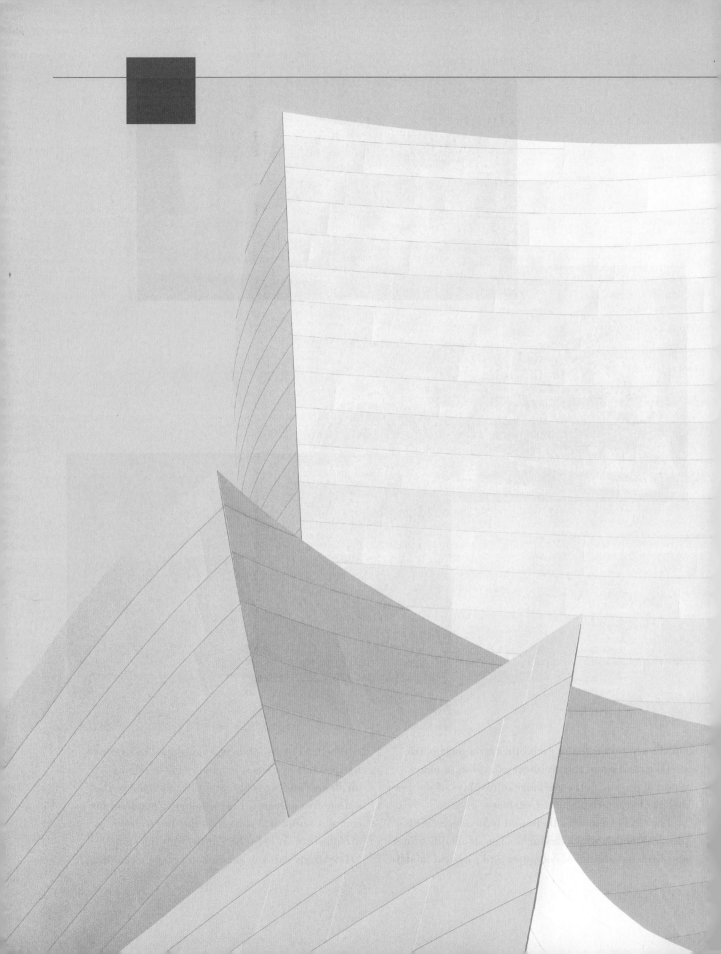

MULTIVARIABLE

Calculus

Concepts and Contexts ◆ 3E

James Stewart
McMASTER UNIVERSITY

© 2003 Tom Bonner

THOMSON
⎯⎯✳⎯⎯ ™
BROOKS/COLE

Australia • Canada • Mexico • Singapore • Spain
United Kingdom • United States

THOMSON

BROOKS/COLE

Publisher □ *Bob Pirtle*

Assistant Editor □ *Stacy Green*

Editorial Assistant □ *Katherine Cook*

Technology Project Manager □ *Earl Perry*

Senior Marketing Manager □ *Karin Sandberg*

Marketing Assistant □ *Erin Mitchell*

Managing Advertising Project Manager □ *Bryan Vann*

Senior Project Manager, Editorial Production □ *Janet Hill*

Senior Art Director □ *Vernon Boes*

Senior Print/Media Buyer □ *Karen Hunt*

Permissions Editor □ *Joohee Lee*

Production Service □ *TECH·arts*

Text Designer □ *TECH·arts*

Copy Editor □ *Kathi Townes*

Compositor □ *Stephanie Kuhns*

Illustrator □ *Brian Betsill*

Art Editor □ *Kathi Townes*

Photo Researchers □ *Denise Davidson, Stephanie Kuhns*

Cover Designer □ *Denise Davidson*

Cover Image □ *Tom Bonner*

Cover Printer □ *CTPS*

Printer □ *CTPS*

Printed in China

3 4 5 6 7 09 08 07

For more information about our products, contact us at:
Thomson Learning Academic Resource Center
1-800-423-0563

For permission to use material from this text or product, submit a request online at **http://www.thomsonrights.com**. Any additional questions about permissions can be submitted by email to **thomsonrights@thomson.com**.

Trademarks
Derive is a registered trademark of Soft Warehouse, Inc.
Journey Through is a trademark used herein under license.
Maple is a registered trademark of Waterloo Maple, Inc.
Mathematica is a registered trademark of Wolfram Research, Inc.
Tools for Enriching is a trademark used herein under license.

Credits continue on page A51.

Library of Congress Control Number: 2004113997

Student Edition ISBN-13: 978-0-534-41004-9
ISBN-10: 0-534-41004-9
Media-Free Version ISBN-13: 978-0-534-41002-5
ISBN-10: 0-534-41002-2

Thomson Higher Education
10 Davis Drive
Belmont, CA 94002-3098
USA

Asia (includes India)
Thomson Learning
5 Shenton Way
#01-01 UIC Building
Singapore 068808

Australia / New Zealand
Thomson Learning Australia
102 Dodds Street
Southbank, Victoria 3006
Australia

Canada
Thomson Nelson
1120 Birchmount Road
Toronto, Ontario M1K 5G4
Canada

UK/Europe/Middle East/Africa
Thomson Learning
High Holborn House
50/51 Bedford Row
London WC1R 4LR
United Kingdom

Latin America
Thomson Learning
Seneca, 53
Colonia Polanco
11560 Mexico
D.F. Mexico

Spain (includes Portugal)
Thomson Paraninfo
Calle Magallanes, 25
28015 Madrid, Spain

Contents

Preface

When the first edition of this book appeared eight years ago, a heated debate about calculus reform was taking place. Such issues as the use of technology, the relevance of rigor, and the role of discovery versus that of drill were causing deep splits in mathematics departments. Since then the rhetoric has calmed down somewhat as reformers and traditionalists have realized that they have a common goal: to enable students to understand and appreciate calculus.

The first and second editions were intended to be a synthesis of reform and traditional approaches to calculus instruction. In this third edition I continue to follow that path by emphasizing conceptual understanding through visual, numerical, and algebraic approaches.

◼◼ What's New in the Third Edition

By way of preparing to write the third edition of this text, I spent a year teaching calculus at the University of Toronto. I listened carefully to my students' questions and my colleagues' suggestions. And as I prepared each lecture I sometimes realized that an additional example was needed, or a sentence could be clarified, or a section could use a few more exercises of a certain type. In addition, I paid attention to the suggestions sent to me by many users and to the comments of the reviewers.

Many hundreds of improvements, large and small, have been incorporated into this edition. Here are some of them.

- ◼ Many examples have been added or changed.

- ◼ Extra steps have been provided in some of the existing examples.

- ◼ The data in examples and exercises have been updated to be more timely.

- ◼ More than 25% of the exercises in each chapter are new. Here are a few of my favorites:

Exercise	Page	Exercise	Page	Exercise	Page
8.4.36	593	8.6.37–38	605	8.9.22	628
10.1.37–38	701	11.4.38	779	11.5.36	787

■ New phrases and margin notes have been added to clarify the exposition.

■ A number of pieces of art have been redrawn.

■ I've also added new problems to the Focus on Problem Solving sections. See, for instance, Problems 14 and 16 on page 635.

■ Two new projects have been added. The project on page 617 shows how computer algebra systems use Taylor series to compute limits, and the project on page 675 shows how computer graphics programmers use clipping planes and hidden line rendering to portray three-dimensional objects on a two-dimensional screen.

■ The CD called *Tools for Enriching Calculus* (TEC) has been completely redesigned and now includes what we call Visuals, brief animations of various figures in the text. In addition there are now Visuals, Modules, and Homework Hints for the multivariable chapters. See the description on page xiii.

■ The symbol $\boxed{V}$ has been placed beside examples (an average of three per section) for which there are videos of instructors explaining the example in more detail. These videos are free to adopters. This material is also included on an *Interactive VideoSkillbuilder CD*. See the description of the Interactive Video Skillbuilder on page xiii.

■ Conscious of the need to control the size of the book, I've put new topics (as well as expanded coverage of some topics already in the book) on the revamped web site **www.stewartcalculus.com** rather than in the text itself. (See the list of additional topics in the description of the web site on page xvii.) As a result, the number of pages in the text is actually a bit less than in the second edition.

Features

Conceptual Exercises
The most important way to foster conceptual understanding is through the problems that we assign. To that end I have devised various types of problems. Some exercise sets begin with requests to explain the meanings of the basic concepts of the section. (See, for instance, the first couple of exercises in Sections 8.2, 11.2, and 11.3. I often use them as a basis for classroom discussions.) Similarly, review sections begin with a Concept Check and a True-False Quiz. Other exercises test conceptual understanding through graphs or tables (see Exercises 8.7.2, 10.2.1–2, 10.3.27–33, 11.1.1–2, 11.1.9–14, 11.3.3–8, 11.6.1–2, 11.7.3–4, 12.1.5–10, 13.1.11–18, 13.2.15–16, and 13.3.1–2).

Graded Exercise Sets
Each exercise set is carefully graded, progressing from basic conceptual exercises and skill-development problems to more challenging problems involving applications and proofs.

Real-World Data
My assistants and I have spent a great deal of time looking in libraries, contacting companies and government agencies, and searching the Internet for interesting real-world data to introduce, motivate, and illustrate the concepts of calculus. As a result, many of the examples and exercises deal with functions defined by such numerical

data or graphs. For instance, functions of two variables are illustrated by a table of values of wave heights (Example 3 in Section 9.6) and by a table of values of the wind-chill index as a function of air temperature and wind speed (Example 1 in Section 11.1). Partial derivatives are introduced in Section 11.3 by examining a column in a table of values of the heat index (perceived air temperature) as a function of the actual temperature and the relative humidity. This example is pursued further in connection with linear approximations (Example 3 in Section 11.4). Directional derivatives are introduced in Section 11.6 by using a temperature contour map to estimate the rate of change of temperature at Reno in the direction of Las Vegas. Double integrals are used to estimate the average snowfall in Colorado on December 24, 1982 (Example 4 in Section 12.1). Vector fields are introduced in Section 13.1 by depictions of actual velocity vector fields showing San Francisco Bay wind patterns.

Projects One way of involving students and making them active learners is to have them work (perhaps in groups) on extended projects that give a feeling of substantial accomplishment when completed. *Applied Projects* involve applications that are designed to appeal to the imagination of students. The project after Section 11.8 uses Lagrange multipliers to determine the masses of the three stages of a rocket so as to minimize the total mass while enabling the rocket to reach a desired velocity. *Discovery Projects* explore aspects of geometry: tetrahedra (after Section 9.4), hyperspheres (after Section 12.7), and intersections of three cylinders (after Section 12.8). The *Laboratory Project* on page 690 uses technology to discover how interesting the shapes of surfaces can be and how these shapes evolve as the parameters change in a family. The *Writing Project* on page 965 explores the historical and physical origins of Green's Theorem and Stokes' Theorem and the interactions of the three men involved. Many additional projects are provided in the *Instructor's Guide*.

Problem Solving Students usually have difficulties with problems for which there is no single well-defined procedure for obtaining the answer. I think nobody has improved very much on George Polya's four-stage problem-solving strategy and, accordingly, I have included a version of his problem-solving principles at the end of Chapter 1. They are applied, both explicitly and implicitly, throughout the book. After the other chapters I have placed sections called *Focus on Problem Solving,* which feature examples of how to tackle challenging calculus problems. In selecting the varied problems for these sections I kept in mind the following advice from David Hilbert: "A mathematical problem should be difficult in order to entice us, yet not inaccessible lest it mock our efforts." When I put these challenging problems on assignments and tests I grade them in a different way. Here I reward a student significantly for ideas toward a solution and for recognizing which problem-solving principles are relevant.

Technology The availability of technology makes it not less important but more important to understand clearly the concepts that underlie the images on the screen. But, when properly used, graphing calculators and computers are powerful tools for discovering and understanding those concepts. I assume that the student has access to either a graphing calculator or a computer algebra system. The icon ⊞ indicates an exercise that definitely requires the use of such technology, but that is not to say that a graphing device can't be used on the other exercises as well. The symbol CAS is reserved for problems in which the full resources of a computer algebra system (like Derive, Maple, Mathematica, or the TI-89/92) are required. But technology doesn't make pencil and paper obsolete. Hand calculation and sketches are often preferable to technology for illustrating and reinforcing some concepts. Both instructors and students need to develop the ability to decide where the hand or the machine is appropriate.

Tools for Enriching™ Calculus The CD-ROM called *TEC* is a companion to the text and is intended to enrich and complement its contents. Developed by Harvey Keynes, Dan Clegg, Hubert Hohn, and myself, *TEC* uses a discovery and exploratory approach. In sections of the book where technology is particularly appropriate, marginal icons direct students to *TEC* Visuals and Modules that provide a laboratory environment in which they can explore the topic in different ways and at different levels. Visuals are animations of figures in the text; Modules are more elaborate activities and include exercises. Instructors can choose to become involved at several different levels, ranging from simply encouraging students to use the Visuals and Modules for independent exploration, to assigning specific exercises from those included with each Module, or to creating additional exercises, labs, and projects that make use of the Visuals and Modules.

TEC also includes *Homework Hints* for representative exercises (usually odd-numbered) in every section of the text, indicated by printing the exercise number in red. These hints are usually presented in the form of questions and try to imitate an effective teaching assistant by functioning as a silent tutor. They are constructed so as not to reveal any more of the actual solution than is minimally necessary to make further progress.

Interactive Video Skilbuilder CD-ROM The *Interactive Video Skillbuilder CD-ROM* contains more than eight hours of video instruction. The problems worked during each video lesson are shown next to the viewing screen so that students can try working them before watching the solution. To help students evaluate their progress, each section contains a ten-question Web quiz (the results of which can be emailed to the instructor) and each chapter contains a chapter test, with answers to each problem.

Web Site: www.stewartcalculus.com This has been renovated and now includes the following.

- Algebra Review

- Lies My Calculator and Computer Told Me

- History of Mathematics, with links to the better historical web sites

- Additional Topics (complete with exercise sets): Trigonometric Integrals, Trigonometric Substitution, Strategy for Integration, Volumes by Cylindrical Shells, Strategy for Testing Series, Fourier Series, Formulas for the Remainder Term in Taylor Series, Linear Differential Equations, Second-Order Linear Differential Equations, Nonhomogeneous Linear Equations, Applications of Second-Order Differential Equations, Using Series to Solve Differential Equations, Rotation of Axes

- Links, for each chapter, to outside Web resources

- Archived Problems (drill exercises that appeared in previous editions, together with their solutions)

- Challenge Problems (some from the Focus on Problem Solving sections of prior editions)

- Downloadable versions of *CalcLabs* for Derive and TI graphing calculators

▨ Content

Chapter 8   Infinite Sequences and Series

Tests for the convergence of series are considered briefly, with intuitive rather than formal justifications. Numerical estimates of sums of series are based on which test was used to prove convergence. The emphasis is on Taylor series and polynomials and their applications to physics. Error estimates include those from graphing devices.

Chapter 9   Vectors and the Geometry of Space

The dot product and cross product of vectors are given geometric definitions, motivated by work and torque, before the algebraic expressions are deduced. To facilitate the discussion of surfaces, functions of two variables and their graphs are introduced here.

Chapter 10   Vector Functions

The calculus of vector functions is used to prove Kepler's First Law of planetary motion, with the proofs of the other laws left as a project. In keeping with the introduction of parametric curves in Chapter 1, parametric surfaces are introduced as soon as possible, namely, in this chapter. I think an early familiarity with such surfaces is desirable, especially with the capability of computers to produce their graphs. Then tangent planes and areas of parametric surfaces can be discussed in Sections 11.4 and 12.6.

Chapter 11   Partial Derivatives

Functions of two or more variables are studied from verbal, numerical, visual, and algebraic points of view. In particular, I introduce partial derivatives by looking at a specific column in a table of values of the heat index (perceived air temperature) as a function of the actual temperature and the relative humidity. Directional derivatives are estimated from contour maps of temperature, pressure, and snowfall.

Chapter 12   Multiple Integrals

Contour maps and the Midpoint Rule are used to estimate the average snowfall and average temperature in given regions. Double and triple integrals are used to compute probabilities, areas of parametric surfaces, volumes of hyperspheres, and the volume of intersection of three cylinders.

Chapter 13   Vector Fields

Vector fields are introduced through pictures of velocity fields showing San Francisco Bay wind patterns. The similarities among the Fundamental Theorem for line integrals, Green's Theorem, Stokes' Theorem, and the Divergence Theorem are emphasized.

▨ Ancillaries

Multivariable Calculus: Concepts and Contexts, Third Edition, is supported by a complete set of ancillaries developed under my direction. Each piece has been designed to enhance student understanding and to facilitate creative instruction. The table on pages xv and xvi lists ancillaries available for instructors and students.

■■ Resources for Instructors

Instructor's Resource CD-ROM
ISBN 0-534-41021-9

Contains Electronic Instructor's Guide, Resource Integration Guide, iLrn Testing, Instructions for iLrn Homework, and Power Point Lecture notes.

 ### Tools for Enriching™ Calculus CD-ROM
by James Stewart and Dan Clegg
ISBN 0-534-40989-X

Completely revised and updated, TEC provides a laboratory environment in which students can explore selected topics. TEC also includes homework hints for representative exercises.

Instructor's Guide
by Douglas Shaw
ISBN 0-534-41030-8

Each section of the main text is discussed from several viewpoints and contains suggested time to allot, points to stress, text discussion topics, core materials for lecture, workshop/discussion suggestions, group work exercises in a form suitable for handout, with solutions, and suggested homework problems. An electronic version is available on the Instructor's Resource CD-ROM.

Complete Solutions Manual, Multivariable
by Dan Clegg
ISBN 0-534-41012-X

Includes worked-out solutions to all exercises in the text.

Printed Test Bank
By William Tomhave & Xueqi Zeng
ISBN 0-534-41031-6

Contains multiple-choice and short-answer test items that key directly to the text.

ILrn Adopter's Fulfillment Folder Kit
ISBN 0-534-41033-2

Featuring full algorithmic generation of problems and free-response mathematics, iLrn allows you to customize exams and track student progress in an accessible, browser-based format, with results flowing automatically into your gradebook! This kit contains the generic Instructor's Guide, Mathematics Instructors User's Guide, Installation CD-ROM (for offline users), a text-specific content CD-ROM, a Quick Start Guide, and a "How do I" quick introduction to widely used functions in iLrn.

Text-Specific Videos
ISBN 0-534-41037-5

Text-specific videotape sets, available at no charge to adopters, consisting of one tape per text chapter. Each tape features a 10- to 20-minute problem-solving lesson for each section of the chapter. Covers both single- and multivariable calculus.

Transparencies, Multivariable
by James Stewart
ISBN 0-534-41015-4

Full-color, large-scale sheets of reproductions of material from the text.

Solutions Builder CD-ROM
ISBN 0534410383

This CD is an electronic version of the complete solutions manual. It provides instructors with an efficient method for creating solution sets to homework or exams. Instructors can easily view, select, and save solution sets that can then be printed or posted.

■■ Resources for Instructors and Students

Stewart Specialty Web Site:
www.stewartcalculus.com

Contents: Algebra Review ▫ Additional Topics ▫ Drill exercises ▫ Problems Plus ▫ Web Links ▫ History of Mathematics ▫ Downloadable versions of CalcLabs for Derive and TI graphing calculators ▫ Maple Projects ▫ Mathematica Projects

iLrn Homework
[http://iLrn.com]
ISBN 0-534-40988-1

iLrn Homework allows instructors to assign machine-gradable homework problems that help students identify where they need additional help. That assistance is available through worked-out solutions that guide students through the steps of problem solving, or via live online tutoring at vMentor. The tutors at this online service will skillfully guide students through a problem, using unique two-way audio and whiteboard features.

■ Electronic items   ■ Printed items

■■ Resources for Instructors and Students

(cont.)

The Brooks/Cole Mathematics Resource Center Web Site
http://mathematics.brookscole.com

When you adopt a Thomson–Brooks/Cole mathematics text, you and your students will have access to a variety of teaching and learning resources. This Web site features everything from book-specific resources to newsgroups. It's a great way to make teaching and learning an interactive and intriguing experience.

WebTutor Advantage™ on WebCT
ISBN 0-534-41028-6

Lecture notes, discussion threads, and quizzes on WebCT.

WebTutor Advantage™ on Blackboard
ISBN 0-534-41039-1

Lecture notes, discussion threads, and quizzes on Blackboard.

■■ Student Resources

Tools for Enriching™ Calculus CD-ROM
by James Stewart and Dan Clegg
ISBN 0-534-40989-X

TEC provides a laboratory environment in which students can explore selected topics. TEC also includes homework hints for representative exercises.

Interactive Video SkillBuilder CD-ROM
ISBN 0-534-41036-7

Think of it as portable office hours! The Interactive Video Skillbuilder CD-ROM contains more than eight hours of video instruction. The problems worked during each video lesson are shown next to the viewing screen so that students can try working them before watching the solution. To help students evaluate their progress, each section contains a ten-question Web quiz (the results of which can be emailed to the instructor) and each chapter contains a chapter test, with answers to each problem.

iLrn Student Resource Kit
ISBN 0-534-39914-2

This helpful kit provides your students with a CD-ROM that contains the plug-ins needed to use the iLrn system and a Student Guide that offers additional assistance for students using iLrn.

Study Guide, Multivariable
by Robert Burton & Dennis Garity
ISBN 0-534-41006-5

Contains key concepts, skills to master, a brief discussion of the ideas of the section, and worked-out examples with tips on how to find the solution.

Student Solutions Manual, Multivariable
by Dan Clegg
ISBN 0-534-41005-7

Provides completely worked-out solutions to all odd-numbered exercises within the text, giving students a way to check their answers and ensure that they took the correct steps to arrive at an answer.

CalcLabs with Maple, Multivariable
by Philip Yasskin and Art Belmonte
ISBN 0-534-41010-3

This comprehensive lab manual will help students learn to effectively use the technology tools available to them. Each lab contains clearly explained exercises and a variety of labs and projects to accompany the text.

Linear Algebra for Calculus
by Konrad J. Heuvers, William P. Francis, John H. Kuisti, Deborah F. Lockhart, Daniel S. Moak, and Gene M. Ortner
ISBN 0-534-25248-6

This comprehensive book, designed to supplement the calculus course, provides an introduction to and review of the basic ideas of linear algebra.

■ Electronic items ■ Printed items

▦ Acknowledgments

I am grateful to the following reviewers for sharing their knowledge and judgment with me. I have learned something from each of them.

Third Edition Reviewers

William Ardis,
Collin County Community College
Jean H. Bevis,
Georgia State University
Martina Bode,
Northwestern University
Paul Wayne Britt,
Louisiana State University
Judith Broadwin,
Jericho High School (retired)
Meghan Anne Burke,
Kennesaw State University
Roxanne M. Byrne,
*University of Colorado
at Denver*
Larry Cannon,
Utah State University
Deborah Troutman Cantrell,
*Chattanooga State Technical
Community College*
Barbara R. Fink,
DeAnza College
Joe W. Fisher,
University of Cincinnati
Richard L. Ford,
California State University Chico
Gerrald Gustave Greivel,
Colorado School of Mines
John R. Griggs,
North Carolina State University
Barbara Bell Grover,
Salt Lake Community College
John William Hagood,
Northern Arizona University
Gary Steven Itzkowitz,
Rowan University

Mohammad A. Kazemi,
*University of North Carolina,
Charlotte*
Kandace Alyson Kling,
Portland Community College
Carrie L. Kyser,
Clackamas Community College
Beth Turner Long,
*Pellissippi State Technical
Community College*
Andre Mathurin,
Bellarmine College Prep
Richard Eugene Mercer,
Wright State University
Laura J. Moore-Mueller,
Green River Community College
Scott L. Mortensen,
Dixie State College
Tejinder Singh Neelon,
California State University San Marcos
Jeanette R. Palmiter,
Portland State University
Dusty Edward Sabo,
Southern Oregon University
Daniel S. Sage,
Louisiana State University
Bernd S.W. Schroeder,
Louisiana Tech University
Jeffrey Scott Scroggs,
North Carolina State University
Linda E. Sundbye,
The Metropolitan State College of Denver
JingLing Wang,
Lansing Community College
Michael B. Ward,
Western Oregon University

Second Edition Reviewers

William Ardis,
Collin County Community College
Judith Broadwin,
Jericho High School
Charles Bu,
Wellesley University
Larry Cannon,
Utah State University
Robert A. Chaffer,
Central Michigan University
Joe W. Fisher,
University of Cincinnati
Barry D. Hughes,
University of Melbourne

Prem K. Kythe,
University of New Orleans
Joyce Riseberg,
Montgomery College
Richard Rochberg,
Washington University
James F. Selgrade,
North Carolina State University
Denise Taunton Reid,
Valdosta State University
Clifton Wingard,
Delta State University
Teri E. Woodington,
Colorado School of Mines

First Edition Reviewers

Neil Berger,
University of Illinois at Chicago
Jay Bourland,
Colorado State University
John Chadam,
University of Pittsburgh
Dan Clegg,
Palomar College
Susan Dean,
DeAnza College
Joseph R. Fiedler,
California State University–Bakersfield
Ronald Freiwald,
Washington University in St. Louis
Frederick Gass,
Miami University
John Gosselin,
University of Georgia
Randall R. Holmes,
Auburn University
Mike Hurley,
Case Western Reserve University
Steve Kahn,
Anne Arundel Community College
Harvey Keynes,
University of Minnesota
Ronald Knill,
Tulane University
Stephen Kokoska,
Bloomsburg University
Kevin Kreider,
University of Akron
James Lang,
*Valencia Community College–
East Campus*

Miroslav Lovrić,
McMaster University
Jim McKinney,
*California State Polytechnic University–
Pomona*
Rennie Mirollo,
Boston College
Bill Moss,
Clemson University
Phil Novinger,
Florida State University
Grace Orzech,
Queen's University
Dan Pritikin,
Miami University
James Reynolds,
Clarion University
Gil Rodriguez,
Los Medanos College
N. Paul Schembari,
East Stroudsburg University
Bettina Schmidt,
Auburn University at Montgomery
William K. Tomhave,
Concordia College
Lorenzo Traldi,
Lafayette College
Tom Tucker,
Colgate University
Stanley Wayment,
Southwest Texas State University
James Wright,
Keuka College

I also thank those who have responded to a survey about attitudes to calculus reform:

Second Edition Respondents

Barbara Bath,
Colorado School of Mines
Paul W. Britt,
Louisiana State University
Maria E. Calzada,
Loyola University–New Orleans
Camille P. Cochrane,
Shelton State Community College
Fred Dodd,
University of South Alabama
Ronald C. Freiwald,
Washington University–St. Louis
Richard Hitt,
University of South Alabama
Tejinder S. Neelon,
California State University San Marcos

Bill Paschke,
University of Kansas
David Patocka,
*Tulsa Community College–
Southeast Campus*
Hernan Rivera,
Texas Lutheran University
David C. Royster,
University of North Carolina–Charlotte
Dr. John Schmeelk,
Virginia Commonwealth University
Jianzhong Wang,
Sam Houston State University
Barak Weiss,
*Ben Gurion University–
Be'er Sheva, Israel*

First Edition Respondents

Irfan Altas,
Charles Sturt University

Robert Burton,
Oregon State University

Bem Cayco,
San Jose State University

James Daly,
University of Colorado

Richard Davis,
Edmonds Community College

Richard DiDio,
LaSalle University

Robert Dieffenbach,
Miami University–Middletown

Helmut Doll,
Bloomsburg University

William Dunham,
Muhlenberg College

David A. Edwards,
The University of Georgia

John Ellison,
Grove City College

James P. Fink,
Gettysburg College

Robert Fontenot,
Whitman College

Laurette Foster,
Prairie View A & M University

Gregory Goodhart,
Columbus State Community College

Daniel Grayson,
*University of Illinois at
Urbana–Champaign*

Raymond Greenwell,
Hofstra University

Murli Gupta,
The George Washington University

Kathy Hann,
*California State University
at Hayward*

Judy Holdener,
United States Air Force Academy

Helmer Junghans,
Montgomery College

Victor Kaftal,
University of Cincinnati

Doug Kuhlmann,
Phillips Academy

David E. Kullman,
Miami University

Carl Leinbach,
Gettysburg College

William L. Lepowsky,
Laney College

Kathryn Lesh,
University of Toledo

Estela Llinas,
University of Pittsburgh at Greensburg

Lou Ann Mahaney,
Tarrant County Junior College–Northeast

John R. Martin,
Tarrant County Junior College

R. J. McKellar,
University of New Brunswick

David Minda,
University of Cincinnati

Brian Mortimer,
Carleton University

Richard Nowakowski,
Dalhousie University

Stephen Ott,
Lexington Community College

Paul Patten,
North Georgia College

Leslie Peek,
Mercer University

Mike Pepe,
Seattle Central Community College

Fred Prydz,
Shoreline Community College

Daniel Russow,
Arizona Western College

Brad Shelton,
University of Oregon

Don Small,
*United States Military Academy–
West Point*

Richard B. Thompson,
The University of Arizona

Alan Tucker,
*State University of New York
at Stony Brook*

George Van Zwalenberg,
Calvin College

Dennis Watson,
Clark College

Paul R. Wenston,
The University of Georgia

Ruth Williams,
University of California–San Diego

In addition, I would like to thank George Bergman, Emile LeBlanc, Martin Erickson, Stuart Goldenberg, Gerald Leibowitz, Larry Peterson, Charles Pugh, Marina Ratner, Peter Rosenthal, and Alan Weinstein for their suggestions; Dan Clegg for his research in libraries and on the Internet; Arnold Good for his treatment of optimization problems with implicit differentiation; Al Shenk and Dennis Zill for permission to use exercises from their calculus texts; COMAP for permission to use project material; George Bergman, David Bleecker, Dan Clegg, John Hagood, Victor Kaftal, Anthony Lam, Jamie Lawson, Ira Rosenholtz, Lowell Smylie, and Larry Wallen for ideas for exercises; Dan Drucker for the roller derby project; Tom Farmer, Fred Gass, John Ramsay, Larry Riddle, V. K. Srinivasan, and Philip Straffin for ideas for projects; and Jeff Cole and Dan Clegg for preparing the answer manuscript. I'm grateful to Jeff Cole for suggesting ways to improve the exercises. Dan Clegg acted as my assistant throughout; he proofread, made suggestions, and contributed some of the new exercises.

In addition, I thank those who have contributed to past editions: Ed Barbeau, Fred Brauer, Andy Bulman-Fleming, Tom DiCiccio, Garret Etgen, Chris Fisher, Gene Hecht, Harvey Keynes, Kevin Kreider, E. L. Koh, Zdislav Kovarik, David Leep, Lothar Redlin, Carl Riehm, Doug Shaw, and Saleem Watson.

I also thank Brian Betsill, Stephanie Kuhns, and Kathi Townes of TECH-arts for their production services, Tom Bonner for the cover image, and the following Brooks/Cole staff: Janet Hill, editorial production project manager; Vernon Boes, art director; Karin Sandberg, Erin Mitchell, and Bryan Vann, marketing team; Earl Perry, technology project manager; Stacy Green, assistant editor; Katherine Cook, editorial assistant; Joohee Lee, permissions editor; Karen Hunt, print/media buyer; and Denise Davidson, cover designer. They have all done an outstanding job.

I have been very fortunate to have worked with some of the best mathematics editors in the business over the past two decades: Ron Munro, Harry Campbell, Craig Barth, Jeremy Hayhurst, Gary Ostedt, and now Bob Pirtle. Bob continues in that tradition of editors who, while offering sound advice and ample assistance, trust my instincts and allow me to write the books that I want to write.

JAMES STEWART

MULTIVARIABLE

Calculus

Concepts and Contexts • 3E

8

Infinite Sequences and Series

Infinite sequences and series were introduced briefly in *A Preview of Calculus* in connection with Zeno's paradoxes and the decimal representation of numbers. Their importance in calculus stems from Newton's idea of representing functions as sums of infinite series. For instance, in finding areas he often integrated a function by first expressing it as a series and then integrating each term of the series. We will pursue his idea in Section 8.7 in order to integrate such functions as e^{-x^2}. (Recall that we have previously been unable to do this.) Many of the functions that arise in mathematical physics and chemistry, such as Bessel functions, are defined as sums of series, so it is important to be familiar with the basic concepts of convergence of infinite sequences and series.

Physicists also use series in another way, as we will see in Section 8.9. In studying fields as diverse as optics, special relativity, and electromagnetism, they analyze phenomena by replacing a function with the first few terms in the series that represents it.

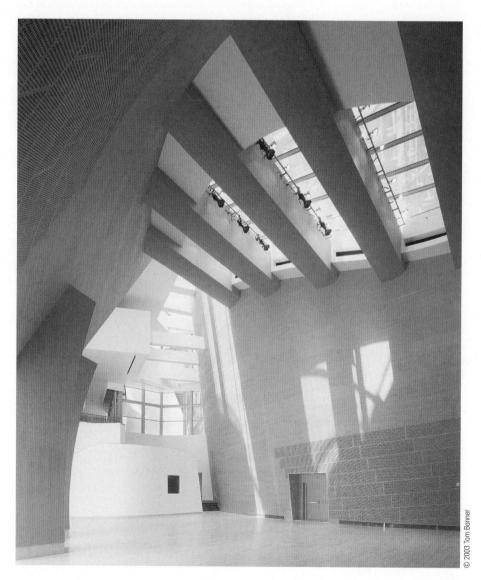

8.1 Sequences

A **sequence** can be thought of as a list of numbers written in a definite order:

$$a_1, \ a_2, \ a_3, \ a_4, \ \ldots, \ a_n, \ldots$$

The number a_1 is called the *first term*, a_2 is the *second term*, and in general a_n is the *nth term*. We will deal exclusively with infinite sequences and so each term a_n will have a successor a_{n+1}.

Notice that for every positive integer n there is a corresponding number a_n and so a sequence can be defined as a function whose domain is the set of positive integers. But we usually write a_n instead of the function notation $f(n)$ for the value of the function at the number n.

NOTATION □ The sequence $\{a_1, a_2, a_3, \ldots\}$ is also denoted by

$$\{a_n\} \qquad \text{or} \qquad \{a_n\}_{n=1}^{\infty}$$

EXAMPLE 1 Some sequences can be defined by giving a formula for the *n*th term. In the following examples we give three descriptions of the sequence: one by using the preceding notation, another by using the defining formula, and a third by writing out the terms of the sequence. Notice that *n* doesn't have to start at 1.

(a) $\left\{ \dfrac{n}{n+1} \right\}_{n=1}^{\infty}$ $a_n = \dfrac{n}{n+1}$ $\left\{ \dfrac{1}{2}, \dfrac{2}{3}, \dfrac{3}{4}, \dfrac{4}{5}, \ldots, \dfrac{n}{n+1}, \ldots \right\}$

(b) $\left\{ \dfrac{(-1)^n(n+1)}{3^n} \right\}$ $a_n = \dfrac{(-1)^n(n+1)}{3^n}$ $\left\{ -\dfrac{2}{3}, \dfrac{3}{9}, -\dfrac{4}{27}, \dfrac{5}{81}, \ldots, \dfrac{(-1)^n(n+1)}{3^n}, \ldots \right\}$

(c) $\left\{ \sqrt{n-3} \right\}_{n=3}^{\infty}$ $a_n = \sqrt{n-3}, \ n \geq 3$ $\left\{ 0, 1, \sqrt{2}, \sqrt{3}, \ldots, \sqrt{n-3}, \ldots \right\}$

(d) $\left\{ \cos \dfrac{n\pi}{6} \right\}_{n=0}^{\infty}$ $a_n = \cos \dfrac{n\pi}{6}, \ n \geq 0$ $\left\{ 1, \dfrac{\sqrt{3}}{2}, \dfrac{1}{2}, 0, \ldots, \cos \dfrac{n\pi}{6}, \ldots \right\}$ ■ ■

$\boxed{V}$ **EXAMPLE 2** Find a formula for the general term a_n of the sequence

$$\left\{ \frac{3}{5}, -\frac{4}{25}, \frac{5}{125}, -\frac{6}{625}, \frac{7}{3125}, \ldots \right\}$$

assuming that the pattern of the first few terms continues.

SOLUTION We are given that

$$a_1 = \frac{3}{5} \qquad a_2 = -\frac{4}{25} \qquad a_3 = \frac{5}{125} \qquad a_4 = -\frac{6}{625} \qquad a_5 = \frac{7}{3125}$$

Notice that the numerators of these fractions start with 3 and increase by 1 whenever we go to the next term. The second term has numerator 4, the third term has numerator 5; in general, the *n*th term will have numerator $n + 2$. The denominators are the powers of 5, so a_n has denominator 5^n. The signs of the terms are alternately positive and negative, so we need to multiply by a power of -1. In Example 1(b) the factor $(-1)^n$ meant we started with a negative term. Here we want to start with a positive term and so we use $(-1)^{n-1}$ or $(-1)^{n+1}$. Therefore,

$$a_n = (-1)^{n-1} \frac{n+2}{5^n}$$

■ ■

EXAMPLE 3 Here are some sequences that don't have a simple defining equation.

(a) The sequence $\{p_n\}$, where p_n is the population of the world as of January 1 in the year n.

(b) If we let a_n be the digit in the nth decimal place of the number e, then $\{a_n\}$ is a well-defined sequence whose first few terms are

$$\{7, 1, 8, 2, 8, 1, 8, 2, 8, 4, 5, \ldots\}$$

(c) The **Fibonacci sequence** $\{f_n\}$ is defined recursively by the conditions

$$f_1 = 1 \qquad f_2 = 1 \qquad f_n = f_{n-1} + f_{n-2} \qquad n \geq 3$$

Each term is the sum of the two preceding terms. The first few terms are

$$\{1, 1, 2, 3, 5, 8, 13, 21, \ldots\}$$

This sequence arose when the 13th-century Italian mathematician known as Fibonacci solved a problem concerning the breeding of rabbits (see Exercise 39).   ■ ■

FIGURE 1

FIGURE 2

A sequence such as the one in Example 1(a), $a_n = n/(n+1)$, can be pictured either by plotting its terms on a number line as in Figure 1 or by plotting its graph as in Figure 2. Note that, since a sequence is a function whose domain is the set of positive integers, its graph consists of isolated points with coordinates

$$(1, a_1) \qquad (2, a_2) \qquad (3, a_3) \qquad \ldots \qquad (n, a_n) \qquad \ldots$$

From Figure 1 or 2 it appears that the terms of the sequence $a_n = n/(n+1)$ are approaching 1 as n becomes large. In fact, the difference

$$1 - \frac{n}{n+1} = \frac{1}{n+1}$$

can be made as small as we like by taking n sufficiently large. We indicate this by writing

$$\lim_{n \to \infty} \frac{n}{n+1} = 1$$

In general, the notation

$$\lim_{n \to \infty} a_n = L$$

means that the terms of the sequence $\{a_n\}$ approach L as n becomes large. Notice that the following definition of the limit of a sequence is very similar to the definition of a limit of a function at infinity given in Section 2.5.

1 Definition A sequence $\{a_n\}$ has the **limit** L and we write

$$\lim_{n\to\infty} a_n = L \qquad \text{or} \qquad a_n \to L \text{ as } n \to \infty$$

if we can make the terms a_n as close to L as we like by taking n sufficiently large. If $\lim_{n\to\infty} a_n$ exists, we say the sequence **converges** (or is **convergent**). Otherwise, we say the sequence **diverges** (or is **divergent**).

■ ■ A more precise definition of the limit of a sequence is given in Appendix D.

Figure 3 illustrates Definition 1 by showing the graphs of two sequences that have the limit L.

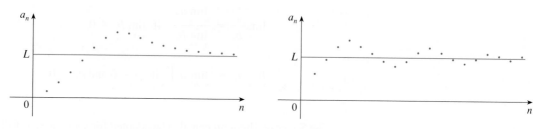

FIGURE 3
Graphs of two sequences with $\lim\limits_{n\to\infty} a_n = L$

If you compare Definition 1 with Definition 2.5.4 you will see that the only difference between $\lim_{n\to\infty} a_n = L$ and $\lim_{x\to\infty} f(x) = L$ is that n is required to be an integer. Thus, we have the following theorem, which is illustrated by Figure 4.

2 Theorem If $\lim_{x\to\infty} f(x) = L$ and $f(n) = a_n$ when n is an integer, then $\lim_{n\to\infty} a_n = L$.

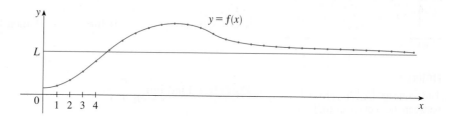

FIGURE 4

In particular, since we know from Section 2.5 that $\lim_{x\to\infty} (1/x^r) = 0$ when $r > 0$, we have

3
$$\lim_{n\to\infty} \frac{1}{n^r} = 0 \qquad \text{if } r > 0$$

If a_n becomes large as n becomes large, we use the notation

$$\lim_{n\to\infty} a_n = \infty$$

In this case the sequence $\{a_n\}$ is divergent, but in a special way. We say that $\{a_n\}$ diverges to ∞.

The Limit Laws given in Section 2.3 also hold for the limits of sequences and their proofs are similar.

Limit Laws for Convergent Sequences

If $\{a_n\}$ and $\{b_n\}$ are convergent sequences and c is a constant, then

$$\lim_{n \to \infty} (a_n + b_n) = \lim_{n \to \infty} a_n + \lim_{n \to \infty} b_n$$

$$\lim_{n \to \infty} (a_n - b_n) = \lim_{n \to \infty} a_n - \lim_{n \to \infty} b_n$$

$$\lim_{n \to \infty} ca_n = c \lim_{n \to \infty} a_n \qquad\qquad \lim_{n \to \infty} c = c$$

$$\lim_{n \to \infty} (a_n b_n) = \lim_{n \to \infty} a_n \cdot \lim_{n \to \infty} b_n$$

$$\lim_{n \to \infty} \frac{a_n}{b_n} = \frac{\lim\limits_{n \to \infty} a_n}{\lim\limits_{n \to \infty} b_n} \quad \text{if } \lim_{n \to \infty} b_n \neq 0$$

$$\lim_{n \to \infty} a_n^p = \left[\lim_{n \to \infty} a_n \right]^p \quad \text{if } p > 0 \text{ and } a_n > 0$$

The Squeeze Theorem can also be adapted for sequences as follows (see Figure 5).

Squeeze Theorem for Sequences

If $a_n \leq b_n \leq c_n$ for $n \geq n_0$ and $\lim\limits_{n \to \infty} a_n = \lim\limits_{n \to \infty} c_n = L$, then $\lim\limits_{n \to \infty} b_n = L$.

Another useful fact about limits of sequences is given by the following theorem, which follows from the Squeeze Theorem because $-|a_n| \leq a_n \leq |a_n|$.

4 Theorem If $\lim\limits_{n \to \infty} |a_n| = 0$, then $\lim\limits_{n \to \infty} a_n = 0$.

FIGURE 5
The sequence $\{b_n\}$ is squeezed between the sequences $\{a_n\}$ and $\{c_n\}$.

EXAMPLE 4 Find $\lim\limits_{n \to \infty} \dfrac{n}{n + 1}$.

SOLUTION The method is similar to the one we used in Section 2.5: Divide numerator and denominator by the highest power of n that occurs in the denominator and then use the Limit Laws.

$$\lim_{n \to \infty} \frac{n}{n + 1} = \lim_{n \to \infty} \frac{1}{1 + \dfrac{1}{n}} = \frac{\lim\limits_{n \to \infty} 1}{\lim\limits_{n \to \infty} 1 + \lim\limits_{n \to \infty} \dfrac{1}{n}}$$

■■ This shows that the guess we made earlier from Figures 1 and 2 was correct.

$$= \frac{1}{1 + 0} = 1$$

Here we used Equation 3 with $r = 1$. ■■

EXAMPLE 5 Calculate $\lim\limits_{n \to \infty} \dfrac{\ln n}{n}$.

SOLUTION Notice that both numerator and denominator approach infinity as $n \to \infty$. We can't apply l'Hospital's Rule directly because it applies not to sequences but to functions of a real variable. However, we can apply l'Hospital's Rule to the related function $f(x) = (\ln x)/x$ and obtain

$$\lim_{x \to \infty} \frac{\ln x}{x} = \lim_{x \to \infty} \frac{1/x}{1} = 0$$

Therefore, by Theorem 2 we have

$$\lim_{n \to \infty} \frac{\ln n}{n} = 0$$

■ ■

EXAMPLE 6 Determine whether the sequence $a_n = (-1)^n$ is convergent or divergent.

SOLUTION If we write out the terms of the sequence, we obtain

$$\{-1, 1, -1, 1, -1, 1, -1, \ldots\}$$

The graph of this sequence is shown in Figure 6. Since the terms oscillate between 1 and -1 infinitely often, a_n does not approach any number. Thus, $\lim_{n \to \infty} (-1)^n$ does not exist; that is, the sequence $\{(-1)^n\}$ is divergent.

■ ■

FIGURE 6

■ ■ The graph of the sequence in Example 7 is shown in Figure 7 and supports the answer.

EXAMPLE 7 Evaluate $\lim_{n \to \infty} \dfrac{(-1)^n}{n}$ if it exists.

SOLUTION

$$\lim_{n \to \infty} \left| \frac{(-1)^n}{n} \right| = \lim_{n \to \infty} \frac{1}{n} = 0$$

Therefore, by Theorem 4,

$$\lim_{n \to \infty} \frac{(-1)^n}{n} = 0$$

■ ■

FIGURE 7

$\boxed{V}$ **EXAMPLE 8** Discuss the convergence of the sequence $a_n = n!/n^n$, where $n! = 1 \cdot 2 \cdot 3 \cdot \cdots \cdot n$.

SOLUTION Both numerator and denominator approach infinity as $n \to \infty$ but here we have no corresponding function for use with l'Hospital's Rule ($x!$ is not defined when x is not an integer). Let's write out a few terms to get a feeling for what happens to a_n as n gets large:

$$a_1 = 1 \qquad a_2 = \frac{1 \cdot 2}{2 \cdot 2} \qquad a_3 = \frac{1 \cdot 2 \cdot 3}{3 \cdot 3 \cdot 3}$$

$\boxed{5}$
$$a_n = \frac{1 \cdot 2 \cdot 3 \cdot \cdots \cdot n}{n \cdot n \cdot n \cdot \cdots \cdot n}$$

It appears from these expressions and the graph in Figure 8 that the terms are decreasing and perhaps approach 0. To confirm this, observe from Equation 5 that

$$a_n = \frac{1}{n} \left(\frac{2 \cdot 3 \cdot \cdots \cdot n}{n \cdot n \cdot \cdots \cdot n} \right)$$

■ ■ CREATING GRAPHS OF SEQUENCES

Some computer algebra systems have special commands that enable us to create sequences and graph them directly. With most graphing calculators, however, sequences can be graphed by using parametric equations. For instance, the sequence in Example 8 can be graphed by entering the parametric equations

$$x = t \qquad y = t!/t^t$$

and graphing in dot mode starting with $t = 1$, setting the t-step equal to 1. The result is shown in Figure 8.

FIGURE 8

Notice that the expression in parentheses is at most 1 because the numerator is less than (or equal to) the denominator. So

$$0 < a_n \leq \frac{1}{n}$$

We know that $1/n \to 0$ as $n \to \infty$. Therefore, $a_n \to 0$ as $n \to \infty$ by the Squeeze Theorem.

■ ■

EXAMPLE 9 For what values of r is the sequence $\{r^n\}$ convergent?

SOLUTION We know from Section 2.5 and the graphs of the exponential functions in Section 1.5 that $\lim_{x\to\infty} a^x = \infty$ for $a > 1$ and $\lim_{x\to\infty} a^x = 0$ for $0 < a < 1$. Therefore, putting $a = r$ and using Theorem 2, we have

$$\lim_{n\to\infty} r^n = \begin{cases} \infty & \text{if } r > 1 \\ 0 & \text{if } 0 < r < 1 \end{cases}$$

For the cases $r = 1$ and $r = 0$ we have

$$\lim_{n\to\infty} 1^n = \lim_{n\to\infty} 1 = 1 \qquad \text{and} \qquad \lim_{n\to\infty} 0^n = \lim_{n\to\infty} 0 = 0$$

If $-1 < r < 0$, then $0 < |r| < 1$, so

$$\lim_{n\to\infty} |r^n| = \lim_{n\to\infty} |r|^n = 0$$

and therefore $\lim_{n\to\infty} r^n = 0$ by Theorem 4. If $r \leq -1$, then $\{r^n\}$ diverges as in Example 6. Figure 9 shows the graphs for various values of r. (The case $r = -1$ is shown in Figure 6.)

FIGURE 9
The sequence $a_n = r^n$

The results of Example 9 are summarized for future use as follows.

---

6 The sequence $\{r^n\}$ is convergent if $-1 < r \leq 1$ and divergent for all other values of r.

$$\lim_{n\to\infty} r^n = \begin{cases} 0 & \text{if } -1 < r < 1 \\ 1 & \text{if } r = 1 \end{cases}$$

Definition A sequence $\{a_n\}$ is called **increasing** if $a_n < a_{n+1}$ for all $n \geq 1$, that is, $a_1 < a_2 < a_3 < \cdots$. It is called **decreasing** if $a_n > a_{n+1}$ for all $n \geq 1$. A sequence is **monotonic** if it is either increasing or decreasing.

EXAMPLE 10 The sequence $\left\{\dfrac{3}{n+5}\right\}$ is decreasing because

■ ■ The right side is smaller because it has a larger denominator.

$$\frac{3}{n+5} > \frac{3}{(n+1)+5} = \frac{3}{n+6}$$

and so $a_n > a_{n+1}$ for all $n \geq 1$.

■ ■

EXAMPLE 11 Show that the sequence $a_n = \dfrac{n}{n^2+1}$ is decreasing.

SOLUTION 1 We must show that $a_{n+1} < a_n$, that is,

$$\frac{n+1}{(n+1)^2+1} < \frac{n}{n^2+1}$$

This inequality is equivalent to the one we get by cross-multiplication:

$$\frac{n+1}{(n+1)^2+1} < \frac{n}{n^2+1} \iff (n+1)(n^2+1) < n[(n+1)^2+1]$$

$$\iff n^3+n^2+n+1 < n^3+2n^2+2n$$

$$\iff 1 < n^2+n$$

Since $n \geq 1$, we know that the inequality $n^2+n > 1$ is true. Therefore, $a_{n+1} < a_n$ and so $\{a_n\}$ is decreasing.

SOLUTION 2 Consider the function $f(x) = \dfrac{x}{x^2+1}$:

$$f'(x) = \frac{x^2+1-2x^2}{(x^2+1)^2} = \frac{1-x^2}{(x^2+1)^2} < 0 \qquad \text{whenever } x^2 > 1$$

Thus, f is decreasing on $(1, \infty)$ and so $f(n) > f(n+1)$. Therefore, $\{a_n\}$ is decreasing.

■ ■

Definition A sequence $\{a_n\}$ is **bounded above** if there is a number M such that

$$a_n \leq M \qquad \text{for all } n \geq 1$$

It is **bounded below** if there is a number m such that

$$m \leq a_n \qquad \text{for all } n \geq 1$$

If it is bounded above and below, then $\{a_n\}$ is a **bounded sequence**.

For instance, the sequence $a_n = n$ is bounded below ($a_n > 0$) but not above. The sequence $a_n = n/(n+1)$ is bounded because $0 < a_n < 1$ for all n.

FIGURE 10

We know that not every bounded sequence is convergent [for instance, the sequence $a_n = (-1)^n$ satisfies $-1 \leq a_n \leq 1$ but is divergent from Example 6] and not every monotonic sequence is convergent ($a_n = n \to \infty$). But if a sequence is both bounded *and* monotonic, then it must be convergent. This fact is stated without proof as Theorem 7, but intuitively you can understand why it is true by looking at Figure 10. If $\{a_n\}$ is increasing and $a_n \leq M$ for all n, then the terms are forced to crowd together and approach some number L.

---

7 Monotonic Sequence Theorem   Every bounded, monotonic sequence is convergent.

EXAMPLE 12 Investigate the sequence $\{a_n\}$ defined by the *recurrence relation*

$$a_1 = 2 \qquad a_{n+1} = \tfrac{1}{2}(a_n + 6) \qquad \text{for } n = 1, 2, 3, \ldots$$

SOLUTION We begin by computing the first several terms:

$$a_1 = 2 \qquad\qquad a_2 = \tfrac{1}{2}(2 + 6) = 4 \qquad\qquad a_3 = \tfrac{1}{2}(4 + 6) = 5$$

$$a_4 = \tfrac{1}{2}(5 + 6) = 5.5 \qquad a_5 = 5.75 \qquad\qquad a_6 = 5.875$$

$$a_7 = 5.9375 \qquad\qquad a_8 = 5.96875 \qquad\qquad a_9 = 5.984375$$

■■ Mathematical induction is often used in dealing with recursive sequences. See page 87 for a discussion of the Principle of Mathematical Induction.

These initial terms suggest that the sequence is increasing and the terms are approaching 6. To confirm that the sequence is increasing, we use mathematical induction to show that $a_{n+1} > a_n$ for all $n \geq 1$. This is true for $n = 1$ because $a_2 = 4 > a_1$. If we assume that it is true for $n = k$, then we have

$$a_{k+1} > a_k$$

so

$$a_{k+1} + 6 > a_k + 6$$

and

$$\tfrac{1}{2}(a_{k+1} + 6) > \tfrac{1}{2}(a_k + 6)$$

Thus

$$a_{k+2} > a_{k+1}$$

We have deduced that $a_{n+1} > a_n$ is true for $n = k + 1$. Therefore, the inequality is true for all n by induction.

Next we verify that $\{a_n\}$ is bounded by showing that $a_n < 6$ for all n. (Since the sequence is increasing, we already know that it has a lower bound: $a_n \geq a_1 = 2$ for all n.) We know that $a_1 < 6$, so the assertion is true for $n = 1$. Suppose it is true for $n = k$. Then

$$a_k < 6$$

so

$$a_k + 6 < 12$$

and

$$\tfrac{1}{2}(a_k + 6) < \tfrac{1}{2}(12) = 6$$

Thus

$$a_{k+1} < 6$$

This shows, by mathematical induction, that $a_n < 6$ for all n.

Since the sequence $\{a_n\}$ is increasing and bounded, the Monotonic Sequence Theorem guarantees that it has a limit. The theorem doesn't tell us what the value of the limit is. But now that we know $L = \lim_{n\to\infty} a_n$ exists, we can use the given recurrence relation to write

$$\lim_{n\to\infty} a_{n+1} = \lim_{n\to\infty} \tfrac{1}{2}(a_n + 6) = \tfrac{1}{2}\left(\lim_{n\to\infty} a_n + 6\right) = \tfrac{1}{2}(L + 6)$$

Since $a_n \to L$, it follows that $a_{n+1} \to L$ too (as $n \to \infty$, $n + 1 \to \infty$ also). So we have

$$L = \tfrac{1}{2}(L + 6)$$

Solving this equation for L, we get $L = 6$, as we predicted. ■■

8.1 Exercises

1. (a) What is a sequence?
(b) What does it mean to say that $\lim_{n\to\infty} a_n = 8$?
(c) What does it mean to say that $\lim_{n\to\infty} a_n = \infty$?

2. (a) What is a convergent sequence? Give two examples.
(b) What is a divergent sequence? Give two examples.

3. List the first six terms of the sequence defined by

$$a_n = \frac{n}{2n + 1}$$

Does the sequence appear to have a limit? If so, find it.

4. List the first nine terms of the sequence $\{\cos(n\pi/3)\}$. Does this sequence appear to have a limit? If so, find it. If not, explain why.

5–8 ■ Find a formula for the general term a_n of the sequence, assuming that the pattern of the first few terms continues.

5. $\left\{1, -\frac{2}{3}, \frac{4}{9}, -\frac{8}{27}, \ldots\right\}$

6. $\left\{-\frac{1}{4}, \frac{2}{9}, -\frac{3}{16}, \frac{4}{25}, \ldots\right\}$

7. $\{2, 7, 12, 17, \ldots\}$

8. $\{5, 1, 5, 1, 5, 1, \ldots\}$

9–28 ■ Determine whether the sequence converges or diverges. If it converges, find the limit.

9. $a_n = \dfrac{3 + 5n^2}{n + n^2}$

10. $a_n = \dfrac{n + 1}{3n - 1}$

11. $a_n = \dfrac{2^n}{3^{n+1}}$

12. $a_n = \dfrac{\sqrt{n}}{1 + \sqrt{n}}$

13. $a_n = \dfrac{(n + 2)!}{n!}$

14. $a_n = \dfrac{n}{1 + \sqrt{n}}$

15. $a_n = \dfrac{(-1)^{n-1}n}{n^2 + 1}$

16. $a_n = \dfrac{(-1)^n n^3}{n^3 + 2n^2 + 1}$

17. $\left\{\dfrac{e^n + e^{-n}}{e^{2n} - 1}\right\}$

18. $a_n = \cos(2/n)$

19. $\{n^2 e^{-n}\}$

20. $\{\arctan 2n\}$

21. $a_n = \dfrac{\cos^2 n}{2^n}$

22. $\{n \cos n\pi\}$

23. $a_n = \left(1 + \dfrac{2}{n}\right)^n$

24. $a_n = \dfrac{\sin 2n}{1 + \sqrt{n}}$

25. $\{0, 1, 0, 0, 1, 0, 0, 0, 1, \ldots\}$

26. $a_n = \dfrac{(\ln n)^2}{n}$

27. $a_n = \ln(2n^2 + 1) - \ln(n^2 + 1)$

28. $a_n = \dfrac{(-3)^n}{n!}$

29–34 ■ Use a graph of the sequence to decide whether the sequence is convergent or divergent. If the sequence is convergent, guess the value of the limit from the graph and then prove your guess. (See the margin note on page 562 for advice on graphing sequences.)

29. $a_n = (-1)^n \dfrac{n + 1}{n}$

30. $a_n = 2 + (-2/\pi)^n$

31. $\left\{\arctan\left(\dfrac{2n}{2n + 1}\right)\right\}$

32. $\left\{\dfrac{\sin n}{\sqrt{n}}\right\}$

33. $a_n = \dfrac{n^3}{n!}$

34. $a_n = \sqrt[n]{3^n + 5^n}$

35. If $\$1000$ is invested at 6% interest, compounded annually, then after n years the investment is worth $a_n = 1000(1.06)^n$ dollars.
(a) Find the first five terms of the sequence $\{a_n\}$.
(b) Is the sequence convergent or divergent? Explain.

36. Find the first 40 terms of the sequence defined by

$$a_{n+1} = \begin{cases} \tfrac{1}{2}a_n & \text{if } a_n \text{ is an even number} \\ 3a_n + 1 & \text{if } a_n \text{ is an odd number} \end{cases}$$

and $a_1 = 11$. Do the same if $a_1 = 25$. Make a conjecture about this type of sequence.

37. (a) Determine whether the sequence defined as follows is convergent or divergent:

$$a_1 = 1 \qquad a_{n+1} = 4 - a_n \quad \text{for } n \geq 1$$

(b) What happens if the first term is $a_1 = 2$?

38. (a) If $\lim_{n \to \infty} a_n = L$, what is the value of $\lim_{n \to \infty} a_{n+1}$?
(b) A sequence $\{a_n\}$ is defined by

$$a_1 = 1 \qquad a_{n+1} = 1/(1 + a_n) \quad \text{for } n \geq 1$$

Find the first ten terms of the sequence correct to five decimal places. Does it appear that the sequence is convergent? If so, estimate the value of the limit to three decimal places.
(c) Assuming that the sequence in part (b) has a limit, use part (a) to find its exact value. Compare with your estimate from part (b).

39. (a) Fibonacci posed the following problem: Suppose that rabbits live forever and that every month each pair produces a new pair which becomes productive at age 2 months. If we start with one newborn pair, how many pairs of rabbits will we have in the nth month? Show that the answer is f_n, where $\{f_n\}$ is the Fibonacci sequence defined in Example 3(c).
(b) Let $a_n = f_{n+1}/f_n$ and show that $a_{n-1} = 1 + 1/a_{n-2}$. Assuming that $\{a_n\}$ is convergent, find its limit.

40. Find the limit of the sequence

$$\left\{ \sqrt{2}, \sqrt{2\sqrt{2}}, \sqrt{2\sqrt{2\sqrt{2}}}, \ldots \right\}$$

41–44 ■ Determine whether the sequence is increasing, decreasing, or not monotonic. Is the sequence bounded?

41. $a_n = \dfrac{1}{2n + 3}$ **42.** $a_n = \dfrac{2n - 3}{3n + 4}$

43. $a_n = \cos(n\pi/2)$ **44.** $a_n = n + \dfrac{1}{n}$

. . . .

45. Suppose you know that $\{a_n\}$ is a decreasing sequence and all its terms lie between the numbers 5 and 8. Explain why the sequence has a limit. What can you say about the value of the limit?

46. A sequence $\{a_n\}$ is given by $a_1 = \sqrt{2}$, $a_{n+1} = \sqrt{2 + a_n}$.
(a) By induction or otherwise, show that $\{a_n\}$ is increasing and bounded above by 3. Apply the Monotonic Sequence Theorem to show that $\lim_{n \to \infty} a_n$ exists.
(b) Find $\lim_{n \to \infty} a_n$.

47. Show that the sequence defined by

$$a_1 = 1 \qquad a_{n+1} = 3 - \dfrac{1}{a_n}$$

is increasing and $a_n < 3$ for all n. Deduce that $\{a_n\}$ is convergent and find its limit.

48. Show that the sequence defined by

$$a_1 = 2 \qquad a_{n+1} = \dfrac{1}{3 - a_n}$$

satisfies $0 < a_n \leq 2$ and is decreasing. Deduce that the sequence is convergent and find its limit.

49. We know that $\lim_{n \to \infty} (0.8)^n = 0$ [from (6) with $r = 0.8$]. Use logarithms to determine how large n has to be so that $(0.8)^n < 0.000001$.

50. (a) Let $a_1 = a$, $a_2 = f(a)$, $a_3 = f(a_2) = f(f(a)), \ldots$, $a_{n+1} = f(a_n)$, where f is a continuous function. If $\lim_{n \to \infty} a_n = L$, show that $f(L) = L$.
(b) Illustrate part (a) by taking $f(x) = \cos x$, $a = 1$, and estimating the value of L to five decimal places.

51. The size of an undisturbed fish population has been modeled by the formula

$$p_{n+1} = \dfrac{bp_n}{a + p_n}$$

where p_n is the fish population after n years and a and b are positive constants that depend on the species and its environment. Suppose that the population in year 0 is $p_0 > 0$.
(a) Show that if $\{p_n\}$ is convergent, then the only possible values for its limit are 0 and $b - a$.
(b) Show that $p_{n+1} < (b/a)p_n$.
(c) Use part (b) to show that if $a > b$, then $\lim_{n \to \infty} p_n = 0$; in other words, the population dies out.
(d) Now assume that $a < b$. Show that if $p_0 < b - a$, then $\{p_n\}$ is increasing and $0 < p_n < b - a$. Show also that if $p_0 > b - a$, then $\{p_n\}$ is decreasing and $p_n > b - a$. Deduce that if $a < b$, then $\lim_{n \to \infty} p_n = b - a$.

52. A sequence is defined recursively by

$$a_1 = 1 \qquad a_{n+1} = 1 + \dfrac{1}{1 + a_n}$$

Find the first eight terms of the sequence $\{a_n\}$. What do you notice about the odd terms and the even terms? By considering the odd and even terms separately, show that $\{a_n\}$ is convergent and deduce that

$$\lim_{n \to \infty} a_n = \sqrt{2}$$

This gives the **continued fraction expansion**

$$\sqrt{2} = 1 + \cfrac{1}{2 + \cfrac{1}{2 + \cdots}}$$

[CAS] **Logistic Sequences**

A sequence that arises in ecology as a model for population growth is defined by the **logistic difference equation**

$$p_{n+1} = kp_n(1 - p_n)$$

where p_n measures the size of the population of the nth generation of a single species. To keep the numbers manageable, p_n is a fraction of the maximal size of the population, so $0 \leq p_n \leq 1$. Notice that the form of this equation is similar to the logistic differential equation in Section 7.5. The discrete model—with sequences instead of continuous functions—is preferable for modeling insect populations, where mating and death occur in a periodic fashion.

An ecologist is interested in predicting the size of the population as time goes on, and asks these questions: Will it stabilize at a limiting value? Will it change in a cyclical fashion? Or will it exhibit random behavior?

Write a program to compute the first n terms of this sequence starting with an initial population p_0, where $0 < p_0 < 1$. Use this program to do the following.

1. Calculate 20 or 30 terms of the sequence for $p_0 = \frac{1}{2}$ and for two values of k such that $1 < k < 3$. Graph the sequences. Do they appear to converge? Repeat for a different value of p_0 between 0 and 1. Does the limit depend on the choice of p_0? Does it depend on the choice of k?

2. Calculate terms of the sequence for a value of k between 3 and 3.4 and plot them. What do you notice about the behavior of the terms?

3. Experiment with values of k between 3.4 and 3.5. What happens to the terms?

4. For values of k between 3.6 and 4, compute and plot at least 100 terms and comment on the behavior of the sequence. What happens if you change p_0 by 0.001? This type of behavior is called *chaotic* and is exhibited by insect populations under certain conditions.

8.2 Series

If we try to add the terms of an infinite sequence $\{a_n\}_{n=1}^{\infty}$ we get an expression of the form

$$\boxed{1} \qquad a_1 + a_2 + a_3 + \cdots + a_n + \cdots$$

which is called an **infinite series** (or just a **series**) and is denoted, for short, by the symbol

$$\sum_{n=1}^{\infty} a_n \qquad \text{or} \qquad \sum a_n$$

But does it make sense to talk about the sum of infinitely many terms?

It would be impossible to find a finite sum for the series

$$1 + 2 + 3 + 4 + 5 + \cdots + n + \cdots$$

because if we start adding the terms we get the cumulative sums 1, 3, 6, 10, 15, 21, ... and, after the nth term, we get $n(n + 1)/2$, which becomes very large as n increases.

However, if we start to add the terms of the series

$$\frac{1}{2} + \frac{1}{4} + \frac{1}{8} + \frac{1}{16} + \frac{1}{32} + \frac{1}{64} + \cdots + \frac{1}{2^n} + \cdots$$

n	Sum of first n terms
1	0.50000000
2	0.75000000
3	0.87500000
4	0.93750000
5	0.96875000
6	0.98437500
7	0.99218750
10	0.99902344
15	0.99996948
20	0.99999905
25	0.99999997

we get $\frac{1}{2}, \frac{3}{4}, \frac{7}{8}, \frac{15}{16}, \frac{31}{32}, \frac{63}{64}, \ldots, 1 - 1/2^n, \ldots$. The table shows that as we add more and more terms, these *partial sums* become closer and closer to 1. (See also Figure 11 in *A Preview of Calculus*, page 7.) In fact, by adding sufficiently many terms of the series we can make the partial sums as close as we like to 1. So it seems reasonable to say that the sum of this infinite series is 1 and to write

$$\sum_{n=1}^{\infty} \frac{1}{2^n} = \frac{1}{2} + \frac{1}{4} + \frac{1}{8} + \frac{1}{16} + \cdots + \frac{1}{2^n} + \cdots = 1$$

We use a similar idea to determine whether or not a general series (1) has a sum. We consider the **partial sums**

$$s_1 = a_1$$

$$s_2 = a_1 + a_2$$

$$s_3 = a_1 + a_2 + a_3$$

$$s_4 = a_1 + a_2 + a_3 + a_4$$

and, in general,

$$s_n = a_1 + a_2 + a_3 + \cdots + a_n = \sum_{i=1}^{n} a_i$$

These partial sums form a new sequence $\{s_n\}$, which may or may not have a limit. If $\lim_{n \to \infty} s_n = s$ exists (as a finite number), then, as in the preceding example, we call it the sum of the infinite series $\Sigma\, a_n$.

> **2 Definition** Given a series $\sum_{n=1}^{\infty} a_n = a_1 + a_2 + a_3 + \cdots$, let s_n denote its nth partial sum:
>
> $$s_n = \sum_{i=1}^{n} a_i = a_1 + a_2 + \cdots + a_n$$
>
> If the sequence $\{s_n\}$ is convergent and $\lim_{n \to \infty} s_n = s$ exists as a real number, then the series $\Sigma\, a_n$ is called **convergent** and we write
>
> $$a_1 + a_2 + \cdots + a_n + \cdots = s \qquad \text{or} \qquad \sum_{n=1}^{\infty} a_n = s$$
>
> The number s is called the **sum** of the series. If the sequence $\{s_n\}$ is divergent, then the series is called **divergent**.

■ ■ Compare with the improper integral

$$\int_1^{\infty} f(x)\, dx = \lim_{t \to \infty} \int_1^t f(x)\, dx$$

To find this integral we integrate from 1 to t and then let $t \to \infty$. For a series, we sum from 1 to n and then let $n \to \infty$.

Thus, the sum of a series is the limit of the sequence of partial sums. So when we write $\sum_{n=1}^{\infty} a_n = s$ we mean that by adding sufficiently many terms of the series we can get as close as we like to the number s. Notice that

$$\sum_{n=1}^{\infty} a_n = \lim_{n \to \infty} \sum_{i=1}^{n} a_i$$

EXAMPLE 1 An important example of an infinite series is the **geometric series**

$$a + ar + ar^2 + ar^3 + \cdots + ar^{n-1} + \cdots = \sum_{n=1}^{\infty} ar^{n-1} \qquad a \neq 0$$

Each term is obtained from the preceding one by multiplying it by the common ratio r. (We have already considered the special case where $a = \frac{1}{2}$ and $r = \frac{1}{2}$.)

If $r = 1$, then $s_n = a + a + \cdots + a = na \to \pm\infty$. Since $\lim_{n \to \infty} s_n$ doesn't exist, the geometric series diverges in this case.

If $r \neq 1$, we have

$$s_n = a + ar + ar^2 + \cdots + ar^{n-1}$$

and

$$rs_n = \qquad ar + ar^2 + \cdots + ar^{n-1} + ar^n$$

Subtracting these equations, we get

$$s_n - rs_n = a - ar^n$$

$$\boxed{3} \qquad s_n = \frac{a(1 - r^n)}{1 - r}$$

If $-1 < r < 1$, we know from (8.1.6) that $r^n \to 0$ as $n \to \infty$, so

$$\lim_{n \to \infty} s_n = \lim_{n \to \infty} \frac{a(1 - r^n)}{1 - r} = \frac{a}{1 - r} - \frac{a}{1 - r} \lim_{n \to \infty} r^n = \frac{a}{1 - r}$$

Thus, when $|r| < 1$ the geometric series is convergent and its sum is $a/(1 - r)$.

If $r \leq -1$ or $r > 1$, the sequence $\{r^n\}$ is divergent by (8.1.6) and so, by Equation 3, $\lim_{n \to \infty} s_n$ does not exist. Therefore, the geometric series diverges in those cases. ■ ■

We summarize the results of Example 1 as follows.

> $\boxed{4}$ The geometric series
>
> $$\sum_{n=1}^{\infty} ar^{n-1} = a + ar + ar^2 + \cdots$$
>
> is convergent if $|r| < 1$ and its sum is
>
> $$\sum_{n=1}^{\infty} ar^{n-1} = \frac{a}{1 - r} \qquad |r| < 1$$
>
> If $|r| \geq 1$, the geometric series is divergent.

$\boxed{V}$ **EXAMPLE 2** Find the sum of the geometric series

$$5 - \frac{10}{3} + \frac{20}{9} - \frac{40}{27} + \cdots$$

SOLUTION The first term is $a = 5$ and the common ratio is $r = -\frac{2}{3}$. Since $|r| = \frac{2}{3} < 1$, the series is convergent by (4) and its sum is

$$5 - \frac{10}{3} + \frac{20}{9} - \frac{40}{27} + \cdots = \frac{5}{1 - \left(-\frac{2}{3}\right)} = \frac{5}{\frac{5}{3}} = 3$$

■ ■

■ ■ Figure 1 provides a geometric demonstration of the result in Example 1. If the triangles are constructed as shown and s is the sum of the series, then, by similar triangles,

$$\frac{s}{a} = \frac{a}{a - ar} \qquad \text{so} \qquad s = \frac{a}{1 - r}$$

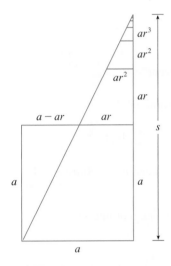

FIGURE 1

■ ■ In words: The sum of a convergent geometric series is

$$\frac{\text{first term}}{1 - \text{common ratio}}$$

■ ■ What do we really mean when we say that the sum of the series in Example 2 is 3? Of course, we can't literally add an infinite number of terms, one by one. But, according to Definition 2, the total sum is the limit of the sequence of partial sums. So, by taking the sum of sufficiently many terms, we can get as close as we like to the number 3. The table shows the first ten partial sums s_n and the graph in Figure 2 shows how the sequence of partial sums approaches 3.

n	s_n
1	5.000000
2	1.666667
3	3.888889
4	2.407407
5	3.395062
6	2.736626
7	3.175583
8	2.882945
9	3.078037
10	2.947975

FIGURE 2

EXAMPLE 3 Is the series $\displaystyle\sum_{n=1}^{\infty} 2^{2n}3^{1-n}$ convergent or divergent?

SOLUTION Let's rewrite the nth term of the series in the form ar^{n-1}:

$$\sum_{n=1}^{\infty} 2^{2n}3^{1-n} = \sum_{n=1}^{\infty} (2^2)^n 3^{-(n-1)} = \sum_{n=1}^{\infty} \frac{4^n}{3^{n-1}} = \sum_{n=1}^{\infty} 4\left(\tfrac{4}{3}\right)^{n-1}$$

We recognize this series as a geometric series with $a = 4$ and $r = \frac{4}{3}$. Since $r > 1$, the series diverges by (4). ■ ■

■ ■ Another way to identify a and r is to write out the first few terms:

$$4 + \tfrac{16}{3} + \tfrac{64}{9} + \cdots$$

$\boxed{\text{V}}$ **EXAMPLE 4** Write the number $2.3\overline{17} = 2.3171717\ldots$ as a ratio of integers.

SOLUTION

$$2.3171717\ldots = 2.3 + \frac{17}{10^3} + \frac{17}{10^5} + \frac{17}{10^7} + \cdots$$

After the first term we have a geometric series with $a = 17/10^3$ and $r = 1/10^2$. Therefore

$$2.3\overline{17} = 2.3 + \frac{\dfrac{17}{10^3}}{1 - \dfrac{1}{10^2}} = 2.3 + \frac{\dfrac{17}{1000}}{\dfrac{99}{100}}$$

$$= \frac{23}{10} + \frac{17}{990} = \frac{1147}{495}$$

■ ■

EXAMPLE 5 Find the sum of the series $\displaystyle\sum_{n=0}^{\infty} x^n$, where $|x| < 1$.

SOLUTION Notice that this series starts with $n = 0$ and so the first term is $x^0 = 1$. (With series, we adopt the convention that $x^0 = 1$ even when $x = 0$.) Thus

$$\sum_{n=0}^{\infty} x^n = 1 + x + x^2 + x^3 + x^4 + \cdots$$

This is a geometric series with $a = 1$ and $r = x$. Since $|r| = |x| < 1$, it converges and (4) gives

Module 8.2 explores a series that depends on an angle θ in a triangle and enables you to see how rapidly the series converges when θ varies.

$$\boxed{5} \qquad \sum_{n=0}^{\infty} x^n = \frac{1}{1-x}$$

■ ■

EXAMPLE 6 Show that the series $\displaystyle\sum_{n=1}^{\infty} \frac{1}{n(n+1)}$ is convergent, and find its sum.

SOLUTION This is not a geometric series, so we go back to the definition of a convergent series and compute the partial sums.

$$s_n = \sum_{i=1}^{n} \frac{1}{i(i+1)} = \frac{1}{1\cdot 2} + \frac{1}{2\cdot 3} + \frac{1}{3\cdot 4} + \cdots + \frac{1}{n(n+1)}$$

We can simplify this expression if we use the partial fraction decomposition

$$\frac{1}{i(i+1)} = \frac{1}{i} - \frac{1}{i+1}$$

(see Section 5.7). Thus, we have

$$s_n = \sum_{i=1}^{n} \frac{1}{i(i+1)} = \sum_{i=1}^{n} \left(\frac{1}{i} - \frac{1}{i+1} \right)$$

$$= \left(1 - \frac{1}{2} \right) + \left(\frac{1}{2} - \frac{1}{3} \right) + \left(\frac{1}{3} - \frac{1}{4} \right) + \cdots + \left(\frac{1}{n} - \frac{1}{n+1} \right)$$

$$= 1 - \frac{1}{n+1}$$

■ ■ Notice that the terms cancel in pairs. This is an example of a **telescoping sum**: Because of all the cancellations, the sum collapses (like a pirate's collapsing telescope) into just two terms.

and so

$$\lim_{n\to\infty} s_n = \lim_{n\to\infty} \left(1 - \frac{1}{n+1} \right) = 1 - 0 = 1$$

■ ■ Figure 3 illustrates Example 6 by showing the graphs of the sequence of terms $a_n = 1/[n(n+1)]$ and the sequence $\{s_n\}$ of partial sums. Notice that $a_n \to 0$ and $s_n \to 1$. See Exercises 46 and 47 for two geometric interpretations of Example 6.

Therefore, the given series is convergent and

$$\sum_{n=1}^{\infty} \frac{1}{n(n+1)} = 1$$

■ ■

FIGURE 3

|V| **EXAMPLE 7** Show that the **harmonic series**

$$\sum_{n=1}^{\infty} \frac{1}{n} = 1 + \frac{1}{2} + \frac{1}{3} + \frac{1}{4} + \cdots$$

is divergent.

SOLUTION For this particular series it's convenient to consider the partial sums s_2, s_4, s_8, s_{16}, s_{32}, ... and show that they become large.

$$s_2 = 1 + \tfrac{1}{2}$$

$$s_4 = 1 + \tfrac{1}{2} + \left(\tfrac{1}{3} + \tfrac{1}{4} \right) > 1 + \tfrac{1}{2} + \left(\tfrac{1}{4} + \tfrac{1}{4} \right) = 1 + \tfrac{2}{2}$$

$$s_8 = 1 + \tfrac{1}{2} + \left(\tfrac{1}{3} + \tfrac{1}{4} \right) + \left(\tfrac{1}{5} + \tfrac{1}{6} + \tfrac{1}{7} + \tfrac{1}{8} \right)$$

$$> 1 + \tfrac{1}{2} + \left(\tfrac{1}{4} + \tfrac{1}{4} \right) + \left(\tfrac{1}{8} + \tfrac{1}{8} + \tfrac{1}{8} + \tfrac{1}{8} \right)$$

$$= 1 + \tfrac{1}{2} + \tfrac{1}{2} + \tfrac{1}{2} = 1 + \tfrac{3}{2}$$

$$s_{16} = 1 + \tfrac{1}{2} + \left(\tfrac{1}{3} + \tfrac{1}{4} \right) + \left(\tfrac{1}{5} + \cdots + \tfrac{1}{8} \right) + \left(\tfrac{1}{9} + \cdots + \tfrac{1}{16} \right)$$

$$> 1 + \tfrac{1}{2} + \left(\tfrac{1}{4} + \tfrac{1}{4} \right) + \left(\tfrac{1}{8} + \cdots + \tfrac{1}{8} \right) + \left(\tfrac{1}{16} + \cdots + \tfrac{1}{16} \right)$$

$$= 1 + \tfrac{1}{2} + \tfrac{1}{2} + \tfrac{1}{2} + \tfrac{1}{2} = 1 + \tfrac{4}{2}$$

Similarly, $s_{32} > 1 + \frac{5}{2}$, $s_{64} > 1 + \frac{6}{2}$, and in general

$$s_{2^n} > 1 + \frac{n}{2}$$

■ ■ The method used in Example 7 for showing that the harmonic series diverges is due to the French scholar Nicole Oresme (1323–1382).

This shows that $s_{2^n} \to \infty$ as $n \to \infty$ and so $\{s_n\}$ is divergent. Therefore, the harmonic series diverges. ■ ■

---

6 **Theorem** If the series $\displaystyle\sum_{n=1}^{\infty} a_n$ is convergent, then $\displaystyle\lim_{n\to\infty} a_n = 0$.

Proof Let $s_n = a_1 + a_2 + \cdots + a_n$. Then $a_n = s_n - s_{n-1}$. Since $\Sigma\, a_n$ is convergent, the sequence $\{s_n\}$ is convergent. Let $\lim_{n\to\infty} s_n = s$. Since $n - 1 \to \infty$ as $n \to \infty$, we also have $\lim_{n\to\infty} s_{n-1} = s$. Therefore

$$\lim_{n\to\infty} a_n = \lim_{n\to\infty} (s_n - s_{n-1}) = \lim_{n\to\infty} s_n - \lim_{n\to\infty} s_{n-1}$$
$$= s - s = 0$$ ■ ■

NOTE 1 ∘ With any *series* $\Sigma\, a_n$ we associate two *sequences:* the sequence $\{s_n\}$ of its partial sums and the sequence $\{a_n\}$ of its terms. If $\Sigma\, a_n$ is convergent, then the limit of the sequence $\{s_n\}$ is s (the sum of the series) and, as Theorem 6 asserts, the limit of the sequence $\{a_n\}$ is 0.

⊘ **NOTE 2** ∘ The converse of Theorem 6 is not true in general. If $\lim_{n\to\infty} a_n = 0$, we cannot conclude that $\Sigma\, a_n$ is convergent. Observe that for the harmonic series $\Sigma\, 1/n$ we have $a_n = 1/n \to 0$ as $n \to \infty$, but we showed in Example 7 that $\Sigma\, 1/n$ is divergent.

---

7 **The Test for Divergence** If $\displaystyle\lim_{n\to\infty} a_n$ does not exist or if $\displaystyle\lim_{n\to\infty} a_n \neq 0$, then the series $\displaystyle\sum_{n=1}^{\infty} a_n$ is divergent.

The Test for Divergence follows from Theorem 6 because, if the series is not divergent, then it is convergent, and so $\lim_{n\to\infty} a_n = 0$.

EXAMPLE 8 Show that the series $\displaystyle\sum_{n=1}^{\infty} \frac{n^2}{5n^2 + 4}$ diverges.

SOLUTION

$$\lim_{n\to\infty} a_n = \lim_{n\to\infty} \frac{n^2}{5n^2 + 4} = \lim_{n\to\infty} \frac{1}{5 + 4/n^2} = \frac{1}{5} \neq 0$$

So the series diverges by the Test for Divergence. ■ ■

NOTE 3 ∘ If we find that $\lim_{n\to\infty} a_n \neq 0$, we know that $\Sigma\, a_n$ is divergent. If we find that $\lim_{n\to\infty} a_n = 0$, we know *nothing* about the convergence or divergence of $\Sigma\, a_n$. Remember the warning in Note 2: If $\lim_{n\to\infty} a_n = 0$, the series $\Sigma\, a_n$ might converge or it might diverge.

---

8 **Theorem** If $\Sigma\, a_n$ and $\Sigma\, b_n$ are convergent series, then so are the series $\Sigma\, ca_n$ (where c is a constant), $\Sigma\, (a_n + b_n)$, and $\Sigma\, (a_n - b_n)$, and

(i) $\displaystyle\sum_{n=1}^{\infty} ca_n = c \sum_{n=1}^{\infty} a_n$
   (ii) $\displaystyle\sum_{n=1}^{\infty} (a_n + b_n) = \sum_{n=1}^{\infty} a_n + \sum_{n=1}^{\infty} b_n$

(iii) $\displaystyle\sum_{n=1}^{\infty} (a_n - b_n) = \sum_{n=1}^{\infty} a_n - \sum_{n=1}^{\infty} b_n$

These properties of convergent series follow from the corresponding Limit Laws for Convergent Sequences in Section 8.1. For instance, here is how part (ii) of Theorem 8 is proved:

Let

$$s_n = \sum_{i=1}^{n} a_i \qquad s = \sum_{n=1}^{\infty} a_n \qquad t_n = \sum_{i=1}^{n} b_i \qquad t = \sum_{n=1}^{\infty} b_n$$

The nth partial sum for the series $\Sigma\, (a_n + b_n)$ is

$$u_n = \sum_{i=1}^{n} (a_i + b_i)$$

and, using Equation 5.2.9, we have

$$\lim_{n\to\infty} u_n = \lim_{n\to\infty} \sum_{i=1}^{n} (a_i + b_i) = \lim_{n\to\infty} \left(\sum_{i=1}^{n} a_i + \sum_{i=1}^{n} b_i \right)$$

$$= \lim_{n\to\infty} \sum_{i=1}^{n} a_i + \lim_{n\to\infty} \sum_{i=1}^{n} b_i$$

$$= \lim_{n\to\infty} s_n + \lim_{n\to\infty} t_n = s + t$$

Therefore, $\Sigma\, (a_n + b_n)$ is convergent and its sum is

$$\sum_{n=1}^{\infty} (a_n + b_n) = s + t = \sum_{n=1}^{\infty} a_n + \sum_{n=1}^{\infty} b_n$$

■ ■

EXAMPLE 9 Find the sum of the series $\displaystyle\sum_{n=1}^{\infty} \left(\frac{3}{n(n+1)} + \frac{1}{2^n} \right)$.

SOLUTION The series $\Sigma\, 1/2^n$ is a geometric series with $a = \frac{1}{2}$ and $r = \frac{1}{2}$, so

$$\sum_{n=1}^{\infty} \frac{1}{2^n} = \frac{\frac{1}{2}}{1 - \frac{1}{2}} = 1$$

In Example 6 we found that

$$\sum_{n=1}^{\infty} \frac{1}{n(n+1)} = 1$$

So, by Theorem 8, the given series is convergent and

$$\sum_{n=1}^{\infty} \left(\frac{3}{n(n+1)} + \frac{1}{2^n} \right) = 3 \sum_{n=1}^{\infty} \frac{1}{n(n+1)} + \sum_{n=1}^{\infty} \frac{1}{2^n}$$

$$= 3 \cdot 1 + 1 = 4$$

■ ■

NOTE 4 ∘ A finite number of terms doesn't affect the convergence or divergence of a series. For instance, suppose that we were able to show that the series

$$\sum_{n=4}^{\infty} \frac{n}{n^3 + 1}$$

is convergent. Since

$$\sum_{n=1}^{\infty} \frac{n}{n^3 + 1} = \frac{1}{2} + \frac{2}{9} + \frac{3}{28} + \sum_{n=4}^{\infty} \frac{n}{n^3 + 1}$$

it follows that the entire series $\sum_{n=1}^{\infty} n/(n^3 + 1)$ is convergent. Similarly, if it is known that the series $\sum_{n=N+1}^{\infty} a_n$ converges, then the full series

$$\sum_{n=1}^{\infty} a_n = \sum_{n=1}^{N} a_n + \sum_{n=N+1}^{\infty} a_n$$

is also convergent.

8.2 Exercises

1. (a) What is the difference between a sequence and a series?
(b) What is a convergent series? What is a divergent series?

2. Explain what it means to say that $\sum_{n=1}^{\infty} a_n = 5$.

3–8 ■ Find at least 10 partial sums of the series. Graph both the sequence of terms and the sequence of partial sums on the same screen. Does it appear that the series is convergent or divergent? If it is convergent, find the sum. If it is divergent, explain why.

3. $\displaystyle\sum_{n=1}^{\infty} \frac{12}{(-5)^n}$

4. $\displaystyle\sum_{n=1}^{\infty} \frac{2n^2 - 1}{n^2 + 1}$

5. $\displaystyle\sum_{n=1}^{\infty} \tan n$

6. $\displaystyle\sum_{n=1}^{\infty} (0.6)^{n-1}$

7. $\displaystyle\sum_{n=1}^{\infty} \left(\frac{1}{n^{1.5}} - \frac{1}{(n+1)^{1.5}} \right)$

8. $\displaystyle\sum_{n=2}^{\infty} \frac{1}{n(n-1)}$

9. Let $a_n = \dfrac{2n}{3n + 1}$.
(a) Determine whether $\{a_n\}$ is convergent.
(b) Determine whether $\sum_{n=1}^{\infty} a_n$ is convergent.

10. (a) Explain the difference between

$$\sum_{i=1}^{n} a_i \quad \text{and} \quad \sum_{j=1}^{n} a_j$$

(b) Explain the difference between

$$\sum_{i=1}^{n} a_i \quad \text{and} \quad \sum_{i=1}^{n} a_j$$

11–16 ■ Determine whether the geometric series is convergent or divergent. If it is convergent, find its sum.

11. $5 - \frac{10}{3} + \frac{20}{9} - \frac{40}{27} + \cdots$

12. $1 + 0.4 + 0.16 + 0.064 + \cdots$

13. $\displaystyle\sum_{n=1}^{\infty} 5\left(\frac{2}{3}\right)^{n-1}$

14. $\displaystyle\sum_{n=1}^{\infty} \frac{(-6)^{n-1}}{5^{n-1}}$

15. $\displaystyle\sum_{n=0}^{\infty} \frac{\pi^n}{3^{n+1}}$

16. $\displaystyle\sum_{n=0}^{\infty} \frac{1}{(\sqrt{2})^n}$

17–26 ■ Determine whether the series is convergent or divergent. If it is convergent, find its sum.

17. $\displaystyle\sum_{n=1}^{\infty} \frac{1}{2n}$

18. $\displaystyle\sum_{n=1}^{\infty} \frac{n+1}{2n - 3}$

19. $\displaystyle\sum_{k=2}^{\infty} \frac{k^2}{k^2 - 1}$

20. $\displaystyle\sum_{k=1}^{\infty} \frac{k(k+2)}{(k+3)^2}$

21. $\displaystyle\sum_{n=1}^{\infty} \frac{1 + 2^n}{3^n}$

22. $\displaystyle\sum_{n=1}^{\infty} \frac{1 + 3^n}{2^n}$

23. $\displaystyle\sum_{n=1}^{\infty} \sqrt[n]{2}$

24. $\displaystyle\sum_{n=1}^{\infty} [(0.8)^{n-1} - (0.3)^n]$

25. $\displaystyle\sum_{n=1}^{\infty} \arctan n$

26. $\displaystyle\sum_{k=1}^{\infty} (\cos 1)^k$

27–30 ■ Determine whether the series is convergent or divergent by expressing s_n as a telescoping sum (as in Example 6). If it is convergent, find its sum.

27. $\displaystyle\sum_{n=2}^{\infty} \frac{2}{n^2 - 1}$

28. $\displaystyle\sum_{n=1}^{\infty} \frac{2}{n^2 + 4n + 3}$

29. $\displaystyle\sum_{n=1}^{\infty} \frac{3}{n(n+3)}$

30. $\displaystyle\sum_{n=1}^{\infty} \ln \frac{n}{n+1}$

31–34 ■ Express the number as a ratio of integers.

31. $0.\overline{2} = 0.2222\ldots$

32. $0.\overline{73} = 0.73737373\ldots$

33. $3.\overline{417} = 3.417417417\ldots$

34. $6.2\overline{54} = 6.2545454\ldots$

35–37 ■ Find the values of x for which the series converges. Find the sum of the series for those values of x.

35. $\displaystyle\sum_{n=1}^{\infty} \frac{x^n}{3^n}$

36. $\displaystyle\sum_{n=0}^{\infty} 2^n(x + 1)^n$

37. $\displaystyle\sum_{n=0}^{\infty} \frac{\cos^n x}{2^n}$

■ ■ ■ ■ ■ ■ ■ ■ ■ ■ ■

38. We have seen that the harmonic series is a divergent series whose terms approach 0. Show that

$$\sum_{n=1}^{\infty} \ln\left(1 + \frac{1}{n}\right)$$

is another series with this property.

[CAS] **39–40** ■ Use the partial fraction command on your CAS to find a convenient expression for the partial sum, and then use this expression to find the sum of the series. Check your answer by using the CAS to sum the series directly.

39. $\displaystyle\sum_{n=1}^{\infty} \frac{1}{(4n + 1)(4n - 3)}$

40. $\displaystyle\sum_{n=1}^{\infty} \frac{n^2 + 3n + 1}{(n^2 + n)^2}$

■ ■ ■ ■ ■ ■ ■ ■ ■ ■ ■

41. If the nth partial sum of a series $\sum_{n=1}^{\infty} a_n$ is

$$s_n = \frac{n - 1}{n + 1}$$

find a_n and $\sum_{n=1}^{\infty} a_n$.

42. If the nth partial sum of a series $\sum_{n=1}^{\infty} a_n$ is $s_n = 3 - n 2^{-n}$, find a_n and $\sum_{n=1}^{\infty} a_n$.

43. When money is spent on goods and services, those that receive the money also spend some of it. The people receiving some of the twice-spent money will spend some of that, and so on. Economists call this chain reaction the *multiplier effect*. In a hypothetical isolated community, the local government begins the process by spending D dollars. Suppose that each recipient of spent money spends $100c\%$ and saves $100s\%$ of the money that he or she receives. The values c and s are called the *marginal propensity to consume* and the *marginal propensity to save* and, of course, $c + s = 1$.
(a) Let S_n be the total spending that has been generated after n transactions. Find an equation for S_n.
(b) Show that $\lim_{n \to \infty} S_n = kD$, where $k = 1/s$. The number k is called the *multiplier*. What is the multiplier if the marginal propensity to consume is 80%?

Note: The federal government uses this principle to justify deficit spending. Banks use this principle to justify lending a large percentage of the money that they receive in deposits.

44. A certain ball has the property that each time it falls from a height h onto a hard, level surface, it rebounds to a height rh, where $0 < r < 1$. Suppose that the ball is dropped from an initial height of H meters.
(a) Assuming that the ball continues to bounce indefinitely, find the total distance that it travels. (Use the fact that the ball falls $\frac{1}{2}gt^2$ meters in t seconds.)

(b) Calculate the total time that the ball travels.
(c) Suppose that each time the ball strikes the surface with velocity v it rebounds with velocity $-kv$, where $0 < k < 1$. How long will it take for the ball to come to rest?

45. What is the value of c if $\displaystyle\sum_{n=2}^{\infty} (1 + c)^{-n} = 2$?

46. Graph the curves $y = x^n$, $0 \le x \le 1$, for $n = 0, 1, 2, 3, 4, \ldots$ on a common screen. By finding the areas between successive curves, give a geometric demonstration of the fact, shown in Example 6, that

$$\sum_{n=1}^{\infty} \frac{1}{n(n + 1)} = 1$$

47. The figure shows two circles C and D of radius 1 that touch at P. T is a common tangent line; C_1 is the circle that touches C, D, and T; C_2 is the circle that touches C, D, and C_1; C_3 is the circle that touches C, D, and C_2. This procedure can be continued indefinitely and produces an infinite sequence of circles $\{C_n\}$. Find an expression for the diameter of C_n and thus provide another geometric demonstration of Example 6.

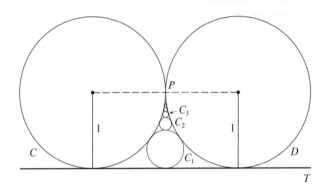

48. A right triangle ABC is given with $\angle A = \theta$ and $|AC| = b$. CD is drawn perpendicular to AB, DE is drawn perpendicular to BC, $EF \perp AB$, and this process is continued indefinitely as shown in the figure. Find the total length of all the perpendiculars

$$|CD| + |DE| + |EF| + |FG| + \cdots$$

in terms of b and θ.

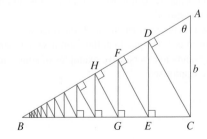

49. What is wrong with the following calculation?

$$0 = 0 + 0 + 0 + \cdots$$
$$= (1 - 1) + (1 - 1) + (1 - 1) + \cdots$$
$$= 1 - 1 + 1 - 1 + 1 - 1 + \cdots$$
$$= 1 + (-1 + 1) + (-1 + 1) + (-1 + 1) + \cdots$$
$$= 1 + 0 + 0 + 0 + \cdots = 1$$

(Guido Ubaldus thought that this proved the existence of God because "something has been created out of nothing.")

50. Suppose that $\sum_{n=1}^{\infty} a_n$ $(a_n \neq 0)$ is known to be a convergent series. Prove that $\sum_{n=1}^{\infty} 1/a_n$ is a divergent series.

51. If Σa_n is convergent and Σb_n is divergent, show that the series $\Sigma (a_n + b_n)$ is divergent. [*Hint:* Argue by contradiction.]

52. If Σa_n and Σb_n are both divergent, is $\Sigma (a_n + b_n)$ necessarily divergent?

53. Suppose that a series Σa_n has positive terms and its partial sums s_n satisfy the inequality $s_n \leq 1000$ for all n. Explain why Σa_n must be convergent.

54. The Fibonacci sequence was defined in Section 8.1 by the equations

$$f_1 = 1, \quad f_2 = 1, \quad f_n = f_{n-1} + f_{n-2} \quad n \geq 3$$

Show that each of the following statements is true.

(a) $\dfrac{1}{f_{n-1} f_{n+1}} = \dfrac{1}{f_{n-1} f_n} - \dfrac{1}{f_n f_{n+1}}$

(b) $\displaystyle\sum_{n=2}^{\infty} \dfrac{1}{f_{n-1} f_{n+1}} = 1$

(c) $\displaystyle\sum_{n=2}^{\infty} \dfrac{f_n}{f_{n-1} f_{n+1}} = 2$

55. The **Cantor set**, named after the German mathematician Georg Cantor (1845–1918), is constructed as follows. We start with the closed interval $[0, 1]$ and remove the open interval $\left(\frac{1}{3}, \frac{2}{3}\right)$. That leaves the two intervals $\left[0, \frac{1}{3}\right]$ and $\left[\frac{2}{3}, 1\right]$ and we remove the open middle third of each. Four intervals remain and again we remove the open middle third of each of them. We continue this procedure indefinitely, at each step removing the open middle third of every interval that remains from the preceding step. The Cantor set consists of the numbers that remain in $[0, 1]$ after all those intervals have been removed.

(a) Show that the total length of all the intervals that are removed is 1. Despite that, the Cantor set contains infinitely many numbers. Give examples of some numbers in the Cantor set.

(b) The **Sierpinski carpet** is a two-dimensional counterpart of the Cantor set. It is constructed by removing the center one-ninth of a square of side 1, then removing the centers of the eight smaller remaining squares, and so on. (The figure shows the first three steps of the construction.) Show that the sum of the areas of the removed squares is 1. This implies that the Sierpinski carpet has area 0.

56. (a) A sequence $\{a_n\}$ is defined recursively by the equation $a_n = \frac{1}{2}(a_{n-1} + a_{n-2})$ for $n \geq 3$, where a_1 and a_2 can be any real numbers. Experiment with various values of a_1 and a_2 and use your calculator to guess the limit of the sequence.

(b) Find $\lim_{n \to \infty} a_n$ in terms of a_1 and a_2 by expressing $a_{n+1} - a_n$ in terms of $a_2 - a_1$ and summing a series.

57. Consider the series

$$\sum_{n=1}^{\infty} \frac{n}{(n + 1)!}$$

(a) Find the partial sums s_1, s_2, s_3, and s_4. Do you recognize the denominators? Use the pattern to guess a formula for s_n.

(b) Use mathematical induction to prove your guess.

(c) Show that the given infinite series is convergent, and find its sum.

58. In the figure there are infinitely many circles approaching the vertices of an equilateral triangle, each circle touching other circles and sides of the triangle. If the triangle has sides of length 1, find the total area occupied by the circles.

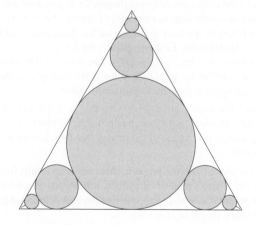

8.3 The Integral and Comparison Tests; Estimating Sums

In general, it is difficult to find the exact sum of a series. We were able to accomplish this for geometric series and the series $\Sigma\, 1/[n(n+1)]$ because in each of those cases we could find a simple formula for the nth partial sum s_n. But usually it isn't easy to compute $\lim_{n\to\infty} s_n$. Therefore, in this section and the next we develop tests that enable us to determine whether a series is convergent or divergent without explicitly finding its sum. In some cases, however, our methods will enable us to find good estimates of the sum.

In this section we deal only with series with positive terms, so the partial sums are increasing. In view of the Monotonic Sequence Theorem, to decide whether a series is convergent or divergent, we need to determine whether the partial sums are bounded or not.

▪▪ Testing with an Integral
▪▪

Let's investigate the series whose terms are the reciprocals of the squares of the positive integers:

$$\sum_{n=1}^{\infty} \frac{1}{n^2} = \frac{1}{1^2} + \frac{1}{2^2} + \frac{1}{3^2} + \frac{1}{4^2} + \frac{1}{5^2} + \cdots$$

n	$s_n = \displaystyle\sum_{i=1}^{n} \frac{1}{i^2}$
5	1.4636
10	1.5498
50	1.6251
100	1.6350
500	1.6429
1000	1.6439
5000	1.6447

There's no simple formula for the sum s_n of the first n terms, but the computer-generated table of values given in the margin suggests that the partial sums are approaching a number near 1.64 as $n \to \infty$ and so it looks as if the series is convergent.

We can confirm this impression with a geometric argument. Figure 1 shows the curve $y = 1/x^2$ and rectangles that lie below the curve. The base of each rectangle is an interval of length 1; the height is equal to the value of the function $y = 1/x^2$ at the right endpoint of the interval. So the sum of the areas of the rectangles is

$$\frac{1}{1^2} + \frac{1}{2^2} + \frac{1}{3^2} + \frac{1}{4^2} + \frac{1}{5^2} + \cdots = \sum_{n=1}^{\infty} \frac{1}{n^2}$$

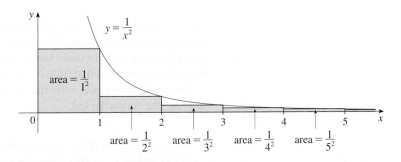

FIGURE 1

If we exclude the first rectangle, the total area of the remaining rectangles is smaller than the area under the curve $y = 1/x^2$ for $x \geqslant 1$, which is the value of the integral $\int_1^{\infty} (1/x^2)\, dx$. In Section 5.10 we discovered that this improper integral is convergent and has value 1. So the picture shows that all the partial sums are less than

$$\frac{1}{1^2} + \int_1^{\infty} \frac{1}{x^2}\, dx = 2$$

Thus, the partial sums are bounded and the series converges. The sum of the series (the limit of the partial sums) is also less than 2:

$$\sum_{n=1}^{\infty} \frac{1}{n^2} = \frac{1}{1^2} + \frac{1}{2^2} + \frac{1}{3^2} + \frac{1}{4^2} + \cdots < 2$$

[The exact sum of this series was found by the Swiss mathematician Leonhard Euler (1707–1783) to be $\pi^2/6$, but the proof of this fact is beyond the scope of this book.]

Now let's look at the series

$$\sum_{n=1}^{\infty} \frac{1}{\sqrt{n}} = \frac{1}{\sqrt{1}} + \frac{1}{\sqrt{2}} + \frac{1}{\sqrt{3}} + \frac{1}{\sqrt{4}} + \frac{1}{\sqrt{5}} + \cdots$$

The table of values of s_n suggests that the partial sums aren't approaching a finite number, so we suspect that the given series may be divergent. Again we use a picture for confirmation. Figure 2 shows the curve $y = 1/\sqrt{x}$, but this time we use rectangles whose tops lie *above* the curve.

n	$s_n = \sum_{i=1}^{n} \dfrac{1}{\sqrt{i}}$
5	3.2317
10	5.0210
50	12.7524
100	18.5896
500	43.2834
1000	61.8010
5000	139.9681

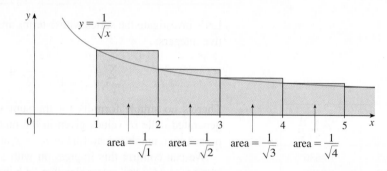

FIGURE 2

The base of each rectangle is an interval of length 1. The height is equal to the value of the function $y = 1/\sqrt{x}$ at the *left* endpoint of the interval. So the sum of the areas of all the rectangles is

$$\frac{1}{\sqrt{1}} + \frac{1}{\sqrt{2}} + \frac{1}{\sqrt{3}} + \frac{1}{\sqrt{4}} + \frac{1}{\sqrt{5}} + \cdots = \sum_{n=1}^{\infty} \frac{1}{\sqrt{n}}$$

This total area is greater than the area under the curve $y = 1/\sqrt{x}$ for $x \geq 1$, which is equal to the integral $\int_1^\infty (1/\sqrt{x}) \, dx$. But we know from Section 5.10 that this improper integral is divergent. In other words, the area under the curve is infinite. So the sum of the series must be infinite, that is, the series is divergent.

The same sort of geometric reasoning that we used for these two series can be used to prove the following test.

The Integral Test Suppose f is a continuous, positive, decreasing function on $[1, \infty)$ and let $a_n = f(n)$. Then the series $\sum_{n=1}^{\infty} a_n$ is convergent if and only if the improper integral $\int_1^\infty f(x) \, dx$ is convergent. In other words:

(a) If $\displaystyle\int_1^\infty f(x) \, dx$ is convergent, then $\displaystyle\sum_{n=1}^{\infty} a_n$ is convergent.

(b) If $\displaystyle\int_1^\infty f(x) \, dx$ is divergent, then $\displaystyle\sum_{n=1}^{\infty} a_n$ is divergent.

NOTE ◦ When we use the Integral Test it is not necessary to start the series or the integral at $n = 1$. For instance, in testing the series

$$\sum_{n=4}^{\infty} \frac{1}{(n-3)^2} \qquad \text{we use} \qquad \int_{4}^{\infty} \frac{1}{(x-3)^2}\, dx$$

Also, it is not necessary that f be always decreasing. What is important is that f be *ultimately* decreasing, that is, decreasing for x larger than some number N. Then $\sum_{n=N}^{\infty} a_n$ is convergent, so $\sum_{n=1}^{\infty} a_n$ is convergent by Note 4 of Section 8.2.

Ⓥ **EXAMPLE 1** Determine whether the series $\sum_{n=1}^{\infty} \dfrac{\ln n}{n}$ converges or diverges.

SOLUTION The function $f(x) = (\ln x)/x$ is positive and continuous for $x > 1$ because the logarithm function is continuous. But it is not obvious whether or not f is decreasing, so we compute its derivative:

$$f'(x) = \frac{x(1/x) - \ln x}{x^2} = \frac{1 - \ln x}{x^2}$$

Thus, $f'(x) < 0$ when $\ln x > 1$, that is, $x > e$. It follows that f is decreasing when $x > e$ and so we can apply the Integral Test:

$$\int_{1}^{\infty} \frac{\ln x}{x}\, dx = \lim_{t \to \infty} \int_{1}^{t} \frac{\ln x}{x}\, dx = \lim_{t \to \infty} \frac{(\ln x)^2}{2} \Bigg]_{1}^{t}$$

$$= \lim_{t \to \infty} \frac{(\ln t)^2}{2} = \infty$$

Since this improper integral is divergent, the series $\sum (\ln n)/n$ is also divergent by the Integral Test. ■ ■

Ⓥ **EXAMPLE 2** For what values of p is the series $\sum_{n=1}^{\infty} \dfrac{1}{n^p}$ convergent?

SOLUTION If $p < 0$, then $\lim_{n \to \infty} (1/n^p) = \infty$. If $p = 0$, then $\lim_{n \to \infty} (1/n^p) = 1$. In either case $\lim_{n \to \infty} (1/n^p) \neq 0$, so the given series diverges by the Test for Divergence [see (8.2.7)].

■ ■ In order to use the Integral Test we need to be able to evaluate $\int_{1}^{\infty} f(x)\, dx$ and therefore we have to be able to find an antiderivative of f. Frequently this is difficult or impossible, so we need other tests for convergence too.

If $p > 0$, then the function $f(x) = 1/x^p$ is clearly continuous, positive, and decreasing on $[1, \infty)$. We found in Chapter 5 [see (5.10.2)] that

$$\int_{1}^{\infty} \frac{1}{x^p}\, dx \qquad \text{converges if } p > 1 \text{ and diverges if } p \leq 1$$

It follows from the Integral Test that the series $\sum 1/n^p$ converges if $p > 1$ and diverges if $0 < p \leq 1$. (For $p = 1$, this series is the harmonic series discussed in Example 7 in Section 8.2.) ■ ■

The series in Example 2 is called the ***p*-series**. It is important in the rest of this chapter, so we summarize the results of Example 2 for future reference as follows.

> $\boxed{1}$ The *p*-series $\sum_{n=1}^{\infty} \dfrac{1}{n^p}$ is convergent if $p > 1$ and divergent if $p \leq 1$.

For instance, the series

$$\sum_{n=1}^{\infty} \frac{1}{n^3} = \frac{1}{1^3} + \frac{1}{2^3} + \frac{1}{3^3} + \frac{1}{4^3} + \cdots$$

is convergent because it is a p-series with $p = 3 > 1$. But the series

$$\sum_{n=1}^{\infty} \frac{1}{n^{1/3}} = \sum_{n=1}^{\infty} \frac{1}{\sqrt[3]{n}} = 1 + \frac{1}{\sqrt[3]{2}} + \frac{1}{\sqrt[3]{3}} + \frac{1}{\sqrt[3]{4}} + \cdots$$

is divergent because it is a p-series with $p = \frac{1}{3} < 1$.

Testing by Comparing

The series

$$\boxed{2} \qquad \sum_{n=1}^{\infty} \frac{1}{2^n + 1}$$

reminds us of the series $\sum_{n=1}^{\infty} 1/2^n$, which is a geometric series with $a = \frac{1}{2}$ and $r = \frac{1}{2}$ and is therefore convergent. Because the series (2) is so similar to a convergent series, we have the feeling that it too must be convergent. Indeed, it is. The inequality

$$\frac{1}{2^n + 1} < \frac{1}{2^n}$$

shows that our given series (2) has smaller terms than those of the geometric series and therefore all its partial sums are also smaller than 1 (the sum of the geometric series). This means that its partial sums form a bounded increasing sequence, which is convergent. It also follows that the sum of the series is less than the sum of the geometric series:

$$\sum_{n=1}^{\infty} \frac{1}{2^n + 1} < 1$$

Similar reasoning can be used to prove the following test, which applies only to series whose terms are positive. The first part says that if we have a series whose terms are *smaller* than those of a known *convergent* series, then our series is also convergent. The second part says that if we start with a series whose terms are *larger* than those of a known *divergent* series, then it too is divergent.

> **The Comparison Test**   Suppose that $\Sigma\, a_n$ and $\Sigma\, b_n$ are series with positive terms.
> (a) If $\Sigma\, b_n$ is convergent and $a_n \leq b_n$ for all n, then $\Sigma\, a_n$ is also convergent.
> (b) If $\Sigma\, b_n$ is divergent and $a_n \geq b_n$ for all n, then $\Sigma\, a_n$ is also divergent.

◼◼ Standard Series for Use with the Comparison Test

In using the Comparison Test we must, of course, have some known series $\Sigma\, b_n$ for the purpose of comparison. Most of the time we use one of these series:

- A p-series $\left[\Sigma\, 1/n^p \text{ converges if } p > 1 \text{ and diverges if } p \leq 1; \text{ see (1)}\right]$
- A geometric series $\left[\Sigma\, ar^{n-1} \text{ converges if } |r| < 1 \text{ and diverges if } |r| \geq 1; \text{ see (8.2.4)}\right]$

$\boxed{V}$ **EXAMPLE 3** Determine whether the series $\displaystyle\sum_{n=1}^{\infty} \frac{5}{2n^2 + 4n + 3}$ converges or diverges.

SOLUTION For large n the dominant term in the denominator is $2n^2$, so we compare the given series with the series $\Sigma\, 5/(2n^2)$. Observe that

$$\frac{5}{2n^2 + 4n + 3} < \frac{5}{2n^2}$$

because the left side has a bigger denominator. (In the notation of the Comparison Test, a_n is the left side and b_n is the right side.) We know that

$$\sum_{n=1}^{\infty} \frac{5}{2n^2} = \frac{5}{2} \sum_{n=1}^{\infty} \frac{1}{n^2}$$

is convergent (p-series with $p = 2 > 1$). Therefore

$$\sum_{n=1}^{\infty} \frac{5}{2n^2 + 4n + 3}$$

is convergent by part (a) of the Comparison Test.   ■ ■

Although the condition $a_n \leq b_n$ or $a_n \geq b_n$ in the Comparison Test is given for all n, we need verify only that it holds for $n \geq N$, where N is some fixed integer, because the convergence of a series is not affected by a finite number of terms. This is illustrated in the next example.

$\boxed{V}$ **EXAMPLE 4** Test the series $\displaystyle\sum_{n=1}^{\infty} \frac{\ln n}{n}$ for convergence or divergence.

SOLUTION We used the Integral Test to test this series in Example 1, but we can also test it by comparing it with the harmonic series. Observe that $\ln n > 1$ for $n \geq 3$ and so

$$\frac{\ln n}{n} > \frac{1}{n} \qquad n \geq 3$$

We know that $\Sigma\, 1/n$ is divergent (p-series with $p = 1$). Thus, the given series is divergent by the Comparison Test.   ■ ■

NOTE ∘ The terms of the series being tested must be smaller than those of a convergent series or larger than those of a divergent series. If the terms are larger than the terms of a convergent series or smaller than those of a divergent series, then the Comparison Test doesn't apply. Consider, for instance, the series

$$\sum_{n=1}^{\infty} \frac{1}{2^n - 1}$$

The inequality

$$\frac{1}{2^n - 1} > \frac{1}{2^n}$$

is useless as far as the Comparison Test is concerned because $\Sigma\, b_n = \Sigma\, \left(\frac{1}{2}\right)^n$ is convergent and $a_n > b_n$. Nonetheless, we have the feeling that $\Sigma\, 1/(2^n - 1)$ ought to be convergent because it is very similar to the convergent geometric series $\Sigma\left(\frac{1}{2}\right)^n$. In such cases the following test can be used.

> **The Limit Comparison Test** Suppose that $\Sigma\, a_n$ and $\Sigma\, b_n$ are series with positive terms. If
>
> $$\lim_{n\to\infty} \frac{a_n}{b_n} = c$$
>
> where c is a finite number and $c > 0$, then either both series converge or both diverge.

Although we won't prove the Limit Comparison Test, it seems reasonable because for large n, $a_n \approx cb_n$.

EXAMPLE 5 Test the series $\displaystyle\sum_{n=1}^{\infty} \frac{1}{2^n - 1}$ for convergence or divergence.

SOLUTION We use the Limit Comparison Test with

$$a_n = \frac{1}{2^n - 1} \qquad b_n = \frac{1}{2^n}$$

and obtain

$$\lim_{n\to\infty} \frac{a_n}{b_n} = \lim_{n\to\infty} \frac{1/(2^n - 1)}{1/2^n} = \lim_{n\to\infty} \frac{2^n}{2^n - 1} = \lim_{n\to\infty} \frac{1}{1 - 1/2^n} = 1 > 0$$

Since this limit exists and $\Sigma\, 1/2^n$ is a convergent geometric series, the given series converges by the Limit Comparison Test. ■ ■

■■ Estimating the Sum of a Series

Suppose we have been able to use the Integral Test to show that a series $\Sigma\, a_n$ is convergent and we now want to find an approximation to the sum s of the series. Of course, any partial sum s_n is an approximation to s because $\lim_{n\to\infty} s_n = s$. But how good is such an approximation? To find out, we need to estimate the size of the **remainder**

$$R_n = s - s_n = a_{n+1} + a_{n+2} + a_{n+3} + \cdots$$

The remainder R_n is the error made when s_n, the sum of the first n terms, is used as an approximation to the total sum.

We use the same notation and ideas as in the Integral Test, assuming that f is decreasing on $[n, \infty)$. Comparing the areas of the rectangles with the area under $y = f(x)$ for $x > n$ in Figure 3, we see that

$$R_n = a_{n+1} + a_{n+2} + \cdots \leq \int_n^{\infty} f(x)\, dx$$

Similarly, we see from Figure 4 that

$$R_n = a_{n+1} + a_{n+2} + \cdots \geq \int_{n+1}^{\infty} f(x)\, dx$$

FIGURE 3

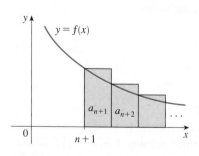

FIGURE 4

So we have proved the following error estimate.

> **3** **Remainder Estimate for the Integral Test** Suppose $f(k) = a_k$, where f is a continuous, positive, decreasing function for $x \geq n$ and $\Sigma\, a_n$ is convergent. If $R_n = s - s_n$, then
>
> $$\int_{n+1}^{\infty} f(x)\, dx \leq R_n \leq \int_{n}^{\infty} f(x)\, dx$$

Ⓥ **EXAMPLE 6**
(a) Approximate the sum of the series $\Sigma\, 1/n^3$ by using the sum of the first 10 terms. Estimate the error involved in this approximation.
(b) How many terms are required to ensure that the sum is accurate to within 0.0005?

SOLUTION In both parts (a) and (b) we need to know $\int_n^{\infty} f(x)\, dx$. With $f(x) = 1/x^3$, which satisfies the conditions of the Integral Test, we have

$$\int_n^{\infty} \frac{1}{x^3}\, dx = \lim_{t \to \infty} \left[-\frac{1}{2x^2} \right]_n^t = \lim_{t \to \infty} \left(-\frac{1}{2t^2} + \frac{1}{2n^2} \right) = \frac{1}{2n^2}$$

(a)
$$\sum_{n=1}^{\infty} \frac{1}{n^3} \approx s_{10} = \frac{1}{1^3} + \frac{1}{2^3} + \frac{1}{3^3} + \cdots + \frac{1}{10^3} \approx 1.1975$$

According to the remainder estimate in (3), we have

$$R_{10} \leq \int_{10}^{\infty} \frac{1}{x^3}\, dx = \frac{1}{2(10)^2} = \frac{1}{200}$$

So the size of the error is at most 0.005.

(b) Accuracy to within 0.0005 means that we have to find a value of n such that $R_n \leq 0.0005$. Since

$$R_n \leq \int_n^{\infty} \frac{1}{x^3}\, dx = \frac{1}{2n^2}$$

we want
$$\frac{1}{2n^2} < 0.0005$$

Solving this inequality, we get

$$n^2 > \frac{1}{0.001} = 1000 \qquad \text{or} \qquad n > \sqrt{1000} \approx 31.6$$

We need 32 terms to ensure accuracy to within 0.0005. ■ ■

If we add s_n to each side of the inequalities in (3), we get

> **4**
> $$s_n + \int_{n+1}^{\infty} f(x)\, dx \leq s \leq s_n + \int_n^{\infty} f(x)\, dx$$

because $s_n + R_n = s$. The inequalities in (4) give a lower bound and an upper bound for s. They provide a more accurate approximation to the sum of the series than the partial sum s_n does.

EXAMPLE 7 Use (4) with $n = 10$ to estimate the sum of the series $\sum_{n=1}^{\infty} \dfrac{1}{n^3}$.

SOLUTION The inequalities in (4) become

$$s_{10} + \int_{11}^{\infty} \frac{1}{x^3}\, dx \leq s \leq s_{10} + \int_{10}^{\infty} \frac{1}{x^3}\, dx$$

From Example 6 we know that

$$\int_{n}^{\infty} \frac{1}{x^3}\, dx = \frac{1}{2n^2}$$

so

$$s_{10} + \frac{1}{2(11)^2} \leq s \leq s_{10} + \frac{1}{2(10)^2}$$

Using $s_{10} \approx 1.197532$, we get

$$1.201664 \leq s \leq 1.202532$$

If we approximate s by the midpoint of this interval, then the error is at most half the length of the interval. So

$$\sum_{n=1}^{\infty} \frac{1}{n^3} \approx 1.2021 \qquad \text{with error} < 0.0005$$ ■ ■

If we compare Example 7 with Example 6, we see that the improved estimate in (4) can be much better than the estimate $s \approx s_n$. To make the error smaller than 0.0005 we had to use 32 terms in Example 6 but only 10 terms in Example 7.

If we have used the Comparison Test to show that a series $\Sigma\, a_n$ converges by comparison with a series $\Sigma\, b_n$, then we may be able to estimate the sum $\Sigma\, a_n$ by comparing remainders, as the following example shows.

V **EXAMPLE 8** Use the sum of the first 100 terms to approximate the sum of the series $\Sigma\, 1/(n^3 + 1)$. Estimate the error involved in this approximation.

SOLUTION Since

$$\frac{1}{n^3 + 1} < \frac{1}{n^3}$$

the given series is convergent by the Comparison Test. The remainder T_n for the comparison series $\Sigma\, 1/n^3$ was estimated in Example 6. There we found that

$$T_n \leq \int_{n}^{\infty} \frac{1}{x^3}\, dx = \frac{1}{2n^2}$$

Therefore, the remainder R_n for the given series satisfies

$$R_n \leq T_n \leq \frac{1}{2n^2}$$

With $n = 100$ we have

$$R_{100} \leq \frac{1}{2(100)^2} = 0.00005$$

Using a programmable calculator or a computer, we find that

$$\sum_{n=1}^{\infty} \frac{1}{n^3 + 1} \approx \sum_{n=1}^{100} \frac{1}{n^3 + 1} \approx 0.6864538$$

with error less than 0.00005.

■ ■

8.3 Exercises

1. Draw a picture to show that

$$\sum_{n=2}^{\infty} \frac{1}{n^{1.3}} < \int_{1}^{\infty} \frac{1}{x^{1.3}} \, dx$$

What can you conclude about the series?

2. Suppose f is a continuous positive decreasing function for $x \geq 1$ and $a_n = f(n)$. By drawing a picture, rank the following three quantities in increasing order:

$$\int_{1}^{6} f(x) \, dx \qquad \sum_{i=1}^{5} a_i \qquad \sum_{i=2}^{6} a_i$$

3. Suppose $\Sigma\, a_n$ and $\Sigma\, b_n$ are series with positive terms and $\Sigma\, b_n$ is known to be convergent.
   (a) If $a_n > b_n$ for all n, what can you say about $\Sigma\, a_n$? Why?
   (b) If $a_n < b_n$ for all n, what can you say about $\Sigma\, a_n$? Why?

4. Suppose $\Sigma\, a_n$ and $\Sigma\, b_n$ are series with positive terms and $\Sigma\, b_n$ is known to be divergent.
   (a) If $a_n > b_n$ for all n, what can you say about $\Sigma\, a_n$? Why?
   (b) If $a_n < b_n$ for all n, what can you say about $\Sigma\, a_n$? Why?

5. It is important to distinguish between

$$\sum_{n=1}^{\infty} n^b \qquad \text{and} \qquad \sum_{n=1}^{\infty} b^n$$

   What name is given to the first series? To the second? For what values of b does the first series converge? For what values of b does the second series converge?

6–8 ■ Use the Integral Test to determine whether the series is convergent or divergent.

6. $\displaystyle\sum_{n=1}^{\infty} \frac{1}{\sqrt[4]{n}}$

7. $\displaystyle\sum_{n=1}^{\infty} \frac{1}{n^4}$

8. $\displaystyle\sum_{n=1}^{\infty} \frac{1}{n^2 + 1}$

■ ■ ■ ■ ■ ■ ■ ■ ■ ■

9–10 ■ Use the Comparison Test to determine whether the series is convergent or divergent.

9. $\displaystyle\sum_{n=1}^{\infty} \frac{1}{n^2 + n + 1}$

10. $\displaystyle\sum_{n=2}^{\infty} \frac{\sqrt{n}}{n - 1}$

■ ■ ■ ■ ■ ■ ■ ■ ■ ■

11–26 ■ Determine whether the series is convergent or divergent.

11. $1 + \dfrac{1}{8} + \dfrac{1}{27} + \dfrac{1}{64} + \dfrac{1}{125} + \cdots$

12. $\displaystyle\sum_{n=1}^{\infty} \left(\frac{5}{n^4} + \frac{4}{n\sqrt{n}} \right)$

13. $\displaystyle\sum_{n=1}^{\infty} ne^{-n}$

14. $\displaystyle\sum_{n=1}^{\infty} \frac{n^2}{n^3 + 1}$

15. $\displaystyle\sum_{n=2}^{\infty} \frac{1}{n \ln n}$

16. $\displaystyle\sum_{n=1}^{\infty} \frac{n^2 - 1}{3n^4 + 1}$

17. $\displaystyle\sum_{n=1}^{\infty} \frac{\cos^2 n}{n^2 + 1}$

18. $\displaystyle\sum_{n=1}^{\infty} \frac{4 + 3^n}{2^n}$

19. $\displaystyle\sum_{n=1}^{\infty} \frac{n - 1}{n4^n}$

20. $\displaystyle\sum_{n=1}^{\infty} \frac{1}{\sqrt{n^3 + 1}}$

21. $\displaystyle\sum_{n=1}^{\infty} \frac{1}{\sqrt{n^2 + 1}}$

22. $\displaystyle\sum_{n=1}^{\infty} \frac{1}{2n + 3}$

23. $\displaystyle\sum_{n=1}^{\infty} \frac{2 + (-1)^n}{n\sqrt{n}}$

24. $\displaystyle\sum_{n=0}^{\infty} \frac{1 + \sin n}{10^n}$

25. $\displaystyle\sum_{n=1}^{\infty} \sin\left(\frac{1}{n} \right)$

26. $\displaystyle\sum_{n=1}^{\infty} \frac{n + 5}{\sqrt[3]{n^7 + n^2}}$

■ ■ ■ ■ ■ ■ ■ ■ ■ ■

27. Find the values of p for which the following series is convergent:

$$\sum_{n=2}^{\infty} \frac{1}{n(\ln n)^p}$$

28. (a) Find the partial sum s_{10} of the series $\sum_{n=1}^{\infty} 1/n^4$. Estimate the error in using s_{10} as an approximation to the sum of the series.
  (b) Use (4) with $n = 10$ to give an improved estimate of the sum.
  (c) Find a value of n so that s_n is within 0.00001 of the sum.

29. (a) Use the sum of the first 10 terms to estimate the sum of the series $\sum_{n=1}^{\infty} 1/n^2$. How good is this estimate?
  (b) Improve this estimate using (4) with $n = 10$.
  (c) Find a value of n that will ensure that the error in the approximation $s \approx s_n$ is less than 0.001.

30. Find the sum of the series $\sum_{n=1}^{\infty} 1/n^5$ correct to three decimal places.

31. Estimate $\sum_{n=1}^{\infty} n^{-3/2}$ to within 0.01.

32. How many terms of the series $\sum_{n=2}^{\infty} 1/[n(\ln n)^2]$ would you need to add to find its sum to within 0.01?

33–34 ■ Use the sum of the first 10 terms to approximate the sum of the series. Estimate the error.

33. $\displaystyle \sum_{n=1}^{\infty} \frac{1}{n^4 + n^2}$

34. $\displaystyle \sum_{n=1}^{\infty} \frac{n}{(n+1)3^n}$

■ ■ ■ ■ ■ ■ ■ ■ ■ ■ ■ ■ ■ ■ ■

35. (a) Use a graph of $y = 1/x$ to show that if s_n is the nth partial sum of the harmonic series, then
$$s_n \leq 1 + \ln n$$
  (b) The harmonic series diverges, but very slowly. Use part (a) to show that the sum of the first million terms is less than 15 and the sum of the first billion terms is less than 22.

36. Show that if we want to approximate the sum of the series $\sum_{n=1}^{\infty} n^{-1.001}$ so that the error is less than 5 in the ninth decimal place, then we need to add more than $10^{11,301}$ terms!

37. The meaning of the decimal representation of a number $0.d_1 d_2 d_3 \ldots$ (where the digit d_i is one of the numbers 0, 1, 2, ..., 9) is that
$$0.d_1 d_2 d_3 d_4 \ldots = \frac{d_1}{10} + \frac{d_2}{10^2} + \frac{d_3}{10^3} + \frac{d_4}{10^4} + \cdots$$
Show that this series always converges.

38. Find all positive values of b for which the series $\sum_{n=1}^{\infty} b^{\ln n}$ converges.

39. If Σa_n is a convergent series with positive terms, is it true that $\Sigma \sin(a_n)$ is also convergent?

40. Show that if $a_n > 0$ and Σa_n is convergent, then $\Sigma \ln(1 + a_n)$ is convergent.

8.4 Other Convergence Tests

The convergence tests that we have looked at so far apply only to series with positive terms. In this section we learn how to deal with series whose terms are not necessarily positive.

■■ Alternating Series

An **alternating series** is a series whose terms are alternately positive and negative. Here are two examples:

$$1 - \frac{1}{2} + \frac{1}{3} - \frac{1}{4} + \frac{1}{5} - \frac{1}{6} + \cdots = \sum_{n=1}^{\infty} (-1)^{n-1} \frac{1}{n}$$

$$-\frac{1}{2} + \frac{2}{3} - \frac{3}{4} + \frac{4}{5} - \frac{5}{6} + \frac{6}{7} - \cdots = \sum_{n=1}^{\infty} (-1)^n \frac{n}{n+1}$$

We see from these examples that the nth term of an alternating series is of the form

$$a_n = (-1)^{n-1} b_n \qquad \text{or} \qquad a_n = (-1)^n b_n$$

where b_n is a positive number. $\left(\text{In fact, } b_n = |a_n|.\right)$

The following test says that if the terms of an alternating series decrease to 0 in absolute value, then the series converges.

The Alternating Series Test If the alternating series

$$\sum_{n=1}^{\infty} (-1)^{n-1} b_n = b_1 - b_2 + b_3 - b_4 + b_5 - b_6 + \cdots \qquad (b_n > 0)$$

satisfies

(i) $b_{n+1} \leq b_n$ for all n

(ii) $\lim_{n \to \infty} b_n = 0$

then the series is convergent.

We won't present a formal proof of this test, but Figure 1 gives a picture of the idea behind the proof.

FIGURE 1

We first plot $s_1 = b_1$ on a number line. To find s_2 we subtract b_2, so s_2 is to the left of s_1. Then to find s_3 we add b_3, so s_3 is to the right of s_2. But, since $b_3 < b_2$, s_3 is to the left of s_1. Continuing in this manner, we see that the partial sums oscillate back and forth. Since $b_n \to 0$, the successive steps are becoming smaller and smaller. The even partial sums s_2, s_4, s_6, ... are increasing and the odd partial sums s_1, s_3, s_5, ... are decreasing. Thus, it seems plausible that both are converging to some number s, which is the sum of the series.

■ ■ Figure 2 illustrates Example 1 by showing the graphs of the terms $a_n = (-1)^{n-1}/n$ and the partial sums s_n. Notice how the values of s_n zigzag across the limiting value, which appears to be about 0.7. In fact, it can be proved that the exact sum of the series is $\ln 2 \approx 0.693$.

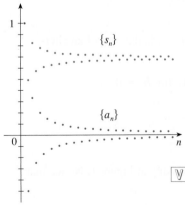

FIGURE 2

[V] **EXAMPLE 1** The alternating harmonic series

$$1 - \frac{1}{2} + \frac{1}{3} - \frac{1}{4} + \cdots = \sum_{n=1}^{\infty} \frac{(-1)^{n-1}}{n}$$

satisfies

(i) $b_{n+1} < b_n$ because $\dfrac{1}{n+1} < \dfrac{1}{n}$

(ii) $\lim_{n \to \infty} b_n = \lim_{n \to \infty} \dfrac{1}{n} = 0$

so the series is convergent by the Alternating Series Test.   ■ ■

[V] **EXAMPLE 2** The series $\displaystyle\sum_{n=1}^{\infty} \frac{(-1)^n 3n}{4n - 1}$ is alternating, but

$$\lim_{n \to \infty} b_n = \lim_{n \to \infty} \frac{3n}{4n - 1} = \lim_{n \to \infty} \frac{3}{4 - \dfrac{1}{n}} = \frac{3}{4}$$

so condition (ii) is not satisfied. Instead, we look at the limit of the nth term of the series:

$$\lim_{n \to \infty} a_n = \lim_{n \to \infty} \frac{(-1)^n 3n}{4n - 1}$$

This limit does not exist, so the series diverges by the Test for Divergence. ■ ■

EXAMPLE 3 Test the series $\sum_{n=1}^{\infty} (-1)^{n+1} \dfrac{n^2}{n^3 + 1}$ for convergence or divergence.

SOLUTION The given series is alternating so we try to verify conditions (i) and (ii) of the Alternating Series Test.

Unlike the situation in Example 1, it is not obvious that the sequence given by $b_n = n^2/(n^3 + 1)$ is decreasing. However, if we consider the related function $f(x) = x^2/(x^3 + 1)$, we find that

$$f'(x) = \frac{x(2 - x^3)}{(x^3 + 1)^2}$$

■ ■ Instead of verifying condition (i) of the Alternating Series Test by computing a derivative, we could verify that $b_{n+1} < b_n$ directly by using the technique of Solution 1 of Example 11 in Section 8.1.

Since we are considering only positive x, we see that $f'(x) < 0$ if $2 - x^3 < 0$, that is, $x > \sqrt[3]{2}$. Thus, f is decreasing on the interval $(\sqrt[3]{2}, \infty)$. This means that $f(n + 1) < f(n)$ and therefore $b_{n+1} < b_n$ when $n \geq 2$. (The inequality $b_2 < b_1$ can be verified directly but all that really matters is that the sequence $\{b_n\}$ is eventually decreasing.)

Condition (ii) is readily verified:

$$\lim_{n \to \infty} b_n = \lim_{n \to \infty} \frac{n^2}{n^3 + 1} = \lim_{n \to \infty} \frac{\dfrac{1}{n}}{1 + \dfrac{1}{n^3}} = 0$$

Thus, the given series is convergent by the Alternating Series Test. ■ ■

The error involved in using the partial sum s_n as an approximation to the total sum s is the remainder $R_n = s - s_n$. The next theorem says that for series that satisfy the conditions of the Alternating Series Test, the size of the error is smaller than b_{n+1}, which is the absolute value of the first neglected term.

Alternating Series Estimation Theorem If $s = \sum (-1)^{n-1} b_n$ is the sum of an alternating series that satisfies

$$\text{(i) } b_{n+1} \leq b_n \qquad \text{and} \qquad \text{(ii) } \lim_{n \to \infty} b_n = 0$$

then

$$|R_n| = |s - s_n| \leq b_{n+1}$$

You can see geometrically why this is true by looking at Figure 1. Notice that $s - s_4 < b_5$, $|s - s_5| < b_6$, and so on.

Ⅴ **EXAMPLE 4** Find the sum of the series $\sum_{n=0}^{\infty} \dfrac{(-1)^n}{n!}$ correct to three decimal places. (By definition, $0! = 1$.)

SOLUTION We first observe that the series is convergent by the Alternating Series Test because

$$\text{(i)} \quad b_{n+1} = \frac{1}{(n+1)!} = \frac{1}{n!(n+1)} < \frac{1}{n!} = b_n$$

$$\text{(ii)} \quad 0 < \frac{1}{n!} < \frac{1}{n} \to 0 \quad \text{so} \quad b_n = \frac{1}{n!} \to 0 \text{ as } n \to \infty$$

To get a feel for how many terms we need to use in our approximation, let's write out the first few terms of the series:

$$s = \frac{1}{0!} - \frac{1}{1!} + \frac{1}{2!} - \frac{1}{3!} + \frac{1}{4!} - \frac{1}{5!} + \frac{1}{6!} - \frac{1}{7!} + \cdots$$

$$= 1 - 1 + \tfrac{1}{2} - \tfrac{1}{6} + \tfrac{1}{24} - \tfrac{1}{120} + \tfrac{1}{720} - \tfrac{1}{5040} + \cdots$$

Notice that

$$b_7 = \tfrac{1}{5040} < \tfrac{1}{5000} = 0.0002$$

and

$$s_6 = 1 - 1 + \tfrac{1}{2} - \tfrac{1}{6} + \tfrac{1}{24} - \tfrac{1}{120} + \tfrac{1}{720} \approx 0.368056$$

By the Alternating Series Estimation Theorem we know that

$$|s - s_6| \le b_7 < 0.0002$$

This error of less than 0.0002 does not affect the third decimal place, so we have

$$s \approx 0.368$$

correct to three decimal places. ■ ■

■ ■ In Section 8.7 we will prove that $e^x = \sum_{n=0}^{\infty} x^n/n!$ for all x, so what we have obtained in Example 4 is actually an approximation to the number e^{-1}.

⊘ **NOTE** ◦ The rule that the error (in using s_n to approximate s) is smaller than the first neglected term is, in general, valid only for alternating series that satisfy the conditions of the Alternating Series Estimation Theorem. The rule does not apply to other types of series.

▰▰ Absolute Convergence

Given any series $\Sigma\, a_n$, we can consider the corresponding series

$$\sum_{n=1}^{\infty} |a_n| = |a_1| + |a_2| + |a_3| + \cdots$$

whose terms are the absolute values of the terms of the original series.

■ ■ We have convergence tests for series with positive terms and for alternating series. But what if the signs of the terms switch back and forth irregularly? We will see in Example 7 that the idea of absolute convergence sometimes helps in such cases.

> **Definition** A series $\Sigma\, a_n$ is called **absolutely convergent** if the series of absolute values $\Sigma\, |a_n|$ is convergent.

Notice that if $\Sigma\, a_n$ is a series with positive terms, then $|a_n| = a_n$ and so absolute convergence is the same as convergence.

EXAMPLE 5 The series

$$\sum_{n=1}^{\infty} \frac{(-1)^{n-1}}{n^2} = 1 - \frac{1}{2^2} + \frac{1}{3^2} - \frac{1}{4^2} + \cdots$$

is absolutely convergent because

$$\sum_{n=1}^{\infty} \left| \frac{(-1)^{n-1}}{n^2} \right| = \sum_{n=1}^{\infty} \frac{1}{n^2} = 1 + \frac{1}{2^2} + \frac{1}{3^2} + \frac{1}{4^2} + \cdots$$

is a convergent p-series ($p = 2$). ■ ■

EXAMPLE 6 We know that the alternating harmonic series

$$\sum_{n=1}^{\infty} \frac{(-1)^{n-1}}{n} = 1 - \frac{1}{2} + \frac{1}{3} - \frac{1}{4} + \cdots$$

is convergent (see Example 1), but it is not absolutely convergent because the corresponding series of absolute values is

$$\sum_{n=1}^{\infty} \left| \frac{(-1)^{n-1}}{n} \right| = \sum_{n=1}^{\infty} \frac{1}{n} = 1 + \frac{1}{2} + \frac{1}{3} + \frac{1}{4} + \cdots$$

which is the harmonic series (p-series with $p = 1$) and is therefore divergent. ■ ■

Example 6 shows that it is possible for a series to be convergent but not absolutely convergent. However, the following theorem shows that absolute convergence implies convergence.

> **1** **Theorem** If a series $\Sigma\, a_n$ is absolutely convergent, then it is convergent.

To see why Theorem 1 is true, observe that the inequality

$$0 \leqslant a_n + |a_n| \leqslant 2|a_n|$$

is true because $|a_n|$ is either a_n or $-a_n$. If $\Sigma\, a_n$ is absolutely convergent, then $\Sigma\, |a_n|$ is convergent, so $\Sigma\, 2|a_n|$ is convergent. Therefore, by the Comparison Test, $\Sigma\, (a_n + |a_n|)$ is convergent. Then

$$\sum a_n = \sum (a_n + |a_n|) - \sum |a_n|$$

is the difference of two convergent series and is therefore convergent. ■ ■

[V] **EXAMPLE 7** Determine whether the series

$$\sum_{n=1}^{\infty} \frac{\cos n}{n^2} = \frac{\cos 1}{1^2} + \frac{\cos 2}{2^2} + \frac{\cos 3}{3^2} + \cdots$$

is convergent or divergent.

SOLUTION This series has both positive and negative terms, but it is not alternating. (The first term is positive, the next three are negative, and the following three are positive. The signs change irregularly.) We can apply the Comparison Test to the

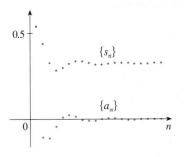

■ ■ Figure 3 shows the graphs of the terms a_n and partial sums s_n of the series in Example 7. Notice that the series is not alternating but has positive and negative terms.

FIGURE 3

series of absolute values

$$\sum_{n=1}^{\infty} \left| \frac{\cos n}{n^2} \right| = \sum_{n=1}^{\infty} \frac{|\cos n|}{n^2}$$

Since $|\cos n| \leq 1$ for all n, we have

$$\frac{|\cos n|}{n^2} \leq \frac{1}{n^2}$$

We know that $\Sigma \, 1/n^2$ is convergent (p-series with $p = 2$) and therefore $\Sigma \, |\cos n|/n^2$ is convergent by the Comparison Test. Thus, the given series $\Sigma \, (\cos n)/n^2$ is absolutely convergent and therefore convergent by Theorem 1. ■ ■

The Ratio Test

The following test is very useful in determining whether a given series is absolutely convergent.

The Ratio Test

(i) If $\displaystyle\lim_{n \to \infty} \left| \frac{a_{n+1}}{a_n} \right| = L < 1$, then the series $\displaystyle\sum_{n=1}^{\infty} a_n$ is absolutely convergent (and therefore convergent).

(ii) If $\displaystyle\lim_{n \to \infty} \left| \frac{a_{n+1}}{a_n} \right| = L > 1$ or $\displaystyle\lim_{n \to \infty} \left| \frac{a_{n+1}}{a_n} \right| = \infty$, then the series $\displaystyle\sum_{n=1}^{\infty} a_n$ is divergent.

(iii) If $\displaystyle\lim_{n \to \infty} \left| \frac{a_{n+1}}{a_n} \right| = 1$, the Ratio Test is inconclusive; that is, no conclusion can be drawn about the convergence or divergence of $\Sigma \, a_n$.

The Ratio Test can be proved by comparing the given series to a geometric series. It's understandable that geometric series are involved because, for those series, the ratio r of consecutive terms is constant and the series converges if $|r| < 1$. In part (i) of the Ratio Test, the ratio of consecutive terms isn't constant but $|a_{n+1}/a_n| \to L$ so, for large n, $|a_{n+1}/a_n|$ is almost constant and the series converges if $L < 1$.

NOTE ◦ Part (iii) of the Ratio Test says that if $\lim_{n \to \infty} |a_{n+1}/a_n| = 1$, the test gives no information. For instance, for the convergent series $\Sigma \, 1/n^2$ we have

$$\left| \frac{a_{n+1}}{a_n} \right| = \frac{\dfrac{1}{(n+1)^2}}{\dfrac{1}{n^2}} = \frac{n^2}{(n+1)^2} = \frac{1}{\left(1 + \dfrac{1}{n}\right)^2} \to 1 \qquad \text{as } n \to \infty$$

whereas for the divergent series $\Sigma \, 1/n$ we have

$$\left| \frac{a_{n+1}}{a_n} \right| = \frac{\dfrac{1}{n+1}}{\dfrac{1}{n}} = \frac{n}{n+1} = \frac{1}{1 + \dfrac{1}{n}} \to 1 \qquad \text{as } n \to \infty$$

Therefore, if $\lim_{n\to\infty} |a_{n+1}/a_n| = 1$, the series $\Sigma\, a_n$ might converge or it might diverge. In this case the Ratio Test fails and we must use some other test.

EXAMPLE 8 Test the series $\displaystyle\sum_{n=1}^{\infty} (-1)^n \frac{n^3}{3^n}$ for absolute convergence.

SOLUTION We use the Ratio Test with $a_n = (-1)^n n^3/3^n$:

$$\left| \frac{a_{n+1}}{a_n} \right| = \left| \frac{\dfrac{(-1)^{n+1}(n+1)^3}{3^{n+1}}}{\dfrac{(-1)^n n^3}{3^n}} \right| = \frac{(n+1)^3}{3^{n+1}} \cdot \frac{3^n}{n^3}$$

$$= \frac{1}{3}\left(\frac{n+1}{n} \right)^3 = \frac{1}{3}\left(1 + \frac{1}{n} \right)^3 \to \frac{1}{3} < 1$$

Thus, by the Ratio Test, the given series is absolutely convergent and therefore convergent. ■ ■

$\boxed{\text{V}}$ **EXAMPLE 9** Test the convergence of the series $\displaystyle\sum_{n=1}^{\infty} \frac{n^n}{n!}$.

SOLUTION Since the terms $a_n = n^n/n!$ are positive, we don't need the absolute value signs.

$$\frac{a_{n+1}}{a_n} = \frac{(n+1)^{n+1}}{(n+1)!} \cdot \frac{n!}{n^n} = \frac{(n+1)(n+1)^n}{(n+1)n!} \cdot \frac{n!}{n^n}$$

$$= \left(\frac{n+1}{n} \right)^n = \left(1 + \frac{1}{n} \right)^n \to e \qquad \text{as } n \to \infty$$

■ ■ Series that involve factorials or other products (including a constant raised to the nth power) are often conveniently tested using the Ratio Test.

(see Equation 3.7.6). Since $e > 1$, the given series is divergent by the Ratio Test. ■ ■

NOTE ◦ Although the Ratio Test works in Example 9, another method is to use the Test for Divergence. Since

■ ■ We now have several tests for convergence of series. So, given a series, how do you know which test to use? For advice, see the web site

www.stewartcalculus.com

Click **Additional Topics** and then on **Strategy for Testing Series**.

$$a_n = \frac{n^n}{n!} = \frac{n \cdot n \cdot n \cdot \cdots \cdot n}{1 \cdot 2 \cdot 3 \cdot \cdots \cdot n} \geq n$$

it follows that a_n does not approach 0 as $n \to \infty$. Therefore, the given series is divergent by the Test for Divergence.

8.4 **Exercises**

1. (a) What is an alternating series?
  (b) Under what conditions does an alternating series converge?
  (c) If these conditions are satisfied, what can you say about the remainder after n terms?

2. What can you say about the series $\Sigma\, a_n$ in each of the following cases?

(a) $\displaystyle\lim_{n\to\infty} \left| \frac{a_{n+1}}{a_n} \right| = 8$ (b) $\displaystyle\lim_{n\to\infty} \left| \frac{a_{n+1}}{a_n} \right| = 0.8$

(c) $\displaystyle\lim_{n\to\infty} \left| \frac{a_{n+1}}{a_n} \right| = 1$

3–8 ■ Test the series for convergence or divergence.

3. $\frac{4}{7} - \frac{4}{8} + \frac{4}{9} - \frac{4}{10} + \frac{4}{11} - \cdots$

4. $-\frac{1}{3} + \frac{2}{4} - \frac{3}{5} + \frac{4}{6} - \frac{5}{7} + \cdots$

5. $\displaystyle\sum_{n=1}^{\infty} \frac{(-1)^{n-1}}{\sqrt{n}}$

6. $\displaystyle\sum_{n=1}^{\infty} (-1)^n \frac{\sqrt{n}}{1 + 2\sqrt{n}}$

7. $\displaystyle\sum_{n=1}^{\infty} (-1)^n \frac{3n-1}{2n+1}$

8. $\displaystyle\sum_{n=1}^{\infty} (-1)^{n-1} \frac{\ln n}{n}$

9. Is the 50th partial sum s_{50} of the alternating series $\sum_{n=1}^{\infty} (-1)^{n-1}/n$ an overestimate or an underestimate of the total sum? Explain.

10. Calculate the first 10 partial sums of the series

$$\sum_{n=1}^{\infty} \frac{(-1)^{n-1}}{n^3}$$

and graph both the sequence of terms and the sequence of partial sums on the same screen. Estimate the error in using the 10th partial sum to approximate the total sum.

11. For what values of p is the following series convergent?

$$\sum_{n=1}^{\infty} \frac{(-1)^{n-1}}{n^p}$$

12–14 ■ Show that the series is convergent. How many terms of the series do we need to add in order to find the sum to the indicated accuracy?

12. $\displaystyle\sum_{n=1}^{\infty} \frac{(-1)^n}{n5^n}$ $(|\text{error}| < 0.0001)$

13. $\displaystyle\sum_{n=1}^{\infty} \frac{(-1)^{n+1}}{n^6}$ $(|\text{error}| < 0.00005)$

14. $\displaystyle\sum_{n=1}^{\infty} (-1)^{n-1} ne^{-n}$ $(|\text{error}| < 0.01)$

15–16 ■ Graph both the sequence of terms and the sequence of partial sums on the same screen. Use the graph to make a rough estimate of the sum of the series. Then use the Alternating Series Estimation Theorem to estimate the sum correct to four decimal places.

15. $\displaystyle\sum_{n=1}^{\infty} \frac{(-1)^{n-1}}{(2n-1)!}$

16. $\displaystyle\sum_{n=0}^{\infty} \frac{(-1)^n}{(2n)!}$

17–18 ■ Approximate the sum of the series correct to four decimal places.

17. $\displaystyle\sum_{n=1}^{\infty} \frac{(-1)^{n-1} n^2}{10^n}$

18. $\displaystyle\sum_{n=1}^{\infty} \frac{(-1)^n}{3^n n!}$

19–30 ■ Determine whether the series is absolutely convergent.

19. $\displaystyle\sum_{n=1}^{\infty} \frac{(-3)^n}{n^3}$

20. $\displaystyle\sum_{n=1}^{\infty} \frac{n^2}{2^n}$

21. $\displaystyle\sum_{n=0}^{\infty} \frac{(-10)^n}{n!}$

22. $\displaystyle\sum_{n=1}^{\infty} (-1)^n \frac{n}{n^2+1}$

23. $\displaystyle\sum_{n=1}^{\infty} \frac{(-1)^{n-1}}{\sqrt{n}}$

24. $\displaystyle\sum_{n=1}^{\infty} (-1)^{n-1} \frac{2^n}{n^4}$

25. $\displaystyle\sum_{n=1}^{\infty} \frac{10^n}{(n+1)4^{2n+1}}$

26. $\displaystyle\sum_{n=1}^{\infty} \frac{\sin 4n}{4^n}$

27. $\displaystyle\sum_{n=1}^{\infty} \frac{(-1)^n \arctan n}{n^2}$

28. $\displaystyle\sum_{n=1}^{\infty} \frac{(-1)^{n+1} 5^{n-1}}{(n+1)^2 4^{n+2}}$

29. $1 - \dfrac{1 \cdot 3}{3!} + \dfrac{1 \cdot 3 \cdot 5}{5!} - \dfrac{1 \cdot 3 \cdot 5 \cdot 7}{7!} + \cdots$

$+ (-1)^{n-1} \dfrac{1 \cdot 3 \cdot 5 \cdot \cdots \cdot (2n-1)}{(2n-1)!} + \cdots$

30. $\dfrac{2}{5} + \dfrac{2 \cdot 6}{5 \cdot 8} + \dfrac{2 \cdot 6 \cdot 10}{5 \cdot 8 \cdot 11} + \dfrac{2 \cdot 6 \cdot 10 \cdot 14}{5 \cdot 8 \cdot 11 \cdot 14} + \cdots$

31. The terms of a series are defined recursively by the equations

$$a_1 = 2 \qquad a_{n+1} = \frac{5n+1}{4n+3} a_n$$

Determine whether $\Sigma\, a_n$ converges or diverges.

32. A series $\Sigma\, a_n$ is defined by the equations

$$a_1 = 1 \qquad a_{n+1} = \frac{2 + \cos n}{\sqrt{n}} a_n$$

Determine whether $\Sigma\, a_n$ converges or diverges.

33. For which of the following series is the Ratio Test inconclusive (that is, it fails to give a definite answer)?

(a) $\displaystyle\sum_{n=1}^{\infty} \frac{1}{n^3}$

(b) $\displaystyle\sum_{n=1}^{\infty} \frac{n}{2^n}$

(c) $\displaystyle\sum_{n=1}^{\infty} \frac{(-3)^{n-1}}{\sqrt{n}}$

(d) $\displaystyle\sum_{n=1}^{\infty} \frac{\sqrt{n}}{1+n^2}$

34. For which positive integers k is the following series convergent?

$$\sum_{n=1}^{\infty} \frac{(n!)^2}{(kn)!}$$

35. (a) Show that $\sum_{n=0}^{\infty} x^n/n!$ converges for all x.
(b) Deduce that $\lim_{n \to \infty} x^n/n! = 0$ for all x.

36. Around 1910, the Indian mathematician Srinivasa Ramanujan discovered the formula

$$\frac{1}{\pi} = \frac{2\sqrt{2}}{9801} \sum_{n=0}^{\infty} \frac{(4n)!(1103 + 26390n)}{(n!)^4 396^{4n}}$$

William Gosper used this series in 1985 to compute the first 17 million digits of π.
(a) Verify that the series is convergent.
(b) How many correct decimal places of π do you get if you use just the first term of the series? What if you use two terms?

8.5 Power Series

A **power series** is a series of the form

$$\boxed{1} \qquad \sum_{n=0}^{\infty} c_n x^n = c_0 + c_1 x + c_2 x^2 + c_3 x^3 + \cdots$$

where x is a variable and the c_n's are constants called the **coefficients** of the series. For each fixed x, the series (1) is a series of constants that we can test for convergence or divergence. A power series may converge for some values of x and diverge for other values of x. The sum of the series is a function

$$f(x) = c_0 + c_1 x + c_2 x^2 + \cdots + c_n x^n + \cdots$$

whose domain is the set of all x for which the series converges. Notice that f resembles a polynomial. The only difference is that f has infinitely many terms.

For instance, if we take $c_n = 1$ for all n, the power series becomes the geometric series

$$\sum_{n=0}^{\infty} x^n = 1 + x + x^2 + \cdots + x^n + \cdots$$

which converges when $-1 < x < 1$ and diverges when $|x| \geq 1$ (see Equation 8.2.5).

More generally, a series of the form

$$\boxed{2} \qquad \sum_{n=0}^{\infty} c_n (x - a)^n = c_0 + c_1 (x - a) + c_2 (x - a)^2 + \cdots$$

is called a **power series in $(x - a)$** or a **power series centered at a** or a **power series about a**. Notice that in writing out the term corresponding to $n = 0$ in Equations 1 and 2 we have adopted the convention that $(x - a)^0 = 1$ even when $x = a$. Notice also that when $x = a$ all of the terms are 0 for $n \geq 1$ and so the power series (2) always converges when $x = a$.

◢ EXAMPLE 1 For what values of x is the series $\displaystyle\sum_{n=0}^{\infty} n! x^n$ convergent?

SOLUTION We use the Ratio Test. If we let a_n, as usual, denote the nth term of the series, then $a_n = n! x^n$. If $x \neq 0$, we have

$$\lim_{n \to \infty} \left| \frac{a_{n+1}}{a_n} \right| = \lim_{n \to \infty} \left| \frac{(n+1)! x^{n+1}}{n! x^n} \right| = \lim_{n \to \infty} (n+1)|x| = \infty$$

By the Ratio Test, the series diverges when $x \neq 0$. Thus, the given series converges only when $x = 0$. ■ ■

◢ EXAMPLE 2 For what values of x does the series $\displaystyle\sum_{n=1}^{\infty} \frac{(x-3)^n}{n}$ converge?

SOLUTION Let $a_n = (x-3)^n/n$. Then

$$\left| \frac{a_{n+1}}{a_n} \right| = \left| \frac{(x-3)^{n+1}}{n+1} \cdot \frac{n}{(x-3)^n} \right|$$

$$= \frac{1}{1 + \dfrac{1}{n}} |x - 3| \to |x - 3| \qquad \text{as } n \to \infty$$

■ ■ TRIGONOMETRIC SERIES

A power series is a series in which each term is a power function. A **trigonometric series**

$$\sum_{n=0}^{\infty} (a_n \cos nx + b_n \sin nx)$$

is a series whose terms are trigonometric functions. This type of series is discussed on the web site

www.stewartcalculus.com

Click on **Additional Topics** and then on **Fourier Series**.

■ ■ Notice that
$$(n+1)! = (n+1)n(n-1) \cdot \cdots \cdot 3 \cdot 2 \cdot 1$$
$$= (n+1)n!$$

By the Ratio Test, the given series is absolutely convergent, and therefore convergent, when $|x - 3| < 1$ and divergent when $|x - 3| > 1$. Now

$$|x - 3| < 1 \iff -1 < x - 3 < 1 \iff 2 < x < 4$$

so the series converges when $2 < x < 4$ and diverges when $x < 2$ or $x > 4$.

The Ratio Test gives no information when $|x - 3| = 1$ so we must consider $x = 2$ and $x = 4$ separately. If we put $x = 4$ in the series, it becomes $\Sigma\, 1/n$, the harmonic series, which is divergent. If $x = 2$, the series is $\Sigma\, (-1)^n/n$, which converges by the Alternating Series Test. Thus, the given power series converges for $2 \leqslant x < 4$. ■ ■

We will see that the main use of a power series is that it provides a way to represent some of the most important functions that arise in mathematics, physics, and chemistry. In particular, the sum of the power series in the next example is called a **Bessel function**, after the German astronomer Friedrich Bessel (1784–1846), and the function given in Exercise 23 is another example of a Bessel function. In fact, these functions first arose when Bessel solved Kepler's equation for describing planetary motion. Since that time, these functions have been applied in many different physical situations, including the temperature distribution in a circular plate and the shape of a vibrating drumhead.

Notice how closely the computer-generated model (which involves Bessel functions and cosine functions) matches the photograph of a vibrating rubber membrane.

EXAMPLE 3 Find the domain of the Bessel function of order 0 defined by

$$J_0(x) = \sum_{n=0}^{\infty} \frac{(-1)^n x^{2n}}{2^{2n}(n!)^2}$$

SOLUTION Let $a_n = (-1)^n x^{2n}/[2^{2n}(n!)^2]$. Then

$$\left| \frac{a_{n+1}}{a_n} \right| = \left| \frac{(-1)^{n+1} x^{2(n+1)}}{2^{2(n+1)}[(n+1)!]^2} \cdot \frac{2^{2n}(n!)^2}{(-1)^n x^{2n}} \right|$$

$$= \frac{x^{2n+2}}{2^{2n+2}(n+1)^2(n!)^2} \cdot \frac{2^{2n}(n!)^2}{x^{2n}}$$

$$= \frac{x^2}{4(n+1)^2} \to 0 < 1 \qquad \text{for all } x$$

Thus, by the Ratio Test, the given series converges for all values of x. In other words, the domain of the Bessel function J_0 is $(-\infty, \infty) = \mathbb{R}$. ■ ■

Recall that the sum of a series is equal to the limit of the sequence of partial sums. So when we define the Bessel function in Example 3 as the sum of a series we mean that, for every real number x,

$$J_0(x) = \lim_{n \to \infty} s_n(x) \qquad \text{where} \qquad s_n(x) = \sum_{i=0}^{n} \frac{(-1)^i x^{2i}}{2^{2i}(i!)^2}$$

The first few partial sums are

$$s_0(x) = 1 \qquad s_1(x) = 1 - \frac{x^2}{4} \qquad s_2(x) = 1 - \frac{x^2}{4} + \frac{x^4}{64}$$

$$s_3(x) = 1 - \frac{x^2}{4} + \frac{x^4}{64} - \frac{x^6}{2304} \qquad s_4(x) = 1 - \frac{x^2}{4} + \frac{x^4}{64} - \frac{x^6}{2304} + \frac{x^8}{147{,}456}$$

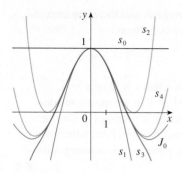

FIGURE 1

Partial sums of the Bessel function J_0

FIGURE 2

Figure 1 shows the graphs of these partial sums, which are polynomials. They are all approximations to the function J_0, but notice that the approximations become better when more terms are included. Figure 2 shows a more complete graph of the Bessel function.

For the power series that we have looked at so far, the set of values of x for which the series is convergent has always turned out to be an interval [a finite interval for the geometric series and the series in Example 2, the infinite interval $(-\infty, \infty)$ in Example 3, and a collapsed interval $[0, 0] = \{0\}$ in Example 1]. The following theorem, which we won't prove, says that this is true in general.

---

3 **Theorem** For a given power series $\sum_{n=0}^{\infty} c_n(x - a)^n$ there are only three possibilities:

(i) The series converges only when $x = a$.

(ii) The series converges for all x.

(iii) There is a positive number R such that the series converges if $|x - a| < R$ and diverges if $|x - a| > R$.

The number R in case (iii) is called the **radius of convergence** of the power series. By convention, the radius of convergence is $R = 0$ in case (i) and $R = \infty$ in case (ii). The **interval of convergence** of a power series is the interval that consists of all values of x for which the series converges. In case (i) the interval consists of just a single point a. In case (ii) the interval is $(-\infty, \infty)$. In case (iii) note that the inequality $|x - a| < R$ can be rewritten as $a - R < x < a + R$. When x is an *endpoint* of the interval, that is, $x = a \pm R$, anything can happen—the series might converge at one or both endpoints or it might diverge at both endpoints. Thus, in case (iii) there are four possibilities for the interval of convergence:

$$(a - R, a + R) \qquad (a - R, a + R] \qquad [a - R, a + R) \qquad [a - R, a + R]$$

The situation is illustrated in Figure 3.

FIGURE 3

We summarize here the radius and interval of convergence for each of the examples already considered in this section.

	Series	Radius of convergence	Interval of convergence
Geometric series	$\sum_{n=0}^{\infty} x^n$	$R = 1$	$(-1, 1)$
Example 1	$\sum_{n=0}^{\infty} n!\, x^n$	$R = 0$	$\{0\}$
Example 2	$\sum_{n=1}^{\infty} \dfrac{(x - 3)^n}{n}$	$R = 1$	$[2, 4)$
Example 3	$\sum_{n=0}^{\infty} \dfrac{(-1)^n x^{2n}}{2^{2n}(n!)^2}$	$R = \infty$	$(-\infty, \infty)$

The Ratio Test can be used to determine the radius of convergence R in most cases. The Ratio Test always fails when x is an endpoint of the interval of convergence, so the endpoints must be checked with some other test.

EXAMPLE 4 Find the radius of convergence and interval of convergence of the series

$$\sum_{n=0}^{\infty} \frac{(-3)^n x^n}{\sqrt{n+1}}$$

SOLUTION Let $a_n = (-3)^n x^n / \sqrt{n+1}$. Then

$$\left| \frac{a_{n+1}}{a_n} \right| = \left| \frac{(-3)^{n+1} x^{n+1}}{\sqrt{n+2}} \cdot \frac{\sqrt{n+1}}{(-3)^n x^n} \right| = \left| -3x \sqrt{\frac{n+1}{n+2}} \right|$$

$$= 3 \sqrt{\frac{1 + (1/n)}{1 + (2/n)}} \, |x| \; \rightarrow \; 3 \, |x| \qquad \text{as } n \rightarrow \infty$$

By the Ratio Test, the given series converges if $3\,|x| < 1$ and diverges if $3\,|x| > 1$. Thus, it converges if $|x| < \frac{1}{3}$ and diverges if $|x| > \frac{1}{3}$. This means that the radius of convergence is $R = \frac{1}{3}$.

We know the series converges in the interval $\left(-\frac{1}{3}, \frac{1}{3}\right)$, but we must now test for convergence at the endpoints of this interval. If $x = -\frac{1}{3}$, the series becomes

$$\sum_{n=0}^{\infty} \frac{(-3)^n \left(-\frac{1}{3}\right)^n}{\sqrt{n+1}} = \sum_{n=0}^{\infty} \frac{1}{\sqrt{n+1}} = \frac{1}{\sqrt{1}} + \frac{1}{\sqrt{2}} + \frac{1}{\sqrt{3}} + \frac{1}{\sqrt{4}} + \cdots$$

which diverges. (Use the Integral Test or simply observe that it is a p-series with $p = \frac{1}{2} < 1$.) If $x = \frac{1}{3}$, the series is

$$\sum_{n=0}^{\infty} \frac{(-3)^n \left(\frac{1}{3}\right)^n}{\sqrt{n+1}} = \sum_{n=0}^{\infty} \frac{(-1)^n}{\sqrt{n+1}}$$

which converges by the Alternating Series Test. Therefore, the given power series converges when $-\frac{1}{3} < x \leq \frac{1}{3}$, so the interval of convergence is $\left(-\frac{1}{3}, \frac{1}{3}\right]$. ■ ■

$\boxed{\text{V}}$ **EXAMPLE 5** Find the radius of convergence and interval of convergence of the series

$$\sum_{n=0}^{\infty} \frac{n(x+2)^n}{3^{n+1}}$$

SOLUTION If $a_n = n(x+2)^n / 3^{n+1}$, then

$$\left| \frac{a_{n+1}}{a_n} \right| = \left| \frac{(n+1)(x+2)^{n+1}}{3^{n+2}} \cdot \frac{3^{n+1}}{n(x+2)^n} \right|$$

$$= \left(1 + \frac{1}{n}\right) \frac{|x+2|}{3} \; \rightarrow \; \frac{|x+2|}{3} \qquad \text{as } n \rightarrow \infty$$

Using the Ratio Test, we see that the series converges if $|x+2|/3 < 1$ and it diverges if $|x+2|/3 > 1$. So it converges if $|x+2| < 3$ and diverges if $|x+2| > 3$. Thus, the radius of convergence is $R = 3$.

The inequality $|x + 2| < 3$ can be written as $-5 < x < 1$, so we test the series at the endpoints -5 and 1. When $x = -5$, the series is

$$\sum_{n=0}^{\infty} \frac{n(-3)^n}{3^{n+1}} = \tfrac{1}{3} \sum_{n=0}^{\infty} (-1)^n n$$

which diverges by the Test for Divergence $[(-1)^n n$ doesn't converge to 0]. When $x = 1$, the series is

$$\sum_{n=0}^{\infty} \frac{n(3)^n}{3^{n+1}} = \tfrac{1}{3} \sum_{n=0}^{\infty} n$$

which also diverges by the Test for Divergence. Thus, the series converges only when $-5 < x < 1$, so the interval of convergence is $(-5, 1)$.   ■ ■

8.5 Exercises

1. What is a power series?

2. (a) What is the radius of convergence of a power series? How do you find it?
   (b) What is the interval of convergence of a power series? How do you find it?

3–18 ■ Find the radius of convergence and interval of convergence of the series.

3. $\displaystyle\sum_{n=1}^{\infty} \frac{x^n}{\sqrt{n}}$

4. $\displaystyle\sum_{n=0}^{\infty} \frac{(-1)^n x^n}{n + 1}$

5. $\displaystyle\sum_{n=1}^{\infty} \frac{(-1)^{n-1} x^n}{n^3}$

6. $\displaystyle\sum_{n=1}^{\infty} \sqrt{n}\,x^n$

7. $\displaystyle\sum_{n=0}^{\infty} \frac{x^n}{n!}$

8. $\displaystyle\sum_{n=1}^{\infty} \frac{x^n}{n3^n}$

9. $\displaystyle\sum_{n=1}^{\infty} \frac{(-2)^n x^n}{\sqrt[4]{n}}$

10. $\displaystyle\sum_{n=1}^{\infty} \frac{x^n}{5^n n^5}$

11. $\displaystyle\sum_{n=2}^{\infty} (-1)^n \frac{x^n}{4^n \ln n}$

12. $\displaystyle\sum_{n=0}^{\infty} (-1)^n \frac{x^{2n}}{(2n)!}$

13. $\displaystyle\sum_{n=1}^{\infty} (-1)^n \frac{(x + 2)^n}{n2^n}$

14. $\displaystyle\sum_{n=1}^{\infty} \frac{(-2)^n}{\sqrt{n}} (x + 3)^n$

15. $\displaystyle\sum_{n=1}^{\infty} \frac{n}{b^n} (x - a)^n, \quad b > 0$

16. $\displaystyle\sum_{n=1}^{\infty} \frac{n(x - 4)^n}{n^3 + 1}$

17. $\displaystyle\sum_{n=1}^{\infty} n!(2x - 1)^n$

18. $\displaystyle\sum_{n=1}^{\infty} \frac{n^2 x^n}{2 \cdot 4 \cdot 6 \cdots \cdots (2n)}$

19. If $\sum_{n=0}^{\infty} c_n 4^n$ is convergent, does it follow that the following series are convergent?

(a) $\displaystyle\sum_{n=0}^{\infty} c_n(-2)^n$

(b) $\displaystyle\sum_{n=0}^{\infty} c_n(-4)^n$

20. Suppose that $\sum_{n=0}^{\infty} c_n x^n$ converges when $x = -4$ and diverges when $x = 6$. What can be said about the convergence or divergence of the following series?

(a) $\displaystyle\sum_{n=0}^{\infty} c_n$

(b) $\displaystyle\sum_{n=0}^{\infty} c_n 8^n$

(c) $\displaystyle\sum_{n=0}^{\infty} c_n(-3)^n$

(d) $\displaystyle\sum_{n=0}^{\infty} (-1)^n c_n 9^n$

21. If k is a positive integer, find the radius of convergence of the series

$$\sum_{n=0}^{\infty} \frac{(n!)^k}{(kn)!} x^n$$

22. Graph the first several partial sums $s_n(x)$ of the series $\sum_{n=0}^{\infty} x^n$, together with the sum function $f(x) = 1/(1 - x)$, on a common screen. On what interval do these partial sums appear to be converging to $f(x)$?

23. The function J_1 defined by

$$J_1(x) = \sum_{n=0}^{\infty} \frac{(-1)^n x^{2n+1}}{n!(n + 1)!2^{2n+1}}$$

is called the *Bessel function of order 1*.
   (a) Find its domain.
   (b) Graph the first several partial sums on a common screen.
   (c) If your CAS has built-in Bessel functions, graph J_1 on the same screen as the partial sums in part (b) and observe how the partial sums approximate J_1.

24. The function A defined by

$$A(x) = 1 + \frac{x^3}{2 \cdot 3} + \frac{x^6}{2 \cdot 3 \cdot 5 \cdot 6} + \frac{x^9}{2 \cdot 3 \cdot 5 \cdot 6 \cdot 8 \cdot 9} + \cdots$$

is called the *Airy function* after the English mathematician and astronomer Sir George Airy (1801–1892).

(a) Find the domain of the Airy function.

(b) Graph the first several partial sums $s_n(x)$ on a common screen.

(c) If your CAS has built-in Airy functions, graph A on the same screen as the partial sums in part (b) and observe how the partial sums approximate A.

25. A function f is defined by

$$f(x) = 1 + 2x + x^2 + 2x^3 + x^4 + \cdots$$

that is, its coefficients are $c_{2n} = 1$ and $c_{2n+1} = 2$ for all

$n \geq 0$. Find the interval of convergence of the series and find an explicit formula for $f(x)$.

26. If $f(x) = \sum_{n=0}^{\infty} c_n x^n$, where $c_{n+4} = c_n$ for all $n \geq 0$, find the interval of convergence of the series and a formula for $f(x)$.

27. Suppose the series $\Sigma\, c_n x^n$ has radius of convergence 2 and the series $\Sigma\, d_n x^n$ has radius of convergence 3. What is the radius of convergence of the series $\Sigma\, (c_n + d_n)x^n$?

28. Suppose that the radius of convergence of the power series $\Sigma\, c_n x^n$ is R. What is the radius of convergence of the power series $\Sigma\, c_n x^{2n}$?

8.6 Representations of Functions as Power Series

In this section we learn how to represent certain types of functions as sums of power series by manipulating geometric series or by differentiating or integrating such a series. You might wonder why we would ever want to express a known function as a sum of infinitely many terms. This strategy is useful for integrating functions that don't have elementary antiderivatives, for solving differential equations, and for approximating functions by polynomials. (Scientists do this to simplify the expressions they deal with; computer scientists do this to represent functions on calculators and computers.)

We start with an equation that we have seen before:

■ ■ A geometric illustration of Equation 1 is shown in Figure 1. Because the sum of a series is the limit of the sequence of partial sums, we have

$$\frac{1}{1-x} = \lim_{n \to \infty} s_n(x)$$

where

$$s_n(x) = 1 + x + x^2 + \cdots + x^n$$

is the nth partial sum. Notice that as n increases, $s_n(x)$ becomes a better approximation to $f(x)$ for $-1 < x < 1$.

$$\boxed{1} \qquad \frac{1}{1-x} = 1 + x + x^2 + x^3 + \cdots = \sum_{n=0}^{\infty} x^n \qquad |x| < 1$$

We first encountered this equation in Example 5 in Section 8.2, where we obtained it by observing that the series is a geometric series with $a = 1$ and $r = x$. But here our point of view is different. We now regard Equation 1 as expressing the function $f(x) = 1/(1-x)$ as a sum of a power series.

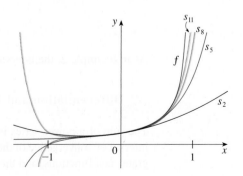

FIGURE 1

$f(x) = \dfrac{1}{1-x}$ and some partial sums

Ⅴ **EXAMPLE 1** Express $1/(1 + x^2)$ as the sum of a power series and find the interval of convergence.

SOLUTION Replacing x by $-x^2$ in Equation 1, we have

$$\frac{1}{1 + x^2} = \frac{1}{1 - (-x^2)} = \sum_{n=0}^{\infty} (-x^2)^n$$

$$= \sum_{n=0}^{\infty} (-1)^n x^{2n} = 1 - x^2 + x^4 - x^6 + x^8 - \cdots$$

Because this is a geometric series, it converges when $|-x^2| < 1$, that is, $x^2 < 1$, or $|x| < 1$. Therefore, the interval of convergence is $(-1, 1)$. (Of course, we could have determined the radius of convergence by applying the Ratio Test, but that much work is unnecessary here.) ■ ■

EXAMPLE 2 Find a power series representation for $1/(x + 2)$.

SOLUTION In order to put this function in the form of the left side of Equation 1 we first factor a 2 from the denominator:

$$\frac{1}{2 + x} = \frac{1}{2\left(1 + \dfrac{x}{2}\right)} = \frac{1}{2\left[1 - \left(-\dfrac{x}{2}\right)\right]}$$

$$= \frac{1}{2} \sum_{n=0}^{\infty} \left(-\frac{x}{2}\right)^n = \sum_{n=0}^{\infty} \frac{(-1)^n}{2^{n+1}} x^n$$

This series converges when $|-x/2| < 1$, that is, $|x| < 2$. So the interval of convergence is $(-2, 2)$. ■ ■

EXAMPLE 3 Find a power series representation of $x^3/(x + 2)$.

SOLUTION Since this function is just x^3 times the function in Example 2, all we have to do is to multiply that series by x^3:

■ ■ It's legitimate to move x^3 across the sigma sign because it doesn't depend on n. [Use Theorem 8.2.8(i) with $c = x^3$.]

$$\frac{x^3}{x + 2} = x^3 \cdot \frac{1}{x + 2} = x^3 \sum_{n=0}^{\infty} \frac{(-1)^n}{2^{n+1}} x^n = \sum_{n=0}^{\infty} \frac{(-1)^n}{2^{n+1}} x^{n+3}$$

$$= \tfrac{1}{2} x^3 - \tfrac{1}{4} x^4 + \tfrac{1}{8} x^5 - \tfrac{1}{16} x^6 + \cdots$$

Another way of writing this series is as follows:

$$\frac{x^3}{x + 2} = \sum_{n=3}^{\infty} \frac{(-1)^{n-1}}{2^{n-2}} x^n$$

As in Example 2, the interval of convergence is $(-2, 2)$. ■ ■

■■
■■ **Differentiation and Integration of Power Series**

The sum of a power series is a function $f(x) = \sum_{n=0}^{\infty} c_n(x - a)^n$ whose domain is the interval of convergence of the series. We would like to be able to differentiate and integrate such functions, and the following theorem (which we won't prove) says that we can do so by differentiating or integrating each individual term in the series, just as we would for a polynomial. This is called **term-by-term differentiation and integration**.

---

2 Theorem If the power series $\sum c_n(x - a)^n$ has radius of convergence $R > 0$, then the function f defined by

$$f(x) = c_0 + c_1(x - a) + c_2(x - a)^2 + \cdots = \sum_{n=0}^{\infty} c_n(x - a)^n$$

is differentiable (and therefore continuous) on the interval $(a - R, a + R)$ and

(i) $f'(x) = c_1 + 2c_2(x - a) + 3c_3(x - a)^2 + \cdots = \sum_{n=1}^{\infty} nc_n(x - a)^{n-1}$

(ii) $\displaystyle\int f(x)\, dx = C + c_0(x - a) + c_1 \frac{(x - a)^2}{2} + c_2 \frac{(x - a)^3}{3} + \cdots$

$$= C + \sum_{n=0}^{\infty} c_n \frac{(x - a)^{n+1}}{n + 1}$$

The radii of convergence of the power series in Equations (i) and (ii) are both R.

■ ■ In part (ii), $\int c_0\, dx = c_0 x + C_1$ is written as $c_0(x - a) + C$, where $C = C_1 + ac_0$, so all the terms of the series have the same form.

NOTE 1 ▫ Equations (i) and (ii) in Theorem 2 can be rewritten in the form

(iii) $\displaystyle\frac{d}{dx}\left[\sum_{n=0}^{\infty} c_n(x - a)^n \right] = \sum_{n=0}^{\infty} \frac{d}{dx}\left[c_n(x - a)^n \right]$

(iv) $\displaystyle\int \left[\sum_{n=0}^{\infty} c_n(x - a)^n \right] dx = \sum_{n=0}^{\infty} \int c_n(x - a)^n\, dx$

■ ■ The idea of differentiating a power series term by term is the basis for a powerful method for solving differential equations. See the web site

www.stewartcalculus.com

Click on **Additional Topics** and then on **Using Series to Solve Differential Equations**.

We know that, for finite sums, the derivative of a sum is the sum of the derivatives and the integral of a sum is the sum of the integrals. Equations (iii) and (iv) assert that the same is true for infinite sums, provided we are dealing with *power series*. (For other types of series of functions the situation is not as simple; see Exercise 34.)

NOTE 2 ▫ Although Theorem 2 says that the radius of convergence remains the same when a power series is differentiated or integrated, this does not mean that the *interval* of convergence remains the same. It may happen that the original series converges at an endpoint, whereas the differentiated series diverges there. (See Exercise 35.)

EXAMPLE 4 In Example 3 in Section 8.5 we saw that the Bessel function

$$J_0(x) = \sum_{n=0}^{\infty} \frac{(-1)^n x^{2n}}{2^{2n}(n!)^2}$$

is defined for all x. Thus, by Theorem 2, J_0 is differentiable for all x and its derivative is found by term-by-term differentiation as follows:

$$J_0'(x) = \sum_{n=0}^{\infty} \frac{d}{dx} \frac{(-1)^n x^{2n}}{2^{2n}(n!)^2} = \sum_{n=1}^{\infty} \frac{(-1)^n 2n x^{2n-1}}{2^{2n}(n!)^2}$$

■ ■

Ⓥ **EXAMPLE 5** Express $1/(1 - x)^2$ as a power series by differentiating Equation 1. What is the radius of convergence?

SOLUTION Differentiating each side of the equation

$$\frac{1}{1 - x} = 1 + x + x^2 + x^3 + \cdots = \sum_{n=0}^{\infty} x^n$$

we get

$$\frac{1}{(1 - x)^2} = 1 + 2x + 3x^2 + \cdots = \sum_{n=1}^{\infty} nx^{n-1}$$

If we wish, we can replace n by $n + 1$ and write the answer as

$$\frac{1}{(1 - x)^2} = \sum_{n=0}^{\infty} (n + 1)x^n$$

According to Theorem 2, the radius of convergence of the differentiated series is the same as the radius of convergence of the original series, namely, $R = 1$. ■ ■

EXAMPLE 6 Find a power series representation for $\ln(1 - x)$ and its radius of convergence.

SOLUTION We notice that, except for a factor of -1, the derivative of this function is $1/(1 - x)$. So we integrate both sides of Equation 1:

$$-\ln(1 - x) = \int \frac{1}{1 - x}\, dx = \int (1 + x + x^2 + \cdots)\, dx$$

$$= x + \frac{x^2}{2} + \frac{x^3}{3} + \cdots + C = \sum_{n=0}^{\infty} \frac{x^{n+1}}{n + 1} + C$$

$$= \sum_{n=1}^{\infty} \frac{x^n}{n} + C \qquad |x| < 1$$

To determine the value of C we put $x = 0$ in this equation and obtain $-\ln(1 - 0) = C$. Thus, $C = 0$ and

$$\ln(1 - x) = -x - \frac{x^2}{2} - \frac{x^3}{3} - \cdots = -\sum_{n=1}^{\infty} \frac{x^n}{n} \qquad |x| < 1$$

The radius of convergence is the same as for the original series: $R = 1$. ■ ■

Notice what happens if we put $x = \frac{1}{2}$ in the result of Example 6. Since $\ln \frac{1}{2} = -\ln 2$, we see that

$$\ln 2 = \frac{1}{2} + \frac{1}{8} + \frac{1}{24} + \frac{1}{64} + \cdots = \sum_{n=1}^{\infty} \frac{1}{n2^n}$$

Ⓥ **EXAMPLE 7** Find a power series representation for $f(x) = \tan^{-1}x$.

SOLUTION We observe that $f'(x) = 1/(1 + x^2)$ and find the required series by integrating the power series for $1/(1 + x^2)$ found in Example 1.

$$\tan^{-1}x = \int \frac{1}{1 + x^2}\, dx = \int (1 - x^2 + x^4 - x^6 + \cdots)\, dx$$

$$= C + x - \frac{x^3}{3} + \frac{x^5}{5} - \frac{x^7}{7} + \cdots$$

■ ■ The power series for $\tan^{-1}x$ obtained in Example 7 is called *Gregory's series* after the Scottish mathematician James Gregory (1638–1675), who had anticipated some of Newton's discoveries. We have shown that Gregory's series is valid when $-1 < x < 1$, but it turns out (although it isn't easy to prove) that it is also valid when $x = \pm 1$. Notice that when $x = 1$ the series becomes

$$\frac{\pi}{4} = 1 - \frac{1}{3} + \frac{1}{5} - \frac{1}{7} + \cdots$$

This beautiful result is known as the Leibniz formula for π.

To find C we put $x = 0$ and obtain $C = \tan^{-1}0 = 0$. Therefore

$$\tan^{-1}x = x - \frac{x^3}{3} + \frac{x^5}{5} - \frac{x^7}{7} + \cdots = \sum_{n=0}^{\infty} (-1)^n \frac{x^{2n+1}}{2n+1}$$

Since the radius of convergence of the series for $1/(1 + x^2)$ is 1, the radius of convergence of this series for $\tan^{-1}x$ is also 1. ■ ■

EXAMPLE 8
(a) Evaluate $\int [1/(1 + x^7)]\,dx$ as a power series.
(b) Use part (a) to approximate $\int_0^{0.5} [1/(1 + x^7)]\,dx$ correct to within 10^{-7}.

SOLUTION
(a) The first step is to express the integrand, $1/(1 + x^7)$, as the sum of a power series. As in Example 1, we start with Equation 1 and replace x by $-x^7$:

$$\frac{1}{1 + x^7} = \frac{1}{1 - (-x^7)} = \sum_{n=0}^{\infty} (-x^7)^n$$

$$= \sum_{n=0}^{\infty} (-1)^n x^{7n} = 1 - x^7 + x^{14} - \cdots$$

■ ■ This example demonstrates one way in which power series representations are useful. Integrating $1/(1 + x^7)$ by hand is incredibly difficult. Different computer algebra systems return different forms of the answer, but they are all extremely complicated. (If you have a CAS, try it yourself.) The infinite series answer that we obtain in Example 8(a) is actually much easier to deal with than the finite answer provided by a CAS.

Now we integrate term by term:

$$\int \frac{1}{1 + x^7}\,dx = \int \sum_{n=0}^{\infty} (-1)^n x^{7n}\,dx = C + \sum_{n=0}^{\infty} (-1)^n \frac{x^{7n+1}}{7n+1}$$

$$= C + x - \frac{x^8}{8} + \frac{x^{15}}{15} - \frac{x^{22}}{22} + \cdots$$

This series converges for $|-x^7| < 1$, that is, for $|x| < 1$.

(b) In applying the Evaluation Theorem it doesn't matter which antiderivative we use, so let's use the antiderivative from part (a) with $C = 0$:

$$\int_0^{0.5} \frac{1}{1 + x^7}\,dx = \left[x - \frac{x^8}{8} + \frac{x^{15}}{15} - \frac{x^{22}}{22} + \cdots \right]_0^{1/2}$$

$$= \frac{1}{2} - \frac{1}{8 \cdot 2^8} + \frac{1}{15 \cdot 2^{15}} - \frac{1}{22 \cdot 2^{22}} + \cdots + \frac{(-1)^n}{(7n+1)2^{7n+1}} + \cdots$$

This infinite series is the exact value of the definite integral, but since it is an alternating series, we can approximate the sum using the Alternating Series Estimation Theorem. If we stop adding after the term with $n = 3$, the error is smaller than the term with $n = 4$:

$$\frac{1}{29 \cdot 2^{29}} \approx 6.4 \times 10^{-11}$$

So we have

$$\int_0^{0.5} \frac{1}{1 + x^7}\,dx \approx \frac{1}{2} - \frac{1}{8 \cdot 2^8} + \frac{1}{15 \cdot 2^{15}} - \frac{1}{22 \cdot 2^{22}} \approx 0.49951374 \quad \blacksquare \blacksquare$$

8.6 Exercises

1. If the radius of convergence of the power series $\sum_{n=0}^{\infty} c_n x^n$ is 10, what is the radius of convergence of the series $\sum_{n=1}^{\infty} n c_n x^{n-1}$? Why?

2. Suppose you know that the series $\sum_{n=0}^{\infty} b_n x^n$ converges for $|x| < 2$. What can you say about the following series? Why?

$$\sum_{n=0}^{\infty} \frac{b_n}{n+1} x^{n+1}$$

3–10 ■ Find a power series representation for the function and determine the interval of convergence.

3. $f(x) = \dfrac{1}{1+x}$

4. $f(x) = \dfrac{3}{1-x^4}$

5. $f(x) = \dfrac{1}{1-x^3}$

6. $f(x) = \dfrac{1}{1+9x^2}$

7. $f(x) = \dfrac{1}{x-5}$

8. $f(x) = \dfrac{x}{4x+1}$

9. $f(x) = \dfrac{x}{9+x^2}$

10. $f(x) = \dfrac{x^2}{a^3-x^3}$

■ ■ ■ ■ ■ ■ ■ ■ ■ ■ ■ ■ ■ ■ ■

11. (a) Use differentiation to find a power series representation for

$$f(x) = \frac{1}{(1+x)^2}$$

What is the radius of convergence?

(b) Use part (a) to find a power series for

$$f(x) = \frac{1}{(1+x)^3}$$

(c) Use part (b) to find a power series for

$$f(x) = \frac{x^2}{(1+x)^3}$$

12. (a) Find a power series representation for $f(x) = \ln(1+x)$. What is the radius of convergence?

(b) Use part (a) to find a power series for $f(x) = x\ln(1+x)$.

(c) Use part (a) to find a power series for $f(x) = \ln(x^2+1)$.

13–16 ■ Find a power series representation for the function and determine the radius of convergence.

13. $f(x) = \ln(5-x)$

14. $f(x) = \dfrac{x^2}{(1-2x)^2}$

15. $f(x) = \dfrac{x^3}{(x-2)^2}$

16. $f(x) = \arctan(x/3)$

■ ■ ■ ■ ■ ■ ■ ■ ■ ■ ■ ■ ■ ■ ■

17–20 ■ Find a power series representation for f, and graph f and several partial sums $s_n(x)$ on the same screen. What happens as n increases?

17. $f(x) = \ln(3+x)$

18. $f(x) = \dfrac{1}{x^2+25}$

19. $f(x) = \ln\left(\dfrac{1+x}{1-x}\right)$

20. $f(x) = \tan^{-1}(2x)$

■ ■ ■ ■ ■ ■ ■ ■ ■ ■ ■ ■ ■ ■ ■

21–24 ■ Evaluate the indefinite integral as a power series. What is the radius of convergence?

21. $\displaystyle\int \frac{t}{1-t^8}\,dt$

22. $\displaystyle\int \frac{\ln(1-t)}{t}\,dt$

23. $\displaystyle\int \frac{x - \tan^{-1}x}{x^3}\,dx$

24. $\displaystyle\int \tan^{-1}(x^2)\,dx$

■ ■ ■ ■ ■ ■ ■ ■ ■ ■ ■ ■ ■ ■ ■

25–28 ■ Use a power series to approximate the definite integral to six decimal places.

25. $\displaystyle\int_0^{0.2} \frac{1}{1+x^5}\,dx$

26. $\displaystyle\int_0^{0.4} \ln(1+x^4)\,dx$

27. $\displaystyle\int_0^{0.1} x\arctan(3x)\,dx$

28. $\displaystyle\int_0^{0.3} \frac{x^2}{1+x^4}\,dx$

■ ■ ■ ■ ■ ■ ■ ■ ■ ■ ■ ■ ■ ■ ■

29. Use the result of Example 6 to compute $\ln 1.1$ correct to five decimal places.

30. Show that the function

$$f(x) = \sum_{n=0}^{\infty} \frac{(-1)^n x^{2n}}{(2n)!}$$

is a solution of the differential equation

$$f''(x) + f(x) = 0$$

31. (a) Show that J_0 (the Bessel function of order 0 given in Example 4) satisfies the differential equation

$$x^2 J_0''(x) + x J_0'(x) + x^2 J_0(x) = 0$$

(b) Evaluate $\int_0^1 J_0(x)\,dx$ correct to three decimal places.

32. The Bessel function of order 1 is defined by

$$J_1(x) = \sum_{n=0}^{\infty} \frac{(-1)^n x^{2n+1}}{n!\,(n+1)!\,2^{2n+1}}$$

(a) Show that J_1 satisfies the differential equation

$$x^2 J_1''(x) + x J_1'(x) + (x^2 - 1)J_1(x) = 0$$

(b) Show that $J_0'(x) = -J_1(x)$.

33. (a) Show that the function

$$f(x) = \sum_{n=0}^{\infty} \frac{x^n}{n!}$$

is a solution of the differential equation

$$f'(x) = f(x)$$

(b) Show that $f(x) = e^x$.

34. Let $f_n(x) = (\sin nx)/n^2$. Show that the series $\Sigma f_n(x)$ converges for all values of x but the series of derivatives $\Sigma f_n'(x)$ diverges when $x = 2n\pi$, n an integer. For what values of x does the series $\Sigma f_n''(x)$ converge?

35. Let

$$f(x) = \sum_{n=1}^{\infty} \frac{x^n}{n^2}$$

Find the intervals of convergence for f, f', and f''.

36. (a) Starting with the geometric series $\Sigma_{n=0}^{\infty} x^n$, find the sum of the series

$$\sum_{n=1}^{\infty} nx^{n-1} \qquad |x| < 1$$

(b) Find the sum of each of the following series.

(i) $\displaystyle\sum_{n=1}^{\infty} nx^n$, $\ |x| < 1$ \qquad (ii) $\displaystyle\sum_{n=1}^{\infty} \frac{n}{2^n}$

(c) Find the sum of each of the following series.

(i) $\displaystyle\sum_{n=2}^{\infty} n(n-1)x^n$, $\ |x| < 1$

(ii) $\displaystyle\sum_{n=2}^{\infty} \frac{n^2 - n}{2^n}$ \qquad (iii) $\displaystyle\sum_{n=1}^{\infty} \frac{n^2}{2^n}$

37. Use the power series for $\tan^{-1} x$ to prove the following expression for π as the sum of an infinite series:

$$\pi = 2\sqrt{3} \sum_{n=0}^{\infty} \frac{(-1)^n}{(2n+1)3^n}$$

38. (a) By completing the square, show that

$$\int_0^{1/2} \frac{dx}{x^2 - x + 1} = \frac{\pi}{3\sqrt{3}}$$

(b) By factoring $x^3 + 1$ as a sum of cubes, rewrite the integral in part (a). Then express $1/(x^3 + 1)$ as the sum of a power series and use it to prove the following formula for π:

$$\pi = \frac{3\sqrt{3}}{4} \sum_{n=0}^{\infty} \frac{(-1)^n}{8^n} \left(\frac{2}{3n+1} + \frac{1}{3n+2} \right)$$

| **8.7** | **Taylor and Maclaurin Series** |

In the preceding section we were able to find power series representations for a certain restricted class of functions. Here we investigate more general problems: Which functions have power series representations? How can we find such representations?

We start by supposing that f is any function that can be represented by a power series.

$$\boxed{1} \qquad f(x) = c_0 + c_1(x - a) + c_2(x - a)^2 + c_3(x - a)^3 + c_4(x - a)^4 + \cdots \qquad |x - a| < R$$

Let's try to determine what the coefficients c_n must be in terms of f. To begin, notice that if we put $x = a$ in Equation 1, then all terms after the first one are 0 and we get

$$f(a) = c_0$$

By Theorem 8.6.2, we can differentiate the series in Equation 1 term by term:

$$\boxed{2} \qquad f'(x) = c_1 + 2c_2(x - a) + 3c_3(x - a)^2 + 4c_4(x - a)^3 + \cdots \qquad |x - a| < R$$

and substitution of $x = a$ in Equation 2 gives

$$f'(a) = c_1$$

Now we differentiate both sides of Equation 2 and obtain

$$\boxed{3} \quad f''(x) = 2c_2 + 2 \cdot 3c_3(x - a) + 3 \cdot 4c_4(x - a)^2 + \cdots \qquad |x - a| < R$$

Again we put $x = a$ in Equation 3. The result is

$$f''(a) = 2c_2$$

Let's apply the procedure one more time. Differentiation of the series in Equation 3 gives

$$\boxed{4} \quad f'''(x) = 2 \cdot 3c_3 + 2 \cdot 3 \cdot 4c_4(x - a) + 3 \cdot 4 \cdot 5c_5(x - a)^2 + \cdots \qquad |x - a| < R$$

and substitution of $x = a$ in Equation 4 gives

$$f'''(a) = 2 \cdot 3c_3 = 3!c_3$$

By now you can see the pattern. If we continue to differentiate and substitute $x = a$, we obtain

$$f^{(n)}(a) = 2 \cdot 3 \cdot 4 \cdot \cdots \cdot nc_n = n!c_n$$

Solving this equation for the nth coefficient c_n, we get

$$c_n = \frac{f^{(n)}(a)}{n!}$$

This formula remains valid even for $n = 0$ if we adopt the conventions that $0! = 1$ and $f^{(0)} = f$. Thus, we have proved the following theorem.

$\boxed{5}$ **Theorem** If f has a power series representation (expansion) at a, that is, if

$$f(x) = \sum_{n=0}^{\infty} c_n(x - a)^n \qquad |x - a| < R$$

then its coefficients are given by the formula

$$c_n = \frac{f^{(n)}(a)}{n!}$$

Substituting this formula for c_n back into the series, we see that *if* f has a power series expansion at a, then it must be of the following form.

$$\boxed{6} \quad f(x) = \sum_{n=0}^{\infty} \frac{f^{(n)}(a)}{n!}(x - a)^n$$

$$= f(a) + \frac{f'(a)}{1!}(x - a) + \frac{f''(a)}{2!}(x - a)^2 + \frac{f'''(a)}{3!}(x - a)^3 + \cdots$$

The series in Equation 6 is called the **Taylor series of the function f at a** (or **about a** or **centered at a**). For the special case $a = 0$ the Taylor series becomes

$$\boxed{7} \qquad f(x) = \sum_{n=0}^{\infty} \frac{f^{(n)}(0)}{n!} x^n = f(0) + \frac{f'(0)}{1!} x + \frac{f''(0)}{2!} x^2 + \cdots$$

This case arises frequently enough that it is given the special name **Maclaurin series**.

NOTE ∘ We have shown that *if f can be represented as a power series about a, then f is equal to the sum of its Taylor series*. But there exist functions that are not equal to the sum of their Taylor series. An example of such a function is given in Exercise 56.

⟨V⟩ **EXAMPLE 1** Find the Maclaurin series of the function $f(x) = e^x$ and its radius of convergence.

SOLUTION If $f(x) = e^x$, then $f^{(n)}(x) = e^x$, so $f^{(n)}(0) = e^0 = 1$ for all n. Therefore, the Taylor series for f at 0 (that is, the Maclaurin series) is

$$\sum_{n=0}^{\infty} \frac{f^{(n)}(0)}{n!} x^n = \sum_{n=0}^{\infty} \frac{x^n}{n!} = 1 + \frac{x}{1!} + \frac{x^2}{2!} + \frac{x^3}{3!} + \cdots$$

To find the radius of convergence we let $a_n = x^n/n!$. Then

$$\left| \frac{a_{n+1}}{a_n} \right| = \left| \frac{x^{n+1}}{(n+1)!} \cdot \frac{n!}{x^n} \right| = \frac{|x|}{n+1} \to 0 < 1$$

so, by the Ratio Test, the series converges for all x and the radius of convergence is $R = \infty$. ■ ■

The conclusion we can draw from Theorem 5 and Example 1 is that *if e^x has a power series expansion at 0*, then

$$e^x = \sum_{n=0}^{\infty} \frac{x^n}{n!}$$

So how can we determine whether e^x *does* have a power series representation?

Let's investigate the more general question: Under what circumstances is a function equal to the sum of its Taylor series? In other words, if f has derivatives of all orders, when is it true that

$$f(x) = \sum_{n=0}^{\infty} \frac{f^{(n)}(a)}{n!} (x - a)^n$$

As with any convergent series, this means that $f(x)$ is the limit of the sequence of partial sums. In the case of the Taylor series, the partial sums are

$$T_n(x) = \sum_{i=0}^{n} \frac{f^{(i)}(a)}{i!} (x - a)^i$$

$$= f(a) + \frac{f'(a)}{1!} (x - a) + \frac{f''(a)}{2!} (x - a)^2 + \cdots + \frac{f^{(n)}(a)}{n!} (x - a)^n$$

■ ■ The Taylor series is named after the English mathematician Brook Taylor (1685–1731) and the Maclaurin series is named in honor of the Scottish mathematician Colin Maclaurin (1698–1746) despite the fact that the Maclaurin series is really just a special case of the Taylor series. But the idea of representing particular functions as sums of power series goes back to Newton, and the general Taylor series was known to the Scottish mathematician James Gregory in 1668 and to the Swiss mathematician John Bernoulli in the 1690s. Taylor was apparently unaware of the work of Gregory and Bernoulli when he published his discoveries on series in 1715 in his book *Methodus incrementorum directa et inversa*. Maclaurin series are named after Colin Maclaurin because he popularized them in his calculus textbook *Treatise of Fluxions* published in 1742.

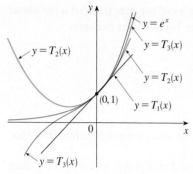

FIGURE 1

■ ■ As n increases, $T_n(x)$ appears to approach e^x in Figure 1. This suggests that e^x is equal to the sum of its Taylor series.

Notice that T_n is a polynomial of degree n called the **nth-degree Taylor polynomial of f at a**. For instance, for the exponential function $f(x) = e^x$, the result of Example 1 shows that the Taylor polynomials at 0 (or Maclaurin polynomials) with $n = 1, 2$, and 3 are

$$T_1(x) = 1 + x \qquad T_2(x) = 1 + x + \frac{x^2}{2!} \qquad T_3(x) = 1 + x + \frac{x^2}{2!} + \frac{x^3}{3!}$$

The graphs of the exponential function and these three Taylor polynomials are drawn in Figure 1.

In general, $f(x)$ is the sum of its Taylor series if

$$f(x) = \lim_{n \to \infty} T_n(x)$$

If we let

$$R_n(x) = f(x) - T_n(x) \qquad \text{so that} \qquad f(x) = T_n(x) + R_n(x)$$

then $R_n(x)$ is called the **remainder** of the Taylor series. If we can somehow show that $\lim_{n \to \infty} R_n(x) = 0$, then it follows that

$$\lim_{n \to \infty} T_n(x) = \lim_{n \to \infty} [f(x) - R_n(x)] = f(x) - \lim_{n \to \infty} R_n(x) = f(x)$$

We have therefore proved the following.

8 Theorem If $f(x) = T_n(x) + R_n(x)$, where T_n is the nth-degree Taylor polynomial of f at a and

$$\lim_{n \to \infty} R_n(x) = 0$$

for $|x - a| < R$, then f is equal to the sum of its Taylor series on the interval $|x - a| < R$.

In trying to show that $\lim_{n \to \infty} R_n(x) = 0$ for a specific function f, we usually use the following fact.

9 Taylor's Inequality If $|f^{(n+1)}(x)| \leq M$ for $|x - a| \leq d$, then the remainder $R_n(x)$ of the Taylor series satisfies the inequality

$$|R_n(x)| \leq \frac{M}{(n + 1)!} |x - a|^{n+1} \qquad \text{for } |x - a| \leq d$$

To see why this is true for $n = 1$, we assume that $|f''(x)| \leq M$. In particular, we have $f''(x) \leq M$, so for $a \leq x \leq a + d$ we have

$$\int_a^x f''(t) \, dt \leq \int_a^x M \, dt$$

An antiderivative of f'' is f', so by the Evaluation Theorem, we have

$$f'(x) - f'(a) \leq M(x - a) \qquad \text{or} \qquad f'(x) \leq f'(a) + M(x - a)$$

■ ■ As alternatives to Taylor's Inequality, we have the following formulas for the remainder term. If $f^{(n+1)}$ is continuous on an interval I and $x \in I$, then

$$R_n(x) = \frac{1}{n!} \int_a^x (x - t)^n f^{(n+1)}(t)\, dt$$

This is called the **integral form of the remainder term**. Another formula, called **Lagrange's form of the remainder term**, states that there is a number z between x and a such that

$$R_n(x) = \frac{f^{(n+1)}(z)}{(n + 1)!}(x - a)^{n+1}$$

This version is an extension of the Mean Value Theorem (which is the case $n = 0$).

Proofs of these formulas, together with discussions of how to use them to solve the examples of Sections 8.7 and 8.9, are given on the web site

www.stewartcalculus.com

Click on **Additional Topics** and then on **Formulas for the Remainder Term in Taylor series**.

Thus

$$\int_a^x f'(t)\, dt \leqslant \int_a^x [f'(a) + M(t - a)]\, dt$$

$$f(x) - f(a) \leqslant f'(a)(x - a) + M\frac{(x - a)^2}{2}$$

$$f(x) - f(a) - f'(a)(x - a) \leqslant \frac{M}{2}(x - a)^2$$

But $R_1(x) = f(x) - T_1(x) = f(x) - f(a) - f'(a)(x - a)$. So

$$R_1(x) \leqslant \frac{M}{2}(x - a)^2$$

A similar argument, using $f''(x) \geqslant -M$, shows that

$$R_1(x) \geqslant -\frac{M}{2}(x - a)^2$$

So

$$|R_1(x)| \leqslant \frac{M}{2}|x - a|^2$$

Although we have assumed that $x > a$, similar calculations show that this inequality is also true for $x < a$.

This proves Taylor's Inequality for the case where $n = 1$. The result for any n is proved in a similar way by integrating $n + 1$ times. (See Exercise 55 for the case $n = 2$.)

NOTE ▫ In Section 8.9 we will explore the use of Taylor's Inequality in approximating functions. Our immediate use of it is in conjunction with Theorem 8.

In applying Theorems 8 and 9 it is often helpful to make use of the following fact.

10
$$\lim_{n \to \infty} \frac{x^n}{n!} = 0 \qquad \text{for every real number } x$$

This is true because we know from Example 1 that the series $\sum x^n/n!$ converges for all x and so its nth term approaches 0.

EXAMPLE 2 Prove that e^x is equal to the sum of its Maclaurin series.

SOLUTION If $f(x) = e^x$, then $f^{(n+1)}(x) = e^x$ for all n. If d is any positive number and $|x| \leqslant d$, then $|f^{(n+1)}(x)| = e^x \leqslant e^d$. So Taylor's Inequality, with $a = 0$ and $M = e^d$, says that

$$|R_n(x)| \leqslant \frac{e^d}{(n + 1)!}|x|^{n+1} \qquad \text{for } |x| \leqslant d$$

Notice that the same constant $M = e^d$ works for every value of n. But, from Equation 10, we have

$$\lim_{n \to \infty} \frac{e^d}{(n + 1)!}|x|^{n+1} = e^d \lim_{n \to \infty} \frac{|x|^{n+1}}{(n + 1)!} = 0$$

It follows from the Squeeze Theorem that $\lim_{n \to \infty} |R_n(x)| = 0$ and therefore $\lim_{n \to \infty} R_n(x) = 0$ for all values of x. By Theorem 8, e^x is equal to the sum of its Maclaurin series, that is,

$$\boxed{11} \qquad e^x = \sum_{n=0}^{\infty} \frac{x^n}{n!} \qquad \text{for all } x$$

■ ■

In particular, if we put $x = 1$ in Equation 11, we obtain the following expression for the number e as a sum of an infinite series:

$$\boxed{12} \qquad e = \sum_{n=0}^{\infty} \frac{1}{n!} = 1 + \frac{1}{1!} + \frac{1}{2!} + \frac{1}{3!} + \cdots$$

■ ■ In 1748 Leonard Euler used Equation 12 to find the value of e correct to 23 digits. In 2003 Shigeru Kondo, again using the series in (12), computed e to more than fifty billion decimal places. The special techniques employed to speed up the computation are explained on the web page

numbers.computation.free.fr

EXAMPLE 3 Find the Taylor series for $f(x) = e^x$ at $a = 2$.

SOLUTION We have $f^{(n)}(2) = e^2$ and so, putting $a = 2$ in the definition of a Taylor series (6), we get

$$\sum_{n=0}^{\infty} \frac{f^{(n)}(2)}{n!} (x - 2)^n = \sum_{n=0}^{\infty} \frac{e^2}{n!} (x - 2)^n$$

Again it can be verified, as in Example 1, that the radius of convergence is $R = \infty$. As in Example 2 we can verify that $\lim_{n \to \infty} R_n(x) = 0$, so

$$\boxed{13} \qquad e^x = \sum_{n=0}^{\infty} \frac{e^2}{n!} (x - 2)^n \qquad \text{for all } x$$

■ ■

We have two power series expansions for e^x, the Maclaurin series in Equation 11 and the Taylor series in Equation 13. The first is better if we are interested in values of x near 0 and the second is better if x is near 2.

EXAMPLE 4 Find the Maclaurin series for $\sin x$ and prove that it represents $\sin x$ for all x.

SOLUTION We arrange our computation in two columns as follows:

$$f(x) = \sin x \qquad\qquad f(0) = 0$$
$$f'(x) = \cos x \qquad\qquad f'(0) = 1$$
$$f''(x) = -\sin x \qquad\qquad f''(0) = 0$$
$$f'''(x) = -\cos x \qquad\qquad f'''(0) = -1$$
$$f^{(4)}(x) = \sin x \qquad\qquad f^{(4)}(0) = 0$$

Since the derivatives repeat in a cycle of four, we can write the Maclaurin series as follows:

$$f(0) + \frac{f'(0)}{1!} x + \frac{f''(0)}{2!} x^2 + \frac{f'''(0)}{3!} x^3 + \cdots$$

$$= x - \frac{x^3}{3!} + \frac{x^5}{5!} - \frac{x^7}{7!} + \cdots = \sum_{n=0}^{\infty} (-1)^n \frac{x^{2n+1}}{(2n + 1)!}$$

▪ ▪ Figure 2 shows the graph of $\sin x$ together with its Taylor (or Maclaurin) polynomials

$$T_1(x) = x$$

$$T_3(x) = x - \frac{x^3}{3!}$$

$$T_5(x) = x - \frac{x^3}{3!} + \frac{x^5}{5!}$$

Notice that, as n increases, $T_n(x)$ becomes a better approximation to $\sin x$.

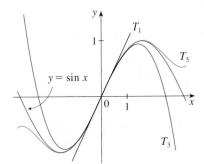

FIGURE 2

Since $f^{(n+1)}(x)$ is $\pm\sin x$ or $\pm\cos x$, we know that $\left| f^{(n+1)}(x) \right| \leq 1$ for all x. So we can take $M = 1$ in Taylor's Inequality:

$$\boxed{14} \qquad \left| R_n(x) \right| \leq \frac{M}{(n+1)!} \left| x^{n+1} \right| = \frac{|x|^{n+1}}{(n+1)!}$$

By Equation 10 the right side of this inequality approaches 0 as $n \to \infty$, so $\left| R_n(x) \right| \to 0$ by the Squeeze Theorem. It follows that $R_n(x) \to 0$ as $n \to \infty$, so $\sin x$ is equal to the sum of its Maclaurin series by Theorem 8. ▪ ▪

We state the result of Example 4 for future reference.

$$\boxed{15} \qquad \sin x = x - \frac{x^3}{3!} + \frac{x^5}{5!} - \frac{x^7}{7!} + \cdots$$

$$= \sum_{n=0}^{\infty} (-1)^n \frac{x^{2n+1}}{(2n+1)!} \qquad \text{for all } x$$

EXAMPLE 5 Find the Maclaurin series for $\cos x$.

SOLUTION We could proceed directly as in Example 4 but it's easier to differentiate the Maclaurin series for $\sin x$ given by Equation 15:

$$\cos x = \frac{d}{dx}(\sin x) = \frac{d}{dx}\left(x - \frac{x^3}{3!} + \frac{x^5}{5!} - \frac{x^7}{7!} + \cdots \right)$$

$$= 1 - \frac{3x^2}{3!} + \frac{5x^4}{5!} - \frac{7x^6}{7!} + \cdots = 1 - \frac{x^2}{2!} + \frac{x^4}{4!} - \frac{x^6}{6!} + \cdots$$

Since the Maclaurin series for $\sin x$ converges for all x, Theorem 8.6.2 tells us that the differentiated series for $\cos x$ also converges for all x. Thus

▪ ▪ The Maclaurin series for e^x, $\sin x$, and $\cos x$ that we found in Examples 2, 4, and 5 were discovered, using different methods, by Newton. These equations are remarkable because they say we know everything about each of these functions if we know all its derivatives at the single number 0.

$$\boxed{16} \qquad \cos x = 1 - \frac{x^2}{2!} + \frac{x^4}{4!} - \frac{x^6}{6!} + \cdots$$

$$= \sum_{n=0}^{\infty} (-1)^n \frac{x^{2n}}{(2n)!} \qquad \text{for all } x$$

EXAMPLE 6 Find the Maclaurin series for the function $f(x) = x\cos x$.

SOLUTION Instead of computing derivatives and substituting in Equation 7, it's easier to multiply the series for $\cos x$ (Equation 16) by x:

$$x\cos x = x \sum_{n=0}^{\infty} (-1)^n \frac{x^{2n}}{(2n)!} = \sum_{n=0}^{\infty} (-1)^n \frac{x^{2n+1}}{(2n)!}$$

EXAMPLE 7 Represent $f(x) = \sin x$ as the sum of its Taylor series centered at $\pi/3$.

SOLUTION Arranging our work in columns, we have

$$f(x) = \sin x \qquad f\left(\frac{\pi}{3}\right) = \frac{\sqrt{3}}{2}$$

$$f'(x) = \cos x \qquad f'\left(\frac{\pi}{3}\right) = \frac{1}{2}$$

$$f''(x) = -\sin x \qquad f''\left(\frac{\pi}{3}\right) = -\frac{\sqrt{3}}{2}$$

$$f'''(x) = -\cos x \qquad f'''\left(\frac{\pi}{3}\right) = -\frac{1}{2}$$

and this pattern repeats indefinitely. Therefore, the Taylor series at $\pi/3$ is

$$f\left(\frac{\pi}{3}\right) + \frac{f'\left(\frac{\pi}{3}\right)}{1!}\left(x - \frac{\pi}{3}\right) + \frac{f''\left(\frac{\pi}{3}\right)}{2!}\left(x - \frac{\pi}{3}\right)^2 + \frac{f'''\left(\frac{\pi}{3}\right)}{3!}\left(x - \frac{\pi}{3}\right)^3 + \cdots$$

$$= \frac{\sqrt{3}}{2} + \frac{1}{2 \cdot 1!}\left(x - \frac{\pi}{3}\right) - \frac{\sqrt{3}}{2 \cdot 2!}\left(x - \frac{\pi}{3}\right)^2 - \frac{1}{2 \cdot 3!}\left(x - \frac{\pi}{3}\right)^3 + \cdots$$

■ ■ We have obtained two different series representations for $\sin x$, the Maclaurin series in Example 4 and the Taylor series in Example 7. It is best to use the Maclaurin series for values of x near 0 and the Taylor series for x near $\pi/3$. Notice that the third Taylor polynomial T_3 in Figure 3 is a good approximation to $\sin x$ near $\pi/3$ but not as good near 0. Compare it with the third Maclaurin polynomial T_3 in Figure 2, where the opposite is true.

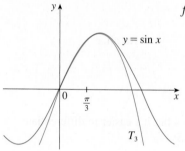

FIGURE 3

The proof that this series represents $\sin x$ for all x is very similar to that in Example 4. [Just replace x by $x - \pi/3$ in (14).] We can write the series in sigma notation if we separate the terms that contain $\sqrt{3}$:

$$\sin x = \sum_{n=0}^{\infty} \frac{(-1)^n \sqrt{3}}{2(2n)!}\left(x - \frac{\pi}{3}\right)^{2n} + \sum_{n=0}^{\infty} \frac{(-1)^n}{2(2n + 1)!}\left(x - \frac{\pi}{3}\right)^{2n+1}$$

■ ■

The power series that we obtained by indirect methods in Examples 5 and 6 and in Section 8.6 are indeed the Taylor or Maclaurin series of the given functions because Theorem 5 asserts that, no matter how we obtain a power series representation $f(x) = \Sigma\, c_n(x - a)^n$, it is always true that $c_n = f^{(n)}(a)/n!$. In other words, the coefficients are uniquely determined.

We collect in the following table, for future reference, some important Maclaurin series that we have derived in this section and the preceding one.

Important Maclaurin series and their intervals of convergence

$\dfrac{1}{1 - x} = \displaystyle\sum_{n=0}^{\infty} x^n = 1 + x + x^2 + x^3 + \cdots$	$(-1, 1)$
$e^x = \displaystyle\sum_{n=0}^{\infty} \frac{x^n}{n!} = 1 + \frac{x}{1!} + \frac{x^2}{2!} + \frac{x^3}{3!} + \cdots$	$(-\infty, \infty)$
$\sin x = \displaystyle\sum_{n=0}^{\infty} (-1)^n \frac{x^{2n+1}}{(2n + 1)!} = x - \frac{x^3}{3!} + \frac{x^5}{5!} - \frac{x^7}{7!} + \cdots$	$(-\infty, \infty)$
$\cos x = \displaystyle\sum_{n=0}^{\infty} (-1)^n \frac{x^{2n}}{(2n)!} = 1 - \frac{x^2}{2!} + \frac{x^4}{4!} - \frac{x^6}{6!} + \cdots$	$(-\infty, \infty)$
$\tan^{-1}x = \displaystyle\sum_{n=0}^{\infty} (-1)^n \frac{x^{2n+1}}{2n + 1} = x - \frac{x^3}{3} + \frac{x^5}{5} - \frac{x^7}{7} + \cdots$	$[-1, 1]$

Module 8.7/8.9 enables you to see how successive Taylor polynomials approach the original function.

One reason that Taylor series are important is that they enable us to integrate functions that we couldn't previously handle. In fact, in the introduction to this chapter we mentioned that Newton often integrated functions by first expressing them as power series and then integrating the series term by term. The function $f(x) = e^{-x^2}$ can't be integrated by techniques discussed so far because its antiderivative is not an elementary function (see Section 5.8). In the following example we use Newton's idea to integrate this function.

EXAMPLE 8

(a) Evaluate $\int e^{-x^2}\, dx$ as an infinite series.

(b) Evaluate $\int_0^1 e^{-x^2}\, dx$ correct to within an error of 0.001.

SOLUTION

(a) First we find the Maclaurin series for $f(x) = e^{-x^2}$. Although it's possible to use the direct method, let's find it simply by replacing x with $-x^2$ in the series for e^x given in the table of Maclaurin series. Thus, for all values of x,

$$e^{-x^2} = \sum_{n=0}^{\infty} \frac{(-x^2)^n}{n!} = \sum_{n=0}^{\infty} (-1)^n \frac{x^{2n}}{n!} = 1 - \frac{x^2}{1!} + \frac{x^4}{2!} - \frac{x^6}{3!} + \cdots$$

Now we integrate term by term:

$$\int e^{-x^2}\, dx = \int \left(1 - \frac{x^2}{1!} + \frac{x^4}{2!} - \frac{x^6}{3!} + \cdots + (-1)^n \frac{x^{2n}}{n!} + \cdots \right) dx$$

$$= C + x - \frac{x^3}{3 \cdot 1!} + \frac{x^5}{5 \cdot 2!} - \frac{x^7}{7 \cdot 3!} + \cdots + (-1)^n \frac{x^{2n+1}}{(2n+1)n!} + \cdots$$

This series converges for all x because the original series for e^{-x^2} converges for all x.

(b) The Evaluation Theorem gives

■ ■ We can take $C = 0$ in the antiderivative in part (a).

$$\int_0^1 e^{-x^2}\, dx = \left[x - \frac{x^3}{3 \cdot 1!} + \frac{x^5}{5 \cdot 2!} - \frac{x^7}{7 \cdot 3!} + \frac{x^9}{9 \cdot 4!} - \cdots \right]_0^1$$

$$= 1 - \tfrac{1}{3} + \tfrac{1}{10} - \tfrac{1}{42} + \tfrac{1}{216} - \cdots$$

$$\approx 1 - \tfrac{1}{3} + \tfrac{1}{10} - \tfrac{1}{42} + \tfrac{1}{216} \approx 0.7475$$

The Alternating Series Estimation Theorem shows that the error involved in this approximation is less than

$$\frac{1}{11 \cdot 5!} = \frac{1}{1320} < 0.001$$

Another use of Taylor series is illustrated in the next example. The limit could be found with l'Hospital's Rule, but instead we use a series.

EXAMPLE 9 Evaluate $\displaystyle\lim_{x \to 0} \frac{e^x - 1 - x}{x^2}$.

SOLUTION Using the Maclaurin series for e^x, we have

◼◼ Some computer algebra systems compute
limits in this way.

$$\lim_{x \to 0} \frac{e^x - 1 - x}{x^2} = \lim_{x \to 0} \frac{\left(1 + \dfrac{x}{1!} + \dfrac{x^2}{2!} + \dfrac{x^3}{3!} + \cdots\right) - 1 - x}{x^2}$$

$$= \lim_{x \to 0} \frac{\dfrac{x^2}{2!} + \dfrac{x^3}{3!} + \dfrac{x^4}{4!} + \cdots}{x^2}$$

$$= \lim_{x \to 0} \left(\frac{1}{2} + \frac{x}{3!} + \frac{x^2}{4!} + \frac{x^3}{5!} + \cdots\right) = \frac{1}{2}$$

because power series are continuous functions. ◼◼

◼◼ **Multiplication and Division of Power Series**

If power series are added or subtracted, they behave like polynomials (Theorem 8.2.8 shows this). In fact, as the following example illustrates, they can also be multiplied and divided like polynomials. We find only the first few terms because the calculations for the later terms become tedious and the initial terms are the most important ones.

EXAMPLE 10 Find the first three nonzero terms in the Maclaurin series for (a) $e^x \sin x$ and (b) $\tan x$.

SOLUTION

(a) Using the Maclaurin series for e^x and $\sin x$ in the table on page 612, we have

$$e^x \sin x = \left(1 + \frac{x}{1!} + \frac{x^2}{2!} + \frac{x^3}{3!} + \cdots\right)\left(x - \frac{x^3}{3!} + \cdots\right)$$

We multiply these expressions, collecting like terms just as for polynomials:

$$
\begin{array}{r}
1 + x + \frac{1}{2}x^2 + \frac{1}{6}x^3 + \cdots \\
x \qquad\qquad - \frac{1}{6}x^3 + \cdots \\
\hline
x + x^2 + \frac{1}{2}x^3 + \frac{1}{6}x^4 + \cdots \\
- \frac{1}{6}x^3 - \frac{1}{6}x^4 - \cdots \\
\hline
x + x^2 + \frac{1}{3}x^3 + \cdots
\end{array}
$$

Thus $\qquad\qquad e^x \sin x = x + x^2 + \frac{1}{3}x^3 + \cdots$

(b) Using the Maclaurin series in the table, we have

$$\tan x = \frac{\sin x}{\cos x} = \frac{x - \dfrac{x^3}{3!} + \dfrac{x^5}{5!} - \cdots}{1 - \dfrac{x^2}{2!} + \dfrac{x^4}{4!} - \cdots}$$

We use a procedure like long division:

$$
\begin{array}{r}
x + \tfrac{1}{3}x^3 + \tfrac{2}{15}x^5 + \cdots \\
1 - \tfrac{1}{2}x^2 + \tfrac{1}{24}x^4 - \cdots \overline{\smash{\big)}\ x - \tfrac{1}{6}x^3 + \tfrac{1}{120}x^5 - \cdots} \\
x - \tfrac{1}{2}x^3 + \tfrac{1}{24}x^5 - \cdots \\
\hline
\tfrac{1}{3}x^3 - \tfrac{1}{30}x^5 + \cdots \\
\tfrac{1}{3}x^3 - \tfrac{1}{6}x^5 + \cdots \\
\hline
\tfrac{2}{15}x^5 + \cdots
\end{array}
$$

Thus $\qquad\qquad \tan x = x + \tfrac{1}{3}x^3 + \tfrac{2}{15}x^5 + \cdots$ ■ ■

Although we have not attempted to justify the formal manipulations used in Example 10, they are legitimate. There is a theorem which states that if both $f(x) = \Sigma\, c_n x^n$ and $g(x) = \Sigma\, b_n x^n$ converge for $|x| < R$ and the series are multiplied as if they were polynomials, then the resulting series also converges for $|x| < R$ and represents $f(x)g(x)$. For division we require $b_0 \neq 0$; the resulting series converges for sufficiently small $|x|$.

8.7 Exercises

1. If $f(x) = \Sigma_{n=0}^{\infty} b_n(x - 5)^n$ for all x, write a formula for b_8.

2. (a) The graph of f is shown. Explain why the series

$$1.6 - 0.8(x - 1) + 0.4(x - 1)^2 - 0.1(x - 1)^3 + \cdots$$

is *not* the Taylor series of f centered at 1.

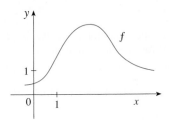

(b) Explain why the series

$$2.8 + 0.5(x - 2) + 1.5(x - 2)^2 - 0.1(x - 2)^3 + \cdots$$

is *not* the Taylor series of f centered at 2.

3. If $f^{(n)}(0) = (n + 1)!$ for $n = 0, 1, 2, \ldots$, find the Maclaurin series for f and its radius of convergence.

4. Find the Taylor series for f centered at 4 if

$$f^{(n)}(4) = \frac{(-1)^n n!}{3^n(n + 1)}$$

What is the radius of convergence of the Taylor series?

5–8 ■ Find the Maclaurin series for $f(x)$ using the definition of a Maclaurin series. [Assume that f has a power series expansion. Do not show that $R_n(x) \to 0$.] Also find the associated radius of convergence.

5. $f(x) = \cos x$ **6.** $f(x) = \sin 2x$

7. $f(x) = e^{5x}$ **8.** $f(x) = xe^x$

■ ■ ■ ■ ■ ■ ■ ■ ■ ■ ■ ■ ■

9–16 ■ Find the Taylor series for $f(x)$ centered at the given value of a. [Assume that f has a power series expansion. Do not show that $R_n(x) \to 0$.]

9. $f(x) = 1 + x + x^2,\quad a = 2$

10. $f(x) = x^3,\quad a = -1$

11. $f(x) = e^x,\quad a = 3$ **12.** $f(x) = \ln x,\quad a = 2$

13. $f(x) = \cos x,\quad a = \pi$ **14.** $f(x) = \sin x,\quad a = \pi/2$

15. $f(x) = 1/\sqrt{x},\quad a = 9$ **16.** $f(x) = x^{-2},\quad a = 1$

■ ■ ■ ■ ■ ■ ■ ■ ■ ■ ■ ■ ■

17. Prove that the series obtained in Exercise 5 represents $\cos x$ for all x.

18. Prove that the series obtained in Exercise 14 represents $\sin x$ for all x.

19–26 ■ Use a Maclaurin series derived in this section to obtain the Maclaurin series for the given function.

19. $f(x) = \cos \pi x$ **20.** $f(x) = e^{-x/2}$

21. $f(x) = x \tan^{-1}x$ **22.** $f(x) = \sin(x^4)$

23. $f(x) = x^2 e^{-x}$ **24.** $f(x) = x \cos 2x$

25. $f(x) = \sin^2 x$ [*Hint:* Use $\sin^2 x = \frac{1}{2}(1 - \cos 2x)$.]

26. $f(x) = \begin{cases} \dfrac{x - \sin x}{x^3} & \text{if } x \neq 0 \\ \dfrac{1}{6} & \text{if } x = 0 \end{cases}$

27–30 ■ Find the Maclaurin series of f (by any method) and its radius of convergence. Graph f and its first few Taylor polynomials on the same screen. What do you notice about the relationship between these polynomials and f?

27. $f(x) = \sqrt{1 + x}$ **28.** $f(x) = e^{-x^2} + \cos x$

29. $f(x) = \cos(x^2)$ **30.** $f(x) = 2^x$

31. Use the Maclaurin series for e^x to calculate $e^{-0.2}$ correct to five decimal places.

32. Use the Maclaurin series for $\sin x$ to compute $\sin 3°$ correct to five decimal places.

33–36 ■ Evaluate the indefinite integral as an infinite series.

33. $\displaystyle\int x \cos(x^3)\, dx$ **34.** $\displaystyle\int \frac{\sin x}{x}\, dx$

35. $\displaystyle\int \sqrt{x^3 + 1}\, dx$ **36.** $\displaystyle\int \frac{e^x - 1}{x}\, dx$

37–40 ■ Use series to approximate the definite integral to within the indicated accuracy.

37. $\displaystyle\int_0^1 x \cos(x^3)\, dx$ (three decimal places)

38. $\displaystyle\int_0^{0.2} [\tan^{-1}(x^3) + \sin(x^3)]\, dx$ (five decimal places)

39. $\displaystyle\int_0^{0.1} \frac{dx}{\sqrt{1 + x^3}}$ ($|\,\text{error}\,| < 10^{-8}$)

40. $\displaystyle\int_0^{0.5} x^2 e^{-x^2}\, dx$ ($|\,\text{error}\,| < 0.001$)

41–43 ■ Use series to evaluate the limit.

41. $\displaystyle\lim_{x \to 0} \frac{x - \tan^{-1} x}{x^3}$ **42.** $\displaystyle\lim_{x \to 0} \frac{1 - \cos x}{1 + x - e^x}$

43. $\displaystyle\lim_{x \to 0} \frac{\sin x - x + \frac{1}{6}x^3}{x^5}$

44. Use the series in Example 10(b) to evaluate

$$\lim_{x \to 0} \frac{\tan x - x}{x^3}$$

We found this limit in Example 4 in Section 4.5 using l'Hospital's Rule three times. Which method do you prefer?

45–48 ■ Use multiplication or division of power series to find the first three nonzero terms in the Maclaurin series for each function.

45. $y = e^{-x^2} \cos x$ **46.** $y = \sec x$

47. $y = \dfrac{x}{\sin x}$ **48.** $y = e^x \ln(1 - x)$

49–54 ■ Find the sum of the series.

49. $\displaystyle\sum_{n=0}^{\infty} (-1)^n \frac{x^{4n}}{n!}$ **50.** $\displaystyle\sum_{n=0}^{\infty} \frac{(-1)^n \pi^{2n}}{6^{2n}(2n)!}$

51. $\displaystyle\sum_{n=0}^{\infty} \frac{(-1)^n \pi^{2n+1}}{4^{2n+1}(2n + 1)!}$ **52.** $\displaystyle\sum_{n=0}^{\infty} \frac{3^n}{5^n n!}$

53. $3 + \dfrac{9}{2!} + \dfrac{27}{3!} + \dfrac{81}{4!} + \cdots$

54. $1 - \ln 2 + \dfrac{(\ln 2)^2}{2!} - \dfrac{(\ln 2)^3}{3!} + \cdots$

55. Prove Taylor's Inequality for $n = 2$, that is, prove that if $|f'''(x)| \leq M$ for $|x - a| \leq d$, then

$$|R_2(x)| \leq \frac{M}{6} |x - a|^3 \quad \text{for } |x - a| \leq d$$

56. (a) Show that the function defined by

$$f(x) = \begin{cases} e^{-1/x^2} & \text{if } x \neq 0 \\ 0 & \text{if } x = 0 \end{cases}$$

is not equal to its Maclaurin series.
(b) Graph the function in part (a) and comment on its behavior near the origin.

[CAS] **An Elusive Limit**

This project deals with the function

$$f(x) = \frac{\sin(\tan x) - \tan(\sin x)}{\arcsin(\arctan x) - \arctan(\arcsin x)}$$

1. Use your computer algebra system to evaluate $f(x)$ for $x = 1, 0.1, 0.01, 0.001$, and 0.0001. Does it appear that f has a limit as $x \to 0$?

2. Use the CAS to graph f near $x = 0$. Does it appear that f has a limit as $x \to 0$?

3. Try to evaluate $\lim_{x \to 0} f(x)$ with l'Hospital's Rule, using the CAS to find derivatives of the numerator and denominator. What do you discover? How many applications of l'Hospital's Rule are required?

4. Evaluate $\lim_{x \to 0} f(x)$ by using the CAS to find sufficiently many terms in the Taylor series of the numerator and denominator. (Use the command `taylor` in Maple or `Series` in Mathematica.)

5. Use the limit command on your CAS to find $\lim_{x \to 0} f(x)$ directly. (Most computer algebra systems use the method of Problem 4 to compute limits.)

6. In view of the answers to Problems 4 and 5, how do you explain the results of Problems 1 and 2?

8.8 The Binomial Series

You may be acquainted with the Binomial Theorem, which states that if a and b are any real numbers and k is a positive integer, then

$$(a + b)^k = a^k + ka^{k-1}b + \frac{k(k-1)}{2!}a^{k-2}b^2 + \frac{k(k-1)(k-2)}{3!}a^{k-3}b^3$$

$$+ \cdots + \frac{k(k-1)(k-2)\cdots(k-n+1)}{n!}a^{k-n}b^n$$

$$+ \cdots + kab^{k-1} + b^k$$

■ ■ For $k = 4$, for instance, the Binomial Theorem says that

$(a + b)^4 = a^4 + 4a^3b + 6a^2b^2$
$\qquad + 4ab^3 + b^4$

The traditional notation for the binomial coefficients is

$$\binom{k}{0} = 1 \qquad \binom{k}{n} = \frac{k(k-1)(k-2)\cdots(k-n+1)}{n!} \qquad n = 1, 2, \ldots, k$$

which enables us to write the Binomial Theorem in the abbreviated form

$$(a + b)^k = \sum_{n=0}^{k} \binom{k}{n} a^{k-n}b^n$$

In particular, if we put $a = 1$ and $b = x$, we get

$$\boxed{1} \qquad (1 + x)^k = \sum_{n=0}^{k} \binom{k}{n} x^n$$

One of Newton's accomplishments was to extend the Binomial Theorem (Equation 1) to the case in which k is no longer a positive integer. (See the Writing Project on page 621.) In this case the expression for $(1 + x)^k$ is no longer a finite sum; it becomes an infinite series. To find this series we compute the Maclaurin series of $(1 + x)^k$ in the usual way:

$$f(x) = (1 + x)^k \qquad\qquad\qquad f(0) = 1$$

$$f'(x) = k(1 + x)^{k-1} \qquad\qquad\qquad f'(0) = k$$

$$f''(x) = k(k - 1)(1 + x)^{k-2} \qquad\qquad f''(0) = k(k - 1)$$

$$f'''(x) = k(k - 1)(k - 2)(1 + x)^{k-3} \qquad f'''(0) = k(k - 1)(k - 2)$$

$$\vdots \qquad\qquad\qquad\qquad\qquad\qquad \vdots$$

$$f^{(n)}(x) = k(k - 1) \cdots (k - n + 1)(1 + x)^{k-n} \qquad f^{(n)}(0) = k(k - 1) \cdots (k - n + 1)$$

Therefore, the Maclaurin series of $f(x) = (1 + x)^k$ is

$$\sum_{n=0}^{\infty} \frac{f^{(n)}(0)}{n!} x^n = \sum_{n=0}^{\infty} \frac{k(k - 1) \cdots (k - n + 1)}{n!} x^n$$

This series is called the **binomial series**. If its nth term is a_n, then

$$\left| \frac{a_{n+1}}{a_n} \right| = \left| \frac{k(k - 1) \cdots (k - n + 1)(k - n)x^{n+1}}{(n + 1)!} \cdot \frac{n!}{k(k - 1) \cdots (k - n + 1)x^n} \right|$$

$$= \frac{|k - n|}{n + 1} |x| = \frac{\left| 1 - \dfrac{k}{n} \right|}{1 + \dfrac{1}{n}} |x| \to |x| \qquad \text{as } n \to \infty$$

Thus, by the Ratio Test, the binomial series converges if $|x| < 1$ and diverges if $|x| > 1$.

The following theorem states that $(1 + x)^k$ is equal to the sum of its Maclaurin series. It is possible to prove this by showing that the remainder term $R_n(x)$ approaches 0, but that turns out to be quite difficult. The proof outlined in Exercise 15 is much easier.

$\boxed{2}$ **The Binomial Series** If k is any real number and $|x| < 1$, then

$$(1 + x)^k = 1 + kx + \frac{k(k - 1)}{2!} x^2 + \frac{k(k - 1)(k - 2)}{3!} x^3 + \cdots$$

$$= \sum_{n=0}^{\infty} \binom{k}{n} x^n$$

where $\displaystyle \binom{k}{n} = \frac{k(k - 1) \cdots (k - n + 1)}{n!} \quad (n \geq 1)$ and $\displaystyle \binom{k}{0} = 1$

Although the binomial series always converges when $|x| < 1$, the question of whether or not it converges at the endpoints, ± 1, depends on the value of k. It turns out that the series converges at 1 if $-1 < k \leq 0$ and at both endpoints if $k \geq 0$. Notice that if k is a positive integer and $n > k$, then the expression for $\binom{k}{n}$ contains a factor $(k - k)$, so $\binom{k}{n} = 0$ for $n > k$. This means that the series terminates and reduces to the ordinary Binomial Theorem (Equation 1) when k is a positive integer.

As we have seen, the binomial series is just a special case of the Maclaurin series; it occurs so frequently that it is worth remembering.

$\boxed{\text{V}}$ **EXAMPLE 1** Expand $\dfrac{1}{(1 + x)^2}$ as a power series.

SOLUTION We use the binomial series with $k = -2$. The binomial coefficient is

$$\binom{-2}{n} = \frac{(-2)(-3)(-4) \cdots (-2 - n + 1)}{n!}$$

$$= \frac{(-1)^n 2 \cdot 3 \cdot 4 \cdot \cdots \cdot n(n + 1)}{n!} = (-1)^n (n + 1)$$

and so, when $|x| < 1$,

$$\frac{1}{(1 + x)^2} = (1 + x)^{-2} = \sum_{n=0}^{\infty} \binom{-2}{n} x^n$$

$$= \sum_{n=0}^{\infty} (-1)^n (n + 1) x^n = 1 - 2x + 3x^2 - 4x^3 + \cdots$$ ■ ■

$\boxed{\text{V}}$ **EXAMPLE 2** Find the Maclaurin series for the function $f(x) = \dfrac{1}{\sqrt{4 - x}}$ and its radius of convergence.

SOLUTION As given, $f(x)$ is not quite of the form $(1 + x)^k$ so we rewrite it as follows:

$$\frac{1}{\sqrt{4 - x}} = \frac{1}{\sqrt{4\left(1 - \dfrac{x}{4}\right)}} = \frac{1}{2\sqrt{1 - \dfrac{x}{4}}} = \frac{1}{2}\left(1 - \frac{x}{4}\right)^{-1/2}$$

Using the binomial series with $k = -\frac{1}{2}$ and with x replaced by $-x/4$, we have

$$\frac{1}{\sqrt{4 - x}} = \frac{1}{2}\left(1 - \frac{x}{4}\right)^{-1/2} = \frac{1}{2} \sum_{n=0}^{\infty} \binom{-\frac{1}{2}}{n} \left(-\frac{x}{4}\right)^n$$

$$= \frac{1}{2}\left[1 + \left(-\frac{1}{2}\right)\left(-\frac{x}{4}\right) + \frac{\left(-\frac{1}{2}\right)\left(-\frac{3}{2}\right)}{2!}\left(-\frac{x}{4}\right)^2 + \frac{\left(-\frac{1}{2}\right)\left(-\frac{3}{2}\right)\left(-\frac{5}{2}\right)}{3!}\left(-\frac{x}{4}\right)^3 \right.$$

$$\left. + \cdots + \frac{\left(-\frac{1}{2}\right)\left(-\frac{3}{2}\right)\left(-\frac{5}{2}\right) \cdots \left(-\frac{1}{2} - n + 1\right)}{n!}\left(-\frac{x}{4}\right)^n + \cdots \right]$$

$$= \frac{1}{2}\left[1 + \frac{1}{8}x + \frac{1 \cdot 3}{2!8^2}x^2 + \frac{1 \cdot 3 \cdot 5}{3!8^3}x^3 + \cdots + \frac{1 \cdot 3 \cdot 5 \cdot \cdots \cdot (2n - 1)}{n!8^n}x^n + \cdots \right]$$

We know from (2) that this series converges when $\left| -x/4 \right| < 1$, that is, $\left| x \right| < 4$, so the radius of convergence is $R = 4$. ■ ■

■ ■ A binomial series is a special case of a Taylor series. Figure 1 shows the graphs of the first three Taylor polynomials computed from the answer to Example 2.

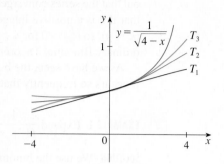

FIGURE 1

8.8 Exercises

1–6 ■ Use the binomial series to expand the function as a power series. State the radius of convergence.

1. $\sqrt{1 + x}$

2. $\dfrac{1}{(1 + x)^4}$

3. $\dfrac{1}{(2 + x)^3}$

4. $(1 - x)^{2/3}$

5. $\dfrac{x}{\sqrt{4 + x^2}}$

6. $\dfrac{x^2}{\sqrt{2 + x}}$

■ ■ ■ ■ ■ ■ ■ ■ ■ ■ ■ ■ ■ ■

7–8 ■ Use the binomial series to expand the function as a Maclaurin series and to find the first three Taylor polynomials T_1, T_2, and T_3. Graph the function and these Taylor polynomials in the interval of convergence.

7. $(1 + 2x)^{3/4}$

8. $\sqrt[3]{1 + 4x}$

■ ■ ■ ■ ■ ■ ■ ■ ■ ■ ■ ■ ■ ■

9. (a) Use the binomial series to expand $1/\sqrt{1 - x^2}$.
  (b) Use part (a) to find the Maclaurin series for $\sin^{-1}x$.

10. (a) Expand $1/\sqrt[4]{1 + x}$ as a power series.
  (b) Use part (a) to estimate $1/\sqrt[4]{1.1}$ correct to three decimal places.

11. (a) Expand $f(x) = x/(1 - x)^2$ as a power series.
  (b) Use part (a) to find the sum of the series

$$\sum_{n=1}^{\infty} \frac{n}{2^n}$$

12. (a) Expand $f(x) = (x + x^2)/(1 - x)^3$ as a power series.
  (b) Use part (a) to find the sum of the series

$$\sum_{n=1}^{\infty} \frac{n^2}{2^n}$$

13. (a) Use the binomial series to find the Maclaurin series of $f(x) = \sqrt{1 + x^2}$.
  (b) Use part (a) to evaluate $f^{(10)}(0)$.

14. (a) Use the binomial series to find the Maclaurin series of $f(x) = 1/\sqrt{1 + x^3}$.
  (b) Use part (a) to evaluate $f^{(9)}(0)$.

15. Use the following steps to prove (2).
  (a) Let $g(x) = \sum_{n=0}^{\infty} \binom{k}{n} x^n$. Differentiate this series to show that

$$g'(x) = \frac{kg(x)}{1 + x} \qquad -1 < x < 1$$

  (b) Let $h(x) = (1 + x)^{-k} g(x)$ and show that $h'(x) = 0$.
  (c) Deduce that $g(x) = (1 + x)^k$.

16. In Exercise 25 in Section 6.3 it was shown that the length of the ellipse $x = a \sin \theta$, $y = b \cos \theta$, where $a > b > 0$, is

$$L = 4a \int_0^{\pi/2} \sqrt{1 - e^2 \sin^2 \theta} \; d\theta$$

where $e = \sqrt{a^2 - b^2}/a$ is the eccentricity of the ellipse. Expand the integrand as a binomial series and use the result of Exercise 36 in Section 5.6 to express L as a series in powers of the eccentricity up to the term in e^6.

How Newton Discovered the Binomial Series

The Binomial Theorem, which gives the expansion of $(a + b)^k$, was known to Chinese mathematicians many centuries before the time of Newton for the case where the exponent k is a positive integer. In 1665, when he was 22, Newton was the first to discover the infinite series expansion of $(a + b)^k$ when k is a fractional exponent (positive or negative). He didn't publish his discovery, but he stated it and gave examples of how to use it in a letter (now called the *epistola prior*) dated June 13, 1676, that he sent to Henry Oldenburg, secretary of the Royal Society of London, to transmit to Leibniz. When Leibniz replied, he asked how Newton had discovered the binomial series. Newton wrote a second letter, the *epistola posterior* of October 24, 1676, in which he explained in great detail how he arrived at his discovery by a very indirect route. He was investigating the areas under the curves $y = (1 - x^2)^{n/2}$ from 0 to x for $n = 0, 1, 2, 3, 4, \ldots$. These are easy to calculate if n is even. By observing patterns and interpolating, Newton was able to guess the answers for odd values of n. Then he realized he could get the same answers by expressing $(1 - x^2)^{n/2}$ as an infinite series.

Write a report on Newton's discovery of the binomial series. Start by giving the statement of the binomial series in Newton's notation (see the *epistola prior* on page 285 of [4] or page 402 of [2]). Explain why Newton's version is equivalent to Theorem 2 on page 618. Then read Newton's *epistola posterior* (page 287 in [4] or page 404 in [2]) and explain the patterns that Newton discovered in the areas under the curves $y = (1 - x^2)^{n/2}$. Show how he was able to guess the areas under the remaining curves and how he verified his answers. Finally, explain how these discoveries led to the binomial series. The books by Edwards [1] and Katz [3] contain commentaries on Newton's letters.

1. C. H. Edwards, *The Historical Development of the Calculus* (New York: Springer-Verlag, 1979), pp. 178–187.

2. John Fauvel and Jeremy Gray, eds., *The History of Mathematics: A Reader* (London: MacMillan Press, 1987).

3. Victor Katz, *A History of Mathematics: An Introduction* (New York: HarperCollins, 1993), pp. 463–466.

4. D. J. Struik, ed., *A Sourcebook in Mathematics, 1200–1800* (Princeton, N.J.: Princeton University Press, 1969).

8.9 Applications of Taylor Polynomials

In this section we explore two types of applications of Taylor polynomials. First we look at how they are used to approximate functions—computer scientists like them because polynomials are the simplest of functions. Then we investigate how physicists and engineers use them in such fields as relativity, optics, blackbody radiation, electric dipoles, the velocity of water waves, and building highways across a desert.

■ Approximating Functions by Polynomials

Suppose that $f(x)$ is equal to the sum of its Taylor series at a:

$$f(x) = \sum_{n=0}^{\infty} \frac{f^{(n)}(a)}{n!} (x - a)^n$$

In Section 8.7 we introduced the notation $T_n(x)$ for the nth partial sum of this series and called it the nth-degree Taylor polynomial of f at a. Thus

$$T_n(x) = \sum_{i=0}^{n} \frac{f^{(i)}(a)}{i!} (x - a)^i$$

$$= f(a) + \frac{f'(a)}{1!} (x - a) + \frac{f''(a)}{2!} (x - a)^2 + \cdots + \frac{f^{(n)}(a)}{n!} (x - a)^n$$

Since f is the sum of its Taylor series, we know that $T_n(x) \to f(x)$ as $n \to \infty$ and so T_n can be used as an approximation to f: $f(x) \approx T_n(x)$.

Notice that the first-degree Taylor polynomial

$$T_1(x) = f(a) + f'(a)(x - a)$$

is the same as the linearization of f at a that we discussed in Section 3.8. Notice also that T_1 and its derivative have the same values at a that f and f' have. In general, it can be shown that the derivatives of T_n at a agree with those of f up to and including derivatives of order n.

To illustrate these ideas let's take another look at the graphs of $y = e^x$ and its first few Taylor polynomials, as shown in Figure 1. The graph of T_1 is the tangent line to $y = e^x$ at $(0, 1)$; this tangent line is the best linear approximation to e^x near $(0, 1)$. The graph of T_2 is the parabola $y = 1 + x + x^2/2$, and the graph of T_3 is the cubic curve $y = 1 + x + x^2/2 + x^3/6$, which is a closer fit to the exponential curve $y = e^x$ than T_2. The next Taylor polynomial T_4 would be an even better approximation, and so on.

The values in the table give a numerical demonstration of the convergence of the Taylor polynomials $T_n(x)$ to the function $y = e^x$. We see that when $x = 0.2$ the convergence is very rapid, but when $x = 3$ it is somewhat slower. In fact, the farther x is from 0, the more slowly $T_n(x)$ converges to e^x.

When using a Taylor polynomial T_n to approximate a function f, we have to ask the questions: How good an approximation is it? How large should we take n to be in order to achieve a desired accuracy? To answer these questions we need to look at the absolute value of the remainder:

$$|R_n(x)| = |f(x) - T_n(x)|$$

There are three possible methods for estimating the size of the error:

1. If a graphing device is available, we can use it to graph $|R_n(x)|$ and thereby estimate the error.

2. If the series happens to be an alternating series, we can use the Alternating Series Estimation Theorem.

3. In all cases we can use Taylor's Inequality (Theorem 8.7.9), which says that if $|f^{(n+1)}(x)| \le M$, then

$$|R_n(x)| \le \frac{M}{(n + 1)!} |x - a|^{n+1}$$

FIGURE 1

x	$x = 0.2$	$x = 3.0$
$T_2(x)$	1.220000	8.500000
$T_4(x)$	1.221400	16.375000
$T_6(x)$	1.221403	19.412500
$T_8(x)$	1.221403	20.009152
$T_{10}(x)$	1.221403	20.079665
e^x	1.221403	20.085537

Ⓥ **EXAMPLE 1**
(a) Approximate the function $f(x) = \sqrt[3]{x}$ by a Taylor polynomial of degree 2 at $a = 8$.
(b) How accurate is this approximation when $7 \le x \le 9$?

SOLUTION

(a)
$$f(x) = \sqrt[3]{x} = x^{1/3} \qquad f(8) = 2$$

$$f'(x) = \tfrac{1}{3}x^{-2/3} \qquad f'(8) = \tfrac{1}{12}$$

$$f''(x) = -\tfrac{2}{9}x^{-5/3} \qquad f''(8) = -\tfrac{1}{144}$$

$$f'''(x) = \tfrac{10}{27}x^{-8/3}$$

Thus, the second-degree Taylor polynomial is

$$T_2(x) = f(8) + \frac{f'(8)}{1!}(x-8) + \frac{f''(8)}{2!}(x-8)^2$$

$$= 2 + \tfrac{1}{12}(x-8) - \tfrac{1}{288}(x-8)^2$$

The desired approximation is

$$\sqrt[3]{x} \approx T_2(x) = 2 + \tfrac{1}{12}(x-8) - \tfrac{1}{288}(x-8)^2$$

(b) The Taylor series is not alternating when $x < 8$, so we can't use the Alternating Series Estimation Theorem in this example. But we can use Taylor's Inequality with $n = 2$ and $a = 8$:

$$|R_2(x)| \leqslant \frac{M}{3!}|x-8|^3$$

where $|f'''(x)| \leqslant M$. Because $x \geqslant 7$, we have $x^{8/3} \geqslant 7^{8/3}$ and so

$$f'''(x) = \frac{10}{27} \cdot \frac{1}{x^{8/3}} \leqslant \frac{10}{27} \cdot \frac{1}{7^{8/3}} < 0.0021$$

Therefore, we can take $M = 0.0021$. Also $7 \leqslant x \leqslant 9$, so $-1 \leqslant x - 8 \leqslant 1$ and $|x - 8| \leqslant 1$. Then Taylor's Inequality gives

$$|R_2(x)| \leqslant \frac{0.0021}{3!} \cdot 1^3 = \frac{0.0021}{6} < 0.0004$$

Thus, if $7 \leqslant x \leqslant 9$, the approximation in part (a) is accurate to within 0.0004.   ■ ■

FIGURE 2

Let's use a graphing device to check the calculation in Example 1. Figure 2 shows that the graphs of $y = \sqrt[3]{x}$ and $y = T_2(x)$ are very close to each other when x is near 8. Figure 3 shows the graph of $|R_2(x)|$ computed from the expression

$$|R_2(x)| = |\sqrt[3]{x} - T_2(x)|$$

We see from this graph that

$$|R_2(x)| < 0.0003$$

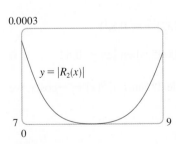

FIGURE 3

when $7 \leqslant x \leqslant 9$. Thus, the error estimate from graphical methods is slightly better than the error estimate from Taylor's Inequality in this case.

$\boxed{\text{V}}$ **EXAMPLE 2**

(a) What is the maximum error possible in using the approximation

$$\sin x \approx x - \frac{x^3}{3!} + \frac{x^5}{5!}$$

when $-0.3 \leqslant x \leqslant 0.3$? Use this approximation to find $\sin 12°$ correct to six decimal places.

(b) For what values of x is this approximation accurate to within 0.00005?

SOLUTION

(a) Notice that the Maclaurin series

$$\sin x = x - \frac{x^3}{3!} + \frac{x^5}{5!} - \frac{x^7}{7!} + \cdots$$

is alternating for all nonzero values of x, and the successive terms decrease in size because $|x| < 1$, so we can use the Alternating Series Estimation Theorem. The error in approximating $\sin x$ by the first three terms of its Maclaurin series is at most

$$\left| \frac{x^7}{7!} \right| = \frac{|x|^7}{5040}$$

If $-0.3 \leqslant x \leqslant 0.3$, then $|x| \leqslant 0.3$, so the error is smaller than

$$\frac{(0.3)^7}{5040} \approx 4.3 \times 10^{-8}$$

To find $\sin 12°$ we first convert to radian measure.

$$\sin 12° = \sin\left(\frac{12\pi}{180} \right) = \sin\left(\frac{\pi}{15} \right)$$

$$\approx \frac{\pi}{15} - \left(\frac{\pi}{15} \right)^3 \frac{1}{3!} + \left(\frac{\pi}{15} \right)^5 \frac{1}{5!}$$

$$\approx 0.20791169$$

Thus, correct to six decimal places, $\sin 12° \approx 0.207912$.

(b) The error will be smaller than 0.00005 if

$$\frac{|x|^7}{5040} < 0.00005$$

Solving this inequality for x, we get

$$|x|^7 < 0.252 \qquad \text{or} \qquad |x| < (0.252)^{1/7} \approx 0.821$$

So the given approximation is accurate to within 0.00005 when $|x| < 0.82$. ■ ■

Module 8.7/8.9 graphically shows the remainders in Taylor polynomial approximations.

What if we use Taylor's Inequality to solve Example 2? Since $f^{(7)}(x) = -\cos x$, we have $|f^{(7)}(x)| \leqslant 1$ and so

$$|R_6(x)| \leqslant \frac{1}{7!} |x|^7$$

So we get the same estimates as with the Alternating Series Estimation Theorem.

FIGURE 4

FIGURE 5

What about graphical methods? Figure 4 shows the graph of

$$| R_6(x) | = | \sin x - (x - \tfrac{1}{6}x^3 + \tfrac{1}{120}x^5) |$$

and we see from it that $| R_6(x) | < 4.3 \times 10^{-8}$ when $| x | \le 0.3$. This is the same estimate that we obtained in Example 2. For part (b) we want $| R_6(x) | < 0.00005$, so we graph both $y = | R_6(x) |$ and $y = 0.00005$ in Figure 5. By placing the cursor on the right intersection point we find that the inequality is satisfied when $| x | < 0.82$. Again this is the same estimate that we obtained in the solution to Example 2.

If we had been asked to approximate $\sin 72°$ instead of $\sin 12°$ in Example 2, it would have been wise to use the Taylor polynomials at $a = \pi/3$ (instead of $a = 0$) because they are better approximations to $\sin x$ for values of x close to $\pi/3$. Notice that $72°$ is close to $60°$ (or $\pi/3$ radians) and the derivatives of $\sin x$ are easy to compute at $\pi/3$.

Figure 6 shows the graphs of the Maclaurin polynomial approximations

$$T_1(x) = x \qquad\qquad T_3(x) = x - \frac{x^3}{3!}$$

$$T_5(x) = x - \frac{x^3}{3!} + \frac{x^5}{5!} \qquad T_7(x) = x - \frac{x^3}{3!} + \frac{x^5}{5!} - \frac{x^7}{7!}$$

to the sine curve. You can see that as n increases, $T_n(x)$ is a good approximation to $\sin x$ on a larger and larger interval.

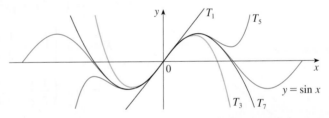

FIGURE 6

One use of the type of calculation done in Examples 1 and 2 occurs in calculators and computers. For instance, when you press the sin or e^x key on your calculator, or when a computer programmer uses a subroutine for a trigonometric or exponential or Bessel function, in many machines a polynomial approximation is calculated. The polynomial is often a Taylor polynomial that has been modified so that the error is spread more evenly throughout an interval.

Applications to Physics

Taylor polynomials are also used frequently in physics. In order to gain insight into an equation, a physicist often simplifies a function by considering only the first two or three terms in its Taylor series. In other words, the physicist uses a Taylor polynomial as an approximation to the function. Taylor's Inequality can then be used to gauge the accuracy of the approximation. The following example shows one way in which this idea is used in special relativity.

▽ **EXAMPLE 3** In Einstein's theory of special relativity the mass of an object moving with velocity v is

$$m = \frac{m_0}{\sqrt{1 - v^2/c^2}}$$

where m_0 is the mass of the object when at rest and c is the speed of light. The kinetic energy of the object is the difference between its total energy and its energy at rest:

$$K = mc^2 - m_0 c^2$$

(a) Show that when v is very small compared with c, this expression for K agrees with classical Newtonian physics: $K = \frac{1}{2} m_0 v^2$.

(b) Use Taylor's Inequality to estimate the difference in these expressions for K when $|v| \leq 100$ m/s.

SOLUTION

(a) Using the expressions given for K and m, we get

$$K = mc^2 - m_0 c^2 = \frac{m_0 c^2}{\sqrt{1 - v^2/c^2}} - m_0 c^2$$

$$= m_0 c^2 \left[\left(1 - \frac{v^2}{c^2} \right)^{-1/2} - 1 \right]$$

■ ■ The upper curve in Figure 7 is the graph of the expression for the kinetic energy K of an object with velocity v in special relativity. The lower curve shows the function used for K in classical Newtonian physics. When v is much smaller than the speed of light, the curves are practically identical.

With $x = -v^2/c^2$, the Maclaurin series for $(1 + x)^{-1/2}$ is most easily computed as a binomial series with $k = -\frac{1}{2}$. (Notice that $|x| < 1$ because $v < c$.) Therefore, we have

$$(1 + x)^{-1/2} = 1 - \frac{1}{2} x + \frac{(-\frac{1}{2})(-\frac{3}{2})}{2!} x^2 + \frac{(-\frac{1}{2})(-\frac{3}{2})(-\frac{5}{2})}{3!} x^3 + \cdots$$

$$= 1 - \frac{1}{2} x + \frac{3}{8} x^2 - \frac{5}{16} x^3 + \cdots$$

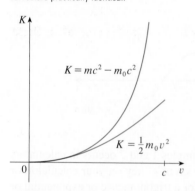

and

$$K = m_0 c^2 \left[\left(1 + \frac{1}{2} \frac{v^2}{c^2} + \frac{3}{8} \frac{v^4}{c^4} + \frac{5}{16} \frac{v^6}{c^6} + \cdots \right) - 1 \right]$$

$$= m_0 c^2 \left(\frac{1}{2} \frac{v^2}{c^2} + \frac{3}{8} \frac{v^4}{c^4} + \frac{5}{16} \frac{v^6}{c^6} + \cdots \right)$$

If v is much smaller than c, then all terms after the first are very small when compared with the first term. If we omit them, we get

$$K \approx m_0 c^2 \left(\frac{1}{2} \frac{v^2}{c^2} \right) = \frac{1}{2} m_0 v^2$$

FIGURE 7

(b) If $x = -v^2/c^2$, $f(x) = m_0 c^2 [(1 + x)^{-1/2} - 1]$, and M is a number such that $|f''(x)| \leq M$, then we can use Taylor's Inequality to write

$$|R_1(x)| \leq \frac{M}{2!} x^2$$

We have $f''(x) = \frac{3}{4} m_0 c^2 (1 + x)^{-5/2}$ and we are given that $|v| \leq 100$ m/s, so

$$|f''(x)| = \frac{3 m_0 c^2}{4(1 - v^2/c^2)^{5/2}} \leq \frac{3 m_0 c^2}{4(1 - 100^2/c^2)^{5/2}} \quad (= M)$$

Thus, with $c = 3 \times 10^8$ m/s,

$$|R_1(x)| \leq \frac{1}{2} \cdot \frac{3 m_0 c^2}{4(1 - 100^2/c^2)^{5/2}} \cdot \frac{100^4}{c^4} < (4.17 \times 10^{-10}) m_0$$

So when $|v| \le 100$ m/s, the magnitude of the error in using the Newtonian expression for kinetic energy is at most $(4.2 \times 10^{-10})m_0$. ■ ■

Another application to physics occurs in optics. Figure 8 is adapted from *Optics*, 4th ed., by Eugene Hecht (San Francisco: Addison Wesley, 2002), page 153. It depicts a wave from the point source S meeting a spherical interface of radius R centered at C. The ray SA is refracted toward P.

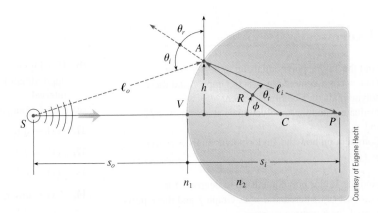

FIGURE 8

Refraction at a spherical interface

Using Fermat's principle that light travels so as to minimize the time taken, Hecht derives the equation

$$\boxed{1} \qquad \frac{n_1}{\ell_o} + \frac{n_2}{\ell_i} = \frac{1}{R}\left(\frac{n_2 s_i}{\ell_i} - \frac{n_1 s_o}{\ell_o}\right)$$

where n_1 and n_2 are indexes of refraction and ℓ_o, ℓ_i, s_o, and s_i are the distances indicated in Figure 8. By the Law of Cosines, applied to triangles ACS and ACP, we have

■ ■ Here we use the identity
$$\cos(\pi - \phi) = -\cos\phi$$

$$\boxed{2} \qquad \ell_o = \sqrt{R^2 + (s_o + R)^2 - 2R(s_o + R)\cos\phi}$$

$$\ell_i = \sqrt{R^2 + (s_i - R)^2 + 2R(s_i - R)\cos\phi}$$

Because Equation 1 is cumbersome to work with, Gauss, in 1841, simplified it by using the linear approximation $\cos\phi \approx 1$ for small values of ϕ. (This amounts to using the Taylor polynomial of degree 1.) Then Equation 1 becomes the following simpler equation [as you are asked to show in Exercise 26(a)]:

$$\boxed{3} \qquad \frac{n_1}{s_o} + \frac{n_2}{s_i} = \frac{n_2 - n_1}{R}$$

The resulting optical theory is known as *Gaussian optics*, or *first-order optics*, and has become the basic theoretical tool used to design lenses.

A more accurate theory is obtained by approximating $\cos\phi$ by its Taylor polynomial of degree 3 (which is the same as the Taylor polynomial of degree 2). This takes into account rays for which ϕ is not so small, that is, rays that strike the surface at greater distances h above the axis. In Exercise 26(b) you are asked to use this approximation to derive the more accurate equation

$$\boxed{4} \quad \frac{n_1}{s_o} + \frac{n_2}{s_i} = \frac{n_2 - n_1}{R} + h^2\left[\frac{n_1}{2s_o}\left(\frac{1}{s_o} + \frac{1}{R}\right)^2 + \frac{n_2}{2s_i}\left(\frac{1}{R} - \frac{1}{s_i}\right)^2\right]$$

The resulting optical theory is known as *third-order optics*.

Other applications of Taylor polynomials to physics and engineering are explored in Exercises 27–30 and in the Applied Project on page 630.

8.9 Exercises

1. (a) Find the Taylor polynomials up to degree 6 for $f(x) = \cos x$ centered at $a = 0$. Graph f and these polynomials on a common screen.
(b) Evaluate f and these polynomials at $x = \pi/4$, $\pi/2$, and π.
(c) Comment on how the Taylor polynomials converge to $f(x)$.

2. (a) Find the Taylor polynomials up to degree 3 for $f(x) = 1/x$ centered at $a = 1$. Graph f and these polynomials on a common screen.
(b) Evaluate f and these polynomials at $x = 0.9$ and 1.3.
(c) Comment on how the Taylor polynomials converge to $f(x)$.

3–8 ■ Find the Taylor polynomial $T_n(x)$ for the function f at the number a. Graph f and T_n on the same screen.

3. $f(x) = \sin x$, $a = \pi/6$, $n = 3$

4. $f(x) = e^x$, $a = 2$, $n = 3$

5. $f(x) = \arcsin x$, $a = 0$, $n = 3$

6. $f(x) = \dfrac{\ln x}{x}$, $a = 1$, $n = 3$

7. $f(x) = xe^{-2x}$, $a = 0$, $n = 3$

8. $f(x) = \sqrt{3 + x^2}$, $a = 1$, $n = 2$

CAS 9–10 ■ Use a computer algebra system to find the Taylor polynomials T_n at $a = 0$ for the given values of n. Then graph these polynomials and f on the same screen.

9. $f(x) = \sec x$, $n = 2, 4, 6, 8$

10. $f(x) = \tan x$, $n = 1, 3, 5, 7, 9$

11–18 ■

(a) Approximate f by a Taylor polynomial with degree n at the number a.

(b) Use Taylor's Inequality to estimate the accuracy of the approximation $f(x) \approx T_n(x)$ when x lies in the given interval.
(c) Check your result in part (b) by graphing $|R_n(x)|$.

11. $f(x) = \sqrt{x}$, $a = 4$, $n = 2$, $4 \le x \le 4.2$

12. $f(x) = x^{-2}$, $a = 1$, $n = 2$, $0.9 \le x \le 1.1$

13. $f(x) = x^{2/3}$, $a = 1$, $n = 3$, $0.8 \le x \le 1.2$

14. $f(x) = \cos x$, $a = \pi/3$, $n = 4$, $0 \le x \le 2\pi/3$

15. $f(x) = e^{x^2}$, $a = 0$, $n = 3$, $0 \le x \le 0.1$

16. $f(x) = \ln(1 + 2x)$, $a = 1$, $n = 3$, $0.5 \le x \le 1.5$

17. $f(x) = x \sin x$, $a = 0$, $n = 4$, $-1 \le x \le 1$

18. $f(x) = x \ln x$, $a = 1$, $n = 3$, $0.5 \le x \le 1.5$

19. Use the information from Exercise 3 to estimate $\sin 35°$ correct to five decimal places.

20. Use the information from Exercise 14 to estimate $\cos 69°$ correct to five decimal places.

21. Use Taylor's Inequality to determine the number of terms of the Maclaurin series for e^x that should be used to estimate $e^{0.1}$ to within 0.00001.

22. Suppose you know that

$$f^{(n)}(4) = \frac{(-1)^n n!}{3^n(n + 1)}$$

and the Taylor series of f centered at 4 converges to $f(x)$ for all x in the interval of convergence. Show that the fifth-degree Taylor polynomial approximates $f(5)$ with error less than 0.0002.

23–24 ■ Use the Alternating Series Estimation Theorem or Taylor's Inequality to estimate the range of values of x for which the given approximation is accurate to within the stated error. Check your answer graphically.

23. $\sin x \approx x - \dfrac{x^3}{6}$ $(|\text{error}| < 0.01)$

24. $\cos x \approx 1 - \dfrac{x^2}{2} + \dfrac{x^4}{24}$ ($|\,\text{error}\,| < 0.005$)

. . . .

25. A car is moving with speed 20 m/s and acceleration 2 m/s^2 at a given instant. Using a second-degree Taylor polynomial, estimate how far the car moves in the next second. Would it be reasonable to use this polynomial to estimate the distance traveled during the next minute?

26. (a) Derive Equation 3 for Gaussian optics from Equation 1 by approximating $\cos\phi$ in Equation 2 by its first-degree Taylor polynomial.
   (b) Show that if $\cos\phi$ is replaced by its third-degree Taylor polynomial in Equation 2, then Equation 1 becomes Equation 4 for third-order optics. [*Hint:* Use the first two terms in the binomial series for ℓ_o^{-1} and ℓ_i^{-1}. Also, use $\phi \approx \sin\phi$.]

27. An electric dipole consists of two electric charges of equal magnitude and opposite sign. If the charges are q and $-q$ and are located at a distance d from each other, then the electric field E at the point P in the figure is

$$E = \frac{q}{D^2} - \frac{q}{(D+d)^2}$$

By expanding this expression for E as a series in powers of d/D, show that E is approximately proportional to $1/D^3$ when P is far away from the dipole.

28. The resistivity ρ of a conducting wire is the reciprocal of the conductivity and is measured in units of ohm-meters (Ω-m). The resistivity of a given metal depends on the temperature according to the equation

$$\rho(t) = \rho_{20}e^{\alpha(t-20)}$$

where t is the temperature in °C. There are tables that list the values of α (called the temperature coefficient) and ρ_{20} (the resistivity at 20°C) for various metals. Except at very low temperatures, the resistivity varies almost linearly with temperature and so it is common to approximate the expression for $\rho(t)$ by its first- or second-degree Taylor polynomial at $t = 20$.
   (a) Find expressions for these linear and quadratic approximations.
   (b) For copper, the tables give $\alpha = 0.0039/°C$ and $\rho_{20} = 1.7 \times 10^{-8}\ \Omega$-m. Graph the resistivity of copper and the linear and quadratic approximations for $-250°C \leqslant t \leqslant 1000°C$.
   (c) For what values of t does the linear approximation agree with the exponential expression to within one percent?

29. If a surveyor measures differences in elevation when making plans for a highway across a desert, corrections must be made for the curvature of the Earth.
   (a) If R is the radius of the Earth and L is the length of the highway, show that the correction is

$$C = R\sec(L/R) - R$$

   (b) Use a Taylor polynomial to show that

$$C \approx \frac{L^2}{2R} + \frac{5L^4}{24R^3}$$

   (c) Compare the corrections given by the formulas in parts (a) and (b) for a highway that is 100 km long. (Take the radius of Earth to be 6370 km.)

30. The period of a pendulum with length L that makes a maximum angle θ_0 with the vertical is

$$T = 4\sqrt{\frac{L}{g}}\int_0^{\pi/2} \frac{dx}{\sqrt{1 - k^2\sin^2 x}}$$

where $k = \sin\left(\frac{1}{2}\theta_0\right)$ and g is the acceleration due to gravity. (In Exercise 32 in Section 5.9 we approximated this integral using Simpson's Rule.)
   (a) Expand the integrand as a binomial series and use the result of Exercise 36 in Section 5.6 to show that

$$T = 2\pi\sqrt{\frac{L}{g}}\left[1 + \frac{1^2}{2^2}k^2 + \frac{1^2 3^2}{2^2 4^2}k^4 + \frac{1^2 3^2 5^2}{2^2 4^2 6^2}k^6 + \cdots\right]$$

If θ_0 is not too large, the approximation $T \approx 2\pi\sqrt{L/g}$, obtained by using only the first term in the series, is often used. A better approximation is obtained by using two terms:

$$T \approx 2\pi\sqrt{\frac{L}{g}}\left(1 + \tfrac{1}{4}k^2\right)$$

   (b) Notice that all the terms in the series after the first one have coefficients that are at most $\frac{1}{4}$. Use this fact to compare this series with a geometric series and show that

$$2\pi\sqrt{\frac{L}{g}}\left(1 + \tfrac{1}{4}k^2\right) \leqslant T \leqslant 2\pi\sqrt{\frac{L}{g}}\frac{4 - 3k^2}{4 - 4k^2}$$

   (c) Use the inequalities in part (b) to estimate the period of a pendulum with $L = 1$ meter and $\theta_0 = 10°$. How does it compare with the estimate $T \approx 2\pi\sqrt{L/g}$? What if $\theta_0 = 42°$?

31. In Section 4.8 we considered Newton's method for approximating a root r of the equation $f(x) = 0$, and from an initial approximation x_1 we obtained successive approximations $x_2, x_3, \ldots$, where

$$x_{n+1} = x_n - \frac{f(x_n)}{f'(x_n)}$$

Use Taylor's Inequality with $n = 1$, $a = x_n$, and $x = r$ to show that if $f''(x)$ exists on an interval I containing r, x_n,

and x_{n+1}, and $\left| f''(x) \right| \leqslant M$, $\left| f'(x) \right| \geqslant K$ for all $x \in I$, then

$$\left| x_{n+1} - r \right| \leqslant \frac{M}{2K} \left| x_n - r \right|^2$$

[This means that if x_n is accurate to d decimal places, then x_{n+1} is accurate to about $2d$ decimal places. More precisely, if the error at stage n is at most 10^{-m}, then the error at stage $n + 1$ is at most $(M/2K)10^{-2m}$.]

© Luke Dodd, Photo Researchers, Inc.

APPLIED PROJECT

Radiation from the Stars

Any object emits radiation when heated. A *blackbody* is a system that absorbs all the radiation that falls on it. For instance, a matte black surface or a large cavity with a small hole in its wall (like a blastfurnace) is a blackbody and emits blackbody radiation. Even the radiation from the Sun is close to being blackbody radiation.

Proposed in the late 19th century, the Rayleigh-Jeans Law expresses the energy density of blackbody radiation of wavelength λ as

$$f(\lambda) = \frac{8\pi kT}{\lambda^4}$$

where λ is measured in meters, T is the temperature in kelvins (K), and k is Boltzmann's constant. The Rayleigh-Jeans Law agrees with experimental measurements for long wavelengths but disagrees drastically for short wavelengths. [The law predicts that $f(\lambda) \to \infty$ as $\lambda \to 0^+$ but experiments have shown that $f(\lambda) \to 0$.] This fact is known as the *ultraviolet catastrophe*.

In 1900 Max Planck found a better model (known now as Planck's Law) for blackbody radiation:

$$f(\lambda) = \frac{8\pi hc\lambda^{-5}}{e^{hc/(\lambda kT)} - 1}$$

where λ is measured in meters, T is the temperature in kelvins, and

$$h = \text{Planck's constant} = 6.6262 \times 10^{-34} \text{ J} \cdot \text{s}$$

$$c = \text{speed of light} = 2.997\,925 \times 10^8 \text{ m/s}$$

$$k = \text{Boltzmann's constant} = 1.3807 \times 10^{-23} \text{ J/K}$$

1. Use l'Hospital's Rule to show that

$$\lim_{\lambda \to 0^+} f(\lambda) = 0 \qquad \text{and} \qquad \lim_{\lambda \to \infty} f(\lambda) = 0$$

for Planck's Law. So, for short wavelengths, this law models blackbody radiation better than the Rayleigh-Jeans Law.

2. Use a Taylor polynomial to show that, for large wavelengths, Planck's Law gives approximately the same values as the Rayleigh-Jeans Law.

3. Graph f as given by both laws on the same screen and comment on the similarities and differences. Use $T = 5700$ K (the temperature of the Sun). (You may want to change from meters to the more convenient unit of micrometers: 1 μm $= 10^{-6}$ m.)

4. Use your graph in Problem 3 to estimate the value of λ for which $f(\lambda)$ is a maximum under Planck's Law.

5. Investigate how the graph of f changes as T varies. (Use Planck's Law.) In particular, graph f for the stars Betelgeuse ($T = 3400$ K), Procyon ($T = 6400$ K), and Sirius ($T = 9200$ K) as well as the Sun. How does the total radiation emitted (the area under the curve) vary with T? Use the graph to comment on why Sirius is known as a blue star and Betelgeuse as a red star.

8 Review

CONCEPT CHECK

1. (a) What is a convergent sequence?
   (b) What is a convergent series?
   (c) What does $\lim_{n\to\infty} a_n = 3$ mean?
   (d) What does $\sum_{n=1}^{\infty} a_n = 3$ mean?

2. (a) What is a bounded sequence?
   (b) What is a monotonic sequence?
   (c) What can you say about a bounded monotonic sequence?

3. (a) What is a geometric series? Under what circumstances is it convergent? What is its sum?
   (b) What is a p-series? Under what circumstances is it convergent?

4. Suppose $\Sigma\, a_n = 3$ and s_n is the nth partial sum of the series. What is $\lim_{n\to\infty} a_n$? What is $\lim_{n\to\infty} s_n$?

5. State the following.
   (a) The Test for Divergence
   (b) The Integral Test
   (c) The Comparison Test
   (d) The Limit Comparison Test
   (e) The Alternating Series Test
   (f) The Ratio Test

6. (a) What is an absolutely convergent series?
   (b) What can you say about such a series?

7. (a) If a series is convergent by the Integral Test, how do you estimate its sum?
   (b) If a series is convergent by the Comparison Test, how do you estimate its sum?

   (c) If a series is convergent by the Alternating Series Test, how do you estimate its sum?

8. (a) Write the general form of a power series.
   (b) What is the radius of convergence of a power series?
   (c) What is the interval of convergence of a power series?

9. Suppose $f(x)$ is the sum of a power series with radius of convergence R.
   (a) How do you differentiate f? What is the radius of convergence of the series for f'?
   (b) How do you integrate f? What is the radius of convergence of the series for $\int f(x)\, dx$?

10. (a) Write an expression for the nth-degree Taylor polynomial of f centered at a.
    (b) Write an expression for the Taylor series of f centered at a.
    (c) Write an expression for the Maclaurin series of f.
    (d) How do you show that $f(x)$ is equal to the sum of its Taylor series?
    (e) State Taylor's Inequality.

11. Write the Maclaurin series and the interval of convergence for each of the following functions.
    (a) $1/(1 - x)$ (b) e^x (c) $\sin x$
    (d) $\cos x$ (e) $\tan^{-1} x$

12. Write the binomial series expansion of $(1 + x)^k$. What is the radius of convergence of this series?

TRUE-FALSE QUIZ

Determine whether the statement is true or false. If it is true, explain why.
If it is false, explain why or give an example that disproves the statement.

1. If $\lim_{n\to\infty} a_n = 0$, then $\Sigma\, a_n$ is convergent.

2. The series $\Sigma_{n=1}^{\infty}\, n^{-\sin 1}$ is convergent.

3. If $\lim_{n\to\infty} a_n = L$, then $\lim_{n\to\infty} a_{2n+1} = L$.

4. If $\Sigma\, c_n 6^n$ is convergent, then $\Sigma\, c_n(-2)^n$ is convergent.

5. If $\Sigma\, c_n 6^n$ is convergent, then $\Sigma\, c_n(-6)^n$ is convergent.

6. If $\Sigma\, c_n x^n$ diverges when $x = 6$, then it diverges when $x = 10$.

7. The Ratio Test can be used to determine whether $\Sigma\, 1/n^3$ converges.

8. The Ratio Test can be used to determine whether $\Sigma\, 1/n!$ converges.

9. If $0 \leq a_n \leq b_n$ and $\Sigma\, b_n$ diverges, then $\Sigma\, a_n$ diverges.

10. $\displaystyle\sum_{n=0}^{\infty} \frac{(-1)^n}{n!} = \frac{1}{e}$

11. If $-1 < \alpha < 1$, then $\lim_{n\to\infty} \alpha^n = 0$.

12. If $\Sigma\, a_n$ is divergent, then $\Sigma\, |a_n|$ is divergent.

13. If $f(x) = 2x - x^2 + \frac{1}{3}x^3 - \cdots$ converges for all x, then $f'''(0) = 2$.

14. If $\{a_n\}$ and $\{b_n\}$ are divergent, then $\{a_n + b_n\}$ is divergent.

15. If $\{a_n\}$ and $\{b_n\}$ are divergent, then $\{a_n b_n\}$ is divergent.

16. If $\{a_n\}$ is decreasing and $a_n > 0$ for all n, then $\{a_n\}$ is convergent.

17. If $a_n > 0$ and $\Sigma\, a_n$ converges, then $\Sigma\, (-1)^n a_n$ converges.

18. If $a_n > 0$ and $\lim_{n\to\infty} (a_{n+1}/a_n) < 1$, then $\lim_{n\to\infty} a_n = 0$.

EXERCISES

1–7 ■ Determine whether the sequence is convergent or divergent. If it is convergent, find its limit.

1. $a_n = \dfrac{2 + n^3}{1 + 2n^3}$

2. $a_n = \dfrac{9^{n+1}}{10^n}$

3. $a_n = \dfrac{n^3}{1 + n^2}$

4. $a_n = \cos(n\pi/2)$

5. $a_n = \dfrac{n \sin n}{n^2 + 1}$

6. $a_n = \dfrac{\ln n}{\sqrt{n}}$

7. $\{(1 + 3/n)^{4n}\}$

■ ■ ■ ■ ■ ■ ■ ■ ■ ■ ■ ■ ■ ■

8. A sequence is defined recursively by the equations $a_1 = 1$, $a_{n+1} = \frac{1}{3}(a_n + 4)$. Show that $\{a_n\}$ is increasing and $a_n < 2$ for all n. Deduce that $\{a_n\}$ is convergent and find its limit.

9–18 ■ Determine whether the series is convergent or divergent.

9. $\displaystyle\sum_{n=1}^{\infty} \frac{n}{n^3 + 1}$

10. $\displaystyle\sum_{n=1}^{\infty} \frac{n^2 + 1}{n^3 + 1}$

11. $\displaystyle\sum_{n=1}^{\infty} \frac{n^3}{5^n}$

12. $\displaystyle\sum_{n=1}^{\infty} \frac{(-1)^n}{\sqrt{n+1}}$

13. $\displaystyle\sum_{n=2}^{\infty} \frac{1}{n\sqrt{\ln n}}$

14. $\displaystyle\sum_{n=1}^{\infty} \ln\!\left(\frac{n}{3n+1}\right)$

15. $\displaystyle\sum_{n=1}^{\infty} (-1)^{n-1}\, \frac{\sqrt{n}}{n+1}$

16. $\displaystyle\sum_{n=1}^{\infty} \frac{\cos 3n}{1 + (1.2)^n}$

17. $\displaystyle\sum_{n=1}^{\infty} \frac{1 \cdot 3 \cdot 5 \cdots (2n-1)}{5^n n!}$

18. $\displaystyle\sum_{n=1}^{\infty} \frac{(-5)^{2n}}{n^2 9^n}$

■ ■ ■ ■ ■ ■ ■ ■ ■ ■ ■ ■ ■ ■

19–22 ■ Find the sum of the series.

19. $\displaystyle\sum_{n=1}^{\infty} \frac{2^{2n+1}}{5^n}$

20. $\displaystyle\sum_{n=0}^{\infty} \frac{(-1)^n \pi^n}{3^{2n}(2n)!}$

21. $\displaystyle\sum_{n=1}^{\infty} [\tan^{-1}(n+1) - \tan^{-1}n]$

22. $1 - e + \dfrac{e^2}{2!} - \dfrac{e^3}{3!} + \dfrac{e^4}{4!} - \cdots$

■ ■ ■ ■ ■ ■ ■ ■ ■ ■ ■ ■ ■ ■

23. Express the repeating decimal $1.2345345345\ldots$ as a fraction.

24. For what values of x does the series $\Sigma_{n=1}^{\infty}\, (\ln x)^n$ converge?

25. Find the sum of the series $\displaystyle\sum_{n=1}^{\infty} \frac{(-1)^{n+1}}{n^5}$ correct to four decimal places.

26. (a) Find the partial sum s_5 of the series $\Sigma_{n=1}^{\infty}\, 1/n^6$ and estimate the error in using it as an approximation to the sum of the series.
(b) Find the sum of this series correct to five decimal places.

27. Use the sum of the first eight terms to approximate the sum of the series $\sum_{n=1}^{\infty} (2 + 5^n)^{-1}$. Estimate the error involved in this approximation.

28. (a) Show that the series $\sum_{n=1}^{\infty} \dfrac{n^n}{(2n)!}$ is convergent.

(b) Deduce that $\lim_{n \to \infty} \dfrac{n^n}{(2n)!} = 0$.

29. Prove that if the series $\sum_{n=1}^{\infty} a_n$ is absolutely convergent, then the series

$$\sum_{n=1}^{\infty} \left(\frac{n + 1}{n} \right) a_n$$

is also absolutely convergent.

30–33 ■ Find the radius of convergence and interval of convergence of the series.

30. $\displaystyle\sum_{n=1}^{\infty} (-1)^n \frac{x^n}{n^2 5^n}$

31. $\displaystyle\sum_{n=1}^{\infty} \frac{(x + 2)^n}{n 4^n}$

32. $\displaystyle\sum_{n=1}^{\infty} \frac{2^n (x - 2)^n}{(n + 2)!}$

33. $\displaystyle\sum_{n=0}^{\infty} \frac{2^n (x - 3)^n}{\sqrt{n + 3}}$

■   ■   ■   ■   ■   ■   ■   ■   ■   ■   ■   ■   ■

34. Find the radius of convergence of the series

$$\sum_{n=1}^{\infty} \frac{(2n)!}{(n!)^2} x^n$$

35. Find the Taylor series of $f(x) = \sin x$ at $a = \pi/6$.

36. Find the Taylor series of $f(x) = \cos x$ at $a = \pi/3$.

37–44 ■ Find the Maclaurin series for f and its radius of convergence. You may use either the direct method (definition of a Maclaurin series) or known series such as geometric series, binomial series, or the Maclaurin series for e^x, $\sin x$, and $\tan^{-1} x$.

37. $f(x) = \dfrac{x^2}{1 + x}$

38. $f(x) = \tan^{-1}(x^2)$

39. $f(x) = \ln(1 - x)$

40. $f(x) = xe^{2x}$

41. $f(x) = \sin(x^4)$

42. $f(x) = 10^x$

43. $f(x) = 1/\sqrt[4]{16 - x}$

44. $f(x) = (1 - 3x)^{-5}$

■   ■   ■   ■   ■   ■   ■   ■   ■   ■   ■   ■   ■

45. Evaluate $\displaystyle\int \frac{e^x}{x} \, dx$ as an infinite series.

46. Use series to approximate $\int_0^1 \sqrt{1 + x^4} \, dx$ correct to two decimal places.

47–48 ■

(a) Approximate f by a Taylor polynomial with degree n at the number a.

(b) Graph f and T_n on a common screen.

(c) Use Taylor's Inequality to estimate the accuracy of the approximation $f(x) \approx T_n(x)$ when x lies in the given interval.

(d) Check your result in part (c) by graphing $|R_n(x)|$.

47. $f(x) = \sqrt{x}$, $a = 1$, $n = 3$, $0.9 \le x \le 1.1$

48. $f(x) = \sec x$, $a = 0$, $n = 2$, $0 \le x \le \pi/6$

■   ■   ■   ■   ■   ■   ■   ■   ■   ■   ■   ■   ■

49. Use series to evaluate the following limit.

$$\lim_{x \to 0} \frac{\sin x - x}{x^3}$$

50. The force due to gravity on an object with mass m at a height h above the surface of the Earth is

$$F = \frac{mgR^2}{(R + h)^2}$$

where R is the radius of the Earth and g is the acceleration due to gravity.

(a) Express F as a series in powers of h/R.

(b) Observe that if we approximate F by the first term in the series, we get the expression $F \approx mg$ that is usually used when h is much smaller than R. Use the Alternating Series Estimation Theorem to estimate the range of values of h for which the approximation $F \approx mg$ is accurate to within 1%. (Use $R = 6400$ km.)

51. (a) Show that $\tan \frac{1}{2} x = \cot \frac{1}{2} x - 2 \cot x$.

(b) Find the sum of the series

$$\sum_{n=1}^{\infty} \frac{1}{2^n} \tan \frac{x}{2^n}$$

52. A function f is defined by

$$f(x) = \lim_{n \to \infty} \frac{x^{2n} - 1}{x^{2n} + 1}$$

Where is f continuous?

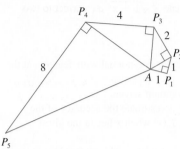

FIGURE FOR PROBLEM 2

1. If $f(x) = \sin(x^3)$, find $f^{(15)}(0)$.

2. Let $\{P_n\}$ be a sequence of points determined as in the figure. Thus $|AP_1| = 1$, $|P_nP_{n+1}| = 2^{n-1}$, and angle AP_nP_{n+1} is a right angle. Find $\lim_{n\to\infty} \angle P_nAP_{n+1}$.

3. To construct the **snowflake curve**, start with an equilateral triangle with sides of length 1. Step 1 in the construction is to divide each side into three equal parts, construct an equilateral triangle on the middle part, and then delete the middle part (see the figure). Step 2 is to repeat Step 1 for each side of the resulting polygon. This process is repeated at each succeeding step. The snowflake curve is the curve that results from repeating this process indefinitely.

   (a) Let s_n, l_n, and p_n represent the number of sides, the length of a side, and the total length of the nth approximating curve (the curve obtained after Step n of the construction), respectively. Find formulas for s_n, l_n, and p_n.

   (b) Show that $p_n \to \infty$ as $n \to \infty$.

   (c) Sum an infinite series to find the area enclosed by the snowflake curve.

   Parts (b) and (c) show that the snowflake curve is infinitely long but encloses only a finite area.

4. Find the sum of the series

$$1 + \frac{1}{2} + \frac{1}{3} + \frac{1}{4} + \frac{1}{6} + \frac{1}{8} + \frac{1}{9} + \frac{1}{12} + \cdots$$

where the terms are the reciprocals of the positive integers whose only prime factors are 2s and 3s.

5. Find the sum of the series $\sum_{n=2}^{\infty} \ln\left(1 - \frac{1}{n^2}\right)$.

FIGURE FOR PROBLEM 8

6. Suppose you have a large supply of books, all the same size, and you stack them at the edge of a table, with each book extending farther beyond the edge of the table than the one beneath it. Show that it is possible to do this so that the top book extends entirely beyond the table. In fact, show that the top book can extend any distance at all beyond the edge of the table if the stack is high enough. Use the following method of stacking: The top book extends half its length beyond the second book. The second book extends a quarter of its length beyond the third. The third extends one-sixth of its length beyond the fourth, and so on. (Try it yourself with a deck of cards.) Consider centers of mass.

7. Let
$$u = 1 + \frac{x^3}{3!} + \frac{x^6}{6!} + \frac{x^9}{9!} + \cdots$$
$$v = x + \frac{x^4}{4!} + \frac{x^7}{7!} + \frac{x^{10}}{10!} + \cdots$$
$$w = \frac{x^2}{2!} + \frac{x^5}{5!} + \frac{x^8}{8!} + \cdots$$

Show that $u^3 + v^3 + w^3 - 3uvw = 1$.

8. If $p > 1$, evaluate the expression

$$\frac{1 + \dfrac{1}{2^p} + \dfrac{1}{3^p} + \dfrac{1}{4^p} + \cdots}{1 - \dfrac{1}{2^p} + \dfrac{1}{3^p} - \dfrac{1}{4^p} + \cdots}$$

9. Suppose that circles of equal diameter are packed tightly in n rows inside an equilateral triangle. (The figure illustrates the case $n = 4$.) If A is the area of the triangle and A_n is the total area occupied by the n rows of circles, show that

$$\lim_{n \to \infty} \frac{A_n}{A} = \frac{\pi}{2\sqrt{3}}$$

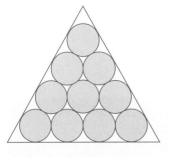

FIGURE FOR PROBLEM 11

10. A sequence $\{a_n\}$ is defined recursively by the equations

$$a_0 = a_1 = 1 \qquad n(n-1)a_n = (n-1)(n-2)a_{n-1} - (n-3)a_{n-2}$$

Find the sum of the series $\sum_{n=0}^{\infty} a_n$.

11. Taking the value of x^x at 0 to be 1 and integrating a series term by term, show that

$$\int_0^1 x^x \, dx = \sum_{n=1}^{\infty} \frac{(-1)^{n-1}}{n^n}$$

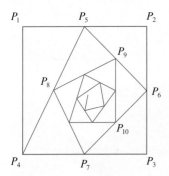

FIGURE FOR PROBLEM 12

12. Starting with the vertices $P_1(0, 1)$, $P_2(1, 1)$, $P_3(1, 0)$, $P_4(0, 0)$ of a square, we construct further points as shown in the figure: P_5 is the midpoint of P_1P_2, P_6 is the midpoint of P_2P_3, P_7 is the midpoint of P_3P_4, and so on. The polygon spiral path $P_1P_2P_3P_4P_5P_6P_7\ldots$ approaches a point P inside the square.
(a) If the coordinates of P_n are (x_n, y_n), show that $\frac{1}{2}x_n + x_{n+1} + x_{n+2} + x_{n+3} = 2$ and find a similar equation for the y-coordinates.
(b) Find the coordinates of P.

13. If $f(x) = \sum_{m=0}^{\infty} c_m x^m$ has positive radius of convergence and $e^{f(x)} = \sum_{n=0}^{\infty} d_n x^n$, show that

$$nd_n = \sum_{i=1}^{n} i c_i d_{n-i} \qquad n \geq 1$$

14. Right-angled triangles are constructed as in the figure. Each triangle has height 1 and its base is the hypotenuse of the preceding triangle. Show that this sequence of triangles makes indefinitely many turns around P by showing that $\sum \theta_n$ is a divergent series.

15. Consider the series whose terms are the reciprocals of the positive integers that can be written in base 10 notation without using the digit 0. Show that this series is convergent and the sum is less than 90.

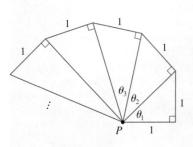

FIGURE FOR PROBLEM 14

16. (a) Show that the Maclaurin series of the function

$$f(x) = \frac{x}{1 - x - x^2} \qquad \text{is} \qquad \sum_{n=1}^{\infty} f_n x^n$$

where f_n is the nth Fibonacci number, that is, $f_1 = 1$, $f_2 = 1$, and $f_n = f_{n-1} + f_{n-2}$ for $n \geq 3$. [*Hint:* Write $x/(1 - x - x^2) = c_0 + c_1 x + c_2 x^2 + \cdots$ and multiply both sides of this equation by $1 - x - x^2$.]
(b) By writing $f(x)$ as a sum of partial fractions and thereby obtaining the Maclaurin series in a different way, find an explicit formula for the nth Fibonacci number.

9

Vectors and the Geometry of Space

In this chapter we introduce vectors and coordinate systems for three-dimensional space. This is the setting for the study of functions of two variables because the graph of such a function is a surface in space. Vectors provide particularly simple descriptions of lines and planes in space as well as velocities and accelerations of objects that move in space.

Three-Dimensional Coordinate Systems

FIGURE 1

Coordinate axes

FIGURE 2

Right-hand rule

To locate a point in a plane, two numbers are necessary. We know that any point in the plane can be represented as an ordered pair (a, b) of real numbers, where a is the x-coordinate and b is the y-coordinate. For this reason, a plane is called two-dimensional. To locate a point in space, three numbers are required. We represent any point in space by an ordered triple (a, b, c) of real numbers.

In order to represent points in space, we first choose a fixed point O (the origin) and three directed lines through O that are perpendicular to each other, called the **coordinate axes** and labeled the x-axis, y-axis, and z-axis. Usually we think of the x- and y-axes as being horizontal and the z-axis as being vertical, and we draw the orientation of the axes as in Figure 1. The direction of the z-axis is determined by the **right-hand rule** as illustrated in Figure 2: If you curl the fingers of your right hand around the z-axis in the direction of a $90°$ counterclockwise rotation from the positive x-axis to the positive y-axis, then your thumb points in the positive direction of the z-axis.

The three coordinate axes determine the three **coordinate planes** illustrated in Figure 3(a). The xy-plane is the plane that contains the x- and y-axes; the yz-plane contains the y- and z-axes; the xz-plane contains the x- and z-axes. These three coordinate planes divide space into eight parts, called **octants**. The **first octant**, in the foreground, is determined by the positive axes.

FIGURE 3   (a) Coordinate planes   (b)

Because many people have some difficulty visualizing diagrams of three-dimensional figures, you may find it helpful to do the following [see Figure 3(b)]. Look at any bottom corner of a room and call the corner the origin. The wall on your left is in the xz-plane, the wall on your right is in the yz-plane, and the floor is in the xy-plane. The x-axis runs along the intersection of the floor and the left wall. The y-axis runs along the intersection of the floor and the right wall. The z-axis runs up from the floor toward the ceiling along the intersection of the two walls. You are situated in the first octant, and you can now imagine seven other rooms situated in the other seven octants (three on the same floor and four on the floor below), all connected by the common corner point O.

Now if P is any point in space, let a be the (directed) distance from the yz-plane to P, let b be the distance from the xz-plane to P, and let c be the distance from the

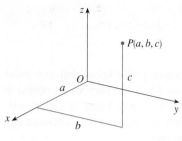

FIGURE 4

xy-plane to P. We represent the point P by the ordered triple (a, b, c) of real numbers and we call a, b, and c the **coordinates** of P; a is the x-coordinate, b is the y-coordinate, and c is the z-coordinate. Thus, to locate the point (a, b, c) we can start at the origin O and move a units along the x-axis, then b units parallel to the y-axis, and then c units parallel to the z-axis as in Figure 4.

The point $P(a, b, c)$ determines a rectangular box as in Figure 5. If we drop a perpendicular from P to the xy-plane, we get a point Q with coordinates $(a, b, 0)$ called the **projection** of P on the xy-plane. Similarly, $R(0, b, c)$ and $S(a, 0, c)$ are the projections of P on the yz-plane and xz-plane, respectively.

As numerical illustrations, the points $(-4, 3, -5)$ and $(3, -2, -6)$ are plotted in Figure 6.

FIGURE 5

FIGURE 6

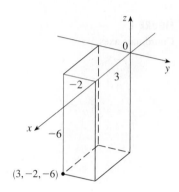

The Cartesian product $\mathbb{R} \times \mathbb{R} \times \mathbb{R} = \{(x, y, z) \mid x, y, z \in \mathbb{R}\}$ is the set of all ordered triples of real numbers and is denoted by $\mathbb{R}^3$. We have given a one-to-one correspondence between points P in space and ordered triples (a, b, c) in $\mathbb{R}^3$. It is called a **three-dimensional rectangular coordinate system**. Notice that, in terms of coordinates, the first octant can be described as the set of points whose coordinates are all positive.

In two-dimensional analytic geometry, the graph of an equation involving x and y is a curve in $\mathbb{R}^2$. In three-dimensional analytic geometry, an equation in x, y, and z represents a *surface* in $\mathbb{R}^3$.

$\boxed{\text{V}}$ **EXAMPLE 1** What surfaces in $\mathbb{R}^3$ are represented by the following equations?
(a) $z = 3$ (b) $y = 5$

SOLUTION
(a) The equation $z = 3$ represents the set $\{(x, y, z) \mid z = 3\}$, which is the set of all points in $\mathbb{R}^3$ whose z-coordinate is 3. This is the horizontal plane that is parallel to the xy-plane and three units above it as in Figure 7(a).

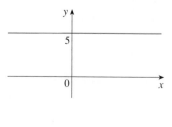

FIGURE 7   (a) $z = 3$, a plane in $\mathbb{R}^3$ (b) $y = 5$, a plane in $\mathbb{R}^3$ (c) $y = 5$, a line in $\mathbb{R}^2$

(b) The equation $y = 5$ represents the set of all points in $\mathbb{R}^3$ whose y-coordinate is 5. This is the vertical plane that is parallel to the xz-plane and five units to the right of it as in Figure 7(b). ■ ■

NOTE ▫ When an equation is given, we must understand from the context whether it represents a curve in $\mathbb{R}^2$ or a surface in $\mathbb{R}^3$. In Example 1, $y = 5$ represents a plane in $\mathbb{R}^3$, but of course $y = 5$ can also represent a line in $\mathbb{R}^2$ if we are dealing with two-dimensional analytic geometry. See Figure 7(b) and (c).

In general, if k is a constant, then $x = k$ represents a plane parallel to the yz-plane, $y = k$ is a plane parallel to the xz-plane, and $z = k$ is a plane parallel to the xy-plane. In Figure 5, the faces of the rectangular box are formed by the three coordinate planes $x = 0$ (the yz-plane), $y = 0$ (the xz-plane), and $z = 0$ (the xy-plane), and the planes $x = a$, $y = b$, and $z = c$.

EXAMPLE 2 Describe and sketch the surface in $\mathbb{R}^3$ represented by the equation $y = x$.

SOLUTION The equation represents the set of all points in $\mathbb{R}^3$ whose x- and y-coordinates are equal, that is, $\{(x, x, z) \mid x \in \mathbb{R}, z \in \mathbb{R}\}$. This is a vertical plane that intersects the xy-plane in the line $y = x$, $z = 0$. The portion of this plane that lies in the first octant is sketched in Figure 8. ■ ■

The familiar formula for the distance between two points in a plane is easily extended to the following three-dimensional formula.

FIGURE 8

The plane $y = x$

> **Distance Formula in Three Dimensions** The distance $|P_1P_2|$ between the points $P_1(x_1, y_1, z_1)$ and $P_2(x_2, y_2, z_2)$ is
>
> $$|P_1P_2| = \sqrt{(x_2 - x_1)^2 + (y_2 - y_1)^2 + (z_2 - z_1)^2}$$

To see why this formula is true, we construct a rectangular box as in Figure 9, where P_1 and P_2 are opposite vertices and the faces of the box are parallel to the coordinate planes. If $A(x_2, y_1, z_1)$ and $B(x_2, y_2, z_1)$ are the vertices of the box indicated in the figure, then

$$|P_1A| = |x_2 - x_1| \qquad |AB| = |y_2 - y_1| \qquad |BP_2| = |z_2 - z_1|$$

Because triangles P_1BP_2 and P_1AB are both right-angled, two applications of the Pythagorean Theorem give

$$|P_1P_2|^2 = |P_1B|^2 + |BP_2|^2$$

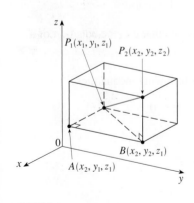

FIGURE 9

and

$$|P_1B|^2 = |P_1A|^2 + |AB|^2$$

Combining these equations, we get

$$|P_1P_2|^2 = |P_1A|^2 + |AB|^2 + |BP_2|^2$$
$$= |x_2 - x_1|^2 + |y_2 - y_1|^2 + |z_2 - z_1|^2$$
$$= (x_2 - x_1)^2 + (y_2 - y_1)^2 + (z_2 - z_1)^2$$

Therefore $\qquad |P_1P_2| = \sqrt{(x_2 - x_1)^2 + (y_2 - y_1)^2 + (z_2 - z_1)^2}$

EXAMPLE 3 The distance from the point $P(2, -1, 7)$ to the point $Q(1, -3, 5)$ is

$$|PQ| = \sqrt{(1 - 2)^2 + (-3 + 1)^2 + (5 - 7)^2} = \sqrt{1 + 4 + 4} = 3 \quad ■ ■$$

$\boxed{V}$ **EXAMPLE 4** Find an equation of a sphere with radius r and center $C(h, k, l)$.

SOLUTION By definition, a sphere is the set of all points $P(x, y, z)$ whose distance from C is r. (See Figure 10.) Thus, P is on the sphere if and only if $|PC| = r$. Squaring both sides, we have $|PC|^2 = r^2$ or

$$(x - h)^2 + (y - k)^2 + (z - l)^2 = r^2 \quad ■ ■$$

The result of Example 4 is worth remembering.

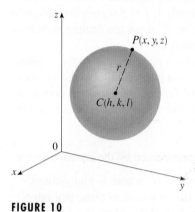

FIGURE 10

> **Equation of a Sphere** An equation of a sphere with center $C(h, k, l)$ and radius r is
>
> $$(x - h)^2 + (y - k)^2 + (z - l)^2 = r^2$$
>
> In particular, if the center is the origin O, then an equation of the sphere is
>
> $$x^2 + y^2 + z^2 = r^2$$

EXAMPLE 5 Show that $x^2 + y^2 + z^2 + 4x - 6y + 2z + 6 = 0$ is the equation of a sphere, and find its center and radius.

SOLUTION We can rewrite the given equation in the form of an equation of a sphere if we complete squares:

$$(x^2 + 4x + 4) + (y^2 - 6y + 9) + (z^2 + 2z + 1) = -6 + 4 + 9 + 1$$
$$(x + 2)^2 + (y - 3)^2 + (z + 1)^2 = 8$$

Comparing this equation with the standard form, we see that it is the equation of a sphere with center $(-2, 3, -1)$ and radius $\sqrt{8} = 2\sqrt{2}$. ■ ■

EXAMPLE 6 What region in $\mathbb{R}^3$ is represented by the following inequalities?

$$1 \leq x^2 + y^2 + z^2 \leq 4 \qquad z \leq 0$$

SOLUTION The inequalities

$$1 \leq x^2 + y^2 + z^2 \leq 4$$

can be rewritten as

$$1 \leq \sqrt{x^2 + y^2 + z^2} \leq 2$$

FIGURE 11

so they represent the points (x, y, z) whose distance from the origin is at least 1 and at most 2. But we are also given that $z \leq 0$, so the points lie on or below the xy-plane. Thus, the given inequalities represent the region that lies between (or on) the spheres $x^2 + y^2 + z^2 = 1$ and $x^2 + y^2 + z^2 = 4$ and beneath (or on) the xy-plane. It is sketched in Figure 11. ■ ■

9.1 Exercises

1. Suppose you start at the origin, move along the x-axis a distance of 4 units in the positive direction, and then move downward a distance of 3 units. What are the coordinates of your position?

2. Sketch the points $(0, 5, 2)$, $(4, 0, -1)$, $(2, 4, 6)$, and $(1, -1, 2)$ on a single set of coordinate axes.

3. Which of the points $P(6, 2, 3)$, $Q(-5, -1, 4)$, and $R(0, 3, 8)$ is closest to the xz-plane? Which point lies in the yz-plane?

4. What are the projections of the point $(2, 3, 5)$ on the xy-, yz-, and xz-planes? Draw a rectangular box with the origin and $(2, 3, 5)$ as opposite vertices and with its faces parallel to the coordinate planes. Label all vertices of the box. Find the length of the diagonal of the box.

5. Describe and sketch the surface in $\mathbb{R}^3$ represented by the equation $x + y = 2$.

6. (a) What does the equation $x = 4$ represent in $\mathbb{R}^2$? What does it represent in $\mathbb{R}^3$? Illustrate with sketches.
(b) What does the equation $y = 3$ represent in $\mathbb{R}^3$? What does $z = 5$ represent? What does the pair of equations $y = 3$, $z = 5$ represent? In other words, describe the set of points (x, y, z) such that $y = 3$ and $z = 5$. Illustrate with a sketch.

7. Find the lengths of the sides of the triangle PQR. Is it a right triangle? Is it an isosceles triangle?
(a) $P(3, -2, -3)$, $Q(7, 0, 1)$, $R(1, 2, 1)$
(b) $P(2, -1, 0)$, $Q(4, 1, 1)$, $R(4, -5, 4)$

8. Find the distance from $(3, 7, -5)$ to each of the following.
(a) The xy-plane   (b) The yz-plane
(c) The xz-plane   (d) The x-axis
(e) The y-axis   (f) The z-axis

9. Determine whether the points lie on straight line.
(a) $A(2, 4, 2)$, $B(3, 7, -2)$, $C(1, 3, 3)$
(b) $D(0, -5, 5)$, $E(1, -2, 4)$, $F(3, 4, 2)$

10. Find an equation of the sphere with center $(2, -6, 4)$ and radius 5. Describe its intersection with each of the coordinate planes.

11. Find an equation of the sphere that passes through the point $(4, 3, -1)$ and has center $(3, 8, 1)$.

12. Find an equation of the sphere that passes through the origin and whose center is $(1, 2, 3)$.

13–14 ■ Show that the equation represents a sphere, and find its center and radius.

13. $x^2 + y^2 + z^2 - 6x + 4y - 2z = 11$

14. $4x^2 + 4y^2 + 4z^2 - 8x + 16y = 1$

15. (a) Prove that the midpoint of the line segment from $P_1(x_1, y_1, z_1)$ to $P_2(x_2, y_2, z_2)$ is
$$\left(\frac{x_1 + x_2}{2}, \frac{y_1 + y_2}{2}, \frac{z_1 + z_2}{2} \right)$$
(b) Find the lengths of the medians of the triangle with vertices $A(1, 2, 3)$, $B(-2, 0, 5)$, and $C(4, 1, 5)$.

16. Find an equation of a sphere if one of its diameters has endpoints $(2, 1, 4)$ and $(4, 3, 10)$.

17. Find equations of the spheres with center $(2, -3, 6)$ that touch (a) the xy-plane, (b) the yz-plane, (c) the xz-plane.

18. Find an equation of the largest sphere with center $(5, 4, 9)$ that is contained in the first octant.

19–28 ■ Describe in words the region of $\mathbb{R}^3$ represented by the equation or inequality.

19. $y = -4$ **20.** $x = 10$

21. $x > 3$ **22.** $y \geqslant 0$

23. $0 \leqslant z \leqslant 6$ **24.** $z^2 = 1$

25. $x^2 + y^2 + z^2 \leqslant 3$ **26.** $x = z$

27. $x^2 + z^2 \leqslant 9$ **28.** $x^2 + y^2 + z^2 > 2z$

29–32 ■ Write inequalities to describe the region.

29. The half-space consisting of all points to the left of the xz-plane

30. The solid rectangular box in the first octant bounded by the planes $x = 1$, $y = 2$, and $z = 3$

31. The region consisting of all points between (but not on) the spheres of radius r and R centered at the origin, where $r < R$

32. The solid upper hemisphere of the sphere of radius 2 centered at the origin

33. The figure shows a line L_1 in space and a second line L_2, which is the projection of L_1 on the xy-plane. (In other

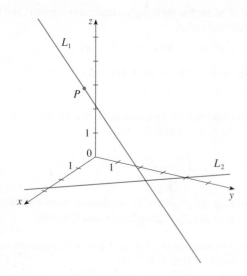

words, the points on L_2 are directly beneath, or above, the points on L_1.)

(a) Find the coordinates of the point P on the line L_1.

(b) Locate on the diagram the points A, B, and C, where the line L_1 intersects the xy-plane, the yz-plane, and the xz-plane, respectively.

34. Consider the points P such that the distance from P to $A(-1, 5, 3)$ is twice the distance from P to $B(6, 2, -2)$. Show that the set of all such points is a sphere, and find its center and radius.

35. Find an equation of the set of all points equidistant from the points $A(-1, 5, 3)$ and $B(6, 2, -2)$. Describe the set.

36. Find the volume of the solid that lies inside both of the spheres

$$x^2 + y^2 + z^2 + 4x - 2y + 4z + 5 = 0$$

and

$$x^2 + y^2 + z^2 = 4$$

9.2 Vectors

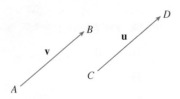

FIGURE 1

Equivalent vectors

The term **vector** is used by scientists to indicate a quantity (such as displacement or velocity or force) that has both magnitude and direction. A vector is often represented by an arrow or a directed line segment. The length of the arrow represents the magnitude of the vector and the arrow points in the direction of the vector. We denote a vector by printing a letter in boldface (**v**) or by putting an arrow above the letter ($\vec{v}$).

For instance, suppose a particle moves along a line segment from point A to point B. The corresponding **displacement vector v**, shown in Figure 1, has **initial point** A (the tail) and **terminal point** B (the tip) and we indicate this by writing $\mathbf{v} = \overrightarrow{AB}$. Notice that the vector $\mathbf{u} = \overrightarrow{CD}$ has the same length and the same direction as **v** even though it is in a different position. We say that **u** and **v** are **equivalent** (or **equal**) and we write $\mathbf{u} = \mathbf{v}$. The **zero vector**, denoted by **0**, has length 0. It is the only vector with no specific direction.

▪▪ Combining Vectors

FIGURE 2

Suppose a particle moves from A to B, so its displacement vector is $\overrightarrow{AB}$. Then the particle changes direction and moves from B to C, with displacement vector $\overrightarrow{BC}$ as in Figure 2. The combined effect of these displacements is that the particle has moved from A to C. The resulting displacement vector $\overrightarrow{AC}$ is called the *sum* of $\overrightarrow{AB}$ and $\overrightarrow{BC}$ and we write

$$\overrightarrow{AC} = \overrightarrow{AB} + \overrightarrow{BC}$$

In general, if we start with vectors **u** and **v**, we first move **v** so that its tail coincides with the tip of **u** and define the sum of **u** and **v** as follows.

> **Definition of Vector Addition** If **u** and **v** are vectors positioned so the initial point of **v** is at the terminal point of **u**, then the **sum u + v** is the vector from the initial point of **u** to the terminal point of **v**.

The definition of vector addition is illustrated in Figure 3. You can see why this definition is sometimes called the **Triangle Law**.

FIGURE 3 The Triangle Law

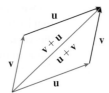

FIGURE 4 The Parallelogram Law

In Figure 4 we start with the same vectors **u** and **v** as in Figure 3 and draw another copy of **v** with the same initial point as **u**. Completing the parallelogram, we see that **u + v = v + u**. This also gives another way to construct the sum: If we place **u** and **v** so they start at the same point, then **u + v** lies along the diagonal of the parallelogram with **u** and **v** as sides. (This is called the **Parallelogram Law**.)

[V] **EXAMPLE 1** Draw the sum of the vectors **a** and **b** shown in Figure 5.

SOLUTION First we translate **b** and place its tail at the tip of **a**, being careful to draw a copy of **b** that has the same length and direction. Then we draw the vector **a + b** [see Figure 6(a)] starting at the initial point of **a** and ending at the terminal point of the copy of **b**.

Alternatively, we could place **b** so it starts where **a** starts and construct **a + b** by the Parallelogram Law as in Figure 6(b).

FIGURE 5

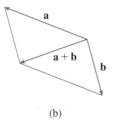

TEC Visual 9.2 shows how the Triangle and Parallelogram Laws work for various vectors **u** and **v**.

FIGURE 6 (a) (b)

It is possible to multiply a vector by a real number c. (In this context we call the real number c a **scalar** to distinguish it from a vector.) For instance, we want $2\mathbf{v}$ to be the same vector as $\mathbf{v} + \mathbf{v}$, which has the same direction as **v** but is twice as long. In general, we multiply a vector by a scalar as follows.

> **Definition of Scalar Multiplication** If c is a scalar and **v** is a vector, then the **scalar multiple** $c\mathbf{v}$ is the vector whose length is $|c|$ times the length of **v** and whose direction is the same as **v** if $c > 0$ and is opposite to **v** if $c < 0$. If $c = 0$ or $\mathbf{v} = \mathbf{0}$, then $c\mathbf{v} = \mathbf{0}$.

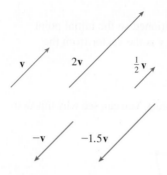

FIGURE 7
Scalar multiples of **v**

This definition is illustrated in Figure 7. We see that real numbers work like scaling factors here; that's why we call them scalars. Notice that two nonzero vectors are **parallel** if they are scalar multiples of one another. In particular, the vector $-\mathbf{v} = (-1)\mathbf{v}$ has the same length as **v** but points in the opposite direction. We call it the **negative** of **v**.

By the **difference** $\mathbf{u} - \mathbf{v}$ of two vectors we mean

$$\mathbf{u} - \mathbf{v} = \mathbf{u} + (-\mathbf{v})$$

So we can construct $\mathbf{u} - \mathbf{v}$ by first drawing the negative of **v**, $-\mathbf{v}$, and then adding it to **u** by the Parallelogram Law as in Figure 8(a). Alternatively, since $\mathbf{v} + (\mathbf{u} - \mathbf{v}) = \mathbf{u}$, the vector $\mathbf{u} - \mathbf{v}$, when added to **v**, gives **u**. So we could construct $\mathbf{u} - \mathbf{v}$ as in Figure 8(b) by means of the Triangle Law.

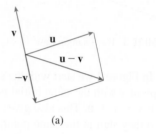

(a)   (b)

FIGURE 8
Drawing $\mathbf{u} - \mathbf{v}$

EXAMPLE 2 If **a** and **b** are the vectors shown in Figure 9, draw $\mathbf{a} - 2\mathbf{b}$.

SOLUTION We first draw the vector $-2\mathbf{b}$ pointing in the direction opposite to **b** and twice as long. We place it with its tail at the tip of **a** and then use the Triangle Law to draw $\mathbf{a} + (-2\mathbf{b})$ as in Figure 10.

FIGURE 9

FIGURE 10   ■ ■

$\mathbf{a} = \langle a_1, a_2 \rangle$

$\mathbf{a} = \langle a_1, a_2, a_3 \rangle$

FIGURE 11

■■ Components
■■

For some purposes it's best to introduce a coordinate system and treat vectors algebraically. If we place the initial point of a vector **a** at the origin of a rectangular coordinate system, then the terminal point of **a** has coordinates of the form (a_1, a_2) or (a_1, a_2, a_3), depending on whether our coordinate system is two- or three-dimensional (see Figure 11). These coordinates are called the **components** of **a** and we write

$$\mathbf{a} = \langle a_1, a_2 \rangle \qquad \text{or} \qquad \mathbf{a} = \langle a_1, a_2, a_3 \rangle$$

We use the notation $\langle a_1, a_2 \rangle$ for the ordered pair that refers to a vector so as not to confuse it with the ordered pair (a_1, a_2) that refers to a point in the plane.

For instance, the vectors shown in Figure 12 are all equivalent to the vector $\overrightarrow{OP} = \langle 3, 2 \rangle$ whose terminal point is $P(3, 2)$. What they have in common is that the terminal point is reached from the initial point by a displacement of three units to the right and two upward. We can think of all these geometric vectors as **representations** of the algebraic vector $\mathbf{a} = \langle 3, 2 \rangle$. The particular representation $\overrightarrow{OP}$ from the origin to the point $P(3, 2)$ is called the **position vector** of the point P.

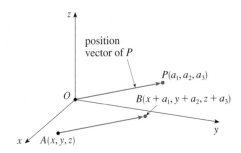

FIGURE 12
Representations of the vector $\mathbf{a} = \langle 3, 2 \rangle$

FIGURE 13
Representations of $\mathbf{a} = \langle a_1, a_2, a_3 \rangle$

In three dimensions, the vector $\mathbf{a} = \overrightarrow{OP} = \langle a_1, a_2, a_3 \rangle$ is the **position vector** of the point $P(a_1, a_2, a_3)$. (See Figure 13.) Let's consider any other representation $\overrightarrow{AB}$ of $\mathbf{a}$, where the initial point is $A(x_1, y_1, z_1)$ and the terminal point is $B(x_2, y_2, z_2)$. Then we must have $x_1 + a_1 = x_2$, $y_1 + a_2 = y_2$, and $z_1 + a_3 = z_2$ and so $a_1 = x_2 - x_1$, $a_2 = y_2 - y_1$, and $a_3 = z_2 - z_1$. Thus, we have the following result.

> **1** Given the points $A(x_1, y_1, z_1)$ and $B(x_2, y_2, z_2)$, the vector $\mathbf{a}$ with representation $\overrightarrow{AB}$ is
> $$\mathbf{a} = \langle x_2 - x_1, y_2 - y_1, z_2 - z_1 \rangle$$

$\boxed{\mathrm{V}}$ **EXAMPLE 3** Find the vector represented by the directed line segment with initial point $A(2, -3, 4)$ and terminal point $B(-2, 1, 1)$.

SOLUTION By (1), the vector corresponding to $\overrightarrow{AB}$ is
$$\mathbf{a} = \langle -2 - 2, 1 - (-3), 1 - 4 \rangle = \langle -4, 4, -3 \rangle \qquad ■ ■$$

The **magnitude** or **length** of the vector $\mathbf{v}$ is the length of any of its representations and is denoted by the symbol $|\mathbf{v}|$ or $\|\mathbf{v}\|$. By using the distance formula to compute the length of a segment OP, we obtain the following formulas.

> The length of the two-dimensional vector $\mathbf{a} = \langle a_1, a_2 \rangle$ is
> $$|\mathbf{a}| = \sqrt{a_1^2 + a_2^2}$$
> The length of the three-dimensional vector $\mathbf{a} = \langle a_1, a_2, a_3 \rangle$ is
> $$|\mathbf{a}| = \sqrt{a_1^2 + a_2^2 + a_3^2}$$

FIGURE 14

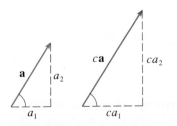

FIGURE 15

How do we add vectors algebraically? Figure 14 shows that if $\mathbf{a} = \langle a_1, a_2 \rangle$ and $\mathbf{b} = \langle b_1, b_2 \rangle$, then the sum is $\mathbf{a} + \mathbf{b} = \langle a_1 + b_1, a_2 + b_2 \rangle$, at least for the case where the components are positive. In other words, *to add algebraic vectors we add their components.* Similarly, *to subtract vectors we subtract components.* From the similar triangles in Figure 15 we see that the components of $c\mathbf{a}$ are ca_1 and ca_2. So *to multiply a vector by a scalar we multiply each component by that scalar.*

If $\mathbf{a} = \langle a_1, a_2 \rangle$ and $\mathbf{b} = \langle b_1, b_2 \rangle$, then

$$\mathbf{a} + \mathbf{b} = \langle a_1 + b_1, a_2 + b_2 \rangle \qquad \mathbf{a} - \mathbf{b} = \langle a_1 - b_1, a_2 - b_2 \rangle$$

$$c\mathbf{a} = \langle ca_1, ca_2 \rangle$$

Similarly, for three-dimensional vectors,

$$\langle a_1, a_2, a_3 \rangle + \langle b_1, b_2, b_3 \rangle = \langle a_1 + b_1, a_2 + b_2, a_3 + b_3 \rangle$$

$$\langle a_1, a_2, a_3 \rangle - \langle b_1, b_2, b_3 \rangle = \langle a_1 - b_1, a_2 - b_2, a_3 - b_3 \rangle$$

$$c\langle a_1, a_2, a_3 \rangle = \langle ca_1, ca_2, ca_3 \rangle$$

$\boxed{\text{V}}$ **EXAMPLE 4** If $\mathbf{a} = \langle 4, 0, 3 \rangle$ and $\mathbf{b} = \langle -2, 1, 5 \rangle$, find $|\mathbf{a}|$ and the vectors $\mathbf{a} + \mathbf{b}$, $\mathbf{a} - \mathbf{b}$, $3\mathbf{b}$, and $2\mathbf{a} + 5\mathbf{b}$.

SOLUTION
$$|\mathbf{a}| = \sqrt{4^2 + 0^2 + 3^2} = \sqrt{25} = 5$$

$$\mathbf{a} + \mathbf{b} = \langle 4, 0, 3 \rangle + \langle -2, 1, 5 \rangle$$
$$= \langle 4 - 2, 0 + 1, 3 + 5 \rangle = \langle 2, 1, 8 \rangle$$

$$\mathbf{a} - \mathbf{b} = \langle 4, 0, 3 \rangle - \langle -2, 1, 5 \rangle$$
$$= \langle 4 - (-2), 0 - 1, 3 - 5 \rangle = \langle 6, -1, -2 \rangle$$

$$3\mathbf{b} = 3\langle -2, 1, 5 \rangle = \langle 3(-2), 3(1), 3(5) \rangle = \langle -6, 3, 15 \rangle$$

$$2\mathbf{a} + 5\mathbf{b} = 2\langle 4, 0, 3 \rangle + 5\langle -2, 1, 5 \rangle$$
$$= \langle 8, 0, 6 \rangle + \langle -10, 5, 25 \rangle = \langle -2, 5, 31 \rangle \qquad ■ ■$$

We denote by V_2 the set of all two-dimensional vectors and by V_3 the set of all three-dimensional vectors. More generally, we will later need to consider the set V_n of all n-dimensional vectors. An n-dimensional vector is an ordered n-tuple:

$$\mathbf{a} = \langle a_1, a_2, \ldots, a_n \rangle$$

where $a_1, a_2, \ldots, a_n$ are real numbers that are called the components of $\mathbf{a}$. Addition and scalar multiplication are defined in terms of components just as for the cases $n = 2$ and $n = 3$.

■ ■ Vectors in n dimensions are used to list various quantities in an organized way. For instance, the components of a six-dimensional vector

$$\mathbf{p} = \langle p_1, p_2, p_3, p_4, p_5, p_6 \rangle$$

might represent the prices of six different ingredients required to make a particular product. Four-dimensional vectors $\langle x, y, z, t \rangle$ are used in relativity theory, where the first three components specify a position in space and the fourth represents time.

Properties of Vectors If $\mathbf{a}$, $\mathbf{b}$, and $\mathbf{c}$ are vectors in V_n and c and d are scalars, then

1. $\mathbf{a} + \mathbf{b} = \mathbf{b} + \mathbf{a}$	**2.** $\mathbf{a} + (\mathbf{b} + \mathbf{c}) = (\mathbf{a} + \mathbf{b}) + \mathbf{c}$
3. $\mathbf{a} + \mathbf{0} = \mathbf{a}$	**4.** $\mathbf{a} + (-\mathbf{a}) = \mathbf{0}$
5. $c(\mathbf{a} + \mathbf{b}) = c\mathbf{a} + c\mathbf{b}$	**6.** $(c + d)\mathbf{a} = c\mathbf{a} + d\mathbf{a}$
7. $(cd)\mathbf{a} = c(d\mathbf{a})$	**8.** $1\mathbf{a} = \mathbf{a}$

These eight properties of vectors can be readily verified either geometrically or algebraically. For instance, Property 1 can be seen from Figure 4 (it's equivalent to the Parallelogram Law) or as follows for the case $n = 2$:

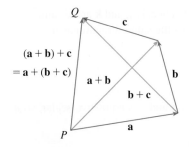

FIGURE 16

$$\mathbf{a} + \mathbf{b} = \langle a_1, a_2 \rangle + \langle b_1, b_2 \rangle = \langle a_1 + b_1, a_2 + b_2 \rangle$$

$$= \langle b_1 + a_1, b_2 + a_2 \rangle = \langle b_1, b_2 \rangle + \langle a_1, a_2 \rangle$$

$$= \mathbf{b} + \mathbf{a}$$

We can see why Property 2 (the associative law) is true by looking at Figure 16 and applying the Triangle Law several times: The vector $\overrightarrow{PQ}$ is obtained either by first constructing $\mathbf{a} + \mathbf{b}$ and then adding $\mathbf{c}$ or by adding $\mathbf{a}$ to the vector $\mathbf{b} + \mathbf{c}$.

Three vectors in V_3 play a special role. Let

$$\mathbf{i} = \langle 1, 0, 0 \rangle \qquad \mathbf{j} = \langle 0, 1, 0 \rangle \qquad \mathbf{k} = \langle 0, 0, 1 \rangle$$

Then $\mathbf{i}$, $\mathbf{j}$, and $\mathbf{k}$ are vectors that have length 1 and point in the directions of the positive x-, y-, and z-axes. Similarly, in two dimensions we define $\mathbf{i} = \langle 1, 0 \rangle$ and $\mathbf{j} = \langle 0, 1 \rangle$. (See Figure 17.)

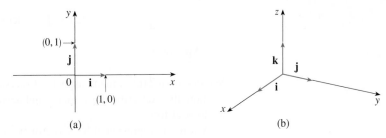

FIGURE 17

Standard basis vectors in V_2 and V_3

If $\mathbf{a} = \langle a_1, a_2, a_3 \rangle$, then we can write

$$\mathbf{a} = \langle a_1, a_2, a_3 \rangle = \langle a_1, 0, 0 \rangle + \langle 0, a_2, 0 \rangle + \langle 0, 0, a_3 \rangle$$

$$= a_1 \langle 1, 0, 0 \rangle + a_2 \langle 0, 1, 0 \rangle + a_3 \langle 0, 0, 1 \rangle$$

2 $\qquad \mathbf{a} = a_1 \mathbf{i} + a_2 \mathbf{j} + a_3 \mathbf{k}$

Thus, any vector in V_3 can be expressed in terms of the **standard basis vectors i, j,** and **k**. For instance,

$$\langle 1, -2, 6 \rangle = \mathbf{i} - 2\mathbf{j} + 6\mathbf{k}$$

Similarly, in two dimensions, we can write

3 $\qquad \mathbf{a} = \langle a_1, a_2 \rangle = a_1 \mathbf{i} + a_2 \mathbf{j}$

See Figure 18 for the geometric interpretation of Equations 3 and 2 and compare with Figure 17.

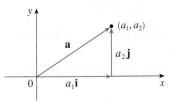

(a) $\mathbf{a} = a_1 \mathbf{i} + a_2 \mathbf{j}$

(b) $\mathbf{a} = a_1 \mathbf{i} + a_2 \mathbf{j} + a_3 \mathbf{k}$

FIGURE 18

EXAMPLE 5 If $\mathbf{a} = \mathbf{i} + 2\mathbf{j} - 3\mathbf{k}$ and $\mathbf{b} = 4\mathbf{i} + 7\mathbf{k}$, express the vector $2\mathbf{a} + 3\mathbf{b}$ in terms of $\mathbf{i}$, $\mathbf{j}$, and $\mathbf{k}$.

SOLUTION Using Properties 1, 2, 5, 6, and 7 of vectors, we have

$$2\mathbf{a} + 3\mathbf{b} = 2(\mathbf{i} + 2\mathbf{j} - 3\mathbf{k}) + 3(4\mathbf{i} + 7\mathbf{k})$$

$$= 2\mathbf{i} + 4\mathbf{j} - 6\mathbf{k} + 12\mathbf{i} + 21\mathbf{k} = 14\mathbf{i} + 4\mathbf{j} + 15\mathbf{k}$$ ■ ■

A **unit vector** is a vector whose length is 1. For instance, **i**, **j**, and **k** are all unit vectors. In general, if $\mathbf{a} \neq \mathbf{0}$, then the unit vector that has the same direction as **a** is

$$\boxed{4} \qquad \mathbf{u} = \frac{1}{|\mathbf{a}|}\mathbf{a} = \frac{\mathbf{a}}{|\mathbf{a}|}$$

In order to verify this, we let $c = 1/|\mathbf{a}|$. Then $\mathbf{u} = c\mathbf{a}$ and c is a positive scalar, so **u** has the same direction as **a**. Also

$$|\mathbf{u}| = |c\mathbf{a}| = |c||\mathbf{a}| = \frac{1}{|\mathbf{a}|}|\mathbf{a}| = 1$$

EXAMPLE 6 Find the unit vector in the direction of the vector $2\mathbf{i} - \mathbf{j} - 2\mathbf{k}$.

SOLUTION The given vector has length

$$|2\mathbf{i} - \mathbf{j} - 2\mathbf{k}| = \sqrt{2^2 + (-1)^2 + (-2)^2} = \sqrt{9} = 3$$

so, by Equation 4, the unit vector with the same direction is

$$\tfrac{1}{3}(2\mathbf{i} - \mathbf{j} - 2\mathbf{k}) = \tfrac{2}{3}\mathbf{i} - \tfrac{1}{3}\mathbf{j} - \tfrac{2}{3}\mathbf{k}$$

■ ■

■■ Applications

Vectors are useful in many aspects of physics and engineering. In Chapter 10 we will see how they describe the velocity and acceleration of objects moving in space. Here we look at forces.

A force is represented by a vector because it has both a magnitude (measured in pounds or newtons) and a direction. If several forces are acting on an object, the **resultant force** experienced by the object is the vector sum of these forces.

EXAMPLE 7 A 100-lb weight hangs from two wires as shown in Figure 19. Find the tensions (forces) $\mathbf{T}_1$ and $\mathbf{T}_2$ in both wires and their magnitudes.

SOLUTION We first express $\mathbf{T}_1$ and $\mathbf{T}_2$ in terms of their horizontal and vertical components. From Figure 20 we see that

$$\boxed{5} \qquad \mathbf{T}_1 = -|\mathbf{T}_1|\cos 50°\,\mathbf{i} + |\mathbf{T}_1|\sin 50°\,\mathbf{j}$$

$$\boxed{6} \qquad \mathbf{T}_2 = |\mathbf{T}_2|\cos 32°\,\mathbf{i} + |\mathbf{T}_2|\sin 32°\,\mathbf{j}$$

The resultant $\mathbf{T}_1 + \mathbf{T}_2$ of the tensions counterbalances the weight **w** and so we must have

$$\mathbf{T}_1 + \mathbf{T}_2 = -\mathbf{w} = 100\,\mathbf{j}$$

Thus

$$\left(-|\mathbf{T}_1|\cos 50° + |\mathbf{T}_2|\cos 32°\right)\mathbf{i} + \left(|\mathbf{T}_1|\sin 50° + |\mathbf{T}_2|\sin 32°\right)\mathbf{j} = 100\,\mathbf{j}$$

Equating components, we get

$$-|\mathbf{T}_1|\cos 50° + |\mathbf{T}_2|\cos 32° = 0$$

$$|\mathbf{T}_1|\sin 50° + |\mathbf{T}_2|\sin 32° = 100$$

Solving the first of these equations for $|\mathbf{T}_2|$ and substituting into the second, we get

$$|\mathbf{T}_1|\sin 50° + \frac{|\mathbf{T}_1|\cos 50°}{\cos 32°}\sin 32° = 100$$

FIGURE 19

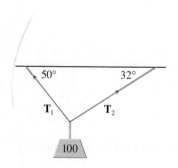

FIGURE 20

So the magnitudes of the tensions are

$$|\mathbf{T}_1| = \frac{100}{\sin 50° + \tan 32° \cos 50°} \approx 85.64 \text{ lb}$$

and

$$|\mathbf{T}_2| = \frac{|\mathbf{T}_1| \cos 50°}{\cos 32°} \approx 64.91 \text{ lb}$$

Substituting these values in (5) and (6), we obtain the tension vectors

$$\mathbf{T}_1 \approx -55.05\,\mathbf{i} + 65.60\,\mathbf{j} \qquad \mathbf{T}_2 \approx 55.05\,\mathbf{i} + 34.40\,\mathbf{j} \qquad ■ ■$$

9.2 Exercises

1. Are the following quantities vectors or scalars? Explain.
  (a) The cost of a theater ticket
  (b) The current in a river
  (c) The initial flight path from Houston to Dallas
  (d) The population of the world

2. What is the relationship between the point $(4, 7)$ and the vector $\langle 4, 7 \rangle$? Illustrate with a sketch.

3. Name all the equal vectors in the parallelogram shown.

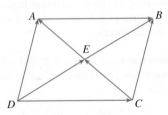

4. Write each combination of vectors as a single vector.
  (a) $\overrightarrow{PQ} + \overrightarrow{QR}$
  (b) $\overrightarrow{RP} + \overrightarrow{PS}$
  (c) $\overrightarrow{QS} - \overrightarrow{PS}$
  (d) $\overrightarrow{RS} + \overrightarrow{SP} + \overrightarrow{PQ}$

5. Copy the vectors in the figure and use them to draw the following vectors.
  (a) $\mathbf{u} + \mathbf{v}$
  (b) $\mathbf{u} - \mathbf{v}$
  (c) $\mathbf{v} + \mathbf{w}$
  (d) $\mathbf{w} + \mathbf{v} + \mathbf{u}$

6. Copy the vectors in the figure and use them to draw the following vectors.
  (a) $\mathbf{a} + \mathbf{b}$
  (b) $\mathbf{a} - \mathbf{b}$
  (c) $2\mathbf{a}$
  (d) $-\frac{1}{2}\mathbf{b}$
  (e) $2\mathbf{a} + \mathbf{b}$
  (f) $\mathbf{b} - 3\mathbf{a}$

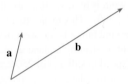

7–10 ■ Find a vector $\mathbf{a}$ with representation given by the directed line segment $\overrightarrow{AB}$. Draw $\overrightarrow{AB}$ and the equivalent representation starting at the origin.

7. $A(2, 3)$, $B(-2, 1)$
8. $A(-2, -2)$, $B(5, 3)$
9. $A(0, 3, 1)$, $B(2, 3, -1)$
10. $A(4, 0, -2)$, $B(4, 2, 1)$

11–14 ■ Find the sum of the given vectors and illustrate geometrically.

11. $\langle 3, -1 \rangle$, $\langle -2, 4 \rangle$
12. $\langle -2, -1 \rangle$, $\langle 5, 7 \rangle$
13. $\langle 0, 1, 2 \rangle$, $\langle 0, 0, -3 \rangle$
14. $\langle -1, 0, 2 \rangle$, $\langle 0, 4, 0 \rangle$

15–18 ■ Find $\mathbf{a} + \mathbf{b}$, $2\mathbf{a} + 3\mathbf{b}$, $|\mathbf{a}|$, and $|\mathbf{a} - \mathbf{b}|$.

15. $\mathbf{a} = \langle 5, -12 \rangle$, $\mathbf{b} = \langle -3, -6 \rangle$
16. $\mathbf{a} = 4\mathbf{i} + \mathbf{j}$, $\mathbf{b} = \mathbf{i} - 2\mathbf{j}$
17. $\mathbf{a} = \mathbf{i} + 2\mathbf{j} - 3\mathbf{k}$, $\mathbf{b} = -2\mathbf{i} - \mathbf{j} + 5\mathbf{k}$
18. $\mathbf{a} = 2\mathbf{i} - 4\mathbf{j} + 4\mathbf{k}$, $\mathbf{b} = 2\mathbf{j} - \mathbf{k}$

19. Find a unit vector with the same direction as $8\mathbf{i} - \mathbf{j} + 4\mathbf{k}$.

20. Find a vector that has the same direction as $\langle -2, 4, 2 \rangle$ but has length 6.

21. If **v** lies in the first quadrant and makes an angle $\pi/3$ with the positive x-axis and $|\mathbf{v}| = 4$, find **v** in component form.

22. If a child pulls a sled through the snow with a force of 50 N exerted at an angle of 38° above the horizontal, find the horizontal and vertical components of the force.

23. Two forces $\mathbf{F}_1$ and $\mathbf{F}_2$ with magnitudes 10 lb and 12 lb act on an object at a point P as shown in the figure. Find the resultant force **F** acting at P as well as its magnitude and its direction. (Indicate the direction by finding the angle θ shown in the figure.)

24. Velocities have both direction and magnitude and thus are vectors. The magnitude of a velocity vector is called *speed*. Suppose that a wind is blowing from the direction N45°W at a speed of 50 km/h. (This means that the direction from which the wind blows is 45° west of the northerly direction.) A pilot is steering a plane in the direction N60°E at an airspeed (speed in still air) of 250 km/h. The *true course*, or *track*, of the plane is the direction of the resultant of the velocity vectors of the plane and the wind. The *ground speed* of the plane is the magnitude of the resultant. Find the true course and the ground speed of the plane.

25. A woman walks due west on the deck of a ship at 3 mi/h. The ship is moving north at a speed of 22 mi/h. Find the speed and direction of the woman relative to the surface of the water.

26. Ropes 3 m and 5 m in length are fastened to a holiday decoration that is suspended over a town square. The decoration has a mass of 5 kg. The ropes, fastened at different heights, make angles of 52° and 40° with the horizontal. Find the tension in each wire and the magnitude of each tension.

27. A clothesline is tied between two poles, 8 m apart. The line is quite taut and has negligible sag. When a wet shirt with a mass of 0.8 kg is hung at the middle of the line, the mid-

point is pulled down 8 cm. Find the tension in each half of the clothesline.

28. The tension **T** at each end of the chain has magnitude 25 N. What is the weight of the chain?

29. (a) Draw the vectors $\mathbf{a} = \langle 3, 2 \rangle$, $\mathbf{b} = \langle 2, -1 \rangle$, and $\mathbf{c} = \langle 7, 1 \rangle$.
   (b) Show, by means of a sketch, that there are scalars s and t such that $\mathbf{c} = s\mathbf{a} + t\mathbf{b}$.
   (c) Use the sketch to estimate the values of s and t.
   (d) Find the exact values of s and t.

30. Suppose that **a** and **b** are nonzero vectors that are not parallel and **c** is any vector in the plane determined by **a** and **b**. Give a geometric argument to show that **c** can be written as $\mathbf{c} = s\mathbf{a} + t\mathbf{b}$ for suitable scalars s and t. Then give an argument using components.

31. Suppose **a** is a three-dimensional unit vector in the first octant that starts at the origin and makes angles of 60° and 72° with the positive x- and y-axes, respectively. Express **a** in terms of its components.

32. Suppose a vector **a** makes angles α, β, and γ with the positive x-, y-, and z-axes, respectively. Find the components of **a** and show that

$$\cos^2\alpha + \cos^2\beta + \cos^2\gamma = 1$$

(The numbers $\cos\alpha$, $\cos\beta$, and $\cos\gamma$ are called the *direction cosines* of **a**.)

33. If $\mathbf{r} = \langle x, y, z \rangle$ and $\mathbf{r}_0 = \langle x_0, y_0, z_0 \rangle$, describe the set of all points (x, y, z) such that $|\mathbf{r} - \mathbf{r}_0| = 1$.

34. If $\mathbf{r} = \langle x, y \rangle$, $\mathbf{r}_1 = \langle x_1, y_1 \rangle$, and $\mathbf{r}_2 = \langle x_2, y_2 \rangle$, describe the set of all points (x, y) such that $|\mathbf{r} - \mathbf{r}_1| + |\mathbf{r} - \mathbf{r}_2| = k$, where $k > |\mathbf{r}_1 - \mathbf{r}_2|$.

35. Figure 16 gives a geometric demonstration of Property 2 of vectors. Use components to give an algebraic proof of this fact for the case $n = 2$.

36. Prove Property 5 of vectors algebraically for the case $n = 3$. Then use similar triangles to give a geometric proof.

37. Use vectors to prove that the line joining the midpoints of two sides of a triangle is parallel to the third side and half its length.

38. Suppose the three coordinate planes are all mirrored and a light ray given by the vector $\mathbf{a} = \langle a_1, a_2, a_3 \rangle$ first strikes the xz-plane, as shown in the figure. Use the fact that the angle of incidence equals the angle of reflection to show that the direction of the reflected ray is given by $\mathbf{b} = \langle a_1, -a_2, a_3 \rangle$. Deduce that, after being reflected by all three mutually perpendicular mirrors, the resulting ray is parallel to the initial ray. (American space scientists used this principle, together with laser beams and an array of corner mirrors on the Moon, to calculate very precisely the distance from the Earth to the Moon.)

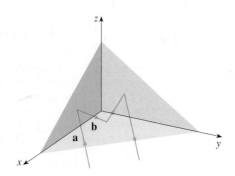

9.3 The Dot Product

So far we have added two vectors and multiplied a vector by a scalar. The question arises: Is it possible to multiply two vectors so that their product is a useful quantity? One such product is the dot product, which we consider in this section. Another is the cross product, which is discussed in the next section.

■ Work and the Dot Product

An example of a situation in physics and engineering where we need to combine two vectors occurs in calculating the work done by a force. In Section 6.5 we defined the work done by a constant force F in moving an object through a distance d as $W = Fd$, but this applies only when the force is directed along the line of motion of the object. Suppose, however, that the constant force is a vector $\mathbf{F} = \overrightarrow{PR}$ pointing in some other direction, as in Figure 1. If the force moves the object from P to Q, then the **displacement vector** is $\mathbf{D} = \overrightarrow{PQ}$. So here we have two vectors: the force $\mathbf{F}$ and the displacement $\mathbf{D}$. The **work** done by $\mathbf{F}$ is defined as the magnitude of the displacement, $|\mathbf{D}|$, multiplied by the magnitude of the applied force in the direction of the motion, which, from Figure 1, is

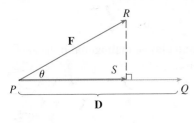

FIGURE 1

$$|\overrightarrow{PS}| = |\mathbf{F}| \cos \theta$$

So the work done by $\mathbf{F}$ is defined to be

$$\boxed{1} \qquad W = |\mathbf{D}| (|\mathbf{F}| \cos \theta) = |\mathbf{F}||\mathbf{D}| \cos \theta$$

Notice that work is a scalar quantity; it has no direction. But its value depends on the angle θ between the force and displacement vectors.

We use the expression in Equation 1 to define the dot product of two vectors even when they don't represent force or displacement.

> **Definition** The **dot product** of two nonzero vectors $\mathbf{a}$ and $\mathbf{b}$ is the number
>
> $$\mathbf{a} \cdot \mathbf{b} = |\mathbf{a}||\mathbf{b}| \cos \theta$$
>
> where θ is the angle between $\mathbf{a}$ and $\mathbf{b}$, $0 \leq \theta \leq \pi$. (So θ is the smaller angle between the vectors when they are drawn with the same initial point.) If either $\mathbf{a}$ or $\mathbf{b}$ is $\mathbf{0}$, we define $\mathbf{a} \cdot \mathbf{b} = 0$.

This product is called the **dot product** because of the dot in the notation $\mathbf{a} \cdot \mathbf{b}$. The result of computing $\mathbf{a} \cdot \mathbf{b}$ is not a vector. It is a real number, that is, a scalar. For this reason, the dot product is sometimes called the **scalar product**.

In the example of finding the work done by a force $\mathbf{F}$ in moving an object through a displacement $\mathbf{D} = \overrightarrow{PQ}$ by calculating $\mathbf{F} \cdot \mathbf{D} = |\mathbf{F}||\mathbf{D}| \cos \theta$, it makes no sense for the angle θ between $\mathbf{F}$ and $\mathbf{D}$ to be $\pi/2$ or larger because movement from P to Q couldn't take place. We make no such restriction in our general definition of $\mathbf{a} \cdot \mathbf{b}$, however, and allow θ to be any angle from 0 to π.

EXAMPLE 1 If the vectors $\mathbf{a}$ and $\mathbf{b}$ have lengths 4 and 6, and the angle between them is $\pi/3$, find $\mathbf{a} \cdot \mathbf{b}$.

SOLUTION According to the definition,

$$\mathbf{a} \cdot \mathbf{b} = |\mathbf{a}||\mathbf{b}| \cos(\pi/3) = 4 \cdot 6 \cdot \tfrac{1}{2} = 12 \qquad ■ ■$$

EXAMPLE 2 A crate is hauled 8 m up a ramp under a constant force of 200 N applied at an angle of 25° to the ramp. Find the work done.

SOLUTION If $\mathbf{F}$ and $\mathbf{D}$ are the force and displacement vectors, as pictured in Figure 2, then the work done is

$$W = \mathbf{F} \cdot \mathbf{D} = |\mathbf{F}||\mathbf{D}| \cos 25°$$

$$= (200)(8) \cos 25° \approx 1450 \text{ N·m} = 1450 \text{ J} \qquad ■ ■$$

FIGURE 2

Two nonzero vectors $\mathbf{a}$ and $\mathbf{b}$ are called **perpendicular** or **orthogonal** if the angle between them is $\theta = \pi/2$. For such vectors we have

$$\mathbf{a} \cdot \mathbf{b} = |\mathbf{a}||\mathbf{b}| \cos(\pi/2) = 0$$

and conversely if $\mathbf{a} \cdot \mathbf{b} = 0$, then $\cos \theta = 0$, so $\theta = \pi/2$. The zero vector $\mathbf{0}$ is considered to be perpendicular to all vectors. Therefore

> $\boxed{2}$ Two vectors $\mathbf{a}$ and $\mathbf{b}$ are orthogonal if and only if $\mathbf{a} \cdot \mathbf{b} = 0$.

Because $\cos \theta > 0$ if $0 \leq \theta < \pi/2$ and $\cos \theta < 0$ if $\pi/2 < \theta \leq \pi$, we see that $\mathbf{a} \cdot \mathbf{b}$ is positive for $\theta < \pi/2$ and negative for $\theta > \pi/2$. We can think of $\mathbf{a} \cdot \mathbf{b}$ as measuring the extent to which $\mathbf{a}$ and $\mathbf{b}$ point in the same direction. The dot product $\mathbf{a} \cdot \mathbf{b}$ is positive if $\mathbf{a}$ and $\mathbf{b}$ point in the same general direction, 0 if they are perpendicular, and negative if they point in generally opposite directions (see Figure 3). In the extreme case where $\mathbf{a}$ and $\mathbf{b}$ point in exactly the same direction, we have $\theta = 0$, so $\cos \theta = 1$ and

$$\mathbf{a} \cdot \mathbf{b} = |\mathbf{a}||\mathbf{b}|$$

If $\mathbf{a}$ and $\mathbf{b}$ point in exactly opposite directions, then $\theta = \pi$ and so $\cos \theta = -1$ and $\mathbf{a} \cdot \mathbf{b} = -|\mathbf{a}||\mathbf{b}|$.

$\mathbf{a} \cdot \mathbf{b} > 0$

$\mathbf{a} \cdot \mathbf{b} = 0$

$\mathbf{a} \cdot \mathbf{b} < 0$

FIGURE 3

Visual 9.3A shows an animation of Figure 3.

■ ■ The Dot Product in Component Form

Suppose we are given two vectors in component form:

$$\mathbf{a} = \langle a_1, a_2, a_3 \rangle \qquad \mathbf{b} = \langle b_1, b_2, b_3 \rangle$$

We want to find a convenient expression for $\mathbf{a} \cdot \mathbf{b}$ in terms of these components. If we apply the Law of Cosines to the triangle in Figure 4, we get

$$|\mathbf{a} - \mathbf{b}|^2 = |\mathbf{a}|^2 + |\mathbf{b}|^2 - 2|\mathbf{a}||\mathbf{b}|\cos\theta$$

$$= |\mathbf{a}|^2 + |\mathbf{b}|^2 - 2\mathbf{a} \cdot \mathbf{b}$$

FIGURE 4

Solving for the dot product, we obtain

$$\mathbf{a} \cdot \mathbf{b} = \tfrac{1}{2}\left(|\mathbf{a}|^2 + |\mathbf{b}|^2 - |\mathbf{a} - \mathbf{b}|^2\right)$$

$$= \tfrac{1}{2}\left[a_1^2 + a_2^2 + a_3^2 + b_1^2 + b_2^2 + b_3^2 - (a_1 - b_1)^2 - (a_2 - b_2)^2 - (a_3 - b_3)^2\right]$$

$$= a_1 b_1 + a_2 b_2 + a_3 b_3$$

The dot product of $\mathbf{a} = \langle a_1, a_2, a_3 \rangle$ and $\mathbf{b} = \langle b_1, b_2, b_3 \rangle$ is

$$\mathbf{a} \cdot \mathbf{b} = a_1 b_1 + a_2 b_2 + a_3 b_3$$

Thus, to find the dot product of $\mathbf{a}$ and $\mathbf{b}$ we multiply corresponding components and add. The dot product of two-dimensional vectors is found in a similar fashion:

$$\langle a_1, a_2 \rangle \cdot \langle b_1, b_2 \rangle = a_1 b_1 + a_2 b_2$$

$\boxed{\text{V}}$ **EXAMPLE 3**
$$\langle 2, 4 \rangle \cdot \langle 3, -1 \rangle = 2(3) + 4(-1) = 2$$

$$\langle -1, 7, 4 \rangle \cdot \left\langle 6, 2, -\tfrac{1}{2} \right\rangle = (-1)(6) + 7(2) + 4\left(-\tfrac{1}{2}\right) = 6$$

$$(\mathbf{i} + 2\mathbf{j} - 3\mathbf{k}) \cdot (2\mathbf{j} - \mathbf{k}) = 1(0) + 2(2) + (-3)(-1) = 7 \qquad\qquad ■\ ■$$

EXAMPLE 4 Show that $2\mathbf{i} + 2\mathbf{j} - \mathbf{k}$ is perpendicular to $5\mathbf{i} - 4\mathbf{j} + 2\mathbf{k}$.

SOLUTION Since

$$(2\mathbf{i} + 2\mathbf{j} - \mathbf{k}) \cdot (5\mathbf{i} - 4\mathbf{j} + 2\mathbf{k}) = 2(5) + 2(-4) + (-1)(2) = 0$$

these vectors are perpendicular by (2). ■ ■

$\boxed{\text{V}}$ **EXAMPLE 5** Find the angle between the vectors $\mathbf{a} = \langle 2, 2, -1 \rangle$ and $\mathbf{b} = \langle 5, -3, 2 \rangle$.

SOLUTION Let θ be the required angle. Since

$$|\mathbf{a}| = \sqrt{2^2 + 2^2 + (-1)^2} = 3 \qquad \text{and} \qquad |\mathbf{b}| = \sqrt{5^2 + (-3)^2 + 2^2} = \sqrt{38}$$

and since

$$\mathbf{a} \cdot \mathbf{b} = 2(5) + 2(-3) + (-1)(2) = 2$$

we have, from the definition of the dot product

$$\cos \theta = \frac{\mathbf{a} \cdot \mathbf{b}}{|\mathbf{a}||\mathbf{b}|} = \frac{2}{3\sqrt{38}}$$

So the angle between **a** and **b** is

$$\theta = \cos^{-1}\left(\frac{2}{3\sqrt{38}}\right) \approx 1.46 \quad (\text{or } 84°)$$

■ ■

EXAMPLE 6 A force is given by a vector $\mathbf{F} = 3\mathbf{i} + 4\mathbf{j} + 5\mathbf{k}$ and moves a particle from the point $P(2, 1, 0)$ to the point $Q(4, 6, 2)$. Find the work done.

SOLUTION The displacement vector is $\mathbf{D} = \overrightarrow{PQ} = \langle 2, 5, 2 \rangle$, so the work done is

$$W = \mathbf{F} \cdot \mathbf{D} = \langle 3, 4, 5 \rangle \cdot \langle 2, 5, 2 \rangle$$
$$= 6 + 20 + 10 = 36$$

If the unit of length is meters and the magnitude of the force is measured in newtons, then the work done is 36 J.

■ ■

The dot product obeys many of the laws that hold for ordinary products of real numbers. These are stated in the following theorem.

Properties of the Dot Product If **a**, **b**, and **c** are vectors in V_3 and c is a scalar, then

1. $\mathbf{a} \cdot \mathbf{a} = |\mathbf{a}|^2$
2. $\mathbf{a} \cdot \mathbf{b} = \mathbf{b} \cdot \mathbf{a}$
3. $\mathbf{a} \cdot (\mathbf{b} + \mathbf{c}) = \mathbf{a} \cdot \mathbf{b} + \mathbf{a} \cdot \mathbf{c}$
4. $(c\mathbf{a}) \cdot \mathbf{b} = c(\mathbf{a} \cdot \mathbf{b}) = \mathbf{a} \cdot (c\mathbf{b})$
5. $\mathbf{0} \cdot \mathbf{a} = 0$

Properties 1, 2, and 5 are immediate consequences of the definition of a dot product. Property 3 is best proved using components:

$$\mathbf{a} \cdot (\mathbf{b} + \mathbf{c}) = \langle a_1, a_2, a_3 \rangle \cdot \langle b_1 + c_1, b_2 + c_2, b_3 + c_3 \rangle$$
$$= a_1(b_1 + c_1) + a_2(b_2 + c_2) + a_3(b_3 + c_3)$$
$$= a_1 b_1 + a_1 c_1 + a_2 b_2 + a_2 c_2 + a_3 b_3 + a_3 c_3$$
$$= (a_1 b_1 + a_2 b_2 + a_3 b_3) + (a_1 c_1 + a_2 c_2 + a_3 c_3)$$
$$= \mathbf{a} \cdot \mathbf{b} + \mathbf{a} \cdot \mathbf{c}$$

The proof of Property 4 is left as Exercise 41.

■■ Projections

Visual 9.3B shows how Figure 5 changes when we vary **a** and **b**.

Figure 5 shows representations $\overrightarrow{PQ}$ and $\overrightarrow{PR}$ of two vectors **a** and **b** with the same initial point P. If S is the foot of the perpendicular from R to the line containing $\overrightarrow{PQ}$, then the vector with representation $\overrightarrow{PS}$ is called the **vector projection** of **b** onto **a** and is

FIGURE 5

Vector projections

FIGURE 6

Scalar projection

denoted by $\text{proj}_a\, \mathbf{b}$. (You can think of it as a shadow of $\mathbf{b}$). The **scalar projection** of $\mathbf{b}$ onto $\mathbf{a}$ (also called the **component of b along a**) is defined to be numerically the length of the vector projection, which is the number $|\mathbf{b}|\cos\theta$, where θ is the angle between $\mathbf{a}$ and $\mathbf{b}$. (See Figure 6.) This is denoted by $\text{comp}_a\, \mathbf{b}$. Observe that it is negative if $\pi/2 < \theta \leq \pi$. (Note that we used the component of the force $\mathbf{F}$ along the displacement $\mathbf{D}$, $\text{comp}_D\, \mathbf{F}$, at the beginning of this section.)

The equation

$$\mathbf{a} \cdot \mathbf{b} = |\mathbf{a}||\mathbf{b}|\cos\theta = |\mathbf{a}|(|\mathbf{b}|\cos\theta)$$

shows that the dot product of $\mathbf{a}$ and $\mathbf{b}$ can be interpreted as the length of $\mathbf{a}$ times the scalar projection of $\mathbf{b}$ onto $\mathbf{a}$. Since

$$|\mathbf{b}|\cos\theta = \frac{\mathbf{a} \cdot \mathbf{b}}{|\mathbf{a}|} = \frac{\mathbf{a}}{|\mathbf{a}|} \cdot \mathbf{b}$$

the component of $\mathbf{b}$ along $\mathbf{a}$ can be computed by taking the dot product of $\mathbf{b}$ with the unit vector in the direction of $\mathbf{a}$. To summarize:

Scalar projection of $\mathbf{b}$ onto $\mathbf{a}$: $\text{comp}_a\, \mathbf{b} = \dfrac{\mathbf{a} \cdot \mathbf{b}}{|\mathbf{a}|}$

Vector projection of $\mathbf{b}$ onto $\mathbf{a}$: $\text{proj}_a\, \mathbf{b} = \left(\dfrac{\mathbf{a} \cdot \mathbf{b}}{|\mathbf{a}|}\right)\dfrac{\mathbf{a}}{|\mathbf{a}|} = \dfrac{\mathbf{a} \cdot \mathbf{b}}{|\mathbf{a}|^2}\,\mathbf{a}$

▽ **EXAMPLE 7** Find the scalar projection and vector projection of $\mathbf{b} = \langle 1, 1, 2\rangle$ onto $\mathbf{a} = \langle -2, 3, 1\rangle$.

SOLUTION Since $|\mathbf{a}| = \sqrt{(-2)^2 + 3^2 + 1^2} = \sqrt{14}$, the scalar projection of $\mathbf{b}$ onto $\mathbf{a}$ is

$$\text{comp}_a\, \mathbf{b} = \frac{\mathbf{a} \cdot \mathbf{b}}{|\mathbf{a}|} = \frac{(-2)(1) + 3(1) + 1(2)}{\sqrt{14}} = \frac{3}{\sqrt{14}}$$

The vector projection is this scalar projection times the unit vector in the direction of $\mathbf{a}$:

$$\text{proj}_a\, \mathbf{b} = \frac{3}{\sqrt{14}}\frac{\mathbf{a}}{|\mathbf{a}|} = \frac{3}{14}\,\mathbf{a} = \left\langle -\frac{3}{7}, \frac{9}{14}, \frac{3}{14}\right\rangle$$

■ ■

At the beginning of this section we saw one use of projections in physics—we used a scalar projection of a force vector in defining work. Other uses of projections occur in three-dimensional geometry. In Exercise 35 you are asked to use a projection to find the distance from a point to a line, and in Section 9.5 we use a projection to find the distance from a point to a plane.

9.3 Exercises

1. Which of the following expressions are meaningful? Which are meaningless? Explain.

(a) $(\mathbf{a} \cdot \mathbf{b}) \cdot \mathbf{c}$

(b) $(\mathbf{a} \cdot \mathbf{b})\mathbf{c}$

(c) $|\mathbf{a}|(\mathbf{b} \cdot \mathbf{c})$

(d) $\mathbf{a} \cdot (\mathbf{b} + \mathbf{c})$

(e) $\mathbf{a} \cdot \mathbf{b} + \mathbf{c}$

(f) $|\mathbf{a}| \cdot (\mathbf{b} + \mathbf{c})$

2. Find the dot product of two vectors if their lengths are 6 and $\frac{1}{3}$ and the angle between them is $\pi/4$.

3–8 ■ Find $\mathbf{a} \cdot \mathbf{b}$.

3. $|\mathbf{a}| = 6$, $\quad |\mathbf{b}| = 5$, $\quad$ the angle between $\mathbf{a}$ and $\mathbf{b}$ is $2\pi/3$

4. $\mathbf{a} = \langle -2, 3 \rangle$, $\quad \mathbf{b} = \langle 0.7, 1.2 \rangle$

5. $\mathbf{a} = \langle 4, 1, \frac{1}{4} \rangle$, $\quad \mathbf{b} = \langle 6, -3, -8 \rangle$

6. $\mathbf{a} = \langle s, 2s, 3s \rangle$, $\quad \mathbf{b} = \langle t, -t, 5t \rangle$

7. $\mathbf{a} = \mathbf{i} - 2\mathbf{j} + 3\mathbf{k}$, $\quad \mathbf{b} = 5\mathbf{i} + 9\mathbf{k}$

8. $\mathbf{a} = 4\mathbf{j} - 3\mathbf{k}$, $\quad \mathbf{b} = 2\mathbf{i} + 4\mathbf{j} + 6\mathbf{k}$

9–10 ■ If $\mathbf{u}$ is a unit vector, find $\mathbf{u} \cdot \mathbf{v}$ and $\mathbf{u} \cdot \mathbf{w}$.

9.

10.

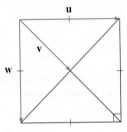

11. (a) Show that $\mathbf{i} \cdot \mathbf{j} = \mathbf{j} \cdot \mathbf{k} = \mathbf{k} \cdot \mathbf{i} = 0$.
(b) Show that $\mathbf{i} \cdot \mathbf{i} = \mathbf{j} \cdot \mathbf{j} = \mathbf{k} \cdot \mathbf{k} = 1$.

12. A street vendor sells a hamburgers, b hot dogs, and c soft drinks on a given day. He charges \$2 for a hamburger, \$1.50 for a hot dog, and \$1 for a soft drink. If $\mathbf{A} = \langle a, b, c \rangle$ and $\mathbf{P} = \langle 2, 1.5, 1 \rangle$, what is the meaning of the dot product $\mathbf{A} \cdot \mathbf{P}$?

13–15 ■ Find the angle between the vectors. (First find an exact expression and then approximate to the nearest degree.)

13. $\mathbf{a} = \langle -8, 6 \rangle$, $\quad \mathbf{b} = \langle \sqrt{7}, 3 \rangle$

14. $\mathbf{a} = \langle 4, 0, 2 \rangle$, $\quad \mathbf{b} = \langle 2, -1, 0 \rangle$

15. $\mathbf{a} = \mathbf{j} + \mathbf{k}$, $\quad \mathbf{b} = \mathbf{i} + 2\mathbf{j} - 3\mathbf{k}$

16. Find, correct to the nearest degree, the three angles of the triangle with vertices $D(0, 1, 1)$, $E(-2, 4, 3)$, and $F(1, 2, -1)$.

17–18 ■ Determine whether the given vectors are orthogonal, parallel, or neither.

17. (a) $\mathbf{a} = \langle -5, 3, 7 \rangle$, $\quad \mathbf{b} = \langle 6, -8, 2 \rangle$
(b) $\mathbf{a} = \langle 4, 6 \rangle$, $\quad \mathbf{b} = \langle -3, 2 \rangle$
(c) $\mathbf{a} = -\mathbf{i} + 2\mathbf{j} + 5\mathbf{k}$, $\quad \mathbf{b} = 3\mathbf{i} + 4\mathbf{j} - \mathbf{k}$
(d) $\mathbf{a} = 2\mathbf{i} + 6\mathbf{j} - 4\mathbf{k}$, $\quad \mathbf{b} = -3\mathbf{i} - 9\mathbf{j} + 6\mathbf{k}$

18. (a) $\mathbf{u} = \langle -3, 9, 6 \rangle$, $\quad \mathbf{v} = \langle 4, -12, -8 \rangle$
(b) $\mathbf{u} = \mathbf{i} - \mathbf{j} + 2\mathbf{k}$, $\quad \mathbf{v} = 2\mathbf{i} - \mathbf{j} + \mathbf{k}$
(c) $\mathbf{u} = \langle a, b, c \rangle$, $\quad \mathbf{v} = \langle -b, a, 0 \rangle$

19. Use vectors to decide whether the triangle with vertices $P(1, -3, -2)$, $Q(2, 0, -4)$, and $R(6, -2, -5)$ is right-angled.

20. For what values of b are the vectors $\langle -6, b, 2 \rangle$ and $\langle b, b^2, b \rangle$ orthogonal?

21. Find a unit vector that is orthogonal to both $\mathbf{i} + \mathbf{j}$ and $\mathbf{i} + \mathbf{k}$.

22. Find two unit vectors that make an angle of $60°$ with $\mathbf{v} = \langle 3, 4 \rangle$.

23–26 ■ Find the scalar and vector projections of $\mathbf{b}$ onto $\mathbf{a}$.

23. $\mathbf{a} = \langle 3, -4 \rangle$, $\quad \mathbf{b} = \langle 5, 0 \rangle$

24. $\mathbf{a} = \langle 1, 2 \rangle$, $\quad \mathbf{b} = \langle -4, 1 \rangle$

25. $\mathbf{a} = \langle 3, 6, -2 \rangle$, $\quad \mathbf{b} = \langle 1, 2, 3 \rangle$

26. $\mathbf{a} = \mathbf{i} + \mathbf{j} + \mathbf{k}$, $\quad \mathbf{b} = \mathbf{i} - \mathbf{j} + \mathbf{k}$

27. Show that the vector $\text{orth}_a \mathbf{b} = \mathbf{b} - \text{proj}_a \mathbf{b}$ is orthogonal to $\mathbf{a}$. (It is called an **orthogonal projection** of $\mathbf{b}$.)

28. For the vectors in Exercise 24, find $\text{orth}_a \mathbf{b}$ and illustrate by drawing the vectors $\mathbf{a}$, $\mathbf{b}$, $\text{proj}_a \mathbf{b}$, and $\text{orth}_a \mathbf{b}$.

29. If $\mathbf{a} = \langle 3, 0, -1 \rangle$, find a vector $\mathbf{b}$ such that $\text{comp}_a \mathbf{b} = 2$.

30. Suppose that $\mathbf{a}$ and $\mathbf{b}$ are nonzero vectors.
(a) Under what circumstances is $\text{comp}_a \mathbf{b} = \text{comp}_b \mathbf{a}$?
(b) Under what circumstances is $\text{proj}_a \mathbf{b} = \text{proj}_b \mathbf{a}$?

31. A constant force with vector representation $\mathbf{F} = 10\mathbf{i} + 18\mathbf{j} - 6\mathbf{k}$ moves an object along a straight line from the point $(2, 3, 0)$ to the point $(4, 9, 15)$. Find the work done if the distance is measured in meters and the magnitude of the force is measured in newtons.

32. Find the work done by a force of 20 lb acting in the direction N50°W in moving an object 4 ft due west.

33. A woman exerts a horizontal force of 25 lb on a crate as she pushes it up a ramp that is 10 ft long and inclined at an angle of $20°$ above the horizontal. Find the work done on the box.

34. A wagon is pulled a distance of 100 m along a horizontal path by a constant force of 50 N. The handle of the wagon is held at an angle of $30°$ above the horizontal. How much work is done?

35. Use a scalar projection to show that the distance from a point $P_1(x_1, y_1)$ to the line $ax + by + c = 0$ is

$$\frac{|ax_1 + by_1 + c|}{\sqrt{a^2 + b^2}}$$

Use this formula to find the distance from the point $(-2, 3)$ to the line $3x - 4y + 5 = 0$.

36. If $\mathbf{r} = \langle x, y, z \rangle$, $\mathbf{a} = \langle a_1, a_2, a_3 \rangle$, and $\mathbf{b} = \langle b_1, b_2, b_3 \rangle$, show that the vector equation $(\mathbf{r} - \mathbf{a}) \cdot (\mathbf{r} - \mathbf{b}) = 0$ represents a sphere, and find its center and radius.

37. Find the angle between a diagonal of a cube and one of its edges.

38. Find the angle between a diagonal of a cube and a diagonal of one of its faces.

39. A molecule of methane, CH_4, is structured with the four hydrogen atoms at the vertices of a regular tetrahedron and the carbon atom at the centroid. The *bond angle* is the angle formed by the H—C—H combination; it is the angle between the lines that join the carbon atom to two of the hydrogen atoms. Show that the bond angle is about 109.5°. [*Hint:* Take the vertices of the tetrahedron to be the points $(1, 0, 0)$, $(0, 1, 0)$, $(0, 0, 1)$, and $(1, 1, 1)$ as shown in the figure. Then the centroid is $\left(\frac{1}{2}, \frac{1}{2}, \frac{1}{2}\right)$.]

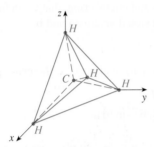

40. If $\mathbf{c} = |\mathbf{a}|\mathbf{b} + |\mathbf{b}|\mathbf{a}$, where $\mathbf{a}$, $\mathbf{b}$, and $\mathbf{c}$ are all nonzero vectors, show that $\mathbf{c}$ bisects the angle between $\mathbf{a}$ and $\mathbf{b}$.

41. Prove Property 4 of the dot product. Use either the definition of a dot product (considering the cases $c > 0$, $c = 0$, and $c < 0$ separately) or the component form.

42. Suppose that all sides of a quadrilateral are equal in length and opposite sides are parallel. Use vector methods to show that the diagonals are perpendicular.

43. Prove the Cauchy-Schwarz Inequality:

$$|\mathbf{a} \cdot \mathbf{b}| \le |\mathbf{a}||\mathbf{b}|$$

44. The Triangle Inequality for vectors is

$$|\mathbf{a} + \mathbf{b}| \le |\mathbf{a}| + |\mathbf{b}|$$

(a) Give a geometric interpretation of the Triangle Inequality.
(b) Use the Cauchy-Schwarz Inequality from Exercise 43 to prove the Triangle Inequality. [*Hint:* Use the fact that $|\mathbf{a} + \mathbf{b}|^2 = (\mathbf{a} + \mathbf{b}) \cdot (\mathbf{a} + \mathbf{b})$ and use Property 3 of the dot product.]

45. The Parallelogram Law states that

$$|\mathbf{a} + \mathbf{b}|^2 + |\mathbf{a} - \mathbf{b}|^2 = 2|\mathbf{a}|^2 + 2|\mathbf{b}|^2$$

(a) Give a geometric interpretation of the Parallelogram Law.
(b) Prove the Parallelogram Law. (See the hint in Exercise 44.)

9.4 | The Cross Product

The **cross product** $\mathbf{a} \times \mathbf{b}$ of two vectors $\mathbf{a}$ and $\mathbf{b}$, unlike the dot product, is a vector. For this reason it is also called the **vector product**. We will see that $\mathbf{a} \times \mathbf{b}$ is useful in geometry because it is perpendicular to both $\mathbf{a}$ and $\mathbf{b}$. But we introduce this product by looking at a situation where it arises in physics and engineering.

FIGURE 1

■■ Torque and the Cross Product

If we tighten a bolt by applying a force to a wrench as in Figure 1, we produce a turning effect called a *torque* τ. The magnitude of the torque depends on two things:

▫ The distance from the axis of the bolt to the point where the force is applied. This is $|\mathbf{r}|$, the length of the position vector $\mathbf{r}$.

▫ The scalar component of the force $\mathbf{F}$ in the direction perpendicular to $\mathbf{r}$. This is the only component that can cause a rotation and, from Figure 2, we see that it is

$$|\mathbf{F}|\sin\theta$$

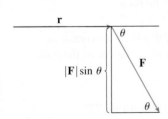

FIGURE 2

where θ is the angle between the vectors $\mathbf{r}$ and $\mathbf{F}$.

We define the magnitude of the torque vector to be the product of these two factors:

$$|\boldsymbol{\tau}| = |\mathbf{r}||\mathbf{F}|\sin\theta$$

The direction is along the axis of rotation. If $\mathbf{n}$ is a unit vector that points in the direction in which a right-threaded bolt moves (see Figure 1), we define the **torque** to be the vector

1 $$\boldsymbol{\tau} = (|\mathbf{r}||\mathbf{F}|\sin\theta)\mathbf{n}$$

We denote this torque vector by $\boldsymbol{\tau} = \mathbf{r} \times \mathbf{F}$ and we call it the *cross product* or *vector product* of $\mathbf{r}$ and $\mathbf{F}$.

The type of expression in Equation 1 occurs so frequently in the study of fluid flow, planetary motion, and other areas of physics and engineering, that we define and study the cross product of *any* pair of three-dimensional vectors $\mathbf{a}$ and $\mathbf{b}$.

Definition If $\mathbf{a}$ and $\mathbf{b}$ are nonzero three-dimensional vectors, the **cross product** of $\mathbf{a}$ and $\mathbf{b}$ is the vector

$$\mathbf{a} \times \mathbf{b} = (|\mathbf{a}||\mathbf{b}|\sin\theta)\mathbf{n}$$

where θ is the angle between $\mathbf{a}$ and $\mathbf{b}$, $0 \le \theta \le \pi$, and $\mathbf{n}$ is a unit vector perpendicular to both $\mathbf{a}$ and $\mathbf{b}$ and whose direction is given by the **right-hand rule**: If the fingers of your right hand curl through the angle θ from $\mathbf{a}$ and $\mathbf{b}$, then your thumb points in the direction of $\mathbf{n}$. (See Figure 3.)

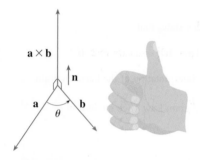

FIGURE 3
The right-hand rule gives
the direction of $\mathbf{a} \times \mathbf{b}$.

If either $\mathbf{a}$ or $\mathbf{b}$ is $\mathbf{0}$, then we define $\mathbf{a} \times \mathbf{b}$ to be $\mathbf{0}$.

Because $\mathbf{a} \times \mathbf{b}$ is a scalar multiple of $\mathbf{n}$, it has the same direction as $\mathbf{n}$ and so

> $\mathbf{a} \times \mathbf{b}$ is orthogonal to both $\mathbf{a}$ and $\mathbf{b}$.

TEC Visual 9.4 shows how $\mathbf{a} \times \mathbf{b}$ changes as $\mathbf{b}$ changes.

Notice that two nonzero vectors $\mathbf{a}$ and $\mathbf{b}$ are parallel if and only if the angle between them is 0 or π. In either case, $\sin\theta = 0$ and so $\mathbf{a} \times \mathbf{b} = \mathbf{0}$.

■ ■ In particular, any vector $\mathbf{a}$ is parallel to itself, so

$$\mathbf{a} \times \mathbf{a} = \mathbf{0}$$

> Two nonzero vectors $\mathbf{a}$ and $\mathbf{b}$ are parallel if and only if $\mathbf{a} \times \mathbf{b} = \mathbf{0}$.

This makes sense in the torque interpretation: If we pull or push the wrench in the direction of its handle (so $\mathbf{F}$ is parallel to $\mathbf{r}$), we produce no torque.

EXAMPLE 1 A bolt is tightened by applying a 40-N force to a 0.25-m wrench as shown in Figure 4. Find the magnitude of the torque about the center of the bolt.

SOLUTION The magnitude of the torque vector is

$$|\boldsymbol{\tau}| = |\mathbf{r} \times \mathbf{F}| = |\mathbf{r}||\mathbf{F}|\sin 75°|\mathbf{n}| = (0.25)(40)\sin 75°$$
$$= 10\sin 75° \approx 9.66 \text{ N·m}$$

FIGURE 4

If the bolt is right-threaded, then the torque vector itself is

$$\boldsymbol{\tau} = |\boldsymbol{\tau}|\,\mathbf{n} \approx 9.66\mathbf{n}$$

where $\mathbf{n}$ is a unit vector directed down into the page.   ■ ■

EXAMPLE 2 Find $\mathbf{i} \times \mathbf{j}$ and $\mathbf{j} \times \mathbf{i}$.

SOLUTION The standard basis vectors $\mathbf{i}$ and $\mathbf{j}$ both have length 1 and the angle between them is $\pi/2$. By the right-hand rule, the unit vector perpendicular to $\mathbf{i}$ and $\mathbf{j}$ is $\mathbf{n} = \mathbf{k}$ (see Figure 5), so

$$\mathbf{i} \times \mathbf{j} = \big(|\mathbf{i}||\mathbf{j}|\sin(\pi/2)\big)\mathbf{k} = \mathbf{k}$$

But if we apply the right-hand rule to the vectors $\mathbf{j}$ and $\mathbf{i}$ (in that order), we see that $\mathbf{n}$ points downward and so $\mathbf{n} = -\mathbf{k}$. Thus

$$\mathbf{j} \times \mathbf{i} = -\mathbf{k}$$ ■ ■

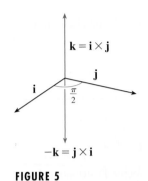

FIGURE 5

From Example 2 we see that

$$\mathbf{i} \times \mathbf{j} \neq \mathbf{j} \times \mathbf{i}$$

so the cross product is not commutative. Similar reasoning shows that

$$\mathbf{j} \times \mathbf{k} = \mathbf{i} \qquad \mathbf{k} \times \mathbf{j} = -\mathbf{i}$$
$$\mathbf{k} \times \mathbf{i} = \mathbf{j} \qquad \mathbf{i} \times \mathbf{k} = -\mathbf{j}$$

In general, the right-hand rule shows that

$$\mathbf{b} \times \mathbf{a} = -\mathbf{a} \times \mathbf{b}$$

Another algebraic law that fails for the cross product is the associative law for multiplication; that is, in general,

$$(\mathbf{a} \times \mathbf{b}) \times \mathbf{c} \neq \mathbf{a} \times (\mathbf{b} \times \mathbf{c})$$

For instance, if $\mathbf{a} = \mathbf{i}$, $\mathbf{b} = \mathbf{i}$, and $\mathbf{c} = \mathbf{j}$, then

$$(\mathbf{i} \times \mathbf{i}) \times \mathbf{j} = \mathbf{0} \times \mathbf{j} = \mathbf{0}$$

whereas

$$\mathbf{i} \times (\mathbf{i} \times \mathbf{j}) = \mathbf{i} \times \mathbf{k} = -\mathbf{j}$$

However, some of the usual laws of algebra *do* hold for cross products:

Properties of the Cross Product If $\mathbf{a}$, $\mathbf{b}$, and $\mathbf{c}$ are vectors and c is a scalar, then

1. $\mathbf{a} \times \mathbf{b} = -\mathbf{b} \times \mathbf{a}$

2. $(c\mathbf{a}) \times \mathbf{b} = c(\mathbf{a} \times \mathbf{b}) = \mathbf{a} \times (c\mathbf{b})$

3. $\mathbf{a} \times (\mathbf{b} + \mathbf{c}) = \mathbf{a} \times \mathbf{b} + \mathbf{a} \times \mathbf{c}$

4. $(\mathbf{a} + \mathbf{b}) \times \mathbf{c} = \mathbf{a} \times \mathbf{c} + \mathbf{b} \times \mathbf{c}$

Property 2 is proved by applying the definition of a cross product to each of the three expressions. Properties 3 and 4 (the Vector Distributive Laws) are more difficult to establish; we won't do so here.

FIGURE 6

A geometric interpretation of the length of the cross product can be seen by looking at Figure 6. If **a** and **b** are represented by directed line segments with the same initial point, then they determine a parallelogram with base $|\mathbf{a}|$, altitude $|\mathbf{b}|\sin\theta$, and area

$$A = |\mathbf{a}|\,(|\mathbf{b}|\sin\theta) = |\mathbf{a}\times\mathbf{b}|$$

> The length of the cross product $\mathbf{a}\times\mathbf{b}$ is equal to the area of the parallelogram determined by **a** and **b**.

The Cross Product in Component Form

Suppose **a** and **b** are given in component form:

$$\mathbf{a} = a_1\mathbf{i} + a_2\mathbf{j} + a_3\mathbf{k} \qquad \mathbf{b} = b_1\mathbf{i} + b_2\mathbf{j} + b_3\mathbf{k}$$

We can express $\mathbf{a}\times\mathbf{b}$ in component form by using the Vector Distributive Laws together with the results from Example 2:

$$\begin{aligned}
\mathbf{a}\times\mathbf{b} &= (a_1\mathbf{i} + a_2\mathbf{j} + a_3\mathbf{k}) \times (b_1\mathbf{i} + b_2\mathbf{j} + b_3\mathbf{k}) \\
&= a_1b_1\mathbf{i}\times\mathbf{i} + a_1b_2\mathbf{i}\times\mathbf{j} + a_1b_3\mathbf{i}\times\mathbf{k} \\
&\quad + a_2b_1\mathbf{j}\times\mathbf{i} + a_2b_2\mathbf{j}\times\mathbf{j} + a_2b_3\mathbf{j}\times\mathbf{k} \\
&\quad + a_3b_1\mathbf{k}\times\mathbf{i} + a_3b_2\mathbf{k}\times\mathbf{j} + a_3b_3\mathbf{k}\times\mathbf{k} \\
&= a_1b_2\mathbf{k} + a_1b_3(-\mathbf{j}) + a_2b_1(-\mathbf{k}) + a_2b_3\mathbf{i} + a_3b_1\mathbf{j} + a_3b_2(-\mathbf{i}) \\
&= (a_2b_3 - a_3b_2)\mathbf{i} + (a_3b_1 - a_1b_3)\mathbf{j} + (a_1b_2 - a_2b_1)\mathbf{k}
\end{aligned}$$

▪ ▪ Note that
$$\mathbf{i}\times\mathbf{i} = 0 \quad \mathbf{j}\times\mathbf{j} = 0 \quad \mathbf{k}\times\mathbf{k} = 0$$

> **2** If $\mathbf{a} = \langle a_1, a_2, a_3 \rangle$ and $\mathbf{b} = \langle b_1, b_2, b_3 \rangle$, then
>
> $$\mathbf{a}\times\mathbf{b} = \langle a_2b_3 - a_3b_2,\ a_3b_1 - a_1b_3,\ a_1b_2 - a_2b_1 \rangle$$

In order to make this expression for $\mathbf{a}\times\mathbf{b}$ easier to remember, we use the notation of determinants. A **determinant of order 2** is defined by

$$\begin{vmatrix} a & b \\ c & d \end{vmatrix} = ad - bc$$

For example,

$$\begin{vmatrix} 2 & 1 \\ -6 & 4 \end{vmatrix} = 2(4) - 1(-6) = 14$$

A **determinant of order 3** can be defined in terms of second-order determinants as follows:

3
$$\begin{vmatrix} a_1 & a_2 & a_3 \\ b_1 & b_2 & b_3 \\ c_1 & c_2 & c_3 \end{vmatrix} = a_1 \begin{vmatrix} b_2 & b_3 \\ c_2 & c_3 \end{vmatrix} - a_2 \begin{vmatrix} b_1 & b_3 \\ c_1 & c_3 \end{vmatrix} + a_3 \begin{vmatrix} b_1 & b_2 \\ c_1 & c_2 \end{vmatrix}$$

Observe that each term on the right side of Equation 3 involves a number a_i in the first row of the determinant, and a_i is multiplied by the second-order determinant obtained from the left side by deleting the row and column in which a_i appears. Notice also the minus sign in the second term. For example,

$$\begin{vmatrix} 1 & 2 & -1 \\ 3 & 0 & 1 \\ -5 & 4 & 2 \end{vmatrix} = 1 \begin{vmatrix} 0 & 1 \\ 4 & 2 \end{vmatrix} - 2 \begin{vmatrix} 3 & 1 \\ -5 & 2 \end{vmatrix} + (-1) \begin{vmatrix} 3 & 0 \\ -5 & 4 \end{vmatrix}$$

$$= 1(0 - 4) - 2(6 + 5) + (-1)(12 - 0) = -38$$

If we now rewrite (2) using second-order determinants and the standard basis vectors **i**, **j**, and **k**, we see that the cross product of $\mathbf{a} = a_1\mathbf{i} + a_2\mathbf{j} + a_3\mathbf{k}$ and $\mathbf{b} = b_1\mathbf{i} + b_2\mathbf{j} + b_3\mathbf{k}$ is

$$\boxed{4} \qquad \mathbf{a} \times \mathbf{b} = \begin{vmatrix} a_2 & a_3 \\ b_2 & b_3 \end{vmatrix} \mathbf{i} - \begin{vmatrix} a_1 & a_3 \\ b_1 & b_3 \end{vmatrix} \mathbf{j} + \begin{vmatrix} a_1 & a_2 \\ b_1 & b_2 \end{vmatrix} \mathbf{k}$$

In view of the similarity between Equations 3 and 4, we often write

$$\boxed{5} \qquad \mathbf{a} \times \mathbf{b} = \begin{vmatrix} \mathbf{i} & \mathbf{j} & \mathbf{k} \\ a_1 & a_2 & a_3 \\ b_1 & b_2 & b_3 \end{vmatrix}$$

Although the first row of the symbolic determinant in Equation 5 consists of vectors, if we expand it as if it were an ordinary determinant using the rule in Equation 3, we obtain Equation 4. The symbolic formula in Equation 5 is probably the easiest way of remembering and computing cross products.

EXAMPLE 3 If $\mathbf{a} = \langle 1, 3, 4 \rangle$ and $\mathbf{b} = \langle 2, 7, -5 \rangle$, then

$$\mathbf{a} \times \mathbf{b} = \begin{vmatrix} \mathbf{i} & \mathbf{j} & \mathbf{k} \\ 1 & 3 & 4 \\ 2 & 7 & -5 \end{vmatrix}$$

$$= \begin{vmatrix} 3 & 4 \\ 7 & -5 \end{vmatrix} \mathbf{i} - \begin{vmatrix} 1 & 4 \\ 2 & -5 \end{vmatrix} \mathbf{j} + \begin{vmatrix} 1 & 3 \\ 2 & 7 \end{vmatrix} \mathbf{k}$$

$$= (-15 - 28)\mathbf{i} - (-5 - 8)\mathbf{j} + (7 - 6)\mathbf{k} = -43\mathbf{i} + 13\mathbf{j} + \mathbf{k}$$

EXAMPLE 4 Find a vector perpendicular to the plane that passes through the points $P(1, 4, 6)$, $Q(-2, 5, -1)$, and $R(1, -1, 1)$.

SOLUTION The vector $\vec{PQ} \times \vec{PR}$ is perpendicular to both $\vec{PQ}$ and $\vec{PR}$ and is therefore perpendicular to the plane through P, Q, and R. We know from (9.2.1) that

$$\vec{PQ} = (-2 - 1)\mathbf{i} + (5 - 4)\mathbf{j} + (-1 - 6)\mathbf{k} = -3\mathbf{i} + \mathbf{j} - 7\mathbf{k}$$

$$\vec{PR} = (1 - 1)\mathbf{i} + (-1 - 4)\mathbf{j} + (1 - 6)\mathbf{k} = -5\mathbf{j} - 5\mathbf{k}$$

We compute the cross product of these vectors:

$$\overrightarrow{PQ} \times \overrightarrow{PR} = \begin{vmatrix} \mathbf{i} & \mathbf{j} & \mathbf{k} \\ -3 & 1 & -7 \\ 0 & -5 & -5 \end{vmatrix}$$

$$= (-5 - 35)\mathbf{i} - (15 - 0)\mathbf{j} + (15 - 0)\mathbf{k} = -40\mathbf{i} - 15\mathbf{j} + 15\mathbf{k}$$

So the vector $\langle -40, -15, 15 \rangle$ is perpendicular to the given plane. Any nonzero scalar multiple of this vector, such as $\langle -8, -3, 3 \rangle$, is also perpendicular to the plane. ■ ■

EXAMPLE 5 Find the area of the triangle with vertices $P(1, 4, 6)$, $Q(-2, 5, -1)$, and $R(1, -1, 1)$.

SOLUTION In Example 4 we computed that $\overrightarrow{PQ} \times \overrightarrow{PR} = \langle -40, -15, 15 \rangle$. The area of the parallelogram with adjacent sides PQ and PR is the length of this cross product:

$$\left| \overrightarrow{PQ} \times \overrightarrow{PR} \right| = \sqrt{(-40)^2 + (-15)^2 + 15^2} = 5\sqrt{82}$$

The area A of the triangle PQR is half the area of this parallelogram, that is, $\frac{5}{2}\sqrt{82}$. ■ ■

■■ Triple Products
■■

FIGURE 7

The product $\mathbf{a} \cdot (\mathbf{b} \times \mathbf{c})$ is called the **scalar triple product** of the vectors $\mathbf{a}$, $\mathbf{b}$, and $\mathbf{c}$. Its geometric significance can be seen by considering the parallelepiped determined by the vectors $\mathbf{a}$, $\mathbf{b}$, and $\mathbf{c}$. (See Figure 7.) The area of the base parallelogram is $A = |\mathbf{b} \times \mathbf{c}|$. If θ is the angle between $\mathbf{a}$ and $\mathbf{b} \times \mathbf{c}$, then the height h of the parallelepiped is $h = |\mathbf{a}||\cos \theta|$. (We must use $|\cos \theta|$ instead of $\cos \theta$ in case $\theta > \pi/2$.) Thus, the volume of the parallelepiped is

$$V = Ah = |\mathbf{b} \times \mathbf{c}||\mathbf{a}||\cos \theta| = |\mathbf{a} \cdot (\mathbf{b} \times \mathbf{c})|$$

Therefore, we have proved the following:

> The volume of the parallelepiped determined by the vectors $\mathbf{a}$, $\mathbf{b}$, and $\mathbf{c}$ is the magnitude of their scalar triple product:
>
> $$V = |\mathbf{a} \cdot (\mathbf{b} \times \mathbf{c})|$$

Instead of thinking of the parallelepiped as having its base parallelogram determined by $\mathbf{b}$ and $\mathbf{c}$, we can think of it with base parallelogram determined by $\mathbf{a}$ and $\mathbf{b}$. In this way, we see that

$$\mathbf{a} \cdot (\mathbf{b} \times \mathbf{c}) = \mathbf{c} \cdot (\mathbf{a} \times \mathbf{b})$$

But the dot product is commutative, so we can write

$$\boxed{6} \qquad \mathbf{a} \cdot (\mathbf{b} \times \mathbf{c}) = (\mathbf{a} \times \mathbf{b}) \cdot \mathbf{c}$$

Suppose that **a**, **b**, and **c** are given in component form:

$$\mathbf{a} = a_1\mathbf{i} + a_2\mathbf{j} + a_3\mathbf{k} \qquad \mathbf{b} = b_1\mathbf{i} + b_2\mathbf{j} + b_3\mathbf{k} \qquad \mathbf{c} = c_1\mathbf{i} + c_2\mathbf{j} + c_3\mathbf{k}$$

Then

$$\mathbf{a} \cdot (\mathbf{b} \times \mathbf{c}) = \mathbf{a} \cdot \left[\begin{vmatrix} b_2 & b_3 \\ c_2 & c_3 \end{vmatrix} \mathbf{i} - \begin{vmatrix} b_1 & b_3 \\ c_1 & c_3 \end{vmatrix} \mathbf{j} + \begin{vmatrix} b_1 & b_2 \\ c_1 & c_2 \end{vmatrix} \mathbf{k} \right]$$

$$= a_1 \begin{vmatrix} b_2 & b_3 \\ c_2 & c_3 \end{vmatrix} - a_2 \begin{vmatrix} b_1 & b_3 \\ c_1 & c_3 \end{vmatrix} + a_3 \begin{vmatrix} b_1 & b_2 \\ c_1 & c_2 \end{vmatrix}$$

This shows that we can write the scalar triple product of **a**, **b**, and **c** as the determinant whose rows are the components of these vectors:

$$\boxed{7} \qquad \mathbf{a} \cdot (\mathbf{b} \times \mathbf{c}) = \begin{vmatrix} a_1 & a_2 & a_3 \\ b_1 & b_2 & b_3 \\ c_1 & c_2 & c_3 \end{vmatrix}$$

Ⓥ **EXAMPLE 6** Use the scalar triple product to show that the vectors $\mathbf{a} = \langle 1, 4, -7 \rangle$, $\mathbf{b} = \langle 2, -1, 4 \rangle$, and $\mathbf{c} = \langle 0, -9, 18 \rangle$ are coplanar; that is, they lie in the same plane.

SOLUTION We use Equation 7 to compute their scalar triple product:

$$\mathbf{a} \cdot (\mathbf{b} \times \mathbf{c}) = \begin{vmatrix} 1 & 4 & -7 \\ 2 & -1 & 4 \\ 0 & -9 & 18 \end{vmatrix}$$

$$= 1 \begin{vmatrix} -1 & 4 \\ -9 & 18 \end{vmatrix} - 4 \begin{vmatrix} 2 & 4 \\ 0 & 18 \end{vmatrix} - 7 \begin{vmatrix} 2 & -1 \\ 0 & -9 \end{vmatrix}$$

$$= 1(18) - 4(36) - 7(-18) = 0$$

Therefore, the volume of the parallelepiped determined by **a**, **b**, and **c** is 0. This means that **a**, **b**, and **c** are coplanar. ■ ■

The product $\mathbf{a} \times (\mathbf{b} \times \mathbf{c})$ is called the **vector triple product** of **a**, **b**, and **c**. The proof of the following formula for the vector triple product is left as Exercise 30.

$$\boxed{8} \qquad \mathbf{a} \times (\mathbf{b} \times \mathbf{c}) = (\mathbf{a} \cdot \mathbf{c})\mathbf{b} - (\mathbf{a} \cdot \mathbf{b})\mathbf{c}$$

Formula 8 will be used to derive Kepler's First Law of planetary motion in Chapter 10.

9.4 Exercises

1. State whether each expression is meaningful. If not, explain why. If so, state whether it is a vector or a scalar.
(a) $\mathbf{a} \cdot (\mathbf{b} \times \mathbf{c})$　　　　(b) $\mathbf{a} \times (\mathbf{b} \cdot \mathbf{c})$
(c) $\mathbf{a} \times (\mathbf{b} \times \mathbf{c})$　　　　(d) $(\mathbf{a} \cdot \mathbf{b}) \times \mathbf{c}$
(e) $(\mathbf{a} \cdot \mathbf{b}) \times (\mathbf{c} \cdot \mathbf{d})$　　(f) $(\mathbf{a} \times \mathbf{b}) \cdot (\mathbf{c} \times \mathbf{d})$

2–3 ■ Find $|\mathbf{u} \times \mathbf{v}|$ and determine whether $\mathbf{u} \times \mathbf{v}$ is directed into the page or out of the page.

2.

3.

4. The figure shows a vector $\mathbf{a}$ in the xy-plane and a vector $\mathbf{b}$ in the direction of $\mathbf{k}$. Their lengths are $|\mathbf{a}| = 3$ and $|\mathbf{b}| = 2$.
(a) Find $|\mathbf{a} \times \mathbf{b}|$.
(b) Use the right-hand rule to decide whether the components of $\mathbf{a} \times \mathbf{b}$ are positive, negative, or 0.

5. A bicycle pedal is pushed by a foot with a 60-N force as shown. The shaft of the pedal is 18 cm long. Find the magnitude of the torque about P.

6. Find the magnitude of the torque about P if a 36-lb force is applied as shown.

7–11 ■ Find the cross product $\mathbf{a} \times \mathbf{b}$ and verify that it is orthogonal to both $\mathbf{a}$ and $\mathbf{b}$.

7. $\mathbf{a} = \langle 1, 2, 0 \rangle$,　$\mathbf{b} = \langle 0, 3, 1 \rangle$

8. $\mathbf{a} = \langle 5, 1, 4 \rangle$,　$\mathbf{b} = \langle -1, 0, 2 \rangle$

9. $\mathbf{a} = \langle t, t^2, t^3 \rangle$,　$\mathbf{b} = \langle 1, 2t, 3t^2 \rangle$

10. $\mathbf{a} = \mathbf{i} + e^t \mathbf{j} + e^{-t} \mathbf{k}$,　$\mathbf{b} = 2\mathbf{i} + e^t \mathbf{j} - e^{-t} \mathbf{k}$

11. $\mathbf{a} = 3\mathbf{i} + 2\mathbf{j} + 4\mathbf{k}$,　$\mathbf{b} = \mathbf{i} - 2\mathbf{j} - 3\mathbf{k}$

12. If $\mathbf{a} = \mathbf{i} - 2\mathbf{k}$ and $\mathbf{b} = \mathbf{j} + \mathbf{k}$, find $\mathbf{a} \times \mathbf{b}$. Sketch $\mathbf{a}$, $\mathbf{b}$, and $\mathbf{a} \times \mathbf{b}$ as vectors starting at the origin.

13. Find two unit vectors orthogonal to both $\langle 2, 0, -3 \rangle$ and $\langle -1, 4, 2 \rangle$.

14. Find two unit vectors orthogonal to both $\mathbf{i} + \mathbf{j} + \mathbf{k}$ and $2\mathbf{i} + \mathbf{k}$.

15. Find the area of the parallelogram with vertices $A(-2, 1)$, $B(0, 4)$, $C(4, 2)$, and $D(2, -1)$.

16. Find the area of the parallelogram with vertices $K(1, 2, 3)$, $L(1, 3, 6)$, $M(3, 8, 6)$, and $N(3, 7, 3)$.

17–18 ■ (a) Find a vector orthogonal to the plane through the points P, Q, and R, and (b) find the area of triangle PQR.

17. $P(0, -2, 0)$,　$Q(4, 1, -2)$,　$R(5, 3, 1)$

18. $P(2, 1, 5)$,　$Q(-1, 3, 4)$,　$R(3, 0, 6)$

19. A wrench 30 cm long lies along the positive y-axis and grips a bolt at the origin. A force is applied in the direction $\langle 0, 3, -4 \rangle$ at the end of the wrench. Find the magnitude of the force needed to supply 100 J of torque to the bolt.

20. Let $\mathbf{v} = 5\mathbf{j}$ and let $\mathbf{u}$ be a vector with length 3 that starts at the origin and rotates in the xy-plane. Find the maximum and minimum values of the length of the vector $\mathbf{u} \times \mathbf{v}$. In what direction does $\mathbf{u} \times \mathbf{v}$ point?

21–22 ■ Find the volume of the parallelepiped determined by the vectors $\mathbf{a}$, $\mathbf{b}$, and $\mathbf{c}$.

21. $\mathbf{a} = \langle 6, 3, -1 \rangle$,　$\mathbf{b} = \langle 0, 1, 2 \rangle$,　$\mathbf{c} = \langle 4, -2, 5 \rangle$

22. $\mathbf{a} = \mathbf{i} + \mathbf{j} - \mathbf{k}$,　$\mathbf{b} = \mathbf{i} - \mathbf{j} + \mathbf{k}$,　$\mathbf{c} = -\mathbf{i} + \mathbf{j} + \mathbf{k}$

23–24 ■ Find the volume of the parallelepiped with adjacent edges PQ, PR, and PS.

23. $P(2, 0, -1)$,　$Q(4, 1, 0)$,　$R(3, -1, 1)$,　$S(2, -2, 2)$

24. $P(3, 0, 1)$,　$Q(-1, 2, 5)$,　$R(5, 1, -1)$,　$S(0, 4, 2)$

25. Use the scalar triple product to verify that the vectors $\mathbf{u} = \mathbf{i} + 5\mathbf{j} - 2\mathbf{k}$, $\mathbf{v} = 3\mathbf{i} - \mathbf{j}$, and $\mathbf{w} = 5\mathbf{i} + 9\mathbf{j} - 4\mathbf{k}$ are coplanar.

26. Use the scalar triple product to determine whether the points $A(1, 3, 2)$, $B(3, -1, 6)$, $C(5, 2, 0)$, and $D(3, 6, -4)$ lie in the same plane.

27. (a) Let P be a point not on the line L that passes through the points Q and R. Show that the distance d from the point P to the line L is

$$d = \frac{|\mathbf{a} \times \mathbf{b}|}{|\mathbf{a}|}$$

where $\mathbf{a} = \overrightarrow{QR}$ and $\mathbf{b} = \overrightarrow{QP}$.

(b) Use the formula in part (a) to find the distance from the point $P(1, 1, 1)$ to the line through $Q(0, 6, 8)$ and $R(-1, 4, 7)$.

28. (a) Let P be a point not on the plane that passes through the points Q, R, and S. Show that the distance d from P to the plane is

$$d = \frac{|\mathbf{a} \cdot (\mathbf{b} \times \mathbf{c})|}{|\mathbf{a} \times \mathbf{b}|}$$

where $\mathbf{a} = \overrightarrow{QR}$, $\mathbf{b} = \overrightarrow{QS}$, and $\mathbf{c} = \overrightarrow{QP}$.

(b) Use the formula in part (a) to find the distance from the point $P(2, 1, 4)$ to the plane through the points $Q(1, 0, 0)$, $R(0, 2, 0)$, and $S(0, 0, 3)$.

29. Prove that $(\mathbf{a} - \mathbf{b}) \times (\mathbf{a} + \mathbf{b}) = 2(\mathbf{a} \times \mathbf{b})$.

30. Prove the following formula for the vector triple product:

$$\mathbf{a} \times (\mathbf{b} \times \mathbf{c}) = (\mathbf{a} \cdot \mathbf{c})\mathbf{b} - (\mathbf{a} \cdot \mathbf{b})\mathbf{c}$$

31. Use Exercise 30 to prove that

$$\mathbf{a} \times (\mathbf{b} \times \mathbf{c}) + \mathbf{b} \times (\mathbf{c} \times \mathbf{a}) + \mathbf{c} \times (\mathbf{a} \times \mathbf{b}) = \mathbf{0}$$

32. Prove that

$$(\mathbf{a} \times \mathbf{b}) \cdot (\mathbf{c} \times \mathbf{d}) = \begin{vmatrix} \mathbf{a} \cdot \mathbf{c} & \mathbf{b} \cdot \mathbf{c} \\ \mathbf{a} \cdot \mathbf{d} & \mathbf{b} \cdot \mathbf{d} \end{vmatrix}$$

33. Suppose that $\mathbf{a} \neq \mathbf{0}$.
(a) If $\mathbf{a} \cdot \mathbf{b} = \mathbf{a} \cdot \mathbf{c}$, does it follow that $\mathbf{b} = \mathbf{c}$?
(b) If $\mathbf{a} \times \mathbf{b} = \mathbf{a} \times \mathbf{c}$, does it follow that $\mathbf{b} = \mathbf{c}$?
(c) If $\mathbf{a} \cdot \mathbf{b} = \mathbf{a} \cdot \mathbf{c}$ and $\mathbf{a} \times \mathbf{b} = \mathbf{a} \times \mathbf{c}$, does it follow that $\mathbf{b} = \mathbf{c}$?

34. If $\mathbf{v}_1$, $\mathbf{v}_2$, and $\mathbf{v}_3$ are noncoplanar vectors, let

$$\mathbf{k}_1 = \frac{\mathbf{v}_2 \times \mathbf{v}_3}{\mathbf{v}_1 \cdot (\mathbf{v}_2 \times \mathbf{v}_3)}$$

$$\mathbf{k}_2 = \frac{\mathbf{v}_3 \times \mathbf{v}_1}{\mathbf{v}_1 \cdot (\mathbf{v}_2 \times \mathbf{v}_3)}$$

$$\mathbf{k}_3 = \frac{\mathbf{v}_1 \times \mathbf{v}_2}{\mathbf{v}_1 \cdot (\mathbf{v}_2 \times \mathbf{v}_3)}$$

(These vectors occur in the study of crystallography. Vectors of the form $n_1\mathbf{v}_1 + n_2\mathbf{v}_2 + n_3\mathbf{v}_3$, where each n_i is an integer, form a *lattice* for a crystal. Vectors written similarly in terms of $\mathbf{k}_1$, $\mathbf{k}_2$, and $\mathbf{k}_3$ form the *reciprocal lattice*.)
(a) Show that $\mathbf{k}_i$ is perpendicular to $\mathbf{v}_j$ if $i \neq j$.
(b) Show that $\mathbf{k}_i \cdot \mathbf{v}_i = 1$ for $i = 1, 2, 3$.
(c) Show that $\mathbf{k}_1 \cdot (\mathbf{k}_2 \times \mathbf{k}_3) = \dfrac{1}{\mathbf{v}_1 \cdot (\mathbf{v}_2 \times \mathbf{v}_3)}$.

DISCOVERY PROJECT

The Geometry of a Tetrahedron

A tetrahedron is a solid with four vertices, P, Q, R, and S, and four triangular faces as shown in the figure.

1. Let $\mathbf{v}_1$, $\mathbf{v}_2$, $\mathbf{v}_3$, and $\mathbf{v}_4$ be vectors with lengths equal to the areas of the faces opposite the vertices P, Q, R, and S, respectively, and directions perpendicular to the respective faces and pointing outward. Show that

$$\mathbf{v}_1 + \mathbf{v}_2 + \mathbf{v}_3 + \mathbf{v}_4 = \mathbf{0}$$

2. The volume V of a tetrahedron is one-third the distance from a vertex to the opposite face, times the area of that face.
(a) Find a formula for the volume of a tetrahedron in terms of the coordinates of its vertices P, Q, R, and S.
(b) Find the volume of the tetrahedron whose vertices are $P(1, 1, 1)$, $Q(1, 2, 3)$, $R(1, 1, 2)$, and $S(3, -1, 2)$.

3. Suppose the tetrahedron in the figure has a trirectangular vertex S. (This means that the three angles at S are all right angles.) Let A, B, and C be the areas of the three faces that meet at S, and let D be the area of the opposite face PQR. Using the result of Problem 1, or otherwise, show that

$$D^2 = A^2 + B^2 + C^2$$

(This is a three-dimensional version of the Pythagorean Theorem.)

9.5 Equations of Lines and Planes

A line in the xy-plane is determined when a point on the line and the direction of the line (its slope or angle of inclination) are given. The equation of the line can then be written using the point-slope form.

Likewise, a line L in three-dimensional space is determined when we know a point $P_0(x_0, y_0, z_0)$ on L and the direction of L. In three dimensions the direction of a line is conveniently described by a vector, so we let $\mathbf{v}$ be a vector parallel to L. Let $P(x, y, z)$ be an arbitrary point on L and let $\mathbf{r}_0$ and $\mathbf{r}$ be the position vectors of P_0 and P (that is, they have representations $\overrightarrow{OP_0}$ and $\overrightarrow{OP}$). If $\mathbf{a}$ is the vector with representation $\overrightarrow{P_0P}$, as in Figure 1, then the Triangle Law for vector addition gives $\mathbf{r} = \mathbf{r}_0 + \mathbf{a}$. But, since $\mathbf{a}$ and $\mathbf{v}$ are parallel vectors, there is a scalar t such that $\mathbf{a} = t\mathbf{v}$. Thus

FIGURE 1

$$\boxed{1} \qquad \boxed{\mathbf{r} = \mathbf{r}_0 + t\mathbf{v}}$$

which is a **vector equation** of L. Each value of the **parameter** t gives the position vector $\mathbf{r}$ of a point on L. In other words, as t varies, the line is traced out by the tip of the vector $\mathbf{r}$. As Figure 2 indicates, positive values of t correspond to points on L that lie on one side of P_0, whereas negative values of t correspond to points that lie on the other side of P_0.

If the vector $\mathbf{v}$ that gives the direction of the line L is written in component form as $\mathbf{v} = \langle a, b, c \rangle$, then we have $t\mathbf{v} = \langle ta, tb, tc \rangle$. We can also write $\mathbf{r} = \langle x, y, z \rangle$ and $\mathbf{r}_0 = \langle x_0, y_0, z_0 \rangle$, so the vector equation (1) becomes

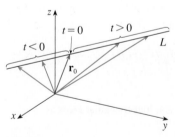

FIGURE 2

$$\langle x, y, z \rangle = \langle x_0 + ta, y_0 + tb, z_0 + tc \rangle$$

Two vectors are equal if and only if corresponding components are equal. Therefore, we have the three scalar equations:

$$\boxed{2} \qquad \boxed{x = x_0 + at \qquad y = y_0 + bt \qquad z = z_0 + ct}$$

where $t \in \mathbb{R}$. These equations are called **parametric equations** of the line L through the point $P_0(x_0, y_0, z_0)$ and parallel to the vector $\mathbf{v} = \langle a, b, c \rangle$. Each value of the parameter t gives a point (x, y, z) on L.

EXAMPLE 1

(a) Find a vector equation and parametric equations for the line that passes through the point $(5, 1, 3)$ and is parallel to the vector $\mathbf{i} + 4\mathbf{j} - 2\mathbf{k}$.

(b) Find two other points on the line.

SOLUTION

(a) Here $\mathbf{r}_0 = \langle 5, 1, 3 \rangle = 5\mathbf{i} + \mathbf{j} + 3\mathbf{k}$ and $\mathbf{v} = \mathbf{i} + 4\mathbf{j} - 2\mathbf{k}$, so the vector equation (1) becomes

$$\mathbf{r} = (5\mathbf{i} + \mathbf{j} + 3\mathbf{k}) + t(\mathbf{i} + 4\mathbf{j} - 2\mathbf{k})$$

or

$$\mathbf{r} = (5 + t)\mathbf{i} + (1 + 4t)\mathbf{j} + (3 - 2t)\mathbf{k}$$

Parametric equations are

$$x = 5 + t \qquad y = 1 + 4t \qquad z = 3 - 2t$$

(b) Choosing the parameter value $t = 1$ gives $x = 6$, $y = 5$, and $z = 1$, so $(6, 5, 1)$ is a point on the line. Similarly, $t = -1$ gives the point $(4, -3, 5)$. ■ ■

■ ■ Figure 3 shows the line L in Example 1 and its relation to the given point and to the vector that gives its direction.

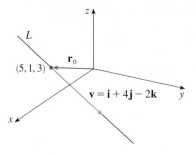

FIGURE 3

The vector equation and parametric equations of a line are not unique. If we change the point or the parameter or choose a different parallel vector, then the equations change. For instance, if, instead of $(5, 1, 3)$, we choose the point $(6, 5, 1)$ in Example 1, then the parametric equations of the line become

$$x = 6 + t \qquad y = 5 + 4t \qquad z = 1 - 2t$$

Or, if we stay with the point $(5, 1, 3)$ but choose the parallel vector $2\mathbf{i} + 8\mathbf{j} - 4\mathbf{k}$, we arrive at the equations

$$x = 5 + 2t \qquad y = 1 + 8t \qquad z = 3 - 4t$$

In general, if a vector $\mathbf{v} = \langle a, b, c \rangle$ is used to describe the direction of a line L, then the numbers a, b, and c are called **direction numbers** of L. Since any vector parallel to $\mathbf{v}$ could also be used, we see that any three numbers proportional to a, b, and c could also be used as a set of direction numbers for L.

Another way of describing a line L is to eliminate the parameter t from Equations 2. If none of a, b, or c is 0, we can solve each of these equations for t, equate the results, and obtain

3

$$\frac{x - x_0}{a} = \frac{y - y_0}{b} = \frac{z - z_0}{c}$$

These equations are called **symmetric equations** of L. Notice that the numbers a, b, and c that appear in the denominators of Equations 3 are direction numbers of L, that is, components of a vector parallel to L. If one of a, b, or c is 0, we can still eliminate t. For instance, if $a = 0$, we could write the equations of L as

$$x = x_0 \qquad \frac{y - y_0}{b} = \frac{z - z_0}{c}$$

This means that L lies in the vertical plane $x = x_0$.

■ ■ Figure 4 shows the line L in Example 2 and the point P where it intersects the xy-plane.

FIGURE 4

EXAMPLE 2

(a) Find parametric equations and symmetric equations of the line that passes through the points $A(2, 4, -3)$ and $B(3, -1, 1)$.

(b) At what point does this line intersect the xy-plane?

SOLUTION

(a) We are not explicitly given a vector parallel to the line, but observe that the vector $\mathbf{v}$ with representation $\overrightarrow{AB}$ is parallel to the line and

$$\mathbf{v} = \langle 3 - 2, -1 - 4, 1 - (-3) \rangle = \langle 1, -5, 4 \rangle$$

Thus, direction numbers are $a = 1$, $b = -5$, and $c = 4$. Taking the point $(2, 4, -3)$ as P_0, we see that parametric equations (2) are

$$x = 2 + t \qquad y = 4 - 5t \qquad z = -3 + 4t$$

and symmetric equations (3) are

$$\frac{x - 2}{1} = \frac{y - 4}{-5} = \frac{z + 3}{4}$$

(b) The line intersects the xy-plane when $z = 0$, so we put $z = 0$ in the symmetric equations and obtain

$$\frac{x - 2}{1} = \frac{y - 4}{-5} = \frac{3}{4}$$

This gives $x = \frac{11}{4}$ and $y = \frac{1}{4}$, so the line intersects the xy-plane at the point $\left(\frac{11}{4}, \frac{1}{4}, 0\right)$.

■ ■

In general, the procedure of Example 2 shows that direction numbers of the line L through the points $P_0(x_0, y_0, z_0)$ and $P_1(x_1, y_1, z_1)$ are $x_1 - x_0$, $y_1 - y_0$, and $z_1 - z_0$ and so symmetric equations of L are

$$\frac{x - x_0}{x_1 - x_0} = \frac{y - y_0}{y_1 - y_0} = \frac{z - z_0}{z_1 - z_0}$$

Often, we need a description, not of an entire line, but of just a line segment. How, for instance, could we describe the line segment AB in Example 2? If we put $t = 0$ in the parametric equations in Example 2(a), we get the point $(2, 4, -3)$ and if we put $t = 1$ we get $(3, -1, 1)$. So the line segment AB is described by the parametric equations

$$x = 2 + t \qquad y = 4 - 5t \qquad z = -3 + 4t \qquad 0 \leqslant t \leqslant 1$$

or by the corresponding vector equation

$$\mathbf{r}(t) = \langle 2 + t, 4 - 5t, -3 + 4t \rangle \qquad 0 \leqslant t \leqslant 1$$

In general, we know from Equation 1 that the vector equation of a line through the (tip of the) vector $\mathbf{r}_0$ in the direction of a vector $\mathbf{v}$ is $\mathbf{r} = \mathbf{r}_0 + t\mathbf{v}$. If the line also passes through (the tip of) $\mathbf{r}_1$, then we can take $\mathbf{v} = \mathbf{r}_1 - \mathbf{r}_0$ and so its vector equation is

$$\mathbf{r} = \mathbf{r}_0 + t(\mathbf{r}_1 - \mathbf{r}_0) = (1 - t)\mathbf{r}_0 + t\mathbf{r}_1$$

The line segment from $\mathbf{r}_0$ to $\mathbf{r}_1$ is given by the parameter interval $0 \leq t \leq 1$.

---

4 The line segment from $\mathbf{r}_0$ to $\mathbf{r}_1$ is given by the vector equation

$$\mathbf{r}(t) = (1 - t)\mathbf{r}_0 + t\mathbf{r}_1 \qquad 0 \leq t \leq 1$$

 The lines L_1 and L_2 in Example 3, shown in Figure 5, are skew lines.

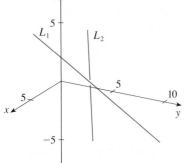

FIGURE 5

⟦V⟧ **EXAMPLE 3** Show that the lines L_1 and L_2 with parametric equations

$$x = 1 + t \qquad y = -2 + 3t \qquad z = 4 - t$$

$$x = 2s \qquad y = 3 + s \qquad z = -3 + 4s$$

are **skew lines**; that is, they do not intersect and are not parallel (and therefore do not lie in the same plane).

SOLUTION The lines are not parallel because the corresponding vectors $\langle 1, 3, -1 \rangle$ and $\langle 2, 1, 4 \rangle$ are not parallel. (Their components are not proportional.) If L_1 and L_2 had a point of intersection, there would be values of t and s such that

$$1 + t = 2s$$

$$-2 + 3t = 3 + s$$

$$4 - t = -3 + 4s$$

But if we solve the first two equations, we get $t = \frac{11}{5}$ and $s = \frac{8}{5}$, and these values don't satisfy the third equation. Therefore, there are no values of t and s that satisfy the three equations, so L_1 and L_2 do not intersect. Thus, L_1 and L_2 are skew lines.

■ ■

▦ Planes

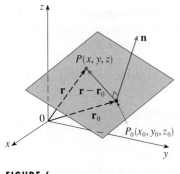

FIGURE 6

Although a line in space is determined by a point and a direction, a plane in space is more difficult to describe. A single vector parallel to a plane is not enough to convey the "direction" of the plane, but a vector perpendicular to the plane does completely specify its direction. Thus, a plane in space is determined by a point $P_0(x_0, y_0, z_0)$ in the plane and a vector $\mathbf{n}$ that is orthogonal to the plane. This orthogonal vector $\mathbf{n}$ is called a **normal vector**. Let $P(x, y, z)$ be an arbitrary point in the plane, and let $\mathbf{r}_0$ and $\mathbf{r}$ be the position vectors of P_0 and P. Then the vector $\mathbf{r} - \mathbf{r}_0$ is represented by $\overrightarrow{P_0P}$. (See Figure 6.) The normal vector $\mathbf{n}$ is orthogonal to every vector in the given plane. In particular, $\mathbf{n}$ is orthogonal to $\mathbf{r} - \mathbf{r}_0$ and so we have

5
$$\mathbf{n} \cdot (\mathbf{r} - \mathbf{r}_0) = 0$$

which can be rewritten as

6
$$\mathbf{n} \cdot \mathbf{r} = \mathbf{n} \cdot \mathbf{r}_0$$

Either Equation 5 or Equation 6 is called a **vector equation of the plane**.

To obtain a scalar equation for the plane, we write $\mathbf{n} = \langle a, b, c \rangle$, $\mathbf{r} = \langle x, y, z \rangle$, and $\mathbf{r}_0 = \langle x_0, y_0, z_0 \rangle$. Then the vector equation (5) becomes

$$\langle a, b, c \rangle \cdot \langle x - x_0, y - y_0, z - z_0 \rangle = 0$$

or

7
$$a(x - x_0) + b(y - y_0) + c(z - z_0) = 0$$

Equation 7 is the **scalar equation of the plane through** $P_0(x_0, y_0, z_0)$ **with normal vector** $\mathbf{n} = \langle a, b, c \rangle$.

Ⓥ **EXAMPLE 4** Find an equation of the plane through the point $(2, 4, -1)$ with normal vector $\mathbf{n} = \langle 2, 3, 4 \rangle$. Find the intercepts and sketch the plane.

SOLUTION Putting $a = 2$, $b = 3$, $c = 4$, $x_0 = 2$, $y_0 = 4$, and $z_0 = -1$ in Equation 7, we see that an equation of the plane is

$$2(x - 2) + 3(y - 4) + 4(z + 1) = 0$$

or

$$2x + 3y + 4z = 12$$

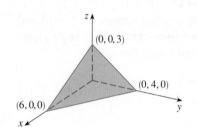

FIGURE 7

To find the x-intercept we set $y = z = 0$ in this equation and obtain $x = 6$. Similarly, the y-intercept is 4 and the z-intercept is 3. This enables us to sketch the portion of the plane that lies in the first octant (see Figure 7). ■ ■

By collecting terms in Equation 7 as we did in Example 4, we can rewrite the equation of a plane as

8
$$ax + by + cz + d = 0$$

where $d = -(ax_0 + by_0 + cz_0)$. Equation 8 is called a **linear equation** in x, y, and z. Conversely, it can be shown that if a, b, and c are not all 0, then the linear equation (8) represents a plane with normal vector $\langle a, b, c \rangle$. (See Exercise 55.)

■ ■ Figure 8 shows the portion of the plane in Example 5 that is enclosed by triangle PQR.

EXAMPLE 5 Find an equation of the plane that passes through the points $P(1, 3, 2)$, $Q(3, -1, 6)$, and $R(5, 2, 0)$.

SOLUTION The vectors $\mathbf{a}$ and $\mathbf{b}$ corresponding to $\overrightarrow{PQ}$ and $\overrightarrow{PR}$ are

$$\mathbf{a} = \langle 2, -4, 4 \rangle \qquad \mathbf{b} = \langle 4, -1, -2 \rangle$$

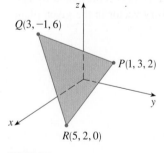

FIGURE 8

Since both $\mathbf{a}$ and $\mathbf{b}$ lie in the plane, their cross product $\mathbf{a} \times \mathbf{b}$ is orthogonal to the plane and can be taken as the normal vector. Thus

$$\mathbf{n} = \mathbf{a} \times \mathbf{b} = \begin{vmatrix} \mathbf{i} & \mathbf{j} & \mathbf{k} \\ 2 & -4 & 4 \\ 4 & -1 & -2 \end{vmatrix} = 12\mathbf{i} + 20\mathbf{j} + 14\mathbf{k}$$

With the point $P(1, 3, 2)$ and the normal vector $\mathbf{n}$, an equation of the plane is

$$12(x - 1) + 20(y - 3) + 14(z - 2) = 0$$

or

$$6x + 10y + 7z = 50$$

■ ■

EXAMPLE 6 Find the point at which the line with parametric equations $x = 2 + 3t$, $y = -4t$, $z = 5 + t$ intersects the plane $4x + 5y - 2z = 18$.

SOLUTION We substitute the expressions for x, y, and z from the parametric equations into the equation of the plane:

$$4(2 + 3t) + 5(-4t) - 2(5 + t) = 18$$

This simplifies to $-10t = 20$, so $t = -2$. Therefore, the point of intersection occurs when the parameter value is $t = -2$. Then $x = 2 + 3(-2) = -4$, $y = -4(-2) = 8$, $z = 5 - 2 = 3$ and so the point of intersection is $(-4, 8, 3)$. ■ ■

FIGURE 9

Two planes are **parallel** if their normal vectors are parallel. For instance, the planes $x + 2y - 3z = 4$ and $2x + 4y - 6z = 3$ are parallel because their normal vectors are $\mathbf{n}_1 = \langle 1, 2, -3 \rangle$ and $\mathbf{n}_2 = \langle 2, 4, -6 \rangle$ and $\mathbf{n}_2 = 2\mathbf{n}_1$. If two planes are not parallel, then they intersect in a straight line and the angle between the two planes is defined as the acute angle between their normal vectors (see angle θ in Figure 9).

EXAMPLE 7
(a) Find the angle between the planes $x + y + z = 1$ and $x - 2y + 3z = 1$.
(b) Find symmetric equations for the line of intersection L of these two planes.

■ ■ Figure 10 shows the planes in Example 7 and their line of intersection L.

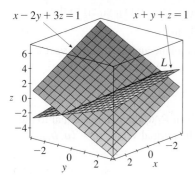

FIGURE 10

SOLUTION
(a) The normal vectors of these planes are

$$\mathbf{n}_1 = \langle 1, 1, 1 \rangle \qquad \mathbf{n}_2 = \langle 1, -2, 3 \rangle$$

and so, if θ is the angle between the planes,

$$\cos \theta = \frac{\mathbf{n}_1 \cdot \mathbf{n}_2}{|\mathbf{n}_1||\mathbf{n}_2|} = \frac{1(1) + 1(-2) + 1(3)}{\sqrt{1 + 1 + 1}\,\sqrt{1 + 4 + 9}} = \frac{2}{\sqrt{42}}$$

$$\theta = \cos^{-1}\left(\frac{2}{\sqrt{42}}\right) \approx 72°$$

(b) We first need to find a point on L. For instance, we can find the point where the line intersects the xy-plane by setting $z = 0$ in the equations of both planes. This gives the equations $x + y = 1$ and $x - 2y = 1$, whose solution is $x = 1$, $y = 0$. So the point $(1, 0, 0)$ lies on L.

Now we observe that, since L lies in both planes, it is perpendicular to both of the normal vectors. Thus, a vector $\mathbf{v}$ parallel to L is given by the cross product

■ ■ Another way to find the line of intersection is to solve the equations of the planes for two of the variables in terms of the third, which can be taken as the parameter.

$$\mathbf{v} = \mathbf{n}_1 \times \mathbf{n}_2 = \begin{vmatrix} \mathbf{i} & \mathbf{j} & \mathbf{k} \\ 1 & 1 & 1 \\ 1 & -2 & 3 \end{vmatrix} = 5\mathbf{i} - 2\mathbf{j} - 3\mathbf{k}$$

and so the symmetric equations of L can be written as

$$\frac{x - 1}{5} = \frac{y}{-2} = \frac{z}{-3}$$

■ ■

NOTE □ Since a linear equation in x, y, and z represents a plane and two non-parallel planes intersect in a line, it follows that two linear equations can represent

FIGURE 11

■ ■ Figure 11 shows how the line L in Example 7 can also be regarded as the line of intersection of planes derived from its symmetric equations.

a line. The points (x, y, z) that satisfy both

$$a_1x + b_1y + c_1z + d_1 = 0 \quad \text{and} \quad a_2x + b_2y + c_2z + d_2 = 0$$

lie on both of these planes, and so the pair of linear equations represents the line of intersection of the planes (if they are not parallel). For instance, in Example 7 the line L was given as the line of intersection of the planes $x + y + z = 1$ and $x - 2y + 3z = 1$. The symmetric equations that we found for L could be written as

$$\frac{x - 1}{5} = \frac{y}{-2} \quad \text{and} \quad \frac{y}{-2} = \frac{z}{-3}$$

which is again a pair of linear equations. They exhibit L as the line of intersection of the planes $(x - 1)/5 = y/(-2)$ and $y/(-2) = z/(-3)$. (See Figure 11.)

In general, when we write the equations of a line in the symmetric form

$$\frac{x - x_0}{a} = \frac{y - y_0}{b} = \frac{z - z_0}{c}$$

we can regard the line as the line of intersection of the two planes

$$\frac{x - x_0}{a} = \frac{y - y_0}{b} \quad \text{and} \quad \frac{y - y_0}{b} = \frac{z - z_0}{c}$$

EXAMPLE 8 Find a formula for the distance D from a point $P_1(x_1, y_1, z_1)$ to the plane $ax + by + cz + d = 0$.

SOLUTION Let $P_0(x_0, y_0, z_0)$ be any point in the given plane and let $\mathbf{b}$ be the vector corresponding to $\overrightarrow{P_0P_1}$. Then

$$\mathbf{b} = \langle x_1 - x_0, y_1 - y_0, z_1 - z_0 \rangle$$

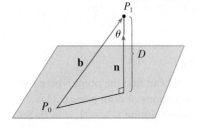

FIGURE 12

From Figure 12 you can see that the distance D from P_1 to the plane is equal to the absolute value of the scalar projection of $\mathbf{b}$ onto the normal vector $\mathbf{n} = \langle a, b, c \rangle$. (See Section 9.3.) Thus

$$D = |\text{comp}_\mathbf{n} \mathbf{b}| = \frac{|\mathbf{n} \cdot \mathbf{b}|}{|\mathbf{n}|}$$

$$= \frac{|a(x_1 - x_0) + b(y_1 - y_0) + c(z_1 - z_0)|}{\sqrt{a^2 + b^2 + c^2}}$$

$$= \frac{|(ax_1 + by_1 + cz_1) - (ax_0 + by_0 + cz_0)|}{\sqrt{a^2 + b^2 + c^2}}$$

Since P_0 lies in the plane, its coordinates satisfy the equation of the plane and so we have $ax_0 + by_0 + cz_0 + d = 0$. Thus, the formula for D can be written as

$$\boxed{\; D = \frac{|ax_1 + by_1 + cz_1 + d|}{\sqrt{a^2 + b^2 + c^2}} \;}$$

9

EXAMPLE 9 Find the distance between the parallel planes $10x + 2y - 2z = 5$ and $5x + y - z = 1$.

SOLUTION First we note that the planes are parallel because their normal vectors $\langle 10, 2, -2 \rangle$ and $\langle 5, 1, -1 \rangle$ are parallel. To find the distance D between the planes, we choose any point on one plane and calculate its distance to the other plane. In particular, if we put $y = z = 0$ in the equation of the first plane, we get $10x = 5$ and so $\left(\frac{1}{2}, 0, 0\right)$ is a point in this plane. By Formula 9, the distance between $\left(\frac{1}{2}, 0, 0\right)$ and the plane $5x + y - z - 1 = 0$ is

$$D = \frac{\left|5\left(\frac{1}{2}\right) + 1(0) - 1(0) - 1\right|}{\sqrt{5^2 + 1^2 + (-1)^2}} = \frac{\frac{3}{2}}{3\sqrt{3}} = \frac{\sqrt{3}}{6}$$

So the distance between the planes is $\sqrt{3}/6$.

EXAMPLE 10 In Example 3 we showed that the lines

$$L_1: \quad x = 1 + t \qquad y = -2 + 3t \qquad z = 4 - t$$
$$L_2: \quad x = 2s \qquad y = 3 + s \qquad z = -3 + 4s$$

are skew. Find the distance between them.

SOLUTION Since the two lines L_1 and L_2 are skew, they can be viewed as lying on two parallel planes P_1 and P_2. The distance between L_1 and L_2 is the same as the distance between P_1 and P_2, which can be computed as in Example 9. The common normal vector to both planes must be orthogonal to both $\mathbf{v}_1 = \langle 1, 3, -1 \rangle$ (the direction of L_1) and $\mathbf{v}_2 = \langle 2, 1, 4 \rangle$ (the direction of L_2). So a normal vector is

$$\mathbf{n} = \mathbf{v}_1 \times \mathbf{v}_2 = \begin{vmatrix} \mathbf{i} & \mathbf{j} & \mathbf{k} \\ 1 & 3 & -1 \\ 2 & 1 & 4 \end{vmatrix} = 13\mathbf{i} - 6\mathbf{j} - 5\mathbf{k}$$

If we put $s = 0$ in the equations of L_2, we get the point $(0, 3, -3)$ on L_2 and so an equation for P_2 is

$$13(x - 0) - 6(y - 3) - 5(z + 3) = 0 \qquad \text{or} \qquad 13x - 6y - 5z + 3 = 0$$

If we now set $t = 0$ in the equations for L_1, we get the point $(1, -2, 4)$ on P_1. So the distance between L_1 and L_2 is the same as the distance from $(1, -2, 4)$ to $13x - 6y - 5z + 3 = 0$. By Formula 9, this distance is

$$D = \frac{\left|13(1) - 6(-2) - 5(4) + 3\right|}{\sqrt{13^2 + (-6)^2 + (-5)^2}} = \frac{8}{\sqrt{230}} \approx 0.53$$

9.5 Exercises

1. Determine whether each statement is true or false.
   (a) Two lines parallel to a third line are parallel.
   (b) Two lines perpendicular to a third line are parallel.
   (c) Two planes parallel to a third plane are parallel.
   (d) Two planes perpendicular to a third plane are parallel.
   (e) Two lines parallel to a plane are parallel.
   (f) Two lines perpendicular to a plane are parallel.
   (g) Two planes parallel to a line are parallel.
   (h) Two planes perpendicular to a line are parallel.
   (i) Two planes either intersect or are parallel.
   (j) Two lines either intersect or are parallel.
   (k) A plane and a line either intersect or are parallel.

2–5 ■ Find a vector equation and parametric equations for the line.

2. The line through the point $(1, 0, -3)$ and parallel to the vector $2\mathbf{i} - 4\mathbf{j} + 5\mathbf{k}$

3. The line through the point $(-2, 4, 10)$ and parallel to the vector $\langle 3, 1, -8 \rangle$

4. The line through the origin and parallel to the line $x = 2t$, $y = 1 - t, z = 4 + 3t$

5. The line through the point $(1, 0, 6)$ and perpendicular to the plane $x + 3y + z = 5$

6–10 ■ Find parametric equations and symmetric equations for the line.

6. The line through the points $(6, 1, -3)$ and $(2, 4, 5)$

7. The line through the points $(0, \frac{1}{2}, 1)$ and $(2, 1, -3)$

8. The line through $(2, 1, 0)$ and perpendicular to both $\mathbf{i} + \mathbf{j}$ and $\mathbf{j} + \mathbf{k}$

9. The line through $(1, -1, 1)$ and parallel to the line $x + 2 = \frac{1}{2}y = z - 3$

10. The line of intersection of the planes $x + y + z = 1$ and $x + z = 0$

11. Is the line through $(-4, -6, 1)$ and $(-2, 0, -3)$ parallel to the line through $(10, 18, 4)$ and $(5, 3, 14)$?

12. Is the line through $(4, 1, -1)$ and $(2, 5, 3)$ perpendicular to the line through $(-3, 2, 0)$ and $(5, 1, 4)$?

13. (a) Find symmetric equations for the line that passes through the point $(0, 2, -1)$ and is parallel to the line with parametric equations $x = 1 + 2t$, $y = 3t$, $z = 5 - 7t$.
(b) Find the points in which the required line in part (a) intersects the coordinate planes.

14. (a) Find parametric equations for the line through $(5, 1, 0)$ that is perpendicular to the plane $2x - y + z = 1$.
(b) In what points does this line intersect the coordinate planes?

15. Find a vector equation for the line segment from $(2, -1, 4)$ to $(4, 6, 1)$.

16. Find parametric equations for the line segment from $(10, 3, 1)$ to $(5, 6, -3)$.

17–20 ■ Determine whether the lines L_1 and L_2 are parallel, skew, or intersecting. If they intersect, find the point of intersection.

17. L_1: $x = -6t$, $y = 1 + 9t$, $z = -3t$
L_2: $x = 1 + 2s$, $y = 4 - 3s$, $z = s$

18. L_1: $x = 1 + 2t$, $y = 3t$, $z = 2 - t$
L_2: $x = -1 + s$, $y = 4 + s$, $z = 1 + 3s$

19. L_1: $\dfrac{x}{1} = \dfrac{y - 1}{2} = \dfrac{z - 2}{3}$

L_2: $\dfrac{x - 3}{-4} = \dfrac{y - 2}{-3} = \dfrac{z - 1}{2}$

20. L_1: $\dfrac{x - 1}{2} = \dfrac{y - 3}{2} = \dfrac{z - 2}{-1}$

L_2: $\dfrac{x - 2}{1} = \dfrac{y - 6}{-1} = \dfrac{z + 2}{3}$

21–30 ■ Find an equation of the plane.

21. The plane through the point $(6, 3, 2)$ and perpendicular to the vector $\langle -2, 1, 5 \rangle$

22. The plane through the point $(4, 0, -3)$ and with normal vector $\mathbf{j} + 2\mathbf{k}$

23. The plane through the origin and parallel to the plane $2x - y + 3z = 1$

24. The plane that contains the line $x = 3 + 2t$, $y = t$, $z = 8 - t$ and is parallel to the plane $2x + 4y + 8z = 17$

25. The plane through the points $(0, 1, 1)$, $(1, 0, 1)$, and $(1, 1, 0)$

26. The plane through the origin and the points $(2, -4, 6)$ and $(5, 1, 3)$

27. The plane that passes through the point $(6, 0, -2)$ and contains the line $x = 4 - 2t$, $y = 3 + 5t$, $z = 7 + 4t$

28. The plane that passes through the point $(1, -1, 1)$ and contains the line with symmetric equations $x = 2y = 3z$

29. The plane that passes through the point $(-1, 2, 1)$ and contains the line of intersection of the planes $x + y - z = 2$ and $2x - y + 3z = 1$

30. The plane that passes through the line of intersection of the planes $x - z = 1$ and $y + 2z = 3$ and is perpendicular to the plane $x + y - 2z = 1$

31. Find the point at which the line $x = 3 - t$, $y = 2 + t$, $z = 5t$ intersects the plane $x - y + 2z = 9$.

32. Where does the line through $(1, 0, 1)$ and $(4, -2, 2)$ intersect the plane $x + y + z = 6$?

33–36 ■ Determine whether the planes are parallel, perpendicular, or neither. If neither, find the angle between them.

33. $x + y + z = 1$, $\quad x - y + z = 1$

34. $2x - 3y + 4z = 5$, $\quad x + 6y + 4z = 3$

35. $x = 4y - 2z$, $\quad 8y = 1 + 2x + 4z$

36. $x + 2y + 2z = 1$, $\quad 2x - y + 2z = 1$

37. (a) Find symmetric equations for the line of intersection of the planes $x + y - z = 2$ and $3x - 4y + 5z = 6$.
(b) Find the angle between these planes.

38. Find an equation for the plane consisting of all points that are equidistant from the points $(-4, 2, 1)$ and $(2, -4, 3)$.

39. Find an equation of the plane with x-intercept a, y-intercept b, and z-intercept c.

40. (a) Find the point at which the given lines intersect:
$$\mathbf{r} = \langle 1, 1, 0 \rangle + t\langle 1, -1, 2 \rangle$$
$$\mathbf{r} = \langle 2, 0, 2 \rangle + s\langle -1, 1, 0 \rangle$$
(b) Find an equation of the plane that contains these lines.

41. Find parametric equations for the line through the point $(0, 1, 2)$ that is parallel to the plane $x + y + z = 2$ and perpendicular to the line $x = 1 + t$, $y = 1 - t$, $z = 2t$.

42. Find parametric equations for the line through the point $(0, 1, 2)$ that is perpendicular to the line $x = 1 + t$, $y = 1 - t$, $z = 2t$ and intersects this line.

43. Which of the following four planes are parallel? Are any of them identical?

P_1: $4x - 2y + 6z = 3$ $\qquad$ P_2: $4x - 2y - 2z = 6$

P_3: $-6x + 3y - 9z = 5$ $\qquad$ P_4: $z = 2x - y - 3$

44. Which of the following four lines are parallel? Are any of them identical?

L_1: $x = 1 + t$, $y = t$, $z = 2 - 5t$

L_2: $x + 1 = y - 2 = 1 - z$

L_3: $x = 1 + t$, $y = 4 + t$, $z = 1 - t$

L_4: $\mathbf{r} = \langle 2, 1, -3 \rangle + t\langle 2, 2, -10 \rangle$

45–46 ■ Use the formula in Exercise 27 in Section 9.4 to find the distance from the point to the given line.

45. $(1, 2, 3)$; $x = 2 + t$, $y = 2 - 3t$, $z = 5t$

46. $(1, 0, -1)$; $x = 5 - t$, $y = 3t$, $z = 1 + 2t$

. . . .

47–48 ■ Find the distance from the point to the given plane.

47. $(2, 8, 5)$, $x - 2y - 2z = 1$

48. $(3, -2, 7)$, $4x - 6y + z = 5$

. . . .

49–50 ■ Find the distance between the given parallel planes.

49. $z = x + 2y + 1$, $3x + 6y - 3z = 4$

50. $3x + 6y - 9z = 4$, $x + 2y - 3z = 1$

. . . .

51. Show that the distance between the parallel planes $ax + by + cz + d_1 = 0$ and $ax + by + cz + d_2 = 0$ is

$$D = \frac{|d_1 - d_2|}{\sqrt{a^2 + b^2 + c^2}}$$

52. Find equations of the planes that are parallel to the plane $x + 2y - 2z = 1$ and two units away from it.

53. Show that the lines with symmetric equations $x = y = z$ and $x + 1 = y/2 = z/3$ are skew, and find the distance between these lines.

54. Find the distance between the skew lines with parametric equations $x = 1 + t$, $y = 1 + 6t$, $z = 2t$, and $x = 1 + 2s$, $y = 5 + 15s$, $z = -2 + 6s$.

55. If a, b, and c are not all 0, show that the equation $ax + by + cz + d = 0$ represents a plane and $\langle a, b, c \rangle$ is a normal vector to the plane.

Hint: Suppose $a \neq 0$ and rewrite the equation in the form

$$a\left(x + \frac{d}{a}\right) + b(y - 0) + c(z - 0) = 0$$

56. Give a geometric description of each family of planes.
(a) $x + y + z = c$ $\qquad$ (b) $x + y + cz = 1$
(c) $y \cos\theta + z \sin\theta = 1$

LABORATORY PROJECT

Putting 3D in Perspective

Computer graphics programmers face the same challenge as the great painters of the past: how to represent a three-dimensional scene as a flat image on a two-dimensional plane (a screen or a canvas). To create the illusion of perspective, in which closer objects appear larger than those farther away, three-dimensional objects in the computer's memory are projected onto a rectangular screen window from a viewpoint where the eye, or camera, is located. The viewing volume—the portion of space that will be visible—is the region contained by the four planes that pass through the viewpoint and an edge of the screen window. If objects in the scene extend beyond these four planes, they must be truncated before pixel data are sent to the screen. These planes are therefore called *clipping planes*.

1. Suppose the screen is represented by a rectangle in the yz-plane with vertices $(0, \pm 400, 0)$ and $(0, \pm 400, 600)$, and the camera is placed at $(1000, 0, 0)$. A line L in the scene passes through the points $(230, -285, 102)$ and $(860, 105, 264)$. At what points should L be clipped by the clipping planes?

2. If the clipped line segment is projected on the screen window, identify the resulting line segment.

3. Use parametric equations to plot the edges of the screen window, the clipped line segment, and its projection on the screen window. Then add sight lines connecting the viewpoint to each end of the clipped segments to verify that the projection is correct.

4. A rectangle with vertices $(621, -147, 206)$, $(563, 31, 242)$, $(657, -111, 86)$, and $(599, 67, 122)$ is added to the scene. The line L intersects this rectangle. To make the rectangle appear opaque, a programmer can use *hidden line rendering,* which removes portions of objects that are behind other objects. Identify the portion of L that should be removed.

9.6 Functions and Surfaces

In this section we take a first look at functions of two variables and their graphs, which are surfaces in three-dimensional space. We will give a much more thorough treatment of such functions in Chapter 11.

■■ **Functions of Two Variables**

The temperature T at a point on the surface of the Earth at any given time depends on the longitude x and latitude y of the point. We can think of T as being a function of the two variables x and y, or as a function of the pair (x, y). We indicate this functional dependence by writing $T = f(x, y)$.

The volume V of a circular cylinder depends on its radius r and its height h. In fact, we know that $V = \pi r^2 h$. We say that V is a function of r and h, and we write $V(r, h) = \pi r^2 h$.

Definition A **function f of two variables** is a rule that assigns to each ordered pair of real numbers (x, y) in a set D a unique real number denoted by $f(x, y)$. The set D is the **domain** of f and its **range** is the set of values that f takes on, that is, $\{f(x, y) \mid (x, y) \in D\}$.

We often write $z = f(x, y)$ to make explicit the value taken on by f at the general point (x, y). The variables x and y are **independent variables** and z is the **dependent variable**. [Compare this with the notation $y = f(x)$ for functions of a single variable.]

The domain is a subset of $\mathbb{R}^2$, the xy-plane. We can think of the domain as the set of all possible inputs and the range as the set of all possible outputs. If a function f is given by a formula and no domain is specified, then the domain of f is understood to be the set of all pairs (x, y) for which the given expression is a well-defined real number.

EXAMPLE 1 If $f(x, y) = 4x^2 + y^2$, then $f(x, y)$ is defined for all possible ordered pairs of real numbers (x, y), so the domain is $\mathbb{R}^2$, the entire xy-plane. The range of f is the set $[0, \infty)$ of all nonnegative real numbers. [Notice that $x^2 \geq 0$ and $y^2 \geq 0$, so $f(x, y) \geq 0$ for all x and y.] ■■

$x + y + 1 = 0$

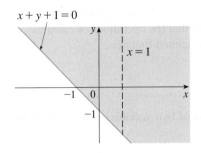

FIGURE 1

Domain of $f(x, y) = \dfrac{\sqrt{x + y + 1}}{x - 1}$

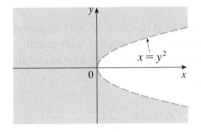

FIGURE 2

Domain of $f(x, y) = x \ln(y^2 - x)$

EXAMPLE 2 Find the domains of the following functions and evaluate $f(3, 2)$.

(a) $f(x, y) = \dfrac{\sqrt{x + y + 1}}{x - 1}$ (b) $f(x, y) = x \ln(y^2 - x)$

SOLUTION

(a) $$f(3, 2) = \frac{\sqrt{3 + 2 + 1}}{3 - 1} = \frac{\sqrt{6}}{2}$$

The expression for f makes sense if the denominator is not 0 and the quantity under the square root sign is nonnegative. So the domain of f is

$$D = \{(x, y) \mid x + y + 1 \geq 0, \ x \neq 1\}$$

The inequality $x + y + 1 \geq 0$, or $y \geq -x - 1$, describes the points that lie on or above the line $y = -x - 1$, while $x \neq 1$ means that the points on the line $x = 1$ must be excluded from the domain. (See Figure 1.)

(b) $$f(3, 2) = 3 \ln(2^2 - 3) = 3 \ln 1 = 0$$

Since $\ln(y^2 - x)$ is defined only when $y^2 - x > 0$, that is, $x < y^2$, the domain of f is $D = \{(x, y) \mid x < y^2\}$. This is the set of points to the left of the parabola $x = y^2$. (See Figure 2.) ■ ■

Not all functions can be represented by explicit formulas. The function in the next example is described verbally and by numerical estimates of its values.

EXAMPLE 3 The wave heights h (in feet) in the open sea depend mainly on the speed v of the wind (in knots) and the length of time t (in hours) that the wind has been blowing at that speed. So h is a function of v and t and we can write $h = f(v, t)$. Observations and measurements have been made by meteorologists and oceanographers and are recorded in Table 1.

TABLE 1

Wave heights (in feet) produced by different wind speeds for various lengths of time

		Duration (hours)						
	$\diagdown \begin{smallmatrix}t\\v\end{smallmatrix}$	5	10	15	20	30	40	50
	10	2	2	2	2	2	2	2
	15	4	4	5	5	5	5	5
Wind	20	5	7	8	8	9	9	9
speed	30	9	13	16	17	18	19	19
(knots)	40	14	21	25	28	31	33	33
	50	19	29	36	40	45	48	50
	60	24	37	47	54	62	67	69

For instance, the table indicates that if the wind has been blowing at 50 knots for 30 hours, then the wave heights are estimated to be 45 ft, so

$$f(50, 30) \approx 45$$

The domain of this function h is given by $v \geqslant 0$ and $t \geqslant 0$. Although there is no exact formula for h in terms of v and t, we will see that the operations of calculus can still be carried out for such an experimentally defined function. ■ ■

▋▋ Graphs

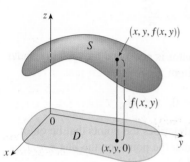

FIGURE 3

One way of visualizing the behavior of a function of two variables is to consider its graph.

> **Definition** If f is a function of two variables with domain D, then the **graph** of f is the set of all points (x, y, z) in $\mathbb{R}^3$ such that $z = f(x, y)$ and (x, y) is in D.

Just as the graph of a function f of one variable is a curve C with equation $y = f(x)$, so the graph of a function f of two variables is a surface S with equation $z = f(x, y)$. We can visualize the graph S of f as lying directly above or below its domain D in the xy-plane (see Figure 3).

EXAMPLE 4 Sketch the graph of the function $f(x, y) = 6 - 3x - 2y$.

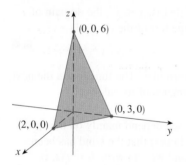

FIGURE 4

SOLUTION The graph of f has the equation $z = 6 - 3x - 2y$, or $3x + 2y + z = 6$, which represents a plane. To graph the plane we first find the intercepts. Putting $y = z = 0$ in the equation, we get $x = 2$ as the x-intercept. Similarly, the y-intercept is 3 and the z-intercept is 6. This helps us sketch the portion of the graph that lies in the first octant in Figure 4. ■ ■

The function in Example 4 is a special case of the function

$$f(x, y) = ax + by + c$$

which is called a **linear function**. The graph of such a function has the equation $z = ax + by + c$, or $ax + by - z + c = 0$, so it is a plane. In much the same way that linear functions of one variable are important in single-variable calculus, we will see that linear functions of two variables play a central role in multivariable calculus.

$\boxed{\text{V}}$ **EXAMPLE 5** Sketch the graph of the function $f(x, y) = x^2$.

SOLUTION Notice that, no matter what value we give y, the value of $f(x, y)$ is always x^2. The equation of the graph is $z = x^2$, which doesn't involve y. This means that any vertical plane with equation $y = k$ (parallel to the xz-plane) intersects the graph in a curve with equation $z = x^2$, that is, a parabola. Figure 5 shows how the graph is formed by taking the parabola $z = x^2$ in the xz-plane and moving it in the direction of the y-axis. So the graph is a surface, called a **parabolic cylinder**, made up of infinitely many shifted copies of the same parabola. ■ ■

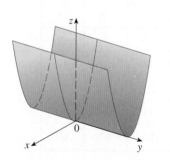

FIGURE 5
The graph of $f(x, y) = x^2$ is the parabolic cylinder $z = x^2$.

In sketching the graphs of functions of two variables, it's often useful to start by determining the shapes of cross-sections (slices) of the graph. For example, if we keep x fixed by putting $x = k$ (a constant) and letting y vary, the result is a function of one

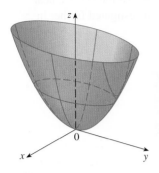

variable $z = f(k, y)$, whose graph is the curve that results when we intersect the surface $z = f(x, y)$ with the vertical plane $x = k$. In a similar fashion we can slice the surface with the vertical plane $y = k$ and look at the curves $z = f(x, k)$. We can also slice with horizontal planes $z = k$. All three types of curves are called **traces** (or cross-sections) of the surface $z = f(x, y)$.

EXAMPLE 6 Use traces to sketch the graph of the function $f(x, y) = 4x^2 + y^2$.

SOLUTION The equation of the graph is $z = 4x^2 + y^2$. If we put $x = 0$, we get $z = y^2$, so the yz-plane intersects the surface in a parabola. If we put $x = k$ (a constant), we get $z = y^2 + 4k^2$. This means that if we slice the graph with any plane parallel to the yz-plane, we obtain a parabola that opens upward. Similarly, if $y = k$, the trace is $z = 4x^2 + k^2$, which is again a parabola that opens upward. If we put $z = k$, we get the horizontal traces $4x^2 + y^2 = k$, which we recognize as a family of ellipses. Knowing the shapes of the traces, we can sketch the graph of f in Figure 6. Because of the elliptical and parabolic traces, the surface $z = 4x^2 + y^2$ is called an **elliptic paraboloid**. ■ ■

FIGURE 6
The graph of $f(x, y) = 4x^2 + y^2$ is the elliptic paraboloid $z = 4x^2 + y^2$. Horizontal traces are ellipses; vertical traces are parabolas.

Ⓥ **EXAMPLE 7** Sketch the graph of $f(x, y) = y^2 - x^2$.

SOLUTION The traces in the vertical planes $x = k$ are the parabolas $z = y^2 - k^2$, which open upward. The traces in $y = k$ are the parabolas $z = -x^2 + k^2$, which open downward. The horizontal traces are $y^2 - x^2 = k$, a family of hyperbolas. We draw the families of traces in Figure 7 and we show how the traces appear when placed in their correct planes in Figure 8.

FIGURE 7
Vertical traces are parabolas; horizontal traces are hyperbolas. All traces are labeled with the value of k.

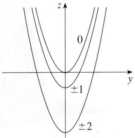

Traces in $x = k$ are $z = y^2 - k^2$

Traces in $y = k$ are $z = -x^2 + k^2$

Traces in $z = k$ are $y^2 - x^2 = k$

FIGURE 8
Traces moved to their correct planes

Traces in $x = k$

Traces in $y = k$

Traces in $z = k$

In Module 9.6A you can investigate how traces determine the shape of a surface.

In Figure 9 we fit together the traces from Figure 8 to form the surface $z = y^2 - x^2$, a **hyperbolic paraboloid**. Notice that the shape of the surface near the origin resembles that of a saddle. This surface will be investigated further in Section 11.7 when we discuss saddle points.

FIGURE 9

The graph of $f(x, y) = y^2 - x^2$ is the hyperbolic paraboloid $z = y^2 - x^2$.

The idea of using traces to draw a surface is employed in three-dimensional graphing software for computers. In most such software, traces in the vertical planes $x = k$ and $y = k$ are drawn for equally spaced values of k and parts of the graph are eliminated using hidden line removal. Figure 10 shows computer-generated graphs of several functions. Notice that we get an especially good picture of a function when rotation is used to give views from different vantage points. In parts (a) and (b) the graph of f is very flat and close to the xy-plane except near the origin; this is because $e^{-x^2 - y^2}$ is very small when x or y is large.

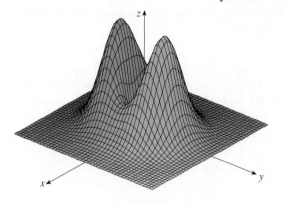

(a) $f(x, y) = (x^2 + 3y^2)e^{-x^2 - y^2}$

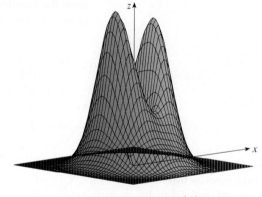

(b) $f(x, y) = (x^2 + 3y^2)e^{-x^2 - y^2}$

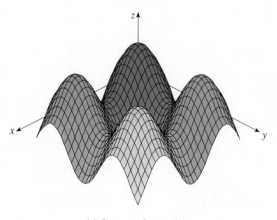

(c) $f(x, y) = \sin x + \sin y$

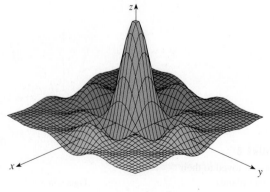

(d) $f(x, y) = \dfrac{\sin x \sin y}{xy}$

FIGURE 10

∷ Quadric Surfaces

The graph of a second-degree equation in three variables x, y, and z is called a **quadric surface**. We have already sketched the quadric surfaces $z = 4x^2 + y^2$ (an elliptic paraboloid) and $z = y^2 - x^2$ (a hyperbolic paraboloid) in Figures 6 and 9. In the next example we investigate a quadric surface called an *ellipsoid*.

EXAMPLE 8 Sketch the quadric surface with equation

$$x^2 + \frac{y^2}{9} + \frac{z^2}{4} = 1$$

SOLUTION The trace in the xy-plane ($z = 0$) is $x^2 + y^2/9 = 1$, which we recognize as an equation of an ellipse. In general, the horizontal trace in the plane $z = k$ is

$$x^2 + \frac{y^2}{9} = 1 - \frac{k^2}{4} \qquad z = k$$

which is an ellipse, provided that $k^2 < 4$, that is, $-2 < k < 2$.

Similarly, the vertical traces are also ellipses:

$$\frac{y^2}{9} + \frac{z^2}{4} = 1 - k^2 \qquad x = k \qquad (\text{if } -1 < k < 1)$$

$$x^2 + \frac{z^2}{4} = 1 - \frac{k^2}{9} \qquad y = k \qquad (\text{if } -3 < k < 3)$$

Figure 11 shows how drawing some traces indicates the shape of the surface. It's called an **ellipsoid** because all of its traces are ellipses. Notice that it is symmetric with respect to each coordinate plane; this is a reflection of the fact that its equation involves only even powers of x, y, and z. ∎

FIGURE 11

The ellipsoid $x^2 + \dfrac{y^2}{9} + \dfrac{z^2}{4} = 1$

The ellipsoid in Example 8 is *not* the graph of a function because some vertical lines (such as the z-axis) intersect it more than once. But the top and bottom halves *are* graphs of functions. In fact, if we solve the equation of the ellipsoid for z, we get

$$z^2 = 4\left(1 - x^2 - \frac{y^2}{9}\right) \qquad z = \pm 2\sqrt{1 - x^2 - \frac{y^2}{9}}$$

So the graphs of the functions

$$f(x, y) = 2\sqrt{1 - x^2 - \frac{y^2}{9}} \qquad \text{and} \qquad g(x, y) = -2\sqrt{1 - x^2 - \frac{y^2}{9}}$$

are the top and bottom halves of the ellipsoid (see Figure 12). The domain of both f and g is the set of all points (x, y) such that

$$1 - x^2 - \frac{y^2}{9} \geq 0 \quad \Longleftrightarrow \quad x^2 + \frac{y^2}{9} \leq 1$$

so the domain is the set of all points that lie on or inside the ellipse $x^2 + y^2/9 = 1$.

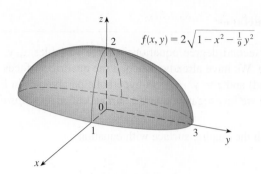

$$f(x, y) = 2\sqrt{1 - x^2 - \tfrac{1}{9}y^2}$$

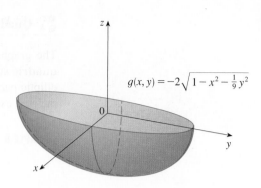

$$g(x, y) = -2\sqrt{1 - x^2 - \tfrac{1}{9}y^2}$$

FIGURE 12

In Module 9.6B you can see how changing a, b, and c in Table 2 affects the shape of the quadric surface.

Table 2 shows computer-drawn graphs of the six basic types of quadric surfaces in standard form. All surfaces are symmetric with respect to the z-axis. If a quadric surface is symmetric about a different axis, its equation changes accordingly.

TABLE 2 Graphs of quadric surfaces

Surface	Equation	Surface	Equation
Ellipsoid	$\dfrac{x^2}{a^2} + \dfrac{y^2}{b^2} + \dfrac{z^2}{c^2} = 1$ All traces are ellipses. If $a = b = c$, the ellipsoid is a sphere.	Cone	$\dfrac{z^2}{c^2} = \dfrac{x^2}{a^2} + \dfrac{y^2}{b^2}$ Horizontal traces are ellipses. Vertical traces in the planes $x = k$ and $y = k$ are hyperbolas if $k \neq 0$ but are pairs of lines if $k = 0$.
Elliptic Paraboloid	$\dfrac{z}{c} = \dfrac{x^2}{a^2} + \dfrac{y^2}{b^2}$ Horizontal traces are ellipses. Vertical traces are parabolas. The variable raised to the first power indicates the axis of the paraboloid.	Hyperboloid of One Sheet	$\dfrac{x^2}{a^2} + \dfrac{y^2}{b^2} - \dfrac{z^2}{c^2} = 1$ Horizontal traces are ellipses. Vertical traces are hyperbolas. The axis of symmetry corresponds to the variable whose coefficient is negative.
Hyperbolic Paraboloid	$\dfrac{z}{c} = \dfrac{x^2}{a^2} - \dfrac{y^2}{b^2}$ Horizontal traces are hyperbolas. Vertical traces are parabolas. The case where $c < 0$ is illustrated.	Hyperboloid of Two Sheets	$-\dfrac{x^2}{a^2} - \dfrac{y^2}{b^2} + \dfrac{z^2}{c^2} = 1$ Horizontal traces in $z = k$ are ellipses if $k > c$ or $k < -c$. Vertical traces are hyperbolas. The two minus signs indicate two sheets.

EXAMPLE 9 Classify the quadric surface $x^2 + 2z^2 - 6x - y + 10 = 0$.

SOLUTION By completing the square we rewrite the equation as

$$y - 1 = (x - 3)^2 + 2z^2$$

Comparing this equation with Table 2, we see that it represents an elliptic paraboloid. Here, however, the axis of the paraboloid is parallel to the y-axis, and it has been shifted so that its vertex is the point $(3, 1, 0)$. The traces in the plane $y = k$ ($k > 1$) are the ellipses

$$(x - 3)^2 + 2z^2 = k - 1 \qquad y = k$$

The trace in the xy-plane is the parabola with equation $y = 1 + (x - 3)^2$, $z = 0$. The paraboloid is sketched in Figure 13. ■ ■

FIGURE 13
$x^2 + 2z^2 - 6x - y + 10 = 0$

9.6 Exercises

1. In Example 3 we considered the function $h = f(v, t)$, where h is the height of waves produced by wind at speed v for a time t. Use Table 1 to answer the following questions.
(a) What is the value of $f(40, 15)$? What is its meaning?
(b) What is the meaning of the function $h = f(30, t)$? Describe the behavior of this function.
(c) What is the meaning of the function $h = f(v, 30)$? Describe the behavior of this function.

2. The figure shows vertical traces for a function $z = f(x, y)$. Which one of the graphs I–IV has these traces? Explain.

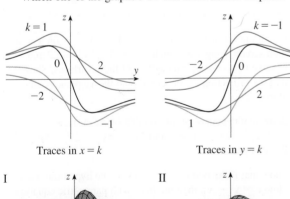

Traces in $x = k$ Traces in $y = k$

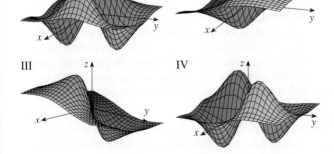

I   II   III   IV

3. Let $f(x, y) = x^2 e^{3xy}$.
(a) Evaluate $f(2, 0)$.
(b) Find the domain of f.
(c) Find the range of f.

4. Let $f(x, y) = \ln(x + y - 1)$.
(a) Evaluate $f(1, 1)$.
(b) Evaluate $f(e, 1)$.
(c) Find and sketch the domain of f.
(d) Find the range of f.

5–8 ■ Find and sketch the domain of the function.

5. $f(x, y) = \dfrac{\sqrt{y - x^2}}{1 - x^2}$

6. $f(x, y) = \sqrt{xy}$

7. $f(x, y) = \sqrt{1 - x^2} - \sqrt{1 - y^2}$

8. $f(x, y) = \sqrt{x^2 + y^2 - 1} + \ln(4 - x^2 - y^2)$

■ ■ ■ ■ ■ ■ ■ ■ ■ ■ ■ ■

9–13 ■ Sketch the graph of the function.

9. $f(x, y) = 3$

10. $f(x, y) = y$

11. $f(x, y) = 6 - 3x - 2y$

12. $f(x, y) = \cos x$

13. $f(x, y) = y^2 + 1$

■ ■ ■ ■ ■ ■ ■ ■ ■ ■ ■ ■

14. (a) Find the traces of the function $f(x, y) = x^2 + y^2$ in the planes $x = k$, $y = k$, and $z = k$. Use these traces to sketch the graph.
(b) Sketch the graph of $g(x, y) = -x^2 - y^2$. How is it related to the graph of f?
(c) Sketch the graph of $h(x, y) = 3 - x^2 - y^2$. How is it related to the graph of g?

15. Match the function with its graph (labeled I–VI) on page 684. Give reasons for your choices.
(a) $f(x, y) = |x| + |y|$ (b) $f(x, y) = |xy|$

(c) $f(x, y) = \dfrac{1}{1 + x^2 + y^2}$ (d) $f(x, y) = (x^2 - y^2)^2$

(e) $f(x, y) = (x - y)^2$ (f) $f(x, y) = \sin(|x| + |y|)$

I

II

III

IV

V

VI

16–18 ■ Use traces to sketch the graph of the function.

16. $f(x, y) = \sqrt{16 - x^2 - 16y^2}$

17. $f(x, y) = \sqrt{4x^2 + y^2}$ **18.** $f(x, y) = x^2 - y^2$

19–20 ■ Use traces to sketch the surface.

19. $y = z^2 - x^2$ **20.** $y = x^2 + z^2$

21–22 ■ Classify the surface by comparing with one of the standard forms in Table 2. Then sketch its graph.

21. $4x^2 + y^2 + 4z^2 - 4y - 24z + 36 = 0$

22. $4y^2 + z^2 - x - 16y - 4z + 20 = 0$

23. (a) What does the equation $x^2 + y^2 = 1$ represent as a curve in $\mathbb{R}^2$?
(b) What does it represent as a surface in $\mathbb{R}^3$?
(c) What does the equation $x^2 + z^2 = 1$ represent?

24. (a) Identify the traces of the surface $z^2 = x^2 + y^2$.
(b) Sketch the surface.
(c) Sketch the graphs of the functions $f(x, y) = \sqrt{x^2 + y^2}$ and $g(x, y) = -\sqrt{x^2 + y^2}$.

25. (a) Find and identify the traces of the quadric surface $x^2 + y^2 - z^2 = 1$ and explain why the graph looks like

the graph of the hyperboloid of one sheet in Table 2.
(b) If we change the equation in part (a) to $x^2 - y^2 + z^2 = 1$, how is the graph affected?
(c) What if we change the equation in part (a) to $x^2 + y^2 + 2y - z^2 = 0$?

26. (a) Find and identify the traces of the quadric surface $-x^2 - y^2 + z^2 = 1$ and explain why the graph looks like the graph of the hyperboloid of two sheets in Table 2.
(b) If the equation in part (a) is changed to $x^2 - y^2 - z^2 = 1$, what happens to the graph? Sketch the new graph.

27–28 ■ Use a computer to graph the function using various domains and viewpoints. Get a printout that gives a good view of the "peaks and valleys." Would you say the function has a maximum value? Can you identify any points on the graph that you might consider to be "local maximum points"? What about "local minimum points"?

27. $f(x, y) = 3x - x^4 - 4y^2 - 10xy$

28. $f(x, y) = xye^{-x^2-y^2}$

29–30 ■ Use a computer to graph the function using various domains and viewpoints. Comment on the limiting behavior of the function. What happens as both x and y become large? What happens as (x, y) approaches the origin?

29. $f(x, y) = \dfrac{x + y}{x^2 + y^2}$ **30.** $f(x, y) = \dfrac{xy}{x^2 + y^2}$

31. Graph the surfaces $z = x^2 + y^2$ and $z = 1 - y^2$ on a common screen using the domain $|x| \le 1.2$, $|y| \le 1.2$ and observe the curve of intersection of these surfaces. Show that the projection of this curve onto the xy-plane is an ellipse.

32. Show that the curve of intersection of the surfaces $x^2 + 2y^2 - z^2 + 3x = 1$ and $2x^2 + 4y^2 - 2z^2 - 5y = 0$ lies in a plane.

33. Show that if the point (a, b, c) lies on the hyperbolic paraboloid $z = y^2 - x^2$, then the lines with parametric equations $x = a + t$, $y = b + t$, $z = c + 2(b - a)t$ and $x = a + t$, $y = b - t$, $z = c - 2(b + a)t$ both lie entirely on this paraboloid. (This shows that the hyperbolic paraboloid is what is called a **ruled surface**; that is, it can be generated by the motion of a straight line. In fact, this exercise shows that through each point on the hyperbolic paraboloid there are two generating lines. The only other quadric surfaces that are ruled surfaces are cylinders, cones, and hyperboloids of one sheet.)

34. Find an equation for the surface consisting of all points P for which the distance from P to the x-axis is twice the distance from P to the yz-plane. Identify the surface.

9.7 Cylindrical and Spherical Coordinates

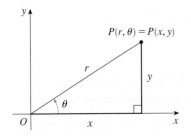

FIGURE 1

In plane geometry the polar coordinate system is used to give a convenient description of certain curves and regions. (See Appendix H.) Figure 1 enables us to recall the connection between polar and Cartesian coordinates. If the point P has Cartesian coordinates (x, y) and polar coordinates (r, θ), then, from the figure,

$$x = r \cos \theta \qquad y = r \sin \theta$$
$$r^2 = x^2 + y^2 \qquad \tan \theta = \frac{y}{x}$$

In three dimensions there are two coordinate systems that are similar to polar coordinates and give convenient descriptions of some commonly occurring surfaces and solids. They will be especially useful in Chapter 12 when we compute volumes and triple integrals.

▪▪ Cylindrical Coordinates

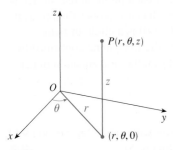

FIGURE 2
The cylindrical coordinates of a point

In the **cylindrical coordinate system**, a point P in three-dimensional space is represented by the ordered triple (r, θ, z), where r and θ are polar coordinates of the projection of P onto the xy-plane and z is the directed distance from the xy-plane to P (see Figure 2).

To convert from cylindrical to rectangular coordinates, we use the equations

1 $\qquad x = r \cos \theta \qquad y = r \sin \theta \qquad z = z$

whereas to convert from rectangular to cylindrical coordinates, we use

2 $\qquad r^2 = x^2 + y^2 \qquad \tan \theta = \frac{y}{x} \qquad z = z$

EXAMPLE 1
(a) Plot the point with cylindrical coordinates $(2, 2\pi/3, 1)$ and find its rectangular coordinates.
(b) Find cylindrical coordinates of the point with rectangular coordinates $(3, -3, -7)$.

SOLUTION
(a) The point with cylindrical coordinates $(2, 2\pi/3, 1)$ is plotted in Figure 3. From Equations 1, its rectangular coordinates are

$$x = 2 \cos \frac{2\pi}{3} = 2\left(-\frac{1}{2}\right) = -1$$

$$y = 2 \sin \frac{2\pi}{3} = 2\left(\frac{\sqrt{3}}{2}\right) = \sqrt{3}$$

$$z = 1$$

Thus, the point is $\left(-1, \sqrt{3}, 1\right)$ in rectangular coordinates.

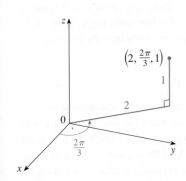

FIGURE 3

(b) From Equations 2 we have

$$r = \sqrt{3^2 + (-3)^2} = 3\sqrt{2}$$

$$\tan \theta = \frac{-3}{3} = -1 \qquad \text{so} \qquad \theta = \frac{7\pi}{4} + 2n\pi$$

$$z = -7$$

Therefore, one set of cylindrical coordinates is $(3\sqrt{2}, 7\pi/4, -7)$. Another is $(3\sqrt{2}, -\pi/4, -7)$. As with polar coordinates, there are infinitely many choices. ■ ■

Cylindrical coordinates are useful in problems that involve symmetry about an axis, and the z-axis is chosen to coincide with this axis of symmetry. For instance, the axis of the circular cylinder with Cartesian equation $x^2 + y^2 = c^2$ is the z-axis. In cylindrical coordinates this cylinder has the very simple equation $r = c$. (See Figure 4.) This is the reason for the name "cylindrical" coordinates.

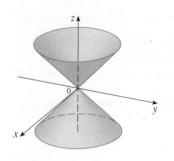

FIGURE 4

$r = c$, a cylinder

V **EXAMPLE 2** Describe the surface whose equation in cylindrical coordinates is $z = r$.

SOLUTION The equation says that the z-value, or height, of each point on the surface is the same as r, the distance from the point to the z-axis. Because θ doesn't appear, it can vary. So any horizontal trace in the plane $z = k$ ($k > 0$) is a circle of radius k. These traces suggest that the surface is a cone. This prediction can be confirmed by converting the equation into rectangular coordinates. From the first equation in (2) we have

$$z^2 = r^2 = x^2 + y^2$$

We recognize the equation $z^2 = x^2 + y^2$ (by comparison with Table 2 in Section 9.6) as being a circular cone whose axis is the z-axis (see Figure 5). ■ ■

EXAMPLE 3 Find an equation in cylindrical coordinates for the ellipsoid $4x^2 + 4y^2 + z^2 = 1$.

SOLUTION Since $r^2 = x^2 + y^2$ from Equations 2, we have

$$z^2 = 1 - 4(x^2 + y^2) = 1 - 4r^2$$

So an equation of the ellipsoid in cylindrical coordinates is $z^2 = 1 - 4r^2$. ■ ■

FIGURE 5

$z = r$, a cone

■ ■ **Spherical Coordinates**
■ ■

The **spherical coordinates** (ρ, θ, ϕ) of a point P in space are shown in Figure 6, where $\rho = |OP|$ is the distance from the origin to P, θ is the same angle as in cylindrical coordinates, and ϕ is the angle between the positive z-axis and the line segment OP. Note that

$$\rho \geqslant 0 \qquad 0 \leqslant \phi \leqslant \pi$$

The spherical coordinate system is especially useful in problems where there is symmetry about a point, and the origin is placed at this point. For example, the sphere with center the origin and radius c has the simple equation $\rho = c$ (see Figure 7); this is the reason for the name "spherical" coordinates. The graph of the equation $\theta = c$ is a vertical half-plane (see Figure 8), and the equation $\phi = c$ represents a half-cone with the z-axis as its axis (see Figure 9).

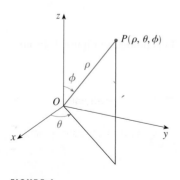

FIGURE 6

The spherical coordinates of a point

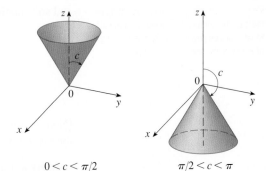

$$0 < c < \pi/2 \qquad \pi/2 < c < \pi$$

FIGURE 7 $\rho = c$, a sphere

FIGURE 8 $\theta = c$, a half-plane

FIGURE 9 $\phi = c$, a half-cone

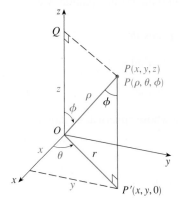

FIGURE 10

The relationship between rectangular and spherical coordinates can be seen from Figure 10. From triangles OPQ and OPP' we have

$$z = \rho \cos \phi \qquad r = \rho \sin \phi$$

But $x = r \cos \theta$ and $y = r \sin \theta$, so to convert from spherical to rectangular coordinates, we use the equations

$$\boxed{3} \qquad x = \rho \sin \phi \cos \theta \qquad y = \rho \sin \phi \sin \theta \qquad z = \rho \cos \phi$$

Also, the distance formula shows that

$$\boxed{4} \qquad \rho^2 = x^2 + y^2 + z^2$$

We use this equation in converting from rectangular to spherical coordinates.

[V] **EXAMPLE 4** The point $(2, \pi/4, \pi/3)$ is given in spherical coordinates. Plot the point and find its rectangular coordinates.

SOLUTION We plot the point in Figure 11. From Equations 3 we have

$$x = \rho \sin \phi \cos \theta = 2 \sin \frac{\pi}{3} \cos \frac{\pi}{4} = 2\left(\frac{\sqrt{3}}{2}\right)\left(\frac{1}{\sqrt{2}}\right) = \sqrt{\frac{3}{2}}$$

$$y = \rho \sin \phi \sin \theta = 2 \sin \frac{\pi}{3} \sin \frac{\pi}{4} = 2\left(\frac{\sqrt{3}}{2}\right)\left(\frac{1}{\sqrt{2}}\right) = \sqrt{\frac{3}{2}}$$

$$z = \rho \cos \phi = 2 \cos \frac{\pi}{3} = 2(\tfrac{1}{2}) = 1$$

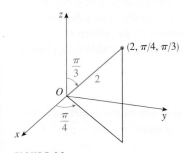

FIGURE 11

Thus, the point $(2, \pi/4, \pi/3)$ is $\left(\sqrt{3/2}, \sqrt{3/2}, 1\right)$ in rectangular coordinates. ■ ■

[V] **EXAMPLE 5** The point $\left(0, 2\sqrt{3}, -2\right)$ is given in rectangular coordinates. Find spherical coordinates for this point.

SOLUTION From Equation 4 we have

$$\rho = \sqrt{x^2 + y^2 + z^2} = \sqrt{0 + 12 + 4} = 4$$

⊘ WARNING: There is not universal agreement on the notation for spherical coordinates. Most books on physics reverse the meanings of θ and ϕ and use r in place of ρ.

and so Equations 3 give

$$\cos\phi = \frac{z}{\rho} = \frac{-2}{4} = -\frac{1}{2} \qquad \phi = \frac{2\pi}{3}$$

$$\cos\theta = \frac{x}{\rho\sin\phi} = 0 \qquad \theta = \frac{\pi}{2}$$

(Note that $\theta \neq 3\pi/2$ because $y = 2\sqrt{3} > 0$.) Therefore, spherical coordinates of the given point are $(4, \pi/2, 2\pi/3)$. ■ ■

EXAMPLE 6 Find an equation in spherical coordinates for the hyperboloid of two sheets with equation $x^2 - y^2 - z^2 = 1$.

SOLUTION Substituting the expressions in Equations 3 into the given equation, we have

$$\rho^2\sin^2\phi\,\cos^2\theta - \rho^2\sin^2\phi\,\sin^2\theta - \rho^2\cos^2\phi = 1$$

$$\rho^2[\sin^2\phi\,(\cos^2\theta - \sin^2\theta) - \cos^2\phi] = 1$$

or $\qquad\qquad \rho^2(\sin^2\phi\,\cos 2\theta - \cos^2\phi) = 1$ ■ ■

TEC In Module 9.7 you can investigate families of surfaces in cylindrical and spherical coordinates.

EXAMPLE 7 Find a rectangular equation for the surface whose spherical equation is $\rho = \sin\theta\,\sin\phi$.

SOLUTION From Equations 4 and 3 we have

$$x^2 + y^2 + z^2 = \rho^2 = \rho\sin\theta\,\sin\phi = y$$

or $\qquad\qquad x^2 + \left(y - \tfrac{1}{2}\right)^2 + z^2 = \tfrac{1}{4}$

which is the equation of a sphere with center $\left(0, \tfrac{1}{2}, 0\right)$ and radius $\tfrac{1}{2}$. ■ ■

EXAMPLE 8 Use a computer to draw a picture of the solid that remains when a hole of radius 3 is drilled through the center of a sphere of radius 4.

■ ■ Most three-dimensional graphing programs can graph surfaces whose equations are given in cylindrical or spherical coordinates. As Example 8 demonstrates, this is often the most convenient way of drawing a solid.

SOLUTION To keep the equations simple, let's choose the coordinate system so that the center of the sphere is at the origin and the axis of the cylinder that forms the hole is the z-axis. We could use either cylindrical or spherical coordinates to describe the solid, but the description is somewhat simpler if we use cylindrical coordinates. Then the equation of the cylinder is $r = 3$ and the equation of the sphere is $x^2 + y^2 + z^2 = 16$, or $r^2 + z^2 = 16$. The points in the solid lie outside the cylinder and inside the sphere, so they satisfy the inequalities

$$3 \leq r \leq \sqrt{16 - z^2}$$

To ensure that the computer graphs only the appropriate parts of these surfaces, we find where they intersect by solving the equations $r = 3$ and $r = \sqrt{16 - z^2}$:

$$\sqrt{16 - z^2} = 3 \quad\Rightarrow\quad 16 - z^2 = 9 \quad\Rightarrow\quad z^2 = 7 \quad\Rightarrow\quad z = \pm\sqrt{7}$$

The solid lies between $z = -\sqrt{7}$ and $z = \sqrt{7}$, so we ask the computer to graph the surfaces with the following equations and domains:

$$r = 3 \qquad\qquad 0 \leq \theta \leq 2\pi \qquad -\sqrt{7} \leq z \leq \sqrt{7}$$

$$r = \sqrt{16 - z^2} \qquad 0 \leq \theta \leq 2\pi \qquad -\sqrt{7} \leq z \leq \sqrt{7}$$

FIGURE 12

The resulting picture, shown in Figure 12, is exactly what we want. ■ ■

9.7 Exercises

1. What are cylindrical coordinates? For what types of surfaces do they provide convenient descriptions?

2. What are spherical coordinates? For what types of surfaces do they provide convenient descriptions?

3-4 ■ Plot the point whose cylindrical coordinates are given. Then find the rectangular coordinates of the point.

3. (a) $(2, \pi/4, 1)$ (b) $(4, -\pi/3, 5)$

4. (a) $(1, \pi, e)$ (b) $(1, 3\pi/2, 2)$

   .   .   .   .

5-6 ■ Change from rectangular to cylindrical coordinates.

5. (a) $(1, -1, 4)$ (b) $\left(-1, -\sqrt{3}, 2\right)$

6. (a) $(3, 3, -2)$ (b) $(3, 4, 5)$

   .   .   .   .

7-8 ■ Plot the point whose spherical coordinates are given. Then find the rectangular coordinates of the point.

7. (a) $(1, 0, 0)$ (b) $(2, \pi/3, \pi/4)$

8. (a) $(5, \pi, \pi/2)$ (b) $(4, 3\pi/4, \pi/3)$

   .   .   .   .

9-10 ■ Change from rectangular to spherical coordinates.

9. (a) $\left(1, \sqrt{3}, 2\sqrt{3}\right)$ (b) $(0, -1, -1)$

10. (a) $\left(0, \sqrt{3}, 1\right)$ (b) $\left(-1, 1, \sqrt{6}\right)$

   .   .   .   .

11-14 ■ Describe in words the surface whose equation is given.

11. $r = 3$ **12.** $\rho = 3$

13. $\phi = \pi/3$ **14.** $\theta = \pi/3$

   .   .   .   .

15-20 ■ Identify the surface whose equation is given.

15. $z = r^2$ **16.** $\rho \sin \phi = 2$

17. $r = 2 \cos \theta$ **18.** $\rho = 2 \cos \phi$

19. $r^2 + z^2 = 25$ **20.** $r^2 - 2z^2 = 4$

   .   .   .   .

21-24 ■ Write the equation (a) in cylindrical coordinates and (b) in spherical coordinates.

21. $z = x^2 + y^2$ **22.** $x^2 + y^2 + z^2 = 2$

23. $x^2 + y^2 = 2y$ **24.** $z = x^2 - y^2$

   .   .   .   .

25-30 ■ Sketch the solid described by the given inequalities.

25. $r^2 \leqslant z \leqslant 2 - r^2$

26. $0 \leqslant \theta \leqslant \pi/2, \quad r \leqslant z \leqslant 2$

27. $\rho \leqslant 2, \quad 0 \leqslant \phi \leqslant \pi/2, \quad 0 \leqslant \theta \leqslant \pi/2$

28. $2 \leqslant \rho \leqslant 3, \quad \pi/2 \leqslant \phi \leqslant \pi$

29. $-\pi/2 \leqslant \theta \leqslant \pi/2, \quad 0 \leqslant \phi \leqslant \pi/6, \quad 0 \leqslant \rho \leqslant \sec \phi$

30. $0 \leqslant \phi \leqslant \pi/3, \quad \rho \leqslant 2$

   .   .   .   .

31. A cylindrical shell is 20 cm long, with inner radius 6 cm and outer radius 7 cm. Write inequalities that describe the shell in an appropriate coordinate system. Explain how you have positioned the coordinate system with respect to the shell.

32. (a) Find inequalities that describe a hollow ball with diameter 30 cm and thickness 0.5 cm. Explain how you have positioned the coordinate system that you have chosen.
     (b) Suppose the ball is cut in half. Write inequalities that describe one of the halves.

33. A solid lies above the cone $z = \sqrt{x^2 + y^2}$ and below the sphere $x^2 + y^2 + z^2 = z$. Write a description of the solid in terms of inequalities involving spherical coordinates.

34. Use a graphing device to draw the solid enclosed by the paraboloids $z = x^2 + y^2$ and $z = 5 - x^2 - y^2$.

35. Use a graphing device to draw a silo consisting of a cylinder with radius 3 and height 10 surmounted by a hemisphere.

36. The latitude and longitude of a point P in the Northern Hemisphere are related to spherical coordinates ρ, θ, ϕ as follows. We take the origin to be the center of the Earth and the positive z-axis to pass through the North Pole. The positive x-axis passes through the point where the prime meridian (the meridian through Greenwich, England) intersects the equator. Then the latitude of P is $\alpha = 90° - \phi°$ and the longitude is $\beta = 360° - \theta°$. Find the great-circle distance from Los Angeles (lat. 34.06° N, long. 118.25° W) to Montréal (lat. 45.50° N, long. 73.60° W). Take the radius of the Earth to be 3960 mi. (A *great circle* is the circle of intersection of a sphere and a plane through the center of the sphere.)

LABORATORY PROJECT

⊞ Families of Surfaces

In this project you will discover the interesting shapes that members of families of surfaces can take. You will also see how the shape of the surface evolves as you vary the constants.

1. Use a computer to investigate the family of functions

$$f(x, y) = (ax^2 + by^2)e^{-x^2-y^2}$$

How does the shape of the graph depend on the numbers a and b?

2. Use a computer to investigate the family of surfaces $z = x^2 + y^2 + cxy$. In particular, you should determine the transitional values of c for which the surface changes from one type of quadric surface to another.

3. Members of the family of surfaces given in spherical coordinates by the equation

$$\rho = 1 + 0.2 \sin m\theta \sin n\phi$$

have been suggested as models for tumors and have been called *bumpy spheres* and *wrinkled spheres*. Use a computer to investigate this family of surfaces, assuming that m and n are positive integers. What roles do the values of m and n play in the shape of the surface?

---

9 Review

CONCEPT CHECK

1. What is the difference between a vector and a scalar?

2. How do you add two vectors geometrically? How do you add them algebraically?

3. If $\mathbf{a}$ is a vector and c is a scalar, how is $c\mathbf{a}$ related to $\mathbf{a}$ geometrically? How do you find $c\mathbf{a}$ algebraically?

4. How do you find the vector from one point to another?

5. How do you find the dot product $\mathbf{a} \cdot \mathbf{b}$ of two vectors if you know their lengths and the angle between them? What if you know their components?

6. How are dot products useful?

7. Write expressions for the scalar and vector projections of $\mathbf{b}$ onto $\mathbf{a}$. Illustrate with diagrams.

8. How do you find the cross product $\mathbf{a} \times \mathbf{b}$ of two vectors if you know their lengths and the angle between them? What if you know their components?

9. How are cross products useful?

10. (a) How do you find the area of the parallelogram determined by $\mathbf{a}$ and $\mathbf{b}$?
(b) How do you find the volume of the parallelepiped determined by $\mathbf{a}$, $\mathbf{b}$, and $\mathbf{c}$?

11. How do you find a vector perpendicular to a plane?

12. How do you find the angle between two intersecting planes?

13. Write a vector equation, parametric equations, and symmetric equations for a line.

14. Write a vector equation and a scalar equation for a plane.

15. (a) How do you tell if two vectors are parallel?
(b) How do you tell if two vectors are perpendicular?
(c) How do you tell if two planes are parallel?

16. (a) Describe a method for determining whether three points P, Q, and R lie on the same line.
(b) Describe a method for determining whether four points P, Q, R, and S lie in the same plane.

17. (a) How do you find the distance from a point to a line?
(b) How do you find the distance from a point to a plane?
(c) How do you find the distance between two lines?

18. How do you sketch the graph of a function of two variables?

19. Write equations in standard form of the six types of quadric surfaces.

20. (a) Write the equations for converting from cylindrical to rectangular coordinates. In what situation would you use cylindrical coordinates?
(b) Write the equations for converting from spherical to rectangular coordinates. In what situation would you use spherical coordinates?

TRUE-FALSE QUIZ

Determine whether the statement is true or false. If it is true, explain why. If it is false, explain why or give an example that disproves the statement.

1. For any vectors $\mathbf{u}$ and $\mathbf{v}$ in V_3, $\mathbf{u} \cdot \mathbf{v} = \mathbf{v} \cdot \mathbf{u}$.

2. For any vectors $\mathbf{u}$ and $\mathbf{v}$ in V_3, $\mathbf{u} \times \mathbf{v} = \mathbf{v} \times \mathbf{u}$.

3. For any vectors $\mathbf{u}$ and $\mathbf{v}$ in V_3, $|\mathbf{u} \times \mathbf{v}| = |\mathbf{v} \times \mathbf{u}|$.

4. For any vectors $\mathbf{u}$ and $\mathbf{v}$ in V_3 and any scalar k,
$k(\mathbf{u} \cdot \mathbf{v}) = (k\mathbf{u}) \cdot \mathbf{v}$.

5. For any vectors $\mathbf{u}$ and $\mathbf{v}$ in V_3 and any scalar k,
$k(\mathbf{u} \times \mathbf{v}) = (k\mathbf{u}) \times \mathbf{v}$.

6. For any vectors $\mathbf{u}$, $\mathbf{v}$, and $\mathbf{w}$ in V_3,
$(\mathbf{u} + \mathbf{v}) \times \mathbf{w} = \mathbf{u} \times \mathbf{w} + \mathbf{v} \times \mathbf{w}$.

7. For any vectors $\mathbf{u}$, $\mathbf{v}$, and $\mathbf{w}$ in V_3,
$\mathbf{u} \cdot (\mathbf{v} \times \mathbf{w}) = (\mathbf{u} \times \mathbf{v}) \cdot \mathbf{w}$.

8. For any vectors $\mathbf{u}$, $\mathbf{v}$, and $\mathbf{w}$ in V_3,
$\mathbf{u} \times (\mathbf{v} \times \mathbf{w}) = (\mathbf{u} \times \mathbf{v}) \times \mathbf{w}$.

9. For any vectors $\mathbf{u}$ and $\mathbf{v}$ in V_3, $(\mathbf{u} \times \mathbf{v}) \cdot \mathbf{u} = 0$.

10. For any vectors $\mathbf{u}$ and $\mathbf{v}$ in V_3, $(\mathbf{u} + \mathbf{v}) \times \mathbf{v} = \mathbf{u} \times \mathbf{v}$.

11. The cross product of two unit vectors is a unit vector.

12. A linear equation $Ax + By + Cz + D = 0$ represents a line in space.

13. The set of points $\{(x, y, z) \mid x^2 + y^2 = 1\}$ is a circle.

14. If $\mathbf{u} = \langle u_1, u_2 \rangle$ and $\mathbf{v} = \langle v_1, v_2 \rangle$, then $\mathbf{u} \cdot \mathbf{v} = \langle u_1 v_1, u_2 v_2 \rangle$.

15. If $\mathbf{u} \cdot \mathbf{v} = 0$, then $\mathbf{u} = \mathbf{0}$ or $\mathbf{v} = \mathbf{0}$.

16. If $\mathbf{u}$ and $\mathbf{v}$ are in V_3, then $|\mathbf{u} \cdot \mathbf{v}| \le |\mathbf{u}||\mathbf{v}|$.

EXERCISES

1. (a) Find an equation of the sphere that passes through the point $(6, -2, 3)$ and has center $(-1, 2, 1)$.
(b) Find the curve in which this sphere intersects the yz-plane.
(c) Find the center and radius of the sphere
$$x^2 + y^2 + z^2 - 8x + 2y + 6z + 1 = 0$$

2. Copy the vectors in the figure and use them to draw each of the following vectors.
(a) $\mathbf{a} + \mathbf{b}$ (b) $\mathbf{a} - \mathbf{b}$
(c) $-\frac{1}{2}\mathbf{a}$ (d) $2\mathbf{a} + \mathbf{b}$

3. If $\mathbf{u}$ and $\mathbf{v}$ are the vectors shown in the figure, find $\mathbf{u} \cdot \mathbf{v}$ and $|\mathbf{u} \times \mathbf{v}|$. Is $\mathbf{u} \times \mathbf{v}$ directed into the page or out of it?

4. Calculate the given quantity if
$$\mathbf{a} = \mathbf{i} + \mathbf{j} - 2\mathbf{k} \qquad \mathbf{b} = 3\mathbf{i} - 2\mathbf{j} + \mathbf{k} \qquad \mathbf{c} = \mathbf{j} - 5\mathbf{k}$$
(a) $2\mathbf{a} + 3\mathbf{b}$ (b) $|\mathbf{b}|$

(c) $\mathbf{a} \cdot \mathbf{b}$ (d) $\mathbf{a} \times \mathbf{b}$
(e) $|\mathbf{b} \times \mathbf{c}|$ (f) $\mathbf{a} \cdot (\mathbf{b} \times \mathbf{c})$
(g) $\mathbf{c} \times \mathbf{c}$ (h) $\mathbf{a} \times (\mathbf{b} \times \mathbf{c})$
(i) $\text{comp}_{\mathbf{a}} \mathbf{b}$ (j) $\text{proj}_{\mathbf{a}} \mathbf{b}$
(k) The angle between $\mathbf{a}$ and $\mathbf{b}$ (correct to the nearest degree)

5. Find the values of x such that the vectors $\langle 3, 2, x \rangle$ and $\langle 2x, 4, x \rangle$ are orthogonal.

6. Find two unit vectors that are orthogonal to both $\mathbf{j} + 2\mathbf{k}$ and $\mathbf{i} - 2\mathbf{j} + 3\mathbf{k}$.

7. Suppose that $\mathbf{u} \cdot (\mathbf{v} \times \mathbf{w}) = 2$. Find
(a) $(\mathbf{u} \times \mathbf{v}) \cdot \mathbf{w}$ (b) $\mathbf{u} \cdot (\mathbf{w} \times \mathbf{v})$
(c) $\mathbf{v} \cdot (\mathbf{u} \times \mathbf{w})$ (d) $(\mathbf{u} \times \mathbf{v}) \cdot \mathbf{v}$

8. Show that if $\mathbf{a}$, $\mathbf{b}$, and $\mathbf{c}$ are in V_3, then
$$(\mathbf{a} \times \mathbf{b}) \cdot [(\mathbf{b} \times \mathbf{c}) \times (\mathbf{c} \times \mathbf{a})] = [\mathbf{a} \cdot (\mathbf{b} \times \mathbf{c})]^2$$

9. Find the acute angle between two diagonals of a cube.

10. Given the points $A(1, 0, 1)$, $B(2, 3, 0)$, $C(-1, 1, 4)$, and $D(0, 3, 2)$, find the volume of the parallelepiped with adjacent edges AB, AC, and AD.

11. (a) Find a vector perpendicular to the plane through the points $A(1, 0, 0)$, $B(2, 0, -1)$, and $C(1, 4, 3)$.
(b) Find the area of triangle ABC.

12. A constant force $\mathbf{F} = 3\mathbf{i} + 5\mathbf{j} + 10\mathbf{k}$ moves an object along the line segment from $(1, 0, 2)$ to $(5, 3, 8)$. Find the work done if the distance is measured in meters and the force in newtons.

13. A boat is pulled onto shore using two ropes, as shown in the diagram. If a force of 255 N is needed, find the magnitude of the force in each rope.

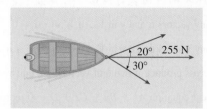

14. Find the magnitude of the torque about P if a 50-N force is applied as shown.

50 N
30°
40 cm
P

15–17 ■ Find parametric equations for the line.

15. The line through $(4, -1, 2)$ and $(1, 1, 5)$

16. The line through $(1, 0, -1)$ and parallel to the line $\frac{1}{3}(x - 4) = \frac{1}{2}y = z + 2$

17. The line through $(-2, 2, 4)$ and perpendicular to the plane $2x - y + 5z = 12$

18–21 ■ Find an equation of the plane.

18. The plane through $(2, 1, 0)$ and parallel to $x + 4y - 3z = 1$

19. The plane through $(3, -1, 1)$, $(4, 0, 2)$, and $(6, 3, 1)$

20. The plane through $(1, 2, -2)$ that contains the line $x = 2t$, $y = 3 - t, z = 1 + 3t$

21. The plane through the line of intersection of the planes $x - z = 1$ and $y + 2z = 3$ and perpendicular to the plane $x + y - 2z = 1$

22. Find the point in which the line with parametric equations $x = 2 - t, y = 1 + 3t, z = 4t$ intersects the plane $2x - y + z = 2$.

23. Determine whether the lines given by the symmetric equations

$$\frac{x - 1}{2} = \frac{y - 2}{3} = \frac{z - 3}{4}$$

and

$$\frac{x + 1}{6} = \frac{y - 3}{-1} = \frac{z + 5}{2}$$

are parallel, skew, or intersecting.

24. (a) Show that the planes $x + y - z = 1$ and $2x - 3y + 4z = 5$ are neither parallel nor perpendicular.
(b) Find, correct to the nearest degree, the angle between these planes.

25. Find the distance between the planes $3x + y - 4z = 2$ and $3x + y - 4z = 24$.

26. Find the distance from the origin to the line $x = 1 + t$, $y = 2 - t, z = -1 + 2t$.

27–28 ■ Find and sketch the domain of the function.

27. $f(x, y) = x \ln(x - y^2)$

28. $f(x, y) = \sqrt{\sin \pi(x^2 + y^2)}$

29–32 ■ Sketch the graph of the function.

29. $f(x, y) = 6 - 2x - 3y$

30. $f(x, y) = \cos y$

31. $f(x, y) = 4 - x^2 - 4y^2$

32. $f(x, y) = \sqrt{4 - x^2 - 4y^2}$

33–36 ■ Identify and sketch the graph of the surface. Include several traces in your sketch.

33. $y^2 + z^2 = 1 - 4x^2$ **34.** $y^2 + z^2 = x$

35. $y^2 + z^2 = 1$ **36.** $y^2 + z^2 = 1 + x^2$

37. The cylindrical coordinates of a point are $(2\sqrt{3}, \pi/3, 2)$. Find the rectangular and spherical coordinates of the point.

38. The rectangular coordinates of a point are $(2, 2, -1)$. Find the cylindrical and spherical coordinates of the point.

39. The spherical coordinates of a point are $(8, \pi/4, \pi/6)$. Find the rectangular and cylindrical coordinates of the point.

40. Identify the surfaces whose equations are given.
(a) $\theta = \pi/4$ (b) $\phi = \pi/4$

41–42 ■ Write the equation in cylindrical coordinates and in spherical coordinates.

41. $x^2 + y^2 + z^2 = 4$ **42.** $x^2 + y^2 = 4$

43. The parabola $z = 4y^2, x = 0$ is rotated about the z-axis. Write an equation of the resulting surface in cylindrical coordinates.

44. Sketch the solid consisting of all points with spherical coordinates (ρ, θ, ϕ) such that $0 \le \theta \le \pi/2, 0 \le \phi \le \pi/6$, and $0 \le \rho \le 2 \cos \phi$.

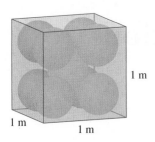

1 m

1 m

1 m

FIGURE FOR PROBLEM 1

1. Each edge of a cubical box has length 1 m. The box contains nine spherical balls with the same radius r. The center of one ball is at the center of the cube and it touches the other eight balls. Each of the other eight balls touches three sides of the box. Thus, the balls are tightly packed in the box. (See the figure.) Find r. (If you have trouble with this problem, read about the problem-solving strategy entitled *Use analogy* on page 86.)

2. Let B be a solid box with length L, width W, and height H. Let S be the set of all points that are a distance at most 1 from some point of B. Express the volume of S in terms of L, W, and H.

3. Let L be the line of intersection of the planes $cx + y + z = c$ and $x - cy + cz = -1$, where c is a real number.
   (a) Find symmetric equations for L.
   (b) As the number c varies, the line L sweeps out a surface S. Find an equation for the curve of intersection of S with the horizontal plane $z = t$ (the trace of S in the plane $z = t$).
   (c) Find the volume of the solid bounded by S and the planes $z = 0$ and $z = 1$.

4. A plane is capable of flying at a speed of 180 km/h in still air. The pilot takes off from an airfield and heads due north according to the plane's compass. After 30 minutes of flight time, the pilot notices that, due to the wind, the plane has actually traveled 80 km at an angle 5° east of north.
   (a) What is the wind velocity?
   (b) In what direction should the pilot have headed to reach the intended destination?

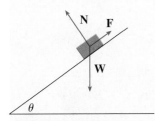

N   F

W

θ

FIGURE FOR PROBLEM 5

5. Suppose a block of mass m is placed on an inclined plane, as shown in the figure. The block's descent down the plane is slowed by friction; if θ is not too large, friction will prevent the block from moving at all. The forces acting on the block are the weight $\mathbf{W}$, where $|\mathbf{W}| = mg$ (g is the acceleration due to gravity); the normal force $\mathbf{N}$ (the normal component of the reactionary force of the plane on the block), where $|\mathbf{N}| = n$; and the force $\mathbf{F}$ due to friction, which acts parallel to the inclined plane, opposing the direction of motion. If the block is at rest and θ is increased, $|\mathbf{F}|$ must also increase until ultimately $|\mathbf{F}|$ reaches its maximum, beyond which the block begins to slide. At this angle θ_s, it has been observed that $|\mathbf{F}|$ is proportional to n. Thus, when $|\mathbf{F}|$ is maximal, we can say that $|\mathbf{F}| = \mu_s n$, where μ_s is called the *coefficient of static friction* and depends on the materials that are in contact.
   (a) Observe that $\mathbf{N} + \mathbf{F} + \mathbf{W} = \mathbf{0}$ and deduce that $\mu_s = \tan \theta_s$.
   (b) Suppose that, for $\theta > \theta_s$, an additional outside force $\mathbf{H}$ is applied to the block, horizontally from the left, and let $|\mathbf{H}| = h$. If h is small, the block may still slide down the plane; if h is large enough, the block will move up the plane. Let h_{min} be the smallest value of h that allows the block to remain motionless (so that $|\mathbf{F}|$ is maximal).

   By choosing the coordinate axes so that $\mathbf{F}$ lies along the x-axis, resolve each force into components parallel and perpendicular to the inclined plane and show that

   $$h_{min} \sin \theta + mg \cos \theta = n \quad \text{and} \quad h_{min} \cos \theta + \mu_s n = mg \sin \theta$$

   (c) Show that
   $$h_{min} = mg \tan(\theta - \theta_s)$$

   Does this equation seem reasonable? Does it make sense for $\theta = \theta_s$? As $\theta \to 90°$? Explain.

   (d) Let h_{max} be the largest value of h that allows the block to remain motionless. (In which direction is $\mathbf{F}$ heading?) Show that

   $$h_{max} = mg \tan(\theta + \theta_s)$$

   Does this equation seem reasonable? Explain.

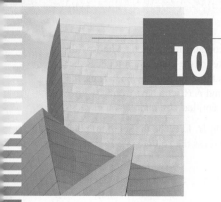

10

Vector Functions

The functions that we have been using so far have been real-valued functions. We now study functions whose values are vectors because such functions are needed to describe curves and surfaces in space. We will also use vector-valued functions to describe the motion of objects through space. In particular, we will use them to derive Kepler's laws of planetary motion.

Walt Disney Concert Hall, CA / Frank O. Gehry & Associates

10.1 Vector Functions and Space Curves

In general, a function is a rule that assigns to each element in the domain an element in the range. A **vector-valued function**, or **vector function**, is simply a function whose domain is a set of real numbers and whose range is a set of vectors. We are most interested in vector functions $\mathbf{r}$ whose values are three-dimensional vectors. This means that for every number t in the domain of $\mathbf{r}$ there is a unique vector in V_3 denoted by $\mathbf{r}(t)$. If $f(t)$, $g(t)$, and $h(t)$ are the components of the vector $\mathbf{r}(t)$, then f, g, and h are real-valued functions called the **component functions** of $\mathbf{r}$ and we can write

$$\mathbf{r}(t) = \langle f(t), g(t), h(t) \rangle = f(t)\,\mathbf{i} + g(t)\,\mathbf{j} + h(t)\,\mathbf{k}$$

We use the letter t to denote the independent variable because it represents time in most applications of vector functions.

EXAMPLE 1 If

$$\mathbf{r}(t) = \langle t^3, \ln(3 - t), \sqrt{t} \rangle$$

then the component functions are

$$f(t) = t^3 \qquad g(t) = \ln(3 - t) \qquad h(t) = \sqrt{t}$$

By our usual convention, the domain of $\mathbf{r}$ consists of all values of t for which the expression for $\mathbf{r}(t)$ is defined. The expressions t^3, $\ln(3 - t)$, and $\sqrt{t}$ are all defined when $3 - t > 0$ and $t \geq 0$. Therefore, the domain of $\mathbf{r}$ is the interval $[0, 3)$. ■ ■

The **limit** of a vector function $\mathbf{r}$ is defined by taking the limits of its component functions as follows.

■ ■ If $\lim_{t \to a} \mathbf{r}(t) = \mathbf{L}$, this definition is equivalent to saying that the length and direction of the vector $\mathbf{r}(t)$ approach the length and direction of the vector $\mathbf{L}$.

> **1** If $\mathbf{r}(t) = \langle f(t), g(t), h(t) \rangle$, then
>
> $$\lim_{t \to a} \mathbf{r}(t) = \left\langle \lim_{t \to a} f(t), \lim_{t \to a} g(t), \lim_{t \to a} h(t) \right\rangle$$
>
> provided the limits of the component functions exist.

Limits of vector functions obey the same rules as limits of real-valued functions (see Exercise 39).

EXAMPLE 2 Find $\lim_{t \to 0} \mathbf{r}(t)$, where $\mathbf{r}(t) = (1 + t^3)\,\mathbf{i} + te^{-t}\,\mathbf{j} + \dfrac{\sin t}{t}\,\mathbf{k}$.

SOLUTION According to Definition 1, the limit of $\mathbf{r}$ is the vector whose components are the limits of the component functions of $\mathbf{r}$:

$$\lim_{t \to 0} \mathbf{r}(t) = \left[\lim_{t \to 0} (1 + t^3) \right] \mathbf{i} + \left[\lim_{t \to 0} te^{-t} \right] \mathbf{j} + \left[\lim_{t \to 0} \frac{\sin t}{t} \right] \mathbf{k}$$

$$= \mathbf{i} + \mathbf{k} \qquad \text{(by Equation 3.4.2)}$$

■ ■

A vector function **r** is **continuous at** a if

$$\lim_{t \to a} \mathbf{r}(t) = \mathbf{r}(a)$$

In view of Definition 1, we see that **r** is continuous at a if and only if its component functions f, g, and h are continuous at a.

There is a close connection between continuous vector functions and space curves. Suppose that f, g, and h are continuous real-valued functions on an interval I. Then the set C of all points (x, y, z) in space, where

$$\boxed{2} \qquad\qquad x = f(t) \qquad y = g(t) \qquad z = h(t)$$

and t varies throughout the interval I, is called a **space curve**. The equations in (2) are called **parametric equations of** C and t is called a **parameter**. We can think of C as being traced out by a moving particle whose position at time t is $(f(t), g(t), h(t))$. If we now consider the vector function $\mathbf{r}(t) = \langle f(t), g(t), h(t)\rangle$, then $\mathbf{r}(t)$ is the position vector of the point $P(f(t), g(t), h(t))$ on C. Thus, any continuous vector function **r** defines a space curve C that is traced out by the tip of the moving vector $\mathbf{r}(t)$, as shown in Figure 1.

FIGURE 1

C is traced out by the tip of a moving position vector $\mathbf{r}(t)$.

 Visual 10.1A shows several curves being traced out by position vectors, including those in Figures 1 and 2.

▽ **EXAMPLE 3** Describe the curve defined by the vector function

$$\mathbf{r}(t) = \langle 1 + t, 2 + 5t, -1 + 6t\rangle$$

SOLUTION The corresponding parametric equations are

$$x = 1 + t \qquad y = 2 + 5t \qquad z = -1 + 6t$$

which we recognize from Equations 9.5.2 as parametric equations of a line passing through the point $(1, 2, -1)$ and parallel to the vector $\langle 1, 5, 6\rangle$. Alternatively, we could observe that the function can be written as $\mathbf{r} = \mathbf{r}_0 + t\mathbf{v}$, where $\mathbf{r}_0 = \langle 1, 2, -1\rangle$ and $\mathbf{v} = \langle 1, 5, 6\rangle$, and this is the vector equation of a line as given by Equation 9.5.1.

■ ■

Plane curves can also be represented in vector notation. For instance, the curve given by the parametric equations $x = t^2 - 2t$ and $y = t + 1$ (see Example 1 in Section 1.7) could also be described by the vector equation

$$\mathbf{r}(t) = \langle t^2 - 2t, t + 1\rangle = (t^2 - 2t)\mathbf{i} + (t + 1)\mathbf{j}$$

where $\mathbf{i} = \langle 1, 0\rangle$ and $\mathbf{j} = \langle 0, 1\rangle$.

▽ **EXAMPLE 4** Sketch the curve whose vector equation is

$$\mathbf{r}(t) = \cos t\,\mathbf{i} + \sin t\,\mathbf{j} + t\,\mathbf{k}$$

SOLUTION The parametric equations for this curve are

$$x = \cos t \qquad y = \sin t \qquad z = t$$

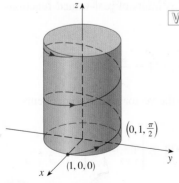

FIGURE 2

Since $x^2 + y^2 = \cos^2 t + \sin^2 t = 1$, the curve must lie on the circular cylinder $x^2 + y^2 = 1$. The point (x, y, z) lies directly above the point $(x, y, 0)$, which moves counterclockwise around the circle $x^2 + y^2 = 1$ in the xy-plane. (See Example 2 in Section 1.7.) Since $z = t$, the curve spirals upward around the cylinder as t increases. The curve, shown in Figure 2, is called a **helix**.

■ ■

FIGURE 3

The corkscrew shape of the helix in Example 4 is familiar from its occurrence in coiled springs. It also occurs in the model of DNA (deoxyribonucleic acid, the genetic material of living cells). In 1953 James Watson and Francis Crick showed that the structure of the DNA molecule is that of two linked, parallel helices that are intertwined as in Figure 3.

In Examples 3 and 4 we were given vector equations of curves and asked for a geometric description or sketch. In the next two examples we are given a geometric description of a curve and are asked to find parametric equations for the curve.

EXAMPLE 5 Find a vector equation and parametric equations for the line segment that joins the point $P(1, 3, -2)$ to the point $Q(2, -1, 3)$.

SOLUTION In Section 9.5 we found a vector equation for the line segment that joins the tip of the vector $\mathbf{r}_0$ to the tip of the vector $\mathbf{r}_1$:

■ ■ Figure 4 shows the line segment PQ in Example 5.

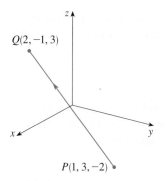

FIGURE 4

$$\mathbf{r}(t) = (1 - t)\mathbf{r}_0 + t\mathbf{r}_1 \qquad 0 \leq t \leq 1$$

(See Equation 9.5.4.) Here we take $\mathbf{r}_0 = \langle 1, 3, -2 \rangle$ and $\mathbf{r}_1 = \langle 2, -1, 3 \rangle$ to obtain a vector equation of the line segment from P to Q:

$$\mathbf{r}(t) = (1 - t)\langle 1, 3, -2 \rangle + t\langle 2, -1, 3 \rangle \qquad 0 \leq t \leq 1$$

or

$$\mathbf{r}(t) = \langle 1 + t, 3 - 4t, -2 + 5t \rangle \qquad 0 \leq t \leq 1$$

The corresponding parametric equations are

$$x = 1 + t \qquad y = 3 - 4t \qquad z = -2 + 5t \qquad 0 \leq t \leq 1 \qquad ■ ■$$

[V] **EXAMPLE 6** Find a vector function that represents the curve of intersection of the cylinder $x^2 + y^2 = 1$ and the plane $y + z = 2$.

SOLUTION Figure 5 shows how the plane and the cylinder intersect, and Figure 6 shows the curve of intersection C, which is an ellipse.

FIGURE 5

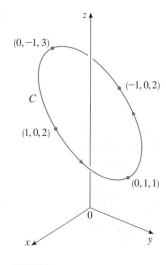

FIGURE 6

The projection of C onto the xy-plane is the circle $x^2 + y^2 = 1$, $z = 0$. So we know from Example 2 in Section 1.7 that we can write

$$x = \cos t \qquad y = \sin t \qquad 0 \le t \le 2\pi$$

From the equation of the plane, we have

$$z = 2 - y = 2 - \sin t$$

So we can write parametric equations for C as

$$x = \cos t \qquad y = \sin t \qquad z = 2 - \sin t \qquad 0 \le t \le 2\pi$$

The corresponding vector equation is

$$\mathbf{r}(t) = \cos t\,\mathbf{i} + \sin t\,\mathbf{j} + (2 - \sin t)\,\mathbf{k} \qquad 0 \le t \le 2\pi$$

This equation is called a *parametrization* of the curve C. The arrows in Figure 6 indicate the direction in which C is traced as the parameter t increases.　■ ■

■■ Using Computers to Draw Space Curves

Space curves are inherently more difficult to draw by hand than plane curves; for an accurate representation we need to use technology. For instance, Figure 7 shows a computer-generated graph of the curve with parametric equations

$$x = (4 + \sin 20t) \cos t \qquad y = (4 + \sin 20t) \sin t \qquad z = \cos 20t$$

It's called a **toroidal spiral** because it lies on a torus. Another interesting curve, the **trefoil knot**, with equations

$$x = (2 + \cos 1.5t) \cos t \qquad y = (2 + \cos 1.5t) \sin t \qquad z = \sin 1.5t$$

is graphed in Figure 8. It wouldn't be easy to plot either of these curves by hand.

Even when a computer is used to draw a space curve, optical illusions make it difficult to get a good impression of what the curve really looks like. (This is especially true in Figure 8. See Exercise 40.) The next example shows how to cope with this problem.

EXAMPLE 7 Use a computer to draw the curve with vector equation $\mathbf{r}(t) = \langle t, t^2, t^3 \rangle$. This curve is called a **twisted cubic**.

SOLUTION We start by using the computer to plot the curve with parametric equations $x = t$, $y = t^2$, $z = t^3$ for $-2 \le t \le 2$. The result is shown in Figure 9(a), but it's hard to see the true nature of the curve from that graph alone. Most three-dimensional computer graphing programs allow the user to enclose a curve or surface in a box instead of displaying the coordinate axes. When we look at the same curve in a box in Figure 9(b), we have a much clearer picture of the curve. We can see that it climbs from a lower corner of the box to the upper corner nearest us, and it twists as it climbs.

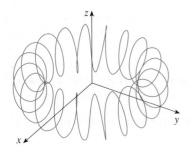

FIGURE 7 A toroidal spiral

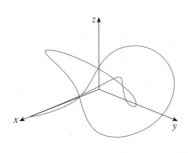

FIGURE 8 A trefoil knot

(a)

(b)

(c)

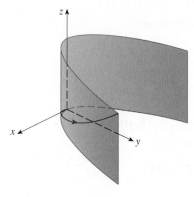

(d)

(e)

(f)

FIGURE 9 Views of the twisted cubic

 In Visual 10.1B you can rotate the box in Figure 9 to see the curve from any viewpoint.

We get an even better idea of the curve when we view it from different vantage points. Part (c) shows the result of rotating the box to give another viewpoint. Parts (d), (e), and (f) show the views we get when we look directly at a face of the box. In particular, part (d) shows the view from directly above the box. It is the projection of the curve on the xy-plane, namely, the parabola $y = x^2$. Part (e) shows the projection on the xz-plane, the cubic curve $z = x^3$. It's now obvious why the given curve is called a twisted cubic. ■ ■

Another method of visualizing a space curve is to draw it on a surface. For instance, the twisted cubic in Example 7 lies on the parabolic cylinder $y = x^2$. (Eliminate the parameter from the first two parametric equations, $x = t$ and $y = t^2$.) Figure 10 shows both the cylinder and the twisted cubic, and we see that the curve moves upward from the origin along the surface of the cylinder. We also used this method in Example 4 to visualize the helix lying on the circular cylinder (see Figure 2).

A third method for visualizing the twisted cubic is to realize that it also lies on the cylinder $z = x^3$. So it can be viewed as the curve of intersection of the cylinders $y = x^2$ and $z = x^3$. (See Figure 11.)

FIGURE 10

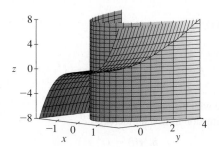

FIGURE 11

■ ■ Some computer algebra systems provide us with a clearer picture of a space curve by enclosing it in a tube. Such a plot enables us to see whether one part of a curve passes in front of or behind another part of the curve. For example, Figure 13 shows the curve of Figure 12(b) as rendered by the `tubeplot` command in Maple.

We have seen that an interesting space curve, the helix, occurs in the model of DNA. Another notable example of a space curve in science is the trajectory of a positively charged particle in orthogonally oriented electric and magnetic fields **E** and **B**. Depending on the initial velocity given the particle at the origin, the path of the particle is either a space curve whose projection on the horizontal plane is the cycloid we studied in Section 1.7 [Figure 12(a)] or a curve whose projection is the trochoid investigated in Exercise 36 in Section 1.7 [Figure 12(b)].

(a) $\mathbf{r}(t) = \langle t - \sin t, 1 - \cos t, t \rangle$

(b) $\mathbf{r}(t) = \langle t - \frac{3}{2} \sin t, 1 - \frac{3}{2} \cos t, t \rangle$

FIGURE 12

Motion of a charged particle in orthogonally oriented electric and magnetic fields

FIGURE 13

For further details concerning the physics involved and animations of the trajectories of the particles, see the following web sites:

□ lompado.uah.edu/Links/CrossedFields.html

□ www.phy.ntnu.edu.tw/java/emField/emField.html

□ www.physics.ucla.edu/plasma-exp/Beam/

10.1 Exercises

1–2 ■ Find the domain of the vector function.

1. $\mathbf{r}(t) = \langle t^2, \sqrt{t-1}, \sqrt{5-t} \rangle$

2. $\mathbf{r}(t) = \dfrac{t-2}{t+2}\,\mathbf{i} + \sin t\,\mathbf{j} + \ln(9 - t^2)\,\mathbf{k}$

■ ■ ■ ■ ■ ■ ■ ■ ■ ■ ■

3–4 ■ Find the limit.

3. $\lim\limits_{t \to 0^+} \langle \cos t, \sin t, t \ln t \rangle$

4. $\lim\limits_{t \to \infty} \left\langle \arctan t, e^{-2t}, \dfrac{\ln t}{t} \right\rangle$

■ ■ ■ ■ ■ ■ ■ ■ ■ ■ ■

5–12 ■ Sketch the curve with the given vector equation. Indicate with an arrow the direction in which t increases.

5. $\mathbf{r}(t) = \langle \sin t, t \rangle$

6. $\mathbf{r}(t) = \langle t^3, t^2 \rangle$

7. $\mathbf{r}(t) = \langle t, \cos 2t, \sin 2t \rangle$

8. $\mathbf{r}(t) = \langle 1 + t, 3t, -t \rangle$

9. $\mathbf{r}(t) = \langle 1, \cos t, 2 \sin t \rangle$

10. $\mathbf{r}(t) = t^2\mathbf{i} + t\mathbf{j} + 2\mathbf{k}$

11. $\mathbf{r}(t) = t^2\mathbf{i} + t^4\mathbf{j} + t^6\mathbf{k}$

12. $\mathbf{r}(t) = \cos t\,\mathbf{i} - \cos t\,\mathbf{j} + \sin t\,\mathbf{k}$

■ ■ ■ ■ ■ ■ ■ ■ ■ ■ ■

13–16 ■ Find a vector equation and parametric equations for the line segment that joins P to Q.

13. $P(0, 0, 0)$, $Q(1, 2, 3)$

14. $P(1, 0, 1)$, $Q(2, 3, 1)$

15. $P(1, -1, 2)$, $Q(4, 1, 7)$

16. $P(-2, 4, 0)$, $Q(6, -1, 2)$

■ ■ ■ ■ ■ ■ ■ ■ ■ ■ ■

17–22 ■ Match the parametric equations with the graphs (labeled I–VI). Give reasons for your choices.

17. $x = \cos 4t$, $y = t$, $z = \sin 4t$

18. $x = t$, $y = t^2$, $z = e^{-t}$

19. $x = t$, $y = 1/(1 + t^2)$, $z = t^2$

20. $x = e^{-t} \cos 10t$, $y = e^{-t} \sin 10t$, $z = e^{-t}$

21. $x = \cos t$, $y = \sin t$, $z = \sin 5t$

22. $x = \cos t$, $y = \sin t$, $z = \ln t$

I

II

III

IV

V

VI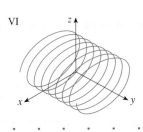

23. Show that the curve with parametric equations $x = t \cos t$, $y = t \sin t$, $z = t$ lies on the cone $z^2 = x^2 + y^2$, and use this fact to help sketch the curve.

24. Show that the curve with parametric equations $x = \sin t$, $y = \cos t$, $z = \sin^2 t$ is the curve of intersection of the surfaces $z = x^2$ and $x^2 + y^2 = 1$. Use this fact to help sketch the curve.

25. At what points does the curve $\mathbf{r}(t) = t\,\mathbf{i} + (2t - t^2)\,\mathbf{k}$ intersect the paraboloid $z = x^2 + y^2$?

26–28 ■ Use a computer to graph the curve with the given vector equation. Make sure you choose a parameter domain and viewpoints that reveal the true nature of the curve.

26. $\mathbf{r}(t) = \langle t^4 - t^2 + 1, t, t^2 \rangle$

27. $\mathbf{r}(t) = \langle t^2, \sqrt{t - 1}, \sqrt{5 - t} \rangle$

28. $\mathbf{r}(t) = \langle \sin t, \sin 2t, \sin 3t \rangle$

29. Graph the curve with parametric equations
$x = (1 + \cos 16t) \cos t$, $y = (1 + \cos 16t) \sin t$,
$z = 1 + \cos 16t$. Explain the appearance of the graph by showing that it lies on a cone.

30. Graph the curve with parametric equations
$$x = \sqrt{1 - 0.25 \cos^2 10t} \, \cos t$$
$$y = \sqrt{1 - 0.25 \cos^2 10t} \, \sin t$$
$$z = 0.5 \cos 10t$$
Explain the appearance of the graph by showing that it lies on a sphere.

31. Show that the curve with parametric equations $x = t^2$, $y = 1 - 3t$, $z = 1 + t^3$ passes through the points $(1, 4, 0)$ and $(9, -8, 28)$ but not through the point $(4, 7, -6)$.

32–34 ■ Find a vector function that represents the curve of intersection of the two surfaces.

32. The cylinder $x^2 + y^2 = 4$ and the surface $z = xy$

33. The cone $z = \sqrt{x^2 + y^2}$ and the plane $z = 1 + y$

34. The paraboloid $z = 4x^2 + y^2$ and the parabolic cylinder $y = x^2$

35. Try to sketch by hand the curve of intersection of the circular cylinder $x^2 + y^2 = 4$ and the parabolic cylinder $z = x^2$. Then find parametric equations for this curve and use these equations and a computer to graph the curve.

36. Try to sketch by hand the curve of intersection of the parabolic cylinder $y = x^2$ and the top half of the ellipsoid $x^2 + 4y^2 + 4z^2 = 16$. Then find parametric equations for this curve and use these equations and a computer to graph the curve.

37. If two objects travel through space along two different curves, it's often important to know whether they will collide. (Will a missile hit its moving target? Will two aircraft collide?) The curves might intersect, but we need to know whether the objects are in the same position *at the same time*. Suppose the trajectories of two particles are given by the vector functions
$$\mathbf{r}_1(t) = \langle t^2, 7t - 12, t^2 \rangle \quad \mathbf{r}_2(t) = \langle 4t - 3, t^2, 5t - 6 \rangle$$
for $t \geq 0$. Do the particles collide?

38. Two particles travel along the space curves
$$\mathbf{r}_1(t) = \langle t, t^2, t^3 \rangle \quad \mathbf{r}_2(t) = \langle 1 + 2t, 1 + 6t, 1 + 14t \rangle$$
Do the particles collide? Do their paths intersect?

39. Suppose $\mathbf{u}$ and $\mathbf{v}$ are vector functions that possess limits as $t \to a$ and let c be a constant. Prove the following properties of limits.

(a) $\lim\limits_{t \to a} [\mathbf{u}(t) + \mathbf{v}(t)] = \lim\limits_{t \to a} \mathbf{u}(t) + \lim\limits_{t \to a} \mathbf{v}(t)$

(b) $\lim\limits_{t \to a} c\mathbf{u}(t) = c \lim\limits_{t \to a} \mathbf{u}(t)$

(c) $\lim\limits_{t \to a} [\mathbf{u}(t) \cdot \mathbf{v}(t)] = \lim\limits_{t \to a} \mathbf{u}(t) \cdot \lim\limits_{t \to a} \mathbf{v}(t)$

(d) $\lim\limits_{t \to a} [\mathbf{u}(t) \times \mathbf{v}(t)] = \lim\limits_{t \to a} \mathbf{u}(t) \times \lim\limits_{t \to a} \mathbf{v}(t)$

40. The view of the trefoil knot shown in Figure 8 is accurate, but it doesn't reveal the whole story. Use the parametric equations

$$x = (2 + \cos 1.5t) \cos t \qquad y = (2 + \cos 1.5t) \sin t$$

$$z = \sin 1.5t$$

to sketch the curve by hand as viewed from above, with gaps indicating where the curve passes over itself. Start by showing that the projection of the curve onto the xy-plane has polar coordinates $r = 2 + \cos 1.5t$ and $\theta = t$, so r varies between 1 and 3. Then show that z has maximum and minimum values when the projection is halfway between $r = 1$ and $r = 3$.

 When you have finished your sketch, use a computer to draw the curve with viewpoint directly above and compare with your sketch. Then use the computer to draw the curve from several other viewpoints. You can get a better impression of the curve if you plot a tube with radius 0.2 around the curve. (Use the `tubeplot` command in Maple.)

10.2 Derivatives and Integrals of Vector Functions

Later in this chapter we are going to use vector functions to describe the motion of planets and other objects through space. Here we prepare the way by developing the calculus of vector functions.

■■ Derivatives

The **derivative r′** of a vector function **r** is defined in much the same way as for real-valued functions:

$$\boxed{1} \qquad \frac{d\mathbf{r}}{dt} = \mathbf{r}'(t) = \lim_{h \to 0} \frac{\mathbf{r}(t + h) - \mathbf{r}(t)}{h}$$

if this limit exists. The geometric significance of this definition is shown in Figure 1. If the points P and Q have position vectors $\mathbf{r}(t)$ and $\mathbf{r}(t + h)$, then $\overrightarrow{PQ}$ represents the vector $\mathbf{r}(t + h) - \mathbf{r}(t)$, which can therefore be regarded as a secant vector. If $h > 0$, the scalar multiple $(1/h)(\mathbf{r}(t + h) - \mathbf{r}(t))$ has the same direction as $\mathbf{r}(t + h) - \mathbf{r}(t)$. As $h \to 0$, it appears that this vector approaches a vector that lies on the tangent line. For this reason, the vector $\mathbf{r}'(t)$ is called the **tangent vector** to the curve defined by **r** at the point P, provided that $\mathbf{r}'(t)$ exists and $\mathbf{r}'(t) \neq \mathbf{0}$. The **tangent line** to C at P is defined to be the line through P parallel to the tangent vector $\mathbf{r}'(t)$. We will also have occasion to consider the **unit tangent vector**, which is

$$\mathbf{T}(t) = \frac{\mathbf{r}'(t)}{|\mathbf{r}'(t)|}$$

Visual 10.2 shows an animation of Figure 1.

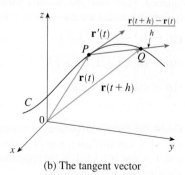

FIGURE 1

(a) The secant vector

(b) The tangent vector

The following theorem gives us a convenient method for computing the derivative of a vector function $\mathbf{r}$: just differentiate each component of $\mathbf{r}$.

> **2 Theorem** If $\mathbf{r}(t) = \langle f(t), g(t), h(t) \rangle = f(t)\,\mathbf{i} + g(t)\,\mathbf{j} + h(t)\,\mathbf{k}$, where f, g, and h are differentiable functions, then
>
> $$\mathbf{r}'(t) = \langle f'(t), g'(t), h'(t) \rangle = f'(t)\,\mathbf{i} + g'(t)\,\mathbf{j} + h'(t)\,\mathbf{k}$$

Proof

$$\mathbf{r}'(t) = \lim_{\Delta t \to 0} \frac{1}{\Delta t} [\mathbf{r}(t + \Delta t) - \mathbf{r}(t)]$$

$$= \lim_{\Delta t \to 0} \frac{1}{\Delta t} [\langle f(t + \Delta t), g(t + \Delta t), h(t + \Delta t) \rangle - \langle f(t), g(t), h(t) \rangle]$$

$$= \lim_{\Delta t \to 0} \left\langle \frac{f(t + \Delta t) - f(t)}{\Delta t}, \frac{g(t + \Delta t) - g(t)}{\Delta t}, \frac{h(t + \Delta t) - h(t)}{\Delta t} \right\rangle$$

$$= \left\langle \lim_{\Delta t \to 0} \frac{f(t + \Delta t) - f(t)}{\Delta t}, \lim_{\Delta t \to 0} \frac{g(t + \Delta t) - g(t)}{\Delta t}, \lim_{\Delta t \to 0} \frac{h(t + \Delta t) - h(t)}{\Delta t} \right\rangle$$

$$= \langle f'(t), g'(t), h'(t) \rangle$$ ■ ■

EXAMPLE 1
(a) Find the derivative of $\mathbf{r}(t) = (1 + t^3)\,\mathbf{i} + te^{-t}\,\mathbf{j} + \sin 2t\,\mathbf{k}$.
(b) Find the unit tangent vector at the point where $t = 0$.

SOLUTION
(a) According to Theorem 2, we differentiate each component of $\mathbf{r}$:

$$\mathbf{r}'(t) = 3t^2\,\mathbf{i} + (1 - t)e^{-t}\,\mathbf{j} + 2\cos 2t\,\mathbf{k}$$

(b) Since $\mathbf{r}(0) = \mathbf{i}$ and $\mathbf{r}'(0) = \mathbf{j} + 2\mathbf{k}$, the unit tangent vector at the point $(1, 0, 0)$ is

$$\mathbf{T}(0) = \frac{\mathbf{r}'(0)}{|\mathbf{r}'(0)|} = \frac{\mathbf{j} + 2\mathbf{k}}{\sqrt{1 + 4}} = \frac{1}{\sqrt{5}}\,\mathbf{j} + \frac{2}{\sqrt{5}}\,\mathbf{k}$$ ■ ■

EXAMPLE 2 For the curve $\mathbf{r}(t) = \sqrt{t}\,\mathbf{i} + (2 - t)\,\mathbf{j}$, find $\mathbf{r}'(t)$ and sketch the position vector $\mathbf{r}(1)$ and the tangent vector $\mathbf{r}'(1)$.

SOLUTION We have

$$\mathbf{r}'(t) = \frac{1}{2\sqrt{t}}\,\mathbf{i} - \mathbf{j} \qquad \text{and} \qquad \mathbf{r}'(1) = \frac{1}{2}\,\mathbf{i} - \mathbf{j}$$

The curve is a plane curve and elimination of the parameter from the equations $x = \sqrt{t}$, $y = 2 - t$ gives $y = 2 - x^2$, $x \geq 0$. In Figure 2 we draw the position vector $\mathbf{r}(1) = \mathbf{i} + \mathbf{j}$ starting at the origin and the tangent vector $\mathbf{r}'(1)$ starting at the corresponding point $(1, 1)$. ■ ■

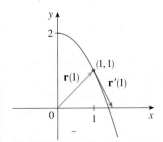

FIGURE 2

V **EXAMPLE 3** Find parametric equations for the tangent line to the helix with parametric equations

$$x = 2 \cos t \qquad y = \sin t \qquad z = t$$

at the point $(0, 1, \pi/2)$.

SOLUTION The vector equation of the helix is $\mathbf{r}(t) = \langle 2 \cos t, \sin t, t \rangle$, so

$$\mathbf{r}'(t) = \langle -2 \sin t, \cos t, 1 \rangle$$

The parameter value corresponding to the point $(0, 1, \pi/2)$ is $t = \pi/2$, so the tangent vector there is $\mathbf{r}'(\pi/2) = \langle -2, 0, 1 \rangle$. The tangent line is the line through $(0, 1, \pi/2)$ parallel to the vector $\langle -2, 0, 1 \rangle$, so by Equations 9.5.2 its parametric equations are

$$x = -2t \qquad y = 1 \qquad z = \frac{\pi}{2} + t$$

■ ■ The helix and the tangent line in Example 3 are shown in Figure 3.

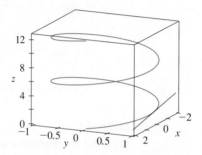

FIGURE 3

■ ■ In Section 10.4 we will see how $\mathbf{r}'(t)$ and $\mathbf{r}''(t)$ can be interpreted as the velocity and acceleration vectors of a particle moving through space with position vector $\mathbf{r}(t)$ at time t.

Just as for real-valued functions, the **second derivative** of a vector function $\mathbf{r}$ is the derivative of $\mathbf{r}'$, that is, $\mathbf{r}'' = (\mathbf{r}')'$. For instance, the second derivative of the function in Example 3 is

$$\mathbf{r}''(t) = \langle -2 \cos t, -\sin t, 0 \rangle$$

A curve given by a vector function $\mathbf{r}(t)$ on an interval I is called **smooth** if $\mathbf{r}'$ is continuous and $\mathbf{r}'(t) \neq \mathbf{0}$ (except possibly at any endpoints of I). For instance, the helix in Example 3 is smooth because $\mathbf{r}'(t)$ is never $\mathbf{0}$.

EXAMPLE 4 Determine whether the semicubical parabola $\mathbf{r}(t) = \langle 1 + t^3, t^2 \rangle$ is smooth.

SOLUTION Since

$$\mathbf{r}'(t) = \langle 3t^2, 2t \rangle$$

we have $\mathbf{r}'(0) = \langle 0, 0 \rangle = \mathbf{0}$ and, therefore, the curve is not smooth. The point that corresponds to $t = 0$ is $(1, 0)$, and we see from the graph in Figure 4 that there is a sharp corner, called a **cusp**, at $(1, 0)$. Any curve with this type of behavior—an abrupt change in direction—is not smooth.

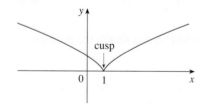

FIGURE 4
The curve $\mathbf{r}(t) = \langle 1 + t^3, t^2 \rangle$ is not smooth.

A curve, such as the semicubical parabola, that is made up of a finite number of smooth pieces is called **piecewise smooth**.

▪▪ Differentiation Rules
▪▪

The next theorem shows that the differentiation formulas for real-valued functions have their counterparts for vector-valued functions.

> **3** **Theorem** Suppose $\mathbf{u}$ and $\mathbf{v}$ are differentiable vector functions, c is a scalar, and f is a real-valued function. Then
>
> 1. $\dfrac{d}{dt}[\mathbf{u}(t) + \mathbf{v}(t)] = \mathbf{u}'(t) + \mathbf{v}'(t)$
>
> 2. $\dfrac{d}{dt}[c\mathbf{u}(t)] = c\mathbf{u}'(t)$
>
> 3. $\dfrac{d}{dt}[f(t)\mathbf{u}(t)] = f'(t)\mathbf{u}(t) + f(t)\mathbf{u}'(t)$
>
> 4. $\dfrac{d}{dt}[\mathbf{u}(t) \cdot \mathbf{v}(t)] = \mathbf{u}'(t) \cdot \mathbf{v}(t) + \mathbf{u}(t) \cdot \mathbf{v}'(t)$
>
> 5. $\dfrac{d}{dt}[\mathbf{u}(t) \times \mathbf{v}(t)] = \mathbf{u}'(t) \times \mathbf{v}(t) + \mathbf{u}(t) \times \mathbf{v}'(t)$
>
> 6. $\dfrac{d}{dt}[\mathbf{u}(f(t))] = f'(t)\mathbf{u}'(f(t))$ (Chain Rule)

This theorem can be proved either directly from Definition 1 or by using Theorem 2 and the corresponding differentiation formulas for real-valued functions. The proof of Formula 4 follows; the remaining proofs are left as exercises.

Proof of Formula 4 Let

$$\mathbf{u}(t) = \langle f_1(t), f_2(t), f_3(t) \rangle \qquad \mathbf{v}(t) = \langle g_1(t), g_2(t), g_3(t) \rangle$$

Then $\mathbf{u}(t) \cdot \mathbf{v}(t) = f_1(t)g_1(t) + f_2(t)g_2(t) + f_3(t)g_3(t) = \displaystyle\sum_{i=1}^{3} f_i(t)g_i(t)$

so the ordinary Product Rule gives

$$\frac{d}{dt}[\mathbf{u}(t) \cdot \mathbf{v}(t)] = \frac{d}{dt} \sum_{i=1}^{3} f_i(t)g_i(t) = \sum_{i=1}^{3} \frac{d}{dt}[f_i(t)g_i(t)]$$

$$= \sum_{i=1}^{3} [f_i'(t)g_i(t) + f_i(t)g_i'(t)]$$

$$= \sum_{i=1}^{3} f_i'(t)g_i(t) + \sum_{i=1}^{3} f_i(t)g_i'(t)$$

$$= \mathbf{u}'(t) \cdot \mathbf{v}(t) + \mathbf{u}(t) \cdot \mathbf{v}'(t)$$

■ ■

V **EXAMPLE 5** Show that if $|\mathbf{r}(t)| = c$ (a constant), then $\mathbf{r}'(t)$ is orthogonal to $\mathbf{r}(t)$ for all t.

SOLUTION Since

$$\mathbf{r}(t) \cdot \mathbf{r}(t) = |\mathbf{r}(t)|^2 = c^2$$

and c^2 is a constant, Formula 4 of Theorem 3 gives

$$0 = \frac{d}{dt}[\mathbf{r}(t) \cdot \mathbf{r}(t)] = \mathbf{r}'(t) \cdot \mathbf{r}(t) + \mathbf{r}(t) \cdot \mathbf{r}'(t) = 2\mathbf{r}'(t) \cdot \mathbf{r}(t)$$

Thus, $\mathbf{r}'(t) \cdot \mathbf{r}(t) = 0$, which says that $\mathbf{r}'(t)$ is orthogonal to $\mathbf{r}(t)$.

Geometrically, this result says that if a curve lies on a sphere with center the origin, then the tangent vector $\mathbf{r}'(t)$ is always perpendicular to the position vector $\mathbf{r}(t)$. ■ ■

■■ Integrals
■■

The **definite integral** of a continuous vector function $\mathbf{r}(t)$ can be defined in much the same way as for real-valued functions except that the integral is a vector. But then we can express the integral of $\mathbf{r}$ in terms of the integrals of its component functions f, g, and h as follows. (We use the notation of Chapter 5.)

$$\int_a^b \mathbf{r}(t)\,dt = \lim_{n\to\infty} \sum_{i=1}^n \mathbf{r}(t_i^*)\,\Delta t$$

$$= \lim_{n\to\infty}\left[\left(\sum_{i=1}^n f(t_i^*)\,\Delta t\right)\mathbf{i} + \left(\sum_{i=1}^n g(t_i^*)\,\Delta t\right)\mathbf{j} + \left(\sum_{i=1}^n h(t_i^*)\,\Delta t\right)\mathbf{k}\right]$$

and so

$$\int_a^b \mathbf{r}(t)\,dt = \left(\int_a^b f(t)\,dt\right)\mathbf{i} + \left(\int_a^b g(t)\,dt\right)\mathbf{j} + \left(\int_a^b h(t)\,dt\right)\mathbf{k}$$

This means that we can evaluate an integral of a vector function by integrating each component function.

We can extend the Fundamental Theorem of Calculus to continuous vector functions as follows:

$$\int_a^b \mathbf{r}(t)\,dt = \mathbf{R}(t)\Big]_a^b = \mathbf{R}(b) - \mathbf{R}(a)$$

where $\mathbf{R}$ is an antiderivative of $\mathbf{r}$, that is, $\mathbf{R}'(t) = \mathbf{r}(t)$. We use the notation $\int \mathbf{r}(t)\,dt$ for indefinite integrals (antiderivatives).

EXAMPLE 6 If $\mathbf{r}(t) = 2\cos t\,\mathbf{i} + \sin t\,\mathbf{j} + 2t\,\mathbf{k}$, then

$$\int \mathbf{r}(t)\,dt = \left(\int 2\cos t\,dt\right)\mathbf{i} + \left(\int \sin t\,dt\right)\mathbf{j} + \left(\int 2t\,dt\right)\mathbf{k}$$

$$= 2\sin t\,\mathbf{i} - \cos t\,\mathbf{j} + t^2\,\mathbf{k} + \mathbf{C}$$

where $\mathbf{C}$ is a vector constant of integration, and

$$\int_0^{\pi/2} \mathbf{r}(t)\,dt = \left[2\sin t\,\mathbf{i} - \cos t\,\mathbf{j} + t^2\,\mathbf{k}\right]_0^{\pi/2} = 2\mathbf{i} + \mathbf{j} + \frac{\pi^2}{4}\mathbf{k}$$

■ ■

10.2 Exercises

1. The figure shows a curve C given by a vector function $\mathbf{r}(t)$.
   (a) Draw the vectors $\mathbf{r}(4.5) - \mathbf{r}(4)$ and $\mathbf{r}(4.2) - \mathbf{r}(4)$.
   (b) Draw the vectors
   $$\frac{\mathbf{r}(4.5) - \mathbf{r}(4)}{0.5} \quad \text{and} \quad \frac{\mathbf{r}(4.2) - \mathbf{r}(4)}{0.2}$$
   (c) Write expressions for $\mathbf{r}'(4)$ and the unit tangent vector $\mathbf{T}(4)$.
   (d) Draw the vector $\mathbf{T}(4)$.

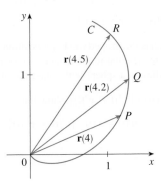

2. (a) Make a large sketch of the curve described by the vector function $\mathbf{r}(t) = \langle t^2, t \rangle$, $0 \le t \le 2$, and draw the vectors $\mathbf{r}(1)$, $\mathbf{r}(1.1)$, and $\mathbf{r}(1.1) - \mathbf{r}(1)$.
   (b) Draw the vector $\mathbf{r}'(1)$ starting at $(1, 1)$ and compare it with the vector
   $$\frac{\mathbf{r}(1.1) - \mathbf{r}(1)}{0.1}$$
   Explain why these vectors are so close to each other in length and direction.

3–8 ■
(a) Sketch the plane curve with the given vector equation.
(b) Find $\mathbf{r}'(t)$.
(c) Sketch the position vector $\mathbf{r}(t)$ and the tangent vector $\mathbf{r}'(t)$ for the given value of t.

3. $\mathbf{r}(t) = \langle t - 2, t^2 + 1 \rangle$, $t = -1$

4. $\mathbf{r}(t) = \langle 1 + t, \sqrt{t} \rangle$, $t = 1$

5. $\mathbf{r}(t) = \sin t\, \mathbf{i} + 2 \cos t\, \mathbf{j}$, $t = \pi/4$

6. $\mathbf{r}(t) = e^t\, \mathbf{i} + e^{-t}\, \mathbf{j}$, $t = 0$

7. $\mathbf{r}(t) = e^t\, \mathbf{i} + e^{3t}\, \mathbf{j}$, $t = 0$

8. $\mathbf{r}(t) = (1 + \cos t)\, \mathbf{i} + (2 + \sin t)\, \mathbf{j}$, $t = \pi/6$

9–14 ■ Find the derivative of the vector function.

9. $\mathbf{r}(t) = \langle t^2, 1 - t, \sqrt{t} \rangle$
10. $\mathbf{r}(t) = \langle \cos 3t, t, \sin 3t \rangle$

11. $\mathbf{r}(t) = e^{t^2}\mathbf{i} - \mathbf{j} + \ln(1 + 3t)\, \mathbf{k}$

12. $\mathbf{r}(t) = at \cos 3t\, \mathbf{i} + b \sin^3 t\, \mathbf{j} + c \cos^3 t\, \mathbf{k}$

13. $\mathbf{r}(t) = \mathbf{a} + t\, \mathbf{b} + t^2\, \mathbf{c}$

14. $\mathbf{r}(t) = t\, \mathbf{a} \times (\mathbf{b} + t\, \mathbf{c})$

15–16 ■ Find the unit tangent vector $\mathbf{T}(t)$ at the point with the given value of the parameter t.

15. $\mathbf{r}(t) = \cos t\, \mathbf{i} + 3t\, \mathbf{j} + 2 \sin 2t\, \mathbf{k}$, $t = 0$

16. $\mathbf{r}(t) = 2 \sin t\, \mathbf{i} + 2 \cos t\, \mathbf{j} + \tan t\, \mathbf{k}$, $t = \pi/4$

17. If $\mathbf{r}(t) = \langle t, t^2, t^3 \rangle$, find $\mathbf{r}'(t)$, $\mathbf{T}(1)$, $\mathbf{r}''(t)$, and $\mathbf{r}'(t) \times \mathbf{r}''(t)$.

18. If $\mathbf{r}(t) = \langle e^{2t}, e^{-2t}, te^{2t} \rangle$, find $\mathbf{T}(0)$, $\mathbf{r}''(0)$, and $\mathbf{r}'(t) \cdot \mathbf{r}''(t)$.

19–22 ■ Find parametric equations for the tangent line to the curve with the given parametric equations at the specified point.

19. $x = t^5$, $y = t^4$, $z = t^3$; $(1, 1, 1)$

20. $x = t^2 - 1$, $y = t^2 + 1$, $z = t + 1$; $(-1, 1, 1)$

21. $x = e^{-t} \cos t$, $y = e^{-t} \sin t$, $z = e^{-t}$; $(1, 0, 1)$

22. $x = \ln t$, $y = 2\sqrt{t}$, $z = t^2$; $(0, 2, 1)$

23–24 ■ Find parametric equations for the tangent line to the curve with the given parametric equations at the specified point. Illustrate by graphing both the curve and the tangent line on a common screen.

23. $x = t$, $y = e^{-t}$, $z = 2t - t^2$; $(0, 1, 0)$

24. $x = 2 \cos t$, $y = 2 \sin t$, $z = 4 \cos 2t$; $(\sqrt{3}, 1, 2)$

25. Determine whether the curve is smooth.
   (a) $\mathbf{r}(t) = \langle t^3, t^4, t^5 \rangle$ (b) $\mathbf{r}(t) = \langle t^3 + t, t^4, t^5 \rangle$
   (c) $\mathbf{r}(t) = \langle \cos^3 t, \sin^3 t \rangle$

26. (a) Find the point of intersection of the tangent lines to the curve $\mathbf{r}(t) = \langle \sin \pi t, 2 \sin \pi t, \cos \pi t \rangle$ at the points where $t = 0$ and $t = 0.5$.
   (b) Illustrate by graphing the curve and both tangent lines.

27. The curves $\mathbf{r}_1(t) = \langle t, t^2, t^3 \rangle$ and $\mathbf{r}_2(t) = \langle \sin t, \sin 2t, t \rangle$ intersect at the origin. Find their angle of intersection correct to the nearest degree.

28. At what point do the curves $\mathbf{r}_1(t) = \langle t, 1 - t, 3 + t^2 \rangle$ and $\mathbf{r}_2(s) = \langle 3 - s, s - 2, s^2 \rangle$ intersect? Find their angle of intersection correct to the nearest degree.

29–34 ■ Evaluate the integral.

29. $\displaystyle \int_0^1 (16t^3\, \mathbf{i} - 9t^2\, \mathbf{j} + 25t^4\, \mathbf{k})\, dt$

30. $\displaystyle \int_0^1 \left(\frac{4}{1 + t^2}\, \mathbf{j} + \frac{2t}{1 + t^2}\, \mathbf{k} \right) dt$

31. $\int_0^{\pi/2} (3 \sin^2 t \cos t \, \mathbf{i} + 3 \sin t \cos^2 t \, \mathbf{j} + 2 \sin t \cos t \, \mathbf{k}) \, dt$

32. $\int_1^2 (t^2 \mathbf{i} + t\sqrt{t-1} \, \mathbf{j} + t \sin \pi t \, \mathbf{k}) \, dt$

33. $\int (e^t \mathbf{i} + 2t \, \mathbf{j} + \ln t \, \mathbf{k}) \, dt$

34. $\int (\cos \pi t \, \mathbf{i} + \sin \pi t \, \mathbf{j} + t \, \mathbf{k}) \, dt$

. . . .

35. Find $\mathbf{r}(t)$ if $\mathbf{r}'(t) = 2t \mathbf{i} + 3t^2 \mathbf{j} + \sqrt{t} \, \mathbf{k}$ and $\mathbf{r}(1) = \mathbf{i} + \mathbf{j}$.

36. Find $\mathbf{r}(t)$ if $\mathbf{r}'(t) = t \mathbf{i} + e^t \mathbf{j} + te^t \mathbf{k}$ and $\mathbf{r}(0) = \mathbf{i} + \mathbf{j} + \mathbf{k}$.

37. Prove Formula 1 of Theorem 3.

38. Prove Formula 3 of Theorem 3.

39. Prove Formula 5 of Theorem 3.

40. Prove Formula 6 of Theorem 3.

41. If $\mathbf{u}(t) = \mathbf{i} - 2t^2 \mathbf{j} + 3t^3 \mathbf{k}$ and $\mathbf{v}(t) = t \mathbf{i} + \cos t \, \mathbf{j} + \sin t \, \mathbf{k}$, find $(d/dt) [\mathbf{u}(t) \cdot \mathbf{v}(t)]$.

42. If $\mathbf{u}$ and $\mathbf{v}$ are the vector functions in Exercise 41, find $(d/dt) [\mathbf{u}(t) \times \mathbf{v}(t)]$.

43. Show that if $\mathbf{r}$ is a vector function such that $\mathbf{r}''$ exists, then

$$\frac{d}{dt} [\mathbf{r}(t) \times \mathbf{r}'(t)] = \mathbf{r}(t) \times \mathbf{r}''(t)$$

44. Find an expression for $\dfrac{d}{dt} [\mathbf{u}(t) \cdot (\mathbf{v}(t) \times \mathbf{w}(t))]$.

45. If $\mathbf{r}(t) \neq \mathbf{0}$, show that $\dfrac{d}{dt} |\mathbf{r}(t)| = \dfrac{1}{|\mathbf{r}(t)|} \mathbf{r}(t) \cdot \mathbf{r}'(t)$.

$\left[\textit{Hint: } |\mathbf{r}(t)|^2 = \mathbf{r}(t) \cdot \mathbf{r}(t) \right]$

46. If a curve has the property that the position vector $\mathbf{r}(t)$ is always perpendicular to the tangent vector $\mathbf{r}'(t)$, show that the curve lies on a sphere with center the origin.

47. If $\mathbf{u}(t) = \mathbf{r}(t) \cdot [\mathbf{r}'(t) \times \mathbf{r}''(t)]$, show that

$$\mathbf{u}'(t) = \mathbf{r}(t) \cdot [\mathbf{r}'(t) \times \mathbf{r}'''(t)]$$

10.3 Arc Length and Curvature

In Section 6.3 we defined the length of a plane curve with parametric equations $x = f(t)$, $y = g(t)$, $a \leq t \leq b$, as the limit of lengths of inscribed polygons and, for the case where f' and g' are continuous, we arrived at the formula

$$\boxed{1} \qquad L = \int_a^b \sqrt{[f'(t)]^2 + [g'(t)]^2} \, dt = \int_a^b \sqrt{\left(\frac{dx}{dt}\right)^2 + \left(\frac{dy}{dt}\right)^2} \, dt$$

The length of a space curve is defined in exactly the same way (see Figure 1). Suppose that the curve has the vector equation $\mathbf{r}(t) = \langle f(t), g(t), h(t) \rangle$, $a \leq t \leq b$, or, equivalently, the parametric equations $x = f(t)$, $y = g(t)$, $z = h(t)$, where f', g', and h' are continuous. If the curve is traversed exactly once as t increases from a to b, then it can be shown that its length is

FIGURE 1
The length of a space curve is the limit of lengths of inscribed polygons.

$$\boxed{2} \qquad L = \int_a^b \sqrt{[f'(t)]^2 + [g'(t)]^2 + [h'(t)]^2} \, dt$$

$$= \int_a^b \sqrt{\left(\frac{dx}{dt}\right)^2 + \left(\frac{dy}{dt}\right)^2 + \left(\frac{dz}{dt}\right)^2} \, dt$$

Notice that both of the arc length formulas (1) and (2) can be put into the more compact form

$$\boxed{3} \qquad L = \int_a^b |\mathbf{r}'(t)| \, dt$$

because, for plane curves $\mathbf{r}(t) = f(t)\,\mathbf{i} + g(t)\,\mathbf{j}$,

$$|\mathbf{r}'(t)| = |f'(t)\,\mathbf{i} + g'(t)\,\mathbf{j}| = \sqrt{[f'(t)]^2 + [g'(t)]^2}$$

whereas, for space curves $\mathbf{r}(t) = f(t)\,\mathbf{i} + g(t)\,\mathbf{j} + h(t)\,\mathbf{k}$,

$$|\mathbf{r}'(t)| = |f'(t)\,\mathbf{i} + g'(t)\,\mathbf{j} + h'(t)\,\mathbf{k}| = \sqrt{[f'(t)]^2 + [g'(t)]^2 + [h'(t)]^2}$$

▽ **EXAMPLE 1** Find the length of the arc of the circular helix with vector equation $\mathbf{r}(t) = \cos t\,\mathbf{i} + \sin t\,\mathbf{j} + t\,\mathbf{k}$ from the point $(1, 0, 0)$ to the point $(1, 0, 2\pi)$.

SOLUTION Since $\mathbf{r}'(t) = -\sin t\,\mathbf{i} + \cos t\,\mathbf{j} + \mathbf{k}$, we have

$$|\mathbf{r}'(t)| = \sqrt{(-\sin t)^2 + \cos^2 t + 1} = \sqrt{2}$$

The arc from $(1, 0, 0)$ to $(1, 0, 2\pi)$ is described by the parameter interval $0 \leqslant t \leqslant 2\pi$ and so, from Formula 3, we have

$$L = \int_0^{2\pi} |\mathbf{r}'(t)|\,dt = \int_0^{2\pi} \sqrt{2}\,dt = 2\sqrt{2}\,\pi$$

■ ■

■ ■ Figure 2 shows the arc of the helix whose length is computed in Example 1.

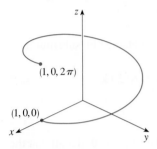

FIGURE 2

A single curve C can be represented by more than one vector function. For instance, the twisted cubic

$$\boxed{4} \qquad \mathbf{r}_1(t) = \langle t, t^2, t^3 \rangle \qquad 1 \leqslant t \leqslant 2$$

could also be represented by the function

$$\boxed{5} \qquad \mathbf{r}_2(u) = \langle e^u, e^{2u}, e^{3u} \rangle \qquad 0 \leqslant u \leqslant \ln 2$$

where the connection between the parameters t and u is given by $t = e^u$. We say that Equations 4 and 5 are **parametrizations** of the curve C. If we were to use Equation 3 to compute the length of C using Equations 4 and 5, we would get the same answer. In general, it can be shown that when Equation 3 is used to compute the length of any piecewise-smooth curve, the arc length is independent of the parametrization that is used.

■ ■ Piecewise-smooth curves were introduced on page 704.

Now we suppose that C is a piecewise-smooth curve given by a vector function $\mathbf{r}(t) = f(t)\,\mathbf{i} + g(t)\,\mathbf{j} + h(t)\,\mathbf{k}$, $a \leqslant t \leqslant b$, and C is traversed exactly once as t increases from a to b. We define its **arc length function** s by

$$\boxed{6} \qquad s(t) = \int_a^t |\mathbf{r}'(u)|\,du = \int_a^t \sqrt{\left(\frac{dx}{du}\right)^2 + \left(\frac{dy}{du}\right)^2 + \left(\frac{dz}{du}\right)^2}\,du$$

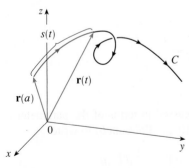

FIGURE 3

Thus, $s(t)$ is the length of the part of C between $\mathbf{r}(a)$ and $\mathbf{r}(t)$. (See Figure 3.) If we differentiate both sides of Equation 6 using Part 1 of the Fundamental Theorem of Calculus, we obtain

$$\boxed{7} \qquad \frac{ds}{dt} = |\mathbf{r}'(t)|$$

It is often useful to **parametrize a curve with respect to arc length** because arc length arises naturally from the shape of the curve and does not depend on a particular coordinate system. If a curve $\mathbf{r}(t)$ is already given in terms of a parameter t and $s(t)$ is the arc length function given by Equation 6, then we may be able to solve for t as a function of s: $t = t(s)$. Then the curve can be reparametrized in terms of s by substituting for t: $\mathbf{r} = \mathbf{r}(t(s))$. Thus, if $s = 3$ for instance, $\mathbf{r}(t(3))$ is the position vector of the point 3 units of length along the curve from its starting point.

EXAMPLE 2 Reparametrize the helix $\mathbf{r}(t) = \cos t\,\mathbf{i} + \sin t\,\mathbf{j} + t\,\mathbf{k}$ with respect to arc length measured from $(1, 0, 0)$ in the direction of increasing t.

SOLUTION The initial point $(1, 0, 0)$ corresponds to the parameter value $t = 0$. From Example 1 we have

$$\frac{ds}{dt} = |\mathbf{r}'(t)| = \sqrt{2}$$

and so

$$s = s(t) = \int_0^t |\mathbf{r}'(u)|\,du = \int_0^t \sqrt{2}\,du = \sqrt{2}\,t$$

Therefore, $t = s/\sqrt{2}$ and the required reparametrization is obtained by substituting for t:

$$\mathbf{r}(t(s)) = \cos(s/\sqrt{2})\,\mathbf{i} + \sin(s/\sqrt{2})\,\mathbf{j} + (s/\sqrt{2})\,\mathbf{k} \qquad ■\ ■$$

■■ Curvature

If C is a smooth curve defined by the vector function $\mathbf{r}$, then $\mathbf{r}'(t) \neq \mathbf{0}$. Recall that the unit tangent vector $\mathbf{T}(t)$ is given by

$$\mathbf{T}(t) = \frac{\mathbf{r}'(t)}{|\mathbf{r}'(t)|}$$

and indicates the direction of the curve. From Figure 4 you can see that $\mathbf{T}(t)$ changes direction very slowly when C is fairly straight, but it changes direction more quickly when C bends or twists more sharply.

The curvature of C at a given point is a measure of how quickly the curve changes direction at that point. Specifically, we define it to be the magnitude of the rate of change of the unit tangent vector with respect to arc length. (We use arc length so that the curvature will be independent of the parametrization.)

FIGURE 4

Unit tangent vectors at equally spaced points on C

Visual 10.3A shows animated unit tangent vectors, like those in Figure 4, for a variety of plane curves and space curves.

> **8 Definition**   The **curvature** of a curve is
>
> $$\kappa = \left|\frac{d\mathbf{T}}{ds}\right|$$
>
> where $\mathbf{T}$ is the unit tangent vector.

The curvature is easier to compute if it is expressed in terms of the parameter t instead of s, so we use the Chain Rule (Theorem 10.2.3, Formula 6) to write

$$\frac{d\mathbf{T}}{dt} = \frac{d\mathbf{T}}{ds}\frac{ds}{dt} \qquad \text{and} \qquad \kappa = \left|\frac{d\mathbf{T}}{ds}\right| = \left|\frac{d\mathbf{T}/dt}{ds/dt}\right|$$

But $ds/dt = |\mathbf{r}'(t)|$ from Equation 7, so

$$\boxed{9 \qquad \kappa(t) = \frac{|\mathbf{T}'(t)|}{|\mathbf{r}'(t)|}}$$

$\boxed{V}$ **EXAMPLE 3** Show that the curvature of a circle of radius a is $1/a$.

SOLUTION We can take the circle to have center the origin, and then a parametrization is

$$\mathbf{r}(t) = a \cos t\, \mathbf{i} + a \sin t\, \mathbf{j}$$

Therefore $\qquad \mathbf{r}'(t) = -a \sin t\, \mathbf{i} + a \cos t\, \mathbf{j} \qquad$ and $\qquad |\mathbf{r}'(t)| = a$

so $\qquad\qquad\qquad \mathbf{T}(t) = \dfrac{\mathbf{r}'(t)}{|\mathbf{r}'(t)|} = -\sin t\, \mathbf{i} + \cos t\, \mathbf{j}$

and $\qquad\qquad\qquad\qquad \mathbf{T}'(t) = -\cos t\, \mathbf{i} - \sin t\, \mathbf{j}$

This gives $|\mathbf{T}'(t)| = 1$, so using Equation 9, we have

$$\kappa(t) = \frac{|\mathbf{T}'(t)|}{|\mathbf{r}'(t)|} = \frac{1}{a}$$ ■ ■

The result of Example 3 shows that small circles have large curvature and large circles have small curvature, in accordance with our intuition. We can see directly from the definition of curvature that the curvature of a straight line is always 0 because the tangent vector is constant.

Although Formula 9 can be used in all cases to compute the curvature, the formula given by the following theorem is often more convenient to apply.

$\boxed{10}$ **Theorem** The curvature of the curve given by the vector function $\mathbf{r}$ is

$$\kappa(t) = \frac{|\mathbf{r}'(t) \times \mathbf{r}''(t)|}{|\mathbf{r}'(t)|^3}$$

Proof Since $\mathbf{T} = \mathbf{r}'/|\mathbf{r}'|$ and $|\mathbf{r}'| = ds/dt$, we have

$$\mathbf{r}' = |\mathbf{r}'|\mathbf{T} = \frac{ds}{dt}\mathbf{T}$$

so the Product Rule (Theorem 10.2.3, Formula 3) gives

$$\mathbf{r}'' = \frac{d^2s}{dt^2}\mathbf{T} + \frac{ds}{dt}\mathbf{T}'$$

Using the fact that $\mathbf{T} \times \mathbf{T} = \mathbf{0}$ (see Section 9.4), we have

$$\mathbf{r}' \times \mathbf{r}'' = \left(\frac{ds}{dt}\right)^2 (\mathbf{T} \times \mathbf{T}')$$

Now $|\mathbf{T}(t)| = 1$ for all t, so $\mathbf{T}$ and $\mathbf{T}'$ are orthogonal by Example 5 in Section 10.2. Therefore, by the definition of a cross product,

$$|\mathbf{r}' \times \mathbf{r}''| = \left(\frac{ds}{dt}\right)^2 |\mathbf{T} \times \mathbf{T}'| = \left(\frac{ds}{dt}\right)^2 |\mathbf{T}||\mathbf{T}'| = \left(\frac{ds}{dt}\right)^2 |\mathbf{T}'|$$

Thus
$$|\mathbf{T}'| = \frac{|\mathbf{r}' \times \mathbf{r}''|}{(ds/dt)^2} = \frac{|\mathbf{r}' \times \mathbf{r}''|}{|\mathbf{r}'|^2}$$

and
$$\kappa = \frac{|\mathbf{T}'|}{|\mathbf{r}'|} = \frac{|\mathbf{r}' \times \mathbf{r}''|}{|\mathbf{r}'|^3}$$

EXAMPLE 4 Find the curvature of the twisted cubic $\mathbf{r}(t) = \langle t, t^2, t^3 \rangle$ at a general point and at $(0, 0, 0)$.

SOLUTION We first compute the required ingredients:

$$\mathbf{r}'(t) = \langle 1, 2t, 3t^2 \rangle \qquad \mathbf{r}''(t) = \langle 0, 2, 6t \rangle$$

$$|\mathbf{r}'(t)| = \sqrt{1 + 4t^2 + 9t^4}$$

$$\mathbf{r}'(t) \times \mathbf{r}''(t) = \begin{vmatrix} \mathbf{i} & \mathbf{j} & \mathbf{k} \\ 1 & 2t & 3t^2 \\ 0 & 2 & 6t \end{vmatrix} = 6t^2\,\mathbf{i} - 6t\,\mathbf{j} + 2\,\mathbf{k}$$

$$|\mathbf{r}'(t) \times \mathbf{r}''(t)| = \sqrt{36t^4 + 36t^2 + 4} = 2\sqrt{9t^4 + 9t^2 + 1}$$

Theorem 10 then gives

$$\kappa(t) = \frac{|\mathbf{r}'(t) \times \mathbf{r}''(t)|}{|\mathbf{r}'(t)|^3} = \frac{2\sqrt{1 + 9t^2 + 9t^4}}{(1 + 4t^2 + 9t^4)^{3/2}}$$

At the origin, where $t = 0$, the curvature is $\kappa(0) = 2$.

For the special case of a plane curve with equation $y = f(x)$, we choose x as the parameter and write $\mathbf{r}(x) = x\,\mathbf{i} + f(x)\,\mathbf{j}$. Then $\mathbf{r}'(x) = \mathbf{i} + f'(x)\,\mathbf{j}$ and $\mathbf{r}''(x) = f''(x)\,\mathbf{j}$. Since $\mathbf{i} \times \mathbf{j} = \mathbf{k}$ and $\mathbf{j} \times \mathbf{j} = \mathbf{0}$, we have $\mathbf{r}'(x) \times \mathbf{r}''(x) = f''(x)\,\mathbf{k}$. We also have $|\mathbf{r}'(x)| = \sqrt{1 + [f'(x)]^2}$ and so, by Theorem 10,

$$\boxed{11 \qquad \kappa(x) = \frac{|f''(x)|}{[1 + (f'(x))^2]^{3/2}}}$$

EXAMPLE 5 Find the curvature of the parabola $y = x^2$ at the points $(0, 0)$, $(1, 1)$, and $(2, 4)$.

SOLUTION Since $y' = 2x$ and $y'' = 2$, Formula 11 gives

$$\kappa(x) = \frac{|y''|}{[1 + (y')^2]^{3/2}} = \frac{2}{(1 + 4x^2)^{3/2}}$$

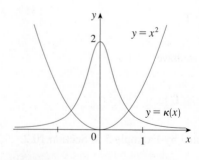

FIGURE 5

The parabola $y = x^2$ and its curvature function

The curvature at $(0, 0)$ is $\kappa(0) = 2$. At $(1, 1)$ it is $\kappa(1) = 2/5^{3/2} \approx 0.18$. At $(2, 4)$ it is $\kappa(2) = 2/17^{3/2} \approx 0.03$. Observe from the expression for $\kappa(x)$ or the graph of κ in Figure 5 that $\kappa(x) \to 0$ as $x \to \pm\infty$. This corresponds to the fact that the parabola appears to become flatter as $x \to \pm\infty$.

▥ The Normal and Binormal Vectors

At a given point on a smooth space curve $\mathbf{r}(t)$, there are many vectors that are orthogonal to the unit tangent vector $\mathbf{T}(t)$. We single out one by observing that, because $|\mathbf{T}(t)| = 1$ for all t, we have $\mathbf{T}(t) \cdot \mathbf{T}'(t) = 0$ by Example 5 in Section 10.2, so $\mathbf{T}'(t)$ is orthogonal to $\mathbf{T}(t)$. Note that $\mathbf{T}'(t)$ is itself not a unit vector. But if $\mathbf{r}'$ is also smooth, we can define the **principal unit normal vector** $\mathbf{N}(t)$ (or simply **unit normal**) as

$$\mathbf{N}(t) = \frac{\mathbf{T}'(t)}{|\mathbf{T}'(t)|}$$

■ ■ We can think of the normal vector as indicating the direction in which the curve is turning at each point.

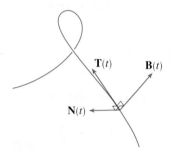

FIGURE 6

The vector $\mathbf{B}(t) = \mathbf{T}(t) \times \mathbf{N}(t)$ is called the **binormal vector**. It is perpendicular to both $\mathbf{T}$ and $\mathbf{N}$ and is also a unit vector. (See Figure 6.)

EXAMPLE 6 Find the unit normal and binormal vectors for the circular helix

$$\mathbf{r}(t) = \cos t\,\mathbf{i} + \sin t\,\mathbf{j} + t\,\mathbf{k}$$

SOLUTION We first compute the ingredients needed for the unit normal vector:

$$\mathbf{r}'(t) = -\sin t\,\mathbf{i} + \cos t\,\mathbf{j} + \mathbf{k} \qquad |\mathbf{r}'(t)| = \sqrt{2}$$

$$\mathbf{T}(t) = \frac{\mathbf{r}'(t)}{|\mathbf{r}'(t)|} = \frac{1}{\sqrt{2}}(-\sin t\,\mathbf{i} + \cos t\,\mathbf{j} + \mathbf{k})$$

$$\mathbf{T}'(t) = \frac{1}{\sqrt{2}}(-\cos t\,\mathbf{i} - \sin t\,\mathbf{j}) \qquad |\mathbf{T}'(t)| = \frac{1}{\sqrt{2}}$$

$$\mathbf{N}(t) = \frac{\mathbf{T}'(t)}{|\mathbf{T}'(t)|} = -\cos t\,\mathbf{i} - \sin t\,\mathbf{j} = \langle -\cos t, -\sin t, 0 \rangle$$

■ ■ Figure 7 illustrates Example 6 by showing the vectors $\mathbf{T}$, $\mathbf{N}$, and $\mathbf{B}$ at two locations on the helix. In general, the vectors $\mathbf{T}$, $\mathbf{N}$, and $\mathbf{B}$, starting at the various points on a curve, form a set of orthogonal vectors, called the **TNB** frame, that moves along the curve as t varies. This **TNB** frame plays an important role in the branch of mathematics known as differential geometry and in its applications to the motion of spacecraft.

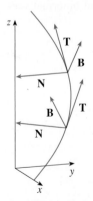

FIGURE 7

This shows that the normal vector at a point on the helix is horizontal and points toward the z-axis. The binormal vector is

$$\mathbf{B}(t) = \mathbf{T}(t) \times \mathbf{N}(t) = \frac{1}{\sqrt{2}}\begin{bmatrix} \mathbf{i} & \mathbf{j} & \mathbf{k} \\ -\sin t & \cos t & 1 \\ -\cos t & -\sin t & 0 \end{bmatrix} = \frac{1}{\sqrt{2}}\langle \sin t, -\cos t, 1 \rangle$$

■ ■

Visual 10.3B shows how the TNB frame moves along several curves.

The plane determined by the normal and binormal vectors $\mathbf{N}$ and $\mathbf{B}$ at a point P on a curve C is called the **normal plane** of C at P. It consists of all lines that are orthogonal to the tangent vector $\mathbf{T}$. The plane determined by the vectors $\mathbf{T}$ and $\mathbf{N}$ is called the **osculating plane** of C at P. The name comes from the Latin *osculum*, meaning "kiss." It is the plane that comes closest to containing the part of the curve near P. (For a plane curve, the osculating plane is simply the plane that contains the curve.)

The circle that lies in the osculating plane of C at P, has the same tangent as C at P, lies on the concave side of C (toward which $\mathbf{N}$ points), and has radius $\rho = 1/\kappa$ (the reciprocal of the curvature) is called the **osculating circle** (or the **circle of curvature**) of C at P. It is the circle that best describes how C behaves near P; it shares the same tangent, normal, and curvature at P.

■ Figure 8 shows the helix and the osculating plane in Example 7.

$z = -x + \frac{\pi}{2}$

FIGURE 8

$y = x^2$

osculating circle

$\frac{1}{2}$

0   1 x

FIGURE 9

Visual 10.3C shows how the osculating circle changes as a point moves along a curve.

⊻ **EXAMPLE 7** Find the equations of the normal plane and osculating plane of the helix in Example 6 at the point $P(0, 1, \pi/2)$.

SOLUTION The normal plane at P has normal vector $\mathbf{r}'(\pi/2) = \langle -1, 0, 1 \rangle$, so an equation is

$$-1(x - 0) + 0(y - 1) + 1\left(z - \frac{\pi}{2}\right) = 0 \qquad \text{or} \qquad z = x + \frac{\pi}{2}$$

The osculating plane at P contains the vectors $\mathbf{T}$ and $\mathbf{N}$, so its normal vector is $\mathbf{T} \times \mathbf{N} = \mathbf{B}$. From Example 6 we have

$$\mathbf{B}(t) = \frac{1}{\sqrt{2}} \langle \sin t, -\cos t, 1 \rangle \qquad \mathbf{B}\left(\frac{\pi}{2}\right) = \left\langle \frac{1}{\sqrt{2}}, 0, \frac{1}{\sqrt{2}} \right\rangle$$

A simpler normal vector is $\langle 1, 0, 1 \rangle$, so an equation of the osculating plane is

$$1(x - 0) + 0(y - 1) + 1\left(z - \frac{\pi}{2}\right) = 0 \qquad \text{or} \qquad z = -x + \frac{\pi}{2} \qquad ■■$$

EXAMPLE 8 Find and graph the osculating circle of the parabola $y = x^2$ at the origin.

SOLUTION From Example 5 the curvature of the parabola at the origin is $\kappa(0) = 2$. So the radius of the osculating circle at the origin is $1/\kappa = \frac{1}{2}$ and its center is $\left(0, \frac{1}{2}\right)$. Its equation is therefore

$$x^2 + \left(y - \tfrac{1}{2}\right)^2 = \tfrac{1}{4}$$

For the graph in Figure 9 we use parametric equations of this circle:

$$x = \tfrac{1}{2} \cos t \qquad y = \tfrac{1}{2} + \tfrac{1}{2} \sin t \qquad ■■$$

We summarize here the formulas for unit tangent, unit normal and binormal vectors, and curvature.

$$\mathbf{T}(t) = \frac{\mathbf{r}'(t)}{|\mathbf{r}'(t)|} \qquad \mathbf{N}(t) = \frac{\mathbf{T}'(t)}{|\mathbf{T}'(t)|} \qquad \mathbf{B}(t) = \mathbf{T}(t) \times \mathbf{N}(t)$$

$$\kappa = \left| \frac{d\mathbf{T}}{ds} \right| = \frac{|\mathbf{T}'(t)|}{|\mathbf{r}'(t)|} = \frac{|\mathbf{r}'(t) \times \mathbf{r}''(t)|}{|\mathbf{r}'(t)|^3}$$

10.3 Exercises

1–4 ■ Find the length of the curve.

1. $\mathbf{r}(t) = \langle 2 \sin t, 5t, 2 \cos t \rangle, \quad -10 \leqslant t \leqslant 10$

2. $\mathbf{r}(t) = \langle t^2, \sin t - t \cos t, \cos t + t \sin t \rangle, \quad 0 \leqslant t \leqslant \pi$

3. $\mathbf{r}(t) = \mathbf{i} + t^2 \mathbf{j} + t^3 \mathbf{k}, \quad 0 \leqslant t \leqslant 1$

4. $\mathbf{r}(t) = 12t \, \mathbf{i} + 8t^{3/2} \mathbf{j} + 3t^2 \mathbf{k}, \quad 0 \leqslant t \leqslant 1$

5. Use Simpson's Rule with $n = 10$ to estimate the length of the arc of the twisted cubic $x = t$, $y = t^2$, $z = t^3$ from the origin to the point $(2, 4, 8)$.

6. Graph the curve with parametric equations $x = \cos t$, $y = \sin 3t$, $z = \sin t$. Find the total length of this curve correct to four decimal places.

7–8 ■ Reparametrize the curve with respect to arc length measured from the point where $t = 0$ in the direction of increasing t.

7. $\mathbf{r}(t) = 2t \, \mathbf{i} + (1 - 3t) \mathbf{j} + (5 + 4t) \mathbf{k}$

8. $\mathbf{r}(t) = e^{2t} \cos 2t \, \mathbf{i} + 2 \mathbf{j} + e^{2t} \sin 2t \, \mathbf{k}$

9. Suppose you start at the point $(0, 0, 3)$ and move 5 units along the curve $x = 3 \sin t$, $y = 4t$, $z = 3 \cos t$ in the positive direction. Where are you now?

10. Reparametrize the curve

$$\mathbf{r}(t) = \left(\frac{2}{t^2 + 1} - 1 \right) \mathbf{i} + \frac{2t}{t^2 + 1} \mathbf{j}$$

with respect to arc length measured from the point $(1, 0)$ in the direction of increasing t. Express the reparametrization in its simplest form. What can you conclude about the curve?

11–14 ■

(a) Find the unit tangent and unit normal vectors $\mathbf{T}(t)$ and $\mathbf{N}(t)$.
(b) Use Formula 9 to find the curvature.

11. $\mathbf{r}(t) = \langle 2 \sin t, 5t, 2 \cos t \rangle$

12. $\mathbf{r}(t) = \langle t^2, \sin t - t \cos t, \cos t + t \sin t \rangle$, $\quad t > 0$

13. $\mathbf{r}(t) = \langle \sqrt{2}\, t, e^t, e^{-t} \rangle$

14. $\mathbf{r}(t) = \langle t, \frac{1}{2}t^2, t^2 \rangle$

15–17 ■ Use Theorem 10 to find the curvature.

15. $\mathbf{r}(t) = t^2 \mathbf{i} + t \mathbf{k}$

16. $\mathbf{r}(t) = t \mathbf{i} + t \mathbf{j} + (1 + t^2) \mathbf{k}$

17. $\mathbf{r}(t) = 3t \mathbf{i} + 4 \sin t \mathbf{j} + 4 \cos t \mathbf{k}$

18. Find the curvature of $\mathbf{r}(t) = \langle e^t \cos t, e^t \sin t, t \rangle$ at the point $(1, 0, 0)$.

19. Find the curvature of $\mathbf{r}(t) = \langle t, t^2, t^3 \rangle$ at the point $(1, 1, 1)$.

20. Graph the curve with parametric equations

$$x = t \qquad y = 4t^{3/2} \qquad z = -t^2$$

and find the curvature at the point $(1, 4, -1)$.

21–23 ■ Use Formula 11 to find the curvature.

21. $y = xe^x$ **22.** $y = \cos x$ **23.** $y = 4x^{5/2}$

24–25 ■ At what point does the curve have maximum curvature? What happens to the curvature as $x \to \infty$?

24. $y = \ln x$ **25.** $y = e^x$

26. Find an equation of a parabola that has curvature 4 at the origin.

27. (a) Is the curvature of the curve C shown in the figure greater at P or at Q? Explain.

(b) Estimate the curvature at P and at Q by sketching the osculating circles at those points.

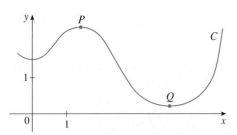

28–29 ■ Use a graphing calculator or computer to graph both the curve and its curvature function $\kappa(x)$ on the same screen. Is the graph of κ what you would expect?

28. $y = x^4 - 2x^2$ **29.** $y = x^{-2}$

30–31 ■ Two graphs, a and b, are shown. One is a curve $y = f(x)$ and the other is the graph of its curvature function $y = \kappa(x)$. Identify each curve and explain your choices.

30. **31.**

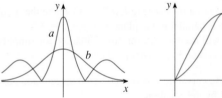

32. (a) Graph the curve $\mathbf{r}(t) = \langle \sin 3t, \sin 2t, \sin 3t \rangle$. At how many points on the curve does it appear that the curvature has a local or absolute maximum?
(b) Use a CAS to find and graph the curvature function. Does this graph confirm your conclusion from part (a)?

33. The graph of $\mathbf{r}(t) = \langle t - \frac{3}{2} \sin t, 1 - \frac{3}{2} \cos t, t \rangle$ is shown in Figure 12(b) in Section 10.1. Where do you think the curvature is largest? Use a CAS to find and graph the curvature function. For which values of t is the curvature largest?

34. Use Theorem 10 to show that the curvature of a plane parametric curve $x = f(t)$, $y = g(t)$ is

$$\kappa = \frac{|\dot{x}\ddot{y} - \dot{y}\ddot{x}|}{[\dot{x}^2 + \dot{y}^2]^{3/2}}$$

where the dots indicate derivatives with respect to t.

35–36 ■ Use the formula in Exercise 34 to find the curvature.

35. $x = e^t \cos t$, $\quad y = e^t \sin t$

36. $x = 1 + t^3$, $\quad y = t + t^2$

37–38 ■ Find the vectors **T**, **N**, and **B** at the given point.

37. $\mathbf{r}(t) = \langle t^2, \frac{2}{3}t^3, t \rangle$, $(1, \frac{2}{3}, 1)$

38. $\mathbf{r}(t) = \langle e^t, e^t \sin t, e^t \cos t \rangle$, $(1, 0, 1)$

· · · · · · · · · · · · ·

39–40 ■ Find equations of the normal plane and osculating plane of the curve at the given point.

39. $x = 2 \sin 3t$, $y = t$, $z = 2 \cos 3t$; $(0, \pi, -2)$

40. $x = t$, $y = t^2$, $z = t^3$; $(1, 1, 1)$

· · · · · · · · · · · · ·

41. Find equations of the osculating circles of the ellipse $9x^2 + 4y^2 = 36$ at the points $(2, 0)$ and $(0, 3)$. Use a graphing calculator or computer to graph the ellipse and both osculating circles on the same screen.

42. Find equations of the osculating circles of the parabola $y = \frac{1}{2}x^2$ at the points $(0, 0)$ and $(1, \frac{1}{2})$. Graph both osculating circles and the parabola on the same screen.

43. At what point on the curve $x = t^3$, $y = 3t$, $z = t^4$ is the normal plane parallel to the plane $6x + 6y - 8z = 1$?

44. Is there a point on the curve in Exercise 43 where the osculating plane is parallel to the plane $x + y + z = 1$? [*Note:* You will need a CAS for differentiating, for simplifying, and for computing a cross product.]

45. Show that the curvature κ is related to the tangent and normal vectors by the equation

$$\frac{d\mathbf{T}}{ds} = \kappa \mathbf{N}$$

46. Show that the curvature of a plane curve is $\kappa = |d\phi/ds|$, where ϕ is the angle between **T** and **i**; that is, ϕ is the angle of inclination of the tangent line.

47. (a) Show that $d\mathbf{B}/ds$ is perpendicular to **B**.
(b) Show that $d\mathbf{B}/ds$ is perpendicular to **T**.
(c) Deduce from parts (a) and (b) that $d\mathbf{B}/ds = -\tau(s)\mathbf{N}$ for some number $\tau(s)$ called the **torsion** of the curve. (The torsion measures the degree of twisting of a curve.)
(d) Show that for a plane curve the torsion is $\tau(s) = 0$.

48. The following formulas, called the **Frenet-Serret formulas**, are of fundamental importance in differential geometry:

1. $d\mathbf{T}/ds = \kappa \mathbf{N}$
2. $d\mathbf{N}/ds = -\kappa \mathbf{T} + \tau \mathbf{B}$
3. $d\mathbf{B}/ds = -\tau \mathbf{N}$

(Formula 1 comes from Exercise 45 and Formula 3 comes from Exercise 47.) Use the fact that $\mathbf{N} = \mathbf{B} \times \mathbf{T}$ to deduce Formula 2 from Formulas 1 and 3.

49. Use the Frenet-Serret formulas to prove each of the following. (Primes denote derivatives with respect to t. Start as in the proof of Theorem 10.)
(a) $\mathbf{r}'' = s''\mathbf{T} + \kappa(s')^2\mathbf{N}$ (b) $\mathbf{r}' \times \mathbf{r}'' = \kappa(s')^3\mathbf{B}$
(c) $\mathbf{r}''' = [s''' - \kappa^2(s')^3]\mathbf{T} + [3\kappa s's'' + \kappa'(s')^2]\mathbf{N}$
$\qquad + \kappa\tau(s')^3\mathbf{B}$

(d) $\tau = \dfrac{(\mathbf{r}' \times \mathbf{r}'') \cdot \mathbf{r}'''}{|\mathbf{r}' \times \mathbf{r}''|^2}$

50. Show that the circular helix

$$\mathbf{r}(t) = \langle a \cos t, a \sin t, bt \rangle$$

where a and b are positive constants, has constant curvature and constant torsion. [Use the result of Exercise 49(d).]

51. The DNA molecule has the shape of a double helix (see Figure 3 on page 697). The radius of each helix is about 10 angstroms (1 Å $= 10^{-8}$ cm). Each helix rises about 34 Å during each complete turn, and there are about 2.9×10^8 complete turns. Estimate the length of each helix.

52. Let's consider the problem of designing a railroad track to make a smooth transition between sections of straight track. Existing track along the negative x-axis is to be joined smoothly to a track along the line $y = 1$ for $x \geqslant 1$.
(a) Find a polynomial $P = P(x)$ of degree 5 such that the function F defined by

$$F(x) = \begin{cases} 0 & \text{if } x \leqslant 0 \\ P(x) & \text{if } 0 < x < 1 \\ 1 & \text{if } x \geqslant 1 \end{cases}$$

is continuous and has continuous slope and continuous curvature.
(b) Use a graphing calculator or computer to draw the graph of F.

10.4 Motion in Space: Velocity and Acceleration

In this section we show how the ideas of tangent and normal vectors and curvature can be used in physics to study the motion of an object, including its velocity and acceleration, along a space curve. In particular, we follow in the footsteps of Newton by using these methods to derive Kepler's First Law of planetary motion.

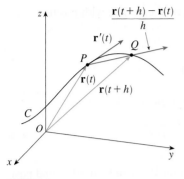

FIGURE 1

Suppose a particle moves through space so that its position vector at time t is $\mathbf{r}(t)$. Notice from Figure 1 that, for small values of h, the vector

$$\boxed{1} \qquad \frac{\mathbf{r}(t + h) - \mathbf{r}(t)}{h}$$

approximates the direction of the particle moving along the curve $\mathbf{r}(t)$. Its magnitude measures the size of the displacement vector per unit time. The vector (1) gives the average velocity over a time interval of length h and its limit is the **velocity vector** $\mathbf{v}(t)$ at time t:

$$\boxed{2} \qquad \mathbf{v}(t) = \lim_{h \to 0} \frac{\mathbf{r}(t + h) - \mathbf{r}(t)}{h} = \mathbf{r}'(t)$$

Thus, the velocity vector is also the tangent vector and points in the direction of the tangent line.

The **speed** of the particle at time t is the magnitude of the velocity vector, that is, $|\mathbf{v}(t)|$. This is appropriate because, from (2) and from Equation 10.3.7, we have

$$|\mathbf{v}(t)| = |\mathbf{r}'(t)| = \frac{ds}{dt} = \text{rate of change of distance with respect to time}$$

As in the case of one-dimensional motion, the **acceleration** of the particle is defined as the derivative of the velocity:

$$\mathbf{a}(t) = \mathbf{v}'(t) = \mathbf{r}''(t)$$

EXAMPLE 1 The position vector of an object moving in a plane is given by $\mathbf{r}(t) = t^3\,\mathbf{i} + t^2\,\mathbf{j}$. Find its velocity, speed, and acceleration when $t = 1$ and illustrate geometrically.

SOLUTION The velocity and acceleration at time t are

$$\mathbf{v}(t) = \mathbf{r}'(t) = 3t^2\,\mathbf{i} + 2t\,\mathbf{j}$$

$$\mathbf{a}(t) = \mathbf{r}''(t) = 6t\,\mathbf{i} + 2\,\mathbf{j}$$

and the speed is

$$|\mathbf{v}(t)| = \sqrt{(3t^2)^2 + (2t)^2} = \sqrt{9t^4 + 4t^2}$$

When $t = 1$, we have

$$\mathbf{v}(1) = 3\,\mathbf{i} + 2\,\mathbf{j} \qquad \mathbf{a}(1) = 6\,\mathbf{i} + 2\,\mathbf{j} \qquad |\mathbf{v}(1)| = \sqrt{13}$$

These velocity and acceleration vectors are shown in Figure 2. ■ ■

FIGURE 2

Visual 10.4 shows animated velocity and acceleration vectors for objects moving along various curves.

■ ■ Figure 3 shows the path of the particle in Example 2 with the velocity and acceleration vectors when $t = 1$.

FIGURE 3

EXAMPLE 2 Find the velocity, acceleration, and speed of a particle with position vector $\mathbf{r}(t) = \langle t^2, e^t, te^t \rangle$.

SOLUTION

$$\mathbf{v}(t) = \mathbf{r}'(t) = \langle 2t, e^t, (1 + t)e^t \rangle$$

$$\mathbf{a}(t) = \mathbf{v}'(t) = \langle 2, e^t, (2 + t)e^t \rangle$$

$$|\mathbf{v}(t)| = \sqrt{4t^2 + e^{2t} + (1 + t)^2 e^{2t}}$$

■ ■

The vector integrals that were introduced in Section 10.2 can be used to find position vectors when velocity or acceleration vectors are known, as in the next example.

�V⟩ **EXAMPLE 3** A moving particle starts at an initial position $\mathbf{r}(0) = \langle 1, 0, 0 \rangle$ with initial velocity $\mathbf{v}(0) = \mathbf{i} - \mathbf{j} + \mathbf{k}$. Its acceleration is $\mathbf{a}(t) = 4t\,\mathbf{i} + 6t\,\mathbf{j} + \mathbf{k}$. Find its velocity and position at time t.

SOLUTION Since $\mathbf{a}(t) = \mathbf{v}'(t)$, we have

$$\mathbf{v}(t) = \int \mathbf{a}(t)\,dt = \int (4t\,\mathbf{i} + 6t\,\mathbf{j} + \mathbf{k})\,dt$$

$$= 2t^2\,\mathbf{i} + 3t^2\,\mathbf{j} + t\,\mathbf{k} + \mathbf{C}$$

To determine the value of the constant vector $\mathbf{C}$, we use the fact that $\mathbf{v}(0) = \mathbf{i} - \mathbf{j} + \mathbf{k}$. The preceding equation gives $\mathbf{v}(0) = \mathbf{C}$, so $\mathbf{C} = \mathbf{i} - \mathbf{j} + \mathbf{k}$ and

$$\mathbf{v}(t) = 2t^2\,\mathbf{i} + 3t^2\,\mathbf{j} + t\,\mathbf{k} + \mathbf{i} - \mathbf{j} + \mathbf{k}$$

$$= (2t^2 + 1)\,\mathbf{i} + (3t^2 - 1)\,\mathbf{j} + (t + 1)\,\mathbf{k}$$

■ ■ The expression for $\mathbf{r}(t)$ that we obtained in Example 3 was used to plot the path of the particle in Figure 4 for $0 \leq t \leq 3$.

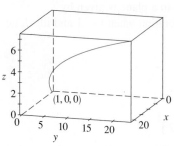

FIGURE 4

Since $\mathbf{v}(t) = \mathbf{r}'(t)$, we have

$$\mathbf{r}(t) = \int \mathbf{v}(t)\,dt$$

$$= \int [(2t^2 + 1)\,\mathbf{i} + (3t^2 - 1)\,\mathbf{j} + (t + 1)\,\mathbf{k}]\,dt$$

$$= \left(\tfrac{2}{3}t^3 + t\right)\mathbf{i} + (t^3 - t)\,\mathbf{j} + \left(\tfrac{1}{2}t^2 + t\right)\mathbf{k} + \mathbf{D}$$

Putting $t = 0$, we find that $\mathbf{D} = \mathbf{r}(0) = \mathbf{i}$, so

$$\mathbf{r}(t) = \left(\tfrac{2}{3}t^3 + t + 1\right)\mathbf{i} + (t^3 - t)\,\mathbf{j} + \left(\tfrac{1}{2}t^2 + t\right)\mathbf{k}$$

■ ■

In general, vector integrals allow us to recover velocity when acceleration is known and position when velocity is known:

$$\mathbf{v}(t) = \mathbf{v}(t_0) + \int_{t_0}^{t} \mathbf{a}(u)\,du \qquad \mathbf{r}(t) = \mathbf{r}(t_0) + \int_{t_0}^{t} \mathbf{v}(u)\,du$$

If the force that acts on a particle is known, then the acceleration can be found from **Newton's Second Law of Motion**. The vector version of this law states that if, at any time t, a force $\mathbf{F}(t)$ acts on an object of mass m producing an acceleration $\mathbf{a}(t)$, then

$$\mathbf{F}(t) = m\mathbf{a}(t)$$

■ ■ The angular speed of the object moving with position P is $\omega = d\theta/dt$, where θ is the angle shown in Figure 5.

EXAMPLE 4 An object with mass m that moves in a circular path with constant angular speed ω has position vector $\mathbf{r}(t) = a \cos \omega t\,\mathbf{i} + a \sin \omega t\,\mathbf{j}$. Find the force acting on the object and show that it is directed toward the origin.

SOLUTION
$$\mathbf{v}(t) = \mathbf{r}'(t) = -a\omega \sin \omega t\,\mathbf{i} + a\omega \cos \omega t\,\mathbf{j}$$

$$\mathbf{a}(t) = \mathbf{v}'(t) = -a\omega^2 \cos \omega t\,\mathbf{i} - a\omega^2 \sin \omega t\,\mathbf{j}$$

Therefore, Newton's Second Law gives the force as

$$\mathbf{F}(t) = m\mathbf{a}(t) = -m\omega^2(a \cos \omega t\,\mathbf{i} + a \sin \omega t\,\mathbf{j})$$

Notice that $\mathbf{F}(t) = -m\omega^2\mathbf{r}(t)$. This shows that the force acts in the direction opposite to the radius vector $\mathbf{r}(t)$ and therefore points toward the origin (see Figure 5). Such a force is called a *centripetal* (center-seeking) force. ■ ■

FIGURE 5

$\boxed{\text{V}}$ **EXAMPLE 5** A projectile is fired with angle of elevation α and initial velocity $\mathbf{v}_0$. (See Figure 6.) Assuming that air resistance is negligible and the only external force is due to gravity, find the position function $\mathbf{r}(t)$ of the projectile. What value of α maximizes the range (the horizontal distance traveled)?

SOLUTION We set up the axes so that the projectile starts at the origin. Since the force due to gravity acts downward, we have

$$\mathbf{F} = m\mathbf{a} = -mg\,\mathbf{j}$$

FIGURE 6

where $g = |\mathbf{a}| \approx 9.8$ m/s^2. Thus

$$\mathbf{a} = -g\,\mathbf{j}$$

Since $\mathbf{v}'(t) = \mathbf{a}$, we have

$$\mathbf{v}(t) = -gt\,\mathbf{j} + \mathbf{C}$$

where $\mathbf{C} = \mathbf{v}(0) = \mathbf{v}_0$. Therefore

$$\mathbf{r}'(t) = \mathbf{v}(t) = -gt\,\mathbf{j} + \mathbf{v}_0$$

Integrating again, we obtain

$$\mathbf{r}(t) = -\tfrac{1}{2}gt^2\,\mathbf{j} + t\,\mathbf{v}_0 + \mathbf{D}$$

But $\mathbf{D} = \mathbf{r}(0) = \mathbf{0}$, so the position vector of the projectile is given by

$$\boxed{3} \qquad \mathbf{r}(t) = -\tfrac{1}{2}gt^2\,\mathbf{j} + t\,\mathbf{v}_0$$

If we write $|\mathbf{v}_0| = v_0$ (the initial speed of the projectile), then

$$\mathbf{v}_0 = v_0 \cos \alpha\,\mathbf{i} + v_0 \sin \alpha\,\mathbf{j}$$

and Equation 3 becomes

$$\mathbf{r}(t) = (v_0 \cos \alpha)t\,\mathbf{i} + \left[(v_0 \sin \alpha)t - \tfrac{1}{2}gt^2\right]\mathbf{j}$$

The parametric equations of the trajectory are therefore

■ ■ If you eliminate t from Equations 4, you will see that y is a quadratic function of x. So the path of the projectile is part of a parabola.

$$\boxed{4} \qquad x = (v_0 \cos \alpha)t \qquad y = (v_0 \sin \alpha)t - \tfrac{1}{2}gt^2$$

The horizontal distance d is the value of x when $y = 0$. Setting $y = 0$, we obtain $t = 0$ or $t = (2v_0 \sin \alpha)/g$. The latter value of t then gives

$$d = x = (v_0 \cos \alpha)\frac{2v_0 \sin \alpha}{g} = \frac{v_0^2(2 \sin \alpha \cos \alpha)}{g} = \frac{v_0^2 \sin 2\alpha}{g}$$

Clearly, d has its maximum value when $\sin 2\alpha = 1$, that is, $\alpha = \pi/4$. ■ ■

$\boxed{\text{V}}$ **EXAMPLE 6** A projectile is fired with muzzle speed 150 m/s and angle of elevation 45° from a position 10 m above ground level. Where does the projectile hit the ground, and with what speed?

SOLUTION If we place the origin at ground level, then the initial position of the projectile is $(0, 10)$ and so we need to adjust Equations 4 by adding 10 to the expression for y. With $v_0 = 150$ m/s, $\alpha = 45°$, and $g = 9.8$ m/s², we have

$$x = 150 \cos(\pi/4)t = 75\sqrt{2}\, t$$

$$y = 10 + 150 \sin(\pi/4)t - \tfrac{1}{2}(9.8)t^2 = 10 + 75\sqrt{2}\, t - 4.9t^2$$

Impact occurs when $y = 0$, that is, $4.9t^2 - 75\sqrt{2}\, t - 10 = 0$. Solving this quadratic equation (and using only the positive value of t), we get

$$t = \frac{75\sqrt{2} + \sqrt{11{,}250 + 196}}{9.8} \approx 21.74$$

Then $x \approx 75\sqrt{2}\,(21.74) \approx 2306$, so the projectile hits the ground about 2306 m away.

The velocity of the projectile is

$$\mathbf{v}(t) = \mathbf{r}'(t) = 75\sqrt{2}\,\mathbf{i} + \left(75\sqrt{2} - 9.8t\right)\mathbf{j}$$

So its speed at impact is

$$|\mathbf{v}(21.74)| = \sqrt{(75\sqrt{2})^2 + (75\sqrt{2} - 9.8 \cdot 21.74)^2} \approx 151 \text{ m/s}$$ ■ ■

■ ■ Tangential and Normal Components of Acceleration

When we study the motion of a particle, it is often useful to resolve the acceleration into two components, one in the direction of the tangent and the other in the direction of the normal. If we write $v = |\mathbf{v}|$ for the speed of the particle, then

$$\mathbf{T}(t) = \frac{\mathbf{r}'(t)}{|\mathbf{r}'(t)|} = \frac{\mathbf{v}(t)}{|\mathbf{v}(t)|} = \frac{\mathbf{v}}{v}$$

and so

$$\mathbf{v} = v\mathbf{T}$$

If we differentiate both sides of this equation with respect to t, we get

$$\boxed{5} \qquad \mathbf{a} = \mathbf{v}' = v'\mathbf{T} + v\mathbf{T}'$$

If we use the expression for the curvature given by Equation 10.3.9, then we have

$$\boxed{6} \qquad \kappa = \frac{|\mathbf{T}'|}{|\mathbf{r}'|} = \frac{|\mathbf{T}'|}{v} \qquad \text{so} \qquad |\mathbf{T}'| = \kappa v$$

The unit normal vector was defined in the preceding section as $\mathbf{N} = \mathbf{T}'/|\mathbf{T}'|$, so (6) gives

$$\mathbf{T}' = |\mathbf{T}'|\mathbf{N} = \kappa v \mathbf{N}$$

and Equation 5 becomes

$$\boxed{7} \qquad \boxed{\mathbf{a} = v'\mathbf{T} + \kappa v^2 \mathbf{N}}$$

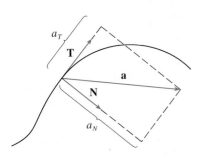

FIGURE 7

Writing a_T and a_N for the tangential and normal components of acceleration, we have

$$\mathbf{a} = a_T \mathbf{T} + a_N \mathbf{N}$$

where

$$\boxed{8} \qquad a_T = v' \qquad \text{and} \qquad a_N = \kappa v^2$$

This resolution is illustrated in Figure 7.

Let's look at what Formula 7 says. The first thing to notice is that the binormal vector $\mathbf{B}$ is absent. No matter how an object moves through space, its acceleration always lies in the plane of $\mathbf{T}$ and $\mathbf{N}$ (the osculating plane). (Recall that $\mathbf{T}$ gives the direction of motion and $\mathbf{N}$ points in the direction the curve is turning.) Next we notice that the tangential component of acceleration is v', the rate of change of speed, and the normal component of acceleration is κv^2, the curvature times the square of the speed. This makes sense if we think of a passenger in a car—a sharp turn in a road means a large value of the curvature κ, so the component of the acceleration perpendicular to the motion is large and the passenger is thrown against a car door. High speed around the turn has the same effect; in fact, if you double your speed, a_N is increased by a factor of 4.

Although we have expressions for the tangential and normal components of acceleration in Equations 8, it's desirable to have expressions that depend only on $\mathbf{r}$, $\mathbf{r}'$, and $\mathbf{r}''$. To this end we take the dot product of $\mathbf{v} = v\mathbf{T}$ with $\mathbf{a}$ as given by Equation 7:

$$\begin{aligned} \mathbf{v} \cdot \mathbf{a} &= v\mathbf{T} \cdot (v'\mathbf{T} + \kappa v^2 \mathbf{N}) \\ &= vv'\mathbf{T} \cdot \mathbf{T} + \kappa v^3 \mathbf{T} \cdot \mathbf{N} \\ &= vv' \qquad\qquad \text{(since } \mathbf{T} \cdot \mathbf{T} = 1 \text{ and } \mathbf{T} \cdot \mathbf{N} = 0\text{)} \end{aligned}$$

Therefore

$$\boxed{9} \qquad a_T = v' = \frac{\mathbf{v} \cdot \mathbf{a}}{v} = \frac{\mathbf{r}'(t) \cdot \mathbf{r}''(t)}{|\mathbf{r}'(t)|}$$

Using the formula for curvature given by Theorem 10.3.10, we have

$$\boxed{10} \qquad a_N = \kappa v^2 = \frac{|\mathbf{r}'(t) \times \mathbf{r}''(t)|}{|\mathbf{r}'(t)|^3} |\mathbf{r}'(t)|^2 = \frac{|\mathbf{r}'(t) \times \mathbf{r}''(t)|}{|\mathbf{r}'(t)|}$$

EXAMPLE 7 A particle moves with position function $\mathbf{r}(t) = \langle t^2, t^2, t^3 \rangle$. Find the tangential and normal components of acceleration.

SOLUTION

$$\mathbf{r}(t) = t^2 \mathbf{i} + t^2 \mathbf{j} + t^3 \mathbf{k}$$

$$\mathbf{r}'(t) = 2t \mathbf{i} + 2t \mathbf{j} + 3t^2 \mathbf{k}$$

$$\mathbf{r}''(t) = 2 \mathbf{i} + 2 \mathbf{j} + 6t \mathbf{k}$$

$$|\mathbf{r}'(t)| = \sqrt{8t^2 + 9t^4}$$

Therefore, Equation 9 gives the tangential component as

$$a_T = \frac{\mathbf{r}'(t) \cdot \mathbf{r}''(t)}{|\mathbf{r}'(t)|} = \frac{8t + 18t^3}{\sqrt{8t^2 + 9t^4}}$$

Since

$$\mathbf{r}'(t) \times \mathbf{r}''(t) = \begin{vmatrix} \mathbf{i} & \mathbf{j} & \mathbf{k} \\ 2t & 2t & 3t^2 \\ 2 & 2 & 6t \end{vmatrix} = 6t^2 \mathbf{i} - 6t^2 \mathbf{j}$$

Equation 10 gives the normal component as

$$a_N = \frac{|\mathbf{r}'(t) \times \mathbf{r}''(t)|}{|\mathbf{r}'(t)|} = \frac{6\sqrt{2}t^2}{\sqrt{8t^2 + 9t^4}}$$

■ ■

▪▪ Kepler's Laws of Planetary Motion

We now describe one of the great accomplishments of calculus by showing how the material of this chapter can be used to prove Kepler's laws of planetary motion. After 20 years of studying the astronomical observations of the Danish astronomer Tycho Brahe, the German mathematician and astronomer Johannes Kepler (1571–1630) formulated the following three laws.

> **Kepler's Laws**
>
> **1.** A planet revolves around the Sun in an elliptical orbit with the Sun at one focus.
>
> **2.** The line joining the Sun to a planet sweeps out equal areas in equal times.
>
> **3.** The square of the period of revolution of a planet is proportional to the cube of the length of the major axis of its orbit.

In his book *Principia Mathematica* of 1687, Sir Isaac Newton was able to show that these three laws are consequences of two of his own laws, the Second Law of Motion and the Law of Universal Gravitation. In what follows we prove Kepler's First Law. The remaining laws are left as exercises (with hints).

Since the gravitational force of the Sun on a planet is so much larger than the forces exerted by other celestial bodies, we can safely ignore all bodies in the universe except the Sun and one planet revolving about it. We use a coordinate system with the Sun at the origin and we let $\mathbf{r} = \mathbf{r}(t)$ be the position vector of the planet. (Equally well, $\mathbf{r}$ could be the position vector of the Moon or a satellite moving around the Earth or a comet moving around a star.) The velocity vector is $\mathbf{v} = \mathbf{r}'$ and the acceleration vector is $\mathbf{a} = \mathbf{r}''$. We use the following laws of Newton:

Second Law of Motion: $\mathbf{F} = m\mathbf{a}$

Law of Gravitation: $\mathbf{F} = -\dfrac{GMm}{r^3}\mathbf{r} = -\dfrac{GMm}{r^2}\mathbf{u}$

where $\mathbf{F}$ is the gravitational force on the planet, m and M are the masses of the planet and the Sun, G is the gravitational constant, $r = |\mathbf{r}|$, and $\mathbf{u} = (1/r)\mathbf{r}$ is the unit vector in the direction of $\mathbf{r}$.

We first show that the planet moves in one plane. By equating the expressions for $\mathbf{F}$ in Newton's two laws, we find that

$$\mathbf{a} = -\frac{GM}{r^3}\mathbf{r}$$

and so $\mathbf{a}$ is parallel to $\mathbf{r}$. It follows that $\mathbf{r} \times \mathbf{a} = \mathbf{0}$. We use Formula 5 in Theorem 10.2.3 to write

$$\frac{d}{dt}(\mathbf{r} \times \mathbf{v}) = \mathbf{r}' \times \mathbf{v} + \mathbf{r} \times \mathbf{v}'$$

$$= \mathbf{v} \times \mathbf{v} + \mathbf{r} \times \mathbf{a} = \mathbf{0} + \mathbf{0} = \mathbf{0}$$

Therefore $\qquad\qquad \mathbf{r} \times \mathbf{v} = \mathbf{h}$

where $\mathbf{h}$ is a constant vector. (We may assume that $\mathbf{h} \neq \mathbf{0}$; that is, $\mathbf{r}$ and $\mathbf{v}$ are not parallel.) This means that the vector $\mathbf{r} = \mathbf{r}(t)$ is perpendicular to $\mathbf{h}$ for all values of t, so the planet always lies in the plane through the origin perpendicular to $\mathbf{h}$. Thus, the orbit of the planet is a plane curve.

To prove Kepler's First Law we rewrite the vector $\mathbf{h}$ as follows:

$$\mathbf{h} = \mathbf{r} \times \mathbf{v} = \mathbf{r} \times \mathbf{r}' = r\mathbf{u} \times (r\mathbf{u})'$$

$$= r\mathbf{u} \times (r\mathbf{u}' + r'\mathbf{u}) = r^2(\mathbf{u} \times \mathbf{u}') + rr'(\mathbf{u} \times \mathbf{u})$$

$$= r^2(\mathbf{u} \times \mathbf{u}')$$

Then

$$\mathbf{a} \times \mathbf{h} = \frac{-GM}{r^2}\mathbf{u} \times (r^2\mathbf{u} \times \mathbf{u}') = -GM\,\mathbf{u} \times (\mathbf{u} \times \mathbf{u}')$$

$$= -GM[(\mathbf{u} \cdot \mathbf{u}')\mathbf{u} - (\mathbf{u} \cdot \mathbf{u})\mathbf{u}'] \qquad \text{(by Formula 9.4.8)}$$

But $\mathbf{u} \cdot \mathbf{u} = |\mathbf{u}|^2 = 1$ and, since $|\mathbf{u}(t)| = 1$, it follows from Example 5 in Section 10.2 that $\mathbf{u} \cdot \mathbf{u}' = 0$. Therefore

$$\mathbf{a} \times \mathbf{h} = GM\,\mathbf{u}'$$

and so $\qquad\qquad (\mathbf{v} \times \mathbf{h})' = \mathbf{v}' \times \mathbf{h} = \mathbf{a} \times \mathbf{h} = GM\,\mathbf{u}'$

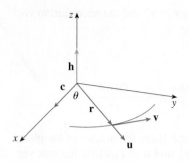

FIGURE 8

Integrating both sides of this equation, we get

$$\boxed{11} \qquad \mathbf{v} \times \mathbf{h} = GM\,\mathbf{u} + \mathbf{c}$$

where $\mathbf{c}$ is a constant vector.

At this point it is convenient to choose the coordinate axes so that the standard basis vector $\mathbf{k}$ points in the direction of the vector $\mathbf{h}$. Then the planet moves in the xy-plane. Since both $\mathbf{v} \times \mathbf{h}$ and $\mathbf{u}$ are perpendicular to $\mathbf{h}$, Equation 11 shows that $\mathbf{c}$ lies in the xy-plane. This means that we can choose the x- and y-axes so that the vector $\mathbf{i}$ lies in the direction of $\mathbf{c}$, as shown in Figure 8.

If θ is the angle between $\mathbf{c}$ and $\mathbf{r}$, then (r, θ) are polar coordinates of the planet. From Equation 11 we have

$$\mathbf{r} \cdot (\mathbf{v} \times \mathbf{h}) = \mathbf{r} \cdot (GM\,\mathbf{u} + \mathbf{c}) = GM\,\mathbf{r} \cdot \mathbf{u} + \mathbf{r} \cdot \mathbf{c}$$

$$= GMr\,\mathbf{u} \cdot \mathbf{u} + |\mathbf{r}||\mathbf{c}|\cos\theta = GMr + rc\cos\theta$$

where $c = |\mathbf{c}|$. Then

$$r = \frac{\mathbf{r} \cdot (\mathbf{v} \times \mathbf{h})}{GM + c\cos\theta} = \frac{1}{GM}\,\frac{\mathbf{r} \cdot (\mathbf{v} \times \mathbf{h})}{1 + e\cos\theta}$$

where $e = c/(GM)$. But

$$\mathbf{r} \cdot (\mathbf{v} \times \mathbf{h}) = (\mathbf{r} \times \mathbf{v}) \cdot \mathbf{h} = \mathbf{h} \cdot \mathbf{h} = |\mathbf{h}|^2 = h^2$$

where $h = |\mathbf{h}|$. So

$$r = \frac{h^2/(GM)}{1 + e\cos\theta} = \frac{eh^2/c}{1 + e\cos\theta}$$

Writing $d = h^2/c$, we obtain the equation

$$\boxed{12} \qquad r = \frac{ed}{1 + e\cos\theta}$$

In Appendix H it is shown that Equation 12 is the polar equation of a conic section with focus at the origin and eccentricity e. We know that the orbit of a planet is a closed curve and so the conic must be an ellipse.

This completes the derivation of Kepler's First Law. We will guide you through the derivation of the Second and Third Laws in the Applied Project on page 727. The proofs of these three laws show that the methods of this chapter provide a powerful tool for describing some of the laws of nature.

10.4 Exercises

1. The table gives coordinates of a particle moving through space along a smooth curve.
   (a) Find the average velocities over the time intervals [0, 1], [0.5, 1], [1, 2], and [1, 1.5].
   (b) Estimate the velocity and speed of the particle at $t = 1$.

t	x	y	z
0	2.7	9.8	3.7
0.5	3.5	7.2	3.3
1.0	4.5	6.0	3.0
1.5	5.9	6.4	2.8
2.0	7.3	7.8	2.7

2. The figure shows the path of a particle that moves with position vector $\mathbf{r}(t)$ at time t.
   (a) Draw a vector that represents the average velocity of the particle over the time interval $2 \le t \le 2.4$.
   (b) Draw a vector that represents the average velocity over the time interval $1.5 \le t \le 2$.
   (c) Write an expression for the velocity vector $\mathbf{v}(2)$.
   (d) Draw an approximation to the vector $\mathbf{v}(2)$ and estimate the speed of the particle at $t = 2$.

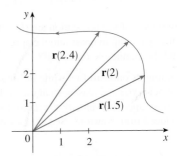

3–8 ■ Find the velocity, acceleration, and speed of a particle with the given position function. Sketch the path of the particle and draw the velocity and acceleration vectors for the specified value of t.

3. $\mathbf{r}(t) = \left\langle -\frac{1}{2}t^2, t \right\rangle$, $t = 2$

4. $\mathbf{r}(t) = \left\langle 2 - t, 4\sqrt{t} \right\rangle$, $t = 1$

5. $\mathbf{r}(t) = 3 \cos t \, \mathbf{i} + 2 \sin t \, \mathbf{j}$, $t = \pi/3$

6. $\mathbf{r}(t) = e^t \mathbf{i} + e^{2t} \mathbf{j}$, $t = 0$

7. $\mathbf{r}(t) = t \mathbf{i} + t^2 \mathbf{j} + 2 \mathbf{k}$, $t = 1$

8. $\mathbf{r}(t) = t \mathbf{i} + 2 \cos t \, \mathbf{j} + \sin t \, \mathbf{k}$, $t = 0$

9–12 ■ Find the velocity, acceleration, and speed of a particle with the given position function.

9. $\mathbf{r}(t) = \langle t^2 + 1, t^3, t^2 - 1 \rangle$

10. $\mathbf{r}(t) = \langle 2 \cos t, 3t, 2 \sin t \rangle$

11. $\mathbf{r}(t) = \sqrt{2}\, t \, \mathbf{i} + e^t \mathbf{j} + e^{-t} \mathbf{k}$

12. $\mathbf{r}(t) = t \sin t \, \mathbf{i} + t \cos t \, \mathbf{j} + t^2 \mathbf{k}$

13–14 ■ Find the velocity and position vectors of a particle that has the given acceleration and the given initial velocity and position.

13. $\mathbf{a}(t) = \mathbf{i} + 2 \mathbf{j}$, $\mathbf{v}(0) = \mathbf{k}$, $\mathbf{r}(0) = \mathbf{i}$

14. $\mathbf{a}(t) = 2 \mathbf{i} + 6t \mathbf{j} + 12t^2 \mathbf{k}$, $\mathbf{v}(0) = \mathbf{i}$, $\mathbf{r}(0) = \mathbf{j} - \mathbf{k}$

15–16 ■

(a) Find the position vector of a particle that has the given acceleration and the specified initial velocity and position.
(b) Use a computer to graph the path of the particle.

15. $\mathbf{a}(t) = 2t \mathbf{i} + \sin t \, \mathbf{j} + \cos 2t \, \mathbf{k}$, $\mathbf{v}(0) = \mathbf{i}$, $\mathbf{r}(0) = \mathbf{j}$

16. $\mathbf{a}(t) = t \mathbf{i} + e^t \mathbf{j} + e^{-t} \mathbf{k}$, $\mathbf{v}(0) = \mathbf{k}$, $\mathbf{r}(0) = \mathbf{j} + \mathbf{k}$

17. The position function of a particle is given by $\mathbf{r}(t) = \langle t^2, 5t, t^2 - 16t \rangle$. When is the speed a minimum?

18. What force is required so that a particle of mass m has the position function $\mathbf{r}(t) = t^3 \mathbf{i} + t^2 \mathbf{j} + t^3 \mathbf{k}$?

19. A force with magnitude 20 N acts directly upward from the xy-plane on an object with mass 4 kg. The object starts at the origin with initial velocity $\mathbf{v}(0) = \mathbf{i} - \mathbf{j}$. Find its position function and its speed at time t.

20. Show that if a particle moves with constant speed, then the velocity and acceleration vectors are orthogonal.

21. A projectile is fired with an initial speed of 500 m/s and angle of elevation 30°. Find (a) the range of the projectile, (b) the maximum height reached, and (c) the speed at impact.

22. Rework Exercise 21 if the projectile is fired from a position 200 m above the ground.

23. A ball is thrown at an angle of 45° to the ground. If the ball lands 90 m away, what was the initial speed of the ball?

24. A gun is fired with angle of elevation 30°. What is the muzzle speed if the maximum height of the shell is 500 m?

25. A gun has muzzle speed 150 m/s. Find two angles of elevation that can be used to hit a target 800 m away.

26. A batter hits a baseball 3 ft above the ground toward the center field fence, which is 10 ft high and 400 ft from home plate. The ball leaves the bat with speed 115 ft/s at an angle 50° above the horizontal. Is it a home run? (In other words, does the ball clear the fence?)

27. A medieval city has the shape of a square and is protected by walls with length 500 m and height 15 m. You are the commander of an attacking army and the closest you can get to the wall is 100 m. Your plan is to set fire to the city by catapulting heated rocks over the wall (with an initial speed of 80 m/s). At what range of angles should you tell your men to set the catapult? (Assume the path of the rocks is perpendicular to the wall.)

28. A ball with mass 0.8 kg is thrown southward into the air with a speed of 30 m/s at an angle of 30° to the ground. A west wind applies a steady force of 4 N to the ball in an easterly direction. Where does the ball land and with what speed?

29. Water traveling along a straight portion of a river normally flows fastest in the middle, and the speed slows to almost zero at the banks. Consider a long straight stretch of river flowing north, with parallel banks 40 m apart. If the maximum water speed is 3 m/s, we can use a quadratic function as a basic model for the rate of water flow x units from the west bank: $f(x) = \frac{3}{400} x(40 - x)$.
  (a) A boat proceeds at a constant speed of 5 m/s from a point A on the west bank while maintaining a heading perpendicular to the bank. How far down the river on the opposite bank will the boat touch shore? Graph the path of the boat.
  (b) Suppose we would like to pilot the boat to land at the point B on the east bank directly opposite A. If we maintain a constant speed of 5 m/s and a constant heading, find the angle at which the boat should head. Then graph the actual path the boat follows. Does the path seem realistic?

30. Another reasonable model for the water speed of the river in Exercise 29 is a sine function: $f(x) = 3 \sin(\pi x/40)$. If a boater would like to cross the river from A to B with constant heading and a constant speed of 5 m/s, determine the angle at which the boat should head.

31–34 ■ Find the tangential and normal components of the acceleration vector.

31. $\mathbf{r}(t) = (3t - t^3)\,\mathbf{i} + 3t^2\,\mathbf{j}$

32. $\mathbf{r}(t) = (1 + t)\,\mathbf{i} + (t^2 - 2t)\,\mathbf{j}$

33. $\mathbf{r}(t) = \cos t\,\mathbf{i} + \sin t\,\mathbf{j} + t\,\mathbf{k}$

34. $\mathbf{r}(t) = t\,\mathbf{i} + t^2\,\mathbf{j} + 3t\,\mathbf{k}$

. . . .

35. The magnitude of the acceleration vector $\mathbf{a}$ is 10 cm/s^2. Use the figure to estimate the tangential and normal components of $\mathbf{a}$.

36. If a particle with mass m moves with position vector $\mathbf{r}(t)$, then its **angular momentum** is defined as $\mathbf{L}(t) = m\mathbf{r}(t) \times \mathbf{v}(t)$ and its **torque** as $\boldsymbol{\tau}(t) = m\mathbf{r}(t) \times \mathbf{a}(t)$. Show that $\mathbf{L}'(t) = \boldsymbol{\tau}(t)$. Deduce that if $\boldsymbol{\tau}(t) = \mathbf{0}$ for all t, then $\mathbf{L}(t)$ is constant. (This is the *law of conservation of angular momentum*.)

37. The position function of a spaceship is

$$\mathbf{r}(t) = (3 + t)\,\mathbf{i} + (2 + \ln t)\,\mathbf{j} + \left(7 - \frac{4}{t^2 + 1}\right)\mathbf{k}$$

and the coordinates of a space station are $(6, 4, 9)$. The captain wants the spaceship to coast into the space station. When should the engines be turned off?

38. A rocket burning its onboard fuel while moving through space has velocity $\mathbf{v}(t)$ and mass $m(t)$ at time t. If the exhaust gases escape with velocity $\mathbf{v}_e$ relative to the rocket, it can be deduced from Newton's Second Law of Motion that

$$m\,\frac{d\mathbf{v}}{dt} = \frac{dm}{dt}\,\mathbf{v}_e$$

  (a) Show that $\mathbf{v}(t) = \mathbf{v}(0) - \ln \dfrac{m(0)}{m(t)}\,\mathbf{v}_e$.
  (b) For the rocket to accelerate in a straight line from rest to twice the speed of its own exhaust gases, what fraction of its initial mass would the rocket have to burn as fuel?

Kepler's Laws

Johannes Kepler stated the following three laws of planetary motion on the basis of masses of data on the positions of the planets at various times.

> **Kepler's Laws**
>
> **1.** A planet revolves around the Sun in an elliptical orbit with the Sun at one focus.
>
> **2.** The line joining the Sun to a planet sweeps out equal areas in equal times.
>
> **3.** The square of the period of revolution of a planet is proportional to the cube of the length of the major axis of its orbit.

Kepler formulated these laws because they fitted the astronomical data. He wasn't able to see why they were true or how they related to each other. But Sir Isaac Newton, in his *Principia Mathematica* of 1687, showed how to deduce Kepler's three laws from two of Newton's own laws, the Second Law of Motion and the Law of Universal Gravitation. In Section 10.4 we proved Kepler's First Law using the calculus of vector functions. In this project we guide you through the proofs of Kepler's Second and Third Laws and explore some of their consequences.

1. Use the following steps to prove Kepler's Second Law. The notation is the same as in the proof of the First Law in Section 10.4. In particular, use polar coordinates so that $\mathbf{r} = (r \cos \theta)\, \mathbf{i} + (r \sin \theta)\, \mathbf{j}$.

   (a) Show that $\mathbf{h} = r^2 \dfrac{d\theta}{dt}\, \mathbf{k}$.

   (b) Deduce that $r^2 \dfrac{d\theta}{dt} = h$.

   (c) If $A = A(t)$ is the area swept out by the radius vector $\mathbf{r} = \mathbf{r}(t)$ in the time interval $[t_0, t]$ as in the figure, show that

   $$\frac{dA}{dt} = \tfrac{1}{2} r^2 \frac{d\theta}{dt}$$

   (d) Deduce that

   $$\frac{dA}{dt} = \tfrac{1}{2} h = \text{constant}$$

   This says that the rate at which A is swept out is constant and proves Kepler's Second Law.

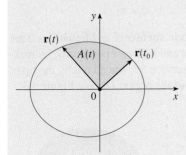

2. Let T be the period of a planet about the Sun; that is, T is the time required for it to travel once around its elliptical orbit. Suppose that the lengths of the major and minor axes of the ellipse are $2a$ and $2b$.

   (a) Use part (d) of Problem 1 to show that $T = 2\pi ab/h$.

   (b) Show that $\dfrac{h^2}{GM} = ed = \dfrac{b^2}{a}$.

   (c) Use parts (a) and (b) to show that $T^2 = \dfrac{4\pi^2}{GM}\, a^3$.

   This proves Kepler's Third Law. [Notice that the proportionality constant $4\pi^2/(GM)$ is independent of the planet.]

3. The period of the Earth's orbit is approximately 365.25 days. Use this fact and Kepler's Third Law to find the length of the major axis of the Earth's orbit. You will need the mass of the Sun, $M = 1.99 \times 10^{30}$ kg, and the gravitational constant, $G = 6.67 \times 10^{-11}$ N·m²/kg².

4. It's possible to place a satellite into orbit about the Earth so that it remains fixed above a given location on the equator. Compute the altitude that is needed for such a satellite. The Earth's mass is 5.98×10^{24} kg; its radius is 6.37×10^6 m. (This orbit is called the Clarke Geosynchronous Orbit after Arthur C. Clarke, who first proposed the idea in 1948. The first such satellite, *Syncom II*, was launched in July 1963.)

10.5 Parametric Surfaces

In Section 9.6 we looked at surfaces that are graphs of functions of two variables. Here we use vector functions to discuss more general surfaces, called *parametric surfaces*.

In much the same way that we describe a space curve by a vector function $\mathbf{r}(t)$ of a single parameter t, we can describe a surface by a vector function $\mathbf{r}(u, v)$ of two parameters u and v. We suppose that

$$\boxed{1} \qquad \mathbf{r}(u, v) = x(u, v)\,\mathbf{i} + y(u, v)\,\mathbf{j} + z(u, v)\,\mathbf{k}$$

is a vector-valued function defined on a region D in the uv-plane. So x, y, and z, the component functions of $\mathbf{r}$, are functions of the two variables u and v with domain D. The set of all points (x, y, z) in $\mathbb{R}^3$ such that

$$\boxed{2} \qquad x = x(u, v) \qquad y = y(u, v) \qquad z = z(u, v)$$

and (u, v) varies throughout D, is called a **parametric surface** S and Equations 2 are called **parametric equations** of S. Each choice of u and v gives a point on S; by making all choices, we get all of S. In other words, the surface S is traced out by the tip of the position vector $\mathbf{r}(u, v)$ as (u, v) moves throughout the region D. (See Figure 1.)

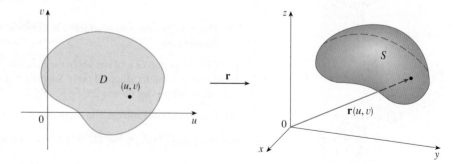

FIGURE 1
A parametric surface

EXAMPLE 1 Identify and sketch the surface with vector equation

$$\mathbf{r}(u, v) = 2\cos u\,\mathbf{i} + v\,\mathbf{j} + 2\sin u\,\mathbf{k}$$

FIGURE 2

FIGURE 3

Visual 10.5 shows animated versions of Figures 4 and 5, with moving grid curves, for several parametric surfaces.

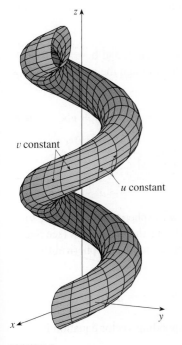

v constant

u constant

FIGURE 5

SOLUTION The parametric equations for this surface are

$$x = 2 \cos u \qquad y = v \qquad z = 2 \sin u$$

So for any point (x, y, z) on the surface, we have

$$x^2 + z^2 = 4 \cos^2 u + 4 \sin^2 u = 4$$

This means that vertical cross-sections parallel to the xz-plane (that is, with y constant) are all circles with radius 2. Since $y = v$ and no restriction is placed on v, the surface is a circular cylinder with radius 2 whose axis is the y-axis (see Figure 2). ■ ■

In Example 1 we placed no restrictions on the parameters u and v and so we got the entire cylinder. If, for instance, we restrict u and v by writing the parameter domain as

$$0 \leqslant u \leqslant \pi/2 \qquad 0 \leqslant v \leqslant 3$$

then $x \geqslant 0, z \geqslant 0, 0 \leqslant y \leqslant 3$, and we get the quarter-cylinder with length 3 illustrated in Figure 3.

If a parametric surface S is given by a vector function $\mathbf{r}(u, v)$, then there are two useful families of curves that lie on S, one family with u constant and the other with v constant. These families correspond to vertical and horizontal lines in the uv-plane. If we keep u constant by putting $u = u_0$, then $\mathbf{r}(u_0, v)$ becomes a vector function of the single parameter v and defines a curve C_1 lying on S. (See Figure 4.)

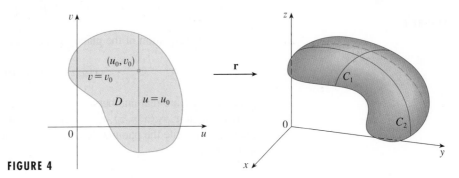

FIGURE 4

Similarly, if we keep v constant by putting $v = v_0$, we get a curve C_2 given by $\mathbf{r}(u, v_0)$ that lies on S. We call these curves **grid curves**. (In Example 1, for instance, the grid curves obtained by letting u be constant are horizontal lines whereas the grid curves with v constant are circles.) In fact, when a computer graphs a parametric surface, it usually depicts the surface by plotting these grid curves, as we see in the following example.

EXAMPLE 2 Use a computer algebra system to graph the surface

$$\mathbf{r}(u, v) = \langle (2 + \sin v) \cos u, (2 + \sin v) \sin u, u + \cos v \rangle$$

Which grid curves have u constant? Which have v constant?

SOLUTION We graph the portion of the surface with parameter domain $0 \leqslant u \leqslant 4\pi$, $0 \leqslant v \leqslant 2\pi$ in Figure 5. It has the appearance of a spiral tube. To identify the grid curves, we write the corresponding parametric equations:

$$x = (2 + \sin v) \cos u \qquad y = (2 + \sin v) \sin u \qquad z = u + \cos v$$

If v is constant, then $\sin v$ and $\cos v$ are constant, so the parametric equations resemble those of the helix in Example 4 in Section 10.1. So the grid curves with v constant are the spiral curves in Figure 5. We deduce that the grid curves with u constant must be the curves that look like circles in the figure. Further evidence for this assertion is that if u is kept constant, $u = u_0$, then the equation $z = u_0 + \cos v$ shows that the z-values vary from $u_0 - 1$ to $u_0 + 1$. ■ ■

In Examples 1 and 2 we were given a vector equation and asked to graph the corresponding parametric surface. In the following examples, however, we are given the more challenging problem of finding a vector function to represent a given surface. In later chapters we will often need to do exactly that.

EXAMPLE 3 Find a vector function that represents the plane that passes through the point P_0 with position vector $\mathbf{r}_0$ and that contains two nonparallel vectors $\mathbf{a}$ and $\mathbf{b}$.

SOLUTION If P is any point in the plane, we can get from P_0 to P by moving a certain distance in the direction of $\mathbf{a}$ and another distance in the direction of $\mathbf{b}$. So there are scalars u and v such that $\overrightarrow{P_0P} = u\mathbf{a} + v\mathbf{b}$. (Figure 6 illustrates how this works, by means of the Parallelogram Law, for the case where u and v are positive. See also Exercise 30 in Section 9.2.) If $\mathbf{r}$ is the position vector of P, then

$$\mathbf{r} = \overrightarrow{OP_0} + \overrightarrow{P_0P} = \mathbf{r}_0 + u\mathbf{a} + v\mathbf{b}$$

FIGURE 6

So the vector equation of the plane can be written as

$$\mathbf{r}(u, v) = \mathbf{r}_0 + u\mathbf{a} + v\mathbf{b}$$

where u and v are real numbers.

If we write $\mathbf{r} = \langle x, y, z \rangle$, $\mathbf{r}_0 = \langle x_0, y_0, z_0 \rangle$, $\mathbf{a} = \langle a_1, a_2, a_3 \rangle$, and $\mathbf{b} = \langle b_1, b_2, b_3 \rangle$, then we can write the parametric equations of the plane through the point (x_0, y_0, z_0) as follows:

$$x = x_0 + ua_1 + vb_1 \qquad y = y_0 + ua_2 + vb_2 \qquad z = z_0 + ua_3 + vb_3 \qquad ■ ■$$

$\boxed{V}$ **EXAMPLE 4** Find a parametric representation of the sphere

$$x^2 + y^2 + z^2 = a^2$$

SOLUTION The sphere has a simple representation $\rho = a$ in spherical coordinates, so let's choose the angles ϕ and θ in spherical coordinates as the parameters (see Section 9.7). Then, putting $\rho = a$ in the equations for conversion from spherical to rectangular coordinates (Equations 9.7.3), we obtain

$$x = a \sin \phi \cos \theta \qquad y = a \sin \phi \sin \theta \qquad z = a \cos \phi$$

as the parametric equations of the sphere. The corresponding vector equation is

$$\mathbf{r}(\phi, \theta) = a \sin \phi \cos \theta \, \mathbf{i} + a \sin \phi \sin \theta \, \mathbf{j} + a \cos \phi \, \mathbf{k}$$

We have $0 \leqslant \phi \leqslant \pi$ and $0 \leqslant \theta \leqslant 2\pi$, so the parameter domain is the rectangle $D = [0, \pi] \times [0, 2\pi]$. The grid curves with ϕ constant are the circles of constant latitude (including the equator). The grid curves with θ constant are the meridians (semicircles), which connect the north and south poles. ■ ■

■ ■ One of the uses of parametric surfaces is in computer graphics. Figure 7 shows the result of trying to graph the sphere $x^2 + y^2 + z^2 = 1$ by solving the equation for z and graphing the top and bottom hemispheres separately. Part of the sphere appears to be missing because of the rectangular grid system used by the computer. The much better picture in Figure 8 was produced by a computer using the parametric equations found in Example 4.

FIGURE 7

FIGURE 8

EXAMPLE 5 Find a parametric representation for the cylinder

$$x^2 + y^2 = 4 \qquad 0 \leqslant z \leqslant 1$$

SOLUTION The cylinder has a simple representation $r = 2$ in cylindrical coordinates, so we choose as parameters θ and z in cylindrical coordinates. Then the parametric equations of the cylinder are

$$x = 2\cos\theta \qquad y = 2\sin\theta \qquad z = z$$

where $0 \leqslant \theta \leqslant 2\pi$ and $0 \leqslant z \leqslant 1$. ■ ■

V $\boxed{V}$ **EXAMPLE 6** Find a vector function that represents the elliptic paraboloid $z = x^2 + 2y^2$.

SOLUTION If we regard x and y as parameters, then the parametric equations are simply

$$x = x \qquad y = y \qquad z = x^2 + 2y^2$$

and the vector equation is

$$\mathbf{r}(x, y) = x\,\mathbf{i} + y\,\mathbf{j} + (x^2 + 2y^2)\,\mathbf{k}$$ ■ ■

In Module 10.5 you can investigate several families of parametric surfaces.

In general, a surface given as the graph of a function of x and y, that is, with an equation of the form $z = f(x, y)$, can always be regarded as a parametric surface by taking x and y as parameters and writing the parametric equations as

$$x = x \qquad y = y \qquad z = f(x, y)$$

Parametric representations (also called parametrizations) of surfaces are not unique. The next example shows two ways to parametrize a cone.

EXAMPLE 7 Find a parametric representation for the surface $z = 2\sqrt{x^2 + y^2}$, that is, the top half of the cone $z^2 = 4x^2 + 4y^2$.

SOLUTION 1 One possible representation is obtained by choosing x and y as parameters:

$$x = x \qquad y = y \qquad z = 2\sqrt{x^2 + y^2}$$

So the vector equation is

$$\mathbf{r}(x, y) = x\,\mathbf{i} + y\,\mathbf{j} + 2\sqrt{x^2 + y^2}\,\mathbf{k}$$

■ ■ For some purposes the parametric representations in Solutions 1 and 2 are equally good, but Solution 2 might be preferable in certain situations. If we are interested only in the part of the cone that lies below the plane $z = 1$, for instance, all we have to do in Solution 2 is change the parameter domain to

$$0 \le r \le \tfrac{1}{2} \qquad 0 \le \theta \le 2\pi$$

SOLUTION 2 Another representation results from choosing as parameters the polar coordinates r and θ. A point (x, y, z) on the cone satisfies $x = r\cos\theta$, $y = r\sin\theta$, and $z = 2\sqrt{x^2 + y^2} = 2r$. So a vector equation for the cone is

$$\mathbf{r}(r, \theta) = r\cos\theta\,\mathbf{i} + r\sin\theta\,\mathbf{j} + 2r\,\mathbf{k}$$

where $r \ge 0$ and $0 \le \theta \le 2\pi$.   ■ ■

▪▪ Surfaces of Revolution

Surfaces of revolution can be represented parametrically and thus graphed using a computer. For instance, let's consider the surface S obtained by rotating the curve $y = f(x)$, $a \le x \le b$, about the x-axis, where $f(x) \ge 0$. Let θ be the angle of rotation as shown in Figure 9. If (x, y, z) is a point on S, then

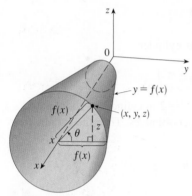

3
$$x = x \qquad y = f(x)\cos\theta \qquad z = f(x)\sin\theta$$

Therefore, we take x and θ as parameters and regard Equations 3 as parametric equations of S. The parameter domain is given by $a \le x \le b$, $0 \le \theta \le 2\pi$.

EXAMPLE 8 Find parametric equations for the surface generated by rotating the curve $y = \sin x$, $0 \le x \le 2\pi$, about the x-axis. Use these equations to graph the surface of revolution.

FIGURE 9

SOLUTION From Equations 3, the parametric equations are

$$x = x \qquad y = \sin x\cos\theta \qquad z = \sin x\sin\theta$$

and the parameter domain is $0 \le x \le 2\pi$, $0 \le \theta \le 2\pi$. Using a computer to plot these equations and rotate the image, we obtain the graph in Figure 10.   ■ ■

We can adapt Equations 3 to represent a surface obtained through revolution about the y- or z-axis. (See Exercise 28.)

FIGURE 10

10.5 Exercises

1–4 ■ Identify the surface with the given vector equation.

1. $\mathbf{r}(u, v) = (u + v)\,\mathbf{i} + (3 - v)\,\mathbf{j} + (1 + 4u + 5v)\,\mathbf{k}$

2. $\mathbf{r}(u, v) = 2\sin u\,\mathbf{i} + 3\cos u\,\mathbf{j} + v\,\mathbf{k}$, $\quad 0 \le v \le 2$

3. $\mathbf{r}(s, t) = \langle s, t, t^2 - s^2 \rangle$

4. $\mathbf{r}(s, t) = \langle s\sin 2t, s^2, s\cos 2t \rangle$

5–10 ■ Use a computer to graph the parametric surface. Get a printout and indicate on it which grid curves have u constant and which have v constant.

5. $\mathbf{r}(u, v) = \langle u^2 + 1, v^3 + 1, u + v \rangle$,
  $-1 \le u \le 1$, $-1 \le v \le 1$

6. $\mathbf{r}(u, v) = \langle u + v, u^2, v^2 \rangle$,
  $-1 \le u \le 1$, $-1 \le v \le 1$

7. $\mathbf{r}(u, v) = \langle \cos^3 u \cos^3 v, \sin^3 u \cos^3 v, \sin^3 v \rangle$,
$0 \le u \le \pi,\ 0 \le v \le 2\pi$

8. $\mathbf{r}(u, v) = \langle \cos u \sin v, \sin u \sin v, \cos v + \ln \tan(v/2) \rangle$,
$0 \le u \le 2\pi,\ 0.1 \le v \le 6.2$

9. $x = \cos u \sin 2v,\quad y = \sin u \sin 2v,\quad z = \sin v$

10. $x = u \sin u \cos v,\quad y = u \cos u \cos v,\quad z = u \sin v$

11–16 ■ Match the equations with the graphs labeled I–VI and give reasons for your answers. Determine which families of grid curves have u constant and which have v constant.

11. $\mathbf{r}(u, v) = \cos v\,\mathbf{i} + \sin v\,\mathbf{j} + u\,\mathbf{k}$

12. $\mathbf{r}(u, v) = u \cos v\,\mathbf{i} + u \sin v\,\mathbf{j} + u\,\mathbf{k}$

13. $\mathbf{r}(u, v) = u \cos v\,\mathbf{i} + u \sin v\,\mathbf{j} + v\,\mathbf{k}$

14. $x = u^3,\quad y = u \sin v,\quad z = u \cos v$

15. $x = (u - \sin u) \cos v,\quad y = (1 - \cos u) \sin v,\quad z = u$

16. $x = (1 - u)(3 + \cos v) \cos 4\pi u$,
$\ y = (1 - u)(3 + \cos v) \sin 4\pi u$,
$\ z = 3u + (1 - u) \sin v$

I

II

III

IV

V

VI

17–24 ■ Find a parametric representation for the surface.

17. The plane that passes through the point $(1, 2, -3)$ and contains the vectors $\mathbf{i} + \mathbf{j} - \mathbf{k}$ and $\mathbf{i} - \mathbf{j} + \mathbf{k}$

18. The lower half of the ellipsoid $2x^2 + 4y^2 + z^2 = 1$

19. The part of the hyperboloid $x^2 + y^2 - z^2 = 1$ that lies to the right of the xz-plane

20. The part of the elliptic paraboloid $x + y^2 + 2z^2 = 4$ that lies in front of the plane $x = 0$

21. The part of the sphere $x^2 + y^2 + z^2 = 4$ that lies above the cone $z = \sqrt{x^2 + y^2}$

22. The part of the sphere $x^2 + y^2 + z^2 = 16$ that lies between the planes $z = -2$ and $z = 2$

23. The part of the cylinder $y^2 + z^2 = 16$ that lies between the planes $x = 0$ and $x = 5$

24. The part of the plane $z = x + 3$ that lies inside the cylinder $x^2 + y^2 = 1$

CAS **25–26** ■ Use a computer algebra system to produce a graph that looks like the given one.

25. **26.**
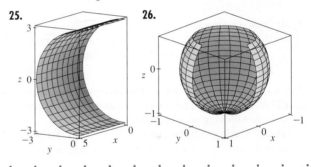

---

27. Find parametric equations for the surface obtained by rotating the curve $y = e^{-x}$, $0 \le x \le 3$, about the x-axis and use them to graph the surface.

28. Find parametric equations for the surface obtained by rotating the curve $x = 4y^2 - y^4$, $-2 \le y \le 2$, about the y-axis and use them to graph the surface.

29. (a) Show that the parametric equations $x = a \sin u \cos v$, $y = b \sin u \sin v$, $z = c \cos u$, $0 \le u \le \pi$, $0 \le v \le 2\pi$, represent an ellipsoid.

(b) Use the parametric equations in part (a) to graph the ellipsoid for the case $a = 1$, $b = 2$, $c = 3$.

30. The surface with parametric equations

$$x = 2 \cos \theta + r \cos(\theta/2)$$

$$y = 2 \sin \theta + r \cos(\theta/2)$$

$$z = r \sin(\theta/2)$$

where $-\frac{1}{2} \le r \le \frac{1}{2}$ and $0 \le \theta \le 2\pi$, is called a **Möbius**

strip. Graph this surface with several viewpoints. What is unusual about it?

31. (a) What happens to the spiral tube in Example 2 (see Figure 5) if we replace cos u by sin u and sin u by cos u?
(b) What happens if we replace cos u by cos $2u$ and sin u by sin $2u$?

32. (a) Find a parametric representation for the torus obtained by rotating about the z-axis the circle in the xz-plane with center $(b, 0, 0)$ and radius $a < b$. [*Hint:* Take as parameters the angles θ and α shown in the figure.]
(b) Use the parametric equations found in part (a) to graph the torus for several values of a and b.

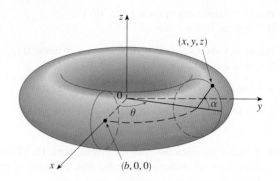

10 Review

CONCEPT CHECK

1. What is a vector function? How do you find its derivative and its integral?

2. What is the connection between vector functions and space curves?

3. (a) What is a smooth curve?
(b) How do you find the tangent vector to a smooth curve at a point? How do you find the tangent line? The unit tangent vector?

4. If **u** and **v** are differentiable vector functions, c is a scalar, and f is a real-valued function, write the rules for differentiating the following vector functions.
(a) $\mathbf{u}(t) + \mathbf{v}(t)$ (b) $c\mathbf{u}(t)$ (c) $f(t)\mathbf{u}(t)$
(d) $\mathbf{u}(t) \cdot \mathbf{v}(t)$ (e) $\mathbf{u}(t) \times \mathbf{v}(t)$ (f) $\mathbf{u}(f(t))$

5. How do you find the length of a space curve given by a vector function $\mathbf{r}(t)$?

6. (a) What is the definition of curvature?
(b) Write a formula for curvature in terms of $\mathbf{r}'(t)$ and $\mathbf{T}'(t)$.
(c) Write a formula for curvature in terms of $\mathbf{r}'(t)$ and $\mathbf{r}''(t)$.
(d) Write a formula for the curvature of a plane curve with equation $y = f(x)$.

7. (a) Write formulas for the unit normal and binormal vectors of a smooth space curve $\mathbf{r}(t)$.
(b) What is the normal plane of a curve at a point? What is the osculating plane? What is the osculating circle?

8. (a) How do you find the velocity, speed, and acceleration of a particle that moves along a space curve?
(b) Write the acceleration in terms of its tangential and normal components.

9. State Kepler's Laws.

10. What is a parametric surface? What are its grid curves?

TRUE-FALSE QUIZ

Determine whether the statement is true or false. If it is true, explain why. If it is false, explain why or give an example that disproves the statement.

1. The curve with vector equation $\mathbf{r}(t) = t^3\mathbf{i} + 2t^3\mathbf{j} + 3t^3\mathbf{k}$ is a line.

2. The curve with vector equation $\mathbf{r}(t) = \langle t, t^3, t^5 \rangle$ is smooth.

3. The curve with vector equation $\mathbf{r}(t) = \langle \cos t, t^2, t^4 \rangle$ is smooth.

4. The derivative of a vector function is obtained by differentiating each component function.

5. If $\mathbf{u}(t)$ and $\mathbf{v}(t)$ are differentiable vector functions, then
$$\frac{d}{dt}[\mathbf{u}(t) \times \mathbf{v}(t)] = \mathbf{u}'(t) \times \mathbf{v}'(t)$$

6. If $\mathbf{r}(t)$ is a differentiable vector function, then
$$\frac{d}{dt}|\mathbf{r}(t)| = |\mathbf{r}'(t)|$$

7. If $\mathbf{T}(t)$ is the unit tangent vector of a smooth curve, then the curvature is $\kappa = |d\mathbf{T}/dt|$.

8. The binormal vector is $\mathbf{B}(t) = \mathbf{N}(t) \times \mathbf{T}(t)$.

9. The osculating circle of a curve C at a point has the same tangent vector, normal vector, and curvature as C at that point.

10. Different parametrizations of the same curve result in identical tangent vectors at a given point on the curve.

EXERCISES

1. (a) Sketch the curve with vector function

$$\mathbf{r}(t) = t\,\mathbf{i} + \cos \pi t\,\mathbf{j} + \sin \pi t\,\mathbf{k} \qquad t \geq 0$$

(b) Find $\mathbf{r}'(t)$ and $\mathbf{r}''(t)$.

2. Let $\mathbf{r}(t) = \langle \sqrt{2-t}, (e^t - 1)/t, \ln(t+1) \rangle$.

(a) Find the domain of $\mathbf{r}$.

(b) Find $\lim_{t \to 0} \mathbf{r}(t)$.

(c) Find $\mathbf{r}'(t)$.

3. Find a vector function that represents the curve of intersection of the cylinder $x^2 + y^2 = 16$ and the plane $x + z = 5$.

4. Find parametric equations for the tangent line to the curve $x = 2 \sin t$, $y = 2 \sin 2t$, $z = 2 \sin 3t$ at the point $(1, \sqrt{3}, 2)$. Graph the curve and the tangent line on a common screen.

5. If $\mathbf{r}(t) = t^2\,\mathbf{i} + t \cos \pi t\,\mathbf{j} + \sin \pi t\,\mathbf{k}$, evaluate $\int_0^1 \mathbf{r}(t)\,dt$.

6. Let C be the curve with equations $x = 2 - t^3$, $y = 2t - 1$, $z = \ln t$. Find (a) the point where C intersects the xz-plane, (b) parametric equations of the tangent line at $(1, 1, 0)$, and (c) an equation of the normal plane to C at $(1, 1, 0)$.

7. Use Simpson's Rule with $n = 6$ to estimate the length of the arc of the curve with equations $x = t^2$, $y = t^3$, $z = t^4$, $0 \leq t \leq 3$.

8. Find the length of the curve $\mathbf{r}(t) = \langle 2t^{3/2}, \cos 2t, \sin 2t \rangle$, $0 \leq t \leq 1$.

9. The helix $\mathbf{r}_1(t) = \cos t\,\mathbf{i} + \sin t\,\mathbf{j} + t\,\mathbf{k}$ intersects the curve $\mathbf{r}_2(t) = (1 + t)\mathbf{i} + t^2\,\mathbf{j} + t^3\,\mathbf{k}$ at the point $(1, 0, 0)$. Find the angle of intersection of these curves.

10. Reparametrize the curve $\mathbf{r}(t) = e^t\,\mathbf{i} + e^t \sin t\,\mathbf{j} + e^t \cos t\,\mathbf{k}$ with respect to arc length measured from the point $(1, 0, 1)$ in the direction of increasing t.

11. For the curve given by $\mathbf{r}(t) = \langle \frac{1}{3}t^3, \frac{1}{2}t^2, t \rangle$, find (a) the unit tangent vector, (b) the unit normal vector, and (c) the curvature.

12. Find the curvature of the ellipse $x = 3 \cos t$, $y = 4 \sin t$ at the points $(3, 0)$ and $(0, 4)$.

13. Find the curvature of the curve $y = x^4$ at the point $(1, 1)$.

14. Find an equation of the osculating circle of the curve $y = x^4 - x^2$ at the origin. Graph both the curve and its osculating circle.

15. Find an equation of the osculating plane of the curve $x = \sin 2t$, $y = t$, $z = \cos 2t$ at the point $(0, \pi, 1)$.

16. The figure shows the curve C traced by a particle with position vector $\mathbf{r}(t)$ at time t.

(a) Draw a vector that represents the average velocity of the particle over the time interval $3 \leq t \leq 3.2$.

(b) Write an expression for the velocity $\mathbf{v}(3)$.

(c) Write an expression for the unit tangent vector $\mathbf{T}(3)$ and draw it.

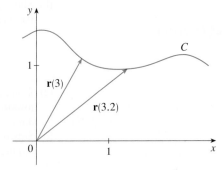

17. A particle moves with position function $\mathbf{r}(t) = t \ln t\,\mathbf{i} + t\,\mathbf{j} + e^{-t}\,\mathbf{k}$. Find the velocity, speed, and acceleration of the particle.

18. A particle starts at the origin with initial velocity $\mathbf{i} - \mathbf{j} + 3\mathbf{k}$. Its acceleration is $\mathbf{a}(t) = 6t\,\mathbf{i} + 12t^2\,\mathbf{j} - 6t\,\mathbf{k}$. Find its position function.

19. An athlete throws a shot at an angle of $45°$ to the horizontal at an initial speed of 43 ft/s. It leaves his hand 7 ft above the ground.

(a) Where is the shot 2 seconds later?

(b) How high does the shot go?

(c) Where does the shot land?

20. Find the tangential and normal components of the acceleration vector of a particle with position function

$$\mathbf{r}(t) = t\,\mathbf{i} + 2t\,\mathbf{j} + t^2\,\mathbf{k}$$

21. Find a parametric representation for the part of the sphere $x^2 + y^2 + z^2 = 4$ that lies between the planes $z = 1$ and $z = -1$.

22. Use a computer to graph the surface with vector equation

$$\mathbf{r}(u, v) = \langle (1 - \cos u) \sin v, u, (u - \sin u) \cos v \rangle$$

Get a printout that gives a good view of the surface and indicate on it which grid curves have u constant and which have v constant.

23. Find the curvature of the curve with parametric equations

$$x = \int_0^t \sin\left(\tfrac{1}{2} \pi \theta^2\right) d\theta \qquad y = \int_0^t \cos\left(\tfrac{1}{2} \pi \theta^2\right) d\theta$$

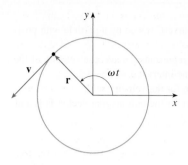

FIGURE FOR PROBLEM 1

1. A particle P moves with constant angular speed ω around a circle whose center is at the origin and whose radius is R. The particle is said to be in *uniform circular motion*. Assume that the motion is counterclockwise and that the particle is at the point $(R, 0)$ when $t = 0$. The position vector at time $t \geqslant 0$ is

$$\mathbf{r}(t) = R \cos \omega t \, \mathbf{i} + R \sin \omega t \, \mathbf{j}$$

(a) Find the velocity vector $\mathbf{v}$ and show that $\mathbf{v} \cdot \mathbf{r} = 0$. Conclude that $\mathbf{v}$ is tangent to the circle and points in the direction of the motion.

(b) Show that the speed $|\mathbf{v}|$ of the particle is the constant ωR. The *period* T of the particle is the time required for one complete revolution. Conclude that

$$T = \frac{2\pi R}{|\mathbf{v}|} = \frac{2\pi}{\omega}$$

(c) Find the acceleration vector $\mathbf{a}$. Show that it is proportional to $\mathbf{r}$ and that it points toward the origin. An acceleration with this property is called a *centripetal acceleration*. Show that the magnitude of the acceleration vector is $|\mathbf{a}| = R\omega^2$.

(d) Suppose that the particle has mass m. Show that the magnitude of the force $\mathbf{F}$ that is required to produce this motion, called a *centripetal force*, is

$$|\mathbf{F}| = \frac{m|\mathbf{v}|^2}{R}$$

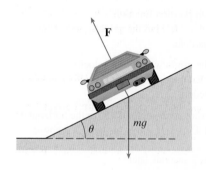

FIGURE FOR PROBLEM 2

2. A circular curve of radius R on a highway is banked at an angle θ so that a car can safely traverse the curve without skidding when there is no friction between the road and the tires. The loss of friction could occur, for example, if the road is covered with a film of water or ice. The rated speed v_R of the curve is the maximum speed that a car can attain without skidding. Suppose a car of mass m is traversing the curve at the rated speed v_R. Two forces are acting on the car: the vertical force, mg, due to the weight of the car, and a force $\mathbf{F}$ exerted by, and normal to, the road. (See the figure.)

The vertical component of $\mathbf{F}$ balances the weight of the car, so that $|\mathbf{F}| \cos \theta = mg$. The horizontal component of $\mathbf{F}$ produces a centripetal force on the car so that, by Newton's Second Law and part (d) of Problem 1,

$$|\mathbf{F}| \sin \theta = \frac{mv_R^2}{R}$$

(a) Show that $v_R^2 = Rg \tan \theta$.

(b) Find the rated speed of a circular curve with radius 400 ft that is banked at an angle of 12°.

(c) Suppose the design engineers want to keep the banking at 12°, but wish to increase the rated speed by 50%. What should the radius of the curve be?

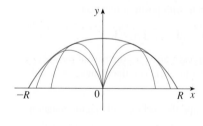

3. A projectile is fired from the origin with angle of elevation α and initial speed v_0. Assuming that air resistance is negligible and that the only force acting on the projectile is gravity, g, we showed in Example 5 in Section 10.4 that the position vector of the projectile is

$$\mathbf{r}(t) = (v_0 \cos \alpha) t \, \mathbf{i} + \left[(v_0 \sin \alpha) t - \tfrac{1}{2} g t^2 \right] \mathbf{j}$$

We also showed that the maximum horizontal distance of the projectile is achieved when $\alpha = 45°$ and in this case the range is $R = v_0^2/g$.

(a) At what angle should the projectile be fired to achieve maximum height and what is the maximum height?

(b) Fix the initial speed v_0 and consider the parabola $x^2 + 2Ry - R^2 = 0$, whose graph is shown in the figure. Show that the projectile can hit any target inside or on the

FIGURE FOR PROBLEM 3

boundary of the region bounded by the parabola and the x-axis, and that it can't hit any target outside this region.

(c) Suppose that the gun is elevated to an angle of inclination α in order to aim at a target that is suspended at a height h directly over a point D units downrange. The target is released at the instant the gun is fired. Show that the projectile always hits the target, regardless of the value v_0, provided the projectile does not hit the ground "before" D.

FIGURE FOR PROBLEM 4

4. (a) A projectile is fired from the origin down an inclined plane that makes an angle θ with the horizontal. The angle of elevation of the gun and the initial speed of the projectile are α and v_0, respectively. Find the position vector of the projectile and the parametric equations of the path of the projectile as functions of the time t. (Ignore air resistance.)

(b) Show that the angle of elevation α that will maximize the downhill range is the angle halfway between the plane and the vertical.

(c) Suppose the projectile is fired up an inclined plane whose angle of inclination is θ. Show that, in order to maximize the (uphill) range, the projectile should be fired in the direction halfway between the plane and the vertical.

(d) In a paper presented in 1686, Edmond Halley summarized the laws of gravity and projectile motion and applied them to gunnery. One problem he posed involved firing a projectile to hit a target a distance R up an inclined plane. Show that the angle at which the projectile should be fired to hit the target but use the least amount of energy is the same as the angle in part (c). (Use the fact that the energy needed to fire the projectile is proportional to the square of the initial speed, so minimizing the energy is equivalent to minimizing the initial speed.)

5. A ball rolls off a table with a speed of 2 ft/s. The table is 3.5 ft high.

(a) Determine the point at which the ball hits the floor and find its speed at the instant of impact.

(b) Find the angle θ between the path of the ball and the vertical line drawn through the point of impact. (See the figure.)

(c) Suppose the ball rebounds from the floor at the same angle with which it hits the floor, but loses 20% of its speed due to energy absorbed by the ball on impact. Where does the ball strike the floor on the second bounce?

FIGURE FOR PROBLEM 5

6. Investigate the shape of the surface with parametric equations

$$x = \sin u \qquad y = \sin v \qquad z = \sin(u + v)$$

Start by graphing the surface from several points of view. Explain the appearance of the graphs by determining the traces in the horizontal planes $z = 0$, $z = \pm 1$, and $z = \pm \frac{1}{2}$.

7. If a projectile is fired with angle of elevation α and initial speed v, then parametric equations for its trajectory are

$$x = (v \cos \alpha)t \qquad y = (v \sin \alpha)t - \tfrac{1}{2}gt^2$$

(See Example 5 in Section 10.4.) We know that the range (horizontal distance traveled) is maximized when $\alpha = 45°$. What value of α maximizes the total distance traveled by the projectile? (State your answer correct to the nearest degree.)

8. A cable has radius r and length L and is wound around a spool with radius R without overlapping. What is the shortest length along the spool that is covered by the cable?

11

Partial Derivatives

Physical quantities often depend on two or more variables. In this chapter
we extend the basic ideas of differential calculus to such functions.

11.1 Functions of Several Variables

In Section 9.6 we discussed functions of two variables and their graphs. Here we study functions of two or more variables from four points of view:

- □ verbally (by a description in words)
- □ numerically (by a table of values)
- □ algebraically (by an explicit formula)
- □ visually (by a graph or level curves)

Recall that a function f of two variables is a rule that assigns to each ordered pair (x, y) of real numbers in its domain a unique real number denoted by $f(x, y)$. In Example 3 in Section 9.6 we looked at the wave heights h in the open sea as a function of the wind speed v and the length of time t that the wind has been blowing at that speed. We presented a table of observed wave heights that represent the function $h = f(v, t)$ numerically. The function in the next example is also described verbally and numerically.

EXAMPLE 1 In regions with severe winter weather, the *wind-chill index* is often used to describe the apparent severity of the cold. This index W is a subjective temperature that depends on the actual temperature T and the wind speed v. So W is a function of T and v, and we can write $W = f(T, v)$. Table 1 records values of W compiled by the U.S. National Weather Service and the Meteorological Service of Canada.

TABLE 1

Wind-chill index as a function of air temperature and wind speed

■ ■ THE NEW WIND-CHILL INDEX

A new wind-chill index was introduced in November of 2001 and is more accurate than the old index at measuring how cold it feels when it's windy. The new index is based on a model of how fast a human face loses heat. It was developed through clinical trials in which volunteers were exposed to a variety of temperatures and wind speeds in a refrigerated wind tunnel.

Wind speed (km/h)

T $\diagdown$ v	5	10	15	20	25	30	40	50	60	70	80
5	4	3	2	1	1	0	−1	−1	−2	−2	−3
0	−2	−3	−4	−5	−6	−6	−7	−8	−9	−9	−10
−5	−7	−9	−11	−12	−12	−13	−14	−15	−16	−16	−17
−10	−13	−15	−17	−18	−19	−20	−21	−22	−23	−23	−24
−15	−19	−21	−23	−24	−25	−26	−27	−29	−30	−30	−31
−20	−24	−27	−29	−30	−32	−33	−34	−35	−36	−37	−38
−25	−30	−33	−35	−37	−38	−39	−41	−42	−43	−44	−45
−30	−36	−39	−41	−43	−44	−46	−48	−49	−50	−51	−52
−35	−41	−45	−48	−49	−51	−52	−54	−56	−57	−58	−60
−40	−47	−51	−54	−56	−57	−59	−61	−63	−64	−65	−67

Actual temperature (°C)

For instance, the table shows that if the temperature is −5°C and the wind speed is 50 km/h, then subjectively it would feel as cold as a temperature of about −15°C with no wind. So

$$f(-5, 50) = -15$$

■ ■

TABLE 2

Year	P	L	K
1899	100	100	100
1900	101	105	107
1901	112	110	114
1902	122	117	122
1903	124	122	131
1904	122	121	138
1905	143	125	149
1906	152	134	163
1907	151	140	176
1908	126	123	185
1909	155	143	198
1910	159	147	208
1911	153	148	216
1912	177	155	226
1913	184	156	236
1914	169	152	244
1915	189	156	266
1916	225	183	298
1917	227	198	335
1918	223	201	366
1919	218	196	387
1920	231	194	407
1921	179	146	417
1922	240	161	431

EXAMPLE 2 In 1928 Charles Cobb and Paul Douglas published a study in which they modeled the growth of the American economy during the period 1899–1922. They considered a simplified view of the economy in which production output is determined by the amount of labor involved and the amount of capital invested. While there are many other factors affecting economic performance, their model proved to be remarkably accurate. The function they used to model production was of the form

$$\boxed{1} \qquad P(L, K) = bL^{\alpha}K^{1-\alpha}$$

where P is the total production (the monetary value of all goods produced in a year), L is the amount of labor (the total number of person-hours worked in a year), and K is the amount of capital invested (the monetary worth of all machinery, equipment, and buildings). In Section 11.3 we will show how the form of Equation 1 follows from certain economic assumptions.

Cobb and Douglas used economic data published by the government to obtain Table 2. They took the year 1899 as a baseline and P, L, and K for 1899 were each assigned the value 100. The values for other years were expressed as percentages of the 1899 figures.

Cobb and Douglas used the method of least squares to fit the data of Table 2 to the function

$$\boxed{2} \qquad P(L, K) = 1.01L^{0.75}K^{0.25}$$

(See Exercise 45 for the details.)

If we use the model given by the function in Equation 2 to compute the production in the years 1910 and 1920, we get the values

$$P(147, 208) = 1.01(147)^{0.75}(208)^{0.25} \approx 161.9$$

$$P(194, 407) = 1.01(194)^{0.75}(407)^{0.25} \approx 235.8$$

which are quite close to the actual values, 159 and 231.

The production function (1) has subsequently been used in many settings, ranging from individual firms to global economic questions. It has become known as the **Cobb-Douglas production function**. ■ ■

The domain of the production function in Example 2 is $\{(L, K) \mid L \geq 0, K \geq 0\}$ because L and K represent labor and capital and are therefore never negative. For a function f given by an algebraic formula, recall that the domain consists of all pairs (x, y) for which the expression for $f(x, y)$ is a well-defined real number.

EXAMPLE 3 Find the domain and range of $g(x, y) = \sqrt{9 - x^2 - y^2}$.

SOLUTION The domain of g is

$$D = \{(x, y) \mid 9 - x^2 - y^2 \geq 0\} = \{(x, y) \mid x^2 + y^2 \leq 9\}$$

which is the disk with center $(0, 0)$ and radius 3 (see Figure 1). The range of g is

$$\left\{z \mid z = \sqrt{9 - x^2 - y^2}, (x, y) \in D\right\}$$

Since z is a positive square root, $z \geq 0$. Also

$$9 - x^2 - y^2 \leq 9 \quad \Rightarrow \quad \sqrt{9 - x^2 - y^2} \leq 3$$

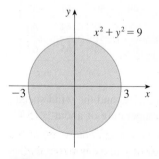

FIGURE 1
Domain of $g(x, y) = \sqrt{9 - x^2 - y^2}$

So the range is

$$\{z \mid 0 \leqslant z \leqslant 3\} = [0, 3]$$ ■ ■

■■ Visual Representations

One way to visualize a function of two variables is through its graph. Recall from Section 9.6 that the graph of f is the surface with equation $z = f(x, y)$.

$\boxed{V}$ **EXAMPLE 4** Sketch the graph of $g(x, y) = \sqrt{9 - x^2 - y^2}$.

SOLUTION The graph has equation $z = \sqrt{9 - x^2 - y^2}$. We square both sides of this equation to obtain $z^2 = 9 - x^2 - y^2$, or $x^2 + y^2 + z^2 = 9$, which we recognize as an equation of the sphere with center the origin and radius 3. But, since $z \geqslant 0$, the graph of g is just the top half of this sphere (see Figure 2). ■ ■

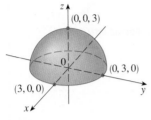

FIGURE 2

Graph of $g(x, y) = \sqrt{9 - x^2 - y^2}$

EXAMPLE 5 Use a computer to draw the graph of the Cobb-Douglas production function $P(L, K) = 1.01 L^{0.75} K^{0.25}$.

SOLUTION Figure 3 shows the graph of P for values of the labor L and capital K that lie between 0 and 300. The computer has drawn the surface by plotting vertical traces. We see from these traces that the value of the production P increases as either L or K increases, as is to be expected.

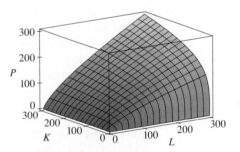

FIGURE 3 ■ ■

Another method for visualizing functions, borrowed from mapmakers, is a contour map on which points of constant elevation are joined to form *contour lines*, or *level curves*.

Definition The **level curves** of a function f of two variables are the curves with equations $f(x, y) = k$, where k is a constant (in the range of f).

A level curve $f(x, y) = k$ is the set of all points in the domain of f at which f takes on a given value k. In other words, it shows where the graph of f has height k.

You can see from Figure 4 the relation between level curves and horizontal traces. The level curves $f(x, y) = k$ are just the traces of the graph of f in the horizontal plane $z = k$ projected down to the xy-plane. So if you draw the level curves of a function and visualize them being lifted up to the surface at the indicated height, then you can mentally piece together a picture of the graph. The surface is steep where the level curves are close together. It is somewhat flatter where they are farther apart.

FIGURE 4

FIGURE 5

Visual 11.1A animates Figure 4 by showing level curves being lifted up to graphs of functions.

One common example of level curves occurs in topographic maps of mountainous regions, such as the map in Figure 5. The level curves are curves of constant elevation above sea level. If you walk along one of these contour lines, you neither ascend nor descend. Another common example is the temperature at locations (x, y) with longitude x and latitude y. Here the level curves are called **isothermals** and join locations with the same temperature. Figure 6 shows a weather map of the world indicating the average January temperatures. The isothermals are the curves that separate the colored bands.

FIGURE 6
World mean sea-level temperatures in January in degrees Celsius

Tarbuck, *Atmosphere: Introduction to Meteorology*, 4th Edition, © 1989. Reprinted by permission of Pearson Education, Inc., Upper Saddle River, NJ.

FIGURE 7

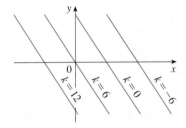

FIGURE 8
Contour map of
$f(x, y) = 6 - 3x - 2y$

EXAMPLE 6 A contour map for a function f is shown in Figure 7. Use it to estimate the values of $f(1, 3)$ and $f(4, 5)$.

SOLUTION The point $(1, 3)$ lies partway between the level curves with z-values 70 and 80. We estimate that

$$f(1, 3) \approx 73$$

Similarly, we estimate that

$$f(4, 5) \approx 56 \qquad ■ ■$$

EXAMPLE 7 Sketch the level curves of the function $f(x, y) = 6 - 3x - 2y$ for the values $k = -6, 0, 6, 12$.

SOLUTION The level curves are

$$6 - 3x - 2y = k \qquad \text{or} \qquad 3x + 2y + (k - 6) = 0$$

This is a family of lines with slope $-\frac{3}{2}$. The four particular level curves with $k = -6, 0, 6$, and 12 are $3x + 2y - 12 = 0$, $3x + 2y - 6 = 0$, $3x + 2y = 0$, and $3x + 2y + 6 = 0$. They are sketched in Figure 8. The level curves are equally spaced parallel lines because the graph of f is a plane (see Figure 4 in Section 9.6). ■ ■

▢V **EXAMPLE 8** Sketch the level curves of the function

$$g(x, y) = \sqrt{9 - x^2 - y^2} \qquad \text{for} \quad k = 0, 1, 2, 3$$

SOLUTION The level curves are

$$\sqrt{9 - x^2 - y^2} = k \qquad \text{or} \qquad x^2 + y^2 = 9 - k^2$$

This is a family of concentric circles with center $(0, 0)$ and radius $\sqrt{9 - k^2}$. The cases $k = 0, 1, 2, 3$ are shown in Figure 9. Try to visualize these level curves lifted up to form a surface and compare with the graph of g (a hemisphere) in Figure 2. (See TEC Visual 11.1A.)

FIGURE 9
Contour map of $g(x, y) = \sqrt{9 - x^2 - y^2}$

■ ■

EXAMPLE 9 Sketch some level curves of the function $h(x, y) = 4x^2 + y^2$.

SOLUTION The level curves are

$$4x^2 + y^2 = k \qquad \text{or} \qquad \frac{x^2}{k/4} + \frac{y^2}{k} = 1$$

which, for $k > 0$, describes a family of ellipses with semiaxes $\sqrt{k}/2$ and $\sqrt{k}$. Figure 10(a) shows a contour map of h drawn by a computer with level curves corresponding to $k = 0.25, 0.5, 0.75, \ldots, 4$. Figure 10(b) shows these level curves lifted up to the graph of h (an elliptic paraboloid) where they become horizontal traces. We see from Figure 10 how the graph of h is put together from the level curves.

Visual 11.1B demonstrates the connection between surfaces and their contour maps.

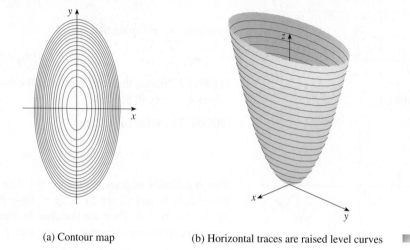

FIGURE 10

The graph of $h(x, y) = 4x^2 + y^2$ is formed by lifting the level curves.

(a) Contour map   (b) Horizontal traces are raised level curves ■ ■

EXAMPLE 10 Plot level curves for the Cobb-Douglas production function of Example 2.

SOLUTION In Figure 11 we use a computer to draw a contour plot for the Cobb-Douglas production function

$$P(L, K) = 1.01 P^{0.75} K^{0.25}$$

Level curves are labeled with the value of the production P. For instance, the level curve labeled 140 shows all values of the labor L and capital investment K that result in a production of $P = 140$. We see that, for a fixed value of P, as L increases K decreases, and vice versa.

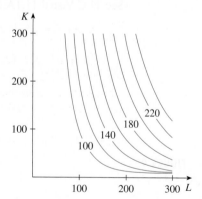

FIGURE 11

For some purposes, a contour map is more useful than a graph. That is certainly true in Example 10. (Compare Figure 11 with Figure 3.) It is also true in estimating function values, as in Example 6.

Figure 12 shows some computer-generated level curves together with the corresponding computer-generated graphs. Notice that the level curves in part (c) crowd together near the origin. That corresponds to the fact that the graph in part (d) is very steep near the origin.

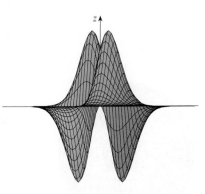

(a) Level curves of $f(x, y) = -xye^{-x^2-y^2}$

(b) Two views of $f(x, y) = -xye^{-x^2-y^2}$

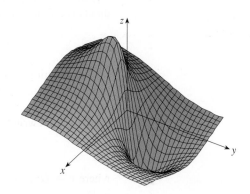

FIGURE 12

(c) Level curves of $f(x, y) = \dfrac{-3y}{x^2 + y^2 + 1}$

(d) $f(x, y) = \dfrac{-3y}{x^2 + y^2 + 1}$

▪▪ Functions of Three or More Variables

A **function of three variables**, f, is a rule that assigns to each ordered triple (x, y, z) in a domain $D \subset \mathbb{R}^3$ a unique real number denoted by $f(x, y, z)$. For instance, the temperature T at a point on the surface of the Earth depends on the longitude x and latitude y of the point and on the time t, so we could write $T = f(x, y, t)$.

EXAMPLE 11 Find the domain of f if

$$f(x, y, z) = \ln(z - y) + xy \sin z$$

SOLUTION The expression for $f(x, y, z)$ is defined as long as $z - y > 0$, so the domain of f is

$$D = \{(x, y, z) \in \mathbb{R}^3 \mid z > y\}$$

This is a **half-space** consisting of all points that lie above the plane $z = y$. ▪▪

It's very difficult to visualize a function f of three variables by its graph, since that would lie in a four-dimensional space. However, we do gain some insight into f by examining its **level surfaces**, which are the surfaces with equations $f(x, y, z) = k$, where k is a constant. If the point (x, y, z) moves along a level surface, the value of $f(x, y, z)$ remains fixed.

EXAMPLE 12 Find the level surfaces of the function

$$f(x, y, z) = x^2 + y^2 + z^2$$

SOLUTION The level surfaces are $x^2 + y^2 + z^2 = k$, where $k \geq 0$. These form a family of concentric spheres with radius $\sqrt{k}$. (See Figure 13.) Thus, as (x, y, z) varies over any sphere with center O, the value of $f(x, y, z)$ remains fixed. ■ ■

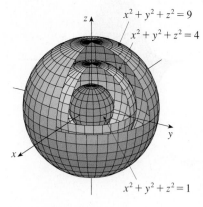

$x^2 + y^2 + z^2 = 9$
$x^2 + y^2 + z^2 = 4$
$x^2 + y^2 + z^2 = 1$

FIGURE 13

Functions of any number of variables can be considered. A **function of n variables** is a rule that assigns a number $z = f(x_1, x_2, \ldots, x_n)$ to an n-tuple $(x_1, x_2, \ldots, x_n)$ of real numbers. We denote by $\mathbb{R}^n$ the set of all such n-tuples. For example, if a company uses n different ingredients in making a food product, c_i is the cost per unit of the ith ingredient, and x_i units of the ith ingredient are used, then the total cost C of the ingredients is a function of the n variables $x_1, x_2, \ldots, x_n$:

$$\boxed{3} \qquad C = f(x_1, x_2, \ldots, x_n) = c_1 x_1 + c_2 x_2 + \cdots + c_n x_n$$

The function f is a real-valued function whose domain is a subset of $\mathbb{R}^n$. Sometimes we will use vector notation in order to write such functions more compactly: If $\mathbf{x} = \langle x_1, x_2, \ldots, x_n \rangle$, we often write $f(\mathbf{x})$ in place of $f(x_1, x_2, \ldots, x_n)$. With this notation we can rewrite the function defined in Equation 3 as

$$f(\mathbf{x}) = \mathbf{c} \cdot \mathbf{x}$$

where $\mathbf{c} = \langle c_1, c_2, \ldots, c_n \rangle$ and $\mathbf{c} \cdot \mathbf{x}$ denotes the dot product of the vectors $\mathbf{c}$ and $\mathbf{x}$ in V_n.

In view of the one-to-one correspondence between points $(x_1, x_2, \ldots, x_n)$ in $\mathbb{R}^n$ and their position vectors $\mathbf{x} = \langle x_1, x_2, \ldots, x_n \rangle$ in V_n, we have three ways of looking at a function f defined on a subset of $\mathbb{R}^n$:

1. As a function of n real variables $x_1, x_2, \ldots, x_n$

2. As a function of a single point variable $(x_1, x_2, \ldots, x_n)$

3. As a function of a single vector variable $\mathbf{x} = \langle x_1, x_2, \ldots, x_n \rangle$

We will see that all three points of view are useful.

11.1 Exercises

1. In Example 1 we considered the function $W = f(T, v)$, where W is the wind-chill index, T is the actual temperature, and v is the wind speed. A numerical representation is given in Table 1.
(a) What is the value of $f(-15, 40)$? What is its meaning?
(b) Describe in words the meaning of the question "For what value of v is $f(-20, v) = -30$?" Then answer the question.

(c) Describe in words the meaning of the question "For what value of T is $f(T, 20) = -49$?" Then answer the question.
(d) What is the meaning of the function $W = f(-5, v)$? Describe the behavior of this function.
(e) What is the meaning of the function $W = f(T, 50)$? Describe the behavior of this function.

2. The *temperature-humidity index I* (or humidex, for short) is the perceived air temperature when the actual temperature is T and the relative humidity is h, so we can write $I = f(T, h)$. The following table of values of I is an excerpt from a table compiled by the National Oceanic and Atmospheric Administration.

TABLE 3 Apparent temperature as a function of temperature and humidity

Relative humidity (%)

T \ h	20	30	40	50	60	70
80	77	78	79	81	82	83
85	82	84	86	88	90	93
90	87	90	93	96	100	106
95	93	96	101	107	114	124
100	99	104	110	120	132	144

Actual temperature (°F)

(a) What is the value of $f(95, 70)$? What is its meaning?
(b) For what value of h is $f(90, h) = 100$?
(c) For what value of T is $f(T, 50) = 88$?
(d) What are the meanings of the functions $I = f(80, h)$ and $I = f(100, h)$? Compare the behavior of these two functions of h.

3. Verify for the Cobb-Douglas production function

$$P(L, K) = 1.01 L^{0.75} K^{0.25}$$

discussed in Example 2 that the production will be doubled if both the amount of labor and the amount of capital are doubled. Is this also true for the general production function $P(L, K) = bL^\alpha K^{1-\alpha}$?

4. The wind-chill index W discussed in Example 1 has been modeled by the following function:

$$W(T, v) = 13.12 + 0.6215T - 11.37v^{0.16} + 0.3965Tv^{0.16}$$

Check to see how closely this model agrees with the values in Table 1 for a few values of T and v.

5. Find and sketch the domain of the function $f(x, y) = \ln(9 - x^2 - 9y^2)$.

6. Find and sketch the domain of the function $f(x, y) = \sqrt{1 + x - y^2}$. What is the range of f?

7. Let $f(x, y, z) = e^{\sqrt{z - x^2 - y^2}}$.
(a) Evaluate $f(2, -1, 6)$.

(b) Find the domain of f.
(c) Find the range of f.

8. Let $g(x, y, z) = \ln(25 - x^2 - y^2 - z^2)$
(a) Evaluate $g(2, -2, 4)$.
(b) Find the domain of g.
(c) Find the range of g.

9. A contour map for a function f is shown. Use it to estimate the values of $f(-3, 3)$ and $f(3, -2)$. What can you say about the shape of the graph?

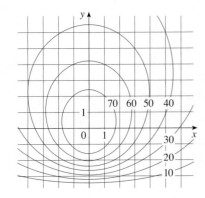

10. Two contour maps are shown. One is for a function f whose graph is a cone. The other is for a function g whose graph is a paraboloid. Which is which, and why?

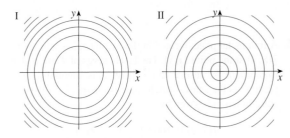

11. Locate the points A and B in the map of Lonesome Mountain (Figure 5). How would you describe the terrain near A? Near B?

12. Make a rough sketch of a contour map for the function whose graph is shown.

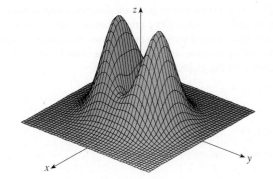

13–14 ■ A contour map of a function is shown. Use it to make a rough sketch of the graph of f.

13.

14.

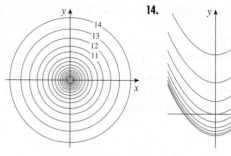

15–22 ■ Draw a contour map of the function showing several level curves.

15. $f(x, y) = (y - 2x)^2$

16. $f(x, y) = x^3 - y$

17. $f(x, y) = y - \ln x$

18. $f(x, y) = e^{y/x}$

19. $f(x, y) = ye^x$

20. $f(x, y) = y \sec x$

21. $f(x, y) = \sqrt{y^2 - x^2}$

22. $f(x, y) = y/(x^2 + y^2)$

23–24 ■ Sketch both a contour map and a graph of the function and compare them.

23. $f(x, y) = x^2 + 9y^2$

24. $f(x, y) = \sqrt{36 - 9x^2 - 4y^2}$

25. A thin metal plate, located in the xy-plane, has temperature $T(x, y)$ at the point (x, y). The level curves of T are called *isothermals* because at all points on an isothermal the temperature is the same. Sketch some isothermals if the temperature function is given by

$$T(x, y) = 100/(1 + x^2 + 2y^2)$$

26. If $V(x, y)$ is the electric potential at a point (x, y) in the xy-plane, then the level curves of V are called *equipotential curves* because at all points on such a curve the electric potential is the same. Sketch some equipotential curves if $V(x, y) = c/\sqrt{r^2 - x^2 - y^2}$, where c is a positive constant.

27–30 ■ Use a computer to graph the function using various domains and viewpoints. Get a printout of one that, in your opinion, gives a good view. If your software also produces level curves, then plot some contour lines of the same function and compare with the graph.

27. $f(x, y) = e^x \cos y$

28. $f(x, y) = (1 - 3x^2 + y^2)e^{1-x^2-y^2}$

29. $f(x, y) = xy^2 - x^3$ (monkey saddle)

30. $f(x, y) = xy^3 - yx^3$ (dog saddle)

31–36 ■ Match the function (a) with its graph (labeled A–F on page 749) and (b) with its contour map (labeled I–VI). Give reasons for your choices.

31. $z = \sin(xy)$

32. $z = e^x \cos y$

33. $z = \sin(x - y)$

34. $z = \sin x - \sin y$

35. $z = (1 - x^2)(1 - y^2)$

36. $z = \dfrac{x - y}{1 + x^2 + y^2}$

37–40 ■ Describe the level surfaces of the function.

37. $f(x, y, z) = x + 3y + 5z$

38. $f(x, y, z) = x^2 + 3y^2 + 5z^2$

39. $f(x, y, z) = x^2 - y^2 + z^2$

40. $f(x, y, z) = x^2 - y^2$

41–42 ■ Describe how the graph of g is obtained from the graph of f.

41. (a) $g(x, y) = f(x, y) + 2$ (b) $g(x, y) = 2f(x, y)$
 (c) $g(x, y) = -f(x, y)$ (d) $g(x, y) = 2 - f(x, y)$

42. (a) $g(x, y) = f(x - 2, y)$ (b) $g(x, y) = f(x, y + 2)$
 (c) $g(x, y) = f(x + 3, y - 4)$

43. Use a computer to investigate the family of functions $f(x, y) = e^{cx^2+y^2}$. How does the shape of the graph depend on c?

44. Graph the functions

$$f(x, y) = \sqrt{x^2 + y^2} \qquad f(x, y) = e^{\sqrt{x^2+y^2}}$$

$$f(x, y) = \ln\sqrt{x^2 + y^2} \qquad f(x, y) = \sin(\sqrt{x^2 + y^2})$$

and $$f(x, y) = \dfrac{1}{\sqrt{x^2 + y^2}}$$

In general, if g is a function of one variable, how is the graph of $f(x, y) = g(\sqrt{x^2 + y^2})$ obtained from the graph of g?

45. (a) Show that, by taking logarithms, the general Cobb-Douglas function $P = bL^\alpha K^{1-\alpha}$ can be expressed as

$$\ln \frac{P}{K} = \ln b + \alpha \ln \frac{L}{K}$$

(b) If we let $x = \ln(L/K)$ and $y = \ln(P/K)$, the equation in part (a) becomes the linear equation $y = \alpha x + \ln b$. Use Table 2 (in Example 2) to make a table of values of $\ln(L/K)$ and $\ln(P/K)$ for the years 1899–1922. Then use a graphing calculator or computer to find the least squares regression line through the points $(\ln(L/K), \ln(P/K))$.

(c) Deduce that the Cobb-Douglas production function is $P = 1.01L^{0.75}K^{0.25}$.

Graphs and Contour Maps for Exercises 31–36

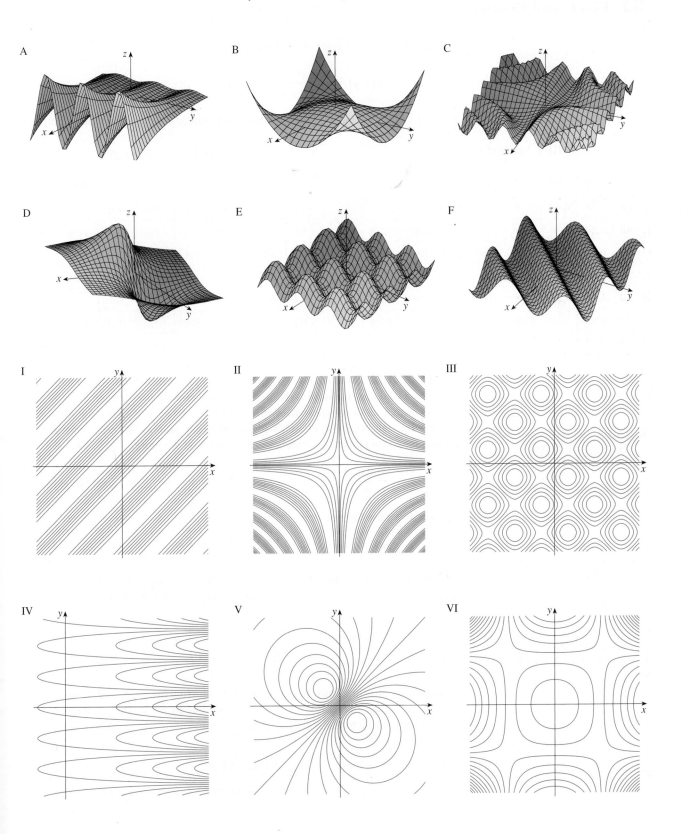

11.2 Limits and Continuity

Let's compare the behavior of the functions

$$f(x, y) = \frac{\sin(x^2 + y^2)}{x^2 + y^2} \quad \text{and} \quad g(x, y) = \frac{x^2 - y^2}{x^2 + y^2}$$

as x and y both approach 0 [and therefore the point (x, y) approaches the origin].

TABLE 1 Values of $f(x, y)$

y \ x	−1.0	−0.5	−0.2	0	0.2	0.5	1.0
−1.0	0.455	0.759	0.829	0.841	0.829	0.759	0.455
−0.5	0.759	0.959	0.986	0.990	0.986	0.959	0.759
−0.2	0.829	0.986	0.999	1.000	0.999	0.986	0.829
0	0.841	0.990	1.000		1.000	0.990	0.841
0.2	0.829	0.986	0.999	1.000	0.999	0.986	0.829
0.5	0.759	0.959	0.986	0.990	0.986	0.959	0.759
1.0	0.455	0.759	0.829	0.841	0.829	0.759	0.455

TABLE 2 Values of $g(x, y)$

y \ x	−1.0	−0.5	−0.2	0	0.2	0.5	1.0
−1.0	0.000	0.600	0.923	1.000	0.923	0.600	0.000
−0.5	−0.600	0.000	0.724	1.000	0.724	0.000	−0.600
−0.2	−0.923	−0.724	0.000	1.000	0.000	−0.724	−0.923
0	−1.000	−1.000	−1.000		−1.000	−1.000	−1.000
0.2	−0.923	−0.724	0.000	1.000	0.000	−0.724	−0.923
0.5	−0.600	0.000	0.724	1.000	0.724	0.000	−0.600
1.0	0.000	0.600	0.923	1.000	0.923	0.600	0.000

Tables 1 and 2 show values of $f(x, y)$ and $g(x, y)$, correct to three decimal places, for points (x, y) near the origin. (Notice that neither function is defined at the origin.) It appears that as (x, y) approaches $(0, 0)$, the values of $f(x, y)$ are approaching 1 whereas the values of $g(x, y)$ aren't approaching any number. It turns out that these guesses based on numerical evidence are correct, and we write

$$\lim_{(x, y) \to (0, 0)} \frac{\sin(x^2 + y^2)}{x^2 + y^2} = 1 \quad \text{and} \quad \lim_{(x, y) \to (0, 0)} \frac{x^2 - y^2}{x^2 + y^2} \quad \text{does not exist}$$

In general, we use the notation

$$\lim_{(x, y) \to (a, b)} f(x, y) = L$$

to indicate that the values of $f(x, y)$ approach the number L as the point (x, y) approaches the point (a, b) along any path that stays within the domain of f.

■ ■ A more precise definition of the limit of a function of two variables is given in Appendix D.

> **1 Definition** We write
> $$\lim_{(x, y) \to (a, b)} f(x, y) = L$$
> and we say that the **limit of $f(x, y)$ as (x, y) approaches (a, b)** is L if we can make the values of $f(x, y)$ as close to L as we like by taking the point (x, y) sufficiently close to the point (a, b), but not equal to (a, b).

Other notations for the limit in Definition 1 are

$$\lim_{\substack{x \to a \\ y \to b}} f(x, y) = L \quad \text{and} \quad f(x, y) \to L \text{ as } (x, y) \to (a, b)$$

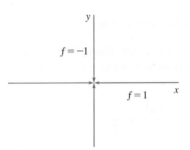

FIGURE 1

For functions of a single variable, when we let x approach a, there are only two possible directions of approach, from the left or from the right. We recall from Chapter 2 that if $\lim_{x \to a^-} f(x) \neq \lim_{x \to a^+} f(x)$, then $\lim_{x \to a} f(x)$ does not exist.

For functions of two variables the situation is not as simple because we can let (x, y) approach (a, b) from an infinite number of directions in any manner whatsoever (see Figure 1) as long as (x, y) stays within the domain of f.

Definition 1 says that the distance between $f(x, y)$ and L can be made arbitrarily small by making the distance from (x, y) to (a, b) sufficiently small (but not 0). The definition refers only to the *distance* between (x, y) and (a, b). It does not refer to the direction of approach. Therefore, if the limit exists, then $f(x, y)$ must approach the same limit no matter how (x, y) approaches (a, b). Thus, if we can find two different paths of approach along which the function $f(x, y)$ has different limits, then it follows that $\lim_{(x, y) \to (a, b)} f(x, y)$ does not exist.

> If $f(x, y) \to L_1$ as $(x, y) \to (a, b)$ along a path C_1 and $f(x, y) \to L_2$ as $(x, y) \to (a, b)$ along a path C_2, where $L_1 \neq L_2$, then $\lim_{(x, y) \to (a, b)} f(x, y)$ does not exist.

$\boxed{V}$ **EXAMPLE 1** Show that $\lim\limits_{(x, y) \to (0, 0)} \dfrac{x^2 - y^2}{x^2 + y^2}$ does not exist.

SOLUTION Let $f(x, y) = (x^2 - y^2)/(x^2 + y^2)$. First let's approach $(0, 0)$ along the x-axis. Then $y = 0$ gives $f(x, 0) = x^2/x^2 = 1$ for all $x \neq 0$, so

$$f(x, y) \to 1 \qquad \text{as} \qquad (x, y) \to (0, 0) \text{ along the } x\text{-axis}$$

We now approach along the y-axis by putting $x = 0$. Then $f(0, y) = \dfrac{-y^2}{y^2} = -1$ for all $y \neq 0$, so

$$f(x, y) \to -1 \qquad \text{as} \qquad (x, y) \to (0, 0) \text{ along the } y\text{-axis}$$

(See Figure 2.) Since f has two different limits along two different lines, the given limit does not exist. (This confirms the conjecture we made on the basis of numerical evidence at the beginning of this section.) ■ ■

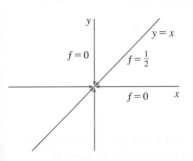

FIGURE 2

EXAMPLE 2 If $f(x, y) = xy/(x^2 + y^2)$, does $\lim\limits_{(x, y) \to (0, 0)} f(x, y)$ exist?

SOLUTION If $y = 0$, then $f(x, 0) = 0/x^2 = 0$. Therefore

$$f(x, y) \to 0 \qquad \text{as} \qquad (x, y) \to (0, 0) \text{ along the } x\text{-axis}$$

If $x = 0$, then $f(0, y) = 0/y^2 = 0$, so

$$f(x, y) \to 0 \qquad \text{as} \qquad (x, y) \to (0, 0) \text{ along the } y\text{-axis}$$

Although we have obtained identical limits along the axes, that does not show that the given limit is 0. Let's now approach $(0, 0)$ along another line, say $y = x$. For all $x \neq 0$,

$$f(x, x) = \frac{x^2}{x^2 + x^2} = \frac{1}{2}$$

Therefore $\qquad f(x, y) \to \frac{1}{2} \qquad \text{as} \qquad (x, y) \to (0, 0)$ along $y = x$

(See Figure 3.) Since we have obtained different limits along different paths, the given limit does not exist. ■ ■

FIGURE 3

Figure 4 sheds some light on Example 2. The ridge that occurs above the line $y = x$ corresponds to the fact that $f(x, y) = \frac{1}{2}$ for all points (x, y) on that line except the origin.

TEC In Visual 11.2 a rotating line on the surface in Figure 4 shows different limits at the origin from different directions.

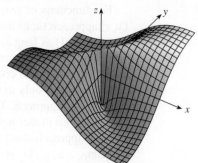

FIGURE 4

$$f(x, y) = \frac{xy}{x^2 + y^2}$$

Ⓥ **EXAMPLE 3** If $f(x, y) = \dfrac{xy^2}{x^2 + y^4}$, does $\displaystyle\lim_{(x, y) \to (0, 0)} f(x, y)$ exist?

SOLUTION With the solution of Example 2 in mind, let's try to save time by letting $(x, y) \to (0, 0)$ along any nonvertical line through the origin. Then $y = mx$, where m is the slope, and

$$f(x, y) = f(x, mx) = \frac{x(mx)^2}{x^2 + (mx)^4} = \frac{m^2x^3}{x^2 + m^4x^4} = \frac{m^2x}{1 + m^4x^2}$$

So $\qquad f(x, y) \to 0 \qquad$ as $\qquad (x, y) \to (0, 0)$ along $y = mx$

■ ■ Figure 5 shows the graph of the function in Example 3. Notice the ridge above the parabola $x = y^2$.

FIGURE 5

Thus, f has the same limiting value along every nonvertical line through the origin. But that does not show that the given limit is 0, for if we now let $(x, y) \to (0, 0)$ along the parabola $x = y^2$, we have

$$f(x, y) = f(y^2, y) = \frac{y^2 \cdot y^2}{(y^2)^2 + y^4} = \frac{y^4}{2y^4} = \frac{1}{2}$$

so $\qquad f(x, y) \to \frac{1}{2} \qquad$ as $\qquad (x, y) \to (0, 0)$ along $x = y^2$

Since different paths lead to different limiting values, the given limit does not exist. ■ ■

Now let's look at limits that *do* exist. Just as for functions of one variable, the calculation of limits for functions of two variables can be greatly simplified by the use of properties of limits. The Limit Laws listed in Section 2.3 can be extended to functions of two variables: The limit of a sum is the sum of the limits, the limit of a product is the product of the limits, and so on. In particular, the following equations are true.

$$\boxed{2} \qquad \lim_{(x, y) \to (a, b)} x = a \qquad \lim_{(x, y) \to (a, b)} y = b \qquad \lim_{(x, y) \to (a, b)} c = c$$

The Squeeze Theorem also holds.

EXAMPLE 4 Find $\displaystyle\lim_{(x, y) \to (0, 0)} \frac{3x^2y}{x^2 + y^2}$ if it exists.

SOLUTION As in Example 3, we could show that the limit along any line through the origin is 0. This doesn't prove that the given limit is 0, but the limits along the parabolas $y = x^2$ and $x = y^2$ also turn out to be 0, so we begin to suspect that the limit does exist and is equal to 0.

To prove it we look at the distance from $f(x, y)$ to 0:

$$\left| \frac{3x^2 y}{x^2 + y^2} - 0 \right| = \left| \frac{3x^2 y}{x^2 + y^2} \right| = \frac{3x^2 |y|}{x^2 + y^2}$$

Notice that $x^2 \leqslant x^2 + y^2$ because $y^2 \geqslant 0$. So

$$\frac{x^2}{x^2 + y^2} \leqslant 1$$

Thus

$$0 \leqslant \frac{3x^2 |y|}{x^2 + y^2} \leqslant 3|y|$$

Now we use the Squeeze Theorem. Since

$$\lim_{(x, y) \to (0, 0)} 0 = 0 \qquad \text{and} \qquad \lim_{(x, y) \to (0, 0)} 3|y| = 0 \qquad \text{[by (2)]}$$

we conclude that

$$\lim_{(x, y) \to (0, 0)} \frac{3x^2 y}{x^2 + y^2} = 0 \qquad\qquad ■\ ■$$

Continuity

Recall that evaluating limits of *continuous* functions of a single variable is easy. It can be accomplished by direct substitution because the defining property of a continuous function is $\lim_{x \to a} f(x) = f(a)$. Continuous functions of two variables are also defined by the direct substitution property.

> **3 Definition** A function f of two variables is called **continuous at** (a, b) if
>
> $$\lim_{(x, y) \to (a, b)} f(x, y) = f(a, b)$$
>
> We say f is **continuous on** D if f is continuous at every point (a, b) in D.

The intuitive meaning of continuity is that if the point (x, y) changes by a small amount, then the value of $f(x, y)$ changes by a small amount. This means that a surface that is the graph of a continuous function has no hole or break.

Using the properties of limits, you can see that sums, differences, products, and quotients of continuous functions are continuous on their domains. Let's use this fact to give examples of continuous functions.

A **polynomial function of two variables** (or polynomial, for short) is a sum of terms of the form $c x^m y^n$, where c is a constant and m and n are nonnegative integers. A **rational function** is a ratio of polynomials. For instance,

$$f(x, y) = x^4 + 5x^3 y^2 + 6xy^4 - 7y + 6$$

is a polynomial, whereas

$$g(x, y) = \frac{2xy + 1}{x^2 + y^2}$$

is a rational function.

The limits in (2) show that the functions $f(x, y) = x$, $g(x, y) = y$, and $h(x, y) = c$ are continuous. Since any polynomial can be built up out of the simple functions f, g, and h by multiplication and addition, it follows that *all polynomials are continuous on* $\mathbb{R}^2$. Likewise, any rational function is continuous on its domain because it is a quotient of continuous functions.

EXAMPLE 5 Evaluate $\lim_{(x, y) \to (1, 2)} (x^2y^3 - x^3y^2 + 3x + 2y)$.

SOLUTION Since $f(x, y) = x^2y^3 - x^3y^2 + 3x + 2y$ is a polynomial, it is continuous everywhere, so we can find the limit by direct substitution:

$$\lim_{(x, y) \to (1, 2)} (x^2y^3 - x^3y^2 + 3x + 2y) = 1^2 \cdot 2^3 - 1^3 \cdot 2^2 + 3 \cdot 1 + 2 \cdot 2 = 11 \quad ■ ■$$

EXAMPLE 6 Where is the function $f(x, y) = \dfrac{x^2 - y^2}{x^2 + y^2}$ continuous?

SOLUTION The function f is discontinuous at $(0, 0)$ because it is not defined there. Since f is a rational function, it is continuous on its domain, which is the set $D = \{(x, y) \mid (x, y) \neq (0, 0)\}$. ■ ■

EXAMPLE 7 Let

$$g(x, y) = \begin{cases} \dfrac{x^2 - y^2}{x^2 + y^2} & \text{if } (x, y) \neq (0, 0) \\ 0 & \text{if } (x, y) = (0, 0) \end{cases}$$

Here g is defined at $(0, 0)$ but g is still discontinuous there because $\lim_{(x, y) \to (0, 0)} g(x, y)$ does not exist (see Example 1). ■ ■

EXAMPLE 8 Let

$$f(x, y) = \begin{cases} \dfrac{3x^2y}{x^2 + y^2} & \text{if } (x, y) \neq (0, 0) \\ 0 & \text{if } (x, y) = (0, 0) \end{cases}$$

We know f is continuous for $(x, y) \neq (0, 0)$ since it is equal to a rational function there. Also, from Example 4, we have

$$\lim_{(x, y) \to (0, 0)} f(x, y) = \lim_{(x, y) \to (0, 0)} \frac{3x^2y}{x^2 + y^2} = 0 = f(0, 0)$$

Therefore, f is continuous at $(0, 0)$, and so it is continuous on $\mathbb{R}^2$. ■ ■

■ ■ Figure 6 shows the graph of the continuous function in Example 8.

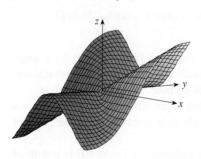

FIGURE 6

Just as for functions of one variable, composition is another way of combining two continuous functions to get a third. In fact, it can be shown that if f is a continuous function of two variables and g is a continuous function of a single variable that is defined on the range of f, then the composite function $h = g \circ f$ defined by $h(x, y) = g(f(x, y))$ is also a continuous function.

EXAMPLE 9 Where is the function $h(x, y) = \arctan(y/x)$ continuous?

SOLUTION The function $f(x, y) = y/x$ is a rational function and therefore continuous except on the line $x = 0$. The function $g(t) = \arctan t$ is continuous everywhere. So the composite function

$$g(f(x, y)) = \arctan(y/x) = h(x, y)$$

is continuous except where $x = 0$. The graph in Figure 7 shows the break in the graph of h above the y-axis. ■ ■

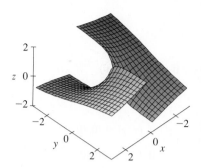

FIGURE 7
The function $h(x, y) = \arctan(y/x)$ is discontinuous where $x = 0$.

Everything that we have done in this section can be extended to functions of three or more variables. The notation

$$\lim_{(x, y, z) \to (a, b, c)} f(x, y, z) = L$$

means that the values of $f(x, y, z)$ approach the number L as the point (x, y, z) approaches the point (a, b, c) along any path in the domain of f. The function f is **continuous** at (a, b, c) if

$$\lim_{(x, y, z) \to (a, b, c)} f(x, y, z) = f(a, b, c)$$

For instance, the function

$$f(x, y, z) = \frac{1}{x^2 + y^2 + z^2 - 1}$$

is a rational function of three variables and so is continuous at every point in $\mathbb{R}^3$ except where $x^2 + y^2 + z^2 = 1$. In other words, it is discontinuous on the sphere with center the origin and radius 1.

11.2 Exercises

1. Suppose that $\lim_{(x, y) \to (3, 1)} f(x, y) = 6$. What can you say about the value of $f(3, 1)$? What if f is continuous?

2. Explain why each function is continuous or discontinuous.
   (a) The outdoor temperature as a function of longitude, latitude, and time
   (b) Elevation (height above sea level) as a function of longitude, latitude, and time
   (c) The cost of a taxi ride as a function of distance traveled and time

3–4 ■ Use a table of numerical values of $f(x, y)$ for (x, y) near the origin to make a conjecture about the value of the limit of $f(x, y)$ as $(x, y) \to (0, 0)$. Then explain why your guess is correct.

3. $f(x, y) = \dfrac{x^2 y^3 + x^3 y^2 - 5}{2 - xy}$

4. $f(x, y) = \dfrac{2xy}{x^2 + 2y^2}$

5–18 ■ Find the limit, if it exists, or show that the limit does not exist.

5. $\lim_{(x, y) \to (5, -2)} (x^5 + 4x^3 y - 5xy^2)$

6. $\lim_{(x, y) \to (6, 3)} xy \cos(x - 2y)$

7. $\lim_{(x, y) \to (0, 0)} \dfrac{y^4}{x^4 + 3y^4}$

8. $\lim_{(x, y) \to (0, 0)} \dfrac{x^2 + \sin^2 y}{2x^2 + y^2}$

9. $\lim_{(x, y) \to (0, 0)} \dfrac{xy \cos y}{3x^2 + y^2}$

10. $\lim_{(x, y) \to (0, 0)} \dfrac{6x^3 y}{2x^4 + y^4}$

11. $\lim_{(x, y) \to (0, 0)} \dfrac{xy}{\sqrt{x^2 + y^2}}$

12. $\lim_{(x, y) \to (0, 0)} \dfrac{x^2 \sin^2 y}{x^2 + 2y^2}$

13. $\lim_{(x, y) \to (0, 0)} \dfrac{2x^2 y}{x^4 + y^2}$

14. $\lim_{(x, y) \to (0, 0)} \dfrac{xy^4}{x^2 + y^8}$

15. $\lim\limits_{(x,y)\to(0,0)} \dfrac{x^2 + y^2}{\sqrt{x^2 + y^2 + 1} - 1}$

16. $\lim\limits_{(x,y)\to(0,0)} \dfrac{x^4 - y^4}{x^2 + y^2}$

17. $\lim\limits_{(x,y,z)\to(0,0,0)} \dfrac{xy + yz^2 + xz^2}{x^2 + y^2 + z^4}$

18. $\lim\limits_{(x,y,z)\to(0,0,0)} \dfrac{x^2 + 2y^2 + 3z^2}{x^2 + y^2 + z^2}$

. . . .

19–20 ■ Use a computer graph of the function to explain why the limit does not exist.

19. $\lim\limits_{(x,y)\to(0,0)} \dfrac{2x^2 + 3xy + 4y^2}{3x^2 + 5y^2}$

20. $\lim\limits_{(x,y)\to(0,0)} \dfrac{xy^3}{x^2 + y^6}$

. . . .

21–22 ■ Find $h(x, y) = g(f(x, y))$ and the set on which h is continuous.

21. $g(t) = t^2 + \sqrt{t}$, $\quad f(x, y) = 2x + 3y - 6$

22. $g(t) = \dfrac{\sqrt{t} - 1}{\sqrt{t} + 1}$, $\quad f(x, y) = x^2 - y$

. . . .

23–24 ■ Graph the function and observe where it is discontinuous. Then use the formula to explain what you have observed.

23. $f(x, y) = e^{1/(x-y)}$

24. $f(x, y) = \dfrac{1}{1 - x^2 - y^2}$

. . . .

25–32 ■ Determine the set of points at which the function is continuous.

25. $F(x, y) = \dfrac{\sin(xy)}{e^x - y^2}$

26. $F(x, y) = \dfrac{x - y}{1 + x^2 + y^2}$

27. $G(x, y) = \ln(x^2 + y^2 - 4)$

28. $F(x, y) = e^{x^2 y} + \sqrt{x + y^2}$

29. $f(x, y, z) = \dfrac{\sqrt{y}}{x^2 - y^2 + z^2}$

30. $f(x, y, z) = \sqrt{x + y + z}$

31. $f(x, y) = \begin{cases} \dfrac{x^2 y^3}{2x^2 + y^2} & \text{if } (x, y) \neq (0, 0) \\ 1 & \text{if } (x, y) = (0, 0) \end{cases}$

32. $f(x, y) = \begin{cases} \dfrac{xy}{x^2 + xy + y^2} & \text{if } (x, y) \neq (0, 0) \\ 0 & \text{if } (x, y) = (0, 0) \end{cases}$

. . . .

33–34 ■ Use polar coordinates to find the limit. [If (r, θ) are polar coordinates of the point (x, y) with $r \geq 0$, note that $r \to 0^+$ as $(x, y) \to (0, 0)$.]

33. $\lim\limits_{(x,y)\to(0,0)} \dfrac{x^3 + y^3}{x^2 + y^2}$

34. $\lim\limits_{(x,y)\to(0,0)} (x^2 + y^2) \ln(x^2 + y^2)$

. . . .

35. Use spherical coordinates to find

$$\lim\limits_{(x,y,z)\to(0,0,0)} \dfrac{xyz}{x^2 + y^2 + z^2}$$

36. At the beginning of this section we considered the function

$$f(x, y) = \dfrac{\sin(x^2 + y^2)}{x^2 + y^2}$$

and guessed that $f(x, y) \to 1$ as $(x, y) \to (0, 0)$ on the basis of numerical evidence. Use polar coordinates to confirm the value of the limit. Then graph the function.

11.3 Partial Derivatives

On a hot day, extreme humidity makes us think the temperature is higher than it really is, whereas in very dry air we perceive the temperature to be lower than the thermometer indicates. The National Weather Service has devised the *heat index* (also called the temperature-humidity index, or humidex, in some countries) to describe the combined effects of temperature and humidity. The heat index I is the perceived air temperature when the actual temperature is T and the relative humidity is H. So I is a function of T and H and we can write $I = f(T, H)$. The following table of values of I is an excerpt from a table compiled by the National Weather Service.

TABLE 1   Heat index I as a function of temperature and humidity

Relative humidity (%)

T \ H	50	55	60	65	70	75	80	85	90
90	96	98	100	103	106	109	112	115	119
92	100	103	105	108	112	115	119	123	128
94	104	107	111	114	118	122	127	132	137
96	109	113	116	121	125	130	135	141	146
98	114	118	123	127	133	138	144	150	157
100	119	124	129	135	141	147	154	161	168

Actual temperature (°F)

If we concentrate on the highlighted column of the table, which corresponds to a relative humidity of $H = 70\%$, we are considering the heat index as a function of the single variable T for a fixed value of H. Let's write $g(T) = f(T, 70)$. Then $g(T)$ describes how the heat index I increases as the actual temperature T increases when the relative humidity is 70%. The derivative of g when $T = 96°F$ is the rate of change of I with respect to T when $T = 96°F$:

$$g'(96) = \lim_{h \to 0} \frac{g(96 + h) - g(96)}{h} = \lim_{h \to 0} \frac{f(96 + h, 70) - f(96, 70)}{h}$$

We can approximate it using the values in Table 1 by taking $h = 2$ and -2:

$$g'(96) \approx \frac{g(98) - g(96)}{2} = \frac{f(98, 70) - f(96, 70)}{2} = \frac{133 - 125}{2} = 4$$

$$g'(96) \approx \frac{g(94) - g(96)}{-2} = \frac{f(94, 70) - f(96, 70)}{-2} = \frac{118 - 125}{-2} = 3.5$$

Averaging these values, we can say that the derivative $g'(96)$ is approximately 3.75. This means that, when the actual temperature is 96°F and the relative humidity is 70%, the apparent temperature (heat index) rises by about 3.75°F for every degree that the actual temperature rises!

Now let's look at the highlighted row in Table 1, which corresponds to a fixed temperature of $T = 96°F$. The numbers in this row are values of the function $G(H) = f(96, H)$, which describes how the heat index increases as the relative humidity H increases when the actual temperature is $T = 96°F$. The derivative of this function when $H = 70\%$ is the rate of change of I with respect to H when $H = 70\%$:

$$G'(70) = \lim_{h \to 0} \frac{G(70 + h) - G(70)}{h} = \lim_{h \to 0} \frac{f(96, 70 + h) - f(96, 70)}{h}$$

By taking $h = 5$ and -5, we approximate $G'(70)$ using the tabular values:

$$G'(70) \approx \frac{G(75) - G(70)}{5} = \frac{f(96, 75) - f(96, 70)}{5} = \frac{130 - 125}{5} = 1$$

$$G'(70) \approx \frac{G(65) - G(70)}{-5} = \frac{f(96, 65) - f(96, 70)}{-5} = \frac{121 - 125}{-5} = 0.8$$

By averaging these values we get the estimate $G'(70) \approx 0.9$. This says that, when the temperature is 96°F and the relative humidity is 70%, the heat index rises about 0.9°F for every percent that the relative humidity rises.

In general, if f is a function of two variables x and y, suppose we let only x vary while keeping y fixed, say $y = b$, where b is a constant. Then we are really considering a function of a single variable x, namely, $g(x) = f(x, b)$. If g has a derivative at a, then we call it the **partial derivative of f with respect to x at (a, b)** and denote it by $f_x(a, b)$. Thus

$$\boxed{1} \qquad f_x(a, b) = g'(a) \qquad \text{where} \qquad g(x) = f(x, b)$$

By the definition of a derivative, we have

$$g'(a) = \lim_{h \to 0} \frac{g(a + h) - g(a)}{h}$$

and so Equation 1 becomes

$$\boxed{2} \qquad f_x(a, b) = \lim_{h \to 0} \frac{f(a + h, b) - f(a, b)}{h}$$

Similarly, the **partial derivative of f with respect to y at (a, b)**, denoted by $f_y(a, b)$, is obtained by keeping x fixed ($x = a$) and finding the ordinary derivative at b of the function $G(y) = f(a, y)$:

$$\boxed{3} \qquad f_y(a, b) = \lim_{h \to 0} \frac{f(a, b + h) - f(a, b)}{h}$$

With this notation for partial derivatives, we can write the rates of change of the heat index I with respect to the actual temperature T and relative humidity H when $T = 96°F$ and $H = 70\%$ as follows:

$$f_T(96, 70) \approx 3.75 \qquad f_H(96, 70) \approx 0.9$$

If we now let the point (a, b) vary in Equations 2 and 3, f_x and f_y become functions of two variables.

4 If f is a function of two variables, its **partial derivatives** are the functions f_x and f_y defined by

$$f_x(x, y) = \lim_{h \to 0} \frac{f(x + h, y) - f(x, y)}{h}$$

$$f_y(x, y) = \lim_{h \to 0} \frac{f(x, y + h) - f(x, y)}{h}$$

There are many alternative notations for partial derivatives. For instance, instead of f_x we can write f_1 or $D_1 f$ (to indicate differentiation with respect to the *first* variable) or $\partial f / \partial x$. But here $\partial f / \partial x$ can't be interpreted as a ratio of differentials.

> **Notations for Partial Derivatives** If $z = f(x, y)$, we write
>
> $$f_x(x, y) = f_x = \frac{\partial f}{\partial x} = \frac{\partial}{\partial x} f(x, y) = \frac{\partial z}{\partial x} = f_1 = D_1 f = D_x f$$
>
> $$f_y(x, y) = f_y = \frac{\partial f}{\partial y} = \frac{\partial}{\partial y} f(x, y) = \frac{\partial z}{\partial y} = f_2 = D_2 f = D_y f$$

To compute partial derivatives, all we have to do is remember from Equation 1 that the partial derivative with respect to x is just the *ordinary* derivative of the function g of a single variable that we get by keeping y fixed. Thus, we have the following rule.

> **Rule for Finding Partial Derivatives of $z = f(x, y)$**
>
> **1.** To find f_x, regard y as a constant and differentiate $f(x, y)$ with respect to x.
>
> **2.** To find f_y, regard x as a constant and differentiate $f(x, y)$ with respect to y.

EXAMPLE 1 If $f(x, y) = x^3 + x^2 y^3 - 2y^2$, find $f_x(2, 1)$ and $f_y(2, 1)$.

SOLUTION Holding y constant and differentiating with respect to x, we get

$$f_x(x, y) = 3x^2 + 2xy^3$$

and so

$$f_x(2, 1) = 3 \cdot 2^2 + 2 \cdot 2 \cdot 1^3 = 16$$

Holding x constant and differentiating with respect to y, we get

$$f_y(x, y) = 3x^2 y^2 - 4y$$

$$f_y(2, 1) = 3 \cdot 2^2 \cdot 1^2 - 4 \cdot 1 = 8$$

■ ■

■■ Interpretations of Partial Derivatives

To give a geometric interpretation of partial derivatives, we recall that the equation $z = f(x, y)$ represents a surface S (the graph of f). If $f(a, b) = c$, then the point $P(a, b, c)$ lies on S. By fixing $y = b$, we are restricting our attention to the curve C_1 in which the vertical plane $y = b$ intersects S. (In other words, C_1 is the trace of S in the plane $y = b$.) Likewise, the vertical plane $x = a$ intersects S in a curve C_2. Both of the curves C_1 and C_2 pass through the point P. (See Figure 1.)

Notice that the curve C_1 is the graph of the function $g(x) = f(x, b)$, so the slope of its tangent T_1 at P is $g'(a) = f_x(a, b)$. The curve C_2 is the graph of the function $G(y) = f(a, y)$, so the slope of its tangent T_2 at P is $G'(b) = f_y(a, b)$.

Thus, the partial derivatives $f_x(a, b)$ and $f_y(a, b)$ can be interpreted geometrically as the slopes of the tangent lines at $P(a, b, c)$ to the traces C_1 and C_2 of S in the planes $y = b$ and $x = a$.

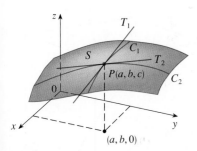

FIGURE 1

The partial derivatives of f at (a, b) are the slopes of the tangents to C_1 and C_2.

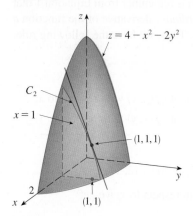

FIGURE 2

As we have seen in the case of the heat index function, partial derivatives can also be interpreted as *rates of change*. If $z = f(x, y)$, then $\partial z/\partial x$ represents the rate of change of z with respect to x when y is fixed. Similarly, $\partial z/\partial y$ represents the rate of change of z with respect to y when x is fixed.

EXAMPLE 2 If $f(x, y) = 4 - x^2 - 2y^2$, find $f_x(1, 1)$ and $f_y(1, 1)$ and interpret these numbers as slopes.

SOLUTION We have

$$f_x(x, y) = -2x \qquad f_y(x, y) = -4y$$

$$f_x(1, 1) = -2 \qquad f_y(1, 1) = -4$$

The graph of f is the paraboloid $z = 4 - x^2 - 2y^2$ and the vertical plane $y = 1$ intersects it in the parabola $z = 2 - x^2$, $y = 1$. (As in the preceding discussion, we label it C_1 in Figure 2.) The slope of the tangent line to this parabola at the point $(1, 1, 1)$ is $f_x(1, 1) = -2$. Similarly, the curve C_2 in which the plane $x = 1$ intersects the paraboloid is the parabola $z = 3 - 2y^2$, $x = 1$, and the slope of the tangent line at $(1, 1, 1)$ is $f_y(1, 1) = -4$. (See Figure 3.) ■ ■

Figure 4 is a computer-drawn counterpart to Figure 2. Part (a) shows the plane $y = 1$ intersecting the surface to form the curve C_1 and part (b) shows C_1 and T_1. [We have used the vector equations $\mathbf{r}(t) = \langle t, 1, 2 - t^2 \rangle$ for C_1 and $\mathbf{r}(t) = \langle 1 + t, 1, 1 - 2t \rangle$ for T_1.] Similarly, Figure 5 corresponds to Figure 3.

FIGURE 4

(a)

(b)

FIGURE 3

FIGURE 5

EXAMPLE 3 If $f(x, y) = \sin\left(\dfrac{x}{1 + y}\right)$, calculate $\dfrac{\partial f}{\partial x}$ and $\dfrac{\partial f}{\partial y}$.

SOLUTION Using the Chain Rule for functions of one variable, we have

$$\frac{\partial f}{\partial x} = \cos\left(\frac{x}{1 + y}\right) \cdot \frac{\partial}{\partial x}\left(\frac{x}{1 + y}\right) = \cos\left(\frac{x}{1 + y}\right) \cdot \frac{1}{1 + y}$$

$$\frac{\partial f}{\partial y} = \cos\left(\frac{x}{1 + y}\right) \cdot \frac{\partial}{\partial y}\left(\frac{x}{1 + y}\right) = -\cos\left(\frac{x}{1 + y}\right) \cdot \frac{x}{(1 + y)^2}$$ ■ ■

 EXAMPLE 4 Find $\partial z/\partial x$ and $\partial z/\partial y$ if z is defined implicitly as a function of x and y by the equation

$$x^3 + y^3 + z^3 + 6xyz = 1$$

■ ■ Some computer algebra systems can plot surfaces defined by implicit equations in three variables. Figure 6 shows such a plot of the surface defined by the equation in Example 4.

SOLUTION To find $\partial z/\partial x$, we differentiate implicitly with respect to x, being careful to treat y as a constant:

$$3x^2 + 3z^2 \frac{\partial z}{\partial x} + 6yz + 6xy \frac{\partial z}{\partial x} = 0$$

Solving this equation for $\partial z/\partial x$, we obtain

$$\frac{\partial z}{\partial x} = -\frac{x^2 + 2yz}{z^2 + 2xy}$$

Similarly, implicit differentiation with respect to y gives

$$\frac{\partial z}{\partial y} = -\frac{y^2 + 2xz}{z^2 + 2xy}$$ ■ ■

FIGURE 6

▪▪ Functions of More than Two Variables

Partial derivatives can also be defined for functions of three or more variables. For example, if f is a function of three variables x, y, and z, then its partial derivative with respect to x is defined as

$$f_x(x, y, z) = \lim_{h \to 0} \frac{f(x + h, y, z) - f(x, y, z)}{h}$$

and it is found by regarding y and z as constants and differentiating $f(x, y, z)$ with respect to x. If $w = f(x, y, z)$, then $f_x = \partial w/\partial x$ can be interpreted as the rate of change of w with respect to x when y and z are held fixed. But we can't interpret it geometrically because the graph of f lies in four-dimensional space.

In general, if u is a function of n variables, $u = f(x_1, x_2, \ldots, x_n)$, its partial derivative with respect to the ith variable x_i is

$$\frac{\partial u}{\partial x_i} = \lim_{h \to 0} \frac{f(x_1, \ldots, x_{i-1}, x_i + h, x_{i+1}, \ldots, x_n) - f(x_1, \ldots, x_i, \ldots, x_n)}{h}$$

and we also write

$$\frac{\partial u}{\partial x_i} = \frac{\partial f}{\partial x_i} = f_{x_i} = f_i = D_i f$$

EXAMPLE 5 Find f_x, f_y, and f_z if $f(x, y, z) = e^{xy} \ln z$.

SOLUTION Holding y and z constant and differentiating with respect to x, we have

$$f_x = ye^{xy} \ln z$$

Similarly, $f_y = xe^{xy} \ln z$ and $f_z = \dfrac{e^{xy}}{z}$ ■ ■

■■ Higher Derivatives

If f is a function of two variables, then its partial derivatives f_x and f_y are also functions of two variables, so we can consider their partial derivatives $(f_x)_x$, $(f_x)_y$, $(f_y)_x$, and $(f_y)_y$, which are called the **second partial derivatives** of f. If $z = f(x, y)$, we use the following notation:

$$(f_x)_x = f_{xx} = f_{11} = \frac{\partial}{\partial x}\left(\frac{\partial f}{\partial x}\right) = \frac{\partial^2 f}{\partial x^2} = \frac{\partial^2 z}{\partial x^2}$$

$$(f_x)_y = f_{xy} = f_{12} = \frac{\partial}{\partial y}\left(\frac{\partial f}{\partial x}\right) = \frac{\partial^2 f}{\partial y \, \partial x} = \frac{\partial^2 z}{\partial y \, \partial x}$$

$$(f_y)_x = f_{yx} = f_{21} = \frac{\partial}{\partial x}\left(\frac{\partial f}{\partial y}\right) = \frac{\partial^2 f}{\partial x \, \partial y} = \frac{\partial^2 z}{\partial x \, \partial y}$$

$$(f_y)_y = f_{yy} = f_{22} = \frac{\partial}{\partial y}\left(\frac{\partial f}{\partial y}\right) = \frac{\partial^2 f}{\partial y^2} = \frac{\partial^2 z}{\partial y^2}$$

Thus, the notation f_{xy} (or $\partial^2 f/\partial y \, \partial x$) means that we first differentiate with respect to x and then with respect to y, whereas in computing f_{yx} the order is reversed.

EXAMPLE 6 Find the second partial derivatives of

$$f(x, y) = x^3 + x^2 y^3 - 2y^2$$

SOLUTION In Example 1 we found that

$$f_x(x, y) = 3x^2 + 2xy^3 \qquad f_y(x, y) = 3x^2 y^2 - 4y$$

Therefore

$$f_{xx} = \frac{\partial}{\partial x}(3x^2 + 2xy^3) = 6x + 2y^3 \qquad f_{xy} = \frac{\partial}{\partial y}(3x^2 + 2xy^3) = 6xy^2$$

$$f_{yx} = \frac{\partial}{\partial x}(3x^2 y^2 - 4y) = 6xy^2 \qquad f_{yy} = \frac{\partial}{\partial y}(3x^2 y^2 - 4y) = 6x^2 y - 4$$

■ ■

f

■ ■ Figure 7 shows the graph of the function f in Example 6 and the graphs of its first- and second-order partial derivatives for $-2 \leqslant x \leqslant 2$, $-2 \leqslant y \leqslant 2$. Notice that these graphs are consistent with our interpretations of f_x and f_y as slopes of tangent lines to traces of the graph of f. For instance, the graph of f decreases if we start at $(0, -2)$ and move in the positive x-direction. This is reflected in the negative values of f_x. You should compare the graphs of f_{yx} and f_{yy} with the graph of f_y to see the relationships.

f_x

f_y

f_{xx}

$f_{xy} = f_{yx}$

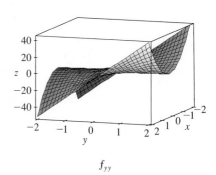

f_{yy}

FIGURE 7

Notice that $f_{xy} = f_{yx}$ in Example 6. This is not just a coincidence. It turns out that the mixed partial derivatives f_{xy} and f_{yx} are equal for most functions that one meets in practice. The following theorem, which was discovered by the French mathematician Alexis Clairaut (1713–1765), gives conditions under which we can assert that $f_{xy} = f_{yx}$. The proof is given in Appendix E.

■ ■ Alexis Clairaut was a child prodigy in mathematics, having read l'Hospital's textbook on calculus when he was ten and presented a paper on geometry to the French Academy of Sciences when he was 13. At the age of 18, Clairaut published *Recherches sur les courbes à double courbure*, which was the first systematic treatise on three-dimensional analytic geometry and included the calculus of space curves.

Clairaut's Theorem Suppose f is defined on a disk D that contains the point (a, b). If the functions f_{xy} and f_{yx} are both continuous on D, then

$$f_{xy}(a, b) = f_{yx}(a, b)$$

Partial derivatives of order 3 or higher can also be defined. For instance,

$$f_{xyy} = (f_{xy})_y = \frac{\partial}{\partial y}\left(\frac{\partial^2 f}{\partial y\, \partial x}\right) = \frac{\partial^3 f}{\partial y^2\, \partial x}$$

and using Clairaut's Theorem it can be shown that $f_{xyy} = f_{yxy} = f_{yyx}$ if these functions are continuous.

EXAMPLE 7 Calculate f_{xxyz} if $f(x, y, z) = \sin(3x + yz)$.

SOLUTION
$$f_x = 3\cos(3x + yz)$$

$$f_{xx} = -9\sin(3x + yz)$$

$$f_{xxy} = -9z\cos(3x + yz)$$

$$f_{xxyz} = -9\cos(3x + yz) + 9yz\sin(3x + yz)$$ ■ ■

Partial Differential Equations

Partial derivatives occur in *partial differential equations* that express certain physical laws. For instance, the partial differential equation

$$\frac{\partial^2 u}{\partial x^2} + \frac{\partial^2 u}{\partial y^2} = 0$$

is called **Laplace's equation** after Pierre Laplace (1749–1827). Solutions of this equation are called **harmonic functions** and play a role in problems of heat conduction, fluid flow, and electric potential.

EXAMPLE 8 Show that the function $u(x, y) = e^x \sin y$ is a solution of Laplace's equation.

SOLUTION
$$u_x = e^x \sin y \qquad\qquad u_y = e^x \cos y$$

$$u_{xx} = e^x \sin y \qquad\qquad u_{yy} = -e^x \sin y$$

$$u_{xx} + u_{yy} = e^x \sin y - e^x \sin y = 0$$

Therefore, u satisfies Laplace's equation. ■ ■

The **wave equation**

$$\frac{\partial^2 u}{\partial t^2} = a^2 \frac{\partial^2 u}{\partial x^2}$$

describes the motion of a waveform, which could be an ocean wave, a sound wave, a light wave, or a wave traveling along a vibrating string. For instance, if $u(x, t)$ represents the displacement of a vibrating violin string at time t and at a distance x from one end of the string (as in Figure 8), then $u(x, t)$ satisfies the wave equation. Here the constant a depends on the density of the string and on the tension in the string.

FIGURE 8

EXAMPLE 9 Verify that the function $u(x, t) = \sin(x - at)$ satisfies the wave equation.

SOLUTION
$$u_x = \cos(x - at) \qquad\qquad u_{xx} = -\sin(x - at)$$

$$u_t = -a\cos(x - at) \qquad\qquad u_{tt} = -a^2 \sin(x - at) = a^2 u_{xx}$$

So u satisfies the wave equation. ■ ■

▪▪ The Cobb-Douglas Production Function

In Example 2 in Section 11.1 we described the work of Cobb and Douglas in modeling the total production P of an economic system as a function of the amount of labor L and the capital investment K. Here we use partial derivatives to show how the particular form of their model follows from certain assumptions they made about the economy.

If the production function is denoted by $P = P(L, K)$, then the partial derivative $\partial P/\partial L$ is the rate at which production changes with respect to the amount of labor. Economists call it the marginal production with respect to labor or the *marginal productivity of labor*. Likewise, the partial derivative $\partial P/\partial K$ is the rate of change of production with respect to capital and is called the *marginal productivity of capital*. In these terms, the assumptions made by Cobb and Douglas can be stated as follows.

   (i) If either labor or capital vanishes, then so will production.

  (ii) The marginal productivity of labor is proportional to the amount of production per unit of labor.

 (iii) The marginal productivity of capital is proportional to the amount of production per unit of capital.

Because the production per unit of labor is P/L, assumption (ii) says that

$$\frac{\partial P}{\partial L} = \alpha \frac{P}{L}$$

for some constant α. If we keep K constant ($K = K_0$), then this partial differential equation becomes an ordinary differential equation:

$$\boxed{5} \qquad \frac{dP}{dL} = \alpha \frac{P}{L}$$

If we solve this separable differential equation by the methods of Section 7.3 (see also Exercise 69), we get

$$\boxed{6} \qquad P(L, K_0) = C_1(K_0)L^\alpha$$

Notice that we have written the constant C_1 as a function of K_0 because it could depend on the value of K_0.

Similarly, assumption (iii) says that

$$\frac{\partial P}{\partial K} = \beta \frac{P}{K}$$

and we can solve this differential equation to get

$$\boxed{7} \qquad P(L_0, K) = C_2(L_0)K^\beta$$

Comparing Equations 6 and 7, we have

$$\boxed{8} \qquad P(L, K) = bL^\alpha K^\beta$$

where b is a constant that is independent of both L and K. Assumption (i) shows that $\alpha > 0$ and $\beta > 0$.

Notice from Equation 8 that if labor and capital are both increased by a factor m, then

$$P(mL, mK) = b(mL)^\alpha(mK)^\beta = m^{\alpha+\beta}bL^\alpha K^\beta = m^{\alpha+\beta}P(L, K)$$

If $\alpha + \beta = 1$, then $P(mL, mK) = mP(L, K)$, which means that production is also increased by a factor of m. That is why Cobb and Douglas assumed that $\alpha + \beta = 1$ and therefore

$$P(L, K) = bL^\alpha K^{1-\alpha}$$

This is the Cobb-Douglas production function that we discussed in Section 11.1.

11.3 Exercises

1. The temperature T at a location in the Northern Hemisphere depends on the longitude x, latitude y, and time t, so we can write $T = f(x, y, t)$. Let's measure time in hours from the beginning of January.
(a) What are the meanings of the partial derivatives $\partial T/\partial x$, $\partial T/\partial y$, and $\partial T/\partial t$?
(b) Honolulu has longitude 158° W and latitude 21° N. Suppose that at 9:00 A.M. on January 1 the wind is blowing hot air to the northeast, so the air to the west and south is warm and the air to the north and east is cooler. Would you expect $f_x(158, 21, 9)$, $f_y(158, 21, 9)$, and $f_t(158, 21, 9)$ to be positive or negative? Explain.

2. At the beginning of this section we discussed the function $I = f(T, H)$, where I is the heat index, T is the temperature, and H is the relative humidity. Use Table 1 to estimate $f_T(92, 60)$ and $f_H(92, 60)$. What are the practical interpretations of these values?

3. The wind-chill index W is the perceived temperature when the actual temperature is T and the wind speed is v, so we can write $W = f(T, v)$. The following table of values is an excerpt from Table 1 in Section 11.1.

Wind speed (km/h)

T \ v	20	30	40	50	60	70
-10	-18	-20	-21	-22	-23	-23
-15	-24	-26	-27	-29	-30	-30
-20	-30	-33	-34	-35	-36	-37
-25	-37	-39	-41	-42	-43	-44

Actual temperature (°C)

(a) Estimate the values of $f_T(-15, 30)$ and $f_v(-15, 30)$. What are the practical interpretations of these values?

(b) In general, what can you say about the signs of $\partial W/\partial T$ and $\partial W/\partial v$?
(c) What appears to be the value of the following limit?

$$\lim_{v \to \infty} \frac{\partial W}{\partial v}$$

4. The wave heights h in the open sea depend on the speed v of the wind and the length of time t that the wind has been blowing at that speed. Values of the function $h = f(v, t)$ are recorded in feet in the following table.

Duration (hours)

v \ t	5	10	15	20	30	40	50
10	2	2	2	2	2	2	2
15	4	4	5	5	5	5	5
20	5	7	8	8	9	9	9
30	9	13	16	17	18	19	19
40	14	21	25	28	31	33	33
50	19	29	36	40	45	48	50
60	24	37	47	54	62	67	69

Wind speed (knots)

(a) What are the meanings of the partial derivatives $\partial h/\partial v$ and $\partial h/\partial t$?
(b) Estimate the values of $f_v(40, 15)$ and $f_t(40, 15)$. What are the practical interpretations of these values?
(c) What appears to be the value of the following limit?

$$\lim_{t \to \infty} \frac{\partial h}{\partial t}$$

5-6 ■ Determine the signs of the partial derivatives for the function f whose graph is shown.

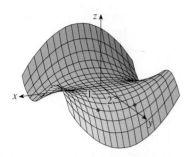

5. (a) $f_x(1, 2)$ (b) $f_y(1, 2)$

6. (a) $f_x(-1, 2)$ (b) $f_y(-1, 2)$

    (c) $f_{xx}(-1, 2)$ (d) $f_{yy}(-1, 2)$

■   ■   ■   ■   ■   ■   ■   ■   ■   ■   ■   ■   ■

7. The following surfaces, labeled a, b, and c, are graphs of a function f and its partial derivatives f_x and f_y. Identify each surface and give reasons for your choices.

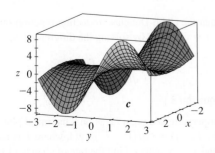

8. A contour map is given for a function f. Use it to estimate $f_x(2, 1)$ and $f_y(2, 1)$.

9. If $f(x, y) = 16 - 4x^2 - y^2$, find $f_x(1, 2)$ and $f_y(1, 2)$ and interpret these numbers as slopes. Illustrate with either hand-drawn sketches or computer plots.

10. If $f(x, y) = \sqrt{4 - x^2 - 4y^2}$, find $f_x(1, 0)$ and $f_y(1, 0)$ and interpret these numbers as slopes. Illustrate with either hand-drawn sketches or computer plots.

11-12 ■ Find f_x and f_y and graph f, f_x, and f_y with domains and viewpoints that enable you to see the relationships between them.

11. $f(x, y) = x^2 + y^2 + x^2 y$ **12.** $f(x, y) = xe^{-x^2-y^2}$

■   ■   ■   ■   ■   ■   ■   ■   ■   ■   ■   ■   ■

13-34 ■ Find the first partial derivatives of the function.

13. $f(x, y) = 3x - 2y^4$

14. $f(x, y) = x^5 + 3x^3y^2 + 3xy^4$

15. $z = xe^{3y}$ **16.** $z = y \ln x$

17. $f(x, y) = \dfrac{x - y}{x + y}$ **18.** $f(x, y) = x^y$

19. $w = \sin \alpha \cos \beta$ **20.** $f(s, t) = st^2/(s^2 + t^2)$

21. $f(r, s) = r \ln(r^2 + s^2)$ **22.** $f(x, t) = \arctan(x\sqrt{t})$

23. $u = te^{w/t}$ **24.** $f(x, y) = \displaystyle\int_y^x \cos(t^2)\, dt$

25. $f(x, y, z) = xy^2z^3 + 3yz$ **26.** $f(x, y, z) = x^2e^{yz}$

27. $w = \ln(x + 2y + 3z)$ **28.** $w = \sqrt{r^2 + s^2 + t^2}$

29. $u = xe^{-t} \sin \theta$ **30.** $u = x^{y/z}$

31. $f(x, y, z, t) = xyz^2 \tan(yt)$ **32.** $f(x, y, z, t) = \dfrac{xy^2}{t + 2z}$

33. $u = \sqrt{x_1^2 + x_2^2 + \cdots + x_n^2}$

34. $u = \sin(x_1 + 2x_2 + \cdots + nx_n)$

■   ■   ■   ■   ■   ■   ■   ■   ■   ■   ■   ■   ■

35-38 ■ Find the indicated partial derivatives.

35. $f(x, y) = \sqrt{x^2 + y^2}$; $f_x(3, 4)$

36. $f(x, y) = \sin(2x + 3y)$; $f_y(-6, 4)$

37. $f(x, y, z) = x/(y + z)$; $f_z(3, 2, 1)$

38. $f(u, v, w) = w \tan(uv)$; $f_v(2, 0, 3)$

39–40 ■ Use the definition of partial derivatives as limits (4) to find $f_x(x, y)$ and $f_y(x, y)$.

39. $f(x, y) = xy^2 - x^3y$

40. $f(x, y) = \dfrac{x}{x + y^2}$

41–44 ■ Use implicit differentiation to find $\partial z/\partial x$ and $\partial z/\partial y$.

41. $x^2 + y^2 + z^2 = 3xyz$

42. $yz = \ln(x + z)$

43. $x - z = \arctan(yz)$

44. $\sin(xyz) = x + 2y + 3z$

45–46 ■ Find $\partial z/\partial x$ and $\partial z/\partial y$.

45. (a) $z = f(x) + g(y)$ (b) $z = f(x + y)$

46. (a) $z = f(x)g(y)$ (b) $z = f(xy)$
    (c) $z = f(x/y)$

47–52 ■ Find all the second partial derivatives.

47. $f(x, y) = x^4 - 3x^2y^3$

48. $f(x, y) = \ln(3x + 5y)$

49. $z = x/(x + y)$

50. $z = y \tan 2x$

51. $u = e^{-s} \sin t$

52. $v = \sqrt{x + y^2}$

53–54 ■ Verify that the conclusion of Clairaut's Theorem holds, that is, $u_{xy} = u_{yx}$.

53. $u = x \sin(x + 2y)$

54. $u = x^4y^2 - 2xy^5$

55–60 ■ Find the indicated partial derivative.

55. $f(x, y) = 3xy^4 + x^3y^2$; f_{xxy}, f_{yyy}

56. $f(x, t) = x^2e^{-ct}$; f_{ttt}, f_{txx}

57. $f(x, y, z) = \cos(4x + 3y + 2z)$; f_{xyz}, f_{yzz}

58. $f(r, s, t) = r \ln(rs^2t^3)$; f_{rss}, f_{rst}

59. $u = e^{r\theta} \sin \theta$; $\dfrac{\partial^3 u}{\partial r^2 \, \partial\theta}$

60. $u = x^ay^bz^c$; $\dfrac{\partial^6 u}{\partial x \, \partial y^2 \, \partial z^3}$

61. Use the table of values of $f(x, y)$ to estimate the values of $f_x(3, 2)$, $f_x(3, 2.2)$, and $f_{xy}(3, 2)$.

x \ y	1.8	2.0	2.2
2.5	12.5	10.2	9.3
3.0	18.1	17.5	15.9
3.5	20.0	22.4	26.1

62. Level curves are shown for a function f. Determine whether the following partial derivatives are positive or negative at the point P.
   (a) f_x (b) f_y (c) f_{xx}
   (d) f_{xy} (e) f_{yy}

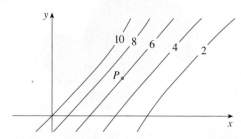

63. Verify that the function $u = e^{-\alpha^2k^2t} \sin kx$ is a solution of the *heat conduction equation* $u_t = \alpha^2 u_{xx}$.

64. Determine whether each of the following functions is a solution of Laplace's equation $u_{xx} + u_{yy} = 0$.
   (a) $u = x^2 + y^2$
   (b) $u = x^2 - y^2$
   (c) $u = x^3 + 3xy^2$
   (d) $u = \ln \sqrt{x^2 + y^2}$
   (e) $u = e^{-x} \cos y - e^{-y} \cos x$

65. Verify that the function $u = 1/\sqrt{x^2 + y^2 + z^2}$ is a solution of the three-dimensional Laplace equation $u_{xx} + u_{yy} + u_{zz} = 0$.

66. Show that each of the following functions is a solution of the wave equation $u_{tt} = a^2 u_{xx}$.
   (a) $u = \sin(kx) \sin(akt)$
   (b) $u = t/(a^2t^2 - x^2)$
   (c) $u = (x - at)^6 + (x + at)^6$
   (d) $u = \sin(x - at) + \ln(x + at)$

67. If f and g are twice differentiable functions of a single variable, show that the function

$$u(x, t) = f(x + at) + g(x - at)$$

is a solution of the wave equation given in Exercise 66.

68. Show that the Cobb-Douglas production function $P = bL^\alpha K^\beta$ satisfies the equation

$$L \frac{\partial P}{\partial L} + K \frac{\partial P}{\partial K} = (\alpha + \beta)P$$

69. Show that the Cobb-Douglas production function satisfies $P(L, K_0) = C_1(K_0)L^\alpha$ by solving the differential equation

$$\frac{dP}{dL} = \alpha \frac{P}{L}$$

(See Equation 5.)

70. The temperature at a point (x, y) on a flat metal plate is given by $T(x, y) = 60/(1 + x^2 + y^2)$, where T is measured in °C and x, y in meters. Find the rate of change of temperature with respect to distance at the point $(2, 1)$ in (a) the x-direction and (b) the y-direction.

71. The total resistance R produced by three conductors with resistances R_1, R_2, R_3 connected in a parallel electrical circuit is given by the formula

$$\frac{1}{R} = \frac{1}{R_1} + \frac{1}{R_2} + \frac{1}{R_3}$$

Find $\partial R/\partial R_1$.

72. The gas law for a fixed mass m of an ideal gas at absolute temperature T, pressure P, and volume V is $PV = mRT$, where R is the gas constant. Show that

$$\frac{\partial P}{\partial V} \frac{\partial V}{\partial T} \frac{\partial T}{\partial P} = -1$$

73. For the ideal gas of Exercise 72, show that

$$T\frac{\partial P}{\partial T} \frac{\partial V}{\partial T} = mR$$

74. The wind-chill index is modeled by the function

$$W = 13.12 + 0.6215T - 11.37v^{0.16} + 0.3965Tv^{0.16}$$

where T is the temperature (°C) and v is the wind speed (km/h). When $T = -15°C$ and $v = 30$ km/h, by how much would you expect the apparent temperature to drop if the actual temperature decreases by $1°C$? What if the wind speed increases by 1 km/h?

75. The kinetic energy of a body with mass m and velocity v is $K = \frac{1}{2}mv^2$. Show that

$$\frac{\partial K}{\partial m} \frac{\partial^2 K}{\partial v^2} = K$$

76. If a, b, c are the sides of a triangle and A, B, C are the opposite angles, find $\partial A/\partial a$, $\partial A/\partial b$, $\partial A/\partial c$ by implicit differentiation of the Law of Cosines.

77. You are told that there is a function f whose partial derivatives are $f_x(x, y) = x + 4y$ and $f_y(x, y) = 3x - y$. Should you believe it?

78. The paraboloid $z = 6 - x - x^2 - 2y^2$ intersects the plane $x = 1$ in a parabola. Find parametric equations for the tangent line to this parabola at the point $(1, 2, -4)$. Use a computer to graph the paraboloid, the parabola, and the tangent line on the same screen.

79. The ellipsoid $4x^2 + 2y^2 + z^2 = 16$ intersects the plane $y = 2$ in an ellipse. Find parametric equations for the tangent line to this ellipse at the point $(1, 2, 2)$.

80. In a study of frost penetration it was found that the temperature T at time t (measured in days) at a depth x (measured in feet) can be modeled by the function

$$T(x, t) = T_0 + T_1e^{-\lambda x}\sin(\omega t - \lambda x)$$

where $\omega = 2\pi/365$ and λ is a positive constant.
(a) Find $\partial T/\partial x$. What is its physical significance?
(b) Find $\partial T/\partial t$. What is its physical significance?
(c) Show that T satisfies the heat equation $T_t = kT_{xx}$ for a certain constant k.
(d) If $\lambda = 0.2$, $T_0 = 0$, and $T_1 = 10$, use a computer to graph $T(x, t)$.
(e) What is the physical significance of the term $-\lambda x$ in the expression $\sin(\omega t - \lambda x)$?

81. If $f(x, y) = x(x^2 + y^2)^{-3/2}e^{\sin(x^2 y)}$, find $f_x(1, 0)$.
[*Hint:* Instead of finding $f_x(x, y)$ first, note that it's easier to use Equation 1 or Equation 2.]

82. If $f(x, y) = \sqrt[3]{x^3 + y^3}$, find $f_x(0, 0)$.

83. Let

$$f(x, y) = \begin{cases} \dfrac{x^3y - xy^3}{x^2 + y^2} & \text{if } (x, y) \neq (0, 0) \\ 0 & \text{if } (x, y) = (0, 0) \end{cases}$$

(a) Use a computer to graph f.
(b) Find $f_x(x, y)$ and $f_y(x, y)$ when $(x, y) \neq (0, 0)$.
(c) Find $f_x(0, 0)$ and $f_y(0, 0)$ using Equations 2 and 3.
(d) Show that $f_{xy}(0, 0) = -1$ and $f_{yx}(0, 0) = 1$.
(e) Does the result of part (d) contradict Clairaut's Theorem? Use graphs of f_{xy} and f_{yx} to illustrate your answer.

11.4 Tangent Planes and Linear Approximations

One of the most important ideas in single-variable calculus is that as we zoom in toward a point on the graph of a differentiable function, the graph becomes indistinguishable from its tangent line and we can approximate the function by a linear function. (See Section 3.8.) Here we develop similar ideas in three dimensions. As we zoom in toward a point on a surface that is the graph of a differentiable function of two variables, the surface looks more and more like a plane (its tangent plane) and we can approximate the function by a linear function of two variables. We also extend the idea of a differential to functions of two or more variables.

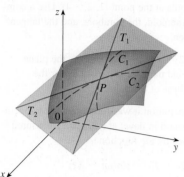

FIGURE 1
The tangent plane contains the
tangent lines T_1 and T_2.

■■ Tangent Planes

Suppose a surface S has equation $z = f(x, y)$, where f has continuous first partial derivatives, and let $P(x_0, y_0, z_0)$ be a point on S. As in the preceding section, let C_1 and C_2 be the curves obtained by intersecting the vertical planes $y = y_0$ and $x = x_0$ with the surface S. Then the point P lies on both C_1 and C_2. Let T_1 and T_2 be the tangent lines to the curves C_1 and C_2 at the point P. Then the **tangent plane** to the surface S at the point P is defined to be the plane that contains both tangent lines T_1 and T_2. (See Figure 1.)

We will see in Section 11.6 that if C is any other curve that lies on the surface S and passes through P, then its tangent line at P also lies in the tangent plane. Therefore, you can think of the tangent plane to S at P as consisting of all possible tangent lines at P to curves that lie on S and pass through P. The tangent plane at P is the plane that most closely approximates the surface S near the point P.

We know from Equation 9.5.7 that any plane passing through the point $P(x_0, y_0, z_0)$ has an equation of the form

$$A(x - x_0) + B(y - y_0) + C(z - z_0) = 0$$

By dividing this equation by C and letting $a = -A/C$ and $b = -B/C$, we can write it in the form

1 $$z - z_0 = a(x - x_0) + b(y - y_0)$$

If Equation 1 represents the tangent plane at P, then its intersection with the plane $y = y_0$ must be the tangent line T_1. Setting $y = y_0$ in Equation 1 gives

$$z - z_0 = a(x - x_0) \qquad y = y_0$$

and we recognize these as the equations (in point-slope form) of a line with slope a. But from Section 11.3 we know that the slope of the tangent T_1 is $f_x(x_0, y_0)$. Therefore, $a = f_x(x_0, y_0)$.

Similarly, putting $x = x_0$ in Equation 1, we get $z - z_0 = b(y - y_0)$, which must represent the tangent line T_2, so $b = f_y(x_0, y_0)$.

■■ Note the similarity between the equation of a tangent plane and the equation of a tangent line:

$$y - y_0 = f'(x_0)(x - x_0)$$

2 Suppose f has continuous partial derivatives. An equation of the tangent plane to the surface $z = f(x, y)$ at the point $P(x_0, y_0, z_0)$ is

$$z - z_0 = f_x(x_0, y_0)(x - x_0) + f_y(x_0, y_0)(y - y_0)$$

V **EXAMPLE 1** Find the tangent plane to the elliptic paraboloid $z = 2x^2 + y^2$ at the point $(1, 1, 3)$.

SOLUTION Let $f(x, y) = 2x^2 + y^2$. Then

$$f_x(x, y) = 4x \qquad f_y(x, y) = 2y$$

$$f_x(1, 1) = 4 \qquad f_y(1, 1) = 2$$

Then (2) gives the equation of the tangent plane at $(1, 1, 3)$ as

$$z - 3 = 4(x - 1) + 2(y - 1)$$

or

$$z = 4x + 2y - 3$$

■ ■

Visual 11.4 shows an animation of Figure 2.

Figure 2(a) shows the elliptic paraboloid and its tangent plane at $(1, 1, 3)$ that we found in Example 1. In parts (b) and (c) we zoom in toward the point $(1, 1, 3)$ by restricting the domain of the function $f(x, y) = 2x^2 + y^2$. Notice that the more we zoom in, the flatter the graph appears and the more it resembles its tangent plane.

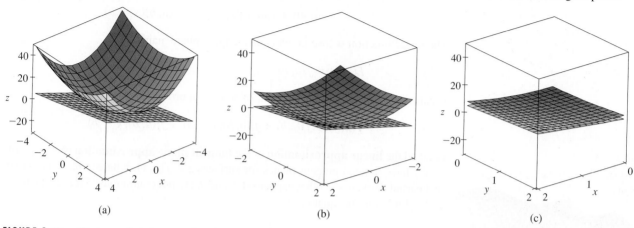

(a)   (b)   (c)

FIGURE 2 The elliptic paraboloid $z = 2x^2 + y^2$ appears to coincide with its tangent plane as we zoom in toward $(1, 1, 3)$.

In Figure 3 we corroborate this impression by zooming in toward the point $(1, 1)$ on a contour map of the function $f(x, y) = 2x^2 + y^2$. Notice that the more we zoom in, the more the level curves look like equally spaced parallel lines, which is characteristic of a plane.

FIGURE 3
Zooming in toward $(1, 1)$
on a contour map of
$f(x, y) = 2x^2 + y^2$

▦ Linear Approximations

In Example 1 we found that an equation of the tangent plane to the graph of the function $f(x, y) = 2x^2 + y^2$ at the point $(1, 1, 3)$ is $z = 4x + 2y - 3$. Therefore, in view of the visual evidence in Figures 2 and 3, the linear function of two variables

$$L(x, y) = 4x + 2y - 3$$

is a good approximation to $f(x, y)$ when (x, y) is near $(1, 1)$. The function L is called the *linearization* of f at $(1, 1)$ and the approximation

$$f(x, y) \approx 4x + 2y - 3$$

is called the *linear approximation* or *tangent plane approximation* of f at $(1, 1)$.
For instance, at the point $(1.1, 0.95)$ the linear approximation gives

$$f(1.1, 0.95) \approx 4(1.1) + 2(0.95) - 3 = 3.3$$

which is quite close to the true value of $f(1.1, 0.95) = 2(1.1)^2 + (0.95)^2 = 3.3225$.

But if we take a point farther away from (1, 1), such as (2, 3), we no longer get a good approximation. In fact, $L(2, 3) = 11$ whereas $f(2, 3) = 17$.

In general, we know from (2) that an equation of the tangent plane to the graph of a function f of two variables at the point $(a, b, f(a, b))$ is

$$z = f(a, b) + f_x(a, b)(x - a) + f_y(a, b)(y - b)$$

The linear function whose graph is this tangent plane, namely

$$\boxed{3} \qquad L(x, y) = f(a, b) + f_x(a, b)(x - a) + f_y(a, b)(y - b)$$

is called the **linearization** of f at (a, b) and the approximation

$$\boxed{4} \qquad f(x, y) \approx f(a, b) + f_x(a, b)(x - a) + f_y(a, b)(y - b)$$

is called the **linear approximation** or the **tangent plane approximation** of f at (a, b).

We have defined tangent planes for surfaces $z = f(x, y)$, where f has continuous first partial derivatives. What happens if f_x and f_y are not continuous? Figure 4 pictures such a function; its equation is

$$f(x, y) = \begin{cases} \dfrac{xy}{x^2 + y^2} & \text{if } (x, y) \neq (0, 0) \\ 0 & \text{if } (x, y) = (0, 0) \end{cases}$$

You can verify (see Exercise 42) that its partial derivatives exist at the origin and, in fact, $f_x(0, 0) = 0$ and $f_y(0, 0) = 0$, but f_x and f_y are not continuous. The linear approximation would be $f(x, y) \approx 0$, but $f(x, y) = \frac{1}{2}$ at all points on the line $y = x$. So a function of two variables can behave badly even though both of its partial derivatives exist. To rule out such behavior, we formulate the idea of a differentiable function of two variables.

Recall that for a function of one variable, $y = f(x)$, if x changes from a to $a + \Delta x$, we defined the increment of y as

$$\Delta y = f(a + \Delta x) - f(a)$$

In Chapter 3 we showed that if f is differentiable at a, then

■ ■ This is Equation 3.5.8.

$$\boxed{5} \qquad \Delta y = f'(a)\,\Delta x + \varepsilon\,\Delta x \qquad \text{where } \varepsilon \to 0 \text{ as } \Delta x \to 0$$

Now consider a function of two variables, $z = f(x, y)$, and suppose x changes from a to $a + \Delta x$ and y changes from b to $b + \Delta y$. Then the corresponding **increment** of z is

$$\boxed{6} \qquad \Delta z = f(a + \Delta x, b + \Delta y) - f(a, b)$$

Thus, the increment Δz represents the change in the value of f when (x, y) changes from (a, b) to $(a + \Delta x, b + \Delta y)$. By analogy with (5) we define the differentiability of a function of two variables as follows.

> $\boxed{7}$ **Definition** If $z = f(x, y)$, then f is **differentiable** at (a, b) if Δz can be expressed in the form
>
> $$\Delta z = f_x(a, b)\,\Delta x + f_y(a, b)\,\Delta y + \varepsilon_1\,\Delta x + \varepsilon_2\,\Delta y$$
>
> where ε_1 and $\varepsilon_2 \to 0$ as $(\Delta x, \Delta y) \to (0, 0)$.

FIGURE 4

$f(x, y) = \dfrac{xy}{x^2 + y^2}$ if $(x, y) \neq (0, 0)$,

$f(0, 0) = 0$

Definition 7 says that a differentiable function is one for which the linear approximation (4) is a good approximation when (x, y) is near (a, b). In other words, the tangent plane approximates the graph of f well near the point of tangency.

It's sometimes hard to use Definition 7 directly to check the differentiability of a function, but the following theorem provides a convenient sufficient condition for differentiability.

■ ■ Theorem 8 is proved in Appendix E.

8 Theorem If the partial derivatives f_x and f_y exist near (a, b) and are continuous at (a, b), then f is differentiable at (a, b).

$\boxed{V}$ **EXAMPLE 2** Show that $f(x, y) = xe^{xy}$ is differentiable at $(1, 0)$ and find its linearization there. Then use it to approximate $f(1.1, -0.1)$.

SOLUTION The partial derivatives are

$$f_x(x, y) = e^{xy} + xye^{xy} \qquad f_y(x, y) = x^2 e^{xy}$$

$$f_x(1, 0) = 1 \qquad\qquad f_y(1, 0) = 1$$

■ ■ Figure 5 shows the graphs of the function f and its linearization L in Example 2.

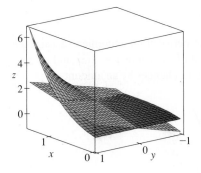

FIGURE 5

Both f_x and f_y are continuous functions, so f is differentiable by Theorem 8. The linearization is

$$L(x, y) = f(1, 0) + f_x(1, 0)(x - 1) + f_y(1, 0)(y - 0)$$

$$= 1 + 1(x - 1) + 1 \cdot y = x + y$$

The corresponding linear approximation is

$$xe^{xy} \approx x + y$$

so

$$f(1.1, -0.1) \approx 1.1 - 0.1 = 1$$

Compare this with the actual value of $f(1.1, -0.1) = 1.1e^{-0.11} \approx 0.98542$. ■ ■

EXAMPLE 3 At the beginning of Section 11.3 we discussed the heat index (perceived temperature) I as a function of the actual temperature T and the relative humidity H and gave the following table of values from the National Weather Service.

Relative humidity (%)

T \\ H	50	55	60	65	70	75	80	85	90
90	96	98	100	103	106	109	112	115	119
92	100	103	105	108	112	115	119	123	128
94	104	107	111	114	118	122	127	132	137
96	109	113	116	121	125	130	135	141	146
98	114	118	123	127	133	138	144	150	157
100	119	124	129	135	141	147	154	161	168

Actual temperature (°F)

Find a linear approximation for the heat index $I = f(T, H)$ when T is near 96°F and H is near 70%. Use it to estimate the heat index when the temperature is 97°F and the relative humidity is 72%.

SOLUTION We read from the table that $f(96, 70) = 125$. In Section 11.3 we used the tabular values to estimate that $f_T(96, 70) \approx 3.75$ and $f_H(96, 70) \approx 0.9$. (See pages 757–58.) So the linear approximation is

$$f(T, H) \approx f(96, 70) + f_T(96, 70)(T - 96) + f_H(96, 70)(H - 70)$$

$$\approx 125 + 3.75(T - 96) + 0.9(H - 70)$$

In particular,

$$f(97, 72) \approx 125 + 3.75(1) + 0.9(2) = 130.55$$

Therefore, when $T = 97°F$ and $H = 72\%$, the heat index is

$$I \approx 131°F$$

■ ■

▪▪ Differentials
▪▪

For a differentiable function of one variable, $y = f(x)$, we define the differential dx to be an independent variable; that is, dx can be given the value of any real number. The differential of y is then defined as

$$\boxed{9} \qquad\qquad dy = f'(x)\, dx$$

(See Section 3.8.) Figure 6 shows the relationship between the increment Δy and the differential dy: Δy represents the change in height of the curve $y = f(x)$ and dy represents the change in height of the tangent line when x changes by an amount $dx = \Delta x$.

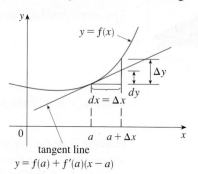

FIGURE 6

For a differentiable function of two variables, $z = f(x, y)$, we define the **differentials** dx and dy to be independent variables; that is, they can be given any values. Then the **differential** dz, also called the **total differential**, is defined by

$$\boxed{10} \qquad dz = f_x(x, y)\, dx + f_y(x, y)\, dy = \frac{\partial z}{\partial x}\, dx + \frac{\partial z}{\partial y}\, dy$$

(Compare with Equation 9.) Sometimes the notation df is used in place of dz.

If we take $dx = \Delta x = x - a$ and $dy = \Delta y = y - b$ in Equation 10, then the differential of z is

$$dz = f_x(a, b)(x - a) + f_y(a, b)(y - b)$$

So, in the notation of differentials, the linear approximation (4) can be written as

$$f(x, y) \approx f(a, b) + dz$$

Figure 7 is the three-dimensional counterpart of Figure 6 and shows the geometric interpretation of the differential dz and the increment Δz: dz represents the change in height of the tangent plane, whereas Δz represents the change in height of the surface $z = f(x, y)$ when (x, y) changes from (a, b) to $(a + \Delta x, b + \Delta y)$.

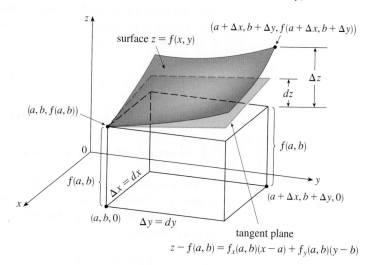

FIGURE 7

$$z - f(a, b) = f_x(a, b)(x - a) + f_y(a, b)(y - b)$$

V | **EXAMPLE 4**

(a) If $z = f(x, y) = x^2 + 3xy - y^2$, find the differential dz.
(b) If x changes from 2 to 2.05 and y changes from 3 to 2.96, compare the values of Δz and dz.

SOLUTION

(a) Definition 10 gives

$$dz = \frac{\partial z}{\partial x}\, dx + \frac{\partial z}{\partial y}\, dy = (2x + 3y)\, dx + (3x - 2y)\, dy$$

■ ■ In Example 4, dz is close to Δz because the tangent plane is a good approximation to the surface $z = x^2 + 3xy - y^2$ near $(2, 3, 13)$. (See Figure 8.)

(b) Putting $x = 2$, $dx = \Delta x = 0.05$, $y = 3$, and $dy = \Delta y = -0.04$, we get

$$dz = [2(2) + 3(3)]0.05 + [3(2) - 2(3)](-0.04)$$
$$= 0.65$$

The increment of z is

$$\Delta z = f(2.05, 2.96) - f(2, 3)$$
$$= [(2.05)^2 + 3(2.05)(2.96) - (2.96)^2] - [2^2 + 3(2)(3) - 3^2]$$
$$= 0.6449$$

FIGURE 8

Notice that $\Delta z \approx dz$ but dz is easier to compute. ■ ■

EXAMPLE 5 The base radius and height of a right circular cone are measured as 10 cm and 25 cm, respectively, with a possible error in measurement of as much as

0.1 cm in each. Use differentials to estimate the maximum error in the calculated volume of the cone.

SOLUTION The volume V of a cone with base radius r and height h is $V = \pi r^2 h/3$. So the differential of V is

$$dV = \frac{\partial V}{\partial r}\,dr + \frac{\partial V}{\partial h}\,dh = \frac{2\pi rh}{3}\,dr + \frac{\pi r^2}{3}\,dh$$

Since each error is at most 0.1 cm, we have $|\Delta r| \leq 0.1$, $|\Delta h| \leq 0.1$. To find the largest error in the volume we take the largest error in the measurement of r and of h. Therefore, we take $dr = 0.1$ and $dh = 0.1$ along with $r = 10$, $h = 25$. This gives

$$dV = \frac{500\pi}{3}(0.1) + \frac{100\pi}{3}(0.1) = 20\pi$$

Thus, the maximum error in the calculated volume is about 20π cm$^3 \approx 63$ cm^3. ■ ■

■■ Functions of Three or More Variables

Linear approximations, differentiability, and differentials can be defined in a similar manner for functions of more than two variables. A differentiable function is defined by an expression similar to the one in Definition 7. For such functions the **linear approximation** is

$$f(x, y, z) \approx f(a, b, c) + f_x(a, b, c)(x - a) + f_y(a, b, c)(y - b) + f_z(a, b, c)(z - c)$$

and the linearization $L(x, y, z)$ is the right side of this expression.

If $w = f(x, y, z)$, then the **increment** of w is

$$\Delta w = f(x + \Delta x, y + \Delta y, z + \Delta z) - f(x, y, z)$$

The **differential** dw is defined in terms of the differentials dx, dy, and dz of the independent variables by

$$dw = \frac{\partial w}{\partial x}\,dx + \frac{\partial w}{\partial y}\,dy + \frac{\partial w}{\partial z}\,dz$$

EXAMPLE 6 The dimensions of a rectangular box are measured to be 75 cm, 60 cm, and 40 cm, and each measurement is correct to within 0.2 cm. Use differentials to estimate the largest possible error when the volume of the box is calculated from these measurements.

SOLUTION If the dimensions of the box are x, y, and z, its volume is $V = xyz$ and so

$$dV = \frac{\partial V}{\partial x}\,dx + \frac{\partial V}{\partial y}\,dy + \frac{\partial V}{\partial z}\,dz = yz\,dx + xz\,dy + xy\,dz$$

We are given that $|\Delta x| \leq 0.2$, $|\Delta y| \leq 0.2$, and $|\Delta z| \leq 0.2$. To find the largest error in the volume, we therefore use $dx = 0.2$, $dy = 0.2$, and $dz = 0.2$ together with $x = 75$, $y = 60$, and $z = 40$:

$$\Delta V \approx dV = (60)(40)(0.2) + (75)(40)(0.2) + (75)(60)(0.2) = 1980$$

Thus, an error of only 0.2 cm in measuring each dimension could lead to an error of as much as 1980 cm^3 in the calculated volume! This may seem like a large error, but it's only about 1% of the volume of the box. ■ ■

▧ Tangent Planes to Parametric Surfaces

Parametric surfaces were introduced in Section 10.5. We now find the tangent plane to a parametric surface S traced out by a vector function

$$\mathbf{r}(u, v) = x(u, v)\,\mathbf{i} + y(u, v)\,\mathbf{j} + z(u, v)\,\mathbf{k}$$

at a point P_0 with position vector $\mathbf{r}(u_0, v_0)$. If we keep u constant by putting $u = u_0$, then $\mathbf{r}(u_0, v)$ becomes a vector function of the single parameter v and defines a grid curve C_1 lying on S. (See Figure 9.) The tangent vector to C_1 at P_0 is obtained by taking the partial derivative of $\mathbf{r}$ with respect to v:

$$\mathbf{r}_v = \frac{\partial x}{\partial v}(u_0, v_0)\,\mathbf{i} + \frac{\partial y}{\partial v}(u_0, v_0)\,\mathbf{j} + \frac{\partial z}{\partial v}(u_0, v_0)\,\mathbf{k}$$

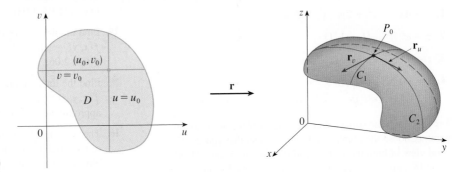

FIGURE 9

Similarly, if we keep v constant by putting $v = v_0$, we get a grid curve C_2 given by $\mathbf{r}(u, v_0)$ that lies on S, and its tangent vector at P_0 is

$$\mathbf{r}_u = \frac{\partial x}{\partial u}(u_0, v_0)\,\mathbf{i} + \frac{\partial y}{\partial u}(u_0, v_0)\,\mathbf{j} + \frac{\partial z}{\partial u}(u_0, v_0)\,\mathbf{k}$$

If $\mathbf{r}_u \times \mathbf{r}_v$ is not $\mathbf{0}$, then the surface S is called **smooth** (it has no "corners"). For a smooth surface, the **tangent plane** is the plane that contains the tangent vectors $\mathbf{r}_u$ and $\mathbf{r}_v$, and the vector $\mathbf{r}_u \times \mathbf{r}_v$ is a normal vector to the tangent plane.

Ⓥ **EXAMPLE 7** Find the tangent plane to the surface with parametric equations $x = u^2$, $y = v^2$, $z = u + 2v$ at the point $(1, 1, 3)$.

■ ■ Figure 10 shows the self-intersecting surface in Example 7 and its tangent plane at $(1, 1, 3)$.

SOLUTION We first compute the tangent vectors:

$$\mathbf{r}_u = \frac{\partial x}{\partial u}\,\mathbf{i} + \frac{\partial y}{\partial u}\,\mathbf{j} + \frac{\partial z}{\partial u}\,\mathbf{k} = 2u\,\mathbf{i} + \mathbf{k}$$

$$\mathbf{r}_v = \frac{\partial x}{\partial v}\,\mathbf{i} + \frac{\partial y}{\partial v}\,\mathbf{j} + \frac{\partial z}{\partial v}\,\mathbf{k} = 2v\,\mathbf{j} + 2\,\mathbf{k}$$

FIGURE 10

Thus, a normal vector to the tangent plane is

$$\mathbf{r}_u \times \mathbf{r}_v = \begin{vmatrix} \mathbf{i} & \mathbf{j} & \mathbf{k} \\ 2u & 0 & 1 \\ 0 & 2v & 2 \end{vmatrix} = -2v\,\mathbf{i} - 4u\,\mathbf{j} + 4uv\,\mathbf{k}$$

Notice that the point $(1, 1, 3)$ corresponds to the parameter values $u = 1$ and $v = 1$, so the normal vector there is

$$-2\,\mathbf{i} - 4\,\mathbf{j} + 4\,\mathbf{k}$$

Therefore, an equation of the tangent plane at $(1, 1, 3)$ is

$$-2(x - 1) - 4(y - 1) + 4(z - 3) = 0$$

or

$$x + 2y - 2z + 3 = 0 \qquad\qquad ■\ ■$$

11.4 Exercises

1–4 ■ Find an equation of the tangent plane to the given surface at the specified point.

1. $z = 4x^2 - y^2 + 2y$, $\quad (-1, 2, 4)$

2. $z = e^{x^2 - y^2}$, $\quad (1, -1, 1)$

3. $z = y \cos(x - y)$, $\quad (2, 2, 2)$

4. $z = y \ln x$, $\quad (1, 4, 0)$

. . . .

5–6 ■ Graph the surface and the tangent plane at the given point. (Choose the domain and viewpoint so that you get a good view of both the surface and the tangent plane.) Then zoom in until the surface and the tangent plane become indistinguishable.

5. $z = x^2 + xy + 3y^2$, $\quad (1, 1, 5)$

6. $z = \arctan(xy^2)$, $\quad (1, 1, \pi/4)$

. . . .

CAS **7–8** ■ Draw the graph of f and its tangent plane at the given point. (Use your computer algebra system both to compute the partial derivatives and to graph the surface and its tangent plane.) Then zoom in until the surface and the tangent plane become indistinguishable.

7. $f(x, y) = \dfrac{xy \sin(x - y)}{1 + x^2 + y^2}$, $\quad (1, 1, 0)$

8. $f(x, y) = e^{-xy/10}\left(\sqrt{x} + \sqrt{y} + \sqrt{xy}\right)$, $\quad (1, 1, 3e^{-0.1})$

. . . .

9–12 ■ Explain why the function is differentiable at the given point. Then find the linearization $L(x, y)$ of the function at that point.

9. $f(x, y) = x\sqrt{y}$, $\quad (1, 4)$

10. $f(x, y) = x/y$, $\quad (6, 3)$

11. $f(x, y) = \tan^{-1}(x + 2y)$, $\quad (1, 0)$

12. $f(x, y) = \sqrt{x + e^{4y}}$, $\quad (3, 0)$

. . . .

13. Find the linear approximation of the function $f(x, y) = \sqrt{20 - x^2 - 7y^2}$ at $(2, 1)$ and use it to approximate $f(1.95, 1.08)$.

14. Find the linear approximation of the function $f(x, y) = \ln(x - 3y)$ at $(7, 2)$ and use it to approximate $f(6.9, 2.06)$. Illustrate by graphing f and the tangent plane.

15. Find the linear approximation of the function $f(x, y, z) = \sqrt{x^2 + y^2 + z^2}$ at $(3, 2, 6)$ and use it to approximate the number $\sqrt{(3.02)^2 + (1.97)^2 + (5.99)^2}$.

16. The wave heights h in the open sea depend on the speed v of the wind and the length of time t that the wind has been blowing at that speed. Values of the function $h = f(v, t)$ are recorded in feet in the following table.

Duration (hours)

v \ t	5	10	15	20	30	40	50
20	5	7	8	8	9	9	9
30	9	13	16	17	18	19	19
40	14	21	25	28	31	33	33
50	19	29	36	40	45	48	50
60	24	37	47	54	62	67	69

Wind speed (knots)

Use the table to find a linear approximation to the wave height function when v is near 40 knots and t is near 20 hours. Then estimate the wave heights when the wind has been blowing for 24 hours at 43 knots.

17. Use the table in Example 3 to find a linear approximation to the heat index function when the temperature is near 94°F and the relative humidity is near 80%. Then estimate the heat index when the temperature is 95°F and the relative humidity is 78%.

18. The wind-chill index W is the perceived temperature when the actual temperature is T and the wind speed is v, so we can write $W = f(T, v)$. The following table of values is an excerpt from Table 1 in Section 11.1.

Wind speed (km/h)

T \ v	20	30	40	50	60	70
-10	-18	-20	-21	-22	-23	-23
-15	-24	-26	-27	-29	-30	-30
-20	-30	-33	-34	-35	-36	-37
-25	-37	-39	-41	-42	-43	-44

Actual temperature (°C)

Use the table to find a linear approximation to the wind-chill index function when T is near $-15°C$ and v is near 50 km/h. Then estimate the wind-chill index when the temperature is $-17°C$ and the wind speed is 55 km/h.

19–22 ■ Find the differential of the function.

19. $z = x^3 \ln(y^2)$ **20.** $u = e^{-t}\sin(s + 2t)$

21. $R = \alpha\beta^2 \cos \gamma$ **22.** $w = xye^{xz}$

23. If $z = 5x^2 + y^2$ and (x, y) changes from $(1, 2)$ to $(1.05, 2.1)$, compare the values of Δz and dz.

24. If $z = x^2 - xy + 3y^2$ and (x, y) changes from $(3, -1)$ to $(2.96, -0.95)$, compare the values of Δz and dz.

25. The length and width of a rectangle are measured as 30 cm and 24 cm, respectively, with an error in measurement of at most 0.1 cm in each. Use differentials to estimate the maximum error in the calculated area of the rectangle.

26. The dimensions of a closed rectangular box are measured as 80 cm, 60 cm, and 50 cm, respectively, with a possible error of 0.2 cm in each dimension. Use differentials to estimate the maximum error in calculating the surface area of the box.

27. Use differentials to estimate the amount of tin in a closed tin can with diameter 8 cm and height 12 cm if the tin is 0.04 cm thick.

28. Use differentials to estimate the amount of metal in a closed cylindrical can that is 10 cm high and 4 cm in diameter if the metal in the top and bottom is 0.1 cm thick and the metal in the sides is 0.05 cm thick.

29. A model for the surface area of a human body is given by $S = 0.1091w^{0.425}h^{0.725}$, where w is the weight (in pounds), h is the height (in inches), and S is measured in square feet. If the errors in measurement of w and h are at most 2%, use differentials to estimate the maximum percentage error in the calculated surface area.

30. The pressure, volume, and temperature of a mole of an ideal gas are related by the equation $PV = 8.31T$, where P is measured in kilopascals, V in liters, and T in kelvins. Use differentials to find the approximate change in the pressure if the volume increases from 12 L to 12.3 L and the temperature decreases from 310 K to 305 K.

31. If R is the total resistance of three resistors, connected in parallel, with resistances R_1, R_2, R_3, then

$$\frac{1}{R} = \frac{1}{R_1} + \frac{1}{R_2} + \frac{1}{R_3}$$

If the resistances are measured in ohms as $R_1 = 25\ \Omega$, $R_2 = 40\ \Omega$, and $R_3 = 50\ \Omega$, with a possible error of 0.5% in each case, estimate the maximum error in the calculated value of R.

32. Four positive numbers, each less than 50, are rounded to the first decimal place and then multiplied together. Use differentials to estimate the maximum possible error in the computed product that might result from the rounding.

33–37 ■ Find an equation of the tangent plane to the given parametric surface at the specified point. If you have software that graphs parametric surfaces, use a computer to graph the surface and the tangent plane.

33. $x = u + v$, $y = 3u^2$, $z = u - v$; $(2, 3, 0)$

34. $x = u^2$, $y = v^2$, $z = uv$; $u = 1, v = 1$

35. $\mathbf{r}(u, v) = u^2\,\mathbf{i} + 2u \sin v\,\mathbf{j} + u \cos v\,\mathbf{k}$; $u = 1, v = 0$

36. $\mathbf{r}(u, v) = uv\,\mathbf{i} + u \sin v\,\mathbf{j} + v \cos u\,\mathbf{k}$; $u = 0, v = \pi$

37. $\mathbf{r}(u, v) = u\,\mathbf{i} + \ln(uv)\,\mathbf{j} + v\,\mathbf{k}$; $u = 1, v = 1$

38. Suppose you need to know an equation of the tangent plane to a surface S at the point $P(2, 1, 3)$. You don't have an equation for S but you know that the curves

$$\mathbf{r}_1(t) = \langle 2 + 3t, 1 - t^2, 3 - 4t + t^2 \rangle$$

$$\mathbf{r}_2(u) = \langle 1 + u^2, 2u^3 - 1, 2u + 1 \rangle$$

both lie on S. Find an equation of the tangent plane at P.

39–40 ■ Show that the function is differentiable by finding values of ε_1 and ε_2 that satisfy Definition 7.

39. $f(x, y) = x^2 + y^2$ **40.** $f(x, y) = xy - 5y^2$

41. Prove that if f is a function of two variables that is differentiable at (a, b), then f is continuous at (a, b).
    Hint: Show that

$$\lim_{(\Delta x,\, \Delta y) \to (0,\, 0)} f(a + \Delta x, b + \Delta y) = f(a, b)$$

42. (a) The function

$$f(x, y) = \begin{cases} \dfrac{xy}{x^2 + y^2} & \text{if } (x, y) \neq (0, 0) \\ 0 & \text{if } (x, y) = (0, 0) \end{cases}$$

was graphed in Figure 4. Show that $f_x(0, 0)$ and $f_y(0, 0)$ both exist but f is not differentiable at $(0, 0)$. [*Hint:* Use the result of Exercise 41.]

   (b) Explain why f_x and f_y are not continuous at $(0, 0)$.

11.5 The Chain Rule

Recall that the Chain Rule for functions of a single variable gives the rule for differentiating a composite function: If $y = f(x)$ and $x = g(t)$, where f and g are differentiable functions, then y is indirectly a differentiable function of t and

$$\boxed{1} \qquad \frac{dy}{dt} = \frac{dy}{dx} \frac{dx}{dt}$$

For functions of more than one variable, the Chain Rule has several versions, each of them giving a rule for differentiating a composite function. The first version (Theorem 2) deals with the case where $z = f(x, y)$ and each of the variables x and y is, in turn, a function of a variable t. This means that z is indirectly a function of t, $z = f(g(t), h(t))$, and the Chain Rule gives a formula for differentiating z as a function of t. We assume that f is differentiable (Definition 11.4.7). Recall that this is the case when f_x and f_y are continuous (Theorem 11.4.8).

$\boxed{2}$ **The Chain Rule (Case 1)** Suppose that $z = f(x, y)$ is a differentiable function of x and y, where $x = g(t)$ and $y = h(t)$ are both differentiable functions of t. Then z is a differentiable function of t and

$$\frac{dz}{dt} = \frac{\partial f}{\partial x} \frac{dx}{dt} + \frac{\partial f}{\partial y} \frac{dy}{dt}$$

Proof A change of Δt in t produces changes of Δx in x and Δy in y. These, in turn, produce a change of Δz in z, and from Definition 11.4.7 we have

$$\Delta z = \frac{\partial f}{\partial x} \Delta x + \frac{\partial f}{\partial y} \Delta y + \varepsilon_1 \Delta x + \varepsilon_2 \Delta y$$

where $\varepsilon_1 \to 0$ and $\varepsilon_2 \to 0$ as $(\Delta x, \Delta y) \to (0, 0)$. [If the functions ε_1 and ε_2 are not defined at $(0, 0)$, we can define them to be 0 there.] Dividing both sides of this equation by Δt, we have

$$\frac{\Delta z}{\Delta t} = \frac{\partial f}{\partial x} \frac{\Delta x}{\Delta t} + \frac{\partial f}{\partial y} \frac{\Delta y}{\Delta t} + \varepsilon_1 \frac{\Delta x}{\Delta t} + \varepsilon_2 \frac{\Delta y}{\Delta t}$$

If we now let $\Delta t \to 0$, then $\Delta x = g(t + \Delta t) - g(t) \to 0$ because g is differentiable and therefore continuous. Similarly, $\Delta y \to 0$. This, in turn, means that $\varepsilon_1 \to 0$ and $\varepsilon_2 \to 0$, so

$$\frac{dz}{dt} = \lim_{\Delta t \to 0} \frac{\Delta z}{\Delta t}$$

$$= \frac{\partial f}{\partial x} \lim_{\Delta t \to 0} \frac{\Delta x}{\Delta t} + \frac{\partial f}{\partial y} \lim_{\Delta t \to 0} \frac{\Delta y}{\Delta t} + \lim_{\Delta t \to 0} \varepsilon_1 \lim_{\Delta t \to 0} \frac{\Delta x}{\Delta t} + \lim_{\Delta t \to 0} \varepsilon_2 \lim_{\Delta t \to 0} \frac{\Delta y}{\Delta t}$$

$$= \frac{\partial f}{\partial x} \frac{dx}{dt} + \frac{\partial f}{\partial y} \frac{dy}{dt} + 0 \cdot \frac{dx}{dt} + 0 \cdot \frac{dy}{dt}$$

$$= \frac{\partial f}{\partial x} \frac{dx}{dt} + \frac{\partial f}{\partial y} \frac{dy}{dt}$$

■ ■

Since we often write $\partial z/\partial x$ in place of $\partial f/\partial x$, we can rewrite the Chain Rule in the form

■■ Notice the similarity to the definition of the differential:

$$dz = \frac{\partial z}{\partial x}\,dx + \frac{\partial z}{\partial y}\,dy$$

$$\frac{dz}{dt} = \frac{\partial z}{\partial x}\frac{dx}{dt} + \frac{\partial z}{\partial y}\frac{dy}{dt}$$

EXAMPLE 1 If $z = x^2y + 3xy^4$, where $x = \sin 2t$ and $y = \cos t$, find dz/dt when $t = 0$.

SOLUTION The Chain Rule gives

$$\frac{dz}{dt} = \frac{\partial z}{\partial x}\frac{dx}{dt} + \frac{\partial z}{\partial y}\frac{dy}{dt}$$

$$= (2xy + 3y^4)(2\cos 2t) + (x^2 + 12xy^3)(-\sin t)$$

It's not necessary to substitute the expressions for x and y in terms of t. We simply observe that when $t = 0$ we have $x = \sin 0 = 0$ and $y = \cos 0 = 1$. Therefore,

$$\frac{dz}{dt}\bigg|_{t=0} = (0 + 3)(2\cos 0) + (0 + 0)(-\sin 0) = 6$$

■ ■

The derivative in Example 1 can be interpreted as the rate of change of z with respect to t as the point (x, y) moves along the curve C with parametric equations $x = \sin 2t$, $y = \cos t$. (See Figure 1.) In particular, when $t = 0$, the point (x, y) is $(0, 1)$ and $dz/dt = 6$ is the rate of increase as we move along the curve C through $(0, 1)$. If, for instance, $z = T(x, y) = x^2y + 3xy^4$ represents the temperature at the point (x, y), then the composite function $z = T(\sin 2t, \cos t)$ represents the temperature at points on C and the derivative dz/dt represents the rate at which the temperature changes along C.

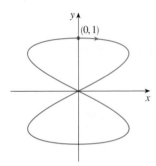

FIGURE 1

The curve $x = \sin 2t$, $y = \cos t$

$\boxed{\text{V}}$ **EXAMPLE 2** The pressure P (in kilopascals), volume V (in liters), and temperature T (in kelvins) of a mole of an ideal gas are related by the equation $PV = 8.31T$. Find the rate at which the pressure is changing when the temperature is 300 K and increasing at a rate of 0.1 K/s and the volume is 100 L and increasing at a rate of 0.2 L/s.

SOLUTION If t represents the time elapsed in seconds, then at the given instant we have $T = 300$, $dT/dt = 0.1$, $V = 100$, $dV/dt = 0.2$. Since

$$P = 8.31\frac{T}{V}$$

the Chain Rule gives

$$\frac{dP}{dt} = \frac{\partial P}{\partial T}\frac{dT}{dt} + \frac{\partial P}{\partial V}\frac{dV}{dt} = \frac{8.31}{V}\frac{dT}{dt} - \frac{8.31T}{V^2}\frac{dV}{dt}$$

$$= \frac{8.31}{100}(0.1) - \frac{8.31(300)}{100^2}(0.2) = -0.04155$$

The pressure is decreasing at a rate of about 0.042 kPa/s. ■ ■

We now consider the situation where $z = f(x, y)$ but each of x and y is a function of two variables s and t: $x = g(s, t)$, $y = h(s, t)$. Then z is indirectly a function of s and t and we wish to find $\partial z/\partial s$ and $\partial z/\partial t$. Recall that in computing $\partial z/\partial t$ we hold s fixed and compute the ordinary derivative of z with respect to t. Therefore, we can apply Theorem 2 to obtain

$$\frac{\partial z}{\partial t} = \frac{\partial z}{\partial x} \frac{\partial x}{\partial t} + \frac{\partial z}{\partial y} \frac{\partial y}{\partial t}$$

A similar argument holds for $\partial z/\partial s$ and so we have proved the following version of the Chain Rule.

---

3 The Chain Rule (Case 2) Suppose that $z = f(x, y)$ is a differentiable function of x and y, where $x = g(s, t)$ and $y = h(s, t)$ are differentiable functions of s and t. Then

$$\frac{\partial z}{\partial s} = \frac{\partial z}{\partial x} \frac{\partial x}{\partial s} + \frac{\partial z}{\partial y} \frac{\partial y}{\partial s} \qquad \frac{\partial z}{\partial t} = \frac{\partial z}{\partial x} \frac{\partial x}{\partial t} + \frac{\partial z}{\partial y} \frac{\partial y}{\partial t}$$

EXAMPLE 3 If $z = e^x \sin y$, where $x = st^2$ and $y = s^2 t$, find $\partial z/\partial s$ and $\partial z/\partial t$.

SOLUTION Applying Case 2 of the Chain Rule, we get

$$\frac{\partial z}{\partial s} = \frac{\partial z}{\partial x} \frac{\partial x}{\partial s} + \frac{\partial z}{\partial y} \frac{\partial y}{\partial s} = (e^x \sin y)(t^2) + (e^x \cos y)(2st)$$

$$= t^2 e^{st^2} \sin(s^2 t) + 2st e^{st^2} \cos(s^2 t)$$

$$\frac{\partial z}{\partial t} = \frac{\partial z}{\partial x} \frac{\partial x}{\partial t} + \frac{\partial z}{\partial y} \frac{\partial y}{\partial t} = (e^x \sin y)(2st) + (e^x \cos y)(s^2)$$

$$= 2st e^{st^2} \sin(s^2 t) + s^2 e^{st^2} \cos(s^2 t) \qquad ■ ■$$

Case 2 of the Chain Rule contains three types of variables: s and t are **independent** variables, x and y are called **intermediate** variables, and z is the **dependent** variable. Notice that Theorem 3 has one term for each intermediate variable and each of these terms resembles the one-dimensional Chain Rule in Equation 1.

To remember the Chain Rule it's helpful to draw the **tree diagram** in Figure 2. We draw branches from the dependent variable z to the intermediate variables x and y to indicate that z is a function of x and y. Then we draw branches from x and y to the independent variables s and t. On each branch we write the corresponding partial derivative. To find $\partial z/\partial s$ we find the product of the partial derivatives along each path from z to s and then add these products:

$$\frac{\partial z}{\partial s} = \frac{\partial z}{\partial x} \frac{\partial x}{\partial s} + \frac{\partial z}{\partial y} \frac{\partial y}{\partial s}$$

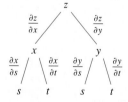

FIGURE 2

Similarly, we find $\partial z/\partial t$ by using the paths from z to t.

Now we consider the general situation in which a dependent variable u is a function of n intermediate variables $x_1, \ldots, x_n$, each of which is, in turn, a function of m

independent variables $t_1, \ldots, t_m$. Notice that there are n terms, one for each intermediate variable. The proof is similar to that of Case 1.

---

4 **The Chain Rule (General Version)** Suppose that u is a differentiable function of the n variables $x_1, x_2, \ldots, x_n$ and each x_j is a differentiable function of the m variables $t_1, t_2, \ldots, t_m$. Then u is a function of $t_1, t_2, \ldots, t_m$ and

$$\frac{\partial u}{\partial t_i} = \frac{\partial u}{\partial x_1}\frac{\partial x_1}{\partial t_i} + \frac{\partial u}{\partial x_2}\frac{\partial x_2}{\partial t_i} + \cdots + \frac{\partial u}{\partial x_n}\frac{\partial x_n}{\partial t_i}$$

for each $i = 1, 2, \ldots, m$.

FIGURE 3

EXAMPLE 4 Write out the Chain Rule for the case where $w = f(x, y, z, t)$ and $x = x(u, v)$, $y = y(u, v)$, $z = z(u, v)$, and $t = t(u, v)$.

SOLUTION We apply Theorem 4 with $n = 4$ and $m = 2$. Figure 3 shows the tree diagram. Although we haven't written the derivatives on the branches, it's understood that if a branch leads from y to u, then the partial derivative for that branch is $\partial y/\partial u$. With the aid of the tree diagram we can now write the required expressions:

$$\frac{\partial w}{\partial u} = \frac{\partial w}{\partial x}\frac{\partial x}{\partial u} + \frac{\partial w}{\partial y}\frac{\partial y}{\partial u} + \frac{\partial w}{\partial z}\frac{\partial z}{\partial u} + \frac{\partial w}{\partial t}\frac{\partial t}{\partial u}$$

$$\frac{\partial w}{\partial v} = \frac{\partial w}{\partial x}\frac{\partial x}{\partial v} + \frac{\partial w}{\partial y}\frac{\partial y}{\partial v} + \frac{\partial w}{\partial z}\frac{\partial z}{\partial v} + \frac{\partial w}{\partial t}\frac{\partial t}{\partial v} \qquad ■ ■$$

FIGURE 4

EXAMPLE 5 If $u = x^4 y + y^2 z^3$, where $x = rse^t$, $y = rs^2 e^{-t}$, and $z = r^2 s \sin t$, find the value of $\partial u/\partial s$ when $r = 2$, $s = 1$, $t = 0$.

SOLUTION With the help of the tree diagram in Figure 4, we have

$$\frac{\partial u}{\partial s} = \frac{\partial u}{\partial x}\frac{\partial x}{\partial s} + \frac{\partial u}{\partial y}\frac{\partial y}{\partial s} + \frac{\partial u}{\partial z}\frac{\partial z}{\partial s}$$

$$= (4x^3 y)(re^t) + (x^4 + 2yz^3)(2rse^{-t}) + (3y^2 z^2)(r^2 \sin t)$$

When $r = 2$, $s = 1$, and $t = 0$, we have $x = 2$, $y = 2$, and $z = 0$, so

$$\frac{\partial u}{\partial s} = (64)(2) + (16)(4) + (0)(0) = 192 \qquad ■ ■$$

EXAMPLE 6 If $g(s, t) = f(s^2 - t^2, t^2 - s^2)$ and f is differentiable, show that g satisfies the equation

$$t\frac{\partial g}{\partial s} + s\frac{\partial g}{\partial t} = 0$$

SOLUTION Let $x = s^2 - t^2$ and $y = t^2 - s^2$. Then $g(s, t) = f(x, y)$ and the Chain Rule gives

$$\frac{\partial g}{\partial s} = \frac{\partial f}{\partial x}\frac{\partial x}{\partial s} + \frac{\partial f}{\partial y}\frac{\partial y}{\partial s} = \frac{\partial f}{\partial x}(2s) + \frac{\partial f}{\partial y}(-2s)$$

$$\frac{\partial g}{\partial t} = \frac{\partial f}{\partial x}\frac{\partial x}{\partial t} + \frac{\partial f}{\partial y}\frac{\partial y}{\partial t} = \frac{\partial f}{\partial x}(-2t) + \frac{\partial f}{\partial y}(2t)$$

Therefore

$$t\frac{\partial g}{\partial s} + s\frac{\partial g}{\partial t} = \left(2st\frac{\partial f}{\partial x} - 2st\frac{\partial f}{\partial y}\right) + \left(-2st\frac{\partial f}{\partial x} + 2st\frac{\partial f}{\partial y}\right) = 0 \quad ■■$$

EXAMPLE 7 If $z = f(x, y)$ has continuous second-order partial derivatives and $x = r^2 + s^2$ and $y = 2rs$, find (a) $\partial z/\partial r$ and (b) $\partial^2 z/\partial r^2$.

SOLUTION

(a) The Chain Rule gives

$$\frac{\partial z}{\partial r} = \frac{\partial z}{\partial x}\frac{\partial x}{\partial r} + \frac{\partial z}{\partial y}\frac{\partial y}{\partial r} = \frac{\partial z}{\partial x}(2r) + \frac{\partial z}{\partial y}(2s)$$

(b) Applying the Product Rule to the expression in part (a), we get

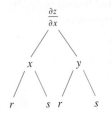

FIGURE 5

$$\boxed{5} \quad \begin{aligned} \frac{\partial^2 z}{\partial r^2} &= \frac{\partial}{\partial r}\left(2r\frac{\partial z}{\partial x} + 2s\frac{\partial z}{\partial y}\right) \\ &= 2\frac{\partial z}{\partial x} + 2r\frac{\partial}{\partial r}\left(\frac{\partial z}{\partial x}\right) + 2s\frac{\partial}{\partial r}\left(\frac{\partial z}{\partial y}\right) \end{aligned}$$

But, using the Chain Rule again (see Figure 5), we have

$$\frac{\partial}{\partial r}\left(\frac{\partial z}{\partial x}\right) = \frac{\partial}{\partial x}\left(\frac{\partial z}{\partial x}\right)\frac{\partial x}{\partial r} + \frac{\partial}{\partial y}\left(\frac{\partial z}{\partial x}\right)\frac{\partial y}{\partial r}$$

$$= \frac{\partial^2 z}{\partial x^2}(2r) + \frac{\partial^2 z}{\partial y\,\partial x}(2s)$$

$$\frac{\partial}{\partial r}\left(\frac{\partial z}{\partial y}\right) = \frac{\partial}{\partial x}\left(\frac{\partial z}{\partial y}\right)\frac{\partial x}{\partial r} + \frac{\partial}{\partial y}\left(\frac{\partial z}{\partial y}\right)\frac{\partial y}{\partial r}$$

$$= \frac{\partial^2 z}{\partial x\,\partial y}(2r) + \frac{\partial^2 z}{\partial y^2}(2s)$$

Putting these expressions into Equation 5 and using the equality of the mixed second-order derivatives, we obtain

$$\begin{aligned} \frac{\partial^2 z}{\partial r^2} &= 2\frac{\partial z}{\partial x} + 2r\left(2r\frac{\partial^2 z}{\partial x^2} + 2s\frac{\partial^2 z}{\partial y\,\partial x}\right) + 2s\left(2r\frac{\partial^2 z}{\partial x\,\partial y} + 2s\frac{\partial^2 z}{\partial y^2}\right) \\ &= 2\frac{\partial z}{\partial x} + 4r^2\frac{\partial^2 z}{\partial x^2} + 8rs\frac{\partial^2 z}{\partial x\,\partial y} + 4s^2\frac{\partial^2 z}{\partial y^2} \quad ■■ \end{aligned}$$

■■ Implicit Differentiation

The Chain Rule can be used to give a more complete description of the process of implicit differentiation that was introduced in Sections 3.6 and 11.3. We suppose that an equation of the form $F(x, y) = 0$ defines y implicitly as a differentiable function of x, that is, $y = f(x)$, where $F(x, f(x)) = 0$ for all x in the domain of f. If F is differentiable, we can apply Case 1 of the Chain Rule to differentiate both sides of the equation $F(x, y) = 0$ with respect to x. Since both x and y are functions of x, we obtain

$$\frac{\partial F}{\partial x}\frac{dx}{dx} + \frac{\partial F}{\partial y}\frac{dy}{dx} = 0$$

But $dx/dx = 1$, so if $\partial F/\partial y \neq 0$ we solve for dy/dx and obtain

$$\boxed{6} \qquad \frac{dy}{dx} = -\frac{\dfrac{\partial F}{\partial x}}{\dfrac{\partial F}{\partial y}} = -\frac{F_x}{F_y}$$

To derive this equation we assumed that $F(x, y) = 0$ defines y implicitly as a function of x. The **Implicit Function Theorem**, proved in advanced calculus, gives conditions under which this assumption is valid. It states that if F is defined on a disk containing (a, b), where $F(a, b) = 0$, $F_y(a, b) \neq 0$, and F_x and F_y are continuous on the disk, then the equation $F(x, y) = 0$ defines y as a function of x near the point (a, b) and the derivative of this function is given by Equation 6.

EXAMPLE 8 Find y' if $x^3 + y^3 = 6xy$.

SOLUTION The given equation can be written as

$$F(x, y) = x^3 + y^3 - 6xy = 0$$

so Equation 6 gives

■ ■ The solution to Example 8 should be compared to the one in Example 2 in Section 3.6.

$$\frac{dy}{dx} = -\frac{F_x}{F_y} = -\frac{3x^2 - 6y}{3y^2 - 6x} = -\frac{x^2 - 2y}{y^2 - 2x}$$

■ ■

Now we suppose that z is given implicitly as a function $z = f(x, y)$ by an equation of the form $F(x, y, z) = 0$. This means that $F(x, y, f(x, y)) = 0$ for all (x, y) in the domain of f. If F and f are differentiable, then we can use the Chain Rule to differentiate the equation $F(x, y, z) = 0$ as follows:

$$\frac{\partial F}{\partial x}\frac{\partial x}{\partial x} + \frac{\partial F}{\partial y}\frac{\partial y}{\partial x} + \frac{\partial F}{\partial z}\frac{\partial z}{\partial x} = 0$$

But

$$\frac{\partial}{\partial x}(x) = 1 \qquad \text{and} \qquad \frac{\partial}{\partial x}(y) = 0$$

so this equation becomes

$$\frac{\partial F}{\partial x} + \frac{\partial F}{\partial z}\frac{\partial z}{\partial x} = 0$$

If $\partial F/\partial z \neq 0$, we solve for $\partial z/\partial x$ and obtain the first formula in Equations 7. The formula for $\partial z/\partial y$ is obtained in a similar manner.

$$\boxed{7} \qquad \frac{\partial z}{\partial x} = -\frac{\dfrac{\partial F}{\partial x}}{\dfrac{\partial F}{\partial z}} \qquad \frac{\partial z}{\partial y} = -\frac{\dfrac{\partial F}{\partial y}}{\dfrac{\partial F}{\partial z}}$$

Again, a version of the **Implicit Function Theorem** gives conditions under which our assumption is valid. If F is defined within a sphere containing (a, b, c), where $F(a, b, c) = 0$, $F_z(a, b, c) \neq 0$, and F_x, F_y, and F_z are continuous inside the sphere, then the equation $F(x, y, z) = 0$ defines z as a function of x and y near the point (a, b, c) and this function is differentiable, with partial derivatives given by (7).

EXAMPLE 9 Find $\dfrac{\partial z}{\partial x}$ and $\dfrac{\partial z}{\partial y}$ if $x^3 + y^3 + z^3 + 6xyz = 1$.

SOLUTION Let $F(x, y, z) = x^3 + y^3 + z^3 + 6xyz - 1$. Then, from Equations 7, we have

■ ■ The solution to Example 9 should be compared to the one in Example 4 in Section 11.3.

$$\frac{\partial z}{\partial x} = -\frac{F_x}{F_z} = -\frac{3x^2 + 6yz}{3z^2 + 6xy} = -\frac{x^2 + 2yz}{z^2 + 2xy}$$

$$\frac{\partial z}{\partial y} = -\frac{F_y}{F_z} = -\frac{3y^2 + 6xz}{3z^2 + 6xy} = -\frac{y^2 + 2xz}{z^2 + 2xy}$$

■ ■

11.5 | **Exercises**

1–4 ■ Use the Chain Rule to find dz/dt or dw/dt.

1. $z = \sin x \cos y$, $\quad x = \pi t$, $\quad y = \sqrt{t}$

2. $z = x \ln(x + 2y)$, $\quad x = \sin t$, $\quad y = \cos t$

3. $w = xe^{y/z}$, $\quad x = t^2$, $\quad y = 1 - t$, $\quad z = 1 + 2t$

4. $w = xy + yz^2$, $\quad x = e^t$, $\quad y = e^t \sin t$, $\quad z = e^t \cos t$

■ ■ ■ ■ ■ ■ ■ ■ ■ ■ ■ ■ ■

5–8 ■ Use the Chain Rule to find $\partial z/\partial s$ and $\partial z/\partial t$.

5. $z = x^2 + xy + y^2$, $\quad x = s + t$, $\quad y = st$

6. $z = x/y$, $\quad x = se^t$, $\quad y = 1 + se^{-t}$

7. $z = e^r \cos \theta$, $\quad r = st$, $\quad \theta = \sqrt{s^2 + t^2}$

8. $z = \sin \alpha \tan \beta$, $\quad \alpha = 3s + t$, $\quad \beta = s - t$

■ ■ ■ ■ ■ ■ ■ ■ ■ ■ ■ ■ ■

9. If $z = f(x, y)$, where f is differentiable, $x = g(t)$, $y = h(t)$, $g(3) = 2$, $g'(3) = 5$, $h(3) = 7$, $h'(3) = -4$, $f_x(2, 7) = 6$, and $f_y(2, 7) = -8$, find dz/dt when $t = 3$.

10. Let $W(s, t) = F(u(s, t), v(s, t))$, where F, u, and v are differentiable, $u(1, 0) = 2$, $u_s(1, 0) = -2$, $u_t(1, 0) = 6$, $v(1, 0) = 3$, $v_s(1, 0) = 5$, $v_t(1, 0) = 4$, $F_u(2, 3) = -1$, and $F_v(2, 3) = 10$. Find $W_s(1, 0)$ and $W_t(1, 0)$.

11. Suppose f is a differentiable function of x and y, and $g(u, v) = f(e^u + \sin v, e^u + \cos v)$. Use the table of values to calculate $g_u(0, 0)$ and $g_v(0, 0)$.

	f	g	f_x	f_y
$(0, 0)$	3	6	4	8
$(1, 2)$	6	3	2	5

12. Suppose f is a differentiable function of x and y, and $g(r, s) = f(2r - s, s^2 - 4r)$. Use the table of values in Exercise 11 to calculate $g_r(1, 2)$ and $g_s(1, 2)$.

13–16 ■ Use a tree diagram to write out the Chain Rule for the given case. Assume all functions are differentiable.

13. $u = f(x, y)$, $\quad$ where $x = x(r, s, t)$, $y = y(r, s, t)$

14. $w = f(x, y, z)$, $\quad$ where $x = x(t, u)$, $y = y(t, u)$, $z = z(t, u)$

15. $v = f(p, q, r)$,
$\quad$ where $p = p(x, y, z)$, $q = q(x, y, z)$, $r = r(x, y, z)$

16. $u = f(s, t)$, $\quad$ where $s = s(w, x, y, z)$, $t = t(w, x, y, z)$

■ ■ ■ ■ ■ ■ ■ ■ ■ ■ ■ ■ ■

17–21 ■ Use the Chain Rule to find the indicated partial derivatives.

17. $z = x^2 + xy^3$, $\quad x = uv^2 + w^3$, $\quad y = u + ve^w$;
$\dfrac{\partial z}{\partial u}$, $\dfrac{\partial z}{\partial v}$, $\dfrac{\partial z}{\partial w}$ $\quad$ when $u = 2$, $v = 1$, $w = 0$

18. $u = \sqrt{r^2 + s^2}$, $\quad r = y + x \cos t$, $\quad s = x + y \sin t$;
$\dfrac{\partial u}{\partial x}$, $\dfrac{\partial u}{\partial y}$, $\dfrac{\partial u}{\partial t}$ $\quad$ when $x = 1$, $y = 2$, $t = 0$

19. $R = \ln(u^2 + v^2 + w^2)$,
$u = x + 2y$, $\quad v = 2x - y$, $\quad w = 2xy$;
$\dfrac{\partial R}{\partial x}$, $\dfrac{\partial R}{\partial y}$ $\quad$ when $x = y = 1$

20. $M = xe^{y - z^2}$, $\quad x = 2uv$, $\quad y = u - v$, $\quad z = u + v$;
$\dfrac{\partial M}{\partial u}$, $\dfrac{\partial M}{\partial v}$ $\quad$ when $u = 3$, $v = -1$

21. $u = x^2 + yz$, $x = pr\cos\theta$, $y = pr\sin\theta$, $z = p + r$;
$\dfrac{\partial u}{\partial p}$, $\dfrac{\partial u}{\partial r}$, $\dfrac{\partial u}{\partial \theta}$ when $p = 2$, $r = 3$, $\theta = 0$

22–24 ■ Use Equation 6 to find dy/dx.

22. $y^5 + x^2y^3 = 1 + ye^{x^2}$ **23.** $\sqrt{xy} = 1 + x^2y$

24. $\sin x + \cos y = \sin x \cos y$

25–28 ■ Use Equations 7 to find $\partial z/\partial x$ and $\partial z/\partial y$.

25. $x^2 + y^2 + z^2 = 3xyz$ **26.** $xyz = \cos(x + y + z)$

27. $x - z = \arctan(yz)$ **28.** $yz = \ln(x + z)$

29. The temperature at a point (x, y) is $T(x, y)$, measured in degrees Celsius. A bug crawls so that its position after t seconds is given by $x = \sqrt{1 + t}$, $y = 2 + \frac{1}{3}t$, where x and y are measured in centimeters. The temperature function satisfies $T_x(2, 3) = 4$ and $T_y(2, 3) = 3$. How fast is the temperature rising on the bug's path after 3 seconds?

30. Wheat production in a given year, W, depends on the average temperature T and the annual rainfall R. Scientists estimate that the average temperature is rising at a rate of $0.15°C/year$ and rainfall is decreasing at a rate of 0.1 cm/year. They also estimate that, at current production levels, $\partial W/\partial T = -2$ and $\partial W/\partial R = 8$.
(a) What is the significance of the signs of these partial derivatives?
(b) Estimate the current rate of change of wheat production, dW/dt.

31. The speed of sound traveling through ocean water with salinity 35 parts per thousand has been modeled by the equation

$$C = 1449.2 + 4.6T - 0.055T^2 + 0.00029T^3 + 0.016D$$

where C is the speed of sound (in meters per second), T is the temperature (in degrees Celsius), and D is the depth below the ocean surface (in meters). A scuba diver began a leisurely dive into the ocean water; the diver's depth and surrounding water temperature over time are recorded in the following graphs. Estimate the rate of change (with respect to time) of the speed of sound through the ocean water experienced by the diver 20 minutes into the dive. What are the units?

32. The radius of a right circular cone is increasing at a rate of 1.8 in/s while its height is decreasing at a rate of 2.5 in/s. At what rate is the volume of the cone changing when the radius is 120 in. and the height is 140 in.?

33. The length ℓ, width w, and height h of a box change with time. At a certain instant the dimensions are $\ell = 1$ m and $w = h = 2$ m, and ℓ and w are increasing at a rate of 2 m/s while h is decreasing at a rate of 3 m/s. At that instant find the rates at which the following quantities are changing.
(a) The volume   (b) The surface area
(c) The length of a diagonal

34. The voltage V in a simple electrical circuit is slowly decreasing as the battery wears out. The resistance R is slowly increasing as the resistor heats up. Use Ohm's Law, $V = IR$, to find how the current I is changing at the moment when $R = 400\ \Omega$, $I = 0.08$ A, $dV/dt = -0.01$ V/s, and $dR/dt = 0.03\ \Omega/s$.

35. The pressure of 1 mole of an ideal gas is increasing at a rate of 0.05 kPa/s and the temperature is increasing at a rate of 0.15 K/s. Use the equation in Example 2 to find the rate of change of the volume when the pressure is 20 kPa and the temperature is 320 K.

36. If a sound with frequency f_s is produced by a source traveling along a line with speed v_s and an observer is traveling with speed v_o along the same line from the opposite direction toward the source, then the frequency of the sound heard by the observer is

$$f_o = \left(\frac{c + v_o}{c - v_s}\right)f_s$$

where c is the speed of sound, about 332 m/s. (This is the **Doppler effect.**) Suppose that, at a particular moment, you are in a train traveling at 34 m/s and accelerating at 1.2 m/s². A train is approaching you from the opposite direction on the other track at 40 m/s, accelerating at 1.4 m/s², and sounds its whistle, which has a frequency of 460 Hz. At that instant, what is the perceived frequency that you hear and how fast is it changing?

37–40 ■ Assume that all the given functions are differentiable.

37. If $z = f(x, y)$, where $x = r\cos\theta$ and $y = r\sin\theta$, (a) find $\partial z/\partial r$ and $\partial z/\partial\theta$ and (b) show that

$$\left(\frac{\partial z}{\partial x}\right)^2 + \left(\frac{\partial z}{\partial y}\right)^2 = \left(\frac{\partial z}{\partial r}\right)^2 + \frac{1}{r^2}\left(\frac{\partial z}{\partial\theta}\right)^2$$

38. If $u = f(x, y)$, where $x = e^s\cos t$ and $y = e^s\sin t$, show that

$$\left(\frac{\partial u}{\partial x}\right)^2 + \left(\frac{\partial u}{\partial y}\right)^2 = e^{-2s}\left[\left(\frac{\partial u}{\partial s}\right)^2 + \left(\frac{\partial u}{\partial t}\right)^2\right]$$

39. If $z = f(x - y)$, show that $\dfrac{\partial z}{\partial x} + \dfrac{\partial z}{\partial y} = 0$.

40. If $z = f(x, y)$, where $x = s + t$ and $y = s - t$, show that

$$\left(\frac{\partial z}{\partial x}\right)^2 - \left(\frac{\partial z}{\partial y}\right)^2 = \frac{\partial z}{\partial s}\frac{\partial z}{\partial t}$$

· · · · · · · · · · · · · · · ·

41–46 ■ Assume that all the given functions have continuous second-order partial derivatives.

41. Show that any function of the form

$$z = f(x + at) + g(x - at)$$

is a solution of the wave equation

$$\frac{\partial^2 z}{\partial t^2} = a^2\frac{\partial^2 z}{\partial x^2}$$

[*Hint:* Let $u = x + at$, $v = x - at$.]

42. If $u = f(x, y)$, where $x = e^s \cos t$ and $y = e^s \sin t$, show that

$$\frac{\partial^2 u}{\partial x^2} + \frac{\partial^2 u}{\partial y^2} = e^{-2s}\left[\frac{\partial^2 u}{\partial s^2} + \frac{\partial^2 u}{\partial t^2}\right]$$

43. If $z = f(x, y)$, where $x = r^2 + s^2$, $y = 2rs$, find $\partial^2 z/\partial r\,\partial s$. (Compare with Example 7.)

44. If $z = f(x, y)$, where $x = r\cos\theta$, $y = r\sin\theta$, find (a) $\partial z/\partial r$, (b) $\partial z/\partial\theta$, and (c) $\partial^2 z/\partial r\,\partial\theta$.

45. If $z = f(x, y)$, where $x = r\cos\theta$, $y = r\sin\theta$, show that

$$\frac{\partial^2 z}{\partial x^2} + \frac{\partial^2 z}{\partial y^2} = \frac{\partial^2 z}{\partial r^2} + \frac{1}{r^2}\frac{\partial^2 z}{\partial\theta^2} + \frac{1}{r}\frac{\partial z}{\partial r}$$

46. Suppose $z = f(x, y)$, where $x = g(s, t)$ and $y = h(s, t)$.
(a) Show that

$$\frac{\partial^2 z}{\partial t^2} = \frac{\partial^2 z}{\partial x^2}\left(\frac{\partial x}{\partial t}\right)^2 + 2\frac{\partial^2 z}{\partial x\,\partial y}\frac{\partial x}{\partial t}\frac{\partial y}{\partial t} + \frac{\partial^2 z}{\partial y^2}\left(\frac{\partial y}{\partial t}\right)^2$$
$$+ \frac{\partial z}{\partial x}\frac{\partial^2 x}{\partial t^2} + \frac{\partial z}{\partial y}\frac{\partial^2 y}{\partial t^2}$$

(b) Find a similar formula for $\partial^2 z/\partial s\,\partial t$.

· · · · · · · · · · · · · · · ·

47. Suppose that the equation $F(x, y, z) = 0$ implicitly defines each of the three variables x, y, and z as functions of the other two: $z = f(x, y)$, $y = g(x, z)$, $x = h(y, z)$. If F is differentiable and F_x, F_y, and F_z are all nonzero, show that

$$\frac{\partial z}{\partial x}\frac{\partial x}{\partial y}\frac{\partial y}{\partial z} = -1$$

11.6 Directional Derivatives and the Gradient Vector

The weather map in Figure 1 shows a contour map of the temperature function $T(x, y)$ for the states of California and Nevada at 3:00 P.M. on a day in October. The level curves, or isothermals, join locations with the same temperature. The partial derivative T_x at a location such as Reno is the rate of change of temperature with respect to distance if we travel east from Reno; T_y is the rate of change of temperature if we travel north. But what if we want to know the rate of change of temperature when we travel southeast (toward Las Vegas), or in some other direction? In this section we intro-

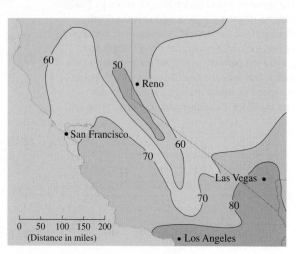

FIGURE 1

duce a type of derivative, called a *directional derivative,* that enables us to find the rate of change of a function of two or more variables in any direction.

■ Directional Derivatives

Recall that if $z = f(x, y)$, then the partial derivatives f_x and f_y are defined as

$$\boxed{1} \qquad f_x(x_0, y_0) = \lim_{h \to 0} \frac{f(x_0 + h, y_0) - f(x_0, y_0)}{h}$$

$$f_y(x_0, y_0) = \lim_{h \to 0} \frac{f(x_0, y_0 + h) - f(x_0, y_0)}{h}$$

and represent the rates of change of z in the x- and y-directions, that is, in the directions of the unit vectors **i** and **j**.

Suppose that we now wish to find the rate of change of z at (x_0, y_0) in the direction of an arbitrary unit vector $\mathbf{u} = \langle a, b \rangle$. (See Figure 2.) To do this we consider the surface S with equation $z = f(x, y)$ (the graph of f) and we let $z_0 = f(x_0, y_0)$. Then the point $P(x_0, y_0, z_0)$ lies on S. The vertical plane that passes through P in the direction of **u** intersects S in a curve C. (See Figure 3.) The slope of the tangent line T to C at the point P is the rate of change of z in the direction of **u**.

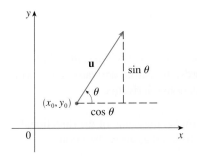

FIGURE 2

A unit vector $\mathbf{u} = \langle a, b \rangle = \langle \cos \theta, \sin \theta \rangle$

TEC Visual 11.6A animates Figure 3 by rotating **u** and therefore T.

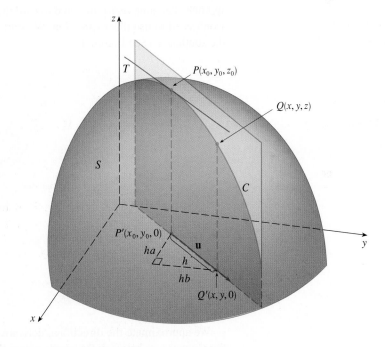

FIGURE 3

If $Q(x, y, z)$ is another point on C and P', Q' are the projections of P, Q on the xy-plane, then the vector $\overrightarrow{P'Q'}$ is parallel to **u** and so

$$\overrightarrow{P'Q'} = h\mathbf{u} = \langle ha, hb \rangle$$

for some scalar h. Therefore, $x - x_0 = ha$, $y - y_0 = hb$, so $x = x_0 + ha$, $y = y_0 + hb$, and

$$\frac{\Delta z}{h} = \frac{z - z_0}{h} = \frac{f(x_0 + ha, y_0 + hb) - f(x_0, y_0)}{h}$$

If we take the limit as $h \to 0$, we obtain the rate of change of z (with respect to distance) in the direction of $\mathbf{u}$, which is called the directional derivative of f in the direction of $\mathbf{u}$.

---

2 **Definition** The **directional derivative** of f at (x_0, y_0) in the direction of a unit vector $\mathbf{u} = \langle a, b \rangle$ is

$$D_{\mathbf{u}} f(x_0, y_0) = \lim_{h \to 0} \frac{f(x_0 + ha, y_0 + hb) - f(x_0, y_0)}{h}$$

if this limit exists.

By comparing Definition 2 with (1), we see that if $\mathbf{u} = \mathbf{i} = \langle 1, 0 \rangle$, then $D_{\mathbf{i}} f = f_x$ and if $\mathbf{u} = \mathbf{j} = \langle 0, 1 \rangle$, then $D_{\mathbf{j}} f = f_y$. In other words, the partial derivatives of f with respect to x and y are just special cases of the directional derivative.

EXAMPLE 1 Use the weather map in Figure 1 to estimate the value of the directional derivative of the temperature function at Reno in the southeasterly direction.

SOLUTION The unit vector directed toward the southeast is $\mathbf{u} = (\mathbf{i} - \mathbf{j})/\sqrt{2}$, but we won't need to use this expression. We start by drawing a line through Reno toward the southeast. (See Figure 4.)

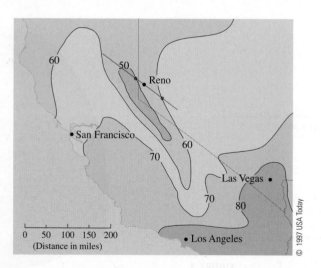

FIGURE 4

We approximate the directional derivative $D_{\mathbf{u}} T$ by the average rate of change of the temperature between the points where this line intersects the isothermals $T = 50$ and $T = 60$. The temperature at the point southeast of Reno is $T = 60°F$ and the temperature at the point northwest of Reno is $T = 50°F$. The distance between these points looks to be about 75 miles. So the rate of change of the temperature in the southeasterly direction is

$$D_{\mathbf{u}} T \approx \frac{60 - 50}{75} = \frac{10}{75} \approx 0.13°F/mi$$

■ ■

When we compute the directional derivative of a function defined by a formula, we generally use the following theorem.

> **3** **Theorem** If f is a differentiable function of x and y, then f has a directional derivative in the direction of any unit vector $\mathbf{u} = \langle a, b \rangle$ and
>
> $$D_{\mathbf{u}} f(x, y) = f_x(x, y)a + f_y(x, y)b$$

Proof If we define a function g of the single variable h by

$$g(h) = f(x_0 + ha, y_0 + hb)$$

then by the definition of a derivative we have

$$\boxed{4} \qquad g'(0) = \lim_{h \to 0} \frac{g(h) - g(0)}{h} = \lim_{h \to 0} \frac{f(x_0 + ha, y_0 + hb) - f(x_0, y_0)}{h}$$

$$= D_{\mathbf{u}} f(x_0, y_0)$$

On the other hand, we can write $g(h) = f(x, y)$, where $x = x_0 + ha$, $y = y_0 + hb$, so the Chain Rule (Theorem 11.5.2) gives

$$g'(h) = \frac{\partial f}{\partial x} \frac{dx}{dh} + \frac{\partial f}{\partial y} \frac{dy}{dh} = f_x(x, y)a + f_y(x, y)b$$

If we now put $h = 0$, then $x = x_0$, $y = y_0$, and

$$\boxed{5} \qquad g'(0) = f_x(x_0, y_0)a + f_y(x_0, y_0)b$$

Comparing Equations 4 and 5, we see that

$$D_{\mathbf{u}} f(x_0, y_0) = f_x(x_0, y_0)a + f_y(x_0, y_0)b \qquad \blacksquare\blacksquare$$

If the unit vector $\mathbf{u}$ makes an angle θ with the positive x-axis (as in Figure 2), then we can write $\mathbf{u} = \langle \cos\theta, \sin\theta \rangle$ and the formula in Theorem 3 becomes

$$\boxed{6} \qquad D_{\mathbf{u}} f(x, y) = f_x(x, y) \cos\theta + f_y(x, y) \sin\theta$$

■ ■ The directional derivative $D_{\mathbf{u}} f(1, 2)$ in Example 2 represents the rate of change of z in the direction of $\mathbf{u}$. This is the slope of the tangent line to the curve of intersection of the surface $z = x^3 - 3xy + 4y^2$ and the vertical plane through $(1, 2, 0)$ in the direction of $\mathbf{u}$ shown in Figure 5.

EXAMPLE 2 Find the directional derivative $D_{\mathbf{u}} f(x, y)$ if

$$f(x, y) = x^3 - 3xy + 4y^2$$

and $\mathbf{u}$ is the unit vector given by angle $\theta = \pi/6$. What is $D_{\mathbf{u}} f(1, 2)$?

SOLUTION Formula 6 gives

$$D_{\mathbf{u}} f(x, y) = f_x(x, y) \cos\frac{\pi}{6} + f_y(x, y) \sin\frac{\pi}{6}$$

$$= (3x^2 - 3y) \frac{\sqrt{3}}{2} + (-3x + 8y)\tfrac{1}{2}$$

$$= \tfrac{1}{2}\left[3\sqrt{3}x^2 - 3x + \left(8 - 3\sqrt{3}\right)y\right]$$

Therefore

$$D_{\mathbf{u}} f(1, 2) = \tfrac{1}{2}\left[3\sqrt{3}(1)^2 - 3(1) + \left(8 - 3\sqrt{3}\right)(2)\right] = \frac{13 - 3\sqrt{3}}{2} \qquad \blacksquare\blacksquare$$

FIGURE 5

▪▪ The Gradient Vector

Notice from Theorem 3 that the directional derivative can be written as the dot product of two vectors:

$$\boxed{7} \qquad D_{\mathbf{u}} f(x, y) = f_x(x, y)a + f_y(x, y)b$$
$$= \langle f_x(x, y), f_y(x, y) \rangle \cdot \langle a, b \rangle$$
$$= \langle f_x(x, y), f_y(x, y) \rangle \cdot \mathbf{u}$$

The first vector in this dot product occurs not only in computing directional derivatives but in many other contexts as well. So we give it a special name (the *gradient* of f) and a special notation (**grad** f or ∇f, which is read "del f").

> **8 Definition** If f is a function of two variables x and y, then the **gradient** of f is the vector function ∇f defined by
>
> $$\nabla f(x, y) = \langle f_x(x, y), f_y(x, y) \rangle = \frac{\partial f}{\partial x} \mathbf{i} + \frac{\partial f}{\partial y} \mathbf{j}$$

EXAMPLE 3 If $f(x, y) = \sin x + e^{xy}$, then

$$\nabla f(x, y) = \langle f_x, f_y \rangle = \langle \cos x + ye^{xy}, xe^{xy} \rangle$$

and
$$\nabla f(0, 1) = \langle 2, 0 \rangle \qquad\qquad ■\ ■$$

With this notation for the gradient vector, we can rewrite the expression (7) for the directional derivative as

$$\boxed{9} \qquad\qquad \boxed{D_{\mathbf{u}} f(x, y) = \nabla f(x, y) \cdot \mathbf{u}}$$

This expresses the directional derivative in the direction of $\mathbf{u}$ as the scalar projection of the gradient vector onto $\mathbf{u}$.

■■ The gradient vector $\nabla f(2, -1)$ in Example 4 is shown in Figure 6 with initial point $(2, -1)$. Also shown is the vector $\mathbf{v}$ that gives the direction of the directional derivative. Both of these vectors are superimposed on a contour plot of the graph of f.

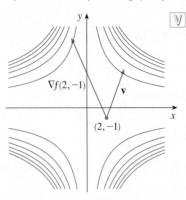

FIGURE 6

$\boxed{\text{V}}$ **EXAMPLE 4** Find the directional derivative of the function $f(x, y) = x^2 y^3 - 4y$ at the point $(2, -1)$ in the direction of the vector $\mathbf{v} = 2\mathbf{i} + 5\mathbf{j}$.

SOLUTION We first compute the gradient vector at $(2, -1)$:

$$\nabla f(x, y) = 2xy^3 \mathbf{i} + (3x^2 y^2 - 4)\mathbf{j}$$
$$\nabla f(2, -1) = -4\mathbf{i} + 8\mathbf{j}$$

Note that $\mathbf{v}$ is not a unit vector, but since $|\mathbf{v}| = \sqrt{29}$, the unit vector in the direction of $\mathbf{v}$ is

$$\mathbf{u} = \frac{\mathbf{v}}{|\mathbf{v}|} = \frac{2}{\sqrt{29}} \mathbf{i} + \frac{5}{\sqrt{29}} \mathbf{j}$$

Therefore, by Equation 9, we have

$$D_{\mathbf{u}}f(2, -1) = \nabla f(2, -1) \cdot \mathbf{u} = (-4\mathbf{i} + 8\mathbf{j}) \cdot \left(\frac{2}{\sqrt{29}}\mathbf{i} + \frac{5}{\sqrt{29}}\mathbf{j}\right)$$

$$= \frac{-4 \cdot 2 + 8 \cdot 5}{\sqrt{29}} = \frac{32}{\sqrt{29}}$$

■■ Functions of Three Variables

For functions of three variables we can define directional derivatives in a similar manner. Again $D_{\mathbf{u}}f(x, y, z)$ can be interpreted as the rate of change of the function in the direction of a unit vector $\mathbf{u}$.

---

10 Definition The **directional derivative** of f at (x_0, y_0, z_0) in the direction of a unit vector $\mathbf{u} = \langle a, b, c \rangle$ is

$$D_{\mathbf{u}}f(x_0, y_0, z_0) = \lim_{h \to 0} \frac{f(x_0 + ha, y_0 + hb, z_0 + hc) - f(x_0, y_0, z_0)}{h}$$

if this limit exists.

If we use vector notation, then we can write both definitions (2 and 10) of the directional derivative in the compact form

$$\boxed{11} \qquad D_{\mathbf{u}}f(\mathbf{x}_0) = \lim_{h \to 0} \frac{f(\mathbf{x}_0 + h\mathbf{u}) - f(\mathbf{x}_0)}{h}$$

where $\mathbf{x}_0 = \langle x_0, y_0 \rangle$ if $n = 2$ and $\mathbf{x}_0 = \langle x_0, y_0, z_0 \rangle$ if $n = 3$. This is reasonable since the vector equation of the line through $\mathbf{x}_0$ in the direction of the vector $\mathbf{u}$ is given by $\mathbf{x} = \mathbf{x}_0 + t\mathbf{u}$ (Equation 9.5.1) and so $f(\mathbf{x}_0 + h\mathbf{u})$ represents the value of f at a point on this line.

If $f(x, y, z)$ is differentiable and $\mathbf{u} = \langle a, b, c \rangle$, then the same method that was used to prove Theorem 3 can be used to show that

$$\boxed{12} \qquad D_{\mathbf{u}}f(x, y, z) = f_x(x, y, z)a + f_y(x, y, z)b + f_z(x, y, z)c$$

For a function f of three variables, the **gradient vector**, denoted by ∇f or **grad** f, is

$$\nabla f(x, y, z) = \langle f_x(x, y, z), f_y(x, y, z), f_z(x, y, z) \rangle$$

or, for short,

$$\boxed{13} \qquad \nabla f = \langle f_x, f_y, f_z \rangle = \frac{\partial f}{\partial x}\mathbf{i} + \frac{\partial f}{\partial y}\mathbf{j} + \frac{\partial f}{\partial z}\mathbf{k}$$

Then, just as with functions of two variables, Formula 12 for the directional derivative can be rewritten as

$$\boxed{14 \qquad D_{\mathbf{u}} f(x, y, z) = \nabla f(x, y, z) \cdot \mathbf{u}}$$

$\boxed{V}$ **EXAMPLE 5** If $f(x, y, z) = x \sin yz$, (a) find the gradient of f and (b) find the directional derivative of f at $(1, 3, 0)$ in the direction of $\mathbf{v} = \mathbf{i} + 2\mathbf{j} - \mathbf{k}$.

SOLUTION
(a) The gradient of f is

$$\nabla f(x, y, z) = \langle f_x(x, y, z), f_y(x, y, z), f_z(x, y, z) \rangle$$

$$= \langle \sin yz, xz \cos yz, xy \cos yz \rangle$$

(b) At $(1, 3, 0)$ we have $\nabla f(1, 3, 0) = \langle 0, 0, 3 \rangle$. The unit vector in the direction of $\mathbf{v} = \mathbf{i} + 2\mathbf{j} - \mathbf{k}$ is

$$\mathbf{u} = \frac{1}{\sqrt{6}} \mathbf{i} + \frac{2}{\sqrt{6}} \mathbf{j} - \frac{1}{\sqrt{6}} \mathbf{k}$$

Therefore, Equation 14 gives

$$D_{\mathbf{u}} f(1, 3, 0) = \nabla f(1, 3, 0) \cdot \mathbf{u}$$

$$= 3\mathbf{k} \cdot \left(\frac{1}{\sqrt{6}} \mathbf{i} + \frac{2}{\sqrt{6}} \mathbf{j} - \frac{1}{\sqrt{6}} \mathbf{k} \right)$$

$$= 3 \left(-\frac{1}{\sqrt{6}} \right) = -\sqrt{\frac{3}{2}} \qquad \blacksquare \blacksquare$$

■ ■ Maximizing the Directional Derivative

Suppose we have a function f of two or three variables and we consider all possible directional derivatives of f at a given point. These give the rates of change of f in all possible directions. We can then ask the questions: In which of these directions does f change fastest and what is the maximum rate of change? The answers are provided by the following theorem.

TEC Visual 11.6B provides visual confirmation of Theorem 15.

$\boxed{15}$ **Theorem** Suppose f is a differentiable function of two or three variables. The maximum value of the directional derivative $D_{\mathbf{u}} f(\mathbf{x})$ is $|\nabla f(\mathbf{x})|$ and it occurs when $\mathbf{u}$ has the same direction as the gradient vector $\nabla f(\mathbf{x})$.

Proof From Equation 9 or 14 we have

$$D_{\mathbf{u}} f = \nabla f \cdot \mathbf{u} = |\nabla f| |\mathbf{u}| \cos \theta = |\nabla f| \cos \theta$$

where θ is the angle between ∇f and $\mathbf{u}$. The maximum value of $\cos \theta$ is 1 and this occurs when $\theta = 0$. Therefore, the maximum value of $D_{\mathbf{u}} f$ is $|\nabla f|$ and it occurs when $\theta = 0$, that is, when $\mathbf{u}$ has the same direction as ∇f. $\qquad \blacksquare \blacksquare$

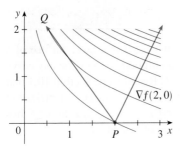

FIGURE 7

■ ■ At $(2, 0)$ the function in Example 6 increases fastest in the direction of the gradient vector $\nabla f(2, 0) = \langle 1, 2 \rangle$. Notice from Figure 7 that this vector appears to be perpendicular to the level curve through $(2, 0)$. Figure 8 shows the graph of f and the gradient vector.

FIGURE 8

EXAMPLE 6

(a) If $f(x, y) = xe^y$, find the rate of change of f at the point $P(2, 0)$ in the direction from P to $Q(\frac{1}{2}, 2)$.

(b) In what direction does f have the maximum rate of change? What is this maximum rate of change?

SOLUTION

(a) We first compute the gradient vector:

$$\nabla f(x, y) = \langle f_x, f_y \rangle = \langle e^y, xe^y \rangle$$

$$\nabla f(2, 0) = \langle 1, 2 \rangle$$

The unit vector in the direction of $\overrightarrow{PQ} = \langle -1.5, 2 \rangle$ is $\mathbf{u} = \langle -\frac{3}{5}, \frac{4}{5} \rangle$, so the rate of change of f in the direction from P to Q is

$$D_{\mathbf{u}} f(2, 0) = \nabla f(2, 0) \cdot \mathbf{u} = \langle 1, 2 \rangle \cdot \langle -\tfrac{3}{5}, \tfrac{4}{5} \rangle$$

$$= 1\left(-\tfrac{3}{5}\right) + 2\left(\tfrac{4}{5}\right) = 1$$

(b) According to Theorem 15, f increases fastest in the direction of the gradient vector $\nabla f(2, 0) = \langle 1, 2 \rangle$. The maximum rate of change is

$$|\nabla f(2, 0)| = |\langle 1, 2 \rangle| = \sqrt{5}$$ ■ ■

EXAMPLE 7 Suppose that the temperature at a point (x, y, z) in space is given by $T(x, y, z) = 80/(1 + x^2 + 2y^2 + 3z^2)$, where T is measured in degrees Celsius and x, y, z in meters. In which direction does the temperature increase fastest at the point $(1, 1, -2)$? What is the maximum rate of increase?

SOLUTION The gradient of T is

$$\nabla T = \frac{\partial T}{\partial x}\mathbf{i} + \frac{\partial T}{\partial y}\mathbf{j} + \frac{\partial T}{\partial z}\mathbf{k}$$

$$= -\frac{160x}{(1 + x^2 + 2y^2 + 3z^2)^2}\mathbf{i} - \frac{320y}{(1 + x^2 + 2y^2 + 3z^2)^2}\mathbf{j} - \frac{480z}{(1 + x^2 + 2y^2 + 3z^2)^2}\mathbf{k}$$

$$= \frac{160}{(1 + x^2 + 2y^2 + 3z^2)^2}(-x\mathbf{i} - 2y\mathbf{j} - 3z\mathbf{k})$$

At the point $(1, 1, -2)$ the gradient vector is

$$\nabla T(1, 1, -2) = \tfrac{160}{256}(-\mathbf{i} - 2\mathbf{j} + 6\mathbf{k}) = \tfrac{5}{8}(-\mathbf{i} - 2\mathbf{j} + 6\mathbf{k})$$

By Theorem 15 the temperature increases fastest in the direction of the gradient vector $\nabla T(1, 1, -2) = \tfrac{5}{8}(-\mathbf{i} - 2\mathbf{j} + 6\mathbf{k})$ or, equivalently, in the direction of $-\mathbf{i} - 2\mathbf{j} + 6\mathbf{k}$ or the unit vector $(-\mathbf{i} - 2\mathbf{j} + 6\mathbf{k})/\sqrt{41}$. The maximum rate of increase is the length of the gradient vector:

$$|\nabla T(1, 1, -2)| = \tfrac{5}{8}|-\mathbf{i} - 2\mathbf{j} + 6\mathbf{k}| = \frac{5\sqrt{41}}{8}$$

Therefore, the maximum rate of increase of temperature is $5\sqrt{41}/8 \approx 4°C/m$. ■ ■

∷ Tangent Planes to Level Surfaces

Suppose S is a surface with equation $F(x, y, z) = k$, that is, it is a level surface of a function F of three variables, and let $P(x_0, y_0, z_0)$ be a point on S. Let C be any curve that lies on the surface S and passes through the point P. Recall from Section 10.1 that the curve C is described by a continuous vector function $\mathbf{r}(t) = \langle x(t), y(t), z(t) \rangle$. Let t_0 be the parameter value corresponding to P; that is, $\mathbf{r}(t_0) = \langle x_0, y_0, z_0 \rangle$. Since C lies on S, any point $(x(t), y(t), z(t))$ must satisfy the equation of S, that is,

$$\boxed{16} \qquad F(x(t), y(t), z(t)) = k$$

If x, y, and z are differentiable functions of t and F is also differentiable, then we can use the Chain Rule to differentiate both sides of Equation 16 as follows:

$$\boxed{17} \qquad \frac{\partial F}{\partial x}\frac{dx}{dt} + \frac{\partial F}{\partial y}\frac{dy}{dt} + \frac{\partial F}{\partial z}\frac{dz}{dt} = 0$$

But, since $\nabla F = \langle F_x, F_y, F_z \rangle$ and $\mathbf{r}'(t) = \langle x'(t), y'(t), z'(t) \rangle$, Equation 17 can be written in terms of a dot product as

$$\nabla F \cdot \mathbf{r}'(t) = 0$$

In particular, when $t = t_0$ we have $\mathbf{r}(t_0) = \langle x_0, y_0, z_0 \rangle$, so

$$\boxed{18} \qquad \nabla F(x_0, y_0, z_0) \cdot \mathbf{r}'(t_0) = 0$$

FIGURE 9

Equation 18 says that *the gradient vector at P, $\nabla F(x_0, y_0, z_0)$, is perpendicular to the tangent vector $\mathbf{r}'(t_0)$ to any curve C on S that passes through P.* (See Figure 9.) If $\nabla F(x_0, y_0, z_0) \neq \mathbf{0}$, it is therefore natural to define the **tangent plane to the level surface** $F(x, y, z) = k$ **at** $P(x_0, y_0, z_0)$ as the plane that passes through P and has normal vector $\nabla F(x_0, y_0, z_0)$. Using the standard equation of a plane (Equation 9.5.7), we can write the equation of this tangent plane as

$$\boxed{19} \quad F_x(x_0, y_0, z_0)(x - x_0) + F_y(x_0, y_0, z_0)(y - y_0) + F_z(x_0, y_0, z_0)(z - z_0) = 0$$

The **normal line** to S at P is the line passing through P and perpendicular to the tangent plane. The direction of the normal line is therefore given by the gradient vector $\nabla F(x_0, y_0, z_0)$ and so, by Equation 9.5.3, its symmetric equations are

$$\boxed{20} \qquad \frac{x - x_0}{F_x(x_0, y_0, z_0)} = \frac{y - y_0}{F_y(x_0, y_0, z_0)} = \frac{z - z_0}{F_z(x_0, y_0, z_0)}$$

In the special case in which the equation of a surface S is of the form $z = f(x, y)$ (that is, S is the graph of a function f of two variables), we can rewrite the equation as

$$F(x, y, z) = f(x, y) - z = 0$$

and regard S as a level surface (with $k = 0$) of F. Then

$$F_x(x_0, y_0, z_0) = f_x(x_0, y_0)$$

$$F_y(x_0, y_0, z_0) = f_y(x_0, y_0)$$

$$F_z(x_0, y_0, z_0) = -1$$

so Equation 19 becomes

$$f_x(x_0, y_0)(x - x_0) + f_y(x_0, y_0)(y - y_0) - (z - z_0) = 0$$

which is equivalent to Equation 11.4.2. Thus, our new, more general, definition of a tangent plane is consistent with the definition that was given for the special case of Section 11.4.

EXAMPLE 8 Find the equations of the tangent plane and normal line at the point $(-2, 1, -3)$ to the ellipsoid

$$\frac{x^2}{4} + y^2 + \frac{z^2}{9} = 3$$

SOLUTION The ellipsoid is the level surface (with $k = 3$) of the function

$$F(x, y, z) = \frac{x^2}{4} + y^2 + \frac{z^2}{9}$$

Therefore, we have

$$F_x(x, y, z) = \frac{x}{2} \qquad F_y(x, y, z) = 2y \qquad F_z(x, y, z) = \frac{2z}{9}$$

$$F_x(-2, 1, -3) = -1 \qquad F_y(-2, 1, -3) = 2 \qquad F_z(-2, 1, -3) = -\tfrac{2}{3}$$

Then Equation 19 gives the equation of the tangent plane at $(-2, 1, -3)$ as

$$-1(x + 2) + 2(y - 1) - \tfrac{2}{3}(z + 3) = 0$$

which simplifies to $3x - 6y + 2z + 18 = 0$.

By Equation 20, symmetric equations of the normal line are

$$\frac{x + 2}{-1} = \frac{y - 1}{2} = \frac{z + 3}{-\tfrac{2}{3}}$$

■ ■ Figure 10 shows the ellipsoid, tangent plane, and normal line in Example 8.

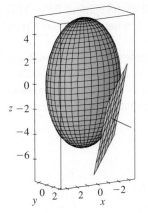

FIGURE 10

▪▪ Significance of the Gradient Vector

We now summarize the ways in which the gradient vector is significant. We first consider a function f of three variables and a point $P(x_0, y_0, z_0)$ in its domain. On the one hand, we know from Theorem 15 that the gradient vector $\nabla f(x_0, y_0, z_0)$ gives the direction of fastest increase of f. On the other hand, we know that $\nabla f(x_0, y_0, z_0)$ is orthogonal to the level surface S of f through P. (Refer to Figure 9.) These two properties are quite compatible intuitively because as we move away from P on the level surface S, the value of f does not change at all. So it seems reasonable that if we move in the perpendicular direction, we get the maximum increase.

In like manner we consider a function f of two variables and a point $P(x_0, y_0)$ in its domain. Again the gradient vector $\nabla f(x_0, y_0)$ gives the direction of fastest increase of f. Also, by considerations similar to our discussion of tangent planes, it can be shown that $\nabla f(x_0, y_0)$ is perpendicular to the level curve $f(x, y) = k$ that passes through P. Again this is intuitively plausible because the values of f remain constant as we move along the curve. (See Figure 11.)

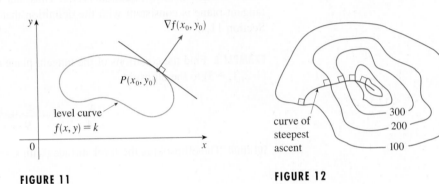

FIGURE 11

FIGURE 12

If we consider a topographical map of a hill and let $f(x, y)$ represent the height above sea level at a point with coordinates (x, y), then a curve of steepest ascent can be drawn as in Figure 12 by making it perpendicular to all of the contour lines. This phenomenon can also be noticed in Figure 5 in Section 11.1, where Lonesome Creek follows a curve of steepest descent.

Computer algebra systems have commands that plot sample gradient vectors. Each gradient vector $\nabla f(a, b)$ is plotted starting at the point (a, b). Figure 13 shows such a plot (called a *gradient vector field*) for the function $f(x, y) = x^2 - y^2$ superimposed on a contour map of f. As expected, the gradient vectors point "uphill" and are perpendicular to the level curves.

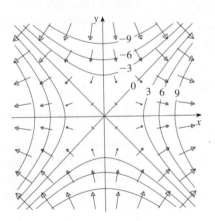

FIGURE 13

11.6 Exercises

1. A contour map of barometric pressure (in millibars) is shown for 7:00 A.M. on September 12, 1960, when Hurricane Donna was raging. Estimate the value of the directional derivative of the pressure function at Raleigh, North Carolina, in the direction of the eye of the hurricane. What are the units of the directional derivative?

2. The contour map shows the average annual snowfall (in inches) near Lake Michigan. Estimate the value of the directional derivative of this snowfall function at Muskegon, Michigan, in the direction of Ludington. What are the units?

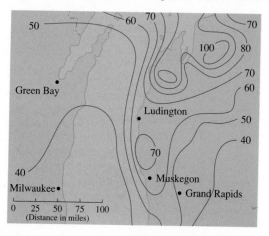

3. A table of values for the wind-chill index $W = f(T, v)$ is given in Exercise 3 on page 766. Use the table to estimate the value of $D_{\mathbf{u}} f(-20, 30)$, where $\mathbf{u} = (\mathbf{i} + \mathbf{j})/\sqrt{2}$.

4–6 ■ Find the directional derivative of f at the given point in the direction indicated by the angle θ.

4. $f(x, y) = x^2 y^3 - y^4$, $(2, 1)$, $\theta = \pi/4$

5. $f(x, y) = \sqrt{5x - 4y}$, $(4, 1)$, $\theta = -\pi/6$

6. $f(x, y) = x \sin(xy)$, $(2, 0)$, $\theta = \pi/3$

7–10 ■
(a) Find the gradient of f.
(b) Evaluate the gradient at the point P.
(c) Find the rate of change of f at P in the direction of the vector $\mathbf{u}$.

7. $f(x, y) = 5xy^2 - 4x^3 y$, $P(1, 2)$, $\mathbf{u} = \langle \frac{5}{13}, \frac{12}{13} \rangle$

8. $f(x, y) = y \ln x$, $P(1, -3)$, $\mathbf{u} = \langle -\frac{4}{5}, \frac{3}{5} \rangle$

9. $f(x, y, z) = xe^{2yz}$, $P(3, 0, 2)$, $\mathbf{u} = \langle \frac{2}{3}, -\frac{2}{3}, \frac{1}{3} \rangle$

10. $f(x, y, z) = \sqrt{x + yz}$, $P(1, 3, 1)$, $\mathbf{u} = \langle \frac{2}{7}, \frac{3}{7}, \frac{6}{7} \rangle$

11–15 ■ Find the directional derivative of the function at the given point in the direction of the vector $\mathbf{v}$.

11. $f(x, y) = 1 + 2x\sqrt{y}$, $(3, 4)$, $\mathbf{v} = \langle 4, -3 \rangle$

12. $f(x, y) = \ln(x^2 + y^2)$, $(2, 1)$, $\mathbf{v} = \langle -1, 2 \rangle$

13. $g(s, t) = s^2 e^t$, $(2, 0)$, $\mathbf{v} = \mathbf{i} + \mathbf{j}$

14. $f(x, y, z) = x/(y + z)$, $(4, 1, 1)$, $\mathbf{v} = \langle 1, 2, 3 \rangle$

15. $g(x, y, z) = (x + 2y + 3z)^{3/2}$, $(1, 1, 2)$, $\mathbf{v} = 2\mathbf{j} - \mathbf{k}$

16. Use the figure to estimate $D_{\mathbf{u}} f(2, 2)$.

17. Find the directional derivative of $f(x, y) = \sqrt{xy}$ at $P(2, 8)$ in the direction of $Q(5, 4)$.

18. Find the directional derivative of $f(x, y, z) = x^2 + y^2 + z^2$ at $P(2, 1, 3)$ in the direction of the origin.

19–22 ■ Find the maximum rate of change of f at the given point and the direction in which it occurs.

19. $f(x, y) = y^2/x$, $(2, 4)$

20. $f(p, q) = qe^{-p} + pe^{-q}$, $(0, 0)$

21. $f(x, y, z) = \ln(xy^2 z^3)$, $(1, -2, -3)$

22. $f(x, y, z) = \tan(x + 2y + 3z)$, $(-5, 1, 1)$

23. (a) Show that a differentiable function f decreases most rapidly at $\mathbf{x}$ in the direction opposite to the gradient vector, that is, in the direction of $-\nabla f(\mathbf{x})$.
   (b) Use the result of part (a) to find the direction in which the function $f(x, y) = x^4 y - x^2 y^3$ decreases fastest at the point $(2, -3)$.

24. Find the directions in which the directional derivative of $f(x, y) = x^2 + \sin xy$ at the point $(1, 0)$ has the value 1.

25. Find all points at which the direction of fastest change of the function $f(x, y) = x^2 + y^2 - 2x - 4y$ is $\mathbf{i} + \mathbf{j}$.

26. Near a buoy, the depth of a lake at the point with coordinates (x, y) is $z = 200 + 0.02x^2 - 0.001y^3$, where x, y, and z are measured in meters. A fisherman in a small boat starts at the point $(80, 60)$ and moves toward the buoy, which is located at $(0, 0)$. Is the water under the boat getting deeper or shallower when he departs? Explain.

27. The temperature T in a metal ball is inversely proportional to the distance from the center of the ball, which we take to be the origin. The temperature at the point $(1, 2, 2)$ is $120°$.
   (a) Find the rate of change of T at $(1, 2, 2)$ in the direction toward the point $(2, 1, 3)$.
   (b) Show that at any point in the ball the direction of greatest increase in temperature is given by a vector that points toward the origin.

28. The temperature at a point (x, y, z) is given by

$$T(x, y, z) = 200 e^{-x^2 - 3y^2 - 9z^2}$$

where T is measured in °C and x, y, z in meters.
   (a) Find the rate of change of temperature at the point $P(2, -1, 2)$ in the direction toward the point $(3, -3, 3)$.
   (b) In which direction does the temperature increase fastest at P?
   (c) Find the maximum rate of increase at P.

29. Suppose that over a certain region of space the electrical potential V is given by $V(x, y, z) = 5x^2 - 3xy + xyz$.
   (a) Find the rate of change of the potential at $P(3, 4, 5)$ in the direction of the vector $\mathbf{v} = \mathbf{i} + \mathbf{j} - \mathbf{k}$.
   (b) In which direction does V change most rapidly at P?
   (c) What is the maximum rate of change at P?

30. Suppose you are climbing a hill whose shape is given by the equation $z = 1000 - 0.005x^2 - 0.01y^2$, where x, y, and z are measured in meters, and you are standing at a point with coordinates $(60, 40, 966)$. The positive x-axis points east and the positive y-axis points north.
   (a) If you walk due south, will you start to ascend or descend? At what rate?
   (b) If you walk northwest, will you start to ascend or descend? At what rate?
   (c) In which direction is the slope largest? What is the rate of ascent in that direction? At what angle above the horizontal does the path in that direction begin?

31. Let f be a function of two variables that has continuous partial derivatives and consider the points $A(1, 3)$, $B(3, 3)$, $C(1, 7)$, and $D(6, 15)$. The directional derivative of f at A in the direction of the vector $\overrightarrow{AB}$ is 3 and the directional derivative at A in the direction of $\overrightarrow{AC}$ is 26. Find the directional derivative of f at A in the direction of the vector $\overrightarrow{AD}$.

32. For the given contour map draw the curves of steepest ascent starting at P and at Q.

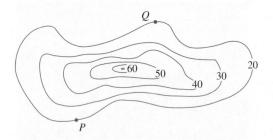

33. Show that the operation of taking the gradient of a function has the given property. Assume that u and v are differentiable functions of x and y and a, b are constants.
   (a) $\nabla(au + bv) = a\,\nabla u + b\,\nabla v$
   (b) $\nabla(uv) = u\,\nabla v + v\,\nabla u$
   (c) $\nabla\left(\dfrac{u}{v}\right) = \dfrac{v\,\nabla u - u\,\nabla v}{v^2}$
   (d) $\nabla u^n = nu^{n-1}\,\nabla u$

34. Sketch the gradient vector $\nabla f(4, 6)$ for the function f whose level curves are shown. Explain how you chose the direction and length of this vector.

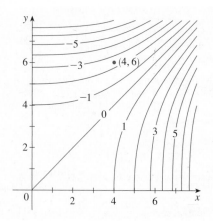

35–38 ■ Find equations of (a) the tangent plane and (b) the normal line to the given surface at the specified point.

35. $x^2 - 2y^2 + z^2 + yz = 2$, $(2, 1, -1)$

36. $x - z = 4 \arctan(yz)$, $(1 + \pi, 1, 1)$

37. $z + 1 = xe^y \cos z$, $(1, 0, 0)$

38. $yz = \ln(x + z)$, $(0, 0, 1)$

. . . .

39–40 ■ Use a computer to graph the surface, the tangent plane, and the normal line on the same screen. Choose the domain carefully so that you avoid extraneous vertical planes. Choose the viewpoint so that you get a good view of all three objects.

39. $xy + yz + zx = 3$, $(1, 1, 1)$

40. $xyz = 6$, $(1, 2, 3)$

. . . .

41. If $f(x, y) = x^2 + 4y^2$, find the gradient vector $\nabla f(2, 1)$ and use it to find the tangent line to the level curve $f(x, y) = 8$ at the point $(2, 1)$. Sketch the level curve, the tangent line, and the gradient vector.

42. If $g(x, y) = x - y^2$, find the gradient vector $\nabla g(3, -1)$ and use it to find the tangent line to the level curve $g(x, y) = 2$ at the point $(3, -1)$. Sketch the level curve, the tangent line, and the gradient vector.

43. Show that the equation of the tangent plane to the ellipsoid $x^2/a^2 + y^2/b^2 + z^2/c^2 = 1$ at the point (x_0, y_0, z_0) can be written as
$$\frac{xx_0}{a^2} + \frac{yy_0}{b^2} + \frac{zz_0}{c^2} = 1$$

44. Find the points on the ellipsoid $x^2 + 2y^2 + 3z^2 = 1$ where the tangent plane is parallel to the plane $3x - y + 3z = 1$.

45. Find the points on the hyperboloid $x^2 - y^2 + 2z^2 = 1$ where the normal line is parallel to the line that joins the points $(3, -1, 0)$ and $(5, 3, 6)$.

46. Show that the ellipsoid $3x^2 + 2y^2 + z^2 = 9$ and the sphere $x^2 + y^2 + z^2 - 8x - 6y - 8z + 24 = 0$ are tangent to each other at the point $(1, 1, 2)$. (This means that they have a common tangent plane at the point.)

47. Show that the sum of the x-, y-, and z-intercepts of any tangent plane to the surface $\sqrt{x} + \sqrt{y} + \sqrt{z} = \sqrt{c}$ is a constant.

48. Show that every normal line to the sphere $x^2 + y^2 + z^2 = r^2$ passes through the center of the sphere.

49. Find parametric equations for the tangent line to the curve of intersection of the paraboloid $z = x^2 + y^2$ and the ellipsoid $4x^2 + y^2 + z^2 = 9$ at the point $(-1, 1, 2)$.

50. (a) The plane $y + z = 3$ intersects the cylinder $x^2 + y^2 = 5$ in an ellipse. Find parametric equations for the tangent line to this ellipse at the point $(1, 2, 1)$.

(b) Graph the cylinder, the plane, and the tangent line on the same screen.

51. (a) Two surfaces are called **orthogonal** at a point of intersection if their normal lines are perpendicular at that point. Show that surfaces with equations $F(x, y, z) = 0$ and $G(x, y, z) = 0$ are orthogonal at a point P where $\nabla F \neq \mathbf{0}$ and $\nabla G \neq \mathbf{0}$ if and only if
$$F_x G_x + F_y G_y + F_z G_z = 0 \quad \text{at } P$$

(b) Use part (a) to show that the surfaces $z^2 = x^2 + y^2$ and $x^2 + y^2 + z^2 = r^2$ are orthogonal at every point of intersection. Can you see why this is true without using calculus?

52. (a) Show that the function $f(x, y) = \sqrt[3]{xy}$ is continuous and the partial derivatives f_x and f_y exist at the origin but the directional derivatives in all other directions do not exist.

(b) Graph f near the origin and comment on how the graph confirms part (a).

53. Suppose that the directional derivatives of $f(x, y)$ are known at a given point in two nonparallel directions given by unit vectors $\mathbf{u}$ and $\mathbf{v}$. Is it possible to find ∇f at this point? If so, how would you do it?

54. Show that if $z = f(x, y)$ is differentiable at $\mathbf{x}_0 = \langle x_0, y_0 \rangle$, then
$$\lim_{\mathbf{x} \to \mathbf{x}_0} \frac{f(\mathbf{x}) - f(\mathbf{x}_0) - \nabla f(\mathbf{x}_0) \cdot (\mathbf{x} - \mathbf{x}_0)}{|\mathbf{x} - \mathbf{x}_0|} = 0$$

[*Hint:* Use Definition 11.4.7 directly.]

11.7 Maximum and Minimum Values

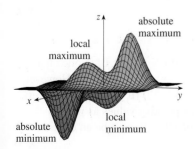

FIGURE 1

As we saw in Chapter 4, one of the main uses of ordinary derivatives is in finding maximum and minimum values. In this section we see how to use partial derivatives to locate maxima and minima of functions of two variables. In particular, in Example 6 we will see how to maximize the volume of a box without a lid if we have a fixed amount of cardboard to work with.

Look at the hills and valleys in the graph of f shown in Figure 1. There are two points (a, b) where f has a *local maximum*, that is, where $f(a, b)$ is larger than nearby values of $f(x, y)$. The larger of these two values is the *absolute maximum*. Likewise, f has two *local minima*, where $f(a, b)$ is smaller than nearby values. The smaller of these two values is the *absolute minimum*.

> **1 Definition** A function of two variables has a **local maximum** at (a, b) if $f(x, y) \leq f(a, b)$ when (x, y) is near (a, b). [This means that $f(x, y) \leq f(a, b)$ for all points (x, y) in some disk with center (a, b).] The number $f(a, b)$ is called a **local maximum value**. If $f(x, y) \geq f(a, b)$ when (x, y) is near (a, b), then $f(a, b)$ is a **local minimum value**.

If the inequalities in Definition 1 hold for *all* points (x, y) in the domain of f, then f has an **absolute maximum** (or **absolute minimum**) at (a, b).

■ ■ Notice that the conclusion of Theorem 2 can be stated in the notation of gradient vectors as $\nabla f(a, b) = \mathbf{0}$.

> **2 Theorem** If f has a local maximum or minimum at (a, b) and the first-order partial derivatives of f exist there, then $f_x(a, b) = 0$ and $f_y(a, b) = 0$.

Proof Let $g(x) = f(x, b)$. If f has a local maximum (or minimum) at (a, b), then g has a local maximum (or minimum) at a, so $g'(a) = 0$ by Fermat's Theorem (see Theorem 4.2.4). But $g'(a) = f_x(a, b)$ (see Equation 11.3.1) and so $f_x(a, b) = 0$. Similarly, by applying Fermat's Theorem to the function $G(y) = f(a, y)$, we obtain $f_y(a, b) = 0$. ■ ■

If we put $f_x(a, b) = 0$ and $f_y(a, b) = 0$ in the equation of a tangent plane (Equation 11.4.2), we get $z = z_0$. Thus, the geometric interpretation of Theorem 2 is that if the graph of f has a tangent plane at a local maximum or minimum, then the tangent plane must be horizontal.

A point (a, b) is called a **critical point** (or *stationary point*) of f if $f_x(a, b) = 0$ and $f_y(a, b) = 0$, or if one of these partial derivatives does not exist. Theorem 2 says that if f has a local maximum or minimum at (a, b), then (a, b) is a critical point of f. However, as in single-variable calculus, not all critical points give rise to maxima or minima. At a critical point, a function could have a local maximum or a local minimum or neither.

EXAMPLE 1 Let $f(x, y) = x^2 + y^2 - 2x - 6y + 14$. Then

$$f_x(x, y) = 2x - 2 \qquad f_y(x, y) = 2y - 6$$

These partial derivatives are equal to 0 when $x = 1$ and $y = 3$, so the only critical point is $(1, 3)$. By completing the square, we find that

$$f(x, y) = 4 + (x - 1)^2 + (y - 3)^2$$

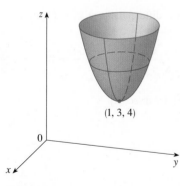

FIGURE 2
$z = x^2 + y^2 - 2x - 6y + 14$

Since $(x - 1)^2 \geq 0$ and $(y - 3)^2 \geq 0$, we have $f(x, y) \geq 4$ for all values of x and y. Therefore, $f(1, 3) = 4$ is a local minimum, and in fact it is the absolute minimum of f. This can be confirmed geometrically from the graph of f, which is the elliptic paraboloid with vertex $(1, 3, 4)$ shown in Figure 2. ■ ■

EXAMPLE 2 Find the extreme values of $f(x, y) = y^2 - x^2$.

SOLUTION Since $f_x = -2x$ and $f_y = 2y$, the only critical point is $(0, 0)$. Notice that for points on the x-axis we have $y = 0$, so $f(x, y) = -x^2 < 0$ (if $x \neq 0$). However, for points on the y-axis we have $x = 0$, so $f(x, y) = y^2 > 0$ (if $y \neq 0$). Thus, every disk with center $(0, 0)$ contains points where f takes positive values as well as points where f takes negative values. Therefore, $f(0, 0) = 0$ can't be an extreme value for f, so f has no extreme value. ■ ■

FIGURE 3
$z = y^2 - x^2$

Example 2 illustrates the fact that a function need not have a maximum or minimum value at a critical point. Figure 3 shows how this is possible. The graph of f is the hyperbolic paraboloid $z = y^2 - x^2$, which has a horizontal tangent plane $(z = 0)$ at the origin. You can see that $f(0, 0) = 0$ is a maximum in the direction of the x-axis but a minimum in the direction of the y-axis. Near the origin the graph has the shape of a saddle and so $(0, 0)$ is called a *saddle point* of f.

We need to be able to determine whether or not a function has an extreme value at a critical point. The following test, which is proved in Appendix E, is analogous to the Second Derivative Test for functions of one variable.

---

3 **Second Derivatives Test** Suppose the second partial derivatives of f are continuous on a disk with center (a, b), and suppose that $f_x(a, b) = 0$ and $f_y(a, b) = 0$ [that is, (a, b) is a critical point of f]. Let

$$D = D(a, b) = f_{xx}(a, b)f_{yy}(a, b) - [f_{xy}(a, b)]^2$$

(a) If $D > 0$ and $f_{xx}(a, b) > 0$, then $f(a, b)$ is a local minimum.

(b) If $D > 0$ and $f_{xx}(a, b) < 0$, then $f(a, b)$ is a local maximum.

(c) If $D < 0$, then $f(a, b)$ is not a local maximum or minimum.

NOTE 1 □ In case (c) the point (a, b) is called a **saddle point** of f and the graph of f crosses its tangent plane at (a, b).

NOTE 2 □ If $D = 0$, the test gives no information: f could have a local maximum or local minimum at (a, b), or (a, b) could be a saddle point of f.

NOTE 3 □ To remember the formula for D it's helpful to write it as a determinant:

$$D = \begin{vmatrix} f_{xx} & f_{xy} \\ f_{yx} & f_{yy} \end{vmatrix} = f_{xx}f_{yy} - (f_{xy})^2$$

$\boxed{V}$ **EXAMPLE 3** Find the local maximum and minimum values and saddle points of $f(x, y) = x^4 + y^4 - 4xy + 1$.

SOLUTION We first locate the critical points:

$$f_x = 4x^3 - 4y \qquad f_y = 4y^3 - 4x$$

Setting these partial derivatives equal to 0, we obtain the equations

$$x^3 - y = 0 \quad \text{and} \quad y^3 - x = 0$$

To solve these equations we substitute $y = x^3$ from the first equation into the second one. This gives

$$0 = x^9 - x = x(x^8 - 1) = x(x^4 - 1)(x^4 + 1) = x(x^2 - 1)(x^2 + 1)(x^4 + 1)$$

so there are three real roots: $x = 0, 1, -1$. The three critical points are $(0, 0)$, $(1, 1)$, and $(-1, -1)$.

FIGURE 4
$z = x^4 + y^4 - 4xy + 1$

Next we calculate the second partial derivatives and $D(x, y)$:

$$f_{xx} = 12x^2 \qquad f_{xy} = -4 \qquad f_{yy} = 12y^2$$

$$D(x, y) = f_{xx}f_{yy} - (f_{xy})^2 = 144x^2y^2 - 16$$

Since $D(0, 0) = -16 < 0$, it follows from case (c) of the Second Derivatives Test that the origin is a saddle point; that is, f has no local maximum or minimum at $(0, 0)$. Since $D(1, 1) = 128 > 0$ and $f_{xx}(1, 1) = 12 > 0$, we see from case (a) of the test that $f(1, 1) = -1$ is a local minimum. Similarly, we have $D(-1, -1) = 128 > 0$ and $f_{xx}(-1, -1) = 12 > 0$, so $f(-1, -1) = -1$ is also a local minimum.

The graph of f is shown in Figure 4. ■ ■

■ ■ A contour map of the function f in Example 3 is shown in Figure 5. The level curves near $(1, 1)$ and $(-1, -1)$ are oval in shape and indicate that as we move away from $(1, 1)$ or $(-1, -1)$ in any direction the values of f are increasing. The level curves near $(0, 0)$, on the other hand, resemble hyperbolas. They reveal that as we move away from the origin (where the value of f is 1), the values of f decrease in some directions but increase in other directions. Thus, the contour map suggests the presence of the minima and saddle point that we found in Example 3.

FIGURE 5

 In Module 11.7 you can use contour maps to estimate the locations of critical points.

EXAMPLE 4 Find and classify the critical points of the function

$$f(x, y) = 10x^2y - 5x^2 - 4y^2 - x^4 - 2y^4$$

Also find the highest point on the graph of f.

SOLUTION The first-order partial derivatives are

$$f_x = 20xy - 10x - 4x^3 \qquad f_y = 10x^2 - 8y - 8y^3$$

So to find the critical points we need to solve the equations

$$\boxed{4} \qquad\qquad 2x(10y - 5 - 2x^2) = 0$$

$$\boxed{5} \qquad\qquad 5x^2 - 4y - 4y^3 = 0$$

From Equation 4 we see that either

$$x = 0 \qquad \text{or} \qquad 10y - 5 - 2x^2 = 0$$

In the first case ($x = 0$), Equation 5 becomes $-4y(1 + y^2) = 0$, so $y = 0$ and we have the critical point $(0, 0)$.

In the second case ($10y - 5 - 2x^2 = 0$), we get

$$x^2 = 5y - 2.5 \qquad \boxed{6}$$

and, putting this in Equation 5, we have $25y - 12.5 - 4y - 4y^3 = 0$. So we have to solve the cubic equation

$$4y^3 - 21y + 12.5 = 0 \qquad \boxed{7}$$

Using a graphing calculator or computer to graph the function

$$g(y) = 4y^3 - 21y + 12.5$$

as in Figure 6, we see that Equation 7 has three real roots. By zooming in, we can find the roots to four decimal places:

$$y \approx -2.5452 \qquad y \approx 0.6468 \qquad y \approx 1.8984$$

FIGURE 6

(Alternatively, we could have used Newton's method or a rootfinder to locate these roots.) From Equation 6, the corresponding x-values are given by

$$x = \pm\sqrt{5y - 2.5}$$

If $y \approx -2.5452$, then x has no corresponding real values. If $y \approx 0.6468$, then $x \approx \pm 0.8567$. If $y \approx 1.8984$, then $x \approx \pm 2.6442$. So we have a total of five critical points, which are analyzed in the following chart. All quantities are rounded to two decimal places.

Critical point	Value of f	f_{xx}	D	Conclusion
$(0, 0)$	0.00	-10.00	80.00	local maximum
$(\pm 2.64, 1.90)$	8.50	-55.93	2488.72	local maximum
$(\pm 0.86, 0.65)$	-1.48	-5.87	-187.64	saddle point

Figures 7 and 8 give two views of the graph of f and we see that the surface opens downward. [This can also be seen from the expression for $f(x, y)$: The dominant terms are $-x^4 - 2y^4$ when $|x|$ and $|y|$ are large.] Comparing the values of f at its local maximum points, we see that the absolute maximum value of f is $f(\pm 2.64, 1.90) \approx 8.50$. In other words, the highest points on the graph of f are $(\pm 2.64, 1.90, 8.50)$.

Visual 11.7 shows several families of surfaces. The surface in Figures 7 and 8 is a member of one of these families.

FIGURE 7

FIGURE 8

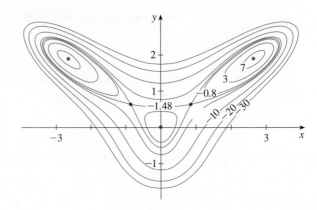

■ ■ The five critical points of the function f in Example 4 are shown in red in the contour map of f in Figure 9.

FIGURE 9

V **EXAMPLE 5** Find the shortest distance from the point $(1, 0, -2)$ to the plane $x + 2y + z = 4$.

SOLUTION The distance from any point (x, y, z) to the point $(1, 0, -2)$ is

$$d = \sqrt{(x - 1)^2 + y^2 + (z + 2)^2}$$

but if (x, y, z) lies on the plane $x + 2y + z = 4$, then $z = 4 - x - 2y$ and so we have $d = \sqrt{(x - 1)^2 + y^2 + (6 - x - 2y)^2}$. We can minimize d by minimizing the simpler expression

$$d^2 = f(x, y) = (x - 1)^2 + y^2 + (6 - x - 2y)^2$$

By solving the equations

$$f_x = 2(x - 1) - 2(6 - x - 2y) = 4x + 4y - 14 = 0$$

$$f_y = 2y - 4(6 - x - 2y) = 4x + 10y - 24 = 0$$

we find that the only critical point is $\left(\frac{11}{6}, \frac{5}{3}\right)$. Since $f_{xx} = 4$, $f_{xy} = 4$, and $f_{yy} = 10$, we have $D(x, y) = f_{xx}f_{yy} - (f_{xy})^2 = 24 > 0$ and $f_{xx} > 0$, so by the Second Derivatives Test f has a local minimum at $\left(\frac{11}{6}, \frac{5}{3}\right)$. Intuitively, we can see that this local minimum is actually an absolute minimum because there must be a point on the given plane that is closest to $(1, 0, -2)$. If $x = \frac{11}{6}$ and $y = \frac{5}{3}$, then

$$d = \sqrt{(x - 1)^2 + y^2 + (6 - x - 2y)^2} = \sqrt{\left(\tfrac{5}{6}\right)^2 + \left(\tfrac{5}{3}\right)^2 + \left(\tfrac{5}{6}\right)^2} = \frac{5\sqrt{6}}{6}$$

■ ■ Example 5 could also be solved using vectors. Compare with the methods of Section 9.5.

The shortest distance from $(1, 0, -2)$ to the plane $x + 2y + z = 4$ is $5\sqrt{6}/6$. ■ ■

V **EXAMPLE 6** A rectangular box without a lid is to be made from 12 m² of cardboard. Find the maximum volume of such a box.

SOLUTION Let the length, width, and height of the box (in meters) be x, y, and z, as shown in Figure 10. Then the volume of the box is

$$V = xyz$$

We can express V as a function of just two variables x and y by using the fact that the area of the four sides and the bottom of the box is

$$2xz + 2yz + xy = 12$$

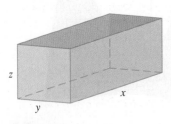

FIGURE 10

Solving this equation for z, we get $z = (12 - xy)/[2(x + y)]$, so the expression for V becomes

$$V = xy \frac{12 - xy}{2(x + y)} = \frac{12xy - x^2y^2}{2(x + y)}$$

We compute the partial derivatives:

$$\frac{\partial V}{\partial x} = \frac{y^2(12 - 2xy - x^2)}{2(x + y)^2} \qquad \frac{\partial V}{\partial y} = \frac{x^2(12 - 2xy - y^2)}{2(x + y)^2}$$

If V is a maximum, then $\partial V/\partial x = \partial V/\partial y = 0$, but $x = 0$ or $y = 0$ gives $V = 0$, so we must solve the equations

$$12 - 2xy - x^2 = 0 \qquad 12 - 2xy - y^2 = 0$$

These imply that $x^2 = y^2$ and so $x = y$. (Note that x and y must both be positive in this problem.) If we put $x = y$ in either equation we get $12 - 3x^2 = 0$, which gives $x = 2$, $y = 2$, and $z = (12 - 2 \cdot 2)/[2(2 + 2)] = 1$.

We could use the Second Derivatives Test to show that this gives a local maximum of V, or we could simply argue from the physical nature of this problem that there must be an absolute maximum volume, which has to occur at a critical point of V, so it must occur when $x = 2$, $y = 2$, $z = 1$. Then $V = 2 \cdot 2 \cdot 1 = 4$, so the maximum volume of the box is 4 m³. ■ ■

▪▪ Absolute Maximum and Minimum Values

For a function f of one variable the Extreme Value Theorem says that if f is continuous on a closed interval $[a, b]$, then f has an absolute minimum value and an absolute maximum value. According to the Closed Interval Method in Section 4.2, we found these by evaluating f not only at the critical numbers but also at the endpoints a and b.

There is a similar situation for functions of two variables. Just as a closed interval contains its endpoints, a **closed set** in $\mathbb{R}^2$ is one that contains all its boundary points. [A boundary point of D is a point (a, b) such that every disk with center (a, b) contains points in D and also points not in D.] For instance, the disk

$$D = \{(x, y) \mid x^2 + y^2 \leq 1\}$$

which consists of all points on and inside the circle $x^2 + y^2 = 1$, is a closed set because it contains all of its boundary points (which are the points on the circle $x^2 + y^2 = 1$). But if even one point on the boundary curve were omitted, the set would not be closed. (See Figure 11.)

A **bounded set** in $\mathbb{R}^2$ is one that is contained within some disk. In other words, it is finite in extent. Then, in terms of closed and bounded sets, we can state the following counterpart of the Extreme Value Theorem in two dimensions.

(a) Closed sets

(b) Sets that are not closed

FIGURE 11

8 Extreme Value Theorem for Functions of Two Variables If f is continuous on a closed, bounded set D in $\mathbb{R}^2$, then f attains an absolute maximum value $f(x_1, y_1)$ and an absolute minimum value $f(x_2, y_2)$ at some points (x_1, y_1) and (x_2, y_2) in D.

To find the extreme values guaranteed by Theorem 8, we note that, by Theorem 2, if f has an extreme value at (x_1, y_1), then (x_1, y_1) is either a critical point of f or a boundary point of D. Thus, we have the following extension of the Closed Interval Method.

---

9 To find the absolute maximum and minimum values of a continuous function f on a closed, bounded set D:

1. Find the values of f at the critical points of f in D.

2. Find the extreme values of f on the boundary of D.

3. The largest of the values from steps 1 and 2 is the absolute maximum value; the smallest of these values is the absolute minimum value.

EXAMPLE 7 Find the absolute maximum and minimum values of the function $f(x, y) = x^2 - 2xy + 2y$ on the rectangle $D = \{(x, y) \mid 0 \leqslant x \leqslant 3, 0 \leqslant y \leqslant 2\}$.

SOLUTION Since f is a polynomial, it is continuous on the closed, bounded rectangle D, so Theorem 8 tells us there is both an absolute maximum and an absolute minimum. According to step 1 in (9), we first find the critical points. These occur when

$$f_x = 2x - 2y = 0 \qquad f_y = -2x + 2 = 0$$

so the only critical point is $(1, 1)$, and the value of f there is $f(1, 1) = 1$.

In step 2 we look at the values of f on the boundary of D, which consists of the four line segments L_1, L_2, L_3, L_4 shown in Figure 12. On L_1 we have $y = 0$ and

$$f(x, 0) = x^2 \qquad 0 \leqslant x \leqslant 3$$

This is an increasing function of x, so its minimum value is $f(0, 0) = 0$ and its maximum value is $f(3, 0) = 9$. On L_2 we have $x = 3$ and

$$f(3, y) = 9 - 4y \qquad 0 \leqslant y \leqslant 2$$

This is a decreasing function of y, so its maximum value is $f(3, 0) = 9$ and its minimum value is $f(3, 2) = 1$. On L_3 we have $y = 2$ and

$$f(x, 2) = x^2 - 4x + 4 \qquad 0 \leqslant x \leqslant 3$$

By the methods of Chapter 4, or simply by observing that $f(x, 2) = (x - 2)^2$, we see that the minimum value of this function is $f(2, 2) = 0$ and the maximum value is $f(0, 2) = 4$. Finally, on L_4 we have $x = 0$ and

$$f(0, y) = 2y \qquad 0 \leqslant y \leqslant 2$$

with maximum value $f(0, 2) = 4$ and minimum value $f(0, 0) = 0$. Thus, on the boundary, the minimum value of f is 0 and the maximum is 9.

In step 3 we compare these values with the value $f(1, 1) = 1$ at the critical point and conclude that the absolute maximum value of f on D is $f(3, 0) = 9$ and the absolute minimum value is $f(0, 0) = f(2, 2) = 0$. Figure 13 shows the graph of f. ■ ■

FIGURE 12

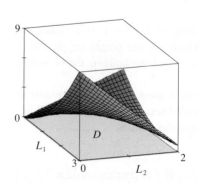

FIGURE 13

$f(x, y) = x^2 - 2xy + 2y$

11.7 | Exercises

1. Suppose $(1, 1)$ is a critical point of a function f with continuous second derivatives. In each case, what can you say about f?
(a) $f_{xx}(1, 1) = 4$, $f_{xy}(1, 1) = 1$, $f_{yy}(1, 1) = 2$
(b) $f_{xx}(1, 1) = 4$, $f_{xy}(1, 1) = 3$, $f_{yy}(1, 1) = 2$

2. Suppose $(0, 2)$ is a critical point of a function g with continuous second derivatives. In each case, what can you say about g?
(a) $g_{xx}(0, 2) = -1$, $g_{xy}(0, 2) = 6$, $g_{yy}(0, 2) = 1$
(b) $g_{xx}(0, 2) = -1$, $g_{xy}(0, 2) = 2$, $g_{yy}(0, 2) = -8$
(c) $g_{xx}(0, 2) = 4$, $g_{xy}(0, 2) = 6$, $g_{yy}(0, 2) = 9$

3–4 ■ Use the level curves in the figure to predict the location of the critical points of f and whether f has a saddle point or a local maximum or minimum at each of those points. Explain your reasoning. Then use the Second Derivatives Test to confirm your predictions.

3. $f(x, y) = 4 + x^3 + y^3 - 3xy$

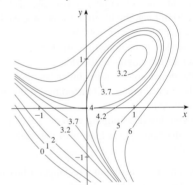

4. $f(x, y) = 3x - x^3 - 2y^2 + y^4$

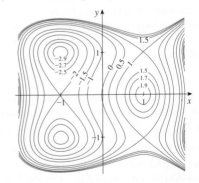

5–16 ■ Find the local maximum and minimum values and saddle point(s) of the function. If you have three-dimensional graphing software, graph the function with a domain and viewpoint that reveal all the important aspects of the function.

5. $f(x, y) = 9 - 2x + 4y - x^2 - 4y^2$

6. $f(x, y) = x^3y + 12x^2 - 8y$

7. $f(x, y) = x^4 + y^4 - 4xy + 2$

8. $f(x, y) = e^{4y - x^2 - y^2}$

9. $f(x, y) = (1 + xy)(x + y)$

10. $f(x, y) = 2x^3 + xy^2 + 5x^2 + y^2$

11. $f(x, y) = e^x \cos y$

12. $f(x, y) = x^2 + y^2 + \dfrac{1}{x^2 y^2}$

13. $f(x, y) = x \sin y$

14. $f(x, y) = (2x - x^2)(2y - y^2)$

15. $f(x, y) = (x^2 + y^2)e^{y^2 - x^2}$

16. $f(x, y) = x^2 y e^{-x^2 - y^2}$

■ ■ ■ ■ ■ ■ ■ ■ ■ ■ ■ ■ ■

17–20 ■ Use a graph and/or level curves to estimate the local maximum and minimum values and saddle point(s) of the function. Then use calculus to find these values precisely.

17. $f(x, y) = 3x^2y + y^3 - 3x^2 - 3y^2 + 2$

18. $f(x, y) = xye^{-x^2 - y^2}$

19. $f(x, y) = \sin x + \sin y + \sin(x + y)$,
$0 \leq x \leq 2\pi$, $0 \leq y \leq 2\pi$

20. $f(x, y) = \sin x + \sin y + \cos(x + y)$,
$0 \leq x \leq \pi/4$, $0 \leq y \leq \pi/4$

■ ■ ■ ■ ■ ■ ■ ■ ■ ■ ■ ■ ■

21–24 ■ Use a graphing device as in Example 4 (or Newton's method or a rootfinder) to find the critical points of f correct to three decimal places. Then classify the critical points and find the highest or lowest points on the graph.

21. $f(x, y) = x^4 - 5x^2 + y^2 + 3x + 2$

22. $f(x, y) = 5 - 10xy - 4x^2 + 3y - y^4$

23. $f(x, y) = 2x + 4x^2 - y^2 + 2xy^2 - x^4 - y^4$

24. $f(x, y) = e^x + y^4 - x^3 + 4 \cos y$

■ ■ ■ ■ ■ ■ ■ ■ ■ ■ ■ ■ ■

25–30 ■ Find the absolute maximum and minimum values of f on the set D.

25. $f(x, y) = 1 + 4x - 5y$, D is the closed triangular region with vertices $(0, 0)$, $(2, 0)$, and $(0, 3)$

26. $f(x, y) = 3 + xy - x - 2y$, D is the closed triangular region with vertices $(1, 0)$, $(5, 0)$, and $(1, 4)$

27. $f(x, y) = x^2 + y^2 + x^2y + 4$,
$D = \{(x, y) \mid |x| \leq 1, |y| \leq 1\}$

28. $f(x, y) = 4x + 6y - x^2 - y^2$,
$D = \{(x, y) \mid 0 \leq x \leq 4, 0 \leq y \leq 5\}$

29. $f(x, y) = x^4 + y^4 - 4xy + 2$,
$D = \{(x, y) \mid 0 \leq x \leq 3, 0 \leq y \leq 2\}$

30. $f(x, y) = xy^2$, $D = \{(x, y) \mid x \geq 0, y \geq 0, x^2 + y^2 \leq 3\}$

31. For functions of one variable it is impossible for a continuous function to have two local maxima and no local minimum. But for functions of two variables such functions exist. Show that the function

$$f(x, y) = -(x^2 - 1)^2 - (x^2y - x - 1)^2$$

has only two critical points, but has local maxima at both of them. Then use a computer to produce a graph with a carefully chosen domain and viewpoint to see how this is possible.

32. If a function of one variable is continuous on an interval and has only one critical number, then a local maximum has to be an absolute maximum. But this is not true for functions of two variables. Show that the function

$$f(x, y) = 3xe^y - x^3 - e^{3y}$$

has exactly one critical point, and that f has a local maximum there that is not an absolute maximum. Then use a computer to produce a graph with a carefully chosen domain and viewpoint to see how this is possible.

33. Find the shortest distance from the point $(2, 1, -1)$ to the plane $x + y - z = 1$.

34. Find the point on the plane $x - y + z = 4$ that is closest to the point $(1, 2, 3)$.

35. Find the points on the cone $z^2 = x^2 + y^2$ that are closest to the point $(4, 2, 0)$.

36. Find the points on the surface $y^2 = 9 + xz$ that are closest to the origin.

37. Find three positive numbers whose sum is 100 and whose product is a maximum.

38. Find three positive numbers x, y, and z whose sum is 100 such that $x^a y^b z^c$ is a maximum.

39. Find the volume of the largest rectangular box with edges parallel to the axes that can be inscribed in the ellipsoid

$$9x^2 + 36y^2 + 4z^2 = 36$$

40. Solve the problem in Exercise 39 for a general ellipsoid

$$\frac{x^2}{a^2} + \frac{y^2}{b^2} + \frac{z^2}{c^2} = 1$$

41. Find the volume of the largest rectangular box in the first octant with three faces in the coordinate planes and one vertex in the plane $x + 2y + 3z = 6$.

42. Find the dimensions of the rectangular box with largest volume if the total surface area is given as 64 cm^2.

43. Find the dimensions of a rectangular box of maximum volume such that the sum of the lengths of its 12 edges is a constant c.

44. The base of an aquarium with given volume V is made of slate and the sides are made of glass. If slate costs five times as much (per unit area) as glass, find the dimensions of the aquarium that minimize the cost of the materials.

45. A cardboard box without a lid is to have a volume of 32,000 cm^3. Find the dimensions that minimize the amount of cardboard used.

46. A rectangular building is being designed to minimize heat loss. The east and west walls lose heat at a rate of 10 units/m^2 per day, the north and south walls at a rate of 8 units/m^2 per day, the floor at a rate of 1 unit/m^2 per day, and the roof at a rate of 5 units/m^2 per day. Each wall must be at least 30 m long, the height must be at least 4 m, and the volume must be exactly 4000 m^3.
(a) Find and sketch the domain of the heat loss as a function of the lengths of the sides.
(b) Find the dimensions that minimize heat loss. (Check both the critical points and the points on the boundary of the domain.)
(c) Could you design a building with even less heat loss if the restrictions on the lengths of the walls were removed?

47. If the length of the diagonal of a rectangular box must be L, what is the largest possible volume?

48. Three alleles (alternative versions of a gene) A, B, and O determine the four blood types A (AA or AO), B (BB or BO), O (OO), and AB. The Hardy-Weinberg Law states that the proportion of individuals in a population who carry two different alleles is

$$P = 2pq + 2pr + 2rq$$

where p, q, and r are the proportions of A, B, and O in the population. Use the fact that $p + q + r = 1$ to show that P is at most $\frac{2}{3}$.

49. Suppose that a scientist has reason to believe that two quantities x and y are related linearly, that is, $y = mx + b$, at least approximately, for some values of m and b. The scientist performs an experiment and collects data in the form of points $(x_1, y_1), (x_2, y_2), \ldots, (x_n, y_n)$, and then plots these points. The points don't lie exactly on a straight line, so the scientist wants to find constants m and b so that the line

$y = mx + b$ "fits" the points as well as possible. (See the figure.)

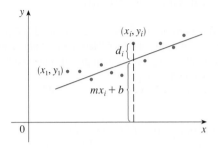

Let $d_i = y_i - (mx_i + b)$ be the vertical deviation of the point (x_i, y_i) from the line. The **method of least squares**

determines m and b so as to minimize $\sum_{i=1}^{n} d_i^2$, the sum of the squares of these deviations. Show that, according to this method, the line of best fit is obtained when

$$m \sum_{i=1}^{n} x_i + bn = \sum_{i=1}^{n} y_i$$

$$m \sum_{i=1}^{n} x_i^2 + b \sum_{i=1}^{n} x_i = \sum_{i=1}^{n} x_i y_i$$

Thus, the line is found by solving these two equations in the two unknowns m and b. (See Section 1.2 for a further discussion and applications of the method of least squares.)

50. Find an equation of the plane that passes through the point $(1, 2, 3)$ and cuts off the smallest volume in the first octant.

Designing a Dumpster

For this project we locate a trash dumpster in order to study its shape and construction. We then attempt to determine the dimensions of a container of similar design that minimize construction cost.

1. First locate a trash dumpster in your area. Carefully study and describe all details of its construction, and determine its volume. Include a sketch of the container.

2. While maintaining the general shape and method of construction, determine the dimensions such a container of the same volume should have in order to minimize the cost of construction. Use the following assumptions in your analysis:

   ■ The sides, back, and front are to be made from 12-gauge (0.1046 inch thick) steel sheets, which cost $0.70 per square foot (including any required cuts or bends).

   ■ The base is to be made from a 10-gauge (0.1345 inch thick) steel sheet, which costs $0.90 per square foot.

   ■ Lids cost approximately $50.00 each, regardless of dimensions.

   ■ Welding costs approximately $0.18 per foot for material and labor combined.

   Give justification of any further assumptions or simplifications made of the details of construction.

3. Describe how any of your assumptions or simplifications may affect the final result.

4. If you were hired as a consultant on this investigation, what would your conclusions be? Would you recommend altering the design of the dumpster? If so, describe the savings that would result.

DISCOVERY PROJECT

Quadratic Approximations and Critical Points

The Taylor polynomial approximation to functions of one variable that we discussed in Chapter 8 can be extended to functions of two or more variables. Here we investigate quadratic approximations to functions of two variables and use them to give insight into the Second Derivatives Test for classifying critical points.

In Section 11.4 we discussed the linearization of a function f of two variables at a point (a, b):

$$L(x, y) = f(a, b) + f_x(a, b)(x - a) + f_y(a, b)(y - b)$$

Recall that the graph of L is the tangent plane to the surface $z = f(x, y)$ at $(a, b, f(a, b))$ and the corresponding linear approximation is $f(x, y) \approx L(x, y)$. The linearization L is also called the **first-degree Taylor polynomial** of f at (a, b).

1. If f has continuous second-order partial derivatives at (a, b), then the **second-degree Taylor polynomial** of f at (a, b) is

$$Q(x, y) = f(a, b) + f_x(a, b)(x - a) + f_y(a, b)(y - b)$$
$$+ \tfrac{1}{2}f_{xx}(a, b)(x - a)^2 + f_{xy}(a, b)(x - a)(y - b) + \tfrac{1}{2}f_{yy}(a, b)(y - b)^2$$

and the approximation $f(x, y) \approx Q(x, y)$ is called the **quadratic approximation** to f at (a, b). Verify that Q has the same first- and second-order partial derivatives as f at (a, b).

2. (a) Find the first- and second-degree Taylor polynomials L and Q of $f(x, y) = e^{-x^2-y^2}$ at $(0, 0)$.
   (b) Graph f, L, and Q. Comment on how well L and Q approximate f.

3. (a) Find the first- and second-degree Taylor polynomials L and Q for $f(x, y) = xe^y$ at $(1, 0)$.
   (b) Compare the values of L, Q, and f at $(0.9, 0.1)$.
   (c) Graph f, L, and Q. Comment on how well L and Q approximate f.

4. In this problem we analyze the behavior of the polynomial $f(x, y) = ax^2 + bxy + cy^2$ (without using the Second Derivatives Test) by identifying the graph as a paraboloid.
   (a) By completing the square, show that if $a \neq 0$, then

$$f(x, y) = ax^2 + bxy + cy^2 = a\left[\left(x + \frac{b}{2a}y\right)^2 + \left(\frac{4ac - b^2}{4a^2}\right)y^2\right]$$

   (b) Let $D = 4ac - b^2$. Show that if $D > 0$ and $a > 0$, then f has a local minimum at $(0, 0)$.
   (c) Show that if $D > 0$ and $a < 0$, then f has a local maximum at $(0, 0)$.
   (d) Show that if $D < 0$, then $(0, 0)$ is a saddle point.

5. (a) Suppose f is any function with continuous second-order partial derivatives such that $f(0, 0) = 0$ and $(0, 0)$ is a critical point of f. Write an expression for the second-degree Taylor polynomial, Q, of f at $(0, 0)$.
   (b) What can you conclude about Q from Problem 4?
   (c) In view of the quadratic approximation $f(x, y) \approx Q(x, y)$, what does part (b) suggest about f?

11.8 Lagrange Multipliers

In Example 6 in Section 11.7 we maximized a volume function $V = xyz$ subject to the constraint $2xz + 2yz + xy = 12$, which expressed the side condition that the surface area was 12 m². In this section we present Lagrange's method for maximizing or minimizing a general function $f(x, y, z)$ subject to a constraint (or side condition) of the form $g(x, y, z) = k$.

It's easier to explain the geometric basis of Lagrange's method for functions of two variables. So we start by trying to find the extreme values of $f(x, y)$ subject to a constraint of the form $g(x, y) = k$. In other words, we seek the extreme values of $f(x, y)$ when the point (x, y) is restricted to lie on the level curve $g(x, y) = k$. Figure 1 shows this curve together with several level curves of f. These have the equations $f(x, y) = c$, where $c = 7, 8, 9, 10, 11$. To maximize $f(x, y)$ subject to $g(x, y) = k$ is to find the largest value of c such that the level curve $f(x, y) = c$ intersects $g(x, y) = k$. It appears from Figure 1 that this happens when these curves just touch each other, that is, when they have a common tangent line. (Otherwise, the value of c could be increased further.) This means that the normal lines at the point (x_0, y_0) where they touch are identical. So the gradient vectors are parallel; that is, $\nabla f(x_0, y_0) = \lambda \nabla g(x_0, y_0)$ for some scalar λ.

This kind of argument also applies to the problem of finding the extreme values of $f(x, y, z)$ subject to the constraint $g(x, y, z) = k$. Thus, the point (x, y, z) is restricted to lie on the level surface S with equation $g(x, y, z) = k$. Instead of the level curves in Figure 1, we consider the level surfaces $f(x, y, z) = c$ and argue that if the maximum value of f is $f(x_0, y_0, z_0) = c$, then the level surface $f(x, y, z) = c$ is tangent to the level surface $g(x, y, z) = k$ and so the corresponding gradient vectors are parallel.

This intuitive argument can be made precise as follows. Suppose that a function f has an extreme value at a point $P(x_0, y_0, z_0)$ on the surface S and let C be a curve with vector equation $\mathbf{r}(t) = \langle x(t), y(t), z(t) \rangle$ that lies on S and passes through P. If t_0 is the parameter value corresponding to the point P, then $\mathbf{r}(t_0) = \langle x_0, y_0, z_0 \rangle$. The composite function $h(t) = f(x(t), y(t), z(t))$ represents the values that f takes on the curve C. Since f has an extreme value at (x_0, y_0, z_0), it follows that h has an extreme value at t_0, so $h'(t_0) = 0$. But if f is differentiable, we can use the Chain Rule to write

$$0 = h'(t_0) = f_x(x_0, y_0, z_0)x'(t_0) + f_y(x_0, y_0, z_0)y'(t_0) + f_z(x_0, y_0, z_0)z'(t_0)$$

$$= \nabla f(x_0, y_0, z_0) \cdot \mathbf{r}'(t_0)$$

This shows that the gradient vector $\nabla f(x_0, y_0, z_0)$ is orthogonal to the tangent vector $\mathbf{r}'(t_0)$ to every such curve C. But we already know from Section 11.6 that the gradient vector of g, $\nabla g(x_0, y_0, z_0)$, is also orthogonal to $\mathbf{r}'(t_0)$. (See Equation 11.6.18.) This means that the gradient vectors $\nabla f(x_0, y_0, z_0)$ and $\nabla g(x_0, y_0, z_0)$ must be parallel. Therefore, if $\nabla g(x_0, y_0, z_0) \neq \mathbf{0}$, there is a number λ such that

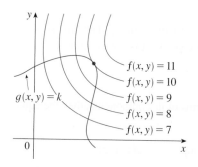

$$f(x, y) = 11$$
$$f(x, y) = 10$$
$$f(x, y) = 9$$
$$f(x, y) = 8$$
$$f(x, y) = 7$$
$$g(x, y) = k$$

FIGURE 1

TEC Visual 11.8 animates Figure 1 for both level curves and level surfaces.

■ ■ Lagrange multipliers are named after the French-Italian mathematician Joseph-Louis Lagrange (1736–1813). See page 279 for a biographical sketch of Lagrange.

1
$$\nabla f(x_0, y_0, z_0) = \lambda \nabla g(x_0, y_0, z_0)$$

The number λ in Equation 1 is called a **Lagrange multiplier**. The procedure based on Equation 1 is as follows.

■ ■ In deriving Lagrange's method we assumed that $\nabla g \neq \mathbf{0}$. In each of our examples you can check that $\nabla g \neq \mathbf{0}$ at all points where $g(x, y, z) = k$.

Method of Lagrange Multipliers To find the maximum and minimum values of $f(x, y, z)$ subject to the constraint $g(x, y, z) = k$ [assuming that these extreme values exist and $\nabla g \neq \mathbf{0}$ on the surface $g(x, y, z) = k$]:

(a) Find all values of x, y, z, and λ such that

$$\nabla f(x, y, z) = \lambda \nabla g(x, y, z)$$

and $$g(x, y, z) = k$$

(b) Evaluate f at all the points (x, y, z) that result from step (a). The largest of these values is the maximum value of f; the smallest is the minimum value of f.

If we write the vector equation $\nabla f = \lambda \nabla g$ in terms of its components, then the equations in step (a) become

$$f_x = \lambda g_x \qquad f_y = \lambda g_y \qquad f_z = \lambda g_z \qquad g(x, y, z) = k$$

This is a system of four equations in the four unknowns x, y, z, and λ, but it is not necessary to find explicit values for λ.

For functions of two variables the method of Lagrange multipliers is similar to the method just described. To find the extreme values of $f(x, y)$ subject to the constraint $g(x, y) = k$, we look for values of x, y, and λ such that

$$\nabla f(x, y) = \lambda \nabla g(x, y) \qquad \text{and} \qquad g(x, y) = k$$

This amounts to solving three equations in three unknowns:

$$f_x = \lambda g_x \qquad f_y = \lambda g_y \qquad g(x, y) = k$$

Our first illustration of Lagrange's method is to reconsider the problem given in Example 6 in Section 11.7.

$\boxed{\text{V}}$ **EXAMPLE 1** A rectangular box without a lid is to be made from 12 m^2 of cardboard. Find the maximum volume of such a box.

SOLUTION As in Example 6 in Section 11.7 we let x, y, and z be the length, width, and height, respectively, of the box in meters. Then we wish to maximize

$$V = xyz$$

subject to the constraint

$$g(x, y, z) = 2xz + 2yz + xy = 12$$

Using the method of Lagrange multipliers, we look for values of x, y, z, and λ such that $\nabla V = \lambda \nabla g$ and $g(x, y, z) = 12$. This gives the equations

$$V_x = \lambda g_x \qquad V_y = \lambda g_y \qquad V_z = \lambda g_z \qquad 2xz + 2yz + xy = 12$$

which become

$\boxed{2}$ $$yz = \lambda(2z + y)$$

$\boxed{3}$ $$xz = \lambda(2z + x)$$

$$\boxed{4} \qquad\qquad xy = \lambda(2x + 2y)$$

$$\boxed{5} \qquad\qquad 2xz + 2yz + xy = 12$$

There are no general rules for solving systems of equations. Sometimes some ingenuity is required. In the present example you might notice that if we multiply (2) by x, (3) by y, and (4) by z, then the left sides of these equations will be identical. Doing this, we have

Another method for solving the system of equations (2–5) is to solve each of Equations 2, 3, and 4 for λ and then to equate the resulting expressions.

$$\boxed{6} \qquad\qquad xyz = \lambda(2xz + xy)$$

$$\boxed{7} \qquad\qquad xyz = \lambda(2yz + xy)$$

$$\boxed{8} \qquad\qquad xyz = \lambda(2xz + 2yz)$$

We observe that $\lambda \neq 0$ because $\lambda = 0$ would imply $yz = xz = xy = 0$ from (2), (3), and (4) and this would contradict (5). Therefore, from (6) and (7) we have

$$2xz + xy = 2yz + xy$$

which gives $xz = yz$. But $z \neq 0$ (since $z = 0$ would give $V = 0$), so $x = y$. From (7) and (8) we have

$$2yz + xy = 2xz + 2yz$$

which gives $2xz = xy$ and so (since $x \neq 0$) $y = 2z$. If we now put $x = y = 2z$ in (5), we get

$$4z^2 + 4z^2 + 4z^2 = 12$$

Since x, y, and z are all positive, we therefore have $z = 1$, $x = 2$, and $y = 2$ as before. ■ ■

$\boxed{\text{V}}$ **EXAMPLE 2** Find the extreme values of the function $f(x, y) = x^2 + 2y^2$ on the circle $x^2 + y^2 = 1$.

SOLUTION We are asked for the extreme values of f subject to the constraint $g(x, y) = x^2 + y^2 = 1$. Using Lagrange multipliers, we solve the equations $\nabla f = \lambda \nabla g$, $g(x, y) = 1$, which can be written as

In geometric terms, Example 2 asks for the highest and lowest points on the curve C in Figure 2 that lies on the paraboloid $z = x^2 + 2y^2$ and directly above the constraint circle $x^2 + y^2 = 1$.

$$f_x = \lambda g_x \qquad f_y = \lambda g_y \qquad g(x, y) = 1$$

or as

$$\boxed{9} \qquad\qquad 2x = 2x\lambda$$

$$\boxed{10} \qquad\qquad 4y = 2y\lambda$$

$$\boxed{11} \qquad\qquad x^2 + y^2 = 1$$

From (9) we have $x = 0$ or $\lambda = 1$. If $x = 0$, then (11) gives $y = \pm 1$. If $\lambda = 1$, then $y = 0$ from (10), so then (11) gives $x = \pm 1$. Therefore, f has possible extreme values at the points $(0, 1)$, $(0, -1)$, $(1, 0)$, and $(-1, 0)$. Evaluating f at these four points, we find that

$$f(0, 1) = 2 \qquad f(0, -1) = 2 \qquad f(1, 0) = 1 \qquad f(-1, 0) = 1$$

Therefore, the maximum value of f on the circle $x^2 + y^2 = 1$ is $f(0, \pm 1) = 2$ and the minimum value is $f(\pm 1, 0) = 1$. Checking with Figure 2, we see that these values look reasonable. ■ ■

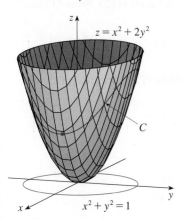

FIGURE 2

■ ■ The geometry behind the use of Lagrange multipliers in Example 2 is shown in Figure 3. The extreme values of $f(x, y) = x^2 + 2y^2$ correspond to the level curves that touch the circle $x^2 + y^2 = 1$.

FIGURE 3

EXAMPLE 3 Find the extreme values of $f(x, y) = x^2 + 2y^2$ on the disk $x^2 + y^2 \leqslant 1$.

SOLUTION According to the procedure in (11.7.9), we compare the values of f at the critical points with values at the points on the boundary. Since $f_x = 2x$ and $f_y = 4y$, the only critical point is $(0, 0)$. We compare the value of f at that point with the extreme values on the boundary from Example 2:

$$f(0, 0) = 0 \qquad f(\pm 1, 0) = 1 \qquad f(0, \pm 1) = 2$$

Therefore, the maximum value of f on the disk $x^2 + y^2 \leqslant 1$ is $f(0, \pm 1) = 2$ and the minimum value is $f(0, 0) = 0$. ■ ■

EXAMPLE 4 Find the points on the sphere $x^2 + y^2 + z^2 = 4$ that are closest to and farthest from the point $(3, 1, -1)$.

SOLUTION The distance from a point (x, y, z) to the point $(3, 1, -1)$ is

$$d = \sqrt{(x - 3)^2 + (y - 1)^2 + (z + 1)^2}$$

but the algebra is simpler if we instead maximize and minimize the square of the distance:

$$d^2 = f(x, y, z) = (x - 3)^2 + (y - 1)^2 + (z + 1)^2$$

The constraint is that the point (x, y, z) lies on the sphere, that is,

$$g(x, y, z) = x^2 + y^2 + z^2 = 4$$

According to the method of Lagrange multipliers, we solve $\nabla f = \lambda \nabla g$, $g = 4$. This gives

$$\boxed{12} \qquad\qquad 2(x - 3) = 2x\lambda$$

$$\boxed{13} \qquad\qquad 2(y - 1) = 2y\lambda$$

$$\boxed{14} \qquad\qquad 2(z + 1) = 2z\lambda$$

$$\boxed{15} \qquad\qquad x^2 + y^2 + z^2 = 4$$

■ ■ Figure 4 shows the sphere and the nearest point P in Example 4. Can you see how to find the coordinates of P without using calculus?

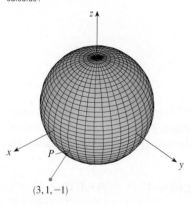

FIGURE 4

The simplest way to solve these equations is to solve for x, y, and z in terms of λ from (12), (13), and (14), and then substitute these values into (15). From (12) we have

$$x - 3 = x\lambda \qquad \text{or} \qquad x(1 - \lambda) = 3 \qquad \text{or} \qquad x = \frac{3}{1 - \lambda}$$

[Note that $1 - \lambda \neq 0$ because $\lambda = 1$ is impossible from (12).] Similarly, (13) and (14) give

$$y = \frac{1}{1 - \lambda} \qquad z = -\frac{1}{1 - \lambda}$$

Therefore, from (15) we have

$$\frac{3^2}{(1 - \lambda)^2} + \frac{1^2}{(1 - \lambda)^2} + \frac{(-1)^2}{(1 - \lambda)^2} = 4$$

which gives $(1 - \lambda)^2 = \frac{11}{4}$, $1 - \lambda = \pm\sqrt{11}/2$, so

$$\lambda = 1 \pm \frac{\sqrt{11}}{2}$$

These values of λ then give the corresponding points (x, y, z):

$$\left(\frac{6}{\sqrt{11}}, \frac{2}{\sqrt{11}}, -\frac{2}{\sqrt{11}} \right) \quad \text{and} \quad \left(-\frac{6}{\sqrt{11}}, -\frac{2}{\sqrt{11}}, \frac{2}{\sqrt{11}} \right)$$

It's easy to see that f has a smaller value at the first of these points, so the closest point is $\left(6/\sqrt{11}, 2/\sqrt{11}, -2/\sqrt{11}\right)$ and the farthest is $\left(-6/\sqrt{11}, -2/\sqrt{11}, 2/\sqrt{11}\right)$.

Two Constraints

Suppose now that we want to find the maximum and minimum values of a function $f(x, y, z)$ subject to two constraints (side conditions) of the form $g(x, y, z) = k$ and $h(x, y, z) = c$. Geometrically, this means that we are looking for the extreme values of f when (x, y, z) is restricted to lie on the curve of intersection C of the level surfaces $g(x, y, z) = k$ and $h(x, y, z) = c$. (See Figure 5.) Suppose f has such an extreme value at a point $P(x_0, y_0, z_0)$. We know from the beginning of this section that ∇f is orthogonal to C there. But we also know that ∇g is orthogonal to $g(x, y, z) = k$ and ∇h is orthogonal to $h(x, y, z) = c$, so ∇g and ∇h are both orthogonal to C. This means that the gradient vector $\nabla f(x_0, y_0, z_0)$ is in the plane determined by $\nabla g(x_0, y_0, z_0)$ and $\nabla h(x_0, y_0, z_0)$. (We assume that these gradient vectors are not zero and not parallel.) So there are numbers λ and μ (called Lagrange multipliers) such that

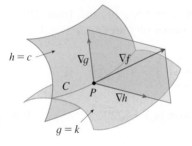

FIGURE 5

$\boxed{16}$
$$\nabla f(x_0, y_0, z_0) = \lambda \, \nabla g(x_0, y_0, z_0) + \mu \, \nabla h(x_0, y_0, z_0)$$

In this case Lagrange's method is to look for extreme values by solving five equations in the five unknowns x, y, z, λ, and μ. These equations are obtained by writing Equation 16 in terms of its components and using the constraint equations:

$$f_x = \lambda g_x + \mu h_x$$
$$f_y = \lambda g_y + \mu h_y$$
$$f_z = \lambda g_z + \mu h_z$$
$$g(x, y, z) = k$$
$$h(x, y, z) = c$$

$\boxed{V}$ **EXAMPLE 5** Find the maximum value of the function $f(x, y, z) = x + 2y + 3z$ on the curve of intersection of the plane $x - y + z = 1$ and the cylinder $x^2 + y^2 = 1$.

SOLUTION We maximize the function $f(x, y, z) = x + 2y + 3z$ subject to the constraints $g(x, y, z) = x - y + z = 1$ and $h(x, y, z) = x^2 + y^2 = 1$. The Lagrange

■ ■ The cylinder $x^2 + y^2 = 1$ intersects the plane $x - y + z = 1$ in an ellipse (Figure 6). Example 5 asks for the maximum value of f when (x, y, z) is restricted to lie on the ellipse.

FIGURE 6

condition is $\nabla f = \lambda \nabla g + \mu \nabla h$, so we solve the equations

$$\boxed{17} \qquad\qquad 1 = \lambda + 2x\mu$$

$$\boxed{18} \qquad\qquad 2 = -\lambda + 2y\mu$$

$$\boxed{19} \qquad\qquad 3 = \lambda$$

$$\boxed{20} \qquad\qquad x - y + z = 1$$

$$\boxed{21} \qquad\qquad x^2 + y^2 = 1$$

Putting $\lambda = 3$ [from (19)] in (17), we get $2x\mu = -2$, so $x = -1/\mu$. Similarly, (18) gives $y = 5/(2\mu)$. Substitution in (21) then gives

$$\frac{1}{\mu^2} + \frac{25}{4\mu^2} = 1$$

and so $\mu^2 = \frac{29}{4}$, $\mu = \pm\sqrt{29}/2$. Then $x = \mp 2/\sqrt{29}$, $y = \pm 5/\sqrt{29}$, and, from (20), $z = 1 - x + y = 1 \pm 7/\sqrt{29}$. The corresponding values of f are

$$\mp\frac{2}{\sqrt{29}} + 2\left(\pm\frac{5}{\sqrt{29}}\right) + 3\left(1 \pm \frac{7}{\sqrt{29}}\right) = 3 \pm \sqrt{29}$$

Therefore, the maximum value of f on the given curve is $3 + \sqrt{29}$. ■ ■

11.8 Exercises

1. Pictured are a contour map of f and a curve with equation $g(x, y) = 8$. Estimate the maximum and minimum values of f subject to the constraint that $g(x, y) = 8$. Explain your reasoning.

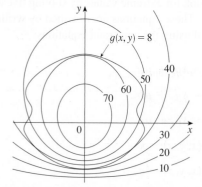

2. (a) Use a graphing calculator or computer to graph the circle $x^2 + y^2 = 1$. On the same screen, graph several curves of the form $x^2 + y = c$ until you find two that just touch the circle. What is the significance of the values of c for these two curves?
(b) Use Lagrange multipliers to find the extreme values of $f(x, y) = x^2 + y$ subject to the constraint $x^2 + y^2 = 1$. Compare your answers with those in part (a).

3–17 ■ Use Lagrange multipliers to find the maximum and minimum values of the function subject to the given constraint(s).

3. $f(x, y) = x^2 + y^2$; $\quad xy = 1$

4. $f(x, y) = 4x + 6y$; $\quad x^2 + y^2 = 13$

5. $f(x, y) = x^2 y$; $\quad x^2 + 2y^2 = 6$

6. $f(x, y) = e^{xy}$; $\quad x^3 + y^3 = 16$

7. $f(x, y, z) = 2x + 6y + 10z$; $\quad x^2 + y^2 + z^2 = 35$

8. $f(x, y, z) = 8x - 4z$; $\quad x^2 + 10y^2 + z^2 = 5$

9. $f(x, y, z) = xyz$; $\quad x^2 + 2y^2 + 3z^2 = 6$

10. $f(x, y, z) = x^2 y^2 z^2$; $\quad x^2 + y^2 + z^2 = 1$

11. $f(x, y, z) = x^2 + y^2 + z^2$; $\quad x^4 + y^4 + z^4 = 1$

12. $f(x, y, z) = x^4 + y^4 + z^4$; $\quad x^2 + y^2 + z^2 = 1$

13. $f(x, y, z, t) = x + y + z + t$; $\quad x^2 + y^2 + z^2 + t^2 = 1$

14. $f(x_1, x_2, \ldots, x_n) = x_1 + x_2 + \cdots + x_n$; $\quad x_1^2 + x_2^2 + \cdots + x_n^2 = 1$

15. $f(x, y, z) = x + 2y$; $\quad x + y + z = 1$, $\quad y^2 + z^2 = 4$

16. $f(x, y, z) = 3x - y - 3z$;
$x + y - z = 0$, $x^2 + 2z^2 = 1$

17. $f(x, y, z) = yz + xy$; $xy = 1$, $y^2 + z^2 = 1$

. . . .

18–19 ■ Find the extreme values of f on the region described by the inequality.

18. $f(x, y) = 2x^2 + 3y^2 - 4x - 5$, $x^2 + y^2 \leq 16$

19. $f(x, y) = e^{-xy}$, $x^2 + 4y^2 \leq 1$

. . . .

CAS 20. (a) If your computer algebra system plots implicitly defined curves, use it to estimate the minimum and maximum values of $f(x, y) = x^3 + y^3 + 3xy$ subject to the constraint $(x - 3)^2 + (y - 3)^2 = 9$ by graphical methods.
(b) Solve the problem in part (a) with the aid of Lagrange multipliers. Use your CAS to solve the equations numerically. Compare your answers with those in part (a).

21. The total production P of a certain product depends on the amount L of labor used and the amount K of capital investment. In Sections 11.1 and 11.3 we discussed how the Cobb-Douglas model $P = bL^\alpha K^{1-\alpha}$ follows from certain economic assumptions, where b and α are positive constants and $\alpha < 1$. If the cost of a unit of labor is m and the cost of a unit of capital is n, and the company can spend only p dollars as its total budget, then maximizing the production P is subject to the constraint $mL + nK = p$. Show that the maximum production occurs when

$$L = \frac{\alpha p}{m} \quad \text{and} \quad K = \frac{(1 - \alpha)p}{n}$$

22. Referring to Exercise 21, we now suppose that the production is fixed at $bL^\alpha K^{1-\alpha} = Q$, where Q is a constant. What values of L and K minimize the cost function $C(L, K) = mL + nK$?

23. Use Lagrange multipliers to prove that the rectangle with maximum area that has a given perimeter p is a square.

24. Use Lagrange multipliers to prove that the triangle with maximum area that has a given perimeter p is equilateral. [*Hint:* Use Heron's formula for the area: $A = \sqrt{s(s - x)(s - y)(s - z)}$, where $s = p/2$ and x, y, z are the lengths of the sides.]

25–37 ■ Use Lagrange multipliers to give an alternate solution to the indicated exercise in Section 11.7.

25. Exercise 33

26. Exercise 34

27. Exercise 35

28. Exercise 36

29. Exercise 37

30. Exercise 38

31. Exercise 39

32. Exercise 40

33. Exercise 41

34. Exercise 42

35. Exercise 43

36. Exercise 44

37. Exercise 47

. . . .

38. Find the maximum and minimum volumes of a rectangular box whose surface area is 1500 cm² and whose total edge length is 200 cm.

39. The plane $x + y + 2z = 2$ intersects the paraboloid $z = x^2 + y^2$ in an ellipse. Find the points on this ellipse that are nearest to and farthest from the origin.

40. The plane $4x - 3y + 8z = 5$ intersects the cone $z^2 = x^2 + y^2$ in an ellipse.
(a) Graph the cone, the plane, and the ellipse.
(b) Use Lagrange multipliers to find the highest and lowest points on the ellipse.

CAS 41–42 ■ Find the maximum and minimum values of f subject to the given constraints. Use a computer algebra system to solve the system of equations that arises in using Lagrange multipliers. (If your CAS finds only one solution, you may need to use additional commands.)

41. $f(x, y, z) = ye^{x-z}$; $9x^2 + 4y^2 + 36z^2 = 36$, $xy + yz = 1$

42. $f(x, y, z) = x + y + z$; $x^2 - y^2 = z$, $x^2 + z^2 = 4$

. . . .

43. (a) Find the maximum value of

$$f(x_1, x_2, \ldots, x_n) = \sqrt[n]{x_1 x_2 \cdots x_n}$$

given that $x_1, x_2, \ldots, x_n$ are positive numbers and $x_1 + x_2 + \cdots + x_n = c$, where c is a constant.
(b) Deduce from part (a) that if $x_1, x_2, \ldots, x_n$ are positive numbers, then

$$\sqrt[n]{x_1 x_2 \cdots x_n} \leq \frac{x_1 + x_2 + \cdots + x_n}{n}$$

This inequality says that the geometric mean of n numbers is no larger than the arithmetic mean of the numbers. Under what circumstances are these two means equal?

44. (a) Maximize $\sum_{i=1}^n x_i y_i$ subject to the constraints $\sum_{i=1}^n x_i^2 = 1$ and $\sum_{i=1}^n y_i^2 = 1$.
(b) Put

$$x_i = \frac{a_i}{\sqrt{\sum a_j^2}} \quad \text{and} \quad y_i = \frac{b_i}{\sqrt{\sum b_j^2}}$$

to show that

$$\sum a_i b_i \leq \sqrt{\sum a_j^2} \sqrt{\sum b_j^2}$$

for any numbers $a_1, \ldots, a_n, b_1, \ldots, b_n$. This inequality is known as the Cauchy-Schwarz Inequality.

Rocket Science

Courtesy of Orbital Sciences Corporation

Many rockets, such as the *Pegasus XL* currently used to launch satellites and the *Saturn V* that first put men on the Moon, are designed to use three stages in their ascent into space. A large first stage initially propels the rocket until its fuel is consumed, at which point the stage is jettisoned to reduce the mass of the rocket. The smaller second and third stages function similarly in order to place the rocket's payload into orbit about the Earth. (With this design, at least two stages are required in order to reach the necessary velocities, and using three stages has proven to be a good compromise between cost and performance.) Our goal here is to determine the individual masses of the three stages to be designed in such a way as to minimize the total mass of the rocket while enabling it to reach a desired velocity.

For a single-stage rocket consuming fuel at a constant rate, the change in velocity resulting from the acceleration of the rocket vehicle has been modeled by

$$\Delta V = -c \ln\left(1 - \frac{(1-S)M_r}{P + M_r}\right)$$

where M_r is the mass of the rocket engine including initial fuel, P is the mass of the payload, S is a *structural factor* determined by the design of the rocket (specifically, it is the ratio of the mass of the rocket vehicle without fuel to the total mass of the rocket with payload), and c is the (constant) speed of exhaust relative to the rocket.

Now consider a rocket with three stages and a payload of mass A. Assume that outside forces are negligible and that c and S remain constant for each stage. If M_i is the mass of the ith stage, we can initially consider the rocket engine to have mass M_1 and its payload to have mass $M_2 + M_3 + A$; the second and third stages can be handled similarly.

1. Show that the velocity attained after all three stages have been jettisoned is given by

$$v_f = c\left[\ln\left(\frac{M_1 + M_2 + M_3 + A}{SM_1 + M_2 + M_3 + A}\right) + \ln\left(\frac{M_2 + M_3 + A}{SM_2 + M_3 + A}\right) + \ln\left(\frac{M_3 + A}{SM_3 + A}\right)\right]$$

2. We wish to minimize the total mass $M = M_1 + M_2 + M_3$ of the rocket engine subject to the constraint that the desired velocity v_f from Problem 1 is attained. The method of Lagrange multipliers is appropriate here, but difficult to implement using the current expressions. To simplify, we define variables N_i so that the constraint equation may be expressed as $v_f = c(\ln N_1 + \ln N_2 + \ln N_3)$. Since M is now difficult to express in terms of the N_i's, we wish to use a simpler function that will be minimized at the same place as M. Show that

$$\frac{M_1 + M_2 + M_3 + A}{M_2 + M_3 + A} = \frac{(1-S)N_1}{1 - SN_1}$$

$$\frac{M_2 + M_3 + A}{M_3 + A} = \frac{(1-S)N_2}{1 - SN_2}$$

$$\frac{M_3 + A}{A} = \frac{(1-S)N_3}{1 - SN_3}$$

and conclude that

$$\frac{M + A}{A} = \frac{(1-S)^3 N_1 N_2 N_3}{(1 - SN_1)(1 - SN_2)(1 - SN_3)}$$

3. Verify that $\ln((M + A)/A)$ is minimized at the same location as M; use Lagrange multipliers and the results of Problem 2 to find expressions for the values of N_i where the minimum occurs subject to the constraint $v_f = c(\ln N_1 + \ln N_2 + \ln N_3)$. [*Hint:* Use properties of logarithms to help simplify the expressions.]

4. Find an expression for the minimum value of M as a function of v_f.

5. If we want to put a three-stage rocket into orbit 100 miles above the Earth's surface, a final velocity of approximately 17,500 mi/h is required. Suppose that each stage is built with a structural factor $S = 0.2$ and an exhaust speed of $c = 6000$ mi/h.
  (a) Find the minimum total mass M of the rocket engines as a function of A.
  (b) Find the mass of each individual stage as a function of A. (They are not equally sized!)

6. The same rocket would require a final velocity of approximately 24,700 mi/h in order to escape Earth's gravity. Find the mass of each individual stage that would minimize the total mass of the rocket engines and allow the rocket to propel a 500-pound probe into deep space.

APPLIED PROJECT

Hydro-Turbine Optimization

The Great Northern Paper Company in Millinocket, Maine, operates a hydroelectric generating station on the Penobscot River. Water is piped from a dam to the power station. The rate at which the water flows through the pipe varies, depending on external conditions.

The power station has three different hydroelectric turbines, each with a known (and unique) power function that gives the amount of electric power generated as a function of the water flow arriving at the turbine. The incoming water can be apportioned in different volumes to each turbine, so the goal is to determine how to distribute water among the turbines to give the maximum total energy production for any rate of flow.

Using experimental evidence and *Bernoulli's equation*, the following quadratic models were determined for the power output of each turbine, along with the allowable flows of operation:

$$KW_1 = (-18.89 + 0.1277Q_1 - 4.08 \cdot 10^{-5}Q_1^2)(170 - 1.6 \cdot 10^{-6}Q_T^2)$$
$$KW_2 = (-24.51 + 0.1358Q_2 - 4.69 \cdot 10^{-5}Q_2^2)(170 - 1.6 \cdot 10^{-6}Q_T^2)$$
$$KW_3 = (-27.02 + 0.1380Q_3 - 3.84 \cdot 10^{-5}Q_3^2)(170 - 1.6 \cdot 10^{-6}Q_T^2)$$

$$250 \le Q_1 \le 1110, \quad 250 \le Q_2 \le 1110, \quad 250 \le Q_3 \le 1225$$

where

$$Q_i = \text{flow through turbine } i \text{ in cubic feet per second}$$
$$KW_i = \text{power generated by turbine } i \text{ in kilowatts}$$
$$Q_T = \text{total flow through the station in cubic feet per second}$$

1. If all three turbines are being used, we wish to determine the flow Q_i to each turbine that will give the maximum total energy production. Our limitations are that the flows must sum to the total incoming flow and the given domain restrictions must be observed. Consequently, use Lagrange multipliers to find the values for the individual flows (as functions of Q_T) that maximize the total energy production $KW_1 + KW_2 + KW_3$ subject to the constraints $Q_1 + Q_2 + Q_3 = Q_T$ and the domain restrictions on each Q_i.

2. For which values of Q_T is your result valid?

3. For an incoming flow of 2500 ft^3/s, determine the distribution to the turbines and verify (by trying some nearby distributions) that your result is indeed a maximum.

4. Until now we assumed that all three turbines are operating; is it possible in some situations that more power could be produced by using only one turbine? Make a graph of the three power functions and use it to help decide if an incoming flow of 1000 ft³/s should be distributed to all three turbines or routed to just one. (If you determine that only one turbine should be used, which one?) What if the flow is only 600 ft³/s?

5. Perhaps for some flow levels it would be advantageous to use two turbines. If the incoming flow is 1500 ft³/s, which two turbines would you recommend using? Use Lagrange multipliers to determine how the flow should be distributed between the two turbines to maximize the energy produced. For this flow, is using two turbines more efficient than using all three?

6. If the incoming flow is 3400 ft³/s, what would you recommend to the company?

11 | Review

1. (a) What is a function of two variables?
   (b) Describe two methods for visualizing a function of two variables. What is the connection between them?

2. What is a function of three variables? How can you visualize such a function?

3. What does
$$\lim_{(x, y) \to (a, b)} f(x, y) = L$$
mean? How can you show that such a limit does not exist?

4. (a) What does it mean to say that f is continuous at (a, b)?
   (b) If f is continuous on $\mathbb{R}^2$, what can you say about its graph?

5. (a) Write expressions for the partial derivatives $f_x(a, b)$ and $f_y(a, b)$ as limits.
   (b) How do you interpret $f_x(a, b)$ and $f_y(a, b)$ geometrically? How do you interpret them as rates of change?
   (c) If $f(x, y)$ is given by a formula, how do you calculate f_x and f_y?

6. What does Clairaut's Theorem say?

7. How do you find a tangent plane to each of the following types of surfaces?
   (a) A graph of a function of two variables, $z = f(x, y)$
   (b) A level surface of a function of three variables, $F(x, y, z) = k$
   (c) A parametric surface given by a vector function $\mathbf{r}(u, v)$

8. Define the linearization of f at (a, b). What is the corresponding linear approximation? What is the geometric interpretation of the linear approximation?

9. (a) What does it mean to say that f is differentiable at (a, b)?
   (b) How do you usually verify that f is differentiable?

10. If $z = f(x, y)$, what are the differentials dx, dy, and dz?

11. State the Chain Rule for the case where $z = f(x, y)$ and x and y are functions of one variable. What if x and y are functions of two variables?

12. If z is defined implicitly as a function of x and y by an equation of the form $F(x, y, z) = 0$, how do you find $\partial z/\partial x$ and $\partial z/\partial y$?

13. (a) Write an expression as a limit for the directional derivative of f at (x_0, y_0) in the direction of a unit vector $\mathbf{u} = \langle a, b \rangle$. How do you interpret it as a rate? How do you interpret it geometrically?
   (b) If f is differentiable, write an expression for $D_{\mathbf{u}} f(x_0, y_0)$ in terms of f_x and f_y.

14. (a) Define the gradient vector ∇f for a function f of two or three variables.
   (b) Express $D_{\mathbf{u}} f$ in terms of ∇f.
   (c) Explain the geometric significance of the gradient.

15. What do the following statements mean?
   (a) f has a local maximum at (a, b).
   (b) f has an absolute maximum at (a, b).
   (c) f has a local minimum at (a, b).
   (d) f has an absolute minimum at (a, b).
   (e) f has a saddle point at (a, b).

16. (a) If f has a local maximum at (a, b), what can you say about its partial derivatives at (a, b)?
   (b) What is a critical point of f?

17. State the Second Derivatives Test.

18. (a) What is a closed set in $\mathbb{R}^2$? What is a bounded set?
   (b) State the Extreme Value Theorem for functions of two variables.
   (c) How do you find the values that the Extreme Value Theorem guarantees?

19. Explain how the method of Lagrange multipliers works in finding the extreme values of $f(x, y, z)$ subject to the constraint $g(x, y, z) = k$. What if there is a second constraint $h(x, y, z) = c$?

TRUE-FALSE QUIZ

Determine whether the statement is true or false. If it is true, explain why. If it is false, explain why or give an example that disproves the statement.

1. $f_y(a, b) = \lim_{y \to b} \dfrac{f(a, y) - f(a, b)}{y - b}$

2. There exists a function f with continuous second-order partial derivatives such that $f_x(x, y) = x + y^2$ and $f_y(x, y) = x - y^2$.

3. $f_{xy} = \dfrac{\partial^2 f}{\partial x\, \partial y}$

4. $D_k f(x, y, z) = f_z(x, y, z)$

5. If $f(x, y) \to L$ as $(x, y) \to (a, b)$ along every straight line through (a, b), then $\lim_{(x, y) \to (a, b)} f(x, y) = L$.

6. If $f_x(a, b)$ and $f_y(a, b)$ both exist, then f is differentiable at (a, b).

7. If f has a local minimum at (a, b) and f is differentiable at (a, b), then $\nabla f(a, b) = \mathbf{0}$.

8. If f is a function, then $\lim_{(x, y) \to (2, 5)} f(x, y) = f(2, 5)$.

9. If $f(x, y) = \ln y$, then $\nabla f(x, y) = 1/y$.

10. If $(2, 1)$ is a critical point of f and
$$f_{xx}(2, 1) f_{yy}(2, 1) < [f_{xy}(2, 1)]^2$$
then f has a saddle point at $(2, 1)$.

11. If $f(x, y) = \sin x + \sin y$, then $-\sqrt{2} \leqslant D_\mathbf{u} f(x, y) \leqslant \sqrt{2}$.

12. If $f(x, y)$ has two local maxima, then f must have a local minimum.

EXERCISES

1–2 ■ Find and sketch the domain of the function.

1. $f(x, y) = \ln(x + y + 1)$

2. $f(x, y) = \sqrt{4 - x^2 - y^2} + \sqrt{1 - x^2}$

• • • • • • • • • • • • • •

3–4 ■ Sketch the graph of the function.

3. $f(x, y) = 1 - y^2$ **4.** $f(x, y) = x^2 + (y - 2)^2$

• • • • • • • • • • • • • •

5–6 ■ Sketch several level curves of the function.

5. $f(x, y) = \sqrt{4x^2 + y^2}$ **6.** $f(x, y) = e^x + y$

• • • • • • • • • • • • • •

7. Make a rough sketch of a contour map for the function whose graph is shown.

8. A contour map of a function f is shown. Use it to make a rough sketch of the graph of f.

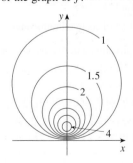

9–10 ■ Evaluate the limit or show that it does not exist.

9. $\displaystyle\lim_{(x, y) \to (1, 1)} \frac{2xy}{x^2 + 2y^2}$ **10.** $\displaystyle\lim_{(x, y) \to (0, 0)} \frac{2xy}{x^2 + 2y^2}$

• • • • • • • • • • • • • •

11. A metal plate is situated in the xy-plane and occupies the rectangle $0 \leqslant x \leqslant 10$, $0 \leqslant y \leqslant 8$, where x and y are measured in meters. The temperature at the point (x, y) in the plate is $T(x, y)$, where T is measured in degrees Celsius. Temperatures at equally spaced points were measured and recorded in the table.
(a) Estimate the values of the partial derivatives $T_x(6, 4)$ and $T_y(6, 4)$. What are the units?
(b) Estimate the value of $D_\mathbf{u} T(6, 4)$, where $\mathbf{u} = (\mathbf{i} + \mathbf{j})/\sqrt{2}$. Interpret your result.
(c) Estimate the value of $T_{xy}(6, 4)$.

x \ y	0	2	4	6	8
0	30	38	45	51	55
2	52	56	60	62	61
4	78	74	72	68	66
6	98	87	80	75	71
8	96	90	86	80	75
10	92	92	91	87	78

12. Find a linear approximation to the temperature function $T(x, y)$ in Exercise 11 near the point $(6, 4)$. Then use it to estimate the temperature at the point $(5, 3.8)$.

13–17 ■ Find the first partial derivatives.

13. $f(x, y) = \sqrt{2x + y^2}$ **14.** $u = e^{-r} \sin 2\theta$

15. $g(u, v) = u \tan^{-1} v$ **16.** $w = \dfrac{x}{y - z}$

17. $T(p, q, r) = p \ln(q + e^r)$

18. The speed of sound traveling through ocean water is a function of temperature, salinity, and pressure. It has been modeled by the function

$$C = 1449.2 + 4.6T - 0.055T^2 + 0.00029T^3$$
$$+ (1.34 - 0.01T)(S - 35) + 0.016D$$

where C is the speed of sound (in meters per second), T is the temperature (in degrees Celsius), S is the salinity (the concentration of salts in parts per thousand, which means the number of grams of dissolved solids per 1000 g of water), and D is the depth below the ocean surface (in meters). Compute $\partial C / \partial T$, $\partial C / \partial S$, and $\partial C / \partial D$ when $T = 10°C$, $S = 35$ parts per thousand, and $D = 100$ m. Explain the physical significance of these partial derivatives.

19–22 ■ Find all second partial derivatives of f.

19. $f(x, y) = 4x^3 - xy^2$ **20.** $z = xe^{-2y}$

21. $f(x, y, z) = x^k y^l z^m$ **22.** $v = r \cos(s + 2t)$

23. If $z = xy + xe^{y/x}$, show that $x\dfrac{\partial z}{\partial x} + y\dfrac{\partial z}{\partial y} = xy + z$.

24. If $z = \sin(x + \sin t)$, show that

$$\frac{\partial z}{\partial x}\frac{\partial^2 z}{\partial x \partial t} = \frac{\partial z}{\partial t}\frac{\partial^2 z}{\partial x^2}$$

25–29 ■ Find equations of (a) the tangent plane and (b) the normal line to the given surface at the specified point.

25. $z = 3x^2 - y^2 + 2x$, $(1, -2, 1)$

26. $z = e^x \cos y$, $(0, 0, 1)$

27. $x^2 + 2y^2 - 3z^2 = 3$, $(2, -1, 1)$

28. $xy + yz + zx = 3$, $(1, 1, 1)$

29. $\mathbf{r}(u, v) = (u + v)\mathbf{i} + u^2\mathbf{j} + v^2\mathbf{k}$, $(3, 4, 1)$

30. Use a computer to graph the surface $z = x^2 + y^4$ and its tangent plane and normal line at $(1, 1, 2)$ on the same screen. Choose the domain and viewpoint so that you get a good view of all three objects.

31. Find the points on the hyperboloid $x^2 + 4y^2 - z^2 = 4$ where the tangent plane is parallel to the plane $2x + 2y + z = 5$.

32. Find du if $u = \ln(1 + se^{2t})$.

33. Find the linear approximation of the function $f(x, y, z) = x^3\sqrt{y^2 + z^2}$ at the point $(2, 3, 4)$ and use it to estimate the number $(1.98)^3\sqrt{(3.01)^2 + (3.97)^2}$.

34. The two legs of a right triangle are measured as 5 m and 12 m with a possible error in measurement of at most 0.2 cm in each. Use differentials to estimate the maximum error in the calculated value of (a) the area of the triangle and (b) the length of the hypotenuse.

35. If $u = x^2 y^3 + z^4$, where $x = p + 3p^2$, $y = pe^p$, and $z = p \sin p$, use the Chain Rule to find du/dp.

36. If $v = x^2 \sin y + ye^{xy}$, where $x = s + 2t$ and $y = st$, use the Chain Rule to find $\partial v / \partial s$ and $\partial v / \partial t$ when $s = 0$ and $t = 1$.

37. Suppose $z = f(x, y)$, where $x = g(s, t)$, $y = h(s, t)$, $g(1, 2) = 3$, $g_s(1, 2) = -1$, $g_t(1, 2) = 4$, $h(1, 2) = 6$, $h_s(1, 2) = -5$, $h_t(1, 2) = 10$, $f_x(3, 6) = 7$, and $f_y(3, 6) = 8$. Find $\partial z / \partial s$ and $\partial z / \partial t$ when $s = 1$ and $t = 2$.

38. Use a tree diagram to write out the Chain Rule for the case where $w = f(t, u, v)$, $t = t(p, q, r, s)$, $u = u(p, q, r, s)$, and $v = v(p, q, r, s)$ are all differentiable functions.

39. If $z = y + f(x^2 - y^2)$, where f is differentiable, show that

$$y\frac{\partial z}{\partial x} + x\frac{\partial z}{\partial y} = x$$

40. The length x of a side of a triangle is increasing at a rate of 3 in/s, the length y of another side is decreasing at a rate of 2 in/s, and the contained angle θ is increasing at a rate of 0.05 radian/s. How fast is the area of the triangle changing when $x = 40$ in, $y = 50$ in, and $\theta = \pi/6$?

41. If $z = f(u, v)$, where $u = xy$, $v = y/x$, and f has continuous second partial derivatives, show that

$$x^2\frac{\partial^2 z}{\partial x^2} - y^2\frac{\partial^2 z}{\partial y^2} = -4uv\frac{\partial^2 z}{\partial u \partial v} + 2v\frac{\partial z}{\partial v}$$

42. If $yz^4 + x^2z^3 = e^{xyz}$, find $\dfrac{\partial z}{\partial x}$ and $\dfrac{\partial z}{\partial y}$.

43. Find the gradient of the function $f(x, y, z) = z^2 e^{x\sqrt{y}}$.

44. (a) When is the directional derivative of f a maximum?
(b) When is it a minimum?
(c) When is it 0?
(d) When is it half of its maximum value?

45–46 ■ Find the directional derivative of f at the given point in the indicated direction.

45. $f(x, y) = 2\sqrt{x} - y^2$, $(1, 5)$,
in the direction toward the point $(4, 1)$

46. $f(x, y, z) = x^2 y + x\sqrt{1 + z}$, $(1, 2, 3)$,
in the direction of $\mathbf{v} = 2\mathbf{i} + \mathbf{j} - 2\mathbf{k}$

47. Find the maximum rate of change of $f(x, y) = x^2 y + \sqrt{y}$ at the point $(2, 1)$. In which direction does it occur?

48. Find the direction in which $f(x, y, z) = ze^{xy}$ increases most rapidly at the point $(0, 1, 2)$. What is the maximum rate of increase?

49. The contour map shows wind speed in knots during Hurricane Andrew on August 24, 1992. Use it to estimate the value of the directional derivative of the wind speed at Homestead, Florida, in the direction of the eye of the hurricane.

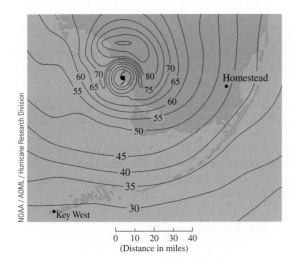

50. Find parametric equations of the tangent line at the point $(-2, 2, 4)$ to the curve of intersection of the surface $z = 2x^2 - y^2$ and the plane $z = 4$.

51–54 ■ Find the local maximum and minimum values and saddle points of the function. If you have three-dimensional graphing software, graph the function with a domain and viewpoint that reveal all the important aspects of the function.

51. $f(x, y) = x^2 - xy + y^2 + 9x - 6y + 10$

52. $f(x, y) = x^3 - 6xy + 8y^3$

53. $f(x, y) = 3xy - x^2 y - xy^2$

54. $f(x, y) = (x^2 + y)e^{y/2}$

. . . .

55–56 ■ Find the absolute maximum and minimum values of f on the set D.

55. $f(x, y) = 4xy^2 - x^2 y^2 - xy^3$; D is the closed triangular region in the xy-plane with vertices $(0, 0)$, $(0, 6)$, and $(6, 0)$

56. $f(x, y) = e^{-x^2-y^2}(x^2 + 2y^2)$; D is the disk $x^2 + y^2 \le 4$

. . . .

57. Use a graph and/or level curves to estimate the local maximum and minimum values and saddle points of $f(x, y) = x^3 - 3x + y^4 - 2y^2$. Then use calculus to find these values precisely.

58. Use a graphing calculator or computer (or Newton's method or a computer algebra system) to find the critical points of $f(x, y) = 12 + 10y - 2x^2 - 8xy - y^4$ correct to three decimal places. Then classify the critical points and find the highest point on the graph.

59–62 ■ Use Lagrange multipliers to find the maximum and minimum values of f subject to the given constraint(s).

59. $f(x, y) = x^2 y$; $\quad x^2 + y^2 = 1$

60. $f(x, y) = \dfrac{1}{x} + \dfrac{1}{y}$; $\quad \dfrac{1}{x^2} + \dfrac{1}{y^2} = 1$

61. $f(x, y, z) = xyz$; $\quad x^2 + y^2 + z^2 = 3$

62. $f(x, y, z) = x^2 + 2y^2 + 3z^2$; $\quad x + y + z = 1, \quad x - y + 2z = 2$

. . . .

63. Find the points on the surface $xy^2 z^3 = 2$ that are closest to the origin.

64. A package in the shape of a rectangular box can be mailed by the U.S. Postal Service if the sum of its length and girth (the perimeter of a cross-section perpendicular to the length) is at most 108 in. Find the dimensions of the package with largest volume that can be mailed.

65. A pentagon is formed by placing an isosceles triangle on a rectangle, as shown in the figure. If the pentagon has fixed perimeter P, find the lengths of the sides of the pentagon that maximize the area of the pentagon.

66. A particle of mass m moves on the surface $z = f(x, y)$. Let $x = x(t)$, $y = y(t)$ be the x- and y-coordinates of the particle at time t.
(a) Find the velocity vector $\mathbf{v}$ and the kinetic energy $K = \frac{1}{2}m|\mathbf{v}|^2$ of the particle.
(b) Determine the acceleration vector $\mathbf{a}$.
(c) Let $z = x^2 + y^2$ and $x(t) = t \cos t$, $y(t) = t \sin t$. Find the velocity vector, the kinetic energy, and the acceleration vector.

1. A rectangle with length L and width W is cut into four smaller rectangles by two lines parallel to the sides. Find the maximum and minimum values of the sum of the squares of the areas of the smaller rectangles.

2. Marine biologists have determined that when a shark detects the presence of blood in the water, it will swim in the direction in which the concentration of the blood increases most rapidly. Based on certain tests, the concentration of blood (in parts per million) at a point $P(x, y)$ on the surface of seawater is approximated by

$$C(x, y) = e^{-(x^2+2y^2)/10^4}$$

where x and y are measured in meters in a rectangular coordinate system with the blood source at the origin.

(a) Identify the level curves of the concentration function and sketch several members of this family together with a path that a shark will follow to the source.

(b) Suppose a shark is at the point (x_0, y_0) when it first detects the presence of blood in the water. Find an equation of the shark's path by setting up and solving a differential equation.

3. A long piece of galvanized sheet metal with width w is to be bent into a symmetric form with three straight sides to make a rain gutter. A cross-section is shown in the figure.

(a) Determine the dimensions that allow the maximum possible flow; that is, find the dimensions that give the maximum possible cross-sectional area.

(b) Would it be better to bend the metal into a gutter with a semicircular cross-section than a three-sided cross-section?

$$w - 2x$$

4. For what values of the number r is the function

$$f(x, y, z) = \begin{cases} \dfrac{(x + y + z)^r}{x^2 + y^2 + z^2} & \text{if } (x, y, z) \neq 0 \\ 0 & \text{if } (x, y, z) = 0 \end{cases}$$

continuous on $\mathbb{R}^3$?

5. Suppose f is a differentiable function of one variable. Show that all tangent planes to the surface $z = xf(y/x)$ intersect in a common point.

6. (a) Newton's method for approximating a root of an equation $f(x) = 0$ (see Section 4.8) can be adapted to approximating a solution of a system of equations $f(x, y) = 0$ and $g(x, y) = 0$. The surfaces $z = f(x, y)$ and $z = g(x, y)$ intersect in a curve that intersects the xy-plane at the point (r, s), which is the solution of the system. If an initial approximation (x_1, y_1) is close to this point, then the tangent planes to the surfaces at (x_1, y_1) intersect in a straight line that intersects the xy-plane in a point (x_2, y_2), which should be closer to (r, s). (Compare with Figure 2 in Section 4.8.) Show that

$$x_2 = x_1 - \frac{fg_y - f_y g}{f_x g_y - f_y g_x} \quad \text{and} \quad y_2 = y_1 - \frac{f_x g - f g_x}{f_x g_y - f_y g_x}$$

where f, g, and their partial derivatives are evaluated at (x_1, y_1). If we continue this procedure, we obtain successive approximations (x_n, y_n).

(b) It was Thomas Simpson (1710–1761) who formulated Newton's method as we know it today and who extended it to functions of two variables as in part (a). (See the biography of Simpson on page 418.) The example that he gave to illustrate the method was to solve the system of equations

$$x^x + y^y = 1000 \qquad x^y + y^x = 100$$

In other words, he found the points of intersection of the curves in the figure. Use the method of part (a) to find the coordinates of the points of intersection correct to six decimal places.

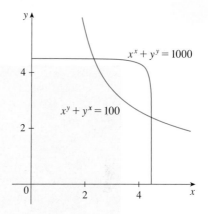

7. (a) Show that when Laplace's equation

$$\frac{\partial^2 u}{\partial x^2} + \frac{\partial^2 u}{\partial y^2} + \frac{\partial^2 u}{\partial z^2} = 0$$

is written in cylindrical coordinates, it becomes

$$\frac{\partial^2 u}{\partial r^2} + \frac{1}{r}\frac{\partial u}{\partial r} + \frac{1}{r^2}\frac{\partial^2 u}{\partial \theta^2} + \frac{\partial^2 u}{\partial z^2} = 0$$

(b) Show that when Laplace's equation is written in spherical coordinates, it becomes

$$\frac{\partial^2 u}{\partial \rho^2} + \frac{2}{\rho}\frac{\partial u}{\partial \rho} + \frac{\cot\phi}{\rho^2}\frac{\partial u}{\partial \phi} + \frac{1}{\rho^2}\frac{\partial^2 u}{\partial \phi^2} + \frac{1}{\rho^2 \sin^2\phi}\frac{\partial^2 u}{\partial \theta^2} = 0$$

8. Among all planes that are tangent to the surface $xy^2z^2 = 1$, find the ones that are farthest from the origin.

9. If the ellipse $x^2/a^2 + y^2/b^2 = 1$ is to enclose the circle $x^2 + y^2 = 2y$, what values of a and b minimize the area of the ellipse?

12

Multiple Integrals

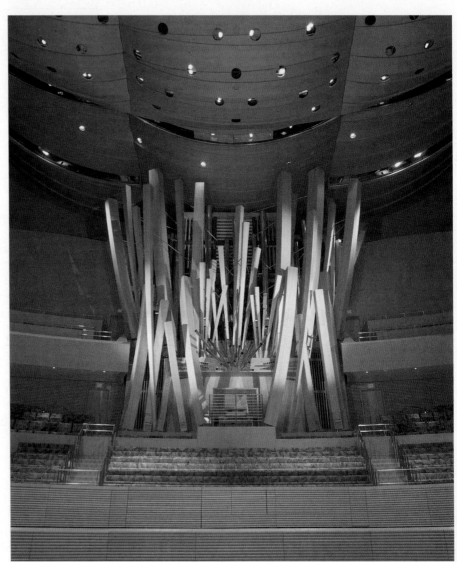

In this chapter we extend the idea of a definite integral to double and triple integrals of functions of two or three variables. These ideas are then used to compute volumes, surface areas, masses, and centroids of more general regions than we were able to consider in Chapter 6. We also use double integrals to calculate probabilities when two random variables are involved.

Double Integrals over Rectangles

In much the same way that our attempt to solve the area problem led to the definition of a definite integral, we now seek to find the volume of a solid and in the process we arrive at the definition of a double integral.

▪▪ Review of the Definite Integral

First let's recall the basic facts concerning definite integrals of functions of a single variable. If $f(x)$ is defined for $a \leq x \leq b$, we start by dividing the interval $[a, b]$ into n subintervals $[x_{i-1}, x_i]$ of equal width $\Delta x = (b - a)/n$ and we choose sample points x_i^* in these subintervals. Then we form the Riemann sum

$$\boxed{1} \qquad \sum_{i=1}^{n} f(x_i^*) \, \Delta x$$

and take the limit of such sums as $n \rightarrow \infty$ to obtain the definite integral of f from a to b:

$$\boxed{2} \qquad \int_a^b f(x) \, dx = \lim_{n \to \infty} \sum_{i=1}^{n} f(x_i^*) \, \Delta x$$

In the special case where $f(x) \geq 0$, the Riemann sum can be interpreted as the sum of the areas of the approximating rectangles in Figure 1, and $\int_a^b f(x) \, dx$ represents the area under the curve $y = f(x)$ from a to b.

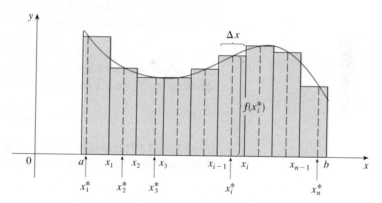

FIGURE 1

▪▪ Volumes and Double Integrals

In a similar manner we consider a function f of two variables defined on a closed rectangle

$$R = [a, b] \times [c, d] = \{(x, y) \in \mathbb{R}^2 \mid a \leq x \leq b, \ c \leq y \leq d\}$$

and we first suppose that $f(x, y) \geq 0$. The graph of f is a surface with equation $z = f(x, y)$. Let S be the solid that lies above R and under the graph of f, that is,

$$S = \{(x, y, z) \in \mathbb{R}^3 \mid 0 \leq z \leq f(x, y), \ (x, y) \in R\}$$

FIGURE 2

(See Figure 2.) Our goal is to find the volume of S.

The first step is to divide the rectangle R into subrectangles. We do this by dividing the interval $[a, b]$ into m subintervals $[x_{i-1}, x_i]$ of equal width $\Delta x = (b - a)/m$ and dividing $[c, d]$ into n subintervals $[y_{j-1}, y_j]$ of equal width $\Delta y = (d - c)/n$. By drawing lines parallel to the coordinate axes through the endpoints of these subintervals as in Figure 3, we form the subrectangles

$$R_{ij} = [x_{i-1}, x_i] \times [y_{j-1}, y_j] = \{(x, y) \mid x_{i-1} \leqslant x \leqslant x_i, \ y_{j-1} \leqslant y \leqslant y_j\}$$

each with area $\Delta A = \Delta x \, \Delta y$.

FIGURE 3
Dividing R into subrectangles

If we choose a **sample point** (x_{ij}^*, y_{ij}^*) in each R_{ij}, then we can approximate the part of S that lies above each R_{ij} by a thin rectangular box (or "column") with base R_{ij} and height $f(x_{ij}^*, y_{ij}^*)$ as shown in Figure 4. (Compare with Figure 1.) The volume of this box is the height of the box times the area of the base rectangle:

$$f(x_{ij}^*, y_{ij}^*) \, \Delta A$$

If we follow this procedure for all the rectangles and add the volumes of the corresponding boxes, we get an approximation to the total volume of S:

$$\boxed{3} \qquad\qquad V \approx \sum_{i=1}^{m} \sum_{j=1}^{n} f(x_{ij}^*, y_{ij}^*) \, \Delta A$$

(See Figure 5.) This double sum means that for each subrectangle we evaluate f at the chosen point and multiply by the area of the subrectangle, and then we add the results.

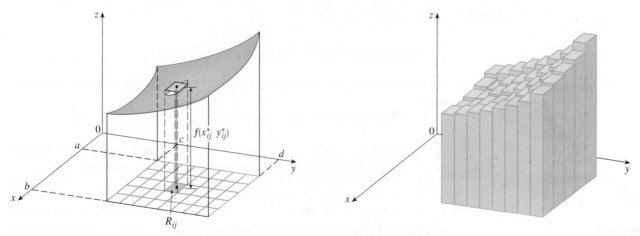

FIGURE 4

FIGURE 5

Our intuition tells us that the approximation given in (3) becomes better as m and n become larger and so we would expect that

$$\boxed{4} \qquad V = \lim_{m, n \to \infty} \sum_{i=1}^{m} \sum_{j=1}^{n} f(x_{ij}^{*}, y_{ij}^{*})\, \Delta A$$

■ ■ The meaning of the double limit in Equation 4 is that we can make the double sum as close as we like to the number V [for any choice of (x_{ij}^{*}, y_{ij}^{*}) in R_{ij}] by taking m and n sufficiently large.

We use the expression in Equation 4 to define the **volume** of the solid S that lies under the graph of f and above the rectangle R. (It can be shown that this definition is consistent with our formula for volume in Section 6.2.)

Limits of the type that appear in Equation 4 occur frequently, not just in finding volumes but in a variety of other situations as well—as we will see in Section 12.5— even when f is not a positive function. So we make the following definition.

■ ■ Notice the similarity between Definition 5 and the definition of a single integral in Equation 2.

> **5 Definition** The **double integral** of f over the rectangle R is
>
> $$\iint_{R} f(x, y)\, dA = \lim_{m, n \to \infty} \sum_{i=1}^{m} \sum_{j=1}^{n} f(x_{ij}^{*}, y_{ij}^{*})\, \Delta A$$
>
> if this limit exists.

It can be proved that the limit in Definition 5 exists if f is a continuous function. (It also exists for some discontinuous functions as long as they are reasonably "well behaved.")

The sample point (x_{ij}^{*}, y_{ij}^{*}) can be chosen to be any point in the subrectangle R_{ij}, but if we choose it to be the upper right-hand corner of R_{ij} [namely (x_i, y_j), see Figure 3], then the expression for the double integral looks simpler:

$$\boxed{6} \qquad \iint_{R} f(x, y)\, dA = \lim_{m, n \to \infty} \sum_{i=1}^{m} \sum_{j=1}^{n} f(x_i, y_j)\, \Delta A$$

By comparing Definitions 4 and 5, we see that a volume can be written as a double integral:

> If $f(x, y) \geqslant 0$, then the volume V of the solid that lies above the rectangle R and below the surface $z = f(x, y)$ is
>
> $$V = \iint_{R} f(x, y)\, dA$$

The sum in Definition 5,

$$\sum_{i=1}^{m} \sum_{j=1}^{n} f(x_{ij}^{*}, y_{ij}^{*})\, \Delta A$$

is called a **double Riemann sum** and is used as an approximation to the value of the double integral. [Notice how similar it is to the Riemann sum in (1) for a function of a single variable.] If f happens to be a *positive* function, then the double Riemann sum represents the sum of volumes of columns, as in Figure 5, and is an approximation to the volume under the graph of f.

FIGURE 6

FIGURE 7

Ⓥ **EXAMPLE 1** Estimate the volume of the solid that lies above the square $R = [0, 2] \times [0, 2]$ and below the elliptic paraboloid $z = 16 - x^2 - 2y^2$. Divide R into four equal squares and choose the sample point to be the upper right corner of each square R_{ij}. Sketch the solid and the approximating rectangular boxes.

SOLUTION The squares are shown in Figure 6. The paraboloid is the graph of $f(x, y) = 16 - x^2 - 2y^2$ and the area of each square is 1. Approximating the volume by the Riemann sum with $m = n = 2$, we have

$$V \approx \sum_{i=1}^{2} \sum_{j=1}^{2} f(x_i, y_j) \, \Delta A$$

$$= f(1, 1) \, \Delta A + f(1, 2) \, \Delta A + f(2, 1) \, \Delta A + f(2, 2) \, \Delta A$$

$$= 13(1) + 7(1) + 10(1) + 4(1) = 34$$

This is the volume of the approximating rectangular boxes shown in Figure 7.   ■ ■

We get better approximations to the volume in Example 1 if we increase the number of squares. Figure 8 shows how the columns start to look more like the actual solid and the corresponding approximations become more accurate when we use 16, 64, and 256 squares. In the next section we will be able to show that the exact volume is 48.

FIGURE 8

The Riemann sum approximations to the volume under $z = 16 - x^2 - 2y^2$ become more accurate as m and n increase.

(a) $m = n = 4$, $V \approx 41.5$ (b) $m = n = 8$, $V \approx 44.875$ (c) $m = n = 16$, $V \approx 46.46875$

Ⓥ **EXAMPLE 2** If $R = \{(x, y) \mid -1 \leq x \leq 1, -2 \leq y \leq 2\}$, evaluate the integral

$$\iint_R \sqrt{1 - x^2} \, dA$$

SOLUTION It would be very difficult to evaluate this integral directly from Definition 5 but, because $\sqrt{1 - x^2} \geq 0$, we can compute the integral by interpreting it as a volume. If $z = \sqrt{1 - x^2}$, then $x^2 + z^2 = 1$ and $z \geq 0$, so the given double integral represents the volume of the solid S that lies below the circular cylinder $x^2 + z^2 = 1$ and above the rectangle R. (See Figure 9.) The volume of S is the area of a semicircle with radius 1 times the length of the cylinder. Thus

$$\iint_R \sqrt{1 - x^2} \, dA = \tfrac{1}{2}\pi(1)^2 \times 4 = 2\pi$$

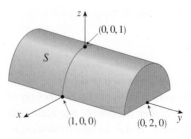

FIGURE 9

■ ■

The Midpoint Rule

The methods that we used for approximating single integrals (the Midpoint Rule, the Trapezoidal Rule, Simpson's Rule) all have counterparts for double integrals. Here we consider only the Midpoint Rule for double integrals. This means that we use a double Riemann sum to approximate the double integral, where the sample point (x_{ij}^*, y_{ij}^*) in R_{ij} is chosen to be the center $(\bar{x}_i, \bar{y}_j)$ of R_{ij}. In other words, $\bar{x}_i$ is the midpoint of $[x_{i-1}, x_i]$ and $\bar{y}_j$ is the midpoint of $[y_{j-1}, y_j]$.

Midpoint Rule for Double Integrals

$$\iint\limits_R f(x, y)\, dA \approx \sum_{i=1}^{m} \sum_{j=1}^{n} f(\bar{x}_i, \bar{y}_j)\, \Delta A$$

where $\bar{x}_i$ is the midpoint of $[x_{i-1}, x_i]$ and $\bar{y}_j$ is the midpoint of $[y_{j-1}, y_j]$.

$\boxed{V}$ **EXAMPLE 3** Use the Midpoint Rule with $m = n = 2$ to estimate the value of the integral $\iint_R (x - 3y^2)\, dA$, where $R = \{(x, y) \mid 0 \le x \le 2,\ 1 \le y \le 2\}$.

SOLUTION In using the Midpoint Rule with $m = n = 2$, we evaluate $f(x, y) = x - 3y^2$ at the centers of the four subrectangles shown in Figure 10. So $\bar{x}_1 = \frac{1}{2}$, $\bar{x}_2 = \frac{3}{2}$, $\bar{y}_1 = \frac{5}{4}$, and $\bar{y}_2 = \frac{7}{4}$. The area of each subrectangle is $\Delta A = \frac{1}{2}$. Thus

$$
\begin{aligned}
\iint\limits_R (x - 3y^2)\, dA &\approx \sum_{i=1}^{2} \sum_{j=1}^{2} f(\bar{x}_i, \bar{y}_j)\, \Delta A \\
&= f(\bar{x}_1, \bar{y}_1)\, \Delta A + f(\bar{x}_1, \bar{y}_2)\, \Delta A + f(\bar{x}_2, \bar{y}_1)\, \Delta A + f(\bar{x}_2, \bar{y}_2)\, \Delta A \\
&= f\!\left(\tfrac{1}{2}, \tfrac{5}{4}\right) \Delta A + f\!\left(\tfrac{1}{2}, \tfrac{7}{4}\right) \Delta A + f\!\left(\tfrac{3}{2}, \tfrac{5}{4}\right) \Delta A + f\!\left(\tfrac{3}{2}, \tfrac{7}{4}\right) \Delta A \\
&= \left(-\tfrac{67}{16}\right)\tfrac{1}{2} + \left(-\tfrac{139}{16}\right)\tfrac{1}{2} + \left(-\tfrac{51}{16}\right)\tfrac{1}{2} + \left(-\tfrac{123}{16}\right)\tfrac{1}{2} \\
&= -\tfrac{95}{8} = -11.875
\end{aligned}
$$

Thus, we have

$$\iint\limits_R (x - 3y^2)\, dA \approx -11.875$$

■ ■

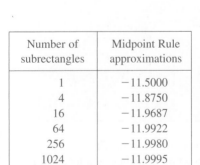

FIGURE 10

NOTE □ In the next section we will develop an efficient method for computing double integrals and then we will see that the exact value of the double integral in Example 3 is -12. (Remember that the interpretation of a double integral as a volume is valid only when the integrand f is a *positive* function. The integrand in Example 3 is not a positive function, so its integral is not a volume. In Examples 2 and 3 in Section 12.2 we will discuss how to interpret integrals of functions that are not always positive in terms of volumes.) If we keep dividing each subrectangle in Figure 10 into four smaller ones with similar shape, we get the Midpoint Rule approximations displayed in the chart in the margin. Notice how these approximations approach the exact value of the double integral, -12.

Number of subrectangles	Midpoint Rule approximations
1	−11.5000
4	−11.8750
16	−11.9687
64	−11.9922
256	−11.9980
1024	−11.9995

Average Value

Recall from Section 6.4 that the average value of a function f of one variable defined on an interval $[a, b]$ is

$$f_{\text{ave}} = \frac{1}{b - a} \int_a^b f(x)\, dx$$

In a similar fashion we define the **average value** of a function f of two variables defined on a rectangle R to be

$$f_{ave} = \frac{1}{A(R)} \iint\limits_R f(x, y)\, dA$$

where $A(R)$ is the area of R.

If $f(x, y) \geqslant 0$, the equation

$$A(R) \times f_{ave} = \iint\limits_R f(x, y)\, dA$$

says that the box with base R and height f_{ave} has the same volume as the solid that lies under the graph of f. [If $z = f(x, y)$ describes a mountainous region and you chop off the tops of the mountains at height f_{ave}, then you can use them to fill in the valleys so that the region becomes completely flat. See Figure 11.]

FIGURE 11

EXAMPLE 4 The contour map in Figure 12 shows the snowfall, in inches, that fell on the state of Colorado on December 24, 1982. (The state is in the shape of a rectangle that measures 388 mi west to east and 276 mi south to north.) Use the contour map to estimate the average snowfall for Colorado as a whole on December 24.

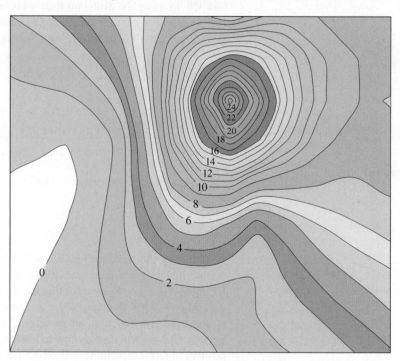

FIGURE 12

SOLUTION Let's place the origin at the southwest corner of the state. Then $0 \leqslant x \leqslant 388$, $0 \leqslant y \leqslant 276$, and $f(x, y)$ is the snowfall, in inches, at a location x miles to the east and y miles to the north of the origin. If R is the rectangle that represents Colorado, then the average snowfall for the state on December 24 was

$$f_{ave} = \frac{1}{A(R)} \iint\limits_R f(x, y)\, dA$$

where $A(R) = 388 \cdot 276$. To estimate the value of this double integral let's use the Midpoint Rule with $m = n = 4$. In other words, we divide R into 16 subrectangles of equal size, as in Figure 13. The area of each subrectangle is

$$\Delta A = \tfrac{1}{16}(388)(276) = 6693 \text{ mi}^2$$

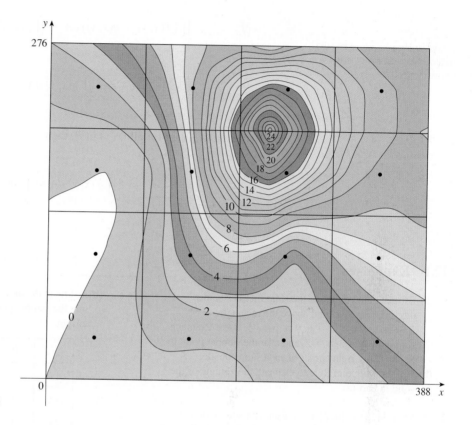

FIGURE 13

Using the contour map to estimate the value of f at the center of each sub-rectangle, we get

$$\iint\limits_R f(x, y)\, dA \approx \sum_{i=1}^{4} \sum_{j=1}^{4} f(\bar{x}_i, \bar{y}_j)\, \Delta A$$

$$\approx \Delta A[0.4 + 1.2 + 1.8 + 3.9 + 0 + 3.9 + 4.0 + 6.5$$
$$+ 0.1 + 6.1 + 16.5 + 8.8 + 1.8 + 8.0 + 16.2 + 9.4]$$
$$= (6693)(88.6)$$

Therefore
$$f_{\text{ave}} \approx \frac{(6693)(88.6)}{(388)(276)} \approx 5.5$$

On December 24, 1982, Colorado received an average of approximately $5\frac{1}{2}$ inches of snow.

■■ Properties of Double Integrals

We list here three properties of double integrals that can be proved in the same manner as in Section 5.2. We assume that all of the integrals exist. Properties 7 and 8 are referred to as the *linearity* of the integral.

■ ■ Double integrals behave this way because the double sums that define them behave this way.

$$\boxed{7} \qquad \iint_R [f(x, y) + g(x, y)] \, dA = \iint_R f(x, y) \, dA + \iint_R g(x, y) \, dA$$

$$\boxed{8} \qquad \iint_R cf(x, y) \, dA = c \iint_R f(x, y) \, dA \qquad \text{where } c \text{ is a constant}$$

If $f(x, y) \geq g(x, y)$ for all (x, y) in R, then

$$\boxed{9} \qquad \iint_R f(x, y) \, dA \geq \iint_R g(x, y) \, dA$$

12.1 Exercises

1. (a) Estimate the volume of the solid that lies below the surface $z = xy$ and above the rectangle $R = \{(x, y) \mid 0 \leq x \leq 6, 0 \leq y \leq 4\}$. Use a Riemann sum with $m = 3$, $n = 2$, and take the sample point to be the upper right corner of each subrectangle.
(b) Use the Midpoint Rule to estimate the volume of the solid in part (a).

2. If $R = [-1, 3] \times [0, 2]$, use a Riemann sum with $m = 4$, $n = 2$ to estimate the value of $\iint_R (y^2 - 2x^2) \, dA$. Take the sample points to be the upper left corners of the subrectangles.

3. (a) Use a Riemann sum with $m = n = 2$ to estimate the value of $\iint_R \sin(x + y) \, dA$, where $R = [0, \pi] \times [0, \pi]$. Take the sample points to be lower left corners.
(b) Use the Midpoint Rule to estimate the integral in part (a).

4. (a) Estimate the volume of the solid that lies below the surface $z = x + 2y^2$ and above the rectangle $R = [0, 2] \times [0, 4]$. Use a Riemann sum with $m = n = 2$ and choose the sample points to be lower right corners.
(b) Use the Midpoint Rule to estimate the volume in part (a).

5. A table of values is given for a function $f(x, y)$ defined on $R = [1, 3] \times [0, 4]$.
(a) Estimate $\iint_R f(x, y) \, dA$ using the Midpoint Rule with $m = n = 2$.

(b) Estimate the double integral with $m = n = 4$ by choosing the sample points to be the points farthest from the origin.

x \ y	0	1	2	3	4
1.0	2	0	−3	−6	−5
1.5	3	1	−4	−8	−6
2.0	4	3	0	−5	−8
2.5	5	5	3	−1	−4
3.0	7	8	6	3	0

6. A 20-ft-by-30-ft swimming pool is filled with water. The depth is measured at 5-ft intervals, starting at one corner of the pool, and the values are recorded in the table. Estimate the volume of water in the pool.

	0	5	10	15	20	25	30
0	2	3	4	6	7	8	8
5	2	3	4	7	8	10	8
10	2	4	6	8	10	12	10
15	2	3	4	5	6	8	7
20	2	2	2	2	2	3	4

7. Let V be the volume of the solid that lies under the graph of $f(x, y) = \sqrt{52 - x^2 - y^2}$ and above the rectangle given by $2 \leq x \leq 4, 2 \leq y \leq 6$. We use the lines $x = 3$ and $y = 4$

to divide R into subrectangles. Let L and U be the Riemann sums computed using lower left corners and upper right corners, respectively. Without calculating the numbers V, L, and U, arrange them in increasing order and explain your reasoning.

8. The figure shows level curves of a function f in the square $R = [0, 2] \times [0, 2]$. Use the Midpoint rule with $m = n = 2$ to estimate $\iint_R f(x, y) \, dA$. How could you improve your estimate?

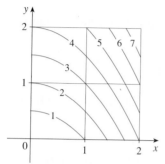

9. A contour map is shown for a function f on the square $R = [0, 4] \times [0, 4]$.
(a) Use the Midpoint Rule with $m = n = 2$ to estimate the value of $\iint_R f(x, y) \, dA$.
(b) Estimate the average value of f.

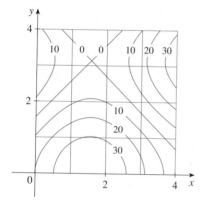

10. The contour map shows the temperature, in degrees Fahrenheit, at 3:00 P.M. on May 1, 1996, in Colorado. (The state measures 388 mi east to west and 276 mi north to south.) Use the Midpoint Rule with $m = n = 4$ to estimate the average temperature in Colorado at that time.

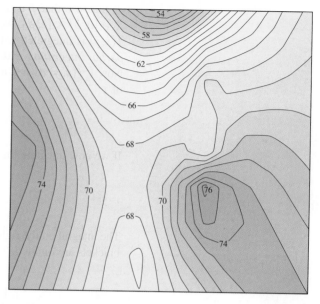

11–13 ■ Evaluate the double integral by first identifying it as the volume of a solid.

11. $\iint_R 3 \, dA, \quad R = \{(x, y) \mid -2 \leqslant x \leqslant 2, 1 \leqslant y \leqslant 6\}$

12. $\iint_R (5 - x) \, dA, \quad R = \{(x, y) \mid 0 \leqslant x \leqslant 5, 0 \leqslant y \leqslant 3\}$

13. $\iint_R (4 - 2y) \, dA, \quad R = [0, 1] \times [0, 1]$

■ ■ ■ ■ ■ ■ ■ ■ ■ ■ ■ ■

14. The integral $\iint_R \sqrt{9 - y^2} \, dA$, where $R = [0, 4] \times [0, 2]$, represents the volume of a solid. Sketch the solid.

15. Use a programmable calculator or computer (or the sum command on a CAS) to estimate

$$\iint_R \sqrt{1 + xe^{-y}} \, dA$$

where $R = [0, 1] \times [0, 1]$. Use the Midpoint Rule with the following numbers of squares of equal size: 1, 4, 16, 64, 256, and 1024.

16. Repeat Exercise 15 for the integral $\iint_R \sin(x + \sqrt{y}) \, dA$.

17. If f is a constant function, $f(x, y) = k$, and $R = [a, b] \times [c, d]$, show that $\iint_R k \, dA = k(b - a)(d - c)$.

18. If $R = [0, 1] \times [0, 1]$, show that $0 \leqslant \iint_R \sin(x + y) \, dA \leqslant 1$.

12.2 Iterated Integrals

Recall that it is usually difficult to evaluate single integrals directly from the definition of an integral, but the Evaluation Theorem (Part 2 of the Fundamental Theorem of Calculus) provides a much easier method. The evaluation of double integrals from first principles is even more difficult, but in this section we see how to express a double integral as an iterated integral, which can then be evaluated by calculating two single integrals.

Suppose that f is a function of two variables that is continuous on the rectangle $R = [a, b] \times [c, d]$. We use the notation $\int_c^d f(x, y)\, dy$ to mean that x is held fixed and $f(x, y)$ is integrated with respect to y from $y = c$ to $y = d$. This procedure is called *partial integration with respect to y*. (Notice its similarity to partial differentiation.) Now $\int_c^d f(x, y)\, dy$ is a number that depends on the value of x, so it defines a function of x:

$$A(x) = \int_c^d f(x, y)\, dy$$

If we now integrate the function A with respect to x from $x = a$ to $x = b$, we get

$$\boxed{1} \qquad \int_a^b A(x)\, dx = \int_a^b \left[\int_c^d f(x, y)\, dy \right] dx$$

The integral on the right side of Equation 1 is called an **iterated integral**. Usually the brackets are omitted. Thus

$$\boxed{2} \qquad \int_a^b \int_c^d f(x, y)\, dy\, dx = \int_a^b \left[\int_c^d f(x, y)\, dy \right] dx$$

means that we first integrate with respect to y from c to d and then with respect to x from a to b.

Similarly, the iterated integral

$$\boxed{3} \qquad \int_c^d \int_a^b f(x, y)\, dx\, dy = \int_c^d \left[\int_a^b f(x, y)\, dx \right] dy$$

means that we first integrate with respect to x (holding y fixed) from $x = a$ to $x = b$ and then we integrate the resulting function of y with respect to y from $y = c$ to $y = d$. Notice that in both Equations 2 and 3 we work *from the inside out*.

EXAMPLE 1 Evaluate the iterated integrals.

(a) $\displaystyle\int_0^3 \int_1^2 x^2 y\, dy\, dx$
(b) $\displaystyle\int_1^2 \int_0^3 x^2 y\, dx\, dy$

SOLUTION

(a) Regarding x as a constant, we obtain

$$\int_1^2 x^2 y\, dy = \left[x^2 \frac{y^2}{2} \right]_{y=1}^{y=2}$$

$$= x^2 \left(\frac{2^2}{2} \right) - x^2 \left(\frac{1^2}{2} \right) = \tfrac{3}{2} x^2$$

Thus, the function A in the preceding discussion is given by $A(x) = \tfrac{3}{2} x^2$ in this example. We now integrate this function of x from 0 to 3:

$$\int_0^3 \int_1^2 x^2 y\, dy\, dx = \int_0^3 \left[\int_1^2 x^2 y\, dy \right] dx$$

$$= \int_0^3 \tfrac{3}{2} x^2\, dx = \frac{x^3}{2} \Bigg]_0^3 = \frac{27}{2} .$$

(b) Here we first integrate with respect to x:

$$\int_1^2 \int_0^3 x^2 y \, dx \, dy = \int_1^2 \left[\int_0^3 x^2 y \, dx \right] dy = \int_1^2 \left[\frac{x^3}{3} y \right]_{x=0}^{x=3} dy$$

$$= \int_1^2 9y \, dy = 9 \frac{y^2}{2} \Big]_1^2 = \frac{27}{2}$$

■ ■

Notice that in Example 1 we obtained the same answer whether we integrated with respect to y or x first. In general, it turns out (see Theorem 4) that the two iterated integrals in Equations 2 and 3 are always equal; that is, the order of integration does not matter. (This is similar to Clairaut's Theorem on the equality of the mixed partial derivatives.)

The following theorem gives a practical method for evaluating a double integral by expressing it as an iterated integral (in either order).

■ ■ Theorem 4 is named after the Italian mathematician Guido Fubini (1879–1943), who proved a very general version of this theorem in 1907. But the version for continuous functions was known to the French mathematician Augustin-Louis Cauchy almost a century earlier.

4 Fubini's Theorem If f is continuous on the rectangle $R = \{(x, y) \mid a \le x \le b, c \le y \le d\}$, then

$$\iint_R f(x, y) \, dA = \int_a^b \int_c^d f(x, y) \, dy \, dx = \int_c^d \int_a^b f(x, y) \, dx \, dy$$

More generally, this is true if we assume that f is bounded on R, f is discontinuous only on a finite number of smooth curves, and the iterated integrals exist.

The proof of Fubini's Theorem is too difficult to include in this book, but we can at least give an intuitive indication of why it is true for the case where $f(x, y) \ge 0$. Recall that if f is positive, then we can interpret the double integral $\iint_R f(x, y) \, dA$ as the volume V of the solid S that lies above R and under the surface $z = f(x, y)$. But we have another formula that we used for volume in Chapter 6, namely,

$$V = \int_a^b A(x) \, dx$$

FIGURE 1

 Visual 12.2 illustrates Fubini's Theorem by showing an animation of Figures 1 and 2.

where $A(x)$ is the area of a cross-section of S in the plane through x perpendicular to the x-axis. From Figure 1 you can see that $A(x)$ is the area under the curve C whose equation is $z = f(x, y)$, where x is held constant and $c \le y \le d$. Therefore

$$A(x) = \int_c^d f(x, y) \, dy$$

and we have

$$\iint_R f(x, y) \, dA = V = \int_a^b A(x) \, dx = \int_a^b \int_c^d f(x, y) \, dy \, dx$$

A similar argument, using cross-sections perpendicular to the y-axis as in Figure 2, shows that

$$\iint_R f(x, y) \, dA = \int_c^d \int_a^b f(x, y) \, dx \, dy$$

FIGURE 2

EXAMPLE 2 Evaluate the double integral $\iint_R (x - 3y^2)\, dA$, where $R = \{(x, y) \mid 0 \leqslant x \leqslant 2,\ 1 \leqslant y \leqslant 2\}$. (Compare with Example 3 in Section 12.1.)

SOLUTION 1 Fubini's Theorem gives

$$\iint_R (x - 3y^2)\, dA = \int_0^2 \int_1^2 (x - 3y^2)\, dy\, dx = \int_0^2 \left[xy - y^3 \right]_{y=1}^{y=2} dx$$

$$= \int_0^2 (x - 7)\, dx = \frac{x^2}{2} - 7x \bigg]_0^2 = -12$$

■ ■ Notice the negative answer in Example 2; nothing is wrong with that. The function f in that example is not a positive function, so its integral doesn't represent a volume. From Figure 3 we see that f is always negative on R, so the value of the integral is the *negative* of the volume that lies *above* the graph of f and *below* R.

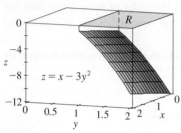

FIGURE 3

SOLUTION 2 Again applying Fubini's Theorem, but this time integrating with respect to x first, we have

$$\iint_R (x - 3y^2)\, dA = \int_1^2 \int_0^2 (x - 3y^2)\, dx\, dy$$

$$= \int_1^2 \left[\frac{x^2}{2} - 3xy^2 \right]_{x=0}^{x=2} dy$$

$$= \int_1^2 (2 - 6y^2)\, dy = 2y - 2y^3 \big]_1^2 = -12$$

■ ■

EXAMPLE 3 Evaluate $\iint_R y \sin(xy)\, dA$, where $R = [1, 2] \times [0, \pi]$.

SOLUTION 1 If we first integrate with respect to x, we get

$$\iint_R y \sin(xy)\, dA = \int_0^\pi \int_1^2 y \sin(xy)\, dx\, dy = \int_0^\pi \left[-\cos(xy) \right]_{x=1}^{x=2} dy$$

$$= \int_0^\pi (-\cos 2y + \cos y)\, dy$$

$$= -\tfrac{1}{2}\sin 2y + \sin y \big]_0^\pi = 0$$

SOLUTION 2 If we reverse the order of integration, we get

$$\iint_R y \sin(xy)\, dA = \int_1^2 \int_0^\pi y \sin(xy)\, dy\, dx$$

To evaluate the inner integral we use integration by parts with

$$u = y \qquad\qquad dv = \sin(xy)\, dy$$

$$du = dy \qquad\qquad v = -\frac{\cos(xy)}{x}$$

■ ■ For a function f that takes on both positive and negative values, $\iint_R f(x, y)\, dA$ is a difference of volumes: $V_1 - V_2$, where V_1 is the volume above R and below the graph of f and V_2 is the volume below R and above the graph. The fact that the integral in Example 3 is 0 means that these two volumes V_1 and V_2 are equal. (See Figure 4.)

and so

$$\int_0^\pi y \sin(xy)\, dy = -\frac{y\cos(xy)}{x} \bigg]_{y=0}^{y=\pi} + \frac{1}{x}\int_0^\pi \cos(xy)\, dy$$

$$= -\frac{\pi\cos \pi x}{x} + \frac{1}{x^2}\left[\sin(xy) \right]_{y=0}^{y=\pi}$$

$$= -\frac{\pi\cos \pi x}{x} + \frac{\sin \pi x}{x^2}$$

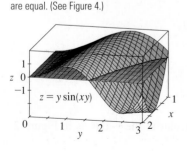

FIGURE 4

If we now integrate the first term by parts with $u = -1/x$ and $dv = \pi \cos \pi x \, dx$, we get $du = dx/x^2$, $v = \sin \pi x$, and

$$\int \left(-\frac{\pi \cos \pi x}{x} \right) dx = -\frac{\sin \pi x}{x} - \int \frac{\sin \pi x}{x^2} \, dx$$

Therefore

$$\int \left(-\frac{\pi \cos \pi x}{x} + \frac{\sin \pi x}{x^2} \right) dx = -\frac{\sin \pi x}{x}$$

and so

$$\int_1^2 \int_0^\pi y \sin(xy) \, dy \, dx = \left[-\frac{\sin \pi x}{x} \right]_1^2$$

$$= -\frac{\sin 2\pi}{2} + \sin \pi = 0$$

■ ■

■ ■ In Example 2, Solutions 1 and 2 are equally straightforward, but in Example 3 the first solution is much easier than the second one. Therefore, when we evaluate double integrals it is wise to choose the order of integration that gives simpler integrals.

$\boxed{\text{V}}$ **EXAMPLE 4** Find the volume of the solid S that is bounded by the elliptic paraboloid $x^2 + 2y^2 + z = 16$, the planes $x = 2$ and $y = 2$, and the three coordinate planes.

SOLUTION We first observe that S is the solid that lies under the surface $z = 16 - x^2 - 2y^2$ and above the square $R = [0, 2] \times [0, 2]$. (See Figure 5.) This solid was considered in Example 1 in Section 12.1, but we are now in a position to evaluate the double integral using Fubini's Theorem. Therefore

$$V = \iint_R (16 - x^2 - 2y^2) \, dA = \int_0^2 \int_0^2 (16 - x^2 - 2y^2) \, dx \, dy$$

$$= \int_0^2 \left[16x - \tfrac{1}{3}x^3 - 2y^2 x \right]_{x=0}^{x=2} dy$$

$$= \int_0^2 \left(\tfrac{88}{3} - 4y^2 \right) dy = \left[\tfrac{88}{3}y - \tfrac{4}{3}y^3 \right]_0^2 = 48$$

■ ■

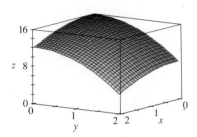

FIGURE 5

In the special case where $f(x, y)$ can be factored as the product of a function of x only and a function of y only, the double integral of f can be written in a particularly simple form. To be specific, suppose that $f(x, y) = g(x)h(y)$ and $R = [a, b] \times [c, d]$. Then Fubini's Theorem gives

$$\iint_R f(x, y) \, dA = \int_c^d \int_a^b g(x)h(y) \, dx \, dy = \int_c^d \left[\int_a^b g(x)h(y) \, dx \right] dy$$

In the inner integral y is a constant, so $h(y)$ is a constant and we can write

$$\int_c^d \left[\int_a^b g(x)h(y) \, dx \right] dy = \int_c^d \left[h(y) \left(\int_a^b g(x) \, dx \right) \right] dy$$

$$= \int_a^b g(x) \, dx \int_c^d h(y) \, dy$$

since $\int_a^b g(x) \, dx$ is a constant. Therefore, in this case, the double integral of f can be written as the product of two single integrals:

$$\boxed{5} \qquad \iint_R g(x)h(y) \, dA = \int_a^b g(x) \, dx \int_c^d h(y) \, dy \qquad \text{where } R = [a, b] \times [c, d]$$

EXAMPLE 5 If $R = [0, \pi/2] \times [0, \pi/2]$, then, by Equation 5,

$$\iint\limits_R \sin x \cos y \, dA = \int_0^{\pi/2} \sin x \, dx \int_0^{\pi/2} \cos y \, dy$$

$$= \big[-\cos x\big]_0^{\pi/2} \big[\sin y\big]_0^{\pi/2} = 1 \cdot 1 = 1 \qquad ■ ■$$

■ ■ The function $f(x, y) = \sin x \cos y$ in Example 5 is positive on R, so the integral represents the volume of the solid that lies above R and below the graph of f shown in Figure 6.

FIGURE 6

12.2 Exercises

1–2 ■ Find $\int_0^3 f(x, y) \, dx$ and $\int_0^4 f(x, y) \, dy$.

1. $f(x, y) = 2x + 3x^2 y$

2. $f(x, y) = \dfrac{y}{x + 2}$

3–12 ■ Calculate the iterated integral.

3. $\int_1^3 \int_0^1 (1 + 4xy) \, dx \, dy$

4. $\int_2^4 \int_{-1}^1 (x^2 + y^2) \, dy \, dx$

5. $\int_0^2 \int_0^{\pi/2} x \sin y \, dy \, dx$

6. $\int_1^4 \int_0^2 (x + \sqrt{y}) \, dx \, dy$

7. $\int_0^2 \int_0^1 (2x + y)^8 \, dx \, dy$

8. $\int_0^1 \int_1^2 \dfrac{xe^x}{y} \, dy \, dx$

9. $\int_1^4 \int_1^2 \left(\dfrac{x}{y} + \dfrac{y}{x}\right) dy \, dx$

10. $\int_1^2 \int_0^1 (x + y)^{-2} \, dx \, dy$

11. $\int_0^{\ln 2} \int_0^{\ln 5} e^{2x-y} \, dx \, dy$

12. $\int_0^1 \int_0^1 xy\sqrt{x^2 + y^2} \, dy \, dx$

13–18 ■ Calculate the double integral.

13. $\iint\limits_R \dfrac{xy^2}{x^2 + 1} \, dA, \quad R = \{(x, y) \mid 0 \leqslant x \leqslant 1, \ -3 \leqslant y \leqslant 3\}$

14. $\iint\limits_R \cos(x + 2y) \, dA,$

$\qquad R = \{(x, y) \mid 0 \leqslant x \leqslant \pi, \ 0 \leqslant y \leqslant \pi/2\}$

15. $\iint\limits_R x \sin(x + y) \, dA, \quad R = [0, \pi/6] \times [0, \pi/3]$

16. $\iint\limits_R \dfrac{1 + x^2}{1 + y^2} \, dA, \quad R = \{(x, y) \mid 0 \leqslant x \leqslant 1, \ 0 \leqslant y \leqslant 1\}$

17. $\iint\limits_R xye^{x^2 y} \, dA, \quad R = [0, 1] \times [0, 2]$

18. $\iint\limits_R \dfrac{x}{1 + xy} \, dA, \quad R = [0, 1] \times [0, 1]$

19–20 ■ Sketch the solid whose volume is given by the iterated integral.

19. $\int_0^1 \int_0^1 (4 - x - 2y) \, dx \, dy$

20. $\int_0^1 \int_0^1 (2 - x^2 - y^2) \, dy \, dx$

21. Find the volume of the solid that lies under the plane $3x + 2y + z = 12$ and above the rectangle $R = \{(x, y) \mid 0 \leqslant x \leqslant 1, \ -2 \leqslant y \leqslant 3\}$.

22. Find the volume of the solid that lies under the hyperbolic paraboloid $z = 4 + x^2 - y^2$ and above the square $R = [-1, 1] \times [0, 2]$.

23. Find the volume of the solid lying under the elliptic paraboloid $x^2/4 + y^2/9 + z = 1$ and above the rectangle $R = [-1, 1] \times [-2, 2]$.

24. Find the volume of the solid enclosed by the surface $z = 1 + e^x \sin y$ and the planes $x = \pm 1$, $y = 0$, $y = \pi$, and $z = 0$.

25. Find the volume of the solid bounded by the surface $z = x\sqrt{x^2 + y}$ and the planes $x = 0$, $x = 1$, $y = 0$, $y = 1$, and $z = 0$.

26. Find the volume of the solid bounded by the elliptic paraboloid $z = 1 + (x - 1)^2 + 4y^2$, the planes $x = 3$ and $y = 2$, and the coordinate planes.

27. Find the volume of the solid in the first octant bounded by the cylinder $z = 9 - y^2$ and the plane $x = 2$.

28. (a) Find the volume of the solid bounded by the surface $z = 6 - xy$ and the planes $x = 2$, $x = -2$, $y = 0$, $y = 3$, and $z = 0$.
(b) Use a computer to draw the solid.

CAS **29.** Use a computer algebra system to find the exact value of the integral $\iint_R x^5 y^3 e^{xy}\, dA$, where $R = [0, 1] \times [0, 1]$. Then use the CAS to draw the solid whose volume is given by the integral.

CAS **30.** Graph the solid that lies between the surfaces $z = e^{-x^2} \cos(x^2 + y^2)$ and $z = 2 - x^2 - y^2$ for $|x| \leq 1$, $|y| \leq 1$. Use a computer algebra system to approximate the volume of this solid correct to four decimal places.

31–32 ■ Find the average value of f over the given rectangle.

31. $f(x, y) = x^2 y$,
R has vertices $(-1, 0)$, $(-1, 5)$, $(1, 5)$, $(1, 0)$

32. $f(x, y) = e^y \sqrt{x + e^y}$, $R = [0, 4] \times [0, 1]$

CAS **33.** Use your CAS to compute the iterated integrals

$$\int_0^1 \int_0^1 \frac{x - y}{(x + y)^3}\, dy\, dx \quad \text{and} \quad \int_0^1 \int_0^1 \frac{x - y}{(x + y)^3}\, dx\, dy$$

Do the answers contradict Fubini's Theorem? Explain what is happening.

34. (a) In what way are the theorems of Fubini and Clairaut similar?
(b) If $f(x, y)$ is continuous on $[a, b] \times [c, d]$ and

$$g(x, y) = \int_a^x \int_c^y f(s, t)\, dt\, ds$$

for $a < x < b$, $c < y < d$, show that $g_{xy} = g_{yx} = f(x, y)$.

12.3 Double Integrals over General Regions

For single integrals, the region over which we integrate is always an interval. But for double integrals, we want to be able to integrate a function f not just over rectangles but also over regions D of more general shape, such as the one illustrated in Figure 1. We suppose that D is a bounded region, which means that D can be enclosed in a rectangular region R as in Figure 2. Then we define a new function F with domain R by

$$\boxed{1} \qquad F(x, y) = \begin{cases} f(x, y) & \text{if } (x, y) \text{ is in } D \\ 0 & \text{if } (x, y) \text{ is in } R \text{ but not in } D \end{cases}$$

FIGURE 1

FIGURE 2

FIGURE 3

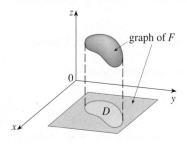

FIGURE 4

If the double integral of F exists over R, then we define the **double integral of f over D** by

$$\boxed{2} \qquad \iint_D f(x, y)\, dA = \iint_R F(x, y)\, dA \qquad \text{where } F \text{ is given by Equation 1}$$

Definition 2 makes sense because R is a rectangle and so $\iint_R F(x, y)\, dA$ has been previously defined in Section 12.1. The procedure that we have used is reasonable because the values of $F(x, y)$ are 0 when (x, y) lies outside D and so they contribute nothing to the integral. This means that it doesn't matter what rectangle R we use as long as it contains D.

In the case where $f(x, y) \geqslant 0$ we can still interpret $\iint_D f(x, y)\, dA$ as the volume of the solid that lies above D and under the surface $z = f(x, y)$ (the graph of f). You can see that this is reasonable by comparing the graphs of f and F in Figures 3 and 4 and remembering that $\iint_R F(x, y)\, dA$ is the volume under the graph of F.

Figure 4 also shows that F is likely to have discontinuities at the boundary points of D. Nonetheless, if f is continuous on D and the boundary curve of D is "well behaved" (in a sense outside the scope of this book), then it can be shown that $\iint_R F(x, y)\, dA$ exists and therefore $\iint_D f(x, y)\, dA$ exists. In particular, this is the case for the following types of regions.

A plane region D is said to be of **type I** if it lies between the graphs of two continuous functions of x, that is,

$$D = \{(x, y) \mid a \leqslant x \leqslant b, \ g_1(x) \leqslant y \leqslant g_2(x)\}$$

where g_1 and g_2 are continuous on $[a, b]$. Some examples of type I regions are shown in Figure 5.

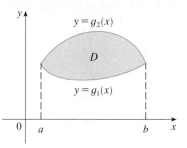

FIGURE 5 Some type I regions

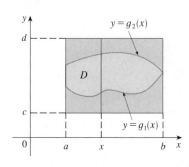

FIGURE 6

In order to evaluate $\iint_D f(x, y)\, dA$ when D is a region of type I, we choose a rectangle $R = [a, b] \times [c, d]$ that contains D, as in Figure 6, and we let F be the function given by Equation 1; that is, F agrees with f on D and F is 0 outside D. Then, by Fubini's Theorem,

$$\iint_D f(x, y)\, dA = \iint_R F(x, y)\, dA = \int_a^b \int_c^d F(x, y)\, dy\, dx$$

Observe that $F(x, y) = 0$ if $y < g_1(x)$ or $y > g_2(x)$ because (x, y) then lies outside D. Therefore

$$\int_c^d F(x, y)\, dy = \int_{g_1(x)}^{g_2(x)} F(x, y)\, dy = \int_{g_1(x)}^{g_2(x)} f(x, y)\, dy$$

because $F(x, y) = f(x, y)$ when $g_1(x) \leq y \leq g_2(x)$. Thus, we have the following formula that enables us to evaluate the double integral as an iterated integral.

3 If f is continuous on a type I region D such that

$$D = \{(x, y) \mid a \leq x \leq b, \ g_1(x) \leq y \leq g_2(x)\}$$

then

$$\iint\limits_D f(x, y) \, dA = \int_a^b \int_{g_1(x)}^{g_2(x)} f(x, y) \, dy \, dx$$

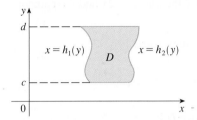

The integral on the right side of (3) is an iterated integral that is similar to the ones we considered in the preceding section, except that in the inner integral we regard x as being constant not only in $f(x, y)$ but also in the limits of integration, $g_1(x)$ and $g_2(x)$.

We also consider plane regions of **type II**, which can be expressed as

4
$$D = \{(x, y) \mid c \leq y \leq d, \ h_1(y) \leq x \leq h_2(y)\}$$

where h_1 and h_2 are continuous. Two such regions are illustrated in Figure 7.

Using the same methods that were used in establishing (3), we can show that

5
$$\iint\limits_D f(x, y) \, dA = \int_c^d \int_{h_1(y)}^{h_2(y)} f(x, y) \, dx \, dy$$

where D is a type II region given by Equation 4.

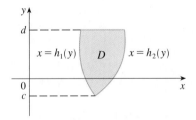

FIGURE 7

Some type II regions

$\boxed{V}$ **EXAMPLE 1** Evaluate $\iint_D (x + 2y) \, dA$, where D is the region bounded by the parabolas $y = 2x^2$ and $y = 1 + x^2$.

SOLUTION The parabolas intersect when $2x^2 = 1 + x^2$, that is, $x^2 = 1$, so $x = \pm 1$. We note that the region D, sketched in Figure 8, is a type I region but not a type II region and we can write

$$D = \{(x, y) \mid -1 \leq x \leq 1, \ 2x^2 \leq y \leq 1 + x^2\}$$

Since the lower boundary is $y = 2x^2$ and the upper boundary is $y = 1 + x^2$, Equation 3 gives

$$\iint\limits_D (x + 2y) \, dA = \int_{-1}^1 \int_{2x^2}^{1+x^2} (x + 2y) \, dy \, dx$$

$$= \int_{-1}^1 \left[xy + y^2 \right]_{y=2x^2}^{y=1+x^2} dx$$

$$= \int_{-1}^1 [x(1 + x^2) + (1 + x^2)^2 - x(2x^2) - (2x^2)^2] \, dx$$

$$= \int_{-1}^1 (-3x^4 - x^3 + 2x^2 + x + 1) \, dx$$

$$= -3\frac{x^5}{5} - \frac{x^4}{4} + 2\frac{x^3}{3} + \frac{x^2}{2} + x \Bigg]_{-1}^1 = \frac{32}{15}$$

FIGURE 8

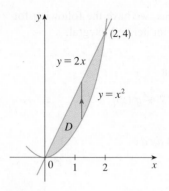

FIGURE 9

D as a type I region

FIGURE 10

D as a type II region

■ ■ Figure 11 shows the solid whose volume is calculated in Example 2. It lies above the *xy*-plane, below the paraboloid $z = x^2 + y^2$, and between the plane $y = 2x$ and the parabolic cylinder $y = x^2$.

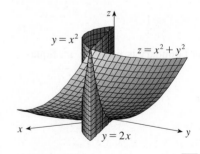

FIGURE 11

NOTE ▫ When we set up a double integral as in Example 1, it is essential to draw a diagram. Often it is helpful to draw a vertical arrow as in Figure 8. Then the limits of integration for the *inner* integral can be read from the diagram as follows: The arrow starts at the lower boundary $y = g_1(x)$, which gives the lower limit in the integral, and the arrow ends at the upper boundary $y = g_2(x)$, which gives the upper limit of integration. For a type II region the arrow is drawn horizontally from the left boundary to the right boundary.

EXAMPLE 2 Find the volume of the solid that lies under the paraboloid $z = x^2 + y^2$ and above the region *D* in the *xy*-plane bounded by the line $y = 2x$ and the parabola $y = x^2$.

SOLUTION 1 From Figure 9 we see that *D* is a type I region and

$$D = \left\{(x, y) \mid 0 \le x \le 2, \ x^2 \le y \le 2x\right\}$$

Therefore, the volume under $z = x^2 + y^2$ and above *D* is

$$V = \iint_D (x^2 + y^2)\, dA = \int_0^2 \int_{x^2}^{2x} (x^2 + y^2)\, dy\, dx$$

$$= \int_0^2 \left[x^2 y + \frac{y^3}{3}\right]_{y=x^2}^{y=2x} dx = \int_0^2 \left[x^2(2x) + \frac{(2x)^3}{3} - x^2 x^2 - \frac{(x^2)^3}{3}\right] dx$$

$$= \int_0^2 \left(-\frac{x^6}{3} - x^4 + \frac{14x^3}{3}\right) dx = -\frac{x^7}{21} - \frac{x^5}{5} + \frac{7x^4}{6}\Big]_0^2 = \frac{216}{35}$$

SOLUTION 2 From Figure 10 we see that *D* can also be written as a type II region:

$$D = \left\{(x, y) \mid 0 \le y \le 4, \ \tfrac{1}{2}y \le x \le \sqrt{y}\right\}$$

Therefore, another expression for *V* is

$$V = \iint_D (x^2 + y^2)\, dA = \int_0^4 \int_{\frac{1}{2}y}^{\sqrt{y}} (x^2 + y^2)\, dx\, dy$$

$$= \int_0^4 \left[\frac{x^3}{3} + y^2 x\right]_{x=\frac{1}{2}y}^{x=\sqrt{y}} dy = \int_0^4 \left(\frac{y^{3/2}}{3} + y^{5/2} - \frac{y^3}{24} - \frac{y^3}{2}\right) dy$$

$$= \tfrac{2}{15}y^{5/2} + \tfrac{2}{7}y^{7/2} - \tfrac{13}{96}y^4\Big]_0^4 = \tfrac{216}{35}$$ ■ ■

$\boxed{\text{V}}$ **EXAMPLE 3** Evaluate $\iint_D xy\, dA$, where *D* is the region bounded by the line $y = x - 1$ and the parabola $y^2 = 2x + 6$.

SOLUTION The region *D* is shown in Figure 12. Again *D* is both type I and type II, but the description of *D* as a type I region is more complicated because the lower boundary consists of two parts. Therefore, we prefer to express *D* as a type II region:

$$D = \left\{(x, y) \mid -2 \le y \le 4, \ \tfrac{1}{2}y^2 - 3 \le x \le y + 1\right\}$$

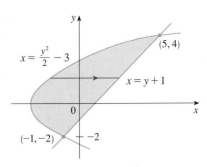

FIGURE 12   (a) D as a type I region   (b) D as a type II region

Then (5) gives

$$\iint_D xy \, dA = \int_{-2}^{4} \int_{\frac{1}{2}y^2-3}^{y+1} xy \, dx \, dy = \int_{-2}^{4} \left[\frac{x^2}{2} y \right]_{x=\frac{1}{2}y^2-3}^{x=y+1} dy$$

$$= \frac{1}{2} \int_{-2}^{4} y \left[(y+1)^2 - \left(\frac{1}{2}y^2 - 3 \right)^2 \right] dy$$

$$= \frac{1}{2} \int_{-2}^{4} \left(-\frac{y^5}{4} + 4y^3 + 2y^2 - 8y \right) dy$$

$$= \frac{1}{2} \left[-\frac{y^6}{24} + y^4 + 2\frac{y^3}{3} - 4y^2 \right]_{-2}^{4} = 36$$

If we had expressed D as a type I region using Figure 12(a), then we would have obtained

$$\iint_D xy \, dA = \int_{-3}^{-1} \int_{-\sqrt{2x+6}}^{\sqrt{2x+6}} xy \, dy \, dx + \int_{-1}^{5} \int_{x-1}^{\sqrt{2x+6}} xy \, dy \, dx$$

but this would have involved more work than the other method.   ■ ■

EXAMPLE 4 Find the volume of the tetrahedron bounded by the planes $x + 2y + z = 2$, $x = 2y$, $x = 0$, and $z = 0$.

SOLUTION In a question such as this, it's wise to draw two diagrams: one of the three-dimensional solid and another of the plane region D over which it lies. Figure 13 shows the tetrahedron T bounded by the coordinate planes $x = 0$, $z = 0$, the vertical plane $x = 2y$, and the plane $x + 2y + z = 2$. Since the plane $x + 2y + z = 2$ intersects the xy-plane (whose equation is $z = 0$) in the line $x + 2y = 2$, we see that T lies above the triangular region D in the xy-plane bounded by the lines $x = 2y$, $x + 2y = 2$, and $x = 0$. (See Figure 14.)

The plane $x + 2y + z = 2$ can be written as $z = 2 - x - 2y$, so the required volume lies under the graph of the function $z = 2 - x - 2y$ and above

$$D = \left\{ (x, y) \mid 0 \leq x \leq 1, \ x/2 \leq y \leq 1 - x/2 \right\}$$

FIGURE 13

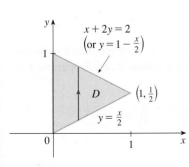

FIGURE 14

Therefore

$$V = \iint\limits_{D} (2 - x - 2y)\, dA = \int_0^1 \int_{x/2}^{1-x/2} (2 - x - 2y)\, dy\, dx$$

$$= \int_0^1 \left[2y - xy - y^2\right]_{y=x/2}^{y=1-x/2} dx$$

$$= \int_0^1 \left[2 - x - x\left(1 - \frac{x}{2}\right) - \left(1 - \frac{x}{2}\right)^2 - x + \frac{x^2}{2} + \frac{x^2}{4}\right] dx$$

$$= \int_0^1 (x^2 - 2x + 1)\, dx = \frac{x^3}{3} - x^2 + x \Big]_0^1 = \frac{1}{3}$$

    ■ ■

$\boxed{V}$ **EXAMPLE 5**   Evaluate the iterated integral $\int_0^1 \int_x^1 \sin(y^2)\, dy\, dx$.

SOLUTION If we try to evaluate the integral as it stands, we are faced with the task of first evaluating $\int \sin(y^2)\, dy$. But it's impossible to do so in finite terms since $\int \sin(y^2)\, dy$ is not an elementary function. (See the end of Section 5.8.) So we must change the order of integration. This is accomplished by first expressing the given iterated integral as a double integral. Using (3) backward, we have

$$\int_0^1 \int_x^1 \sin(y^2)\, dy\, dx = \iint\limits_{D} \sin(y^2)\, dA$$

where

$$D = \left\{(x, y) \mid 0 \leqslant x \leqslant 1,\ x \leqslant y \leqslant 1\right\}$$

We sketch this region D in Figure 15. Then from Figure 16 we see that an alternative description of D is

$$D = \left\{(x, y) \mid 0 \leqslant y \leqslant 1,\ 0 \leqslant x \leqslant y\right\}$$

This enables us to use (5) to express the double integral as an iterated integral in the reverse order:

$$\int_0^1 \int_x^1 \sin(y^2)\, dy\, dx = \iint\limits_{D} \sin(y^2)\, dA$$

$$= \int_0^1 \int_0^y \sin(y^2)\, dx\, dy = \int_0^1 \left[x \sin(y^2)\right]_{x=0}^{x=y} dy$$

$$= \int_0^1 y \sin(y^2)\, dy = -\tfrac{1}{2} \cos(y^2)\big]_0^1$$

$$= \tfrac{1}{2}(1 - \cos 1)$$

    ■ ■

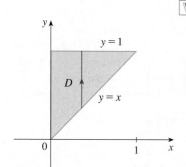

FIGURE 15

D as a type I region

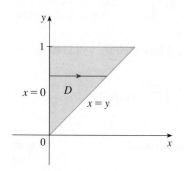

FIGURE 16

D as a type II region

■ ■ **Properties of Double Integrals**

We assume that all of the following integrals exist. The first three properties of double integrals over a region D follow immediately from Definition 2 and Properties 7, 8, and 9 in Section 12.1.

$\boxed{6}$

$$\iint\limits_{D} [f(x, y) + g(x, y)]\, dA = \iint\limits_{D} f(x, y)\, dA + \iint\limits_{D} g(x, y)\, dA$$

$\boxed{7}$

$$\iint\limits_{D} c f(x, y)\, dA = c \iint\limits_{D} f(x, y)\, dA$$

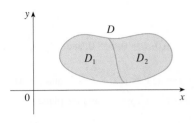

FIGURE 17

If $f(x, y) \geq g(x, y)$ for all (x, y) in D, then

$$\boxed{8} \qquad \iint_D f(x, y) \, dA \geq \iint_D g(x, y) \, dA$$

The next property of double integrals is similar to the property of single integrals given by the equation $\int_a^b f(x) \, dx = \int_a^c f(x) \, dx + \int_c^b f(x) \, dx$.

If $D = D_1 \cup D_2$, where D_1 and D_2 don't overlap except perhaps on their boundaries (see Figure 17), then

$$\boxed{9} \qquad \iint_D f(x, y) \, dA = \iint_{D_1} f(x, y) \, dA + \iint_{D_2} f(x, y) \, dA$$

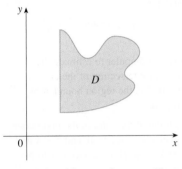

(a) D is neither type I nor type II.

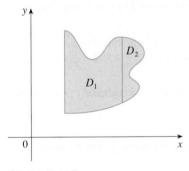

(b) $D = D_1 \cup D_2$,
  D_1 is type I, D_2 is type II.

FIGURE 18

Property 9 can be used to evaluate double integrals over regions D that are neither type I nor type II but can be expressed as a union of regions of type I or type II. Figure 18 illustrates this procedure. (See Exercises 45 and 46.)

The next property of integrals says that if we integrate the constant function $f(x, y) = 1$ over a region D, we get the area of D:

$$\boxed{10} \qquad \iint_D 1 \, dA = A(D)$$

Figure 19 illustrates why Equation 10 is true: A solid cylinder whose base is D and whose height is 1 has volume $A(D) \cdot 1 = A(D)$, but we know that we can also write its volume as $\iint_D 1 \, dA$.

Finally, we can combine Properties 7, 8, and 10 to prove the following property. (See Exercise 49.)

$\boxed{11}$ If $m \leq f(x, y) \leq M$ for all (x, y) in D, then

$$mA(D) \leq \iint_D f(x, y) \, dA \leq MA(D)$$

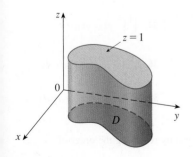

FIGURE 19
Cylinder with base D and height 1

EXAMPLE 6 Use Property 11 to estimate the integral $\iint_D e^{\sin x \cos y} \, dA$, where D is the disk with center the origin and radius 2.

SOLUTION Since $-1 \leq \sin x \leq 1$ and $-1 \leq \cos y \leq 1$, we have $-1 \leq \sin x \cos y \leq 1$ and therefore

$$e^{-1} \leq e^{\sin x \cos y} \leq e^1 = e$$

Thus, using $m = e^{-1} = 1/e$, $M = e$, and $A(D) = \pi(2)^2$ in Property 11, we obtain

$$\frac{4\pi}{e} \leq \iint_D e^{\sin x \cos y} \, dA \leq 4\pi e$$

■ ■

12.3 Exercises

1–6 ■ Evaluate the iterated integral.

1. $\int_0^1 \int_0^{x^2} (x + 2y)\, dy\, dx$

2. $\int_1^2 \int_y^2 xy\, dx\, dy$

3. $\int_0^1 \int_y^{e^y} \sqrt{x}\, dx\, dy$

4. $\int_0^1 \int_x^{2-x} (x^2 - y)\, dy\, dx$

5. $\int_0^{\pi/2} \int_0^{\cos\theta} e^{\sin\theta}\, dr\, d\theta$

6. $\int_0^1 \int_0^v \sqrt{1 - v^2}\, du\, dv$

- - - - - - - - - - - - - - - - - -

7–16 ■ Evaluate the double integral.

7. $\iint_D x^3 y^2\, dA$, $D = \{(x, y) \mid 0 \leq x \leq 2, \ -x \leq y \leq x\}$

8. $\iint_D \frac{4y}{x^3 + 2}\, dA$, $D = \{(x, y) \mid 1 \leq x \leq 2, \ 0 \leq y \leq 2x\}$

9. $\iint_D \frac{2y}{x^2 + 1}\, dA$, $D = \{(x, y) \mid 0 \leq x \leq 1, \ 0 \leq y \leq \sqrt{x}\}$

10. $\iint_D e^{y^2}\, dA$, $D = \{(x, y) \mid 0 \leq y \leq 1, \ 0 \leq x \leq y\}$

11. $\iint_D x \cos y\, dA$, D is bounded by $y = 0$, $y = x^2$, $x = 1$

12. $\iint_D (x + y)\, dA$, D is bounded by $y = \sqrt{x}$ and $y = x^2$

13. $\iint_D y^3\, dA$,
D is the triangular region with vertices $(0, 2)$, $(1, 1)$, $(3, 2)$

14. $\iint_D xy^2\, dA$, D is enclosed by $x = 0$ and $x = \sqrt{1 - y^2}$

15. $\iint_D (2x - y)\, dA$,
D is bounded by the circle with center the origin and radius 2

16. $\iint_D 2xy\, dA$, D is the triangular region with vertices $(0, 0)$,
$(1, 2)$, and $(0, 3)$

- - - - - - - - - - - - - - - - - -

17–26 ■ Find the volume of the given solid.

17. Under the plane $x + 2y - z = 0$ and above the region bounded by $y = x$ and $y = x^4$

18. Under the surface $z = 2x + y^2$ and above the region bounded by $x = y^2$ and $x = y^3$

19. Under the surface $z = xy$ and above the triangle with vertices $(1, 1)$, $(4, 1)$, and $(1, 2)$

20. Enclosed by the paraboloid $z = x^2 + 3y^2$ and the planes $x = 0$, $y = 1$, $y = x$, $z = 0$

21. Bounded by the planes $x = 0$, $y = 0$, $z = 0$, and $x + y + z = 1$

22. Bounded by the planes $z = x$, $y = x$, $x + y = 2$, and $z = 0$

23. Enclosed by the cylinders $z = x^2$, $y = x^2$ and the planes $z = 0$, $y = 4$

24. Bounded by the cylinder $y^2 + z^2 = 4$ and the planes $x = 2y$, $x = 0$, $z = 0$ in the first octant

25. Bounded by the cylinder $x^2 + y^2 = 1$ and the planes $y = z$, $x = 0$, $z = 0$ in the first octant

26. Bounded by the cylinders $x^2 + y^2 = r^2$ and $y^2 + z^2 = r^2$

- - - - - - - - - - - - - - - - - -

27. Use a graphing calculator or computer to estimate the x-coordinates of the points of intersection of the curves $y = x^4$ and $y = 3x - x^2$. If D is the region bounded by these curves, estimate $\iint_D x\, dA$.

28. Find the approximate volume of the solid in the first octant that is bounded by the planes $y = x$, $z = 0$, and $z = x$ and the cylinder $y = \cos x$. (Use a graphing device to estimate the points of intersection.)

29–30 ■ Find the volume of the solid by subtracting two volumes.

29. The solid enclosed by the parabolic cylinders $y = 1 - x^2$, $y = x^2 - 1$ and the planes $x + y + z = 2$, $2x + 2y - z + 10 = 0$

30. The solid enclosed by the parabolic cylinder $y = x^2$ and the planes $z = 3y$, $z = 2 + y$

- - - - - - - - - - - - - - - - - -

[CAS] **31–32** ■ Use a computer algebra system to find the exact volume of the solid.

31. Enclosed by $z = 1 - x^2 - y^2$ and $z = 0$

32. Enclosed by $z = x^2 + y^2$ and $z = 2y$

- - - - - - - - - - - - - - - - - -

33–38 ■ Sketch the region of integration and change the order of integration.

33. $\int_0^4 \int_0^{\sqrt{x}} f(x, y)\, dy\, dx$

34. $\int_0^1 \int_{4x}^4 f(x, y)\, dy\, dx$

35. $\int_0^3 \int_{-\sqrt{9-y^2}}^{\sqrt{9-y^2}} f(x, y)\, dx\, dy$

36. $\int_0^3 \int_0^{\sqrt{9-y}} f(x, y)\, dx\, dy$

37. $\int_1^2 \int_0^{\ln x} f(x, y)\, dy\, dx$

38. $\int_0^1 \int_{\arctan x}^{\pi/4} f(x, y)\, dy\, dx$

- - - - - - - - - - - - - - - - - -

39–44 ■ Evaluate the integral by reversing the order of integration.

39. $\int_0^1 \int_{3y}^3 e^{x^2}\, dx\, dy$

40. $\int_0^1 \int_{\sqrt{y}}^1 \sqrt{x^3 + 1}\, dx\, dy$

41. $\int_0^3 \int_{y^2}^9 y \cos(x^2) \, dx \, dy$

42. $\int_0^1 \int_{x^2}^1 x^3 \sin(y^3) \, dy \, dx$

43. $\int_0^1 \int_{\arcsin y}^{\pi/2} \cos x \sqrt{1 + \cos^2 x} \, dx \, dy$

44. $\int_0^8 \int_{\sqrt[3]{y}}^2 e^{x^4} \, dx \, dy$

- - - - - - - - - - - - -

45–46 ■ Express D as a union of regions of type I or type II and evaluate the integral.

45. $\iint_D x^2 \, dA$ **46.** $\iint_D xy \, dA$

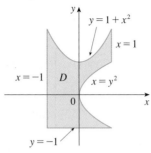

- - - - - - - - - - - - -

47–48 ■ Use Property 11 to estimate the value of the integral.

47. $\iint_D \sqrt{x^3 + y^3} \, dA$, $D = [0, 1] \times [0, 1]$

48. $\iint_D e^{x^2 + y^2} \, dA$,

D is the disk with center the origin and radius $\frac{1}{2}$

- - - - - - - - - - - - -

49. Prove Property 11.

50. In evaluating a double integral over a region D, a sum of iterated integrals was obtained as follows:

$$\iint_D f(x, y) \, dA = \int_0^1 \int_0^{2y} f(x, y) \, dx \, dy + \int_1^3 \int_0^{3-y} f(x, y) \, dx \, dy$$

Sketch the region D and express the double integral as an iterated integral with reversed order of integration.

51. Evaluate $\iint_D (x^2 \tan x + y^3 + 4) \, dA$, where
$D = \{(x, y) \mid x^2 + y^2 \leq 2\}$.
[*Hint:* Exploit the fact that D is symmetric with respect to both axes.]

52. Use symmetry to evaluate $\iint_D (2 - 3x + 4y) \, dA$, where D is the region bounded by the square with vertices $(\pm 5, 0)$ and $(0, \pm 5)$.

53. Compute $\iint_D \sqrt{1 - x^2 - y^2} \, dA$, where D is the disk $x^2 + y^2 \leq 1$, by first identifying the integral as the volume of a solid.

54. Graph the solid bounded by the plane $x + y + z = 1$ and the paraboloid $z = 4 - x^2 - y^2$ and find its exact volume. (Use your CAS to do the graphing, to find the equations of the boundary curves of the region of integration, and to evaluate the double integral.) [CAS]

- - - - - - - - - - - - -

12.4 Double Integrals in Polar Coordinates

■ ■ See Appendix H for information about polar coordinates.

Suppose that we want to evaluate a double integral $\iint_R f(x, y) \, dA$, where R is one of the regions shown in Figure 1. In either case the description of R in terms of rectangular coordinates is rather complicated but R is easily described using polar coordinates.

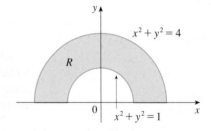

FIGURE 1   (a) $R = \{(r, \theta) \mid 0 \leq r \leq 1, 0 \leq \theta \leq 2\pi\}$ (b) $R = \{(r, \theta) \mid 1 \leq r \leq 2, 0 \leq \theta \leq \pi\}$

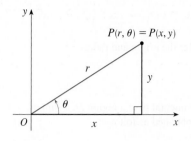

FIGURE 2

Recall from Figure 2 that the polar coordinates (r, θ) of a point are related to the rectangular coordinates (x, y) by the equations

$$r^2 = x^2 + y^2 \qquad x = r\cos\theta \qquad y = r\sin\theta$$

The regions in Figure 1 are special cases of a **polar rectangle**

$$R = \{(r, \theta) \mid a \leq r \leq b, \alpha \leq \theta \leq \beta\}$$

which is shown in Figure 3. In order to compute the double integral $\iint_R f(x, y)\, dA$, where R is a polar rectangle, we divide the interval $[a, b]$ into m subintervals $[r_{i-1}, r_i]$ of equal width $\Delta r = (b - a)/m$ and we divide the interval $[\alpha, \beta]$ into n subintervals $[\theta_{j-1}, \theta_j]$ of equal width $\Delta\theta = (\beta - \alpha)/n$. Then the circles $r = r_i$ and the rays $\theta = \theta_j$ divide the polar rectangle R into the small polar rectangles shown in Figure 4.

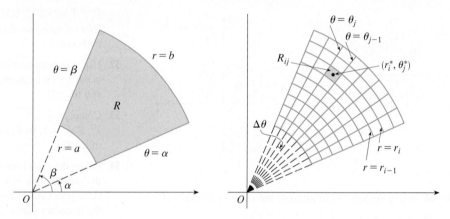

FIGURE 3 Polar rectangle

FIGURE 4 Dividing R into polar subrectangles

The "center" of the polar subrectangle

$$R_{ij} = \{(r, \theta) \mid r_{i-1} \leq r \leq r_i, \theta_{j-1} \leq \theta \leq \theta_j\}$$

has polar coordinates

$$r_i^* = \tfrac{1}{2}(r_{i-1} + r_i) \qquad \theta_j^* = \tfrac{1}{2}(\theta_{j-1} + \theta_j)$$

We compute the area of R_{ij} using the fact that the area of a sector of a circle with radius r and central angle θ is $\tfrac{1}{2}r^2\theta$. Subtracting the areas of two such sectors, each of which has central angle $\Delta\theta = \theta_j - \theta_{j-1}$, we find that the area of R_{ij} is

$$\Delta A_i = \tfrac{1}{2}r_i^2\,\Delta\theta - \tfrac{1}{2}r_{i-1}^2\,\Delta\theta = \tfrac{1}{2}(r_i^2 - r_{i-1}^2)\,\Delta\theta$$

$$= \tfrac{1}{2}(r_i + r_{i-1})(r_i - r_{i-1})\,\Delta\theta = r_i^*\,\Delta r\,\Delta\theta$$

Although we have defined the double integral $\iint_R f(x, y)\, dA$ in terms of ordinary rectangles, it can be shown that, for continuous functions f, we always obtain the same answer using polar rectangles. The rectangular coordinates of the center of R_{ij} are $(r_i^*\cos\theta_j^*, r_i^*\sin\theta_j^*)$, so a typical Riemann sum is

$$\boxed{1} \quad \sum_{i=1}^{m}\sum_{j=1}^{n} f(r_i^*\cos\theta_j^*, r_i^*\sin\theta_j^*)\,\Delta A_i = \sum_{i=1}^{m}\sum_{j=1}^{n} f(r_i^*\cos\theta_j^*, r_i^*\sin\theta_j^*)\, r_i^*\,\Delta r\,\Delta\theta$$

If we write $g(r, \theta) = rf(r\cos\theta, r\sin\theta)$, then the Riemann sum in Equation 1 can be written as

$$\sum_{i=1}^{m}\sum_{j=1}^{n} g(r_i^*, \theta_j^*)\,\Delta r\,\Delta\theta$$

which is a Riemann sum for the double integral

$$\int_{\alpha}^{\beta}\int_{a}^{b} g(r, \theta)\,dr\,d\theta$$

Therefore, we have

$$\iint_{R} f(x, y)\,dA = \lim_{m, n \to \infty}\sum_{i=1}^{m}\sum_{j=1}^{n} f(r_i^*\cos\theta_j^*, r_i^*\sin\theta_j^*)\,\Delta A_i$$

$$= \lim_{m, n \to \infty}\sum_{i=1}^{m}\sum_{j=1}^{n} g(r_i^*, \theta_j^*)\,\Delta r\,\Delta\theta = \int_{\alpha}^{\beta}\int_{a}^{b} g(r, \theta)\,dr\,d\theta$$

$$= \int_{\alpha}^{\beta}\int_{a}^{b} f(r\cos\theta, r\sin\theta)\,r\,dr\,d\theta$$

> **2** **Change to Polar Coordinates in a Double Integral** If f is continuous on a polar rectangle R given by $0 \le a \le r \le b$, $\alpha \le \theta \le \beta$, where $0 \le \beta - \alpha \le 2\pi$, then
>
> $$\iint_{R} f(x, y)\,dA = \int_{\alpha}^{\beta}\int_{a}^{b} f(r\cos\theta, r\sin\theta)\,r\,dr\,d\theta$$

The formula in (2) says that we convert from rectangular to polar coordinates in a double integral by writing $x = r\cos\theta$ and $y = r\sin\theta$, using the appropriate limits of integration for r and θ, and replacing dA by $r\,dr\,d\theta$. Be careful not to forget the additional factor r on the right side of Formula 2. A classical method for remembering this is shown in Figure 5, where the "infinitesimal" polar rectangle can be thought of as an ordinary rectangle with dimensions $r\,d\theta$ and dr and therefore has "area" $dA = r\,dr\,d\theta$.

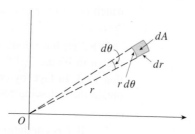

FIGURE 5

EXAMPLE 1 Evaluate $\iint_{R} (3x + 4y^2)\,dA$, where R is the region in the upper half-plane bounded by the circles $x^2 + y^2 = 1$ and $x^2 + y^2 = 4$.

SOLUTION The region R can be described as

$$R = \{(x, y) \mid y \ge 0,\ 1 \le x^2 + y^2 \le 4\}$$

It is the half-ring shown in Figure 1(b), and in polar coordinates it is given by $1 \le r \le 2, 0 \le \theta \le \pi$. Therefore, by Formula 2,

$$\iint\limits_R (3x + 4y^2) \, dA = \int_0^\pi \int_1^2 (3r \cos \theta + 4r^2 \sin^2\theta) \, r \, dr \, d\theta$$

$$= \int_0^\pi \int_1^2 (3r^2 \cos \theta + 4r^3 \sin^2\theta) \, dr \, d\theta$$

$$= \int_0^\pi \left[r^3 \cos \theta + r^4 \sin^2\theta \right]_{r=1}^{r=2} d\theta = \int_0^\pi (7 \cos \theta + 15 \sin^2\theta) \, d\theta$$

$$= \int_0^\pi \left[7 \cos \theta + \tfrac{15}{2}(1 - \cos 2\theta) \right] d\theta$$

$$= 7 \sin \theta + \frac{15\theta}{2} - \frac{15}{4} \sin 2\theta \Big]_0^\pi = \frac{15\pi}{2} \quad ■■$$

■■ Here we use the trigonometric identity

$$\sin^2\theta = \tfrac{1}{2}(1 - \cos 2\theta)$$

as discussed in Section 5.7. Alternatively, we could have used Formula 63 in the Table of Integrals:

$$\int \sin^2 u \, du = \tfrac{1}{2}u - \tfrac{1}{4}\sin 2u + C$$

|V| **EXAMPLE 2** Find the volume of the solid bounded by the plane $z = 0$ and the paraboloid $z = 1 - x^2 - y^2$.

SOLUTION If we put $z = 0$ in the equation of the paraboloid, we get $x^2 + y^2 = 1$. This means that the plane intersects the paraboloid in the circle $x^2 + y^2 = 1$, so the solid lies under the paraboloid and above the circular disc D given by $x^2 + y^2 \le 1$ [see Figures 6 and 1(a)]. In polar coordinates D is given by $0 \le r \le 1, 0 \le \theta \le 2\pi$. Since $1 - x^2 - y^2 = 1 - r^2$, the volume is

$$V = \iint\limits_D (1 - x^2 - y^2) \, dA = \int_0^{2\pi} \int_0^1 (1 - r^2) \, r \, dr \, d\theta$$

$$= \int_0^{2\pi} d\theta \int_0^1 (r - r^3) \, dr = 2\pi \left[\frac{r^2}{2} - \frac{r^4}{4} \right]_0^1 = \frac{\pi}{2}$$

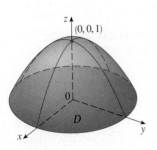

FIGURE 6

If we had used rectangular coordinates instead of polar coordinates, then we would have obtained

$$V = \iint\limits_D (1 - x^2 - y^2) \, dA = \int_{-1}^1 \int_{-\sqrt{1-x^2}}^{\sqrt{1-x^2}} (1 - x^2 - y^2) \, dy \, dx$$

which is not easy to evaluate because it involves finding $\int (1 - x^2)^{3/2} \, dx$. ■■

What we have done so far can be extended to the more complicated type of region shown in Figure 7. It's similar to the type II rectangular regions considered in Section 12.3. In fact, by combining Formula 2 in this section with Formula 12.3.5, we obtain the following formula.

3 If f is continuous on a polar region of the form

$$D = \{(r, \theta) \mid \alpha \le \theta \le \beta, \ h_1(\theta) \le r \le h_2(\theta)\}$$

then

$$\iint\limits_D f(x, y) \, dA = \int_\alpha^\beta \int_{h_1(\theta)}^{h_2(\theta)} f(r \cos \theta, r \sin \theta) \, r \, dr \, d\theta$$

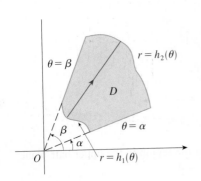

FIGURE 7
$D = \{(r, \theta) \mid \alpha \le \theta \le \beta, h_1(\theta) \le r \le h_2(\theta)\}$

In particular, taking $f(x, y) = 1$, $h_1(\theta) = 0$, and $h_2(\theta) = h(\theta)$ in this formula, we see that the area of the region D bounded by $\theta = \alpha$, $\theta = \beta$, and $r = h(\theta)$ is

$$A(D) = \iint\limits_{D} 1\, dA = \int_{\alpha}^{\beta} \int_{0}^{h(\theta)} r\, dr\, d\theta$$

$$= \int_{\alpha}^{\beta} \left[\frac{r^2}{2} \right]_{0}^{h(\theta)} d\theta = \int_{\alpha}^{\beta} \tfrac{1}{2} [h(\theta)]^2\, d\theta$$

and this agrees with Formula 3 in Appendix H.2.

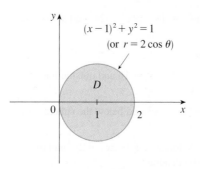

EXAMPLE 3 Find the volume of the solid that lies under the paraboloid $z = x^2 + y^2$, above the xy-plane, and inside the cylinder $x^2 + y^2 = 2x$.

SOLUTION The solid lies above the disk D whose boundary circle has equation $x^2 + y^2 = 2x$ or, after completing the square,

$$(x - 1)^2 + y^2 = 1$$

(See Figures 8 and 9.) In polar coordinates we have $x^2 + y^2 = r^2$ and $x = r \cos \theta$, so the boundary circle becomes $r^2 = 2r \cos \theta$, or $r = 2 \cos \theta$. Thus, the disk D is given by

$$D = \{(r, \theta) \mid -\pi/2 \leqslant \theta \leqslant \pi/2,\ 0 \leqslant r \leqslant 2 \cos \theta\}$$

and, by Formula 3, we have

$$V = \iint\limits_{D} (x^2 + y^2)\, dA = \int_{-\pi/2}^{\pi/2} \int_{0}^{2 \cos \theta} r^2\, r\, dr\, d\theta$$

$$= \int_{-\pi/2}^{\pi/2} \left[\frac{r^4}{4} \right]_{0}^{2 \cos \theta} d\theta = 4 \int_{-\pi/2}^{\pi/2} \cos^4\theta\, d\theta$$

$$= 8 \int_{0}^{\pi/2} \cos^4\theta\, d\theta$$

Using Formula 74 in the Table of Integrals with $n = 4$, we get

$$V = 8 \int_{0}^{\pi/2} \cos^4\theta\, d\theta = 8\left(\tfrac{1}{4} \cos^3\theta\, \sin\theta \Big]_{0}^{\pi/2} + \tfrac{3}{4} \int_{0}^{\pi/2} \cos^2\theta\, d\theta \right)$$

$$= 6 \int_{0}^{\pi/2} \cos^2\theta\, d\theta$$

Now we use Formula 64 in the Table of Integrals:

$$V = 6 \int_{0}^{\pi/2} \cos^2\theta\, d\theta = 6\left[\tfrac{1}{2}\theta + \tfrac{1}{4} \sin 2\theta \right]_{0}^{\pi/2}$$

$$= 6 \cdot \frac{1}{2} \cdot \frac{\pi}{2} = \frac{3\pi}{2}$$

FIGURE 8

FIGURE 9

■ ■ Instead of using tables, we could have used the identity $\cos^2\theta = \tfrac{1}{2}(1 + \cos 2\theta)$ twice.

12.4   Exercises

1–6 ■ A region R is shown. Decide whether to use polar coordinates or rectangular coordinates and write $\iint_R f(x, y)\, dA$ as an iterated integral, where f is an arbitrary continuous function on R.

1.

2.

3.

4.

5.

6.

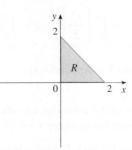

7–8 ■ Sketch the region whose area is given by the integral and evaluate the integral.

7. $\int_{\pi}^{2\pi} \int_{4}^{7} r\, dr\, d\theta$

8. $\int_{0}^{\pi/2} \int_{0}^{4\cos\theta} r\, dr\, d\theta$

9–14 ■ Evaluate the given integral by changing to polar coordinates.

9. $\iint_D xy\, dA$,
where D is the disk with center the origin and radius 3

10. $\iint_R (x + y)\, dA$,   where R is the region that lies to the left of the y-axis between the circles $x^2 + y^2 = 1$ and $x^2 + y^2 = 4$

11. $\iint_R \cos(x^2 + y^2)\, dA$,   where R is the region that lies above the x-axis within the circle $x^2 + y^2 = 9$

12. $\iint_R \sqrt{4 - x^2 - y^2}\, dA$,
where $R = \{(x, y) \mid x^2 + y^2 \leq 4,\ x \geq 0\}$

13. $\iint_R \arctan(y/x)\, dA$,
where $R = \{(x, y) \mid 1 \leq x^2 + y^2 \leq 4,\ 0 \leq y \leq x\}$

14. $\iint_R ye^x\, dA$,   where R is the region in the first quadrant enclosed by the circle $x^2 + y^2 = 25$

15–21 ■ Use polar coordinates to find the volume of the given solid.

15. Under the cone $z = \sqrt{x^2 + y^2}$ and above the disk $x^2 + y^2 \leq 4$

16. Below the paraboloid $z = 18 - 2x^2 - 2y^2$ and above the xy-plane

17. A sphere of radius a

18. Inside the sphere $x^2 + y^2 + z^2 = 16$ and outside the cylinder $x^2 + y^2 = 4$

19. Above the cone $z = \sqrt{x^2 + y^2}$ and below the sphere $x^2 + y^2 + z^2 = 1$

20. Bounded by the paraboloid $z = 1 + 2x^2 + 2y^2$ and the plane $z = 7$ in the first quadrant

21. Inside both the cylinder $x^2 + y^2 = 4$ and the ellipsoid $4x^2 + 4y^2 + z^2 = 64$

22. (a) A cylindrical drill with radius r_1 is used to bore a hole through the center of a sphere of radius r_2. Find the volume of the ring-shaped solid that remains.
(b) Express the volume in part (a) in terms of the height h of the ring. Notice that the volume depends only on h, not on r_1 or r_2.

23–24 ■ Use a double integral to find the area of the region.

23. One loop of the rose $r = \cos 3\theta$

24. The region enclosed by the curve $r = 4 + 3\cos\theta$

25–28 ■ Evaluate the iterated integral by converting to polar coordinates.

25. $\int_{-3}^{3} \int_{0}^{\sqrt{9-x^2}} \sin(x^2 + y^2)\, dy\, dx$

26. $\int_{0}^{a} \int_{-\sqrt{a^2-y^2}}^{0} x^2 y\, dx\, dy$

27. $\int_{0}^{1} \int_{y}^{\sqrt{2-y^2}} (x + y)\, dx\, dy$

28. $\int_{0}^{2} \int_{0}^{\sqrt{2x-x^2}} \sqrt{x^2 + y^2}\, dy\, dx$

29. A swimming pool is circular with a 40-ft diameter. The depth is constant along east-west lines and increases linearly from 2 ft at the south end to 7 ft at the north end. Find the volume of water in the pool.

30. An agricultural sprinkler distributes water in a circular pattern of radius 100 ft. It supplies water to a depth of e^{-r} feet per hour at a distance of r feet from the sprinkler.
(a) What is the total amount of water supplied per hour to the region inside the circle of radius R centered at the sprinkler?
(b) Determine an expression for the average amount of water per hour per square foot supplied to the region inside the circle of radius R.

31. Use polar coordinates to combine the sum

$$\int_{1/\sqrt{2}}^{1}\int_{\sqrt{1-x^2}}^{x} xy\,dy\,dx + \int_{1}^{\sqrt{2}}\int_{0}^{x} xy\,dy\,dx + \int_{\sqrt{2}}^{2}\int_{0}^{\sqrt{4-x^2}} xy\,dy\,dx$$

into one double integral. Then evaluate the double integral.

32. (a) We define the improper integral (over the entire plane $\mathbb{R}^2$)

$$I = \iint_{\mathbb{R}^2} e^{-(x^2+y^2)}\,dA = \int_{-\infty}^{\infty}\int_{-\infty}^{\infty} e^{-(x^2+y^2)}\,dy\,dx$$
$$= \lim_{a\to\infty} \iint_{D_a} e^{-(x^2+y^2)}\,dA$$

where D_a is the disk with radius a and center the origin.

Show that

$$\int_{-\infty}^{\infty}\int_{-\infty}^{\infty} e^{-(x^2+y^2)}\,dA = \pi$$

(b) An equivalent definition of the improper integral in part (a) is

$$\iint_{\mathbb{R}^2} e^{-(x^2+y^2)}\,dA = \lim_{a\to\infty} \iint_{S_a} e^{-(x^2+y^2)}\,dA$$

where S_a is the square with vertices $(\pm a, \pm a)$. Use this to show that

$$\int_{-\infty}^{\infty} e^{-x^2}\,dx \int_{-\infty}^{\infty} e^{-y^2}\,dy = \pi$$

(c) Deduce that

$$\int_{-\infty}^{\infty} e^{-x^2}\,dx = \sqrt{\pi}$$

(d) By making the change of variable $t = \sqrt{2}\,x$, show that

$$\int_{-\infty}^{\infty} e^{-x^2/2}\,dx = \sqrt{2\pi}$$

(This is a fundamental result for probability and statistics.)

33. Use the result of Exercise 32 part (c) to evaluate the following integrals.
(a) $\int_{0}^{\infty} x^2 e^{-x^2}\,dx$ (b) $\int_{0}^{\infty} \sqrt{x}\, e^{-x}\,dx$

12.5 Applications of Double Integrals

We have already seen one application of double integrals: computing volumes. Another geometric application is finding areas of surfaces and this will be done in the next section. In this section we explore physical applications such as computing mass, electric charge, center of mass, and moment of inertia. We will see that these physical ideas are also important when applied to probability density functions of two random variables.

Density and Mass

In Chapter 6 we were able to use single integrals to compute moments and the center of mass of a thin plate or lamina with constant density. But now, equipped with the double integral, we can consider a lamina with variable density. Suppose the lamina occupies a region D of the xy-plane and its **density** (in units of mass per unit area) at a point (x, y) in D is given by $\rho(x, y)$, where ρ is a continuous function on D. This means that

$$\rho(x, y) = \lim \frac{\Delta m}{\Delta A}$$

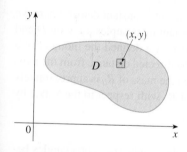

FIGURE 1

where Δm and ΔA are the mass and area of a small rectangle that contains (x, y) and the limit is taken as the dimensions of the rectangle approach 0. (See Figure 1.)

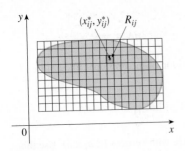

FIGURE 2

To find the total mass m of the lamina we divide a rectangle R containing D into subrectangles R_{ij} of the same size (as in Figure 2) and consider $\rho(x, y)$ to be 0 outside D. If we choose a point (x_{ij}^*, y_{ij}^*) in R_{ij}, then the mass of the part of the lamina that occupies R_{ij} is approximately $\rho(x_{ij}^*, y_{ij}^*) \, \Delta A$, where ΔA is the area of R_{ij}. If we add all such masses, we get an approximation to the total mass:

$$m \approx \sum_{i=1}^{k} \sum_{j=1}^{l} \rho(x_{ij}^*, y_{ij}^*) \, \Delta A$$

If we now increase the number of subrectangles, we obtain the total mass m of the lamina as the limiting value of the approximations:

$$\boxed{1} \qquad m = \lim_{k, l \to \infty} \sum_{i=1}^{k} \sum_{j=1}^{l} \rho(x_{ij}^*, y_{ij}^*) \, \Delta A = \iint_{D} \rho(x, y) \, dA$$

Physicists also consider other types of density that can be treated in the same manner. For example, if an electric charge is distributed over a region D and the charge density (in units of charge per unit area) is given by $\sigma(x, y)$ at a point (x, y) in D, then the total charge Q is given by

$$\boxed{2} \qquad Q = \iint_{D} \sigma(x, y) \, dA$$

EXAMPLE 1 Charge is distributed over the triangular region D in Figure 3 so that the charge density at (x, y) is $\sigma(x, y) = xy$, measured in coulombs per square meter (C/m^2). Find the total charge.

SOLUTION From Equation 2 and Figure 3 we have

$$Q = \iint_{D} \sigma(x, y) \, dA = \int_{0}^{1} \int_{1-x}^{1} xy \, dy \, dx$$

$$= \int_{0}^{1} \left[x \frac{y^2}{2} \right]_{y=1-x}^{y=1} dx = \int_{0}^{1} \frac{x}{2} [1^2 - (1 - x)^2] \, dx$$

$$= \frac{1}{2} \int_{0}^{1} (2x^2 - x^3) \, dx = \frac{1}{2} \left[\frac{2x^3}{3} - \frac{x^4}{4} \right]_{0}^{1} = \frac{5}{24}$$

Thus, the total charge is $\frac{5}{24}$ C. ■ ■

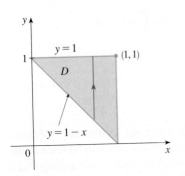

FIGURE 3

■■ Moments and Centers of Mass

In Section 6.5 we found the center of mass of a lamina with constant density; here we consider a lamina with variable density. Suppose the lamina occupies a region D and has density function $\rho(x, y)$. Recall from Chapter 6 that we defined the moment of a particle about an axis as the product of its mass and its directed distance from the axis. We divide D into small rectangles as in Figure 2. Then the mass of R_{ij} is approximately $\rho(x_{ij}^*, y_{ij}^*) \, \Delta A$, so we can approximate the moment of R_{ij} with respect to the x-axis by

$$[\rho(x_{ij}^*, y_{ij}^*) \, \Delta A] y_{ij}^*$$

If we now add these quantities and take the limit as the number of subrectangles becomes large, we obtain the **moment** of the entire lamina **about the x-axis**:

$$3 \qquad M_x = \lim_{m, n \to \infty} \sum_{i=1}^{m} \sum_{j=1}^{n} y_{ij}^* \, \rho(x_{ij}^*, y_{ij}^*) \, \Delta A = \iint_D y\rho(x, y) \, dA$$

Similarly, the **moment about the y-axis** is

$$4 \qquad M_y = \lim_{m, n \to \infty} \sum_{i=1}^{m} \sum_{j=1}^{n} x_{ij}^* \, \rho(x_{ij}^*, y_{ij}^*) \, \Delta A = \iint_D x\rho(x, y) \, dA$$

FIGURE 4

As before, we define the center of mass $(\bar{x}, \bar{y})$ so that $m\bar{x} = M_y$ and $m\bar{y} = M_x$. The physical significance is that the lamina behaves as if its entire mass is concentrated at its center of mass. Thus, the lamina balances horizontally when supported at its center of mass (see Figure 4).

> **5** The coordinates $(\bar{x}, \bar{y})$ of the center of mass of a lamina occupying the region D and having density function $\rho(x, y)$ are
>
> $$\bar{x} = \frac{M_y}{m} = \frac{1}{m} \iint_D x\rho(x, y) \, dA \qquad \bar{y} = \frac{M_x}{m} = \frac{1}{m} \iint_D y\rho(x, y) \, dA$$
>
> where the mass m is given by
>
> $$m = \iint_D \rho(x, y) \, dA$$

$\boxed{V}$ **EXAMPLE 2** Find the mass and center of mass of a triangular lamina with vertices $(0, 0)$, $(1, 0)$, and $(0, 2)$ if the density function is $\rho(x, y) = 1 + 3x + y$.

SOLUTION The triangle is shown in Figure 5. (Note that the equation of the upper boundary is $y = 2 - 2x$.) The mass of the lamina is

$$m = \iint_D \rho(x, y) \, dA = \int_0^1 \int_0^{2-2x} (1 + 3x + y) \, dy \, dx$$

$$= \int_0^1 \left[y + 3xy + \frac{y^2}{2} \right]_{y=0}^{y=2-2x} dx$$

$$= 4 \int_0^1 (1 - x^2) \, dx = 4 \left[x - \frac{x^3}{3} \right]_0^1 = \frac{8}{3}$$

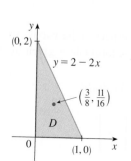

FIGURE 5

Then the formulas in (5) give

$$\bar{x} = \frac{1}{m} \iint_D x\rho(x, y) \, dA = \frac{3}{8} \int_0^1 \int_0^{2-2x} (x + 3x^2 + xy) \, dy \, dx$$

$$= \frac{3}{8} \int_0^1 \left[xy + 3x^2y + x\frac{y^2}{2} \right]_{y=0}^{y=2-2x} dx$$

$$= \frac{3}{2} \int_0^1 (x - x^3) \, dx = \frac{3}{2} \left[\frac{x^2}{2} - \frac{x^4}{4} \right]_0^1 = \frac{3}{8}$$

$$\bar{y} = \frac{1}{m} \iint_D y\rho(x, y) \, dA = \tfrac{3}{8} \int_0^1 \int_0^{2-2x} (y + 3xy + y^2) \, dy \, dx$$

$$= \frac{3}{8} \int_0^1 \left[\frac{y^2}{2} + 3x\frac{y^2}{2} + \frac{y^3}{3} \right]_{y=0}^{y=2-2x} dx = \tfrac{1}{4} \int_0^1 (7 - 9x - 3x^2 + 5x^3) \, dx$$

$$= \frac{1}{4} \left[7x - 9\frac{x^2}{2} - x^3 + 5\frac{x^4}{4} \right]_0^1 = \frac{11}{16}$$

The center of mass is at the point $\left(\frac{3}{8}, \frac{11}{16} \right)$. ▪ ▪

EXAMPLE 3 The density at any point on a semicircular lamina is proportional to the distance from the center of the circle. Find the center of mass of the lamina.

SOLUTION Let's place the lamina as the upper half of the circle $x^2 + y^2 = a^2$. (See Figure 6.) Then the distance from a point (x, y) to the center of the circle (the origin) is $\sqrt{x^2 + y^2}$. Therefore, the density function is

$$\rho(x, y) = K\sqrt{x^2 + y^2}$$

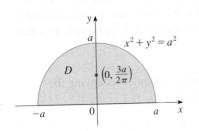

FIGURE 6

where K is some constant. Both the density function and the shape of the lamina suggest that we convert to polar coordinates. Then $\sqrt{x^2 + y^2} = r$ and the region D is given by $0 \le r \le a$, $0 \le \theta \le \pi$. Thus, the mass of the lamina is

$$m = \iint_D \rho(x, y) \, dA = \iint_D K\sqrt{x^2 + y^2} \, dA$$

$$= \int_0^\pi \int_0^a (Kr) \, r \, dr \, d\theta = K \int_0^\pi d\theta \int_0^a r^2 \, dr$$

$$= K\pi \frac{r^3}{3} \Big]_0^a = \frac{K\pi a^3}{3}$$

Both the lamina and the density function are symmetric with respect to the y-axis, so the center of mass must lie on the y-axis, that is, $\bar{x} = 0$. The y-coordinate is given by

$$\bar{y} = \frac{1}{m} \iint_D y\rho(x, y) \, dA = \frac{3}{K\pi a^3} \int_0^\pi \int_0^a r \sin\theta \, (Kr) \, r \, dr \, d\theta$$

$$= \frac{3}{\pi a^3} \int_0^\pi \sin\theta \, d\theta \int_0^a r^3 \, dr = \frac{3}{\pi a^3} \left[-\cos\theta \right]_0^\pi \left[\frac{r^4}{4} \right]_0^a$$

$$= \frac{3}{\pi a^3} \frac{2a^4}{4} = \frac{3a}{2\pi}$$

▪ ▪ Compare the location of the center of mass in Example 3 with Example 7 in Section 6.5 where we found that the center of mass of a lamina with the same shape but uniform density is located at the point $(0, 4a/(3\pi))$.

Therefore, the center of mass is located at the point $(0, 3a/(2\pi))$. ▪ ▪

▪▪ Moment of Inertia

The **moment of inertia** (also called the **second moment**) of a particle of mass m about an axis is defined to be mr^2, where r is the distance from the particle to the axis. We extend this concept to a lamina with density function $\rho(x, y)$ and occupying a region D by proceeding as we did for ordinary moments. We divide D into small rectangles, approximate the moment of inertia of each subrectangle about the x-axis, and take the limit of the sum as the number of subrectangles becomes large. The result is the **moment of inertia** of the lamina **about the x-axis**:

$$\boxed{6} \qquad I_x = \lim_{m, n \to \infty} \sum_{i=1}^{m} \sum_{j=1}^{n} (y_{ij}^*)^2 \rho(x_{ij}^*, y_{ij}^*) \, \Delta A = \iint\limits_{D} y^2 \rho(x, y) \, dA$$

Similarly, the **moment of inertia about the y-axis** is

$$\boxed{7} \qquad I_y = \lim_{m, n \to \infty} \sum_{i=1}^{m} \sum_{j=1}^{n} (x_{ij}^*)^2 \rho(x_{ij}^*, y_{ij}^*) \, \Delta A = \iint\limits_{D} x^2 \rho(x, y) \, dA$$

It is also of interest to consider the **moment of inertia about the origin**, also called the **polar moment of inertia**:

$$\boxed{8} \qquad I_0 = \lim_{m, n \to \infty} \sum_{i=1}^{m} \sum_{j=1}^{n} \left[(x_{ij}^*)^2 + (y_{ij}^*)^2 \right] \rho(x_{ij}^*, y_{ij}^*) \, \Delta A = \iint\limits_{D} (x^2 + y^2) \rho(x, y) \, dA$$

Note that $I_0 = I_x + I_y$.

$\boxed{V}$ **EXAMPLE 4** Find the moments of inertia I_x, I_y, and I_0 of a homogeneous disk D with density $\rho(x, y) = \rho$, center the origin, and radius a.

SOLUTION The boundary of D is the circle $x^2 + y^2 = a^2$ and in polar coordinates D is described by $0 \leqslant \theta \leqslant 2\pi$, $0 \leqslant r \leqslant a$. Let's compute I_0 first:

$$I_0 = \iint\limits_{D} (x^2 + y^2) \rho \, dA = \rho \int_0^{2\pi} \int_0^a r^2 \, r \, dr \, d\theta$$

$$= \rho \int_0^{2\pi} d\theta \int_0^a r^3 \, dr = 2\pi\rho \left[\frac{r^4}{4} \right]_0^a = \frac{\pi\rho a^4}{2}$$

Instead of computing I_x and I_y directly, we use the facts that $I_x + I_y = I_0$ and $I_x = I_y$ (from the symmetry of the problem). Thus

$$I_x = I_y = \frac{I_0}{2} = \frac{\pi\rho a^4}{4} \qquad \blacksquare\blacksquare$$

In Example 4 notice that the mass of the disk is

$$m = \text{density} \times \text{area} = \rho(\pi a^2)$$

so the moment of inertia of the disk about the origin (like a wheel about its axle) can be written as

$$I_0 = \frac{\pi\rho a^4}{2} = \tfrac{1}{2}(\rho\pi a^2)a^2 = \tfrac{1}{2}ma^2$$

Thus, if we increase the mass or the radius of the disk, we thereby increase the moment of inertia. In general, the moment of inertia plays much the same role in rotational motion that mass plays in linear motion. The moment of inertia of a wheel is what makes it difficult to start or stop the rotation of the wheel, just as the mass of a car is what makes it difficult to start or stop the motion of the car.

Probability

In Section 6.7 we considered the *probability density function* f of a continuous random variable X. This means that $f(x) \geq 0$ for all x, $\int_{-\infty}^{\infty} f(x)\, dx = 1$, and the probability that X lies between a and b is found by integrating f from a to b:

$$P(a \leq X \leq b) = \int_a^b f(x)\, dx$$

Now we consider a pair of continuous random variables X and Y, such as the lifetimes of two components of a machine or the height and weight of an adult female chosen at random. The **joint density function** of X and Y is a function f of two variables such that the probability that (X, Y) lies in a region D is

$$P((X, Y) \in D) = \iint_D f(x, y)\, dA$$

In particular, if the region is a rectangle, the probability that X lies between a and b and Y lies between c and d is

$$P(a \leq X \leq b,\ c \leq Y \leq d) = \int_a^b \int_c^d f(x, y)\, dy\, dx$$

(See Figure 7.)

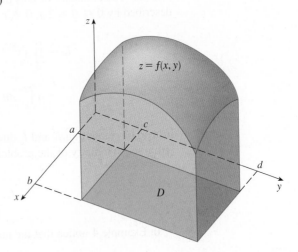

FIGURE 7

The probability that X lies between a and b and Y lies between c and d is the volume that lies above the rectangle $D = [a, b] \times [c, d]$ and below the graph of the joint density function.

Because probabilities aren't negative and are measured on a scale from 0 to 1, the joint density function has the following properties:

$$f(x, y) \geq 0 \qquad \iint_{\mathbb{R}^2} f(x, y)\, dA = 1$$

As in Exercise 32 in Section 12.4, the double integral over $\mathbb{R}^2$ is an improper integral defined as the limit of double integrals over expanding circles or squares and we can write

$$\iint_{\mathbb{R}^2} f(x, y) \, dA = \int_{-\infty}^{\infty} \int_{-\infty}^{\infty} f(x, y) \, dx \, dy = 1$$

EXAMPLE 5 If the joint density function for X and Y is given by

$$f(x, y) = \begin{cases} C(x + 2y) & \text{if } 0 \leqslant x \leqslant 10, \ 0 \leqslant y \leqslant 10 \\ 0 & \text{otherwise} \end{cases}$$

find the value of the constant C. Then find $P(X \leqslant 7, Y \geqslant 2)$.

SOLUTION We find the value of C by ensuring that the double integral of f is equal to 1. Because $f(x, y) = 0$ outside the rectangle $[0, 10] \times [0, 10]$, we have

$$\int_{-\infty}^{\infty} \int_{-\infty}^{\infty} f(x, y) \, dy \, dx = \int_{0}^{10} \int_{0}^{10} C(x + 2y) \, dy \, dx = C \int_{0}^{10} \left[xy + y^2 \right]_{y=0}^{y=10} \, dx$$

$$= C \int_{0}^{10} (10x + 100) \, dx = 1500C$$

Therefore, $1500C = 1$ and so $C = \frac{1}{1500}$.

Now we can compute the probability that X is at most 7 and Y is at least 2:

$$P(X \leqslant 7, Y \geqslant 2) = \int_{-\infty}^{7} \int_{2}^{\infty} f(x, y) \, dy \, dx = \int_{0}^{7} \int_{2}^{10} \tfrac{1}{1500}(x + 2y) \, dy \, dx$$

$$= \tfrac{1}{1500} \int_{0}^{7} \left[xy + y^2 \right]_{y=2}^{y=10} \, dx = \tfrac{1}{1500} \int_{0}^{7} (8x + 96) \, dx$$

$$= \tfrac{868}{1500} \approx 0.5787$$

Suppose X is a random variable with probability density function $f_1(x)$ and Y is a random variable with density function $f_2(y)$. Then X and Y are called **independent random variables** if their joint density function is the product of their individual density functions:

$$f(x, y) = f_1(x) f_2(y)$$

In Section 6.7 we modeled waiting times by using exponential density functions

$$f(t) = \begin{cases} 0 & \text{if } t < 0 \\ \mu^{-1} e^{-t/\mu} & \text{if } t \geqslant 0 \end{cases}$$

where μ is the mean waiting time. In the next example we consider a situation with two independent waiting times.

EXAMPLE 6 The manager of a movie theater determines that the average time movie-goers wait in line to buy a ticket for this week's film is 10 minutes and the average time they wait to buy popcorn is 5 minutes. Assuming that the waiting times are independent, find the probability that a moviegoer waits a total of less than 20 minutes before taking his or her seat.

SOLUTION Assuming that both the waiting time X for the ticket purchase and the waiting time Y in the refreshment line are modeled by exponential probability density functions, we can write the individual density functions as

$$f_1(x) = \begin{cases} 0 & \text{if } x < 0 \\ \frac{1}{10}e^{-x/10} & \text{if } x \geq 0 \end{cases} \qquad f_2(y) = \begin{cases} 0 & \text{if } y < 0 \\ \frac{1}{5}e^{-y/5} & \text{if } y \geq 0 \end{cases}$$

Since X and Y are independent, the joint density function is the product:

$$f(x, y) = f_1(x)f_2(y) = \begin{cases} \frac{1}{50}e^{-x/10}e^{-y/5} & \text{if } x \geq 0, \ y \geq 0 \\ 0 & \text{otherwise} \end{cases}$$

We are asked for the probability that $X + Y < 20$:

$$P(X + Y < 20) = P((X, Y) \in D)$$

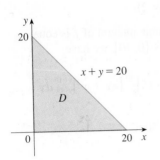

FIGURE 8

where D is the triangular region shown in Figure 8. Thus

$$P(X + Y < 20) = \iint\limits_D f(x, y) \, dA = \int_0^{20} \int_0^{20-x} \tfrac{1}{50}e^{-x/10}e^{-y/5} \, dy \, dx$$

$$= \tfrac{1}{50} \int_0^{20} \left[e^{-x/10}(-5)e^{-y/5} \right]_{y=0}^{y=20-x} dx$$

$$= \tfrac{1}{10} \int_0^{20} e^{-x/10}\left(1 - e^{(x-20)/5}\right) dx$$

$$= \tfrac{1}{10} \int_0^{20} \left(e^{-x/10} - e^{-4}e^{x/10}\right) dx$$

$$= 1 + e^{-4} - 2e^{-2} \approx 0.7476$$

This means that about 75% of the moviegoers wait less than 20 minutes before taking their seats. ■ ■

▪▪ Expected Values

Recall from Section 6.7 that if X is a random variable with probability density function f, then its *mean* is

$$\mu = \int_{-\infty}^{\infty} x f(x) \, dx$$

Now if X and Y are random variables with joint density function f, we define the **X-mean** and **Y-mean**, also called the **expected values** of X and Y, to be

9
$$\mu_1 = \iint\limits_{\mathbb{R}^2} x f(x, y) \, dA \qquad \mu_2 = \iint\limits_{\mathbb{R}^2} y f(x, y) \, dA$$

Notice how closely the expressions for μ_1 and μ_2 in (9) resemble the moments M_x and M_y of a lamina with density function ρ in Equations 3 and 4. In fact, we can think of probability as being like continuously distributed mass. We calculate probability the way we calculate mass—by integrating a density function. And because the total "probability mass" is 1, the expressions for $\bar{x}$ and $\bar{y}$ in (5) show that we can think of the expected values of X and Y, μ_1 and μ_2, as the coordinates of the "center of mass" of the probability distribution.

In the next example we deal with normal distributions. As in Section 6.7, a single random variable is *normally distributed* if its probability density function is of the form

$$f(x) = \frac{1}{\sigma\sqrt{2\pi}} e^{-(x-\mu)^2/(2\sigma^2)}$$

where μ is the mean and σ is the standard deviation.

EXAMPLE 7 A factory produces (cylindrically shaped) roller bearings that are sold as having diameter 4.0 cm and length 6.0 cm. In fact, the diameters X are normally distributed with mean 4.0 cm and standard deviation 0.01 cm while the lengths Y are normally distributed with mean 6.0 cm and standard deviation 0.01 cm. Assuming that X and Y are independent, write the joint density function and graph it. Find the probability that a bearing randomly chosen from the production line has either length or diameter that differs from the mean by more than 0.02 cm.

SOLUTION We are given that X and Y are normally distributed with $\mu_1 = 4.0$, $\mu_2 = 6.0$, and $\sigma_1 = \sigma_2 = 0.01$. So the individual density functions for X and Y are

$$f_1(x) = \frac{1}{0.01\sqrt{2\pi}} e^{-(x-4)^2/0.0002} \qquad f_2(y) = \frac{1}{0.01\sqrt{2\pi}} e^{-(y-6)^2/0.0002}$$

Since X and Y are independent, the joint density function is the product:

$$f(x, y) = f_1(x)f_2(y) = \frac{1}{0.0002\pi} e^{-(x-4)^2/0.0002} e^{-(y-6)^2/0.0002}$$

$$= \frac{5000}{\pi} e^{-5000[(x-4)^2+(y-6)^2]}$$

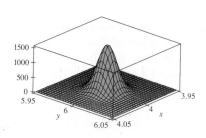

FIGURE 9

Graph of the bivariate normal joint density function in Example 7

A graph of this function is shown in Figure 9.

Let's first calculate the probability that both X and Y differ from their means by less than 0.02 cm. Using a calculator or computer to estimate the integral, we have

$$P(3.98 < X < 4.02, 5.98 < Y < 6.02) = \int_{3.98}^{4.02} \int_{5.98}^{6.02} f(x, y) \, dy \, dx$$

$$= \frac{5000}{\pi} \int_{3.98}^{4.02} \int_{5.98}^{6.02} e^{-5000[(x-4)^2+(y-6)^2]} \, dy \, dx$$

$$\approx 0.91$$

Then the probability that either X or Y differs from its mean by more than 0.02 cm is approximately

$$1 - 0.91 = 0.09$$

■ ■

12.5 Exercises

1. Electric charge is distributed over the rectangle $1 \leqslant x \leqslant 3$, $0 \leqslant y \leqslant 2$ so that the charge density at (x, y) is $\sigma(x, y) = 2xy + y^2$ (measured in coulombs per square meter). Find the total charge on the rectangle.

2. Electric charge is distributed over the disk $x^2 + y^2 \leqslant 4$ so that the charge density at (x, y) is $\sigma(x, y) = x + y + x^2 + y^2$ (measured in coulombs per square meter). Find the total charge on the disk.

3–10 ■ Find the mass and center of mass of the lamina that occupies the region D and has the given density function ρ.

3. $D = \{(x, y) \mid 0 \leqslant x \leqslant 2, -1 \leqslant y \leqslant 1\};$ $\rho(x, y) = xy^2$

4. $D = \{(x, y) \mid 0 \leqslant x \leqslant a, 0 \leqslant y \leqslant b\};$ $\rho(x, y) = cxy$

5. D is the triangular region with vertices $(0, 0)$, $(2, 1)$, $(0, 3)$; $\rho(x, y) = x + y$

6. D is the triangular region with vertices $(0, 0)$, $(1, 1)$, $(4, 0)$; $\rho(x, y) = x$

7. D is bounded by $y = e^x$, $y = 0$, $x = 0$, and $x = 1$; $\rho(x, y) = y$

8. D is bounded by $y = \sqrt{x}$, $y = 0$, and $x = 1$; $\rho(x, y) = x$

9. D is bounded by the parabola $x = y^2$ and the line $y = x - 2$; $\rho(x, y) = 3$

10. $D = \{(x, y) \mid 0 \leqslant y \leqslant \cos x, 0 \leqslant x \leqslant \pi/2\};$ $\rho(x, y) = x$

11. A lamina occupies the part of the disk $x^2 + y^2 \leqslant 1$ in the first quadrant. Find its center of mass if the density at any point is proportional to its distance from the x-axis.

12. Find the center of mass of the lamina in Exercise 11 if the density at any point is proportional to the square of its distance from the origin.

13. Find the center of mass of a lamina in the shape of an isosceles right triangle with equal sides of length a if the density at any point is proportional to the square of the distance from the vertex opposite the hypotenuse.

14. A lamina occupies the region inside the circle $x^2 + y^2 = 2y$ but outside the circle $x^2 + y^2 = 1$. Find the center of mass if the density at any point is inversely proportional to its distance from the origin.

15. Find the moments of inertia I_x, I_y, I_0 for the lamina of Exercise 7.

16. Find the moments of inertia I_x, I_y, I_0 for the lamina of Exercise 12.

17. Find the moments of inertia I_x, I_y, I_0 for the lamina of Exercise 13.

18. Consider a square fan blade with sides of length 2 and the lower left corner placed at the origin. If the density of the

blade is $\rho(x, y) = 1 + 0.1x$, is it more difficult to rotate the blade about the x-axis or the y-axis?

CAS **19–20** ■ Use a computer algebra system to find the mass, center of mass, and moments of inertia of the lamina that occupies the region D and has the given density function.

19. $D = \{(x, y) \mid 0 \leqslant y \leqslant \sin x, 0 \leqslant x \leqslant \pi\};$ $\rho(x, y) = xy$

20. D is enclosed by the cardioid $r = 1 + \cos \theta$; $\rho(x, y) = \sqrt{x^2 + y^2}$

21. The joint density function for a pair of random variables X and Y is
$$f(x, y) = \begin{cases} Cx(1 + y) & \text{if } 0 \leqslant x \leqslant 1, 0 \leqslant y \leqslant 2 \\ 0 & \text{otherwise} \end{cases}$$

(a) Find the value of the constant C.
(b) Find $P(X \leqslant 1, Y \leqslant 1)$.
(c) Find $P(X + Y \leqslant 1)$.

22. (a) Verify that
$$f(x, y) = \begin{cases} 4xy & \text{if } 0 \leqslant x \leqslant 1, 0 \leqslant y \leqslant 1 \\ 0 & \text{otherwise} \end{cases}$$

is a joint density function.
(b) If X and Y are random variables whose joint density function is the function f in part (a), find
(i) $P\left(X \geqslant \frac{1}{2}\right)$ (ii) $P\left(X \geqslant \frac{1}{2}, Y \leqslant \frac{1}{2}\right)$
(c) Find the expected values of X and Y.

23. Suppose X and Y are random variables with joint density function
$$f(x, y) = \begin{cases} 0.1e^{-(0.5x + 0.2y)} & \text{if } x \geqslant 0, y \geqslant 0 \\ 0 & \text{otherwise} \end{cases}$$

(a) Verify that f is indeed a joint density function.
(b) Find the following probabilities.
(i) $P(Y \geqslant 1)$ (ii) $P(X \leqslant 2, Y \leqslant 4)$
(c) Find the expected values of X and Y.

24. (a) A lamp has two bulbs of a type with an average lifetime of 1000 hours. Assuming that we can model the probability of failure of these bulbs by an exponential density function with mean $\mu = 1000$, find the probability that both of the lamp's bulbs fail within 1000 hours.
(b) Another lamp has just one bulb of the same type as in part (a). If one bulb burns out and is replaced by a bulb of the same type, find the probability that the two bulbs fail within a total of 1000 hours.

CAS **25.** Suppose that X and Y are independent random variables, where X is normally distributed with mean 45 and standard deviation 0.5 and Y is normally distributed with mean 20 and standard deviation 0.1.
(a) Find $P(40 \leqslant X \leqslant 50, 20 \leqslant Y \leqslant 25)$.
(b) Find $P(4(X - 45)^2 + 100(Y - 20)^2 \leqslant 2)$.

26. Xavier and Yolanda both have classes that end at noon and they agree to meet every day after class. They arrive at the coffee shop independently. Xavier's arrival time is X and Yolanda's arrival time is Y, where X and Y are measured in minutes after noon. The individual density functions are

$$f_1(x) = \begin{cases} e^{-x} & \text{if } x \geq 0 \\ 0 & \text{if } x < 0 \end{cases} \qquad f_2(y) = \begin{cases} \frac{1}{50}y & \text{if } 0 \leq y \leq 10 \\ 0 & \text{otherwise} \end{cases}$$

(Xavier arrives sometime after noon and is more likely to arrive promptly than late. Yolanda always arrives by 12:10 P.M. and is more likely to arrive late than promptly.) After Yolanda arrives, she'll wait for up to half an hour for Xavier, but he won't wait for her. Find the probability that they meet.

27. When studying the spread of an epidemic, we assume that the probability that an infected individual will spread the disease to an uninfected individual is a function of the distance between them. Consider a circular city of radius 10 mi in which the population is uniformly distributed. For an uninfected individual at a fixed point $A(x_0, y_0)$, assume that the probability function is given by

$$f(P) = \tfrac{1}{20}[20 - d(P, A)]$$

where $d(P, A)$ denotes the distance between P and A.

(a) Suppose the exposure of a person to the disease is the sum of the probabilities of catching the disease from all members of the population. Assume that the infected people are uniformly distributed throughout the city, with k infected individuals per square mile. Find a double integral that represents the exposure of a person residing at A.

(b) Evaluate the integral for the case in which A is the center of the city and for the case in which A is located on the edge of the city. Where would you prefer to live?

12.6 Surface Area

In this section we apply double integrals to the problem of computing the area of a surface. We start by finding a formula for the area of a parametric surface and then, as a special case, we deduce a formula for the surface area of the graph of a function of two variables.

We recall from Section 10.5 that a parametric surface S is defined by a vector-valued function of two parameters

1 $$\mathbf{r}(u, v) = x(u, v)\,\mathbf{i} + y(u, v)\,\mathbf{j} + z(u, v)\,\mathbf{k}$$

or, equivalently, by parametric equations

$$x = x(u, v) \qquad y = y(u, v) \qquad z = z(u, v)$$

where (u, v) varies throughout a region D in the uv-plane.

We will find the area of S by dividing S into patches and approximating the area of each patch by the area of a piece of a tangent plane. So first let's recall from Section 11.4 how to find tangent planes to parametric surfaces.

Let P_0 be a point on S with position vector $\mathbf{r}(u_0, v_0)$. If we keep u constant by putting $u = u_0$, then $\mathbf{r}(u_0, v)$ becomes a vector function of the single parameter v and defines a grid curve C_1 lying on S. (See Figure 1.) The tangent vector to C_1 at P_0 is obtained by taking the partial derivative of $\mathbf{r}$ with respect to v:

2 $$\mathbf{r}_v = \frac{\partial x}{\partial v}(u_0, v_0)\,\mathbf{i} + \frac{\partial y}{\partial v}(u_0, v_0)\,\mathbf{j} + \frac{\partial z}{\partial v}(u_0, v_0)\,\mathbf{k}$$

Similarly, if we keep v constant by putting $v = v_0$, we get a grid curve C_2 given by $\mathbf{r}(u, v_0)$ that lies on S, and its tangent vector at P_0 is

3 $$\mathbf{r}_u = \frac{\partial x}{\partial u}(u_0, v_0)\,\mathbf{i} + \frac{\partial y}{\partial u}(u_0, v_0)\,\mathbf{j} + \frac{\partial z}{\partial u}(u_0, v_0)\,\mathbf{k}$$

FIGURE 1

If the **normal vector** $\mathbf{r}_u \times \mathbf{r}_v$ is not $\mathbf{0}$, then the surface S is called **smooth**. (It has no "corners"). In this case the tangent plane to S at P_0 exists and can be found using the normal vector.

Now we define the surface area of a general parametric surface given by Equation 1. For simplicity we start by considering a surface whose parameter domain D is a rectangle, and we divide it into subrectangles R_{ij}. Let's choose (u_i^*, v_j^*) to be the lower left corner of R_{ij}. (See Figure 2.) The part S_{ij} of the surface S that corresponds to R_{ij} is called a *patch* and has the point P_{ij} with position vector $\mathbf{r}(u_i^*, v_j^*)$ as one of its corners. Let

$$\mathbf{r}_u^* = \mathbf{r}_u(u_i^*, v_j^*) \qquad \text{and} \qquad \mathbf{r}_v^* = \mathbf{r}_v(u_i^*, v_j^*)$$

be the tangent vectors at P_{ij} as given by Equations 3 and 2.

FIGURE 2

The image of the subrectangle R_{ij} is the patch S_{ij}.

Figure 3(a) shows how the two edges of the patch that meet at P_{ij} can be approximated by vectors. These vectors, in turn, can be approximated by the vectors $\Delta u \, \mathbf{r}_u^*$ and $\Delta v \, \mathbf{r}_v^*$ because partial derivatives can be approximated by difference quotients. So we approximate S_{ij} by the parallelogram determined by the vectors $\Delta u \, \mathbf{r}_u^*$ and $\Delta v \, \mathbf{r}_v^*$. This parallelogram is shown in Figure 3(b) and lies in the tangent plane to S at P_{ij}. The area of this parallelogram is

$$\left| (\Delta u \, \mathbf{r}_u^*) \times (\Delta v \, \mathbf{r}_v^*) \right| = \left| \mathbf{r}_u^* \times \mathbf{r}_v^* \right| \Delta u \, \Delta v$$

and so an approximation to the area of S is

$$\sum_{i=1}^{m} \sum_{j=1}^{n} \left| \mathbf{r}_u^* \times \mathbf{r}_v^* \right| \Delta u \, \Delta v$$

Our intuition tells us that this approximation gets better as we increase the number of subrectangles, and we recognize the double sum as a Riemann sum for the double integral $\iint_D \left| \mathbf{r}_u \times \mathbf{r}_v \right| du \, dv$. This motivates the following definition.

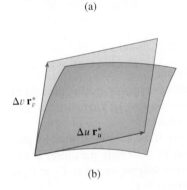

(a)

(b)

FIGURE 3

Approximating a patch by a parallelogram

> **4** **Definition** If a smooth parametric surface S is given by the equation
>
> $$\mathbf{r}(u, v) = x(u, v) \, \mathbf{i} + y(u, v) \, \mathbf{j} + z(u, v) \, \mathbf{k} \qquad (u, v) \in D$$
>
> and S is covered just once as (u, v) ranges throughout the parameter domain D, then the **surface area** of S is
>
> $$A(S) = \iint\limits_D \left| \mathbf{r}_u \times \mathbf{r}_v \right| dA$$
>
> where $\qquad \mathbf{r}_u = \dfrac{\partial x}{\partial u} \, \mathbf{i} + \dfrac{\partial y}{\partial u} \, \mathbf{j} + \dfrac{\partial z}{\partial u} \, \mathbf{k} \qquad \mathbf{r}_v = \dfrac{\partial x}{\partial v} \, \mathbf{i} + \dfrac{\partial y}{\partial v} \, \mathbf{j} + \dfrac{\partial z}{\partial v} \, \mathbf{k}$

EXAMPLE 1 Find the surface area of a sphere of radius a.

SOLUTION In Example 4 in Section 10.5 we found the parametric representation

$$x = a \sin \phi \cos \theta \qquad y = a \sin \phi \sin \theta \qquad z = a \cos \phi$$

where the parameter domain is

$$D = \left\{ (\phi, \theta) \mid 0 \leqslant \phi \leqslant \pi, \ 0 \leqslant \theta \leqslant 2\pi \right\}$$

We first compute the cross product of the tangent vectors:

$$\mathbf{r}_\phi \times \mathbf{r}_\theta = \begin{vmatrix} \mathbf{i} & \mathbf{j} & \mathbf{k} \\ \dfrac{\partial x}{\partial \phi} & \dfrac{\partial y}{\partial \phi} & \dfrac{\partial z}{\partial \phi} \\ \dfrac{\partial x}{\partial \theta} & \dfrac{\partial y}{\partial \theta} & \dfrac{\partial z}{\partial \theta} \end{vmatrix} = \begin{vmatrix} \mathbf{i} & \mathbf{j} & \mathbf{k} \\ a \cos \phi \cos \theta & a \cos \phi \sin \theta & -a \sin \phi \\ -a \sin \phi \sin \theta & a \sin \phi \cos \theta & 0 \end{vmatrix}$$

$$= a^2 \sin^2 \phi \, \cos \theta \, \mathbf{i} + a^2 \sin^2 \phi \, \sin \theta \, \mathbf{j} + a^2 \sin \phi \, \cos \phi \, \mathbf{k}$$

Thus

$$\left| \mathbf{r}_\phi \times \mathbf{r}_\theta \right| = \sqrt{a^4 \sin^4 \phi \, \cos^2 \theta + a^4 \sin^4 \phi \, \sin^2 \theta + a^4 \sin^2 \phi \, \cos^2 \phi}$$

$$= \sqrt{a^4 \sin^4 \phi + a^4 \sin^2 \phi \, \cos^2 \phi} = a^2 \sqrt{\sin^2 \phi} = a^2 \sin \phi$$

since $\sin \phi \geqslant 0$ for $0 \leqslant \phi \leqslant \pi$. Therefore, by Definition 4, the area of the sphere is

$$A = \iint_D \left| \mathbf{r}_\phi \times \mathbf{r}_\theta \right| dA = \int_0^{2\pi} \int_0^\pi a^2 \sin \phi \, d\phi \, d\theta$$

$$= a^2 \int_0^{2\pi} d\theta \int_0^\pi \sin \phi \, d\phi = a^2 (2\pi) 2 = 4\pi a^2$$

■■

▪▪ Surface Area of a Graph

For the special case of a surface S with equation $z = f(x, y)$, where (x, y) lies in D and f has continuous partial derivatives, we take x and y as parameters. The parametric equations are

$$x = x \qquad y = y \qquad z = f(x, y)$$

so

$$\mathbf{r}_x = \mathbf{i} + \left(\dfrac{\partial f}{\partial x} \right) \mathbf{k} \qquad \mathbf{r}_y = \mathbf{j} + \left(\dfrac{\partial f}{\partial y} \right) \mathbf{k}$$

and

$$\boxed{5} \qquad \mathbf{r}_x \times \mathbf{r}_y = \begin{vmatrix} \mathbf{i} & \mathbf{j} & \mathbf{k} \\ 1 & 0 & \dfrac{\partial f}{\partial x} \\ 0 & 1 & \dfrac{\partial f}{\partial y} \end{vmatrix} = -\dfrac{\partial f}{\partial x} \mathbf{i} - \dfrac{\partial f}{\partial y} \mathbf{j} + \mathbf{k}$$

■ ■ Notice the similarity between the surface area formula in Equation 6 and the arc length formula

$$L = \int_a^b \sqrt{1 + \left(\frac{dy}{dx}\right)^2}\, dx$$

from Section 6.3.

Thus, the surface area formula in Definition 4 becomes

6
$$A(S) = \iint_D \sqrt{1 + \left(\frac{\partial z}{\partial x}\right)^2 + \left(\frac{\partial z}{\partial y}\right)^2}\, dA$$

▼ **EXAMPLE 2** Find the area of the part of the paraboloid $z = x^2 + y^2$ that lies under the plane $z = 9$.

SOLUTION The plane intersects the paraboloid in the circle $x^2 + y^2 = 9$, $z = 9$. Therefore, the given surface lies above the disk D with center the origin and radius 3. (See Figure 4.) Using Formula 6, we have

$$A = \iint_D \sqrt{1 + \left(\frac{\partial z}{\partial x}\right)^2 + \left(\frac{\partial z}{\partial y}\right)^2}\, dA = \iint_D \sqrt{1 + (2x)^2 + (2y)^2}\, dA$$

$$= \iint_D \sqrt{1 + 4(x^2 + y^2)}\, dA$$

Converting to polar coordinates, we obtain

$$A = \int_0^{2\pi} \int_0^3 \sqrt{1 + 4r^2}\, r\, dr\, d\theta = \int_0^{2\pi} d\theta \int_0^3 r\sqrt{1 + 4r^2}\, dr$$

$$= 2\pi\left(\tfrac{1}{8}\right)\tfrac{2}{3}(1 + 4r^2)^{3/2}\big]_0^3 = \frac{\pi}{6}\left(37\sqrt{37} - 1\right)$$ ■ ■

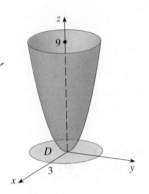

FIGURE 4

A common type of surface is a **surface of revolution** S obtained by rotating the curve $y = f(x)$, $a \leqslant x \leqslant b$, about the x-axis, where $f(x) \geqslant 0$ and f' is continuous. In Exercise 25 you are asked to use a parametric representation of S and Definition 4 to prove the following formula for the area of a surface of revolution:

7
$$A = 2\pi \int_a^b f(x)\sqrt{1 + [f'(x)]^2}\, dx$$

12.6 Exercises

1–12 ■ Find the area of the surface.

1. The part of the plane $z = 2 + 3x + 4y$ that lies above the rectangle $[0, 5] \times [1, 4]$

2. The part of the plane $2x + 5y + z = 10$ that lies inside the cylinder $x^2 + y^2 = 9$

3. The part of the plane $3x + 2y + z = 6$ that lies in the first octant

4. The part of the plane with vector equation $\mathbf{r}(u, v) = \langle 1 + v, u - 2v, 3 - 5u + v \rangle$ that is given by $0 \leqslant u \leqslant 1, 0 \leqslant v \leqslant 1$

5. The part of the hyperbolic paraboloid $z = y^2 - x^2$ that lies between the cylinders $x^2 + y^2 = 1$ and $x^2 + y^2 = 4$

6. The part of the surface $z = 1 + 3x + 2y^2$ that lies above the triangle with vertices $(0, 0)$, $(0, 1)$, and $(2, 1)$

7. The surface with parametric equations $x = u^2$, $y = uv$, $z = \tfrac{1}{2}v^2$, $0 \leqslant u \leqslant 1, 0 \leqslant v \leqslant 2$

8. The helicoid (or spiral ramp) with vector equation $\mathbf{r}(u, v) = u \cos v\,\mathbf{i} + u \sin v\,\mathbf{j} + v\,\mathbf{k}$, $0 \leqslant u \leqslant 1, 0 \leqslant v \leqslant \pi$

9. The part of the surface $y = 4x + z^2$ that lies between the planes $x = 0$, $x = 1$, $z = 0$, and $z = 1$

10. The part of the paraboloid $x = y^2 + z^2$ that lies inside the cylinder $y^2 + z^2 = 9$

11. The part of the surface $z = xy$ that lies within the cylinder $x^2 + y^2 = 1$

12. The surface $z = \frac{2}{3}(x^{3/2} + y^{3/2})$, $0 \le x \le 1$, $0 \le y \le 1$

- - - - - - - - - - - - - - - - -

13–14 ■ Find the area of the surface correct to four decimal places by expressing the area in terms of a single integral and using your calculator to estimate the integral.

13. The part of the surface $z = e^{-x^2-y^2}$ that lies above the disk $x^2 + y^2 \le 4$

14. The part of the surface $z = \cos(x^2 + y^2)$ that lies inside the cylinder $x^2 + y^2 = 1$

- - - - - - - - - - - - - - - - -

15. (a) Use the Midpoint Rule for double integrals (see Section 12.1) with six squares to estimate the area of the surface $z = 1/(1 + x^2 + y^2)$, $0 \le x \le 6$, $0 \le y \le 4$.

CAS (b) Use a computer algebra system to approximate the surface area in part (a) to four decimal places. Compare with the answer to part (a).

16. (a) Use the Midpoint Rule for double integrals with $m = n = 2$ to estimate the area of the surface $z = xy + x^2 + y^2$, $0 \le x \le 2$, $0 \le y \le 2$.

CAS (b) Use a computer algebra system to approximate the surface area in part (a) to four decimal places. Compare with the answer to part (a).

CAS **17.** Find the area of the surface with vector equation $\mathbf{r}(u, v) = \langle \cos^3 u \cos^3 v, \sin^3 u \cos^3 v, \sin^3 v \rangle$, $0 \le u \le \pi$, $0 \le v \le 2\pi$. State your answer correct to four decimal places.

CAS **18.** Find, to four decimal places, the area of the part of the surface $z = (1 + x^2)/(1 + y^2)$ that lies above the square $|x| + |y| \le 1$. Illustrate by graphing this part of the surface.

CAS **19.** Find the exact area of the surface $z = 1 + 2x + 3y + 4y^2$, $1 \le x \le 4$, $0 \le y \le 1$.

20. (a) Set up, but do not evaluate, a double integral for the area of the surface with parametric equations $x = au \cos v$, $y = bu \sin v$, $z = u^2$, $0 \le u \le 2$, $0 \le v \le 2\pi$.
(b) Eliminate the parameters to show that the surface is an elliptic paraboloid and set up another double integral for the surface area.
(c) Use the parametric equations in part (a) with $a = 2$ and $b = 3$ to graph the surface.
CAS (d) For the case $a = 2$, $b = 3$, use a computer algebra system to find the surface area correct to four decimal places.

21. (a) Show that the parametric equations $x = a \sin u \cos v$, $y = b \sin u \sin v$, $z = c \cos u$, $0 \le u \le \pi$, $0 \le v \le 2\pi$, represent an ellipsoid.
(b) Use the parametric equations in part (a) to graph the ellipsoid for the case $a = 1$, $b = 2$, $c = 3$.
(c) Set up, but do not evaluate, a double integral for the surface area of the ellipsoid in part (b).

22. (a) Show that the parametric equations $x = a \cosh u \cos v$, $y = b \cosh u \sin v$, $z = c \sinh u$, represent a hyperboloid of one sheet.
(b) Use the parametric equations in part (a) to graph the hyperboloid for the case $a = 1$, $b = 2$, $c = 3$.
(c) Set up, but do not evaluate, a double integral for the surface area of the part of the hyperboloid in part (b) that lies between the planes $z = -3$ and $z = 3$.

23. Find the area of the part of the sphere $x^2 + y^2 + z^2 = 4z$ that lies inside the paraboloid $z = x^2 + y^2$.

24. The figure shows the surface created when the cylinder $y^2 + z^2 = 1$ intersects the cylinder $x^2 + z^2 = 1$. Find the area of this surface.

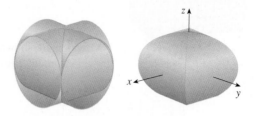

25. Use Definition 4 and the parametric equations for a surface of revolution (see Equations 10.5.3) to derive Formula 7.

26–27 ■ Use Formula 7 to find the area of the surface obtained by rotating the given curve about the x-axis.

26. $y = x^3$, $0 \le x \le 2$ **27.** $y = \sqrt{x}$, $4 \le x \le 9$

- - - - - - - - - - - - - - - - -

28. The figure shows the torus obtained by rotating about the z-axis the circle in the xz-plane with center $(b, 0, 0)$ and radius $a < b$. Parametric equations for the torus are

$$x = b \cos \theta + a \cos \alpha \cos \theta$$

$$y = b \sin \theta + a \cos \alpha \sin \theta$$

$$z = a \sin \alpha$$

where θ and α are the angles shown in the figure. Find the surface area of the torus.

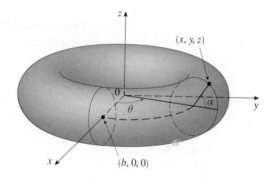

12.7 Triple Integrals

Just as we defined single integrals for functions of one variable and double integrals for functions of two variables, so we can define triple integrals for functions of three variables. Let's first deal with the simplest case where f is defined on a rectangular box:

$$\boxed{1} \qquad B = \{(x, y, z) \mid a \leqslant x \leqslant b,\ c \leqslant y \leqslant d,\ r \leqslant z \leqslant s\}$$

The first step is to divide B into sub-boxes. We do this by dividing the interval $[a, b]$ into l subintervals $[x_{i-1}, x_i]$ of equal width Δx, dividing $[c, d]$ into m subintervals of width Δy, and dividing $[r, s]$ into n subintervals of width Δz. The planes through the endpoints of these subintervals parallel to the coordinate planes divide the box B into lmn sub-boxes

$$B_{ijk} = [x_{i-1}, x_i] \times [y_{j-1}, y_j] \times [z_{k-1}, z_k]$$

which are shown in Figure 1. Each sub-box has volume $\Delta V = \Delta x\, \Delta y\, \Delta z$.

Then we form the **triple Riemann sum**

$$\boxed{2} \qquad \sum_{i=1}^{l} \sum_{j=1}^{m} \sum_{k=1}^{n} f(x_{ijk}^*, y_{ijk}^*, z_{ijk}^*)\, \Delta V$$

where the sample point $(x_{ijk}^*, y_{ijk}^*, z_{ijk}^*)$ is in B_{ijk}. By analogy with the definition of a double integral (12.1.5), we define the triple integral as the limit of the triple Riemann sums in (2).

> **3 Definition** The **triple integral** of f over the box B is
>
> $$\iiint\limits_{B} f(x, y, z)\, dV = \lim_{l, m, n \to \infty} \sum_{i=1}^{l} \sum_{j=1}^{m} \sum_{k=1}^{n} f(x_{ijk}^*, y_{ijk}^*, z_{ijk}^*)\, \Delta V$$
>
> if this limit exists.

FIGURE 1

Again, the triple integral always exists if f is continuous. We can choose the sample point to be any point in the sub-box, but if we choose it to be the point (x_i, y_j, z_k) we get a simpler-looking expression for the triple integral:

$$\iiint\limits_{B} f(x, y, z)\, dV = \lim_{l, m, n \to \infty} \sum_{i=1}^{l} \sum_{j=1}^{m} \sum_{k=1}^{n} f(x_i, y_j, z_k)\, \Delta V$$

Just as for double integrals, the practical method for evaluating triple integrals is to express them as iterated integrals as follows.

> **4 Fubini's Theorem for Triple Integrals** If f is continuous on the rectangular box $B = [a, b] \times [c, d] \times [r, s]$, then
>
> $$\iiint\limits_{B} f(x, y, z)\, dV = \int_{r}^{s} \int_{c}^{d} \int_{a}^{b} f(x, y, z)\, dx\, dy\, dz$$

The iterated integral on the right side of Fubini's Theorem means that we integrate first with respect to x (keeping y and z fixed), then we integrate with respect to y (keeping z fixed), and finally we integrate with respect to z. There are five other possible orders in which we can integrate, all of which give the same value. For instance, if we integrate with respect to y, then z, and then x, we have

$$\iiint_B f(x, y, z)\, dV = \int_a^b \int_r^s \int_c^d f(x, y, z)\, dy\, dz\, dx$$

▽ **EXAMPLE 1** Evaluate the triple integral $\iiint_B xyz^2\, dV$, where B is the rectangular box given by

$$B = \{(x, y, z) \mid 0 \le x \le 1,\ -1 \le y \le 2,\ 0 \le z \le 3\}$$

SOLUTION We could use any of the six possible orders of integration. If we choose to integrate with respect to x, then y, and then z, we obtain

$$\iiint_B xyz^2\, dV = \int_0^3 \int_{-1}^2 \int_0^1 xyz^2\, dx\, dy\, dz = \int_0^3 \int_{-1}^2 \left[\frac{x^2 yz^2}{2} \right]_{x=0}^{x=1} dy\, dz$$

$$= \int_0^3 \int_{-1}^2 \frac{yz^2}{2}\, dy\, dz = \int_0^3 \left[\frac{y^2 z^2}{4} \right]_{y=-1}^{y=2} dz$$

$$= \int_0^3 \frac{3z^2}{4}\, dz = \frac{z^3}{4} \Big]_0^3 = \frac{27}{4}$$

■ ■

Now we define the **triple integral over a general bounded region E** in three-dimensional space (a solid) by much the same procedure that we used for double integrals (12.3.2). We enclose E in a box B of the type given by Equation 1. Then we define a function F so that it agrees with f on E but is 0 for points in B that are outside E. By definition,

$$\iiint_E f(x, y, z)\, dV = \iiint_B F(x, y, z)\, dV$$

This integral exists if f is continuous and the boundary of E is "reasonably smooth." The triple integral has essentially the same properties as the double integral (Properties 6–9 in Section 12.3).

We restrict our attention to continuous functions f and to certain simple types of regions. A solid region E is said to be of **type 1** if it lies between the graphs of two continuous functions of x and y, that is,

$$\boxed{5} \qquad E = \{(x, y, z) \mid (x, y) \in D,\ u_1(x, y) \le z \le u_2(x, y)\}$$

where D is the projection of E onto the xy-plane as shown in Figure 2. Notice that the upper boundary of the solid E is the surface with equation $z = u_2(x, y)$, while the lower boundary is the surface $z = u_1(x, y)$.

By the same sort of argument that led to (12.3.3), it can be shown that if E is a type 1 region given by Equation 5, then

$$\boxed{6} \qquad \iiint_E f(x, y, z)\, dV = \iint_D \left[\int_{u_1(x, y)}^{u_2(x, y)} f(x, y, z)\, dz \right] dA$$

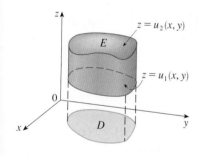

FIGURE 2

A type 1 solid region

FIGURE 3
A type 1 solid region

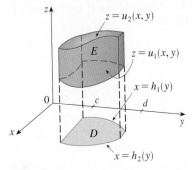

FIGURE 4
Another type 1 solid region

FIGURE 5

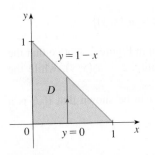

FIGURE 6

The meaning of the inner integral on the right side of Equation 6 is that x and y are held fixed, and therefore $u_1(x, y)$ and $u_2(x, y)$ are regarded as constants, while $f(x, y, z)$ is integrated with respect to z.

In particular, if the projection D of E onto the xy-plane is a type I plane region (as in Figure 3), then

$$E = \{(x, y, z) \mid a \leq x \leq b,\ g_1(x) \leq y \leq g_2(x),\ u_1(x, y) \leq z \leq u_2(x, y)\}$$

and Equation 6 becomes

7
$$\iiint\limits_{E} f(x, y, z)\, dV = \int_a^b \int_{g_1(x)}^{g_2(x)} \int_{u_1(x, y)}^{u_2(x, y)} f(x, y, z)\, dz\, dy\, dx$$

If, on the other hand, D is a type II plane region (as in Figure 4), then

$$E = \{(x, y, z) \mid c \leq y \leq d,\ h_1(y) \leq x \leq h_2(y),\ u_1(x, y) \leq z \leq u_2(x, y)\}$$

and Equation 6 becomes

8
$$\iiint\limits_{E} f(x, y, z)\, dV = \int_c^d \int_{h_1(y)}^{h_2(y)} \int_{u_1(x, y)}^{u_2(x, y)} f(x, y, z)\, dz\, dx\, dy$$

EXAMPLE 2 Evaluate $\iiint_E z\, dV$, where E is the solid tetrahedron bounded by the four planes $x = 0$, $y = 0$, $z = 0$, and $x + y + z = 1$.

SOLUTION When we set up a triple integral it's wise to draw *two* diagrams: one of the solid region E (see Figure 5) and one of its projection D on the xy-plane (see Figure 6). The lower boundary of the tetrahedron is the plane $z = 0$ and the upper boundary is the plane $x + y + z = 1$ (or $z = 1 - x - y$), so we use $u_1(x, y) = 0$ and $u_2(x, y) = 1 - x - y$ in Formula 7. Notice that the planes $x + y + z = 1$ and $z = 0$ intersect in the line $x + y = 1$ (or $y = 1 - x$) in the xy-plane. So the projection of E is the triangular region shown in Figure 6, and we have

9 $\quad E = \{(x, y, z) \mid 0 \leq x \leq 1,\ 0 \leq y \leq 1 - x,\ 0 \leq z \leq 1 - x - y\}$

This description of E as a type 1 region enables us to evaluate the integral as follows:

$$\iiint\limits_{E} z\, dV = \int_0^1 \int_0^{1-x} \int_0^{1-x-y} z\, dz\, dy\, dx = \int_0^1 \int_0^{1-x} \left[\frac{z^2}{2}\right]_{z=0}^{z=1-x-y} dy\, dx$$

$$= \frac{1}{2} \int_0^1 \int_0^{1-x} (1 - x - y)^2\, dy\, dx$$

$$= \frac{1}{2} \int_0^1 \left[-\frac{(1 - x - y)^3}{3}\right]_{y=0}^{y=1-x} dx$$

$$= \frac{1}{6} \int_0^1 (1 - x)^3\, dx = \frac{1}{6}\left[-\frac{(1 - x)^4}{4}\right]_0^1 = \frac{1}{24}$$

■ ■

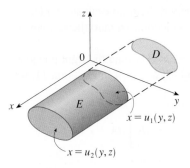

FIGURE 7
A type 2 region

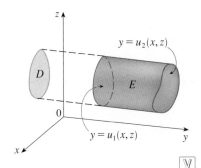

FIGURE 8
A type 3 region

Visual 12.7 illustrates how solid regions (including the one in Figure 9) project onto coordinate planes.

A solid region E is of **type 2** if it is of the form

$$E = \{(x, y, z) \mid (y, z) \in D,\ u_1(y, z) \leqslant x \leqslant u_2(y, z)\}$$

where, this time, D is the projection of E onto the yz-plane (see Figure 7). The back surface is $x = u_1(y, z)$, the front surface is $x = u_2(y, z)$, and we have

$$\boxed{10} \qquad \iiint_E f(x, y, z)\, dV = \iint_D \left[\int_{u_1(y, z)}^{u_2(y, z)} f(x, y, z)\, dx \right] dA$$

Finally, a **type 3** region is of the form

$$E = \{(x, y, z) \mid (x, z) \in D,\ u_1(x, z) \leqslant y \leqslant u_2(x, z)\}$$

where D is the projection of E onto the xz-plane, $y = u_1(x, z)$ is the left surface, and $y = u_2(x, z)$ is the right surface (see Figure 8). For this type of region we have

$$\boxed{11} \qquad \iiint_E f(x, y, z)\, dV = \iint_D \left[\int_{u_1(x, z)}^{u_2(x, z)} f(x, y, z)\, dy \right] dA$$

In each of Equations 10 and 11 there may be two possible expressions for the integral depending on whether D is a type I or type II plane region (and corresponding to Equations 7 and 8).

EXAMPLE 3 Evaluate $\iiint_E \sqrt{x^2 + z^2}\, dV$, where E is the region bounded by the paraboloid $y = x^2 + z^2$ and the plane $y = 4$.

SOLUTION The solid E is shown in Figure 9. If we regard it as a type 1 region, then we need to consider its projection D_1 onto the xy-plane, which is the parabolic region in Figure 10. (The trace of $y = x^2 + z^2$ in the plane $z = 0$ is the parabola $y = x^2$.)

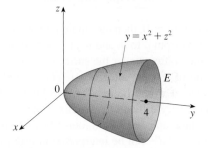

FIGURE 9
Region of integration

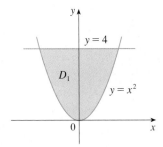

FIGURE 10
Projection on xy-plane

From $y = x^2 + z^2$ we obtain $z = \pm\sqrt{y - x^2}$, so the lower boundary surface of E is $z = -\sqrt{y - x^2}$ and the upper surface is $z = \sqrt{y - x^2}$. Therefore, the description of E as a type 1 region is

$$E = \{(x, y, z) \mid -2 \leqslant x \leqslant 2,\ x^2 \leqslant y \leqslant 4,\ -\sqrt{y - x^2} \leqslant z \leqslant \sqrt{y - x^2}\}$$

and so we obtain

$$\iiint_E \sqrt{x^2 + z^2}\, dV = \int_{-2}^{2} \int_{x^2}^{4} \int_{-\sqrt{y - x^2}}^{\sqrt{y - x^2}} \sqrt{x^2 + z^2}\, dz\, dy\, dx$$

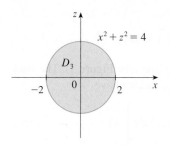

FIGURE 11
Projection on xz-plane

Although this expression is correct, it is extremely difficult to evaluate. So let's instead consider E as a type 3 region. As such, its projection D_3 onto the xz-plane is the disk $x^2 + z^2 \le 4$ shown in Figure 11.

Then the left boundary of E is the paraboloid $y = x^2 + z^2$ and the right boundary is the plane $y = 4$, so taking $u_1(x, z) = x^2 + z^2$ and $u_2(x, z) = 4$ in Equation 11, we have

$$\iiint_E \sqrt{x^2 + z^2}\, dV = \iint_{D_3} \left[\int_{x^2+z^2}^4 \sqrt{x^2 + z^2}\, dy \right] dA$$

$$= \iint_{D_3} (4 - x^2 - z^2)\sqrt{x^2 + z^2}\, dA$$

Although this integral could be written as

$$\int_{-2}^2 \int_{-\sqrt{4-x^2}}^{\sqrt{4-x^2}} (4 - x^2 - z^2)\sqrt{x^2 + z^2}\, dz\, dx$$

⊘ The most difficult step in evaluating a triple integral is setting up an expression for the region of integration (such as Equation 9 in Example 2). Remember that the limits of integration in the inner integral contain at most two variables, the limits of integration in the middle integral contain at most one variable, and the limits of integration in the outer integral must be constants.

it's easier to convert to polar coordinates in the xz-plane: $x = r \cos \theta$, $z = r \sin \theta$. This gives

$$\iiint_E \sqrt{x^2 + z^2}\, dV = \iint_{D_3} (4 - x^2 - z^2)\sqrt{x^2 + z^2}\, dA$$

$$= \int_0^{2\pi} \int_0^2 (4 - r^2) r\, r\, dr\, d\theta = \int_0^{2\pi} d\theta \int_0^2 (4r^2 - r^4)\, dr$$

$$= 2\pi \left[\frac{4r^3}{3} - \frac{r^5}{5} \right]_0^2 = \frac{128\pi}{15}$$ ◼ ◼

◼◼ Applications of Triple Integrals

Recall that if $f(x) \ge 0$, then the single integral $\int_a^b f(x)\, dx$ represents the area under the curve $y = f(x)$ from a to b, and if $f(x, y) \ge 0$, then the double integral $\iint_D f(x, y)\, dA$ represents the volume under the surface $z = f(x, y)$ and above D. The corresponding interpretation of a triple integral $\iiint_E f(x, y, z)\, dV$, where $f(x, y, z) \ge 0$, is not very useful because it would be the "hypervolume" of a four-dimensional object and, of course, that is very difficult to visualize. (Remember that E is just the *domain* of the function f; the graph of f lies in four-dimensional space.) Nonetheless, the triple integral $\iiint_E f(x, y, z)\, dV$ can be interpreted in different ways in different physical situations, depending on the physical interpretations of x, y, z and $f(x, y, z)$.

Let's begin with the special case where $f(x, y, z) = 1$ for all points in E. Then the triple integral does represent the volume of E:

12

$$V(E) = \iiint_E dV$$

For example, you can see this in the case of a type 1 region by putting $f(x, y, z) = 1$ in Formula 6:

$$\iiint_E 1\, dV = \iint_D \left[\int_{u_1(x, y)}^{u_2(x, y)} dz \right] dA = \iint_D [u_2(x, y) - u_1(x, y)]\, dA$$

and from Section 12.3 we know this represents the volume that lies between the surfaces $z = u_1(x, y)$ and $z = u_2(x, y)$.

EXAMPLE 4 Use a triple integral to find the volume of the tetrahedron T bounded by the planes $x + 2y + z = 2$, $x = 2y$, $x = 0$, and $z = 0$.

SOLUTION The tetrahedron T and its projection D on the xy-plane are shown in Figures 12 and 13. The lower boundary of T is the plane $z = 0$ and the upper boundary is the plane $x + 2y + z = 2$, that is, $z = 2 - x - 2y$. Therefore, we have

$$V(T) = \iiint_T dV = \int_0^1 \int_{x/2}^{1-x/2} \int_0^{2-x-2y} dz\, dy\, dx$$

$$= \int_0^1 \int_{x/2}^{1-x/2} (2 - x - 2y)\, dy\, dx = \tfrac{1}{3}$$

by the same calculation as in Example 4 in Section 12.3.

(Notice that it is not necessary to use triple integrals to compute volumes. They simply give an alternative method for setting up the calculation.) ■ ■

FIGURE 12

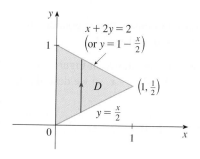

FIGURE 13

All the applications of double integrals in Section 12.5 can be immediately extended to triple integrals. For example, if the density function of a solid object that occupies the region E is $\rho(x, y, z)$, in units of mass per unit volume, at any given point (x, y, z), then its **mass** is

$$\boxed{13} \qquad m = \iiint_E \rho(x, y, z)\, dV$$

and its **moments** about the three coordinate planes are

$$\boxed{14} \qquad M_{yz} = \iiint_E x\rho(x, y, z)\, dV \qquad M_{xz} = \iiint_E y\rho(x, y, z)\, dV$$

$$M_{xy} = \iiint_E z\rho(x, y, z)\, dV$$

The **center of mass** is located at the point $(\bar{x}, \bar{y}, \bar{z})$, where

$$\boxed{15} \qquad \bar{x} = \frac{M_{yz}}{m} \qquad \bar{y} = \frac{M_{xz}}{m} \qquad \bar{z} = \frac{M_{xy}}{m}$$

If the density is constant, the center of mass of the solid is called the **centroid** of E. The **moments of inertia** about the three coordinate axes are

$$\boxed{16} \quad I_x = \iiint_E (y^2 + z^2)\rho(x, y, z)\, dV \qquad I_y = \iiint_E (x^2 + z^2)\rho(x, y, z)\, dV$$

$$I_z = \iiint_E (x^2 + y^2)\rho(x, y, z)\, dV$$

As in Section 12.5, the total **electric charge** on a solid object occupying a region

E and having charge density $\sigma(x, y, z)$ is

$$Q = \iiint_E \sigma(x, y, z) \, dV$$

If we have three continuous random variables X, Y, and Z, their **joint density function** is a function of three variables such that the probability that (X, Y, Z) lies in E is

$$P((X, Y, Z) \in E) = \iiint_E f(x, y, z) \, dV$$

In particular,

$$P(a \leq X \leq b, \ c \leq Y \leq d, \ r \leq Z \leq s) = \int_a^b \int_c^d \int_r^s f(x, y, z) \, dz \, dy \, dx$$

The joint density function satisfies

$$f(x, y, z) \geq 0 \qquad \int_{-\infty}^{\infty} \int_{-\infty}^{\infty} \int_{-\infty}^{\infty} f(x, y, z) \, dz \, dy \, dx = 1$$

$\boxed{\text{V}}$ **EXAMPLE 5** Find the center of mass of a solid of constant density that is bounded by the parabolic cylinder $x = y^2$ and the planes $x = z$, $z = 0$, and $x = 1$.

SOLUTION The solid E and its projection onto the xy-plane are shown in Figure 14. The lower and upper surfaces of E are the planes $z = 0$ and $z = x$, so we describe E as a type 1 region:

$$E = \left\{ (x, y, z) \mid -1 \leq y \leq 1, \ y^2 \leq x \leq 1, \ 0 \leq z \leq x \right\}$$

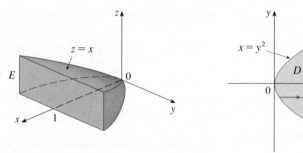

FIGURE 14

Then, if the density is $\rho(x, y, z) = \rho$, the mass is

$$m = \iiint_E \rho \, dV = \int_{-1}^1 \int_{y^2}^1 \int_0^x \rho \, dz \, dx \, dy$$

$$= \rho \int_{-1}^1 \int_{y^2}^1 x \, dx \, dy = \rho \int_{-1}^1 \left[\frac{x^2}{2} \right]_{x=y^2}^{x=1} dy$$

$$= \frac{\rho}{2} \int_{-1}^1 (1 - y^4) \, dy = \rho \int_0^1 (1 - y^4) \, dy$$

$$= \rho \left[y - \frac{y^5}{5} \right]_0^1 = \frac{4\rho}{5}$$

Because of the symmetry of E and ρ about the xz-plane, we can immediately say that $M_{xz} = 0$ and, therefore, $\bar{y} = 0$. The other moments are

$$M_{yz} = \iiint_E x\rho \, dV = \int_{-1}^1 \int_{y^2}^1 \int_0^x x\rho \, dz \, dx \, dy$$

$$= \rho \int_{-1}^1 \int_{y^2}^1 x^2 \, dx \, dy = \rho \int_{-1}^1 \left[\frac{x^3}{3} \right]_{x=y^2}^{x=1} dy$$

$$= \frac{2\rho}{3} \int_0^1 (1 - y^6) \, dy = \frac{2\rho}{3} \left[y - \frac{y^7}{7} \right]_0^1 = \frac{4\rho}{7}$$

$$M_{xy} = \iiint_E z\rho \, dV = \int_{-1}^1 \int_{y^2}^1 \int_0^x z\rho \, dz \, dx \, dy$$

$$= \rho \int_{-1}^1 \int_{y^2}^1 \left[\frac{z^2}{2} \right]_{z=0}^{z=x} dx \, dy = \frac{\rho}{2} \int_{-1}^1 \int_{y^2}^1 x^2 \, dx \, dy$$

$$= \frac{\rho}{3} \int_0^1 (1 - y^6) \, dy = \frac{2\rho}{7}$$

Therefore, the center of mass is

$$(\bar{x}, \bar{y}, \bar{z}) = \left(\frac{M_{yz}}{m}, \frac{M_{xz}}{m}, \frac{M_{xy}}{m} \right) = \left(\tfrac{5}{7}, 0, \tfrac{5}{14} \right)$$

■ ■

12.7 Exercises

1. Evaluate the integral in Example 1, integrating first with respect to z, then x, and then y.

2. Evaluate the integral $\iiint_E (xz - y^3) \, dV$, where
$$E = \{(x, y, z) \mid -1 \leqslant x \leqslant 1, 0 \leqslant y \leqslant 2, 0 \leqslant z \leqslant 1\}$$
using three different orders of integration.

3–6 ■ Evaluate the iterated integral.

3. $\int_0^1 \int_0^z \int_0^{x+z} 6xz \, dy \, dx \, dz$

4. $\int_0^1 \int_x^{2x} \int_0^y 2xyz \, dz \, dy \, dx$

5. $\int_0^3 \int_0^1 \int_0^{\sqrt{1-z^2}} ze^y \, dx \, dz \, dy$

6. $\int_0^1 \int_0^z \int_0^y ze^{-y^2} \, dx \, dy \, dz$

. . . .

7–16 ■ Evaluate the triple integral.

7. $\iiint_E 2x \, dV$, where
$$E = \{(x, y, z) \mid 0 \leqslant y \leqslant 2, 0 \leqslant x \leqslant \sqrt{4 - y^2}, 0 \leqslant z \leqslant y\}$$

8. $\iiint_E yz \cos(x^5) \, dV$, where
$$E = \{(x, y, z) \mid 0 \leqslant x \leqslant 1, 0 \leqslant y \leqslant x, x \leqslant z \leqslant 2x\}$$

9. $\iiint_E 6xy \, dV$, where E lies under the plane $z = 1 + x + y$ and above the region in the xy-plane bounded by the curves $y = \sqrt{x}$, $y = 0$, and $x = 1$

10. $\iiint_E y \, dV$, where E is bounded by the planes $x = 0$, $y = 0$, $z = 0$, and $2x + 2y + z = 4$

11. $\iiint_E xy \, dV$, where E is the solid tetrahedron with vertices $(0, 0, 0)$, $(1, 0, 0)$, $(0, 2, 0)$, and $(0, 0, 3)$

12. $\iiint_E xz \, dV$, where E is the solid tetrahedron with vertices $(0, 0, 0)$, $(0, 1, 0)$, $(1, 1, 0)$, and $(0, 1, 1)$

13. $\iiint_E x^2 e^y \, dV$, where E is bounded by the parabolic cylinder $z = 1 - y^2$ and the planes $z = 0$, $x = 1$, and $x = -1$

14. $\iiint_E (x + 2y) \, dV$, where E is bounded by the parabolic cylinder $y = x^2$ and the planes $x = z$, $x = y$, and $z = 0$

15. $\iiint_E x \, dV$, where E is bounded by the paraboloid $x = 4y^2 + 4z^2$ and the plane $x = 4$

16. $\iiint_E z \, dV$, where E is bounded by the cylinder $y^2 + z^2 = 9$ and the planes $x = 0$, $y = 3x$, and $z = 0$ in the first octant

. . . .

17–20 ■ Use a triple integral to find the volume of the given solid.

17. The tetrahedron enclosed by the coordinate planes and the plane $2x + y + z = 4$

18. The solid bounded by the cylinder $y = x^2$ and the planes $z = 0$, $z = 4$, and $y = 9$

19. The solid enclosed by the cylinder $x^2 + y^2 = 9$ and the planes $y + z = 5$ and $z = 1$

20. The solid enclosed by the paraboloid $x = y^2 + z^2$ and the plane $x = 16$

. . . .

21. (a) Express the volume of the wedge in the first octant that is cut from the cylinder $y^2 + z^2 = 1$ by the planes $y = x$ and $x = 1$ as a triple integral.

(CAS) (b) Use either the Table of Integrals (on Reference Pages 6–10) or a computer algebra system to find the exact value of the triple integral in part (a).

22. (a) In the **Midpoint Rule for triple integrals** we use a triple Riemann sum to approximate a triple integral over a box B, where $f(x, y, z)$ is evaluated at the center $(\bar{x}_i, \bar{y}_j, \bar{z}_k)$ of the box B_{ijk}. Use the Midpoint Rule to estimate $\iiint_B \sqrt{x^2 + y^2 + z^2}\, dV$, where B is the cube defined by $0 \le x \le 4$, $0 \le y \le 4$, $0 \le z \le 4$. Divide B into eight cubes of equal size.

(CAS) (b) Use a computer algebra system to approximate the integral in part (a) correct to the nearest integer. Compare with the answer to part (a).

23–24 ■ Use the Midpoint Rule for triple integrals (Exercise 22) to estimate the value of the integral. Divide B into eight sub-boxes of equal size.

23. $\iiint_B \dfrac{1}{\ln(1 + x + y + z)}\, dV$, where
$B = \{(x, y, z) \mid 0 \le x \le 4,\ 0 \le y \le 8,\ 0 \le z \le 4\}$

24. $\iiint_B \sin(xy^2 z^3)\, dV$, where
$B = \{(x, y, z) \mid 0 \le x \le 4,\ 0 \le y \le 2,\ 0 \le z \le 1\}$

. . . .

25–26 ■ Sketch the solid whose volume is given by the iterated integral.

25. $\displaystyle\int_0^1 \int_0^{1-x} \int_0^{2-2z} dy\, dz\, dx$

26. $\displaystyle\int_0^2 \int_0^{2-y} \int_0^{4-y^2} dx\, dz\, dy$

. . . .

27–30 ■ Express the integral $\iiint_E f(x, y, z)\, dV$ as an iterated integral in six different ways, where E is the solid bounded by the given surfaces.

27. $x^2 + z^2 = 4,\quad y = 0,\quad y = 6$

28. $z = 0,\quad x = 0,\quad y = 2,\quad z = y - 2x$

29. $z = 0,\quad z = y,\quad x^2 = 1 - y$

30. $9x^2 + 4y^2 + z^2 = 1$

. . . .

31. The figure shows the region of integration for the integral

$$\int_0^1 \int_{\sqrt{x}}^1 \int_0^{1-y} f(x, y, z)\, dz\, dy\, dx$$

Rewrite this integral as an equivalent iterated integral in the five other orders.

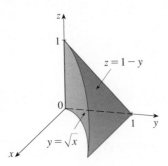

32. The figure shows the region of integration for the integral

$$\int_0^1 \int_0^{1-x^2} \int_0^{1-x} f(x, y, z)\, dy\, dz\, dx$$

Rewrite this integral as an equivalent iterated integral in the five other orders.

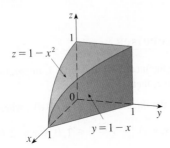

33–34 ■ Write five other iterated integrals that are equal to the given iterated integral.

33. $\displaystyle\int_0^1 \int_y^1 \int_0^y f(x, y, z)\, dz\, dx\, dy$

34. $\displaystyle\int_0^1 \int_0^{x^2} \int_0^y f(x, y, z)\, dz\, dy\, dx$

. . . .

35–38 ■ Find the mass and center of mass of the solid E with the given density function ρ.

35. E is the solid of Exercise 9; $\rho(x, y, z) = 2$

36. E is bounded by the parabolic cylinder $z = 1 - y^2$ and the planes $x + z = 1$, $x = 0$, and $z = 0$; $\rho(x, y, z) = 4$

37. E is the cube given by $0 \le x \le a$, $0 \le y \le a$, $0 \le z \le a$; $\rho(x, y, z) = x^2 + y^2 + z^2$

38. E is the tetrahedron bounded by the planes $x = 0$, $y = 0$, $z = 0$, $x + y + z = 1$; $\rho(x, y, z) = y$

39–40 ■ Set up, but do not evaluate, integral expressions for (a) the mass, (b) the center of mass, and (c) the moment of inertia about the z-axis.

39. The solid of Exercise 19; $\rho(x, y, z) = \sqrt{x^2 + y^2}$

40. The hemisphere $x^2 + y^2 + z^2 \leqslant 1$, $z \geqslant 0$;
$\rho(x, y, z) = \sqrt{x^2 + y^2 + z^2}$

41. Let E be the solid in the first octant bounded by the cylinder $x^2 + y^2 = 1$ and the planes $y = z$, $x = 0$, and $z = 0$ with the density function $\rho(x, y, z) = 1 + x + y + z$. Use a computer algebra system to find the exact values of the following quantities for E.
(a) The mass
(b) The center of mass
(c) The moment of inertia about the z-axis

42. If E is the solid of Exercise 16 with density function $\rho(x, y, z) = x^2 + y^2$, find the following quantities, correct to three decimal places.
(a) The mass
(b) The center of mass
(c) The moment of inertia about the z-axis

43. Find the moments of inertia for a cube of constant density k and side length L if one vertex is located at the origin and three edges lie along the coordinate axes.

44. Find the moments of inertia for a rectangular brick with dimensions a, b, and c, mass M, and constant density if the center of the brick is situated at the origin and the edges are parallel to the coordinate axes.

45. The joint density function for random variables X, Y, and Z is $f(x, y, z) = Cxyz$ if $0 \leqslant x \leqslant 2$, $0 \leqslant y \leqslant 2$, $0 \leqslant z \leqslant 2$, and $f(x, y, z) = 0$ otherwise.
(a) Find the value of the constant C.
(b) Find $P(X \leqslant 1, Y \leqslant 1, Z \leqslant 1)$.
(c) Find $P(X + Y + Z \leqslant 1)$.

46. Suppose X, Y, and Z are random variables with joint density function $f(x, y, z) = Ce^{-(0.5x+0.2y+0.1z)}$ if $x \geqslant 0$, $y \geqslant 0$, $z \geqslant 0$, and $f(x, y, z) = 0$ otherwise.
(a) Find the value of the constant C.
(b) Find $P(X \leqslant 1, Y \leqslant 1)$.
(c) Find $P(X \leqslant 1, Y \leqslant 1, Z \leqslant 1)$.

47–48 ■ The **average value** of a function $f(x, y, z)$ over a solid region E is defined to be

$$f_{ave} = \frac{1}{V(E)} \iiint_E f(x, y, z) \, dV$$

where $V(E)$ is the volume of E. For instance, if ρ is a density function, then ρ_{ave} is the average density of E.

47. Find the average value of the function $f(x, y, z) = xyz$ over the cube with side length L that lies in the first octant with one vertex at the origin and edges parallel to the coordinate axes.

48. Find the average value of the function $f(x, y, z) = x^2 z + y^2 z$ over the region enclosed by the paraboloid $z = 1 - x^2 - y^2$ and the plane $z = 0$.

49. Find the region E for which the triple integral

$$\iiint_E (1 - x^2 - 2y^2 - 3z^2) \, dV$$

is a maximum.

DISCOVERY PROJECT

Volumes of Hyperspheres

In this project we find formulas for the volume enclosed by a hypersphere in n-dimensional space.

1. Use a double integral and the trigonometric substitution $y = r \sin \theta$, together with Formula 64 in the Table of Integrals, to find the area of a circle with radius r.

2. Use a triple integral and trigonometric substitution to find the volume of a sphere with radius r.

3. Use a quadruple integral to find the hypervolume enclosed by the hypersphere $x^2 + y^2 + z^2 + w^2 = r^2$ in $\mathbb{R}^4$. (Use only trigonometric substitution and the reduction formulas for $\int \sin^n x \, dx$ or $\int \cos^n x \, dx$.)

4. Use an n-tuple integral to find the volume enclosed by a hypersphere of radius r in n-dimensional space $\mathbb{R}^n$. [*Hint:* The formulas are different for n even and n odd.]

12.8 Triple Integrals in Cylindrical and Spherical Coordinates

We saw in Section 12.4 that some double integrals are easier to evaluate using polar coordinates. In this section we see that some triple integrals are easier to evaluate using cylindrical or spherical coordinates.

■ Cylindrical Coordinates

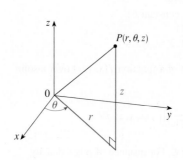

FIGURE 1

Recall from Section 9.7 that the cylindrical coordinates of a point P are (r, θ, z), where r, θ, and z are shown in Figure 1. Suppose that E is a type 1 region whose projection D on the xy-plane is conveniently described in polar coordinates (see Figure 2). In particular, suppose that f is continuous and

$$E = \{(x, y, z) \mid (x, y) \in D, \ u_1(x, y) \leq z \leq u_2(x, y)\}$$

where D is given in polar coordinates by

$$D = \{(r, \theta) \mid \alpha \leq \theta \leq \beta, \ h_1(\theta) \leq r \leq h_2(\theta)\}$$

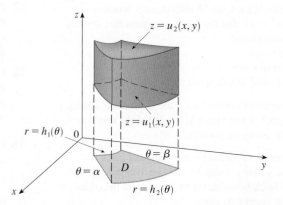

FIGURE 2

We know from Equation 12.7.6 that

$$\boxed{1} \qquad \iiint_E f(x, y, z) \, dV = \iint_D \left[\int_{u_1(x, y)}^{u_2(x, y)} f(x, y, z) \, dz \right] dA$$

But we also know how to evaluate double integrals in polar coordinates. In fact, combining Equation 1 with Equation 12.4.3, we obtain

$$\boxed{2} \qquad \iiint_E f(x, y, z) \, dV = \int_\alpha^\beta \int_{h_1(\theta)}^{h_2(\theta)} \int_{u_1(r \cos \theta, \, r \sin \theta)}^{u_2(r \cos \theta, \, r \sin \theta)} f(r \cos \theta, r \sin \theta, z) \, r \, dz \, dr \, d\theta$$

FIGURE 3

Volume element in cylindrical coordinates: $dV = r \, dz \, dr \, d\theta$

Formula 2 is the **formula for triple integration in cylindrical coordinates**. It says that we convert a triple integral from rectangular to cylindrical coordinates by writing $x = r \cos \theta$, $y = r \sin \theta$, leaving z as it is, using the appropriate limits of integration for z, r, and θ, and replacing dV by $r \, dz \, dr \, d\theta$. (Figure 3 shows how to remember this.)

It is worthwhile to use this formula when E is a solid region easily described in cylindrical coordinates, and especially when the function $f(x, y, z)$ involves the expression $x^2 + y^2$.

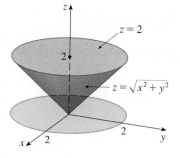

FIGURE 4

▣ **EXAMPLE 1** A solid E lies within the cylinder $x^2 + y^2 = 1$, below the plane $z = 4$, and above the paraboloid $z = 1 - x^2 - y^2$. (See Figure 4.) The density at any point is proportional to its distance from the axis of the cylinder. Find the mass of E.

SOLUTION In cylindrical coordinates the cylinder is $r = 1$ and the paraboloid is $z = 1 - r^2$, so we can write

$$E = \{(r, \theta, z) \mid 0 \le \theta \le 2\pi, \ 0 \le r \le 1, \ 1 - r^2 \le z \le 4\}$$

Since the density at (x, y, z) is proportional to the distance from the z-axis, the density function is

$$f(x, y, z) = K\sqrt{x^2 + y^2} = Kr$$

where K is the proportionality constant. Therefore, from Formula 12.7.13, the mass of E is

$$m = \iiint_E K\sqrt{x^2 + y^2} \, dV = \int_0^{2\pi} \int_0^1 \int_{1-r^2}^4 (Kr) \, r \, dz \, dr \, d\theta$$

$$= \int_0^{2\pi} \int_0^1 Kr^2 [4 - (1 - r^2)] \, dr \, d\theta = K \int_0^{2\pi} d\theta \int_0^1 (3r^2 + r^4) \, dr$$

$$= 2\pi K \left[r^3 + \frac{r^5}{5} \right]_0^1 = \frac{12\pi K}{5}$$

■■

EXAMPLE 2 Evaluate $\int_{-2}^{2} \int_{-\sqrt{4-x^2}}^{\sqrt{4-x^2}} \int_{\sqrt{x^2+y^2}}^{2} (x^2 + y^2) \, dz \, dy \, dx$.

SOLUTION This iterated integral is a triple integral over the solid region

$$E = \left\{(x, y, z) \mid -2 \le x \le 2, \ -\sqrt{4 - x^2} \le y \le \sqrt{4 - x^2}, \ \sqrt{x^2 + y^2} \le z \le 2 \right\}$$

and the projection of E onto the xy-plane is the disk $x^2 + y^2 \le 4$. The lower surface of E is the cone $z = \sqrt{x^2 + y^2}$ and its upper surface is the plane $z = 2$. (See Figure 5.) This region has a much simpler description in cylindrical coordinates:

$$E = \{(r, \theta, z) \mid 0 \le \theta \le 2\pi, \ 0 \le r \le 2, \ r \le z \le 2\}$$

Therefore, we have

$$\int_{-2}^{2} \int_{-\sqrt{4-x^2}}^{\sqrt{4-x^2}} \int_{\sqrt{x^2+y^2}}^{2} (x^2 + y^2) \, dz \, dy \, dx = \iiint_E (x^2 + y^2) \, dV$$

$$= \int_0^{2\pi} \int_0^2 \int_r^2 r^2 \, r \, dz \, dr \, d\theta$$

$$= \int_0^{2\pi} d\theta \int_0^2 r^3(2 - r) \, dr$$

$$= 2\pi \left[\tfrac{1}{2}r^4 - \tfrac{1}{5}r^5 \right]_0^2 = \tfrac{16}{5}\pi$$

■■

FIGURE 5

◫ Spherical Coordinates

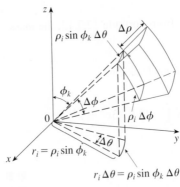

FIGURE 6
Spherical coordinates of P

In Section 9.7 we defined the spherical coordinates (ρ, θ, ϕ) of a point (see Figure 6) and we demonstrated the following relationships between rectangular coordinates and spherical coordinates:

$$\boxed{3} \qquad x = \rho \sin \phi \cos \theta \qquad y = \rho \sin \phi \sin \theta \qquad z = \rho \cos \phi$$

In this coordinate system the counterpart of a rectangular box is a **spherical wedge**

$$E = \{(\rho, \theta, \phi) \mid a \le \rho \le b, \ \alpha \le \theta \le \beta, \ c \le \phi \le d\}$$

where $a \ge 0$ and $\beta - \alpha \le 2\pi$. Although we defined triple integrals by dividing solids into small boxes, it can be shown that dividing a solid into small spherical wedges always gives the same result. So we divide E into smaller spherical wedges E_{ijk} by means of equally spaced spheres $\rho = \rho_i$, half-planes $\theta = \theta_j$, and half-cones $\phi = \phi_k$. Figure 7 shows that E_{ijk} is approximately a rectangular box with dimensions $\Delta\rho$, $\rho_i \Delta\phi$ (arc of a circle with radius ρ_i, angle $\Delta\phi$), and $\rho_i \sin \phi_k \Delta\theta$ (arc of a circle with radius $\rho_i \sin \phi_k$, angle $\Delta\theta$). So an approximation to the volume of E_{ijk} is given by

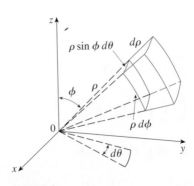

FIGURE 7

$$(\Delta\rho) \times (\rho_i \Delta\phi) \times (\rho_i \sin \phi_k \Delta\theta) = \rho_i^2 \sin \phi_k \, \Delta\rho \, \Delta\theta \, \Delta\phi$$

Thus, an approximation to a typical triple Riemann sum is

$$\sum_{i=1}^{l} \sum_{j=1}^{m} \sum_{k=1}^{n} f(\rho_i \sin \phi_k \cos \theta_j, \ \rho_i \sin \phi_k \sin \theta_j, \ \rho_i \cos \phi_k) \rho_i^2 \sin \phi_k \, \Delta\rho \, \Delta\theta \, \Delta\phi$$

But this sum is a Riemann sum for the function

$$F(\rho, \theta, \phi) = f(\rho \sin \phi \cos \theta, \ \rho \sin \phi \sin \theta, \ \rho \cos \phi) \cdot \rho^2 \sin \phi$$

Consequently, the following **formula for triple integration in spherical coordinates** is plausible.

$$\boxed{4} \qquad \iiint_E f(x, y, z) \, dV$$

$$= \int_c^d \int_\alpha^\beta \int_a^b f(\rho \sin \phi \cos \theta, \ \rho \sin \phi \sin \theta, \ \rho \cos \phi) \, \rho^2 \sin \phi \, d\rho \, d\theta \, d\phi$$

where E is a spherical wedge given by

$$E = \{(\rho, \theta, \phi) \mid a \le \rho \le b, \ \alpha \le \theta \le \beta, \ c \le \phi \le d\}$$

FIGURE 8
Volume element in spherical coordinates: $dV = \rho^2 \sin \phi \, d\rho \, d\theta \, d\phi$

Formula 4 says that we convert a triple integral from rectangular coordinates to spherical coordinates by writing

$$x = \rho \sin \phi \cos \theta \qquad y = \rho \sin \phi \sin \theta \qquad z = \rho \cos \phi$$

using the appropriate limits of integration, and replacing dV by $\rho^2 \sin \phi \, d\rho \, d\theta \, d\phi$. This is illustrated in Figure 8.

This formula can be extended to include more general spherical regions such as

$$E = \left\{(\rho, \theta, \phi) \mid \alpha \le \theta \le \beta,\ c \le \phi \le d,\ g_1(\theta, \phi) \le \rho \le g_2(\theta, \phi)\right\}$$

In this case the formula is the same as in (4) except that the limits of integration for ρ are $g_1(\theta, \phi)$ and $g_2(\theta, \phi)$.

Usually, spherical coordinates are used in triple integrals when surfaces such as cones and spheres form the boundary of the region of integration.

V **EXAMPLE 3** Evaluate $\iiint_B e^{(x^2+y^2+z^2)^{3/2}}\, dV$, where B is the unit ball:

$$B = \left\{(x, y, z) \mid x^2 + y^2 + z^2 \le 1\right\}$$

SOLUTION Since the boundary of B is a sphere, we use spherical coordinates:

$$B = \left\{(\rho, \theta, \phi) \mid 0 \le \rho \le 1,\ 0 \le \theta \le 2\pi,\ 0 \le \phi \le \pi\right\}$$

In addition, spherical coordinates are appropriate because

$$x^2 + y^2 + z^2 = \rho^2$$

Thus, (4) gives

$$\iiint_B e^{(x^2+y^2+z^2)^{3/2}}\, dV = \int_0^\pi \int_0^{2\pi} \int_0^1 e^{(\rho^2)^{3/2}} \rho^2 \sin\phi\, d\rho\, d\theta\, d\phi$$

$$= \int_0^\pi \sin\phi\, d\phi \int_0^{2\pi} d\theta \int_0^1 \rho^2 e^{\rho^3}\, d\rho$$

$$= \left[-\cos\phi\right]_0^\pi (2\pi) \left[\tfrac{1}{3} e^{\rho^3}\right]_0^1 = \tfrac{4}{3}\pi(e - 1)$$ ■ ■

NOTE □ It would have been extremely awkward to evaluate the integral in Example 3 without spherical coordinates. In rectangular coordinates the iterated integral would have been

$$\int_{-1}^1 \int_{-\sqrt{1-x^2}}^{\sqrt{1-x^2}} \int_{-\sqrt{1-x^2-y^2}}^{\sqrt{1-x^2-y^2}} e^{(x^2+y^2+z^2)^{3/2}}\, dz\, dy\, dx$$

V **EXAMPLE 4** Use spherical coordinates to find the volume of the solid that lies above the cone $z = \sqrt{x^2 + y^2}$ and below the sphere $x^2 + y^2 + z^2 = z$. (See Figure 9.)

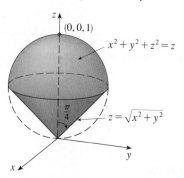

FIGURE 9

SOLUTION Notice that the sphere passes through the origin and has center $\left(0, 0, \tfrac{1}{2}\right)$. We write the equation of the sphere in spherical coordinates as

$$\rho^2 = \rho\cos\phi \qquad \text{or} \qquad \rho = \cos\phi$$

■ ■ Figure 10 gives another look (this time drawn by Maple) at the solid of Example 4.

FIGURE 10

 Visual 12.8 shows an animation of Figure 11.

The equation of the cone can be written as

$$\rho \cos \phi = \sqrt{\rho^2 \sin^2\phi \, \cos^2\theta + \rho^2 \sin^2\phi \, \sin^2\theta} = \rho \sin \phi$$

This gives $\sin \phi = \cos \phi$, or $\phi = \pi/4$. Therefore, the description of the solid E in spherical coordinates is

$$E = \{(\rho, \theta, \phi) \mid 0 \leq \theta \leq 2\pi, \ 0 \leq \phi \leq \pi/4, \ 0 \leq \rho \leq \cos \phi\}$$

Figure 11 shows how E is swept out if we integrate first with respect to ρ, then ϕ, and then θ. The volume of E is

$$V(E) = \iiint_E dV = \int_0^{2\pi} \int_0^{\pi/4} \int_0^{\cos \phi} \rho^2 \sin \phi \, d\rho \, d\phi \, d\theta$$

$$= \int_0^{2\pi} d\theta \int_0^{\pi/4} \sin \phi \left[\frac{\rho^3}{3}\right]_{\rho=0}^{\rho=\cos\phi} d\phi$$

$$= \frac{2\pi}{3} \int_0^{\pi/4} \sin \phi \cos^3\phi \, d\phi = \frac{2\pi}{3} \left[-\frac{\cos^4\phi}{4}\right]_0^{\pi/4} = \frac{\pi}{8}$$

ρ varies from 0 to $\cos \phi$ while ϕ and θ are constant.

ϕ varies from 0 to $\pi/4$ while θ is constant.

θ varies from 0 to 2π.

FIGURE 11

12.8 Exercises

1–4 ■ Sketch the solid whose volume is given by the integral and evaluate the integral.

1. $\int_0^4 \int_0^{2\pi} \int_r^4 r \, dz \, d\theta \, dr$

2. $\int_0^{\pi/2} \int_0^2 \int_0^{9-r^2} r \, dz \, dr \, d\theta$

3. $\int_0^{\pi/6} \int_0^{\pi/2} \int_0^3 \rho^2 \sin \phi \, d\rho \, d\theta \, d\phi$

4. $\int_0^{2\pi} \int_{\pi/2}^{\pi} \int_1^2 \rho^2 \sin \phi \, d\rho \, d\phi \, d\theta$

5–6 ■ Set up the triple integral of an arbitrary continuous function $f(x, y, z)$ in cylindrical or spherical coordinates over the solid shown.

5.

6.

7–16 ■ Use cylindrical coordinates.

7. Evaluate $\iiint_E \sqrt{x^2 + y^2} \, dV$, where E is the region that lies inside the cylinder $x^2 + y^2 = 16$ and between the planes $z = -5$ and $z = 4$.

8. Evaluate $\iiint_E (x^3 + xy^2) \, dV$, where E is the solid in the first octant that lies beneath the paraboloid $z = 1 - x^2 - y^2$.

9. Evaluate $\iiint_E e^z \, dV$, where E is enclosed by the paraboloid $z = 1 + x^2 + y^2$, the cylinder $x^2 + y^2 = 5$, and the xy-plane.

10. Evaluate $\iiint_E x \, dV$, where E is enclosed by the planes $z = 0$ and $z = x + y + 5$ and by the cylinders $x^2 + y^2 = 4$ and $x^2 + y^2 = 9$.

11. Evaluate $\iiint_E x^2 \, dV$, where E is the solid that lies within the cylinder $x^2 + y^2 = 1$, above the plane $z = 0$, and below the cone $z^2 = 4x^2 + 4y^2$.

12. Find the volume of the solid that lies within both the cylinder $x^2 + y^2 = 1$ and the sphere $x^2 + y^2 + z^2 = 4$.

13. (a) Find the volume of the region E bounded by the paraboloids $z = x^2 + y^2$ and $z = 36 - 3x^2 - 3y^2$.
(b) Find the centroid of E (the center of mass in the case where the density is constant).

14. (a) Find the volume of the solid that the cylinder $r = a \cos \theta$ cuts out of the sphere of radius a centered at the origin.
(b) Illustrate the solid of part (a) by graphing the sphere and the cylinder on the same screen.

15. Find the mass and center of mass of the solid S bounded by the paraboloid $z = 4x^2 + 4y^2$ and the plane $z = a$ $(a > 0)$ if S has constant density K.

16. Find the mass of a ball B given by $x^2 + y^2 + z^2 \leq a^2$ if the density at any point is proportional to its distance from the z-axis.

17–26 ■ Use spherical coordinates.

17. Evaluate $\iiint_B (x^2 + y^2 + z^2) \, dV$, where B is the unit ball $x^2 + y^2 + z^2 \leq 1$.

18. Evaluate $\iiint_H (x^2 + y^2) \, dV$, where H is the hemispherical region that lies above the xy-plane and below the sphere $x^2 + y^2 + z^2 = 1$.

19. Evaluate $\iiint_E z \, dV$, where E lies between the spheres $x^2 + y^2 + z^2 = 1$ and $x^2 + y^2 + z^2 = 4$ in the first octant.

20. Evaluate $\iiint_E e^{\sqrt{x^2+y^2+z^2}} \, dV$, where E is enclosed by the sphere $x^2 + y^2 + z^2 = 9$ in the first octant.

21. Evaluate $\iiint_E x^2 \, dV$, where E is bounded by the xz-plane and the hemispheres $y = \sqrt{9 - x^2 - z^2}$ and $y = \sqrt{16 - x^2 - z^2}$.

22. Find the volume of the solid that lies within the sphere $x^2 + y^2 + z^2 = 4$, above the xy-plane, and below the cone $z = \sqrt{x^2 + y^2}$.

23. (a) Find the volume of the solid that lies above the cone $\phi = \pi/3$ and below the sphere $\rho = 4 \cos \phi$.
(b) Find the centroid of the solid in part (a).

24. Let H be a solid hemisphere of radius a whose density at any point is proportional to its distance from the center of the base.
(a) Find the mass of H.
(b) Find the center of mass of H.
(c) Find the moment of inertia of H about its axis.

25. (a) Find the centroid of a solid homogeneous hemisphere of radius a.
(b) Find the moment of inertia of the solid in part (a) about a diameter of its base.

26. Find the mass and center of mass of a solid hemisphere of radius a if the density at any point is proportional to its distance from the base.

27–30 ■ Use cylindrical or spherical coordinates, whichever seems more appropriate.

27. Find the volume and centroid of the solid E that lies above the cone $z = \sqrt{x^2 + y^2}$ and below the sphere $x^2 + y^2 + z^2 = 1$.

28. Find the volume of the smaller wedge cut from a sphere of radius a by two planes that intersect along a diameter at an angle of $\pi/6$.

CAS 29. Evaluate $\iiint_E z \, dV$, where E lies above the paraboloid $z = x^2 + y^2$ and below the plane $z = 2y$. Use either the Table of Integrals (on Reference Pages 6–10) or a computer algebra system to evaluate the integral.

30. (a) Find the volume enclosed by the torus $\rho = \sin \phi$.
(b) Use a computer to draw the torus.

31–32 ■ Evaluate the integral by changing to cylindrical coordinates.

31. $\int_{-2}^{2} \int_{-\sqrt{4-y^2}}^{\sqrt{4-y^2}} \int_{\sqrt{x^2+y^2}}^{2} xz \, dz \, dx \, dy$

32. $\int_{-3}^{3} \int_{0}^{\sqrt{9-x^2}} \int_{0}^{9-x^2-y^2} \sqrt{x^2 + y^2} \, dz \, dy \, dx$

33–34 ■ Evaluate the integral by changing to spherical coordinates.

33. $\int_{0}^{1} \int_{0}^{\sqrt{1-x^2}} \int_{\sqrt{x^2+y^2}}^{\sqrt{2-x^2-y^2}} xy \, dz \, dy \, dx$

34. $\int_{-a}^{a} \int_{-\sqrt{a^2-y^2}}^{\sqrt{a^2-y^2}} \int_{-\sqrt{a^2-x^2-y^2}}^{\sqrt{a^2-x^2-y^2}} (x^2z + y^2z + z^3) \, dz \, dx \, dy$

35. In the Laboratory Project on page 690 we investigated the family of surfaces $\rho = 1 + \frac{1}{5} \sin m\theta \sin n\phi$ that have been used as models for tumors. The "bumpy sphere" with $m = 6$ and $n = 5$ is shown. Use a computer algebra system to find the volume it encloses.

36. Show that

$$\int_{-\infty}^{\infty} \int_{-\infty}^{\infty} \int_{-\infty}^{\infty} \sqrt{x^2 + y^2 + z^2}\, e^{-(x^2+y^2+z^2)}\, dx\, dy\, dz = 2\pi$$

(The improper triple integral is defined as the limit of a triple integral over a solid sphere as the radius of the sphere increases indefinitely.)

37. When studying the formation of mountain ranges, geologists estimate the amount of work required to lift a mountain from sea level. Consider a mountain that is essentially in the shape of a right circular cone. Suppose that the weight density of the material in the vicinity of a point P is $g(P)$ and the height is $h(P)$.

(a) Find a definite integral that represents the total work done in forming the mountain.

(b) Assume that Mount Fuji in Japan is in the shape of a right circular cone with radius 62,000 ft, height 12,400 ft, and density a constant 200 lb/ft³. How much work was done in forming Mount Fuji if the land was initially at sea level?

APPLIED PROJECT

Roller Derby

Suppose that a solid ball (a marble), a hollow ball (a squash ball), a solid cylinder (a steel bar), and a hollow cylinder (a lead pipe) roll down a slope. Which of these objects reaches the bottom first? (Make a guess before proceeding.)

To answer this question we consider a ball or cylinder with mass m, radius r, and moment of inertia I (about the axis of rotation). If the vertical drop is h, then the potential energy at the top is mgh. Suppose the object reaches the bottom with velocity v and angular velocity ω, so $v = \omega r$. The kinetic energy at the bottom consists of two parts: $\frac{1}{2}mv^2$ from translation (moving down the slope) and $\frac{1}{2}I\omega^2$ from rotation. If we assume that energy loss from rolling friction is negligible, then conservation of energy gives

$$mgh = \tfrac{1}{2}mv^2 + \tfrac{1}{2}I\omega^2$$

1. Show that

$$v^2 = \frac{2gh}{1 + I^*} \qquad \text{where } I^* = \frac{I}{mr^2}$$

2. If $y(t)$ is the vertical distance traveled at time t, then the same reasoning as used in Problem 1 shows that $v^2 = 2gy/(1 + I^*)$ at any time t. Use this result to show that y satisfies the differential equation

$$\frac{dy}{dt} = \sqrt{\frac{2g}{1 + I^*}}\, (\sin \alpha) \sqrt{y}$$

where α is the angle of inclination of the plane.

3. By solving the differential equation in Problem 2, show that the total travel time is

$$T = \sqrt{\frac{2h(1 + I^*)}{g \sin^2\alpha}}$$

This shows that the object with the smallest value of I^* wins the race.

4. Show that $I^* = \frac{1}{2}$ for a solid cylinder and $I^* = 1$ for a hollow cylinder.

5. Calculate I^* for a partly hollow ball with inner radius a and outer radius r. Express your answer in terms of $b = a/r$. What happens as $a \to 0$ and as $a \to r$?

6. Show that $I^* = \frac{2}{5}$ for a solid ball and $I^* = \frac{2}{3}$ for a hollow ball. Thus, the objects finish in the following order: solid ball, solid cylinder, hollow ball, hollow cylinder.

DISCOVERY PROJECT

The Intersection of Three Cylinders

The figure shows the solid enclosed by three circular cylinders with the same diameter that intersect at right angles. In this project we compute its volume and determine how its shape changes if the cylinders have different diameters.

1. Sketch carefully the solid enclosed by the three cylinders $x^2 + y^2 = 1$, $x^2 + z^2 = 1$, and $y^2 + z^2 = 1$. Indicate the positions of the coordinate axes and label the faces with the equations of the corresponding cylinders.

2. Find the volume of the solid in Problem 1.

CAS **3.** Use a computer algebra system to draw the edges of the solid.

4. What happens to the solid in Problem 1 if the radius of the first cylinder is different from 1? Illustrate with a hand-drawn sketch or a computer graph.

5. If the first cylinder is $x^2 + y^2 = a^2$, where $a < 1$, set up, but do not evaluate, a double integral for the volume of the solid. What if $a > 1$?

12.9 Change of Variables in Multiple Integrals

In one-dimensional calculus we often use a change of variable (a substitution) to simplify an integral. By reversing the roles of x and u, we can write the Substitution Rule (5.5.5) as

$$\boxed{1} \qquad \int_a^b f(x)\, dx = \int_c^d f(g(u))g'(u)\, du$$

where $x = g(u)$ and $a = g(c)$, $b = g(d)$. Another way of writing Formula 1 is as follows:

$$\boxed{2} \qquad \int_a^b f(x)\, dx = \int_c^d f(x(u))\, \frac{dx}{du}\, du$$

A change of variables can also be useful in double integrals. We have already seen one example of this: conversion to polar coordinates. The new variables r and θ are related to the old variables x and y by the equations

$$x = r \cos \theta \qquad y = r \sin \theta$$

and the change of variables formula (12.4.2) can be written as

$$\iint_R f(x, y)\, dA = \iint_S f(r \cos \theta, r \sin \theta)\, r\, dr\, d\theta$$

where S is the region in the $r\theta$-plane that corresponds to the region R in the xy-plane.

More generally, we consider a change of variables that is given by a **transformation** T from the uv-plane to the xy-plane:

$$T(u, v) = (x, y)$$

where x and y are related to u and v by the equations

$$\boxed{3} \qquad x = g(u, v) \qquad y = h(u, v)$$

or, as we sometimes write,

$$x = x(u, v) \qquad y = y(u, v)$$

We usually assume that T is a C^1 **transformation**, which means that g and h have continuous first-order partial derivatives.

A transformation T is really just a function whose domain and range are both subsets of $\mathbb{R}^2$. If $T(u_1, v_1) = (x_1, y_1)$, then the point (x_1, y_1) is called the **image** of the point (u_1, v_1). If no two points have the same image, T is called **one-to-one**. Figure 1 shows the effect of a transformation T on a region S in the uv-plane. T transforms S into a region R in the xy-plane called the **image of S**, consisting of the images of all points in S.

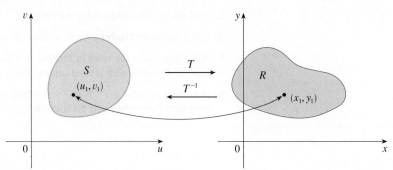

FIGURE 1

If T is a one-to-one transformation, then it has an **inverse transformation** T^{-1} from the xy-plane to the uv-plane and it may be possible to solve Equations 3 for u and v in terms of x and y:

$$u = G(x, y) \qquad v = H(x, y)$$

$\boxed{V}$ **EXAMPLE 1** A transformation is defined by the equations

$$x = u^2 - v^2 \qquad y = 2uv$$

Find the image of the square $S = \{(u, v) \mid 0 \leqslant u \leqslant 1, \ 0 \leqslant v \leqslant 1\}$.

SOLUTION The transformation maps the boundary of S into the boundary of the image. So we begin by finding the images of the sides of S. The first side, S_1, is given by $v = 0$ ($0 \leqslant u \leqslant 1$). (See Figure 2.) From the given equations we have $x = u^2$, $y = 0$, and so $0 \leqslant x \leqslant 1$. Thus, S_1 is mapped into the line segment from $(0, 0)$ to $(1, 0)$ in the xy-plane. The second side, S_2, is $u = 1$ ($0 \leqslant v \leqslant 1$) and, putting $u = 1$ in the given equations, we get

$$x = 1 - v^2 \qquad y = 2v$$

Eliminating v, we obtain

$$\boxed{4} \qquad\qquad\qquad x = 1 - \frac{y^2}{4} \qquad 0 \leqslant x \leqslant 1$$

which is part of a parabola. Similarly, S_3 is given by $v = 1$ ($0 \leqslant u \leqslant 1$), whose image is the parabolic arc

$$\boxed{5} \qquad\qquad\qquad x = \frac{y^2}{4} - 1 \qquad -1 \leqslant x \leqslant 0$$

Finally, S_4 is given by $u = 0$ ($0 \leqslant v \leqslant 1$) whose image is $x = -v^2$, $y = 0$, that is, $-1 \leqslant x \leqslant 0$. (Notice that as we move around the square in the counterclockwise direction, we also move around the parabolic region in the counterclockwise direction.) The image of S is the region R (shown in Figure 2) bounded by the x-axis and the parabolas given by Equations 4 and 5. ■ ■

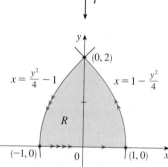

FIGURE 2

Now let's see how a change of variables affects a double integral. We start with a small rectangle S in the uv-plane whose lower left corner is the point (u_0, v_0) and whose dimensions are Δu and Δv. (See Figure 3.)

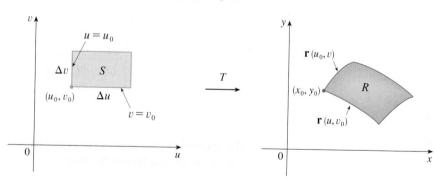

FIGURE 3

The image of S is a region R in the xy-plane, one of whose boundary points is $(x_0, y_0) = T(u_0, v_0)$. The vector

$$\mathbf{r}(u, v) = g(u, v)\,\mathbf{i} + h(u, v)\,\mathbf{j}$$

is the position vector of the image of the point (u, v). The equation of the lower side of S is $v = v_0$, whose image curve is given by the vector function $\mathbf{r}(u, v_0)$. The tangent vector at (x_0, y_0) to this image curve is

$$\mathbf{r}_u = g_u(u_0, v_0)\mathbf{i} + h_u(u_0, v_0)\mathbf{j} = \frac{\partial x}{\partial u}\mathbf{i} + \frac{\partial y}{\partial u}\mathbf{j}$$

Similarly, the tangent vector at (x_0, y_0) to the image curve of the left side of S (namely, $u = u_0$) is

$$\mathbf{r}_v = g_v(u_0, v_0)\mathbf{i} + h_v(u_0, v_0)\mathbf{j} = \frac{\partial x}{\partial v}\mathbf{i} + \frac{\partial y}{\partial v}\mathbf{j}$$

We can approximate the image region $R = T(S)$ by a parallelogram determined by the secant vectors

$$\mathbf{a} = \mathbf{r}(u_0 + \Delta u, v_0) - \mathbf{r}(u_0, v_0) \qquad \mathbf{b} = \mathbf{r}(u_0, v_0 + \Delta v) - \mathbf{r}(u_0, v_0)$$

shown in Figure 4. But

$$\mathbf{r}_u = \lim_{\Delta u \to 0} \frac{\mathbf{r}(u_0 + \Delta u, v_0) - \mathbf{r}(u_0, v_0)}{\Delta u}$$

and so

$$\mathbf{r}(u_0 + \Delta u, v_0) - \mathbf{r}(u_0, v_0) \approx \Delta u\, \mathbf{r}_u$$

Similarly

$$\mathbf{r}(u_0, v_0 + \Delta v) - \mathbf{r}(u_0, v_0) \approx \Delta v\, \mathbf{r}_v$$

This means that we can approximate R by a parallelogram determined by the vectors $\Delta u\, \mathbf{r}_u$ and $\Delta v\, \mathbf{r}_v$. (See Figure 5.) Therefore, we can approximate the area of R by the area of this parallelogram, which, from Section 9.4, is

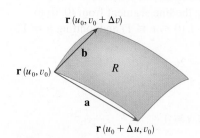

r$(u_0, v_0 + \Delta v)$

b

R

r(u_0, v_0)

a

r$(u_0 + \Delta u, v_0)$

FIGURE 4

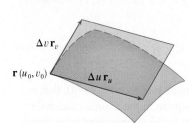

$\Delta v\, \mathbf{r}_v$

r(u_0, v_0)

$\Delta u\, \mathbf{r}_u$

FIGURE 5

$\boxed{6}$

$$\left| (\Delta u\, \mathbf{r}_u) \times (\Delta v\, \mathbf{r}_v) \right| = \left| \mathbf{r}_u \times \mathbf{r}_v \right| \Delta u\, \Delta v$$

Computing the cross product, we obtain

$$\mathbf{r}_u \times \mathbf{r}_v = \begin{vmatrix} \mathbf{i} & \mathbf{j} & \mathbf{k} \\ \dfrac{\partial x}{\partial u} & \dfrac{\partial y}{\partial u} & 0 \\ \dfrac{\partial x}{\partial v} & \dfrac{\partial y}{\partial v} & 0 \end{vmatrix} = \begin{vmatrix} \dfrac{\partial x}{\partial u} & \dfrac{\partial y}{\partial u} \\ \dfrac{\partial x}{\partial v} & \dfrac{\partial y}{\partial v} \end{vmatrix} \mathbf{k} = \begin{vmatrix} \dfrac{\partial x}{\partial u} & \dfrac{\partial x}{\partial v} \\ \dfrac{\partial y}{\partial u} & \dfrac{\partial y}{\partial v} \end{vmatrix} \mathbf{k}$$

The determinant that arises in this calculation is called the *Jacobian* of the transformation and is given a special notation.

■■ The Jacobian is named after the German mathematician Carl Gustav Jacob Jacobi (1804–1851). Although the French mathematician Cauchy first used these special determinants involving partial derivatives, Jacobi developed them into a method for evaluating multiple integrals.

$\boxed{7}$ **Definition** The **Jacobian** of the transformation T given by $x = g(u, v)$ and $y = h(u, v)$ is

$$\frac{\partial(x, y)}{\partial(u, v)} = \begin{vmatrix} \dfrac{\partial x}{\partial u} & \dfrac{\partial x}{\partial v} \\ \dfrac{\partial y}{\partial u} & \dfrac{\partial y}{\partial v} \end{vmatrix} = \frac{\partial x}{\partial u}\frac{\partial y}{\partial v} - \frac{\partial x}{\partial v}\frac{\partial y}{\partial u}$$

With this notation we can use Equation 6 to give an approximation to the area ΔA of R:

$$\boxed{8} \qquad \Delta A \approx \left| \frac{\partial(x, y)}{\partial(u, v)} \right| \Delta u \, \Delta v$$

where the Jacobian is evaluated at (u_0, v_0).

Next we divide a region S in the uv-plane into rectangles S_{ij} and call their images in the xy-plane R_{ij}. (See Figure 6.)

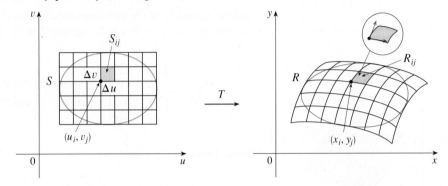

FIGURE 6

Applying the approximation (8) to each R_{ij}, we approximate the double integral of f over R as follows:

$$\iint\limits_{R} f(x, y) \, dA \approx \sum_{i=1}^{m} \sum_{j=1}^{n} f(x_i, y_j) \, \Delta A$$

$$\approx \sum_{i=1}^{m} \sum_{j=1}^{n} f(g(u_i, v_j), h(u_i, v_j)) \left| \frac{\partial(x, y)}{\partial(u, v)} \right| \Delta u \, \Delta v$$

where the Jacobian is evaluated at (u_i, v_j). Notice that this double sum is a Riemann sum for the integral

$$\iint\limits_{S} f(g(u, v), h(u, v)) \left| \frac{\partial(x, y)}{\partial(u, v)} \right| du \, dv$$

The foregoing argument suggests that the following theorem is true. (A full proof is given in books on advanced calculus.)

$\boxed{9}$ **Change of Variables in a Double Integral**   Suppose that T is a C^1 transformation whose Jacobian is nonzero and that maps a region S in the uv-plane onto a region R in the xy-plane. Suppose that f is continuous on R and that R and S are type I or type II plane regions. Suppose also that T is one-to-one, except perhaps on the boundary of S. Then

$$\iint\limits_{R} f(x, y) \, dA = \iint\limits_{S} f(x(u, v), y(u, v)) \left| \frac{\partial(x, y)}{\partial(u, v)} \right| du \, dv$$

Theorem 9 says that we change from an integral in x and y to an integral in u and v by expressing x and y in terms of u and v and writing

$$dA = \left| \frac{\partial(x, y)}{\partial(u, v)} \right| du \, dv$$

Notice the similarity between Theorem 9 and the one-dimensional formula in Equation 2. Instead of the derivative dx/du, we have the absolute value of the Jacobian, that is, $|\partial(x, y)/\partial(u, v)|$.

As a first illustration of Theorem 9, we show that the formula for integration in polar coordinates is just a special case. Here the transformation T from the $r\theta$-plane to the xy-plane is given by

$$x = g(r, \theta) = r \cos \theta \qquad y = h(r, \theta) = r \sin \theta$$

and the geometry of the transformation is shown in Figure 7. T maps an ordinary rectangle in the $r\theta$-plane to a polar rectangle in the xy-plane. The Jacobian of T is

$$\frac{\partial(x, y)}{\partial(r, \theta)} = \begin{vmatrix} \dfrac{\partial x}{\partial r} & \dfrac{\partial x}{\partial \theta} \\ \dfrac{\partial y}{\partial r} & \dfrac{\partial y}{\partial \theta} \end{vmatrix} = \begin{vmatrix} \cos \theta & -r \sin \theta \\ \sin \theta & r \cos \theta \end{vmatrix} = r \cos^2\theta + r \sin^2\theta = r > 0$$

Thus, Theorem 9 gives

$$\iint\limits_{R} f(x, y)\, dx\, dy = \iint\limits_{S} f(r \cos \theta, r \sin \theta) \left| \frac{\partial(x, y)}{\partial(r, \theta)} \right| dr\, d\theta$$

$$= \int_{\alpha}^{\beta} \int_{a}^{b} f(r \cos \theta, r \sin \theta)\, r\, dr\, d\theta$$

which is the same as Formula 12.4.2.

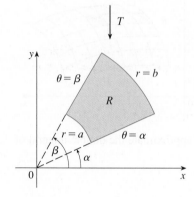

FIGURE 7
The polar coordinate transformation

$\boxed{V}$ **EXAMPLE 2** Use the change of variables $x = u^2 - v^2$, $y = 2uv$ to evaluate the integral $\iint_R y\, dA$, where R is the region bounded by the x-axis and the parabolas $y^2 = 4 - 4x$ and $y^2 = 4 + 4x$, $y \geq 0$.

SOLUTION The region R is pictured in Figure 2. In Example 1 we discovered that $T(S) = R$, where S is the square $[0, 1] \times [0, 1]$. Indeed, the reason for making the change of variables to evaluate the integral is that S is a much simpler region than R. First we need to compute the Jacobian:

$$\frac{\partial(x, y)}{\partial(u, v)} = \begin{vmatrix} \dfrac{\partial x}{\partial u} & \dfrac{\partial x}{\partial v} \\ \dfrac{\partial y}{\partial u} & \dfrac{\partial y}{\partial v} \end{vmatrix} = \begin{vmatrix} 2u & -2v \\ 2v & 2u \end{vmatrix} = 4u^2 + 4v^2 > 0$$

Therefore, by Theorem 9,

$$\iint\limits_{R} y\, dA = \iint\limits_{S} 2uv \left| \frac{\partial(x, y)}{\partial(u, v)} \right| dA = \int_{0}^{1} \int_{0}^{1} (2uv)4(u^2 + v^2)\, du\, dv$$

$$= 8 \int_{0}^{1} \int_{0}^{1} (u^3 v + uv^3)\, du\, dv = 8 \int_{0}^{1} \left[\tfrac{1}{4} u^4 v + \tfrac{1}{2} u^2 v^3 \right]_{u=0}^{u=1} dv$$

$$= \int_{0}^{1} (2v + 4v^3)\, dv = \left[v^2 + v^4 \right]_{0}^{1} = 2$$

■ ■

NOTE ▫ Example 2 was not a very difficult problem to solve because we were given a suitable change of variables. If we are not supplied with a transformation, then the first step is to think of an appropriate change of variables. If $f(x, y)$ is difficult to integrate, then the form of $f(x, y)$ may suggest a transformation. If the region of integration R is awkward, then the transformation should be chosen so that the corresponding region S in the uv-plane has a convenient description.

EXAMPLE 3 Evaluate the integral $\iint_R e^{(x+y)/(x-y)} \, dA$, where R is the trapezoidal region with vertices $(1, 0)$, $(2, 0)$, $(0, -2)$, and $(0, -1)$.

SOLUTION Since it isn't easy to integrate $e^{(x+y)/(x-y)}$, we make a change of variables suggested by the form of this function:

$$\boxed{10} \qquad\qquad u = x + y \qquad v = x - y$$

These equations define a transformation T^{-1} from the xy-plane to the uv-plane. Theorem 9 talks about a transformation T from the uv-plane to the xy-plane. It is obtained by solving Equations 10 for x and y:

$$\boxed{11} \qquad\qquad x = \tfrac{1}{2}(u + v) \qquad y = \tfrac{1}{2}(u - v)$$

The Jacobian of T is

$$\frac{\partial(x, y)}{\partial(u, v)} = \begin{vmatrix} \dfrac{\partial x}{\partial u} & \dfrac{\partial x}{\partial v} \\[2mm] \dfrac{\partial y}{\partial u} & \dfrac{\partial y}{\partial v} \end{vmatrix} = \begin{vmatrix} \tfrac{1}{2} & \tfrac{1}{2} \\[1mm] \tfrac{1}{2} & -\tfrac{1}{2} \end{vmatrix} = -\tfrac{1}{2}$$

To find the region S in the uv-plane corresponding to R, we note that the sides of R lie on the lines

$$y = 0 \qquad x - y = 2 \qquad x = 0 \qquad x - y = 1$$

and, from either Equations 10 or Equations 11, the image lines in the uv-plane are

$$u = v \qquad v = 2 \qquad u = -v \qquad v = 1$$

Thus, the region S is the trapezoidal region with vertices $(1, 1)$, $(2, 2)$, $(-2, 2)$, and $(-1, 1)$ shown in Figure 8. Since

$$S = \left\{ (u, v) \mid 1 \leqslant v \leqslant 2, \; -v \leqslant u \leqslant v \right\}$$

Theorem 9 gives

$$\iint_R e^{(x+y)/(x-y)} \, dA = \iint_S e^{u/v} \left| \frac{\partial(x, y)}{\partial(u, v)} \right| du \, dv$$

$$= \int_1^2 \int_{-v}^v e^{u/v} \left(\tfrac{1}{2} \right) du \, dv = \tfrac{1}{2} \int_1^2 \left[v e^{u/v} \right]_{u=-v}^{u=v} dv$$

$$= \tfrac{1}{2} \int_1^2 (e - e^{-1}) v \, dv = \tfrac{3}{4}(e - e^{-1})$$

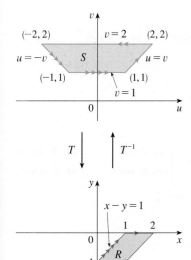

FIGURE 8

■■ Triple Integrals

There is a similar change of variables formula for triple integrals. Let T be a transformation that maps a region S in uvw-space onto a region R in xyz-space by means of the equations

$$x = g(u, v, w) \qquad y = h(u, v, w) \qquad z = k(u, v, w)$$

The **Jacobian** of T is the following 3×3 determinant:

$$\boxed{12} \qquad \frac{\partial(x, y, z)}{\partial(u, v, w)} = \begin{vmatrix} \dfrac{\partial x}{\partial u} & \dfrac{\partial x}{\partial v} & \dfrac{\partial x}{\partial w} \\[2mm] \dfrac{\partial y}{\partial u} & \dfrac{\partial y}{\partial v} & \dfrac{\partial y}{\partial w} \\[2mm] \dfrac{\partial z}{\partial u} & \dfrac{\partial z}{\partial v} & \dfrac{\partial z}{\partial w} \end{vmatrix}$$

Under hypotheses similar to those in Theorem 9, we have the following formula for triple integrals:

$$\boxed{13} \qquad \iiint\limits_{R} f(x, y, z)\, dV$$

$$= \iiint\limits_{S} f(x(u, v, w), y(u, v, w), z(u, v, w)) \left| \frac{\partial(x, y, z)}{\partial(u, v, w)} \right| du\, dv\, dw$$

▽ **EXAMPLE 4** Use Formula 13 to derive the formula for triple integration in spherical coordinates.

SOLUTION Here the change of variables is given by

$$x = \rho \sin \phi \cos \theta \qquad y = \rho \sin \phi \sin \theta \qquad z = \rho \cos \phi$$

We compute the Jacobian as follows:

$$\frac{\partial(x, y, z)}{\partial(\rho, \theta, \phi)} = \begin{vmatrix} \sin \phi \cos \theta & -\rho \sin \phi \sin \theta & \rho \cos \phi \cos \theta \\ \sin \phi \sin \theta & \rho \sin \phi \cos \theta & \rho \cos \phi \sin \theta \\ \cos \phi & 0 & -\rho \sin \phi \end{vmatrix}$$

$$= \cos \phi \begin{vmatrix} -\rho \sin \phi \sin \theta & \rho \cos \phi \cos \theta \\ \rho \sin \phi \cos \theta & \rho \cos \phi \sin \theta \end{vmatrix} - \rho \sin \phi \begin{vmatrix} \sin \phi \cos \theta & -\rho \sin \phi \sin \theta \\ \sin \phi \sin \theta & \rho \sin \phi \cos \theta \end{vmatrix}$$

$$= \cos \phi \, (-\rho^2 \sin \phi \cos \phi \sin^2\theta - \rho^2 \sin \phi \cos \phi \cos^2\theta)$$

$$\qquad - \rho \sin \phi \, (\rho \sin^2\phi \cos^2\theta + \rho \sin^2\phi \sin^2\theta)$$

$$= -\rho^2 \sin \phi \cos^2\phi - \rho^2 \sin \phi \sin^2\phi = -\rho^2 \sin \phi$$

Since $0 \leq \phi \leq \pi$, we have $\sin \phi \geq 0$. Therefore

$$\left| \frac{\partial(x, y, z)}{\partial(\rho, \theta, \phi)} \right| = |{-\rho^2 \sin \phi}| = \rho^2 \sin \phi$$

and Formula 13 gives

$$\iiint_R f(x, y, z)\, dV = \iiint_S f(\rho \sin \phi \cos \theta, \rho \sin \phi \sin \theta, \rho \cos \phi)\, \rho^2 \sin \phi \, d\rho \, d\theta \, d\phi$$

which is equivalent to Formula 12.8.4.

■ ■

12.9 Exercises

1–6 ■ Find the Jacobian of the transformation.

1. $x = u + 4v, \quad y = 3u - 2v$

2. $x = u^2 - v^2, \quad y = u^2 + v^2$

3. $x = \dfrac{u}{u + v}, \quad y = \dfrac{v}{u - v}$

4. $x = \alpha \sin \beta, \quad y = \alpha \cos \beta$

5. $x = uv, \quad y = vw, \quad z = uw$

6. $x = e^{u-v}, \quad y = e^{u+v}, \quad z = e^{u+v+w}$

7–10 ■ Find the image of the set S under the given transformation.

7. $S = \{(u, v) \mid 0 \leqslant u \leqslant 3, \ 0 \leqslant v \leqslant 2\}$;
$x = 2u + 3v, \ y = u - v$

8. S is the square bounded by the lines $u = 0$, $u = 1$, $v = 0$, $v = 1$; $\quad x = v, \ y = u(1 + v^2)$

9. S is the triangular region with vertices $(0, 0)$, $(1, 1)$, $(0, 1)$; $x = u^2, \ y = v$

10. S is the disk given by $u^2 + v^2 \leqslant 1$; $\quad x = au, \ y = bv$

11–16 ■ Use the given transformation to evaluate the integral.

11. $\iint_R (x - 3y)\, dA$, where R is the triangular region with vertices $(0, 0)$, $(2, 1)$, and $(1, 2)$; $\quad x = 2u + v, \ y = u + 2v$

12. $\iint_R (4x + 8y)\, dA$, where R is the parallelogram with vertices $(-1, 3)$, $(1, -3)$, $(3, -1)$, and $(1, 5)$; $x = \frac{1}{4}(u + v), \ y = \frac{1}{4}(v - 3u)$

13. $\iint_R x^2\, dA$, where R is the region bounded by the ellipse $9x^2 + 4y^2 = 36$; $\quad x = 2u, \ y = 3v$

14. $\iint_R (x^2 - xy + y^2)\, dA$, where R is the region bounded by the ellipse $x^2 - xy + y^2 = 2$; $x = \sqrt{2}\, u - \sqrt{2/3}\, v, \ y = \sqrt{2}\, u + \sqrt{2/3}\, v$

15. $\iint_R xy\, dA$, where R is the region in the first quadrant bounded by the lines $y = x$ and $y = 3x$ and the hyperbolas $xy = 1$, $xy = 3$; $\quad x = u/v, \ y = v$

16. $\iint_R y^2\, dA$, where R is the region bounded by the curves $xy = 1$, $xy = 2$, $xy^2 = 1$, $xy^2 = 2$; $\quad u = xy, \ v = xy^2$. Illustrate by using a graphing calculator or computer to draw R.

■ ■ ■ ■ ■ ■ ■ ■ ■ ■ ■ ■

17. (a) Evaluate $\iiint_E dV$, where E is the solid enclosed by the ellipsoid $x^2/a^2 + y^2/b^2 + z^2/c^2 = 1$. Use the transformation $x = au, \ y = bv, \ z = cw$.

(b) The Earth is not a perfect sphere; rotation has resulted in flattening at the poles. So the shape can be approximated by an ellipsoid with $a = b = 6378$ km and $c = 6356$ km. Use part (a) to estimate the volume of the Earth.

18. Evaluate $\iiint_E x^2 y\, dV$, where E is the solid of Exercise 17(a).

19–23 ■ Evaluate the integral by making an appropriate change of variables.

19. $\iint_R \dfrac{x - 2y}{3x - y}\, dA$, where R is the parallelogram enclosed by the lines $x - 2y = 0$, $x - 2y = 4$, $3x - y = 1$, and $3x - y = 8$

20. $\iint_R (x + y)e^{x^2 - y^2}\, dA$, where R is the rectangle enclosed by the lines $x - y = 0$, $x - y = 2$, $x + y = 0$, and $x + y = 3$

21. $\iint_R \cos\left(\dfrac{y - x}{y + x}\right)\, dA$, where R is the trapezoidal region with vertices $(1, 0)$, $(2, 0)$, $(0, 2)$, and $(0, 1)$

22. $\iint_R \sin(9x^2 + 4y^2)\, dA$, where R is the region in the first quadrant bounded by the ellipse $9x^2 + 4y^2 = 1$

23. $\iint_R e^{x+y}\, dA$, where R is given by the inequality $|x| + |y| \leqslant 1$

■ ■ ■ ■ ■ ■ ■ ■ ■ ■ ■ ■

24. Let f be continuous on $[0, 1]$ and let R be the triangular region with vertices $(0, 0)$, $(1, 0)$, and $(0, 1)$. Show that

$$\iint_R f(x + y)\, dA = \int_0^1 uf(u)\, du$$

12 Review

CONCEPT CHECK

1. Suppose f is a continuous function defined on a rectangle $R = [a, b] \times [c, d]$.
   (a) Write an expression for a double Riemann sum of f. If $f(x, y) \geq 0$, what does the sum represent?
   (b) Write the definition of $\iint_R f(x, y)\, dA$ as a limit.
   (c) What is the geometric interpretation of $\iint_R f(x, y)\, dA$ if $f(x, y) \geq 0$? What if f takes on both positive and negative values?
   (d) How do you evaluate $\iint_R f(x, y)\, dA$?
   (e) What does the Midpoint Rule for double integrals say?
   (f) Write an expression for the average value of f.

2. (a) How do you define $\iint_D f(x, y)\, dA$ if D is a bounded region that is not a rectangle?
   (b) What is a type I region? How do you evaluate $\iint_D f(x, y)\, dA$ if D is a type I region?
   (c) What is a type II region? How do you evaluate $\iint_D f(x, y)\, dA$ if D is a type II region?
   (d) What properties do double integrals have?

3. How do you change from rectangular coordinates to polar coordinates in a double integral? Why would you want to make the change?

4. If a lamina occupies a plane region D and has density function $\rho(x, y)$, write expressions for each of the following in terms of double integrals.
   (a) The mass
   (b) The moments about the axes
   (c) The center of mass
   (d) The moments of inertia about the axes and the origin

5. Let f be a joint density function of a pair of continuous random variables X and Y.
   (a) Write a double integral for the probability that X lies between a and b and Y lies between c and d.
   (b) What properties does f possess?
   (c) What are the expected values of X and Y?

6. Write an expression for the area of a surface S for each of the following cases.
   (a) S is a parametric surface given by a vector function $\mathbf{r}(u, v), (u, v) \in D$.
   (b) S has the equation $z = f(x, y), (x, y) \in D$.
   (c) S is the surface of revolution obtained by rotating the curve $y = f(x), a \leq x \leq b$, about the x-axis.

7. (a) Write the definition of the triple integral of f over a rectangular box B.
   (b) How do you evaluate $\iiint_B f(x, y, z)\, dV$?
   (c) How do you define $\iiint_E f(x, y, z)\, dV$ if E is a bounded solid region that is not a box?
   (d) What is a type 1 solid region? How do you evaluate $\iiint_E f(x, y, z)\, dV$ if E is such a region?
   (e) What is a type 2 solid region? How do you evaluate $\iiint_E f(x, y, z)\, dV$ if E is such a region?
   (f) What is a type 3 solid region? How do you evaluate $\iiint_E f(x, y, z)\, dV$ if E is such a region?

8. Suppose a solid object occupies the region E and has density function $\rho(x, y, z)$. Write expressions for each of the following.
   (a) The mass
   (b) The moments about the coordinate planes
   (c) The coordinates of the center of mass
   (d) The moments of inertia about the axes

9. (a) How do you change from rectangular coordinates to cylindrical coordinates in a triple integral?
   (b) How do you change from rectangular coordinates to spherical coordinates in a triple integral?
   (c) In what situations would you change to cylindrical or spherical coordinates?

10. (a) If a transformation T is given by $x = g(u, v)$, $y = h(u, v)$, what is the Jacobian of T?
    (b) How do you change variables in a double integral?
    (c) How do you change variables in a triple integral?

Determine whether the statement is true or false. If it is true, explain why. If it is false, explain why or give an example that disproves the statement.

1. $\int_{-1}^{2} \int_{0}^{6} x^2 \sin(x-y)\, dx\, dy = \int_{0}^{6} \int_{-1}^{2} x^2 \sin(x-y)\, dy\, dx$

2. $\int_{0}^{1} \int_{0}^{x} \sqrt{x+y^2}\, dy\, dx = \int_{0}^{x} \int_{0}^{1} \sqrt{x+y^2}\, dx\, dy$

3. $\int_{1}^{2} \int_{3}^{4} x^2 e^y\, dy\, dx = \int_{1}^{2} x^2\, dx \int_{3}^{4} e^y\, dy$

4. $\int_{-1}^{1} \int_{0}^{1} e^{x^2+y^2} \sin y\, dx\, dy = 0$

5. If D is the disk given by $x^2 + y^2 \leq 4$, then

$$\iint_{D} \sqrt{4 - x^2 - y^2}\, dA = \frac{16}{3}\pi$$

6. $\int_{1}^{4} \int_{0}^{1} \left(x^2 + \sqrt{y}\right) \sin(x^2 y^2)\, dx\, dy \leq 9$

7. The integral

$$\int_{0}^{2\pi} \int_{0}^{2} \int_{r}^{2} dz\, dr\, d\theta$$

represents the volume enclosed by the cone $z = \sqrt{x^2 + y^2}$ and the plane $z = 2$.

8. The integral $\iiint_{E} kr^3\, dz\, dr\, d\theta$ represents the moment of inertia about the z-axis of a solid E with constant density k.

---

1. A contour map is shown for a function f on the square $R = [0, 3] \times [0, 3]$. Use a Riemann sum with nine terms to estimate the value of $\iint_{R} f(x, y)\, dA$. Take the sample points to be the upper right corners of the squares.

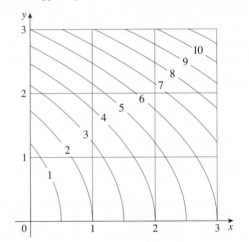

2. Use the Midpoint Rule to estimate the integral in Exercise 1.

3–8 ■ Calculate the iterated integral.

3. $\int_{1}^{2} \int_{0}^{2} (y + 2xe^y)\, dx\, dy$

4. $\int_{0}^{1} \int_{0}^{1} ye^{xy}\, dx\, dy$

5. $\int_{0}^{1} \int_{0}^{x} \cos(x^2)\, dy\, dx$

6. $\int_{0}^{1} \int_{x}^{e^x} 3xy^2\, dy\, dx$

7. $\int_{0}^{\pi} \int_{0}^{1} \int_{0}^{\sqrt{1-y^2}} y \sin x\, dz\, dy\, dx$

8. $\int_{0}^{1} \int_{0}^{y} \int_{x}^{1} 6xyz\, dz\, dx\, dy$

9–10 ■ Write $\iint_{R} f(x, y)\, dA$ as an iterated integral, where R is the region shown and f is an arbitrary continuous function on R.

9.

10.
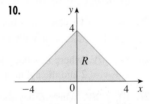

11. Describe the region whose area is given by the integral

$$\int_{0}^{\pi/2} \int_{0}^{\sin 2\theta} r\, dr\, d\theta$$

12. Describe the solid whose volume is given by the integral

$$\int_{0}^{\pi/2} \int_{0}^{\pi/2} \int_{1}^{2} \rho^2 \sin \phi\, d\rho\, d\phi\, d\theta$$

and evaluate the integral.

13–14 ■ Calculate the iterated integral by first reversing the order of integration.

13. $\int_{0}^{1} \int_{x}^{1} \cos(y^2)\, dy\, dx$

14. $\int_{0}^{1} \int_{\sqrt{y}}^{1} \frac{ye^{x^2}}{x^3}\, dx\, dy$

15–28 ■ Calculate the value of the multiple integral.

15. $\iint_{R} ye^{xy}\, dA$, where $R = \{(x, y) \mid 0 \leq x \leq 2,\ 0 \leq y \leq 3\}$

16. $\iint_{D} xy\, dA$, where $D = \{(x, y) \mid 0 \leq y \leq 1,\ y^2 \leq x \leq y + 2\}$

17. $\iint\limits_{D} \dfrac{y}{1 + x^2}\, dA$,

where D is bounded by $y = \sqrt{x}$, $y = 0$, $x = 1$

18. $\iint\limits_{D} \dfrac{1}{1 + x^2}\, dA$, where D is the triangular region with

vertices $(0, 0)$, $(1, 1)$, and $(0, 1)$

19. $\iint_{D} y\, dA$, where D is the region in the first quadrant bounded by the parabolas $x = y^2$ and $x = 8 - y^2$

20. $\iint_{D} y\, dA$, where D is the region in the first quadrant that lies above the hyperbola $xy = 1$ and the line $y = x$ and below the line $y = 2$

21. $\iint_{D} (x^2 + y^2)^{3/2}\, dA$, where D is the region in the first quadrant bounded by the lines $y = 0$ and $y = \sqrt{3}x$ and the circle $x^2 + y^2 = 9$

22. $\iint_{D} x\, dA$, where D is the region in the first quadrant that lies between the circles $x^2 + y^2 = 1$ and $x^2 + y^2 = 2$

23. $\iiint_{E} xy\, dV$, where
$E = \{(x, y, z)\,|\,0 \leqslant x \leqslant 3,\ 0 \leqslant y \leqslant x,\ 0 \leqslant z \leqslant x + y\}$

24. $\iiint_{T} xy\, dV$, where T is the solid tetrahedron with vertices $(0, 0, 0)$, $(\tfrac{1}{3}, 0, 0)$, $(0, 1, 0)$, and $(0, 0, 1)$

25. $\iiint_{E} y^2 z^2\, dV$, where E is bounded by the paraboloid $x = 1 - y^2 - z^2$ and the plane $x = 0$

26. $\iiint_{E} z\, dV$, where E is bounded by the planes $y = 0$, $z = 0$, $x + y = 2$ and the cylinder $y^2 + z^2 = 1$ in the first octant

27. $\iiint_{E} yz\, dV$, where E lies above the plane $z = 0$, below the plane $z = y$, and inside the cylinder $x^2 + y^2 = 4$

28. $\iiint_{H} z^3\sqrt{x^2 + y^2 + z^2}\, dV$, where H is the solid hemisphere that lies above the xy-plane and has center the origin and radius 1

29–34 ■ Find the volume of the given solid.

29. Under the paraboloid $z = x^2 + 4y^2$ and above the rectangle $R = [0, 2] \times [1, 4]$

30. Under the surface $z = x^2 y$ and above the triangle in the xy-plane with vertices $(1, 0)$, $(2, 1)$, and $(4, 0)$

31. The solid tetrahedron with vertices $(0, 0, 0)$, $(0, 0, 1)$, $(0, 2, 0)$, and $(2, 2, 0)$

32. Bounded by the cylinder $x^2 + y^2 = 4$ and the planes $z = 0$ and $y + z = 3$

33. One of the wedges cut from the cylinder $x^2 + 9y^2 = a^2$ by the planes $z = 0$ and $z = mx$

34. Above the paraboloid $z = x^2 + y^2$ and below the half-cone $z = \sqrt{x^2 + y^2}$

35. Consider a lamina that occupies the region D bounded by the parabola $x = 1 - y^2$ and the coordinate axes in the first quadrant with density function $\rho(x, y) = y$.
(a) Find the mass of the lamina.
(b) Find the center of mass.
(c) Find the moments of inertia and radii of gyration about the x- and y-axes.

36. A lamina occupies the part of the disk $x^2 + y^2 \leqslant a^2$ that lies in the first quadrant.
(a) Find the centroid of the lamina.
(b) Find the center of mass of the lamina if the density function is $\rho(x, y) = xy^2$.

37. (a) Find the centroid of a right circular cone with height h and base radius a. (Place the cone so that its base is in the xy-plane with center the origin and its axis along the positive z-axis.)
(b) Find the moment of inertia of the cone about its axis (the z-axis).

38. (a) Set up, but don't evaluate, an integral for the surface area of the parametric surface given by the vector function $\mathbf{r}(u, v) = v^2\mathbf{i} - uv\mathbf{j} + u^2\mathbf{k}$, $0 \leqslant u \leqslant 3$, $-3 \leqslant v \leqslant 3$.
CAS (b) Use a computer algebra system to approximate the surface area correct to four significant digits.

39. Find the area of the part of the surface $z = x^2 + y$ that lies above the triangle with vertices $(0, 0)$, $(1, 0)$, and $(0, 2)$.

CAS 40. Graph the surface $z = x \sin y$, $-3 \leqslant x \leqslant 3$, $-\pi \leqslant y \leqslant \pi$, and find its surface area correct to four decimal places.

41. Use polar coordinates to evaluate
$$\int_{0}^{3}\int_{-\sqrt{9-x^2}}^{\sqrt{9-x^2}} (x^3 + xy^2)\, dy\, dx$$

42. Use spherical coordinates to evaluate
$$\int_{-2}^{2}\int_{0}^{\sqrt{4-y^2}}\int_{-\sqrt{4-x^2-y^2}}^{\sqrt{4-x^2-y^2}} y^2\sqrt{x^2 + y^2 + z^2}\, dz\, dx\, dy$$

43. If D is the region bounded by the curves $y = 1 - x^2$ and $y = e^x$, find the approximate value of the integral $\iint_{D} y^2\, dA$. (Use a graphing device to estimate the points of intersection of the curves.)

CAS 44. Find the center of mass of the solid tetrahedron with vertices $(0, 0, 0)$, $(1, 0, 0)$, $(0, 2, 0)$, $(0, 0, 3)$ and density function $\rho(x, y, z) = x^2 + y^2 + z^2$.

45. The joint density function for random variables X and Y is
$$f(x, y) = \begin{cases} C(x + y) & \text{if } 0 \leqslant x \leqslant 3,\ 0 \leqslant y \leqslant 2 \\ 0 & \text{otherwise} \end{cases}$$

(a) Find the value of the constant C.
(b) Find $P(X \leqslant 2, Y \geqslant 1)$.
(c) Find $P(X + Y \leqslant 1)$.

46. A lamp has three bulbs, each of a type with average lifetime 800 hours. If we model the probability of failure of the bulbs by an exponential density function with mean 800, find the probability that all three bulbs fail within a total of 1000 hours.

47. Rewrite the integral

$$\int_{-1}^{1} \int_{x^2}^{1} \int_{0}^{1-y} f(x, y, z) \, dz \, dy \, dx$$

as an iterated integral in the order $dx \, dy \, dz$.

48. Give five other iterated integrals that are equal to

$$\int_{0}^{2} \int_{0}^{y^3} \int_{0}^{y^2} f(x, y, z) \, dz \, dx \, dy$$

49. Use the transformation $u = x - y$, $v = x + y$ to evaluate $\iint_R (x - y)/(x + y) \, dA$, where R is the square with vertices $(0, 2)$, $(1, 1)$, $(2, 2)$, and $(1, 3)$.

50. Use the transformation $x = u^2$, $y = v^2$, $z = w^2$ to find the volume of the region bounded by the surface $\sqrt{x} + \sqrt{y} + \sqrt{z} = 1$ and the coordinate planes.

51. Use the change of variables formula and an appropriate transformation to evaluate $\iint_R xy \, dA$, where R is the square with vertices $(0, 0)$, $(1, 1)$, $(2, 0)$, and $(1, -1)$.

52. (a) Evaluate $\displaystyle\iint_D \frac{1}{(x^2 + y^2)^{n/2}} \, dA$, where n is an integer and D is the region bounded by the circles with center the origin and radii r and R, $0 < r < R$.

    (b) For what values of n does the integral in part (a) have a limit as $r \to 0^+$?

    (c) Find $\displaystyle\iiint_E \frac{1}{(x^2 + y^2 + z^2)^{n/2}} \, dV$, where E is the region bounded by the spheres with center the origin and radii r and R, $0 < r < R$.

    (d) For what values of n does the integral in part (c) have a limit as $r \to 0^+$?

1. If $[\![x]\!]$ denotes the greatest integer in x, evaluate the integral

$$\iint_R [\![x + y]\!] \, dA$$

where $R = \{(x, y) \mid 1 \leqslant x \leqslant 3, \; 2 \leqslant y \leqslant 5\}$.

2. Evaluate the integral

$$\int_0^1 \int_0^1 e^{\max\{x^2, y^2\}} \, dy \, dx$$

where $\max\{x^2, y^2\}$ means the larger of the numbers x^2 and y^2.

3. Find the average value of the function $f(x) = \int_x^1 \cos(t^2) \, dt$ on the interval $[0, 1]$.

4. If $\mathbf{a}$, $\mathbf{b}$, and $\mathbf{c}$ are constant vectors, $\mathbf{r}$ is the position vector $x\mathbf{i} + y\mathbf{j} + z\mathbf{k}$, and E is given by the inequalities $0 \leqslant \mathbf{a} \cdot \mathbf{r} \leqslant \alpha$, $0 \leqslant \mathbf{b} \cdot \mathbf{r} \leqslant \beta$, $0 \leqslant \mathbf{c} \cdot \mathbf{r} \leqslant \gamma$, show that

$$\iiint_E (\mathbf{a} \cdot \mathbf{r})(\mathbf{b} \cdot \mathbf{r})(\mathbf{c} \cdot \mathbf{r}) \, dV = \frac{(\alpha \beta \gamma)^2}{8|\mathbf{a} \cdot (\mathbf{b} \times \mathbf{c})|}$$

5. The double integral $\int_0^1 \int_0^1 \dfrac{1}{1 - xy} \, dx \, dy$ is an improper integral and could be defined as the limit of double integrals over the rectangle $[0, t] \times [0, t]$ as $t \to 1^-$. But if we expand the integrand as a geometric series, we can express the integral as the sum of an infinite series. Show that

$$\int_0^1 \int_0^1 \frac{1}{1 - xy} \, dx \, dy = \sum_{n=1}^{\infty} \frac{1}{n^2}$$

6. Leonhard Euler was able to find the exact sum of the series in Problem 5. In 1736 he proved that

$$\sum_{n=1}^{\infty} \frac{1}{n^2} = \frac{\pi^2}{6}$$

In this problem we ask you to prove this fact by evaluating the double integral in Problem 5. Start by making the change of variables

$$x = \frac{u - v}{\sqrt{2}} \qquad y = \frac{u + v}{\sqrt{2}}$$

This gives a rotation about the origin through the angle $\pi/4$. You will need to sketch the corresponding region in the uv-plane.

[*Hint:* If, in evaluating the integral, you encounter either of the expressions $(1 - \sin \theta)/\cos \theta$ or $(\cos \theta)/(1 + \sin \theta)$, you might like to use the identity $\cos \theta = \sin((\pi/2) - \theta)$ and the corresponding identity for $\sin \theta$.]

7. (a) Show that

$$\int_0^1 \int_0^1 \int_0^1 \frac{1}{1 - xyz} \, dx \, dy \, dz = \sum_{n=1}^{\infty} \frac{1}{n^3}$$

(Nobody has ever been able to find the exact value of the sum of this series.)

(b) Show that

$$\int_0^1 \int_0^1 \int_0^1 \frac{1}{1 + xyz} \, dx \, dy \, dz = \sum_{n=1}^{\infty} \frac{(-1)^{n-1}}{n^3}$$

Use this equation to evaluate the triple integral correct to two decimal places.

8. Show that

$$\int_0^\infty \frac{\arctan \pi x - \arctan x}{x} \, dx = \frac{\pi}{2} \ln \pi$$

by first expressing the integral as an iterated integral.

9. If f is continuous, show that

$$\int_0^x \int_0^y \int_0^z f(t) \, dt \, dz \, dy = \frac{1}{2} \int_0^x (x - t)^2 f(t) \, dt$$

10. (a) A lamina has constant density ρ and takes the shape of a disk with center the origin and radius R. Use Newton's Law of Gravitation (see page 723) to show that the magnitude of the force of attraction that the lamina exerts on a body with mass m located at the point $(0, 0, d)$ on the positive z-axis is

$$F = 2\pi Gm\rho d \left(\frac{1}{d} - \frac{1}{\sqrt{R^2 + d^2}} \right)$$

[*Hint:* Divide the disk as in Figure 4 in Section 12.4 and first compute the vertical component of the force exerted by the polar subrectangle R_{ij}.]

(b) Show that the magnitude of the force of attraction of a lamina with density ρ that occupies an entire plane on an object with mass m located at a distance d from the plane is

$$F = 2\pi Gm\rho$$

Notice that this expression does not depend on d.

13

Vector Calculus

In this chapter we study the calculus of vector fields. (These are functions that assign vectors to points in space.) In particular we define line integrals (which can be used to find the work done by a force field in moving an object along a curve). Then we define surface integrals (which can be used to find the rate of fluid flow across a surface). The connections between these new types of integrals and the single, double, and triple integrals that we have already met are given by the higher-dimensional versions of the Fundamental Theorem of Calculus: Green's Theorem, Stokes' Theorem, and the Divergence Theorem.

13.1 Vector Fields

The vectors in Figure 1 are air velocity vectors that indicate the wind speed and direction at points 10 m above the surface elevation in the San Francisco Bay area. We see at a glance from the largest arrows in part (a) that the greatest wind speeds at that time occurred as the winds entered the bay across the Golden Gate Bridge. Part (b) shows the very different wind pattern at a later date. Associated with every point in the air we can imagine a wind velocity vector. This is an example of a *velocity vector field*.

(a) 12:00 P.M., June 11, 2002

(b) 4:00 P.M., June 30, 2002

FIGURE 1 Velocity vector fields showing San Francisco Bay wind patterns

Other examples of velocity vector fields are illustrated in Figure 2: ocean currents and flow past an airfoil.

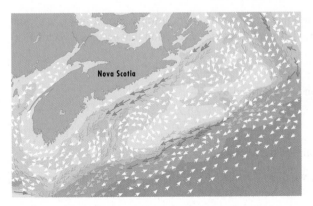

(a) Ocean currents off the coast of Nova Scotia

(b) Airflow past an inclined airfoil

FIGURE 2 Velocity vector fields

Another type of vector field, called a *force field*, associates a force vector with each point in a region. An example is the gravitational force field that we will look at in Example 4.

In general, a vector field is a function whose domain is a set of points in $\mathbb{R}^2$ (or $\mathbb{R}^3$) and whose range is a set of vectors in V_2 (or V_3).

---

1 **Definition** Let D be a set in $\mathbb{R}^2$ (a plane region). A **vector field on** $\mathbb{R}^2$ is a function $\mathbf{F}$ that assigns to each point (x, y) in D a two-dimensional vector $\mathbf{F}(x, y)$.

The best way to picture a vector field is to draw the arrow representing the vector $\mathbf{F}(x, y)$ starting at the point (x, y). Of course, it's impossible to do this for all points (x, y), but we can gain a reasonable impression of $\mathbf{F}$ by doing it for a few representative points in D as in Figure 3. Since $\mathbf{F}(x, y)$ is a two-dimensional vector, we can write it in terms of its **component functions** P and Q as follows:

$$\mathbf{F}(x, y) = P(x, y)\, \mathbf{i} + Q(x, y)\, \mathbf{j} = \langle P(x, y), Q(x, y)\rangle$$

or, for short,
$$\mathbf{F} = P\,\mathbf{i} + Q\,\mathbf{j}$$

Notice that P and Q are scalar functions of two variables and are sometimes called **scalar fields** to distinguish them from vector fields.

FIGURE 3
Vector field on $\mathbb{R}^2$

---

2 **Definition** Let E be a subset of $\mathbb{R}^3$. A **vector field on** $\mathbb{R}^3$ is a function $\mathbf{F}$ that assigns to each point (x, y, z) in E a three-dimensional vector $\mathbf{F}(x, y, z)$.

A vector field $\mathbf{F}$ on $\mathbb{R}^3$ is pictured in Figure 4. We can express it in terms of its component functions P, Q, and R as

$$\mathbf{F}(x, y, z) = P(x, y, z)\, \mathbf{i} + Q(x, y, z)\, \mathbf{j} + R(x, y, z)\, \mathbf{k}$$

As with the vector functions in Section 10.1, we can define continuity of vector fields and show that $\mathbf{F}$ is continuous if and only if its component functions P, Q, and R are continuous.

We sometimes identify a point (x, y, z) with its position vector $\mathbf{x} = \langle x, y, z\rangle$ and write $\mathbf{F}(\mathbf{x})$ instead of $\mathbf{F}(x, y, z)$. Then $\mathbf{F}$ becomes a function that assigns a vector $\mathbf{F}(\mathbf{x})$ to a vector $\mathbf{x}$.

FIGURE 4
Vector field on $\mathbb{R}^3$

[V] **EXAMPLE 1** A vector field on $\mathbb{R}^2$ is defined by

$$\mathbf{F}(x, y) = -y\,\mathbf{i} + x\,\mathbf{j}$$

Describe $\mathbf{F}$ by sketching some of the vectors $\mathbf{F}(x, y)$ as in Figure 3.

SOLUTION Since $\mathbf{F}(1, 0) = \mathbf{j}$, we draw the vector $\mathbf{j} = \langle 0, 1\rangle$ starting at the point $(1, 0)$ in Figure 5. Since $\mathbf{F}(0, 1) = -\mathbf{i}$, we draw the vector $\langle -1, 0\rangle$ with starting point $(0, 1)$. Continuing in this way, we calculate several other representative values of

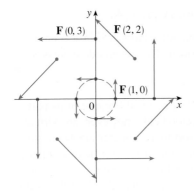

FIGURE 5
$\mathbf{F}(x, y) = -y\,\mathbf{i} + x\,\mathbf{j}$

$\mathbf{F}(x, y)$ in the table and draw the corresponding vectors to represent the vector field in Figure 5.

(x, y)	$\mathbf{F}(x, y)$	(x, y)	$\mathbf{F}(x, y)$
$(1, 0)$	$\langle 0, 1 \rangle$	$(-1, 0)$	$\langle 0, -1 \rangle$
$(2, 2)$	$\langle -2, 2 \rangle$	$(-2, -2)$	$\langle 2, -2 \rangle$
$(3, 0)$	$\langle 0, 3 \rangle$	$(-3, 0)$	$\langle 0, -3 \rangle$
$(0, 1)$	$\langle -1, 0 \rangle$	$(0, -1)$	$\langle 1, 0 \rangle$
$(-2, 2)$	$\langle -2, -2 \rangle$	$(2, -2)$	$\langle 2, 2 \rangle$
$(0, 3)$	$\langle -3, 0 \rangle$	$(0, -3)$	$\langle 3, 0 \rangle$

It appears from Figure 5 that each arrow is tangent to a circle with center the origin. To confirm this, we take the dot product of the position vector $\mathbf{x} = x\,\mathbf{i} + y\,\mathbf{j}$ with the vector $\mathbf{F}(\mathbf{x}) = \mathbf{F}(x, y)$:

$$\mathbf{x} \cdot \mathbf{F}(\mathbf{x}) = (x\,\mathbf{i} + y\,\mathbf{j}) \cdot (-y\,\mathbf{i} + x\,\mathbf{j})$$
$$= -xy + yx = 0$$

This shows that $\mathbf{F}(x, y)$ is perpendicular to the position vector $\langle x, y \rangle$ and is therefore tangent to a circle with center the origin and radius $|\mathbf{x}| = \sqrt{x^2 + y^2}$. Notice also that

$$|\mathbf{F}(x, y)| = \sqrt{(-y)^2 + x^2} = \sqrt{x^2 + y^2} = |\mathbf{x}|$$

so the magnitude of the vector $\mathbf{F}(x, y)$ is equal to the radius of the circle. ■ ■

Some computer algebra systems are capable of plotting vector fields in two or three dimensions. They give a better impression of the vector field than is possible by hand because the computer can plot a large number of representative vectors. Figure 6 shows a computer plot of the vector field in Example 1; Figures 7 and 8 show two other vector fields. Notice that the computer scales the lengths of the vectors so they are not too long and yet are proportional to their true lengths.

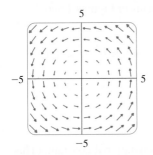

FIGURE 6
$\mathbf{F}(x, y) = \langle -y, x \rangle$

FIGURE 7
$\mathbf{F}(x, y) = \langle y, \sin x \rangle$

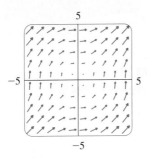

FIGURE 8
$\mathbf{F}(x, y) = \langle \ln(1 + y^2), \ln(1 + x^2) \rangle$

FIGURE 9
$\mathbf{F}(x, y, z) = z\,\mathbf{k}$

Ⓥ **EXAMPLE 2** Sketch the vector field on $\mathbb{R}^3$ given by $\mathbf{F}(x, y, z) = z\,\mathbf{k}$.

SOLUTION The sketch is shown in Figure 9. Notice that all vectors are vertical and point upward above the xy-plane or downward below it. The magnitude increases with the distance from the xy-plane. ■ ■

We were able to draw the vector field in Example 2 by hand because of its particularly simple formula. Most three-dimensional vector fields, however, are virtually impossible to sketch by hand and so we need to resort to a computer algebra system. Examples are shown in Figures 10, 11, and 12. Notice that the vector fields in Figures 10 and 11 have similar formulas, but all the vectors in Figure 11 point in the general direction of the negative y-axis because their y-components are all -2. If the vector field in Figure 12 represents a velocity field, then a particle would be swept upward and would spiral around the z-axis in the clockwise direction as viewed from above.

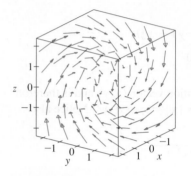

FIGURE 10
$\mathbf{F}(x, y, z) = y\,\mathbf{i} + z\,\mathbf{j} + x\,\mathbf{k}$

FIGURE 11
$\mathbf{F}(x, y, z) = y\,\mathbf{i} - 2\,\mathbf{j} + x\,\mathbf{k}$

FIGURE 12
$\mathbf{F}(x, y, z) = \dfrac{y}{z}\,\mathbf{i} - \dfrac{x}{z}\,\mathbf{j} + \dfrac{z}{4}\,\mathbf{k}$

 In Visual 13.1 you can rotate the vector fields in Figures 10–12 as well as additional fields.

FIGURE 13
Velocity field in fluid flow

EXAMPLE 3 Imagine a fluid flowing steadily along a pipe and let $\mathbf{V}(x, y, z)$ be the velocity vector at a point (x, y, z). Then $\mathbf{V}$ assigns a vector to each point (x, y, z) in a certain domain E (the interior of the pipe) and so $\mathbf{V}$ is a vector field on $\mathbb{R}^3$ called a **velocity field**. A possible velocity field is illustrated in Figure 13. The speed at any given point is indicated by the length of the arrow.

Velocity fields also occur in other areas of physics. For instance, the vector field in Example 1 could be used as the velocity field describing the counterclockwise rotation of a wheel. We have seen other examples of velocity fields in Figures 1 and 2. ■ ■

EXAMPLE 4 Newton's Law of Gravitation states that the magnitude of the gravitational force between two objects with masses m and M is

$$|\mathbf{F}| = \frac{mMG}{r^2}$$

where r is the distance between the objects and G is the gravitational constant. (This is an example of an inverse square law.) Let's assume that the object with mass M is located at the origin in $\mathbb{R}^3$. (For instance, M could be the mass of the Earth and the origin would be at its center.) Let the position vector of the object with mass m be $\mathbf{x} = \langle x, y, z \rangle$. Then $r = |\mathbf{x}|$, so $r^2 = |\mathbf{x}|^2$. The gravitational force exerted on this second object acts toward the origin, and the unit vector in this direction is

$$-\frac{\mathbf{x}}{|\mathbf{x}|}$$

Therefore, the gravitational force acting on the object at $\mathbf{x} = \langle x, y, z \rangle$ is

$$\boxed{3} \qquad \mathbf{F}(\mathbf{x}) = -\frac{mMG}{|\mathbf{x}|^3}\,\mathbf{x}$$

[Physicists often use the notation $\mathbf{r}$ instead of $\mathbf{x}$ for the position vector, so you may see Formula 3 written in the form $\mathbf{F} = -(mMG/r^3)\mathbf{r}$.] The function given by Equation 3 is an example of a vector field, called the **gravitational field**, because it associates a vector [the force $\mathbf{F}(\mathbf{x})$] with every point $\mathbf{x}$ in space.

Formula 3 is a compact way of writing the gravitational field, but we can also write it in terms of its component functions by using the facts that $\mathbf{x} = x\,\mathbf{i} + y\,\mathbf{j} + z\,\mathbf{k}$ and $|\mathbf{x}| = \sqrt{x^2 + y^2 + z^2}$:

$$\mathbf{F}(x, y, z) = \frac{-mMGx}{(x^2 + y^2 + z^2)^{3/2}}\,\mathbf{i} + \frac{-mMGy}{(x^2 + y^2 + z^2)^{3/2}}\,\mathbf{j} + \frac{-mMGz}{(x^2 + y^2 + z^2)^{3/2}}\,\mathbf{k}$$

FIGURE 14
Gravitational force field

The gravitational field $\mathbf{F}$ is pictured in Figure 14. ■ ■

EXAMPLE 5 Suppose an electric charge Q is located at the origin. According to Coulomb's Law, the electric force $\mathbf{F}(\mathbf{x})$ exerted by this charge on a charge q located at a point (x, y, z) with position vector $\mathbf{x} = \langle x, y, z \rangle$ is

$$\boxed{4} \qquad \mathbf{F}(\mathbf{x}) = \frac{\varepsilon q Q}{|\mathbf{x}|^3}\,\mathbf{x}$$

where ε is a constant (that depends on the units used). For like charges, we have $qQ > 0$ and the force is repulsive; for unlike charges, we have $qQ < 0$ and the force is attractive. Notice the similarity between Formulas 3 and 4. Both vector fields are examples of **force fields**.

Instead of considering the electric force $\mathbf{F}$, physicists often consider the force per unit charge:

$$\mathbf{E}(\mathbf{x}) = \frac{1}{q}\,\mathbf{F}(\mathbf{x}) = \frac{\varepsilon Q}{|\mathbf{x}|^3}\,\mathbf{x}$$

Then $\mathbf{E}$ is a vector field on $\mathbb{R}^3$ called the **electric field** of Q. ■ ■

▪ Gradient Fields

If f is a scalar function of two variables, recall from Section 11.6 that its gradient ∇f (or grad f) is defined by

$$\nabla f(x, y) = f_x(x, y)\,\mathbf{i} + f_y(x, y)\,\mathbf{j}$$

Therefore, ∇f is really a vector field on $\mathbb{R}^2$ and is called a **gradient vector field**. Likewise, if f is a scalar function of three variables, its gradient is a vector field on $\mathbb{R}^3$ given by

$$\nabla f(x, y, z) = f_x(x, y, z)\,\mathbf{i} + f_y(x, y, z)\,\mathbf{j} + f_z(x, y, z)\,\mathbf{k}$$

$\boxed{V}$ **EXAMPLE 6** Find the gradient vector field of $f(x, y) = x^2y - y^3$. Plot the gradient vector field together with a contour map of f. How are they related?

SOLUTION The gradient vector field is given by

$$\nabla f(x, y) = \frac{\partial f}{\partial x}\mathbf{i} + \frac{\partial f}{\partial y}\mathbf{j} = 2xy\,\mathbf{i} + (x^2 - 3y^2)\mathbf{j}$$

Figure 15 shows a contour map of f with the gradient vector field. Notice that the gradient vectors are perpendicular to the level curves, as we would expect from Section 11.6. Notice also that the gradient vectors are long where the level curves are close to each other and short where they are farther apart. That's because the length of the gradient vector is the value of the directional derivative of f and closely spaced level curves indicate a steep graph. ■ ■

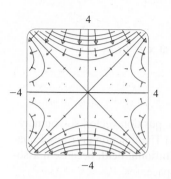

FIGURE 15

A vector field $\mathbf{F}$ is called a **conservative vector field** if it is the gradient of some scalar function, that is, if there exists a function f such that $\mathbf{F} = \nabla f$. In this situation f is called a **potential function** for $\mathbf{F}$.

Not all vector fields are conservative, but such fields do arise frequently in physics. For example, the gravitational field $\mathbf{F}$ in Example 4 is conservative because if we define

$$f(x, y, z) = \frac{mMG}{\sqrt{x^2 + y^2 + z^2}}$$

then

$$\nabla f(x, y, z) = \frac{\partial f}{\partial x}\mathbf{i} + \frac{\partial f}{\partial y}\mathbf{j} + \frac{\partial f}{\partial z}\mathbf{k}$$

$$= \frac{-mMGx}{(x^2 + y^2 + z^2)^{3/2}}\mathbf{i} + \frac{-mMGy}{(x^2 + y^2 + z^2)^{3/2}}\mathbf{j} + \frac{-mMGz}{(x^2 + y^2 + z^2)^{3/2}}\mathbf{k}$$

$$= \mathbf{F}(x, y, z)$$

In Sections 13.3 and 13.5 we will learn how to tell whether or not a given vector field is conservative.

13.1 Exercises

1–10 ■ Sketch the vector field $\mathbf{F}$ by drawing a diagram like Figure 5 or Figure 9.

1. $\mathbf{F}(x, y) = \frac{1}{2}(\mathbf{i} + \mathbf{j})$

2. $\mathbf{F}(x, y) = \mathbf{i} + x\mathbf{j}$

3. $\mathbf{F}(x, y) = y\mathbf{i} + \frac{1}{2}\mathbf{j}$

4. $\mathbf{F}(x, y) = (x - y)\mathbf{i} + x\mathbf{j}$

5. $\mathbf{F}(x, y) = \dfrac{y\mathbf{i} + x\mathbf{j}}{\sqrt{x^2 + y^2}}$

6. $\mathbf{F}(x, y) = \dfrac{y\mathbf{i} - x\mathbf{j}}{\sqrt{x^2 + y^2}}$

7. $\mathbf{F}(x, y, z) = \mathbf{k}$

8. $\mathbf{F}(x, y, z) = -y\mathbf{k}$

9. $\mathbf{F}(x, y, z) = x\mathbf{k}$

10. $\mathbf{F}(x, y, z) = \mathbf{j} - \mathbf{i}$

11–14 ■ Match the vector fields $\mathbf{F}$ with the plots labeled I–IV. Give reasons for your choices.

11. $\mathbf{F}(x, y) = \langle y, x \rangle$

12. $\mathbf{F}(x, y) = \langle 1, \sin y \rangle$

13. $\mathbf{F}(x, y) = \langle x - 2, x + 1 \rangle$

14. $\mathbf{F}(x, y) = \langle y, 1/x \rangle$

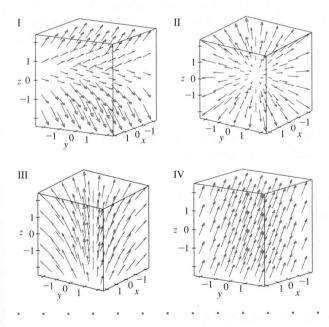

Mathematica), use it to plot

$$\mathbf{F}(x, y) = (y^2 - 2xy)\,\mathbf{i} + (3xy - 6x^2)\,\mathbf{j}$$

Explain the appearance by finding the set of points (x, y) such that $\mathbf{F}(x, y) = \mathbf{0}$.

20. Let $\mathbf{F}(\mathbf{x}) = (r^2 - 2r)\mathbf{x}$, where $\mathbf{x} = \langle x, y \rangle$ and $r = |\mathbf{x}|$. Use a CAS to plot this vector field in various domains until you can see what is happening. Describe the appearance of the plot and explain it by finding the points where $\mathbf{F}(\mathbf{x}) = \mathbf{0}$.

21–24 ■ Find the gradient vector field of f.

21. $f(x, y) = \ln(x + 2y)$
22. $f(x, y) = x^\alpha e^{-\beta x}$

23. $f(x, y, z) = \sqrt{x^2 + y^2 + z^2}$
24. $f(x, y, z) = x \cos(y/z)$

25–26 ■ Find the gradient vector field ∇f of f and sketch it.

25. $f(x, y) = xy - 2x$
26. $f(x, y) = \frac{1}{4}(x + y)^2$

27–28 ■ Plot the gradient vector field of f together with a contour map of f. Explain how they are related to each other.

27. $f(x, y) = \sin x + \sin y$
28. $f(x, y) = \sin(x + y)$

29–32 ■ Match the functions f with the plots of their gradient vector fields (labeled I–IV). Give reasons for your choices.

29. $f(x, y) = xy$
30. $f(x, y) = x^2 - y^2$

31. $f(x, y) = x^2 + y^2$
32. $f(x, y) = \sqrt{x^2 + y^2}$

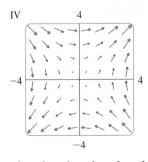

15–18 ■ Match the vector fields $\mathbf{F}$ on $\mathbb{R}^3$ with the plots labeled I–IV. Give reasons for your choices.

15. $\mathbf{F}(x, y, z) = \mathbf{i} + 2\mathbf{j} + 3\mathbf{k}$

16. $\mathbf{F}(x, y, z) = \mathbf{i} + 2\mathbf{j} + z\mathbf{k}$

17. $\mathbf{F}(x, y, z) = x\mathbf{i} + y\mathbf{j} + 3\mathbf{k}$

18. $\mathbf{F}(x, y, z) = x\mathbf{i} + y\mathbf{j} + z\mathbf{k}$

19. If you have a CAS that plots vector fields (the command is fieldplot in Maple and PlotVectorField in

33. A particle moves in a velocity field $\mathbf{V}(x, y) = \langle x^2, x + y^2 \rangle$. If it is at position $(2, 1)$ at time $t = 3$, estimate its location at time $t = 3.01$.

34. At time $t = 1$, a particle is located at position $(1, 3)$. If it moves in a velocity field $\mathbf{F}(x, y) = \langle xy - 2, y^2 - 10 \rangle$, find its approximate location at time $t = 1.05$.

35. The **flow lines** (or **streamlines**) of a vector field are the paths followed by a particle whose velocity field is the given vector field. Thus, the vectors in a vector field are tangent to the flow lines.

(a) Use a sketch of the vector field $\mathbf{F}(x, y) = x\,\mathbf{i} - y\,\mathbf{j}$ to draw some flow lines. From your sketches, can you guess the equations of the flow lines?

(b) If parametric equations of a flow line are $x = x(t)$, $y = y(t)$, explain why these functions satisfy the differ-

ential equations $dx/dt = x$ and $dy/dt = -y$. Then solve the differential equations to find an equation of the flow line that passes through the point $(1, 1)$.

36. (a) Sketch the vector field $\mathbf{F}(x, y) = \mathbf{i} + x\,\mathbf{j}$ and then sketch some flow lines. What shape do these flow lines appear to have?

(b) If parametric equations of the flow lines are $x = x(t)$, $y = y(t)$, what differential equations do these functions satisfy? Deduce that $dy/dx = x$.

(c) If a particle starts at the origin in the velocity field given by $\mathbf{F}$, find an equation of the path it follows.

13.2 Line Integrals

In this section we define an integral that is similar to a single integral except that instead of integrating over an interval $[a, b]$, we integrate over a curve C. Such integrals are called *line integrals,* although "curve integrals" would be better terminology. They were invented in the early 19th century to solve problems involving fluid flow, forces, electricity, and magnetism.

We start with a plane curve C given by the parametric equations

$$\boxed{1} \qquad x = x(t) \qquad y = y(t) \qquad a \leqslant t \leqslant b$$

or, equivalently, by the vector equation $\mathbf{r}(t) = x(t)\,\mathbf{i} + y(t)\,\mathbf{j}$, and we assume that C is a smooth curve. [This means that $\mathbf{r}'$ is continuous and $\mathbf{r}'(t) \neq \mathbf{0}$. See Section 10.2.] If we divide the parameter interval $[a, b]$ into n subintervals $[t_{i-1}, t_i]$ of equal width and we let $x_i = x(t_i)$ and $y_i = y(t_i)$, then the corresponding points $P_i(x_i, y_i)$ divide C into n subarcs with lengths $\Delta s_1, \Delta s_2, \ldots, \Delta s_n$. (See Figure 1.) We choose any point $P_i^*(x_i^*, y_i^*)$ in the ith subarc. (This corresponds to a point t_i^* in $[t_{i-1}, t_i]$.) Now if f is any function of two variables whose domain includes the curve C, we evaluate f at the point (x_i^*, y_i^*), multiply by the length Δs_i of the subarc, and form the sum

$$\sum_{i=1}^{n} f(x_i^*, y_i^*)\,\Delta s_i$$

which is similar to a Riemann sum. Then we take the limit of these sums and make the following definition by analogy with a single integral.

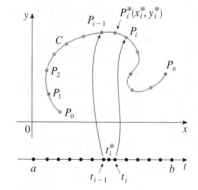

FIGURE 1

---

2 **Definition** If f is defined on a smooth curve C given by Equations 1, then the **line integral of f along C** is

$$\int_C f(x, y)\,ds = \lim_{n \to \infty} \sum_{i=1}^{n} f(x_i^*, y_i^*)\,\Delta s_i$$

if this limit exists.

In Section 6.3 we found that the length of C is

$$L = \int_a^b \sqrt{\left(\frac{dx}{dt}\right)^2 + \left(\frac{dy}{dt}\right)^2}\,dt$$

A similar type of argument can be used to show that if f is a continuous function, then the limit in Definition 2 always exists and the following formula can be used to evaluate the line integral:

$$\boxed{3} \qquad \int_C f(x, y) \, ds = \int_a^b f(x(t), y(t)) \sqrt{\left(\frac{dx}{dt}\right)^2 + \left(\frac{dy}{dt}\right)^2} \, dt$$

The value of the line integral does not depend on the parametrization of the curve, provided that the curve is traversed exactly once as t increases from a to b.

If $s(t)$ is the length of C between $\mathbf{r}(a)$ and $\mathbf{r}(t)$, then

$$\frac{ds}{dt} = \sqrt{\left(\frac{dx}{dt}\right)^2 + \left(\frac{dy}{dt}\right)^2}$$

So the way to remember Formula 3 is to express everything in terms of the parameter t: Use the parametric equations to express x and y in terms of t and write ds as

$$ds = \sqrt{\left(\frac{dx}{dt}\right)^2 + \left(\frac{dy}{dt}\right)^2} \, dt$$

In the special case where C is the line segment that joins $(a, 0)$ to $(b, 0)$, using x as the parameter, we can write the parametric equations of C as follows: $x = x$, $y = 0$, $a \le x \le b$. Formula 3 then becomes

$$\int_C f(x, y) \, ds = \int_a^b f(x, 0) \, dx$$

and so the line integral reduces to an ordinary single integral in this case.

Just as for an ordinary single integral, we can interpret the line integral of a *positive* function as an area. In fact, if $f(x, y) \ge 0$, $\int_C f(x, y) \, ds$ represents the area of one side of the "fence" or "curtain" in Figure 2, whose base is C and whose height above the point (x, y) is $f(x, y)$.

EXAMPLE 1 Evaluate $\int_C (2 + x^2 y) \, ds$, where C is the upper half of the unit circle $x^2 + y^2 = 1$.

SOLUTION In order to use Formula 3 we first need parametric equations to represent C. Recall that the unit circle can be parametrized by means of the equations

$$x = \cos t \qquad y = \sin t$$

and the upper half of the circle is described by the parameter interval $0 \le t \le \pi$. (See Figure 3.) Therefore, Formula 3 gives

$$\int_C (2 + x^2 y) \, ds = \int_0^\pi (2 + \cos^2 t \sin t) \sqrt{\left(\frac{dx}{dt}\right)^2 + \left(\frac{dy}{dt}\right)^2} \, dt$$

$$= \int_0^\pi (2 + \cos^2 t \sin t) \sqrt{\sin^2 t + \cos^2 t} \, dt$$

$$= \int_0^\pi (2 + \cos^2 t \sin t) \, dt = \left[2t - \frac{\cos^3 t}{3} \right]_0^\pi$$

$$= 2\pi + \tfrac{2}{3}$$

■ ■ The arc length function s is discussed in Section 10.3.

FIGURE 2

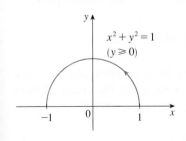

$x^2 + y^2 = 1$
$(y \ge 0)$

FIGURE 3

FIGURE 4

A piecewise-smooth curve

FIGURE 5

$C = C_1 \cup C_2$

Suppose now that C is a **piecewise-smooth curve**; that is, C is a union of a finite number of smooth curves $C_1, C_2, \ldots, C_n$, where, as illustrated in Figure 4, the initial point of C_{i+1} is the terminal point of C_i. Then we define the integral of f along C as the sum of the integrals of f along each of the smooth pieces of C:

$$\int_C f(x, y)\, ds = \int_{C_1} f(x, y)\, ds + \int_{C_2} f(x, y)\, ds + \cdots + \int_{C_n} f(x, y)\, ds$$

EXAMPLE 2 Evaluate $\int_C 2x\, ds$, where C consists of the arc C_1 of the parabola $y = x^2$ from $(0, 0)$ to $(1, 1)$ followed by the vertical line segment C_2 from $(1, 1)$ to $(1, 2)$.

SOLUTION The curve C is shown in Figure 5. C_1 is the graph of a function of x, so we can choose x as the parameter and the equations for C_1 become

$$x = x \qquad y = x^2 \qquad 0 \leqslant x \leqslant 1$$

Therefore

$$\int_{C_1} 2x\, ds = \int_0^1 2x \sqrt{\left(\frac{dx}{dx}\right)^2 + \left(\frac{dy}{dx}\right)^2}\, dx$$

$$= \int_0^1 2x\sqrt{1 + 4x^2}\, dx = \tfrac{1}{4} \cdot \tfrac{2}{3}(1 + 4x^2)^{3/2}\Big]_0^1 = \frac{5\sqrt{5} - 1}{6}$$

On C_2 we choose y as the parameter, so the equations of C_2 are

$$x = 1 \qquad y = y \qquad 1 \leqslant y \leqslant 2$$

and

$$\int_{C_2} 2x\, ds = \int_1^2 2(1)\sqrt{\left(\frac{dx}{dy}\right)^2 + \left(\frac{dy}{dy}\right)^2}\, dy = \int_1^2 2\, dy = 2$$

Thus

$$\int_C 2x\, ds = \int_{C_1} 2x\, ds + \int_{C_2} 2x\, ds = \frac{5\sqrt{5} - 1}{6} + 2 \qquad ■ ■$$

Any physical interpretation of a line integral $\int_C f(x, y)\, ds$ depends on the physical interpretation of the function f. Suppose that $\rho(x, y)$ represents the linear density at a point (x, y) of a thin wire shaped like a curve C. Then the mass of the part of the wire from P_{i-1} to P_i in Figure 1 is approximately $\rho(x_i^*, y_i^*)\,\Delta s_i$ and so the total mass of the wire is approximately $\Sigma\, \rho(x_i^*, y_i^*)\,\Delta s_i$. By taking more and more points on the curve, we obtain the **mass** m of the wire as the limiting value of these approximations:

$$m = \lim_{n \to \infty} \sum_{i=1}^n \rho(x_i^*, y_i^*)\,\Delta s_i = \int_C \rho(x, y)\, ds$$

[For example, if $f(x, y) = 2 + x^2 y$ represents the density of a semicircular wire, then the integral in Example 1 would represent the mass of the wire.] The **center of mass** of the wire with density function ρ is located at the point $(\bar{x}, \bar{y})$, where

4

$$\bar{x} = \frac{1}{m} \int_C x\rho(x, y)\, ds \qquad \bar{y} = \frac{1}{m} \int_C y\rho(x, y)\, ds$$

Other physical interpretations of line integrals will be discussed later in this chapter.

⟨V⟩ **EXAMPLE 3** A wire takes the shape of the semicircle $x^2 + y^2 = 1$, $y \geqslant 0$, and is thicker near its base than near the top. Find the center of mass of the wire if the linear density at any point is proportional to its distance from the line $y = 1$.

SOLUTION As in Example 1 we use the parametrization $x = \cos t$, $y = \sin t$, $0 \leqslant t \leqslant \pi$, and find that $ds = dt$. The linear density is

$$\rho(x, y) = k(1 - y)$$

where k is a constant, and so the mass of the wire is

$$m = \int_C k(1 - y) \, ds = \int_0^\pi k(1 - \sin t) \, dt$$
$$= k\big[t + \cos t\big]_0^\pi = k(\pi - 2)$$

From Equations 4 we have

$$\bar{y} = \frac{1}{m} \int_C y\rho(x, y) \, ds = \frac{1}{k(\pi - 2)} \int_C yk(1 - y) \, ds$$

$$= \frac{1}{\pi - 2} \int_0^\pi (\sin t - \sin^2 t) \, dt = \frac{1}{\pi - 2} \left[-\cos t - \tfrac{1}{2}t + \tfrac{1}{4}\sin 2t \right]_0^\pi$$

$$= \frac{4 - \pi}{2(\pi - 2)}$$

By symmetry we see that $\bar{x} = 0$, so the center of mass is

$$\left(0, \frac{4 - \pi}{2(\pi - 2)} \right) \approx (0, 0.38)$$

See Figure 6. ■ ■

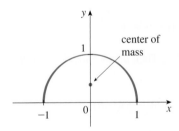

center of mass

FIGURE 6

Two other line integrals are obtained by replacing Δs_i by either $\Delta x_i = x_i - x_{i-1}$ or $\Delta y_i = y_i - y_{i-1}$ in Definition 2. They are called the **line integrals of f along C with respect to x and y**:

$$\boxed{5} \qquad \int_C f(x, y) \, dx = \lim_{n \to \infty} \sum_{i=1}^n f(x_i^*, y_i^*) \, \Delta x_i$$

$$\boxed{6} \qquad \int_C f(x, y) \, dy = \lim_{n \to \infty} \sum_{i=1}^n f(x_i^*, y_i^*) \, \Delta y_i$$

When we want to distinguish the original line integral $\int_C f(x, y) \, ds$ from those in Equations 5 and 6, we call it the **line integral with respect to arc length**.

The following formulas say that line integrals with respect to x and y can also be evaluated by expressing everything in terms of t: $x = x(t)$, $y = y(t)$, $dx = x'(t) \, dt$, $dy = y'(t) \, dt$.

$$\boxed{7} \qquad \int_C f(x, y) \, dx = \int_a^b f(x(t), y(t)) \, x'(t) \, dt$$

$$\int_C f(x, y) \, dy = \int_a^b f(x(t), y(t)) \, y'(t) \, dt$$

It frequently happens that line integrals with respect to x and y occur together. When this happens, it's customary to abbreviate by writing

$$\int_C P(x, y)\, dx + \int_C Q(x, y)\, dy = \int_C P(x, y)\, dx + Q(x, y)\, dy$$

When we are setting up a line integral, sometimes the most difficult thing is to think of a parametric representation for a curve whose geometric description is given. In particular, we often need to parametrize a line segment, so it's useful to remember that a vector representation of the line segment that starts at $\mathbf{r}_0$ and ends at $\mathbf{r}_1$ is given by

$$\boxed{8} \qquad \boxed{\quad \mathbf{r}(t) = (1 - t)\mathbf{r}_0 + t\mathbf{r}_1 \qquad 0 \leqslant t \leqslant 1 \quad}$$

(See Equation 9.5.4.)

$\boxed{V}$ **EXAMPLE 4** Evaluate $\int_C y^2\, dx + x\, dy$, where (a) $C = C_1$ is the line segment from $(-5, -3)$ to $(0, 2)$ and (b) $C = C_2$ is the arc of the parabola $x = 4 - y^2$ from $(-5, -3)$ to $(0, 2)$. (See Figure 7.)

SOLUTION
(a) A parametric representation for the line segment is

$$x = 5t - 5 \qquad y = 5t - 3 \qquad 0 \leqslant t \leqslant 1$$

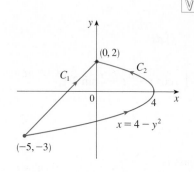

FIGURE 7

(Use Equation 8 with $\mathbf{r}_0 = \langle -5, -3 \rangle$ and $\mathbf{r}_1 = \langle 0, 2 \rangle$.) Then $dx = 5\, dt$, $dy = 5\, dt$, and Formulas 7 give

$$\int_{C_1} y^2\, dx + x\, dy = \int_0^1 (5t - 3)^2 (5\, dt) + (5t - 5)(5\, dt)$$

$$= 5 \int_0^1 (25t^2 - 25t + 4)\, dt$$

$$= 5 \left[\frac{25t^3}{3} - \frac{25t^2}{2} + 4t \right]_0^1 = -\frac{5}{6}$$

(b) Since the parabola is given as a function of y, let's take y as the parameter and write C_2 as

$$x = 4 - y^2 \qquad y = y \qquad -3 \leqslant y \leqslant 2$$

Then $dx = -2y\, dy$ and by Formulas 7 we have

$$\int_{C_2} y^2\, dx + x\, dy = \int_{-3}^2 y^2(-2y)\, dy + (4 - y^2)\, dy$$

$$= \int_{-3}^2 (-2y^3 - y^2 + 4)\, dy$$

$$= \left[-\frac{y^4}{2} - \frac{y^3}{3} + 4y \right]_{-3}^2 = 40\frac{5}{6} \qquad ■\ ■$$

Notice that we got different answers in parts (a) and (b) of Example 4 even though the two curves had the same endpoints. Thus, in general, the value of a line integral depends not just on the endpoints of the curve but also on the path. (But see Section 13.3 for conditions under which the integral is independent of the path.)

Notice also that the answers in Example 4 depend on the direction, or orientation, of the curve. If $-C_1$ denotes the line segment from $(0, 2)$ to $(-5, -3)$, you can verify, using the parametrization

$$x = -5t \qquad y = 2 - 5t \qquad 0 \leqslant t \leqslant 1$$

that

$$\int_{-C_1} y^2 \, dx + x \, dy = \tfrac{5}{6}$$

In general, a given parametrization $x = x(t)$, $y = y(t)$, $a \leqslant t \leqslant b$, determines an **orientation** of a curve C, with the positive direction corresponding to increasing values of the parameter t. (See Figure 8, where the initial point A corresponds to the parameter value a and the terminal point B corresponds to $t = b$.)

If $-C$ denotes the curve consisting of the same points as C but with the opposite orientation (from initial point B to terminal point A in Figure 8), then we have

$$\int_{-C} f(x, y) \, dx = -\int_{C} f(x, y) \, dx \qquad \int_{-C} f(x, y) \, dy = -\int_{C} f(x, y) \, dy$$

But if we integrate with respect to arc length, the value of the line integral does *not* change when we reverse the orientation of the curve:

$$\int_{-C} f(x, y) \, ds = \int_{C} f(x, y) \, ds$$

This is because Δs_i is always positive, whereas Δx_i and Δy_i change sign when we reverse the orientation of C.

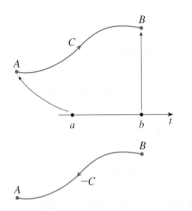

FIGURE 8

▪▪ Line Integrals in Space

We now suppose that C is a smooth space curve given by the parametric equations

$$x = x(t) \qquad y = y(t) \qquad z = z(t) \qquad a \leqslant t \leqslant b$$

or by a vector equation $\mathbf{r}(t) = x(t)\,\mathbf{i} + y(t)\,\mathbf{j} + z(t)\,\mathbf{k}$. If f is a function of three variables that is continuous on some region containing C, then we define the **line integral of f along C** (with respect to arc length) in a manner similar to that for plane curves:

$$\int_{C} f(x, y, z) \, ds = \lim_{n \to \infty} \sum_{i=1}^{n} f(x_i^*, y_i^*, z_i^*) \, \Delta s_i$$

We evaluate it using a formula similar to Formula 3:

$$\boxed{9} \quad \int_{C} f(x, y, z) \, ds = \int_{a}^{b} f(x(t), y(t), z(t)) \sqrt{\left(\frac{dx}{dt}\right)^2 + \left(\frac{dy}{dt}\right)^2 + \left(\frac{dz}{dt}\right)^2} \, dt$$

Observe that the integrals in both Formulas 3 and 9 can be written in the more compact vector notation

$$\int_{a}^{b} f(\mathbf{r}(t)) \, |\, \mathbf{r}'(t)\,| \, dt$$

For the special case $f(x, y, z) = 1$, we get

$$\int_C ds = \int_a^b |\mathbf{r}'(t)| \, dt = L$$

where L is the length of the curve C (see Formula 10.3.3).

Line integrals along C with respect to x, y, and z can also be defined. For example,

$$\int_C f(x, y, z) \, dz = \lim_{n \to \infty} \sum_{i=1}^{n} f(x_i^*, y_i^*, z_i^*) \, \Delta z_i$$

$$= \int_a^b f(x(t), y(t), z(t)) \, z'(t) \, dt$$

Therefore, as with line integrals in the plane, we evaluate integrals of the form

$$\boxed{10} \qquad \int_C P(x, y, z) \, dx + Q(x, y, z) \, dy + R(x, y, z) \, dz$$

by expressing everything (x, y, z, dx, dy, dz) in terms of the parameter t.

$\boxed{V}$ **EXAMPLE 5** Evaluate $\int_C y \sin z \, ds$, where C is the circular helix given by the equations $x = \cos t$, $y = \sin t$, $z = t$, $0 \le t \le 2\pi$. (See Figure 9.)

SOLUTION Formula 9 gives

$$\int_C y \sin z \, ds = \int_0^{2\pi} (\sin t) \sin t \sqrt{\left(\frac{dx}{dt}\right)^2 + \left(\frac{dy}{dt}\right)^2 + \left(\frac{dz}{dt}\right)^2} \, dt$$

$$= \int_0^{2\pi} \sin^2 t \sqrt{\sin^2 t + \cos^2 t + 1} \, dt$$

$$= \sqrt{2} \int_0^{2\pi} \tfrac{1}{2}(1 - \cos 2t) \, dt = \frac{\sqrt{2}}{2} \left[t - \tfrac{1}{2} \sin 2t \right]_0^{2\pi} = \sqrt{2}\,\pi \qquad ∎$$

FIGURE 9

EXAMPLE 6 Evaluate $\int_C y \, dx + z \, dy + x \, dz$, where C consists of the line segment C_1 from $(2, 0, 0)$ to $(3, 4, 5)$ followed by the vertical line segment C_2 from $(3, 4, 5)$ to $(3, 4, 0)$.

SOLUTION The curve C is shown in Figure 10. Using Equation 8, we write C_1 as

$$\mathbf{r}(t) = (1 - t)\langle 2, 0, 0 \rangle + t\langle 3, 4, 5 \rangle = \langle 2 + t, 4t, 5t \rangle$$

or, in parametric form, as

$$x = 2 + t \qquad y = 4t \qquad z = 5t \qquad 0 \le t \le 1$$

Thus

$$\int_{C_1} y \, dx + z \, dy + x \, dz = \int_0^1 (4t) \, dt + (5t)4 \, dt + (2 + t)5 \, dt$$

$$= \int_0^1 (10 + 29t) \, dt = 10t + 29 \frac{t^2}{2} \Bigg]_0^1 = 24.5$$

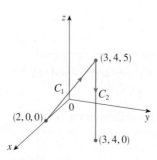

FIGURE 10

Likewise, C_2 can be written in the form

$$\mathbf{r}(t) = (1 - t)\langle 3, 4, 5 \rangle + t\langle 3, 4, 0 \rangle = \langle 3, 4, 5 - 5t \rangle$$

or $\qquad\qquad x = 3 \qquad y = 4 \qquad z = 5 - 5t \qquad 0 \le t \le 1$

Then $dx = 0 = dy$, so

$$\int_{C_2} y \, dx + z \, dy + x \, dz = \int_0^1 3(-5) \, dt = -15$$

Adding the values of these integrals, we obtain

$$\int_C y \, dx + z \, dy + x \, dz = 24.5 - 15 = 9.5$$

■ ■

■■ Line Integrals of Vector Fields

Recall from Section 6.5 that the work done by a variable force $f(x)$ in moving a particle from a to b along the x-axis is $W = \int_a^b f(x) \, dx$. Then in Section 9.3 we found that the work done by a constant force $\mathbf{F}$ in moving an object from a point P to another point Q in space is $W = \mathbf{F} \cdot \mathbf{D}$, where $\mathbf{D} = \overrightarrow{PQ}$ is the displacement vector.

Now suppose that $\mathbf{F} = P \mathbf{i} + Q \mathbf{j} + R \mathbf{k}$ is a continuous force field on $\mathbb{R}^3$, such as the gravitational field of Example 4 in Section 13.1 or the electric force field of Example 5 in Section 13.1. (A force field on $\mathbb{R}^2$ could be regarded as a special case where $R = 0$ and P and Q depend only on x and y.) We wish to compute the work done by this force in moving a particle along a smooth curve C.

We divide C into subarcs $P_{i-1}P_i$ with lengths Δs_i by dividing the parameter interval $[a, b]$ into subintervals of equal width. (See Figure 1 for the two-dimensional case or Figure 11 for the three-dimensional case.) Choose a point $P_i^*(x_i^*, y_i^*, z_i^*)$ on the ith subarc corresponding to the parameter value t_i^*. If Δs_i is small, then as the particle moves from P_{i-1} to P_i along the curve, it proceeds approximately in the direction of $\mathbf{T}(t_i^*)$, the unit tangent vector at P_i^*. Thus, the work done by the force $\mathbf{F}$ in moving the particle from P_{i-1} to P_i is approximately

$$\mathbf{F}(x_i^*, y_i^*, z_i^*) \cdot [\Delta s_i \, \mathbf{T}(t_i^*)] = [\mathbf{F}(x_i^*, y_i^*, z_i^*) \cdot \mathbf{T}(t_i^*)] \, \Delta s_i$$

and the total work done in moving the particle along C is approximately

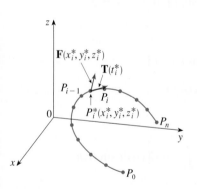

FIGURE 11

$$\boxed{11} \qquad \sum_{i=1}^n [\mathbf{F}(x_i^*, y_i^*, z_i^*) \cdot \mathbf{T}(x_i^*, y_i^*, z_i^*)] \, \Delta s_i$$

where $\mathbf{T}(x, y, z)$ is the unit tangent vector at the point (x, y, z) on C. Intuitively, we see that these approximations ought to become better as n becomes larger. Therefore, we define the **work** W done by the force field $\mathbf{F}$ as the limit of the Riemann sums in (11), namely,

$$\boxed{12} \qquad W = \int_C \mathbf{F}(x, y, z) \cdot \mathbf{T}(x, y, z) \, ds = \int_C \mathbf{F} \cdot \mathbf{T} \, ds$$

Equation 12 says that *work is the line integral with respect to arc length of the tangential component of the force.*

If the curve C is given by the vector equation $\mathbf{r}(t) = x(t)\,\mathbf{i} + y(t)\,\mathbf{j} + z(t)\,\mathbf{k}$, then $\mathbf{T}(t) = \mathbf{r}'(t)/|\mathbf{r}'(t)|$, so using Equation 9 we can rewrite Equation 12 in the form

$$W = \int_a^b \left[\mathbf{F}(\mathbf{r}(t)) \cdot \frac{\mathbf{r}'(t)}{|\mathbf{r}'(t)|} \right] |\mathbf{r}'(t)|\, dt = \int_a^b \mathbf{F}(\mathbf{r}(t)) \cdot \mathbf{r}'(t)\, dt$$

This integral is often abbreviated as $\int_C \mathbf{F} \cdot d\mathbf{r}$ and occurs in other areas of physics as well. Therefore, we make the following definition for the line integral of *any* continuous vector field.

> **13 Definition** Let $\mathbf{F}$ be a continuous vector field defined on a smooth curve C given by a vector function $\mathbf{r}(t)$, $a \le t \le b$. Then the **line integral of F along C** is
>
> $$\int_C \mathbf{F} \cdot d\mathbf{r} = \int_a^b \mathbf{F}(\mathbf{r}(t)) \cdot \mathbf{r}'(t)\, dt = \int_C \mathbf{F} \cdot \mathbf{T}\, ds$$

When using Definition 13, remember that $\mathbf{F}(\mathbf{r}(t))$ is just an abbreviation for $\mathbf{F}(x(t), y(t), z(t))$, so we evaluate $\mathbf{F}(\mathbf{r}(t))$ simply by putting $x = x(t)$, $y = y(t)$, and $z = z(t)$ in the expression for $\mathbf{F}(x, y, z)$. Notice also that we can formally write $d\mathbf{r} = \mathbf{r}'(t)\, dt$.

■ ■ Figure 12 shows the force field and the curve in Example 7. The work done is negative because the field impedes movement along the curve.

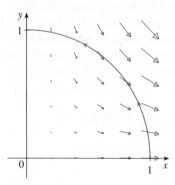

FIGURE 12

EXAMPLE 7 Find the work done by the force field $\mathbf{F}(x, y) = x^2\,\mathbf{i} - xy\,\mathbf{j}$ in moving a particle along the quarter-circle $\mathbf{r}(t) = \cos t\,\mathbf{i} + \sin t\,\mathbf{j}$, $0 \le t \le \pi/2$.

SOLUTION Since $x = \cos t$ and $y = \sin t$, we have

$$\mathbf{F}(\mathbf{r}(t)) = \cos^2 t\,\mathbf{i} - \cos t \sin t\,\mathbf{j}$$

and

$$\mathbf{r}'(t) = -\sin t\,\mathbf{i} + \cos t\,\mathbf{j}$$

Therefore, the work done is

$$\int_C \mathbf{F} \cdot d\mathbf{r} = \int_0^{\pi/2} \mathbf{F}(\mathbf{r}(t)) \cdot \mathbf{r}'(t)\, dt = \int_0^{\pi/2} (-2\cos^2 t \sin t)\, dt$$

$$= 2\,\frac{\cos^3 t}{3}\bigg]_0^{\pi/2} = -\frac{2}{3}$$

■ ■

NOTE ▫ Even though $\int_C \mathbf{F} \cdot d\mathbf{r} = \int_C \mathbf{F} \cdot \mathbf{T}\, ds$ and integrals with respect to arc length are unchanged when orientation is reversed, it is still true that

$$\int_{-C} \mathbf{F} \cdot d\mathbf{r} = -\int_C \mathbf{F} \cdot d\mathbf{r}$$

because the unit tangent vector $\mathbf{T}$ is replaced by its negative when C is replaced by $-C$.

EXAMPLE 8 Evaluate $\int_C \mathbf{F} \cdot d\mathbf{r}$, where $\mathbf{F}(x, y, z) = xy\,\mathbf{i} + yz\,\mathbf{j} + zx\,\mathbf{k}$ and C is the twisted cubic given by

$$x = t \qquad y = t^2 \qquad z = t^3 \qquad 0 \le t \le 1$$

■ ■ Figure 13 shows the twisted cubic C in Example 8 and some typical vectors acting at three points on C.

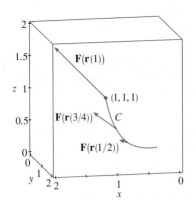

FIGURE 13

SOLUTION We have

$$\mathbf{r}(t) = t\,\mathbf{i} + t^2\,\mathbf{j} + t^3\,\mathbf{k}$$

$$\mathbf{r}'(t) = \mathbf{i} + 2t\,\mathbf{j} + 3t^2\,\mathbf{k}$$

$$\mathbf{F}(\mathbf{r}(t)) = t^3\,\mathbf{i} + t^5\,\mathbf{j} + t^4\,\mathbf{k}$$

Thus

$$\int_C \mathbf{F} \cdot d\mathbf{r} = \int_0^1 \mathbf{F}(\mathbf{r}(t)) \cdot \mathbf{r}'(t)\, dt$$

$$= \int_0^1 (t^3 + 5t^6)\, dt = \frac{t^4}{4} + \frac{5t^7}{7} \bigg]_0^1 = \frac{27}{28}$$

■ ■

Finally, we note the connection between line integrals of vector fields and line integrals of scalar fields. Suppose the vector field $\mathbf{F}$ on $\mathbb{R}^3$ is given in component form by the equation $\mathbf{F} = P\,\mathbf{i} + Q\,\mathbf{j} + R\,\mathbf{k}$. We use Definition 13 to compute its line integral along C:

$$\int_C \mathbf{F} \cdot d\mathbf{r} = \int_a^b \mathbf{F}(\mathbf{r}(t)) \cdot \mathbf{r}'(t)\, dt$$

$$= \int_a^b (P\,\mathbf{i} + Q\,\mathbf{j} + R\,\mathbf{k}) \cdot (x'(t)\,\mathbf{i} + y'(t)\,\mathbf{j} + z'(t)\,\mathbf{k})\, dt$$

$$= \int_a^b [P(x(t), y(t), z(t))\, x'(t) + Q(x(t), y(t), z(t))\, y'(t) + R(x(t), y(t), z(t))\, z'(t)]\, dt$$

But this last integral is precisely the line integral in (10). Therefore, we have

$$\int_C \mathbf{F} \cdot d\mathbf{r} = \int_C P\, dx + Q\, dy + R\, dz \qquad \text{where } \mathbf{F} = P\,\mathbf{i} + Q\,\mathbf{j} + R\,\mathbf{k}$$

For example, the integral $\int_C y\, dx + z\, dy + x\, dz$ in Example 6 could be expressed as $\int_C \mathbf{F} \cdot d\mathbf{r}$ where

$$\mathbf{F}(x, y, z) = y\,\mathbf{i} + z\,\mathbf{j} + x\,\mathbf{k}$$

13.2 Exercises

1–14 ■ Evaluate the line integral, where C is the given curve.

1. $\int_C y\, ds$, $C: x = t^2$, $y = t$, $0 \le t \le 2$

2. $\int_C (y/x)\, ds$, $C: x = t^4$, $y = t^3$, $\frac{1}{2} \le t \le 1$

3. $\int_C xy^4\, ds$,
C is the right half of the circle $x^2 + y^2 = 16$

4. $\int_C xe^y\, dx$,
C is the arc of the curve $x = e^y$ from $(1, 0)$ to $(e, 1)$

5. $\int_C xy\, dx + (x - y)\, dy$, C consists of line segments from $(0, 0)$ to $(2, 0)$ and from $(2, 0)$ to $(3, 2)$

6. $\int_C \sin x\, dx + \cos y\, dy$, C consists of the top half of the circle $x^2 + y^2 = 1$ from $(1, 0)$ to $(-1, 0)$ and the line segment from $(-1, 0)$ to $(-2, 3)$

7. $\int_C xy^3\, ds$,
$C: x = 4 \sin t$, $y = 4 \cos t$, $z = 3t$, $0 \le t \le \pi/2$

8. $\int_C x^2 z\, ds$, C is the line segment from $(0, 6, -1)$ to $(4, 1, 5)$

9. $\int_C xe^{yz}\, ds$, C is the line segment from $(0, 0, 0)$ to $(1, 2, 3)$

10. $\int_C (2x + 9z)\, ds$, $C: x = t$, $y = t^2$, $z = t^3$, $0 \le t \le 1$

11. $\int_C x^2 y \sqrt{z}\, dz$, $C: x = t^3$, $y = t$, $z = t^2$, $0 \le t \le 1$

12. $\int_C z\,dx + x\,dy + y\,dz$,
  $C: x = t^2,\ y = t^3,\ z = t^2,\ 0 \le t \le 1$

13. $\int_C (x + yz)\,dx + 2x\,dy + xyz\,dz$, C consists of line
  segments from $(1, 0, 1)$ to $(2, 3, 1)$ and from $(2, 3, 1)$ to
  $(2, 5, 2)$

14. $\int_C x^2\,dx + y^2\,dy + z^2\,dz$, C consists of line segments
  from $(0, 0, 0)$ to $(1, 2, -1)$ and from $(1, 2, -1)$ to $(3, 2, 0)$

15. Let **F** be the vector field shown in the figure.
  (a) If C_1 is the vertical line segment from $(-3, -3)$ to
    $(-3, 3)$, determine whether $\int_{C_1} \mathbf{F} \cdot d\mathbf{r}$ is positive, nega-
    tive, or zero.
  (b) If C_2 is the counterclockwise-oriented circle with radius
    3 and center the origin, determine whether $\int_{C_2} \mathbf{F} \cdot d\mathbf{r}$ is
    positive, negative, or zero.

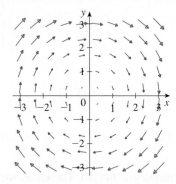

16. The figure shows a vector field **F** and two curves C_1 and C_2.
  Are the line integrals of **F** over C_1 and C_2 positive, negative,
  or zero? Explain.

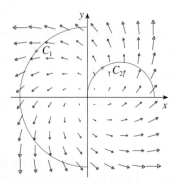

17–20 ■ Evaluate the line integral $\int_C \mathbf{F} \cdot d\mathbf{r}$, where C is given
by the vector function $\mathbf{r}(t)$.

17. $\mathbf{F}(x, y) = x^2 y^3\,\mathbf{i} - y\sqrt{x}\,\mathbf{j}$,
  $\mathbf{r}(t) = t^2\,\mathbf{i} - t^3\,\mathbf{j}$, $0 \le t \le 1$

18. $\mathbf{F}(x, y, z) = yz\,\mathbf{i} + xz\,\mathbf{j} + xy\,\mathbf{k}$,
  $\mathbf{r}(t) = t\,\mathbf{i} + t^2\,\mathbf{j} + t^3\,\mathbf{k}$, $0 \le t \le 2$

19. $\mathbf{F}(x, y, z) = \sin x\,\mathbf{i} + \cos y\,\mathbf{j} + xz\,\mathbf{k}$,
  $\mathbf{r}(t) = t^3\,\mathbf{i} - t^2\,\mathbf{j} + t\,\mathbf{k}$, $0 \le t \le 1$

20. $\mathbf{F}(x, y, z) = z\,\mathbf{i} + y\,\mathbf{j} - x\,\mathbf{k}$,
  $\mathbf{r}(t) = t\,\mathbf{i} + \sin t\,\mathbf{j} + \cos t\,\mathbf{k}$, $0 \le t \le \pi$

CAS **21–22** ■ Use a graph of the vector field **F** and the curve C to
guess whether the line integral of **F** over C is positive, negative,
or zero. Then evaluate the line integral.

21. $\mathbf{F}(x, y) = (x - y)\,\mathbf{i} + xy\,\mathbf{j}$,
  C is the arc of the circle $x^2 + y^2 = 4$ traversed counterclockwise from $(2, 0)$ to $(0, -2)$

22. $\mathbf{F}(x, y) = \dfrac{x}{\sqrt{x^2 + y^2}}\,\mathbf{i} + \dfrac{y}{\sqrt{x^2 + y^2}}\,\mathbf{j}$,
  C is the parabola $y = 1 + x^2$ from $(-1, 2)$ to $(1, 2)$

23. (a) Evaluate the line integral $\int_C \mathbf{F} \cdot d\mathbf{r}$, where
    $\mathbf{F}(x, y) = e^{x-1}\,\mathbf{i} + xy\,\mathbf{j}$ and C is given by
    $\mathbf{r}(t) = t^2\,\mathbf{i} + t^3\,\mathbf{j}, 0 \le t \le 1$.
  (b) Illustrate part (a) by using a graphing calculator or com-
    puter to graph C and the vectors from the vector field
    corresponding to $t = 0, 1/\sqrt{2}$, and 1 (as in Figure 13).

24. (a) Evaluate the line integral $\int_C \mathbf{F} \cdot d\mathbf{r}$, where
    $\mathbf{F}(x, y, z) = x\,\mathbf{i} - z\,\mathbf{j} + y\,\mathbf{k}$ and C is given by
    $\mathbf{r}(t) = 2t\,\mathbf{i} + 3t\,\mathbf{j} - t^2\,\mathbf{k}, -1 \le t \le 1$.
  (b) Illustrate part (a) by using a computer to graph C and
    the vectors from the vector field corresponding to
    $t = \pm 1$ and $\pm\frac{1}{2}$ (as in Figure 13).

CAS **25.** Find the exact value of $\int_C x^3 y^5\,ds$, where C is the part of the
  astroid $x = \cos^3 t, y = \sin^3 t$ in the first quadrant.

26. (a) Find the work done by the force field
    $\mathbf{F}(x, y) = x^2\,\mathbf{i} + xy\,\mathbf{j}$ on a particle that moves once
    around the circle $x^2 + y^2 = 4$ oriented in the
    counterclockwise direction.
  CAS (b) Use a computer algebra system to graph the force field
    and circle on the same screen. Use the graph to explain
    your answer to part (a).

27. A thin wire is bent into the shape of a semicircle
  $x^2 + y^2 = 4, x \ge 0$. If the linear density is a constant k,
  find the mass and center of mass of the wire.

28. Find the mass and center of mass of a thin wire in the shape
  of a quarter-circle $x^2 + y^2 = r^2, x \ge 0, y \ge 0$, if the density function is $\rho(x, y) = x + y$.

29. (a) Write the formulas similar to Equations 4 for the center
    of mass $(\bar{x}, \bar{y}, \bar{z})$ of a thin wire with density function
    $\rho(x, y, z)$ in the shape of a space curve C.
  (b) Find the center of mass of a wire in the shape of the
    helix $x = 2\sin t, y = 2\cos t, z = 3t, 0 \le t \le 2\pi$, if the
    density is a constant k.

30. Find the mass and center of mass of a wire in the shape of the helix $x = t$, $y = \cos t$, $z = \sin t$, $0 \leqslant t \leqslant 2\pi$, if the density at any point is equal to the square of the distance from the origin.

31. If a wire with linear density $\rho(x, y)$ lies along a plane curve C, its **moments of inertia** about the x- and y-axes are defined as

$$I_x = \int_C y^2 \rho(x, y)\, ds$$

$$I_y = \int_C x^2 \rho(x, y)\, ds$$

Find the moments of inertia for the wire in Example 3.

32. If a wire with linear density $\rho(x, y, z)$ lies along a space curve C, its **moments of inertia** about the x-, y-, and z-axes are defined as

$$I_x = \int_C (y^2 + z^2) \rho(x, y, z)\, ds$$

$$I_y = \int_C (x^2 + z^2) \rho(x, y, z)\, ds$$

$$I_z = \int_C (x^2 + y^2) \rho(x, y, z)\, ds$$

Find the moments of inertia for the wire in Exercise 29.

33. Find the work done by the force field $\mathbf{F}(x, y) = x\,\mathbf{i} + (y + 2)\,\mathbf{j}$ in moving an object along an arch of the cycloid $\mathbf{r}(t) = (t - \sin t)\,\mathbf{i} + (1 - \cos t)\,\mathbf{j}$, $0 \leqslant t \leqslant 2\pi$.

34. Find the work done by the force field $\mathbf{F}(x, y) = x \sin y\,\mathbf{i} + y\,\mathbf{j}$ on a particle that moves along the parabola $y = x^2$ from $(-1, 1)$ to $(2, 4)$.

35. Find the work done by the force field $\mathbf{F}(x, y, z) = \langle y + z, x + z, x + y \rangle$ on a particle that moves along the line segment from $(1, 0, 0)$ to $(3, 4, 2)$.

36. The force exerted by an electric charge at the origin on a charged particle at a point (x, y, z) with position vector $\mathbf{r} = \langle x, y, z \rangle$ is $\mathbf{F}(\mathbf{r}) = K\mathbf{r}/|\mathbf{r}|^3$ where K is a constant. (See Example 5 in Section 13.1.) Find the work done as the particle moves along a straight line from $(2, 0, 0)$ to $(2, 1, 5)$.

37. A 160-lb man carries a 25-lb can of paint up a helical staircase that encircles a silo with a radius of 20 ft. If the silo is 90 ft high and the man makes exactly three complete revolutions, how much work is done by the man against gravity in climbing to the top?

38. Suppose there is a hole in the can of paint in Exercise 37 and 9 lb of paint leaks steadily out of the can during the man's ascent. How much work is done?

39. (a) Show that a constant force field does zero work on a particle that moves once uniformly around the circle $x^2 + y^2 = 1$.

(b) Is this also true for a force field $\mathbf{F}(\mathbf{x}) = k\mathbf{x}$, where k is a constant and $\mathbf{x} = \langle x, y \rangle$?

40. The base of a circular fence with radius 10 m is given by $x = 10 \cos t$, $y = 10 \sin t$. The height of the fence at position (x, y) is given by the function $h(x, y) = 4 + 0.01(x^2 - y^2)$, so the height varies from 3 m to 5 m. Suppose that 1 L of paint covers 100 m². Sketch the fence and determine how much paint you will need if you paint both sides of the fence.

41. An object moves along the curve C shown in the figure from $(1, 2)$ to $(9, 8)$. The lengths of the vectors in the force field $\mathbf{F}$ are measured in newtons by the scales on the axes. Estimate the work done by $\mathbf{F}$ on the object.

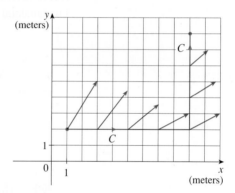

42. Experiments show that a steady current I in a long wire produces a magnetic field $\mathbf{B}$ that is tangent to any circle that lies in the plane perpendicular to the wire and whose center is the axis of the wire (as in the figure). *Ampère's Law* relates the electric current to its magnetic effects and states that

$$\int_C \mathbf{B} \cdot d\mathbf{r} = \mu_0 I$$

where I is the net current that passes through any surface bounded by a closed curve C and μ_0 is a constant called the permeability of free space. By taking C to be a circle with radius r, show that the magnitude $B = |\mathbf{B}|$ of the magnetic field at a distance r from the center of the wire is

$$B = \frac{\mu_0 I}{2\pi r}$$

13.3 The Fundamental Theorem for Line Integrals

Recall from Section 5.4 that Part 2 of the Fundamental Theorem of Calculus can be written as

$$\boxed{1} \qquad \int_a^b F'(x)\,dx = F(b) - F(a)$$

where F' is continuous on $[a, b]$. We also called Equation 1 the Net Change Theorem: The integral of a rate of change is the net change.

If we think of the gradient vector ∇f of a function f of two or three variables as a sort of derivative of f, then the following theorem can be regarded as a version of the Fundamental Theorem for line integrals.

> **2 Theorem** Let C be a smooth curve given by the vector function $\mathbf{r}(t)$, $a \leq t \leq b$. Let f be a differentiable function of two or three variables whose gradient vector ∇f is continuous on C. Then
>
> $$\int_C \nabla f \cdot d\mathbf{r} = f(\mathbf{r}(b)) - f(\mathbf{r}(a))$$

NOTE ▫ Theorem 2 says that we can evaluate the line integral of a conservative vector field (the gradient vector field of the potential function f) simply by knowing the value of f at the endpoints of C. In fact, Theorem 2 says that the line integral of ∇f is the net change in f. If f is a function of two variables and C is a plane curve with initial point $A(x_1, y_1)$ and terminal point $B(x_2, y_2)$, as in Figure 1, then Theorem 2 becomes

$$\int_C \nabla f \cdot d\mathbf{r} = f(x_2, y_2) - f(x_1, y_1)$$

If f is a function of three variables and C is a space curve joining the point $A(x_1, y_1, z_1)$ to the point $B(x_2, y_2, z_2)$, then we have

$$\int_C \nabla f \cdot d\mathbf{r} = f(x_2, y_2, z_2) - f(x_1, y_1, z_1)$$

Let's prove Theorem 2 for this case.

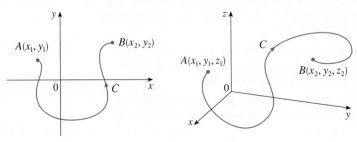

FIGURE 1

Proof of Theorem 2   Using Definition 13.2.13, we have

$$\int_C \nabla f \cdot d\mathbf{r} = \int_a^b \nabla f(\mathbf{r}(t)) \cdot \mathbf{r}'(t)\, dt$$

$$= \int_a^b \left(\frac{\partial f}{\partial x} \frac{dx}{dt} + \frac{\partial f}{\partial y} \frac{dy}{dt} + \frac{\partial f}{\partial z} \frac{dz}{dt} \right) dt$$

$$= \int_a^b \frac{d}{dt} f(\mathbf{r}(t))\, dt \qquad\qquad \text{(by the Chain Rule)}$$

$$= f(\mathbf{r}(b)) - f(\mathbf{r}(a))$$

The last step follows from the Fundamental Theorem of Calculus (Equation 1). ■ ■

Although we have proved Theorem 2 for smooth curves, it is also true for piecewise-smooth curves. This can be seen by subdividing C into a finite number of smooth curves and adding the resulting integrals.

EXAMPLE 1   Find the work done by the gravitational field

$$\mathbf{F}(\mathbf{x}) = -\frac{mMG}{|\mathbf{x}|^3}\, \mathbf{x}$$

in moving a particle with mass m from the point $(3, 4, 12)$ to the point $(2, 2, 0)$ along a piecewise-smooth curve C. (See Example 4 in Section 13.1.)

SOLUTION   From Section 13.1 we know that $\mathbf{F}$ is a conservative vector field and, in fact, $\mathbf{F} = \nabla f$, where

$$f(x, y, z) = \frac{mMG}{\sqrt{x^2 + y^2 + z^2}}$$

Therefore, by Theorem 2, the work done is

$$W = \int_C \mathbf{F} \cdot d\mathbf{r} = \int_C \nabla f \cdot d\mathbf{r}$$

$$= f(2, 2, 0) - f(3, 4, 12)$$

$$= \frac{mMG}{\sqrt{2^2 + 2^2}} - \frac{mMG}{\sqrt{3^2 + 4^2 + 12^2}} = mMG\left(\frac{1}{2\sqrt{2}} - \frac{1}{13} \right) \qquad ■ ■$$

■■ Independence of Path

Suppose C_1 and C_2 are two piecewise-smooth curves (which are called **paths**) that have the same initial point A and terminal point B. We know from Example 4 in Section 13.2 that, in general, $\int_{C_1} \mathbf{F} \cdot d\mathbf{r} \neq \int_{C_2} \mathbf{F} \cdot d\mathbf{r}$. But one implication of Theorem 2 is that

$$\int_{C_1} \nabla f \cdot d\mathbf{r} = \int_{C_2} \nabla f \cdot d\mathbf{r}$$

whenever ∇f is continuous. In other words, the line integral of a *conservative* vector field depends only on the initial point and terminal point of a curve.

In general, if $\mathbf{F}$ is a continuous vector field with domain D, we say that the line integral $\int_C \mathbf{F} \cdot d\mathbf{r}$ is **independent of path** if $\int_{C_1} \mathbf{F} \cdot d\mathbf{r} = \int_{C_2} \mathbf{F} \cdot d\mathbf{r}$ for any two paths

FIGURE 2
A closed curve

FIGURE 3

C_1 and C_2 in D that have the same initial and terminal points. With this terminology we can say that *line integrals of conservative vector fields are independent of path.*

A curve is called **closed** if its terminal point coincides with its initial point, that is, $\mathbf{r}(b) = \mathbf{r}(a)$. (See Figure 2.) If $\int_C \mathbf{F} \cdot d\mathbf{r}$ is independent of path in D and C is any closed path in D, we can choose any two points A and B on C and regard C as being composed of the path C_1 from A to B followed by the path C_2 from B to A. (See Figure 3.) Then

$$\int_C \mathbf{F} \cdot d\mathbf{r} = \int_{C_1} \mathbf{F} \cdot d\mathbf{r} + \int_{C_2} \mathbf{F} \cdot d\mathbf{r} = \int_{C_1} \mathbf{F} \cdot d\mathbf{r} - \int_{-C_2} \mathbf{F} \cdot d\mathbf{r} = 0$$

since C_1 and $-C_2$ have the same initial and terminal points.

Conversely, if it is true that $\int_C \mathbf{F} \cdot d\mathbf{r} = 0$ whenever C is a closed path in D, then we demonstrate independence of path as follows. Take any two paths C_1 and C_2 from A to B in D and define C to be the curve consisting of C_1 followed by $-C_2$. Then

$$0 = \int_C \mathbf{F} \cdot d\mathbf{r} = \int_{C_1} \mathbf{F} \cdot d\mathbf{r} + \int_{-C_2} \mathbf{F} \cdot d\mathbf{r} = \int_{C_1} \mathbf{F} \cdot d\mathbf{r} - \int_{C_2} \mathbf{F} \cdot d\mathbf{r}$$

and so $\int_{C_1} \mathbf{F} \cdot d\mathbf{r} = \int_{C_2} \mathbf{F} \cdot d\mathbf{r}$. Thus, we have proved the following theorem.

---

3 Theorem $\int_C \mathbf{F} \cdot d\mathbf{r}$ is independent of path in D if and only if $\int_C \mathbf{F} \cdot d\mathbf{r} = 0$ for every closed path C in D.

Since we know that the line integral of any conservative vector field $\mathbf{F}$ is independent of path, it follows that $\int_C \mathbf{F} \cdot d\mathbf{r} = 0$ for any closed path. The physical interpretation is that the work done by a conservative force field (such as the gravitational or electric field in Section 13.1) as it moves an object around a closed path is 0.

The following theorem says that the *only* vector fields that are independent of path are conservative. It is stated and proved for plane curves, but there is a similar version for space curves. We assume that D is **open**, which means that for every point P in D there is a disk with center P that lies entirely in D. (So D doesn't contain any of its boundary points.) In addition, we assume that D is **connected**. This means that any two points in D can be joined by a path that lies in D.

---

4 Theorem Suppose $\mathbf{F}$ is a vector field that is continuous on an open connected region D. If $\int_C \mathbf{F} \cdot d\mathbf{r}$ is independent of path in D, then $\mathbf{F}$ is a conservative vector field on D; that is, there exists a function f such that $\nabla f = \mathbf{F}$.

Proof Let $A(a, b)$ be a fixed point in D. We construct the desired potential function f by defining

$$f(x, y) = \int_{(a, b)}^{(x, y)} \mathbf{F} \cdot d\mathbf{r}$$

for any point (x, y) in D. Since $\int_C \mathbf{F} \cdot d\mathbf{r}$ is independent of path, it does not matter which path C from (a, b) to (x, y) is used to evaluate $f(x, y)$. Since D is open, there exists a disk contained in D with center (x, y). Choose any point (x_1, y) in the disk with $x_1 < x$ and let C consist of any path C_1 from (a, b) to (x_1, y) followed by the horizontal line segment C_2 from (x_1, y) to (x, y). (See Figure 4.) Then

$$f(x, y) = \int_{C_1} \mathbf{F} \cdot d\mathbf{r} + \int_{C_2} \mathbf{F} \cdot d\mathbf{r} = \int_{(a, b)}^{(x_1, y)} \mathbf{F} \cdot d\mathbf{r} + \int_{C_2} \mathbf{F} \cdot d\mathbf{r}$$

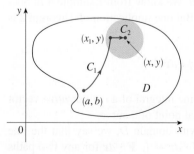

FIGURE 4

Notice that the first of these integrals does not depend on x, so

$$\frac{\partial}{\partial x} f(x, y) = 0 + \frac{\partial}{\partial x} \int_{C_2} \mathbf{F} \cdot d\mathbf{r}$$

If we write $\mathbf{F} = P \mathbf{i} + Q \mathbf{j}$, then

$$\int_{C_2} \mathbf{F} \cdot d\mathbf{r} = \int_{C_2} P \, dx + Q \, dy$$

On C_2, y is constant, so $dy = 0$. Using t as the parameter, where $x_1 \le t \le x$, we have

$$\frac{\partial}{\partial x} f(x, y) = \frac{\partial}{\partial x} \int_{C_2} P \, dx + Q \, dy = \frac{\partial}{\partial x} \int_{x_1}^{x} P(t, y) \, dt = P(x, y)$$

by Part 1 of the Fundamental Theorem of Calculus (see Section 5.4). A similar argument, using a vertical line segment (see Figure 5), shows that

$$\frac{\partial}{\partial y} f(x, y) = \frac{\partial}{\partial y} \int_{C_2} P \, dx + Q \, dy = \frac{\partial}{\partial y} \int_{y_1}^{y} Q(x, t) \, dt = Q(x, y)$$

Thus

$$\mathbf{F} = P \mathbf{i} + Q \mathbf{j} = \frac{\partial f}{\partial x} \mathbf{i} + \frac{\partial f}{\partial y} \mathbf{j} = \nabla f$$

which says that $\mathbf{F}$ is conservative.　■ ■

The question remains: How is it possible to determine whether or not a vector field $\mathbf{F}$ is conservative? Suppose it is known that $\mathbf{F} = P \mathbf{i} + Q \mathbf{j}$ is conservative, where P and Q have continuous first-order partial derivatives. Then there is a function f such that $\mathbf{F} = \nabla f$, that is,

$$P = \frac{\partial f}{\partial x} \qquad \text{and} \qquad Q = \frac{\partial f}{\partial y}$$

Therefore, by Clairaut's Theorem,

$$\frac{\partial P}{\partial y} = \frac{\partial^2 f}{\partial y \, \partial x} = \frac{\partial^2 f}{\partial x \, \partial y} = \frac{\partial Q}{\partial x}$$

> **5 Theorem** If $\mathbf{F}(x, y) = P(x, y) \mathbf{i} + Q(x, y) \mathbf{j}$ is a conservative vector field, where P and Q have continuous first-order partial derivatives on a domain D, then throughout D we have
>
> $$\frac{\partial P}{\partial y} = \frac{\partial Q}{\partial x}$$

The converse of Theorem 5 is true only for a special type of region. To explain this, we first need the concept of a **simple curve**, which is a curve that doesn't intersect itself anywhere between its endpoints. [See Figure 6; $\mathbf{r}(a) = \mathbf{r}(b)$ for a simple closed curve, but $\mathbf{r}(t_1) \ne \mathbf{r}(t_2)$ when $a < t_1 < t_2 < b$.]

In Theorem 4 we needed an open connected region. For the next theorem we need a stronger condition. A **simply-connected region** in the plane is a connected region

FIGURE 5

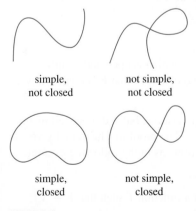

simple,　　　not simple,
not closed　　not closed

simple,　　　not simple,
closed　　　　closed

FIGURE 6
Types of curves

simply-connected region

regions that are not simply-connected

FIGURE 7

D such that every simple closed curve in D encloses only points that are in D. Notice from Figure 7 that, intuitively speaking, a simply-connected region contains no hole and can't consist of two separate pieces.

In terms of simply-connected regions we can now state a partial converse to Theorem 5 that gives a convenient method for verifying that a vector field on $\mathbb{R}^2$ is conservative. The proof will be sketched in the next section as a consequence of Green's Theorem.

---

6 **Theorem**   Let $\mathbf{F} = P\,\mathbf{i} + Q\,\mathbf{j}$ be a vector field on an open simply-connected region D. Suppose that P and Q have continuous first-order derivatives and

$$\frac{\partial P}{\partial y} = \frac{\partial Q}{\partial x} \qquad \text{throughout } D$$

Then $\mathbf{F}$ is conservative.

FIGURE 8

■ ■ Figures 8 and 9 show the vector fields in Examples 2 and 3, respectively. The vectors in Figure 8 that start on the closed curve C all appear to point in roughly the same direction as C. So it looks as if $\int_C \mathbf{F} \cdot d\mathbf{r} > 0$ and therefore $\mathbf{F}$ is not conservative. The calculation in Example 2 confirms this impression. Some of the vectors near the curves C_1 and C_2 in Figure 9 point in approximately the same direction as the curves, whereas others point in the opposite direction. So it appears plausible that line integrals around all closed paths are 0. Example 3 shows that $\mathbf{F}$ is indeed conservative.

FIGURE 9

$\boxed{\text{V}}$ **EXAMPLE 2** Determine whether or not the vector field

$$\mathbf{F}(x, y) = (x - y)\,\mathbf{i} + (x - 2)\,\mathbf{j}$$

is conservative.

SOLUTION Let $P(x, y) = x - y$ and $Q(x, y) = x - 2$. Then

$$\frac{\partial P}{\partial y} = -1 \qquad \frac{\partial Q}{\partial x} = 1$$

Since $\partial P/\partial y \neq \partial Q/\partial x$, $\mathbf{F}$ is not conservative by Theorem 5.   ■ ■

$\boxed{\text{V}}$ **EXAMPLE 3** Determine whether or not the vector field

$$\mathbf{F}(x, y) = (3 + 2xy)\,\mathbf{i} + (x^2 - 3y^2)\,\mathbf{j}$$

is conservative.

SOLUTION Let $P(x, y) = 3 + 2xy$ and $Q(x, y) = x^2 - 3y^2$. Then

$$\frac{\partial P}{\partial y} = 2x = \frac{\partial Q}{\partial x}$$

Also, the domain of $\mathbf{F}$ is the entire plane $(D = \mathbb{R}^2)$, which is open and simply-connected. Therefore, we can apply Theorem 6 and conclude that $\mathbf{F}$ is conservative.   ■ ■

In Example 3, Theorem 6 told us that $\mathbf{F}$ is conservative, but it did not tell us how to find the (potential) function f such that $\mathbf{F} = \nabla f$. The proof of Theorem 4 gives us a clue as to how to find f. We use "partial integration" as in the following example.

EXAMPLE 4
(a) If $\mathbf{F}(x, y) = (3 + 2xy)\,\mathbf{i} + (x^2 - 3y^2)\,\mathbf{j}$, find a function f such that $\mathbf{F} = \nabla f$.
(b) Evaluate the line integral $\int_C \mathbf{F} \cdot d\mathbf{r}$, where C is the curve given by
$\mathbf{r}(t) = e^t \sin t\,\mathbf{i} + e^t \cos t\,\mathbf{j}, 0 \leq t \leq \pi$.

SOLUTION
(a) From Example 3 we know that $\mathbf{F}$ is conservative and so there exists a function f with $\nabla f = \mathbf{F}$, that is,

$$\boxed{7} \qquad\qquad f_x(x, y) = 3 + 2xy$$

$$\boxed{8} \qquad\qquad f_y(x, y) = x^2 - 3y^2$$

Integrating (7) with respect to x, we obtain

$$\boxed{9} \qquad\qquad f(x, y) = 3x + x^2 y + g(y)$$

Notice that the constant of integration is a constant with respect to x, that is, a function of y, which we have called $g(y)$. Next we differentiate both sides of (9) with respect to y:

$$\boxed{10} \qquad\qquad f_y(x, y) = x^2 + g'(y)$$

Comparing (8) and (10), we see that

$$g'(y) = -3y^2$$

Integrating with respect to y, we have

$$g(y) = -y^3 + K$$

where K is a constant. Putting this in (9), we have

$$f(x, y) = 3x + x^2 y - y^3 + K$$

as the desired potential function.

(b) To use Theorem 2 all we have to know are the initial and terminal points of C, namely, $\mathbf{r}(0) = (0, 1)$ and $\mathbf{r}(\pi) = (0, -e^\pi)$. In the expression for $f(x, y)$ in part (a), any value of the constant K will do, so let's choose $K = 0$. Then we have

$$\int_C \mathbf{F} \cdot d\mathbf{r} = \int_C \nabla f \cdot d\mathbf{r} = f(0, -e^\pi) - f(0, 1)$$

$$= e^{3\pi} - (-1) = e^{3\pi} + 1$$

This method is much shorter than the straightforward method for evaluating line integrals that we learned in Section 13.2. ■ ■

A criterion for determining whether or not a vector field $\mathbf{F}$ on $\mathbb{R}^3$ is conservative is given in Section 13.5. Meanwhile, the next example shows that the technique for finding the potential function is much the same as for vector fields on $\mathbb{R}^2$.

$\boxed{\text{V}}$ **EXAMPLE 5** If $\mathbf{F}(x, y, z) = y^2 \mathbf{i} + (2xy + e^{3z})\mathbf{j} + 3ye^{3z}\mathbf{k}$, find a function f such that $\nabla f = \mathbf{F}$.

SOLUTION If there is such a function f, then

$$\boxed{11} \qquad\qquad f_x(x, y, z) = y^2$$

$$\boxed{12} \qquad\qquad f_y(x, y, z) = 2xy + e^{3z}$$

$$\boxed{13} \qquad\qquad f_z(x, y, z) = 3ye^{3z}$$

Integrating (11) with respect to x, we get

$$\boxed{14} \qquad f(x, y, z) = xy^2 + g(y, z)$$

where $g(y, z)$ is a constant with respect to x. Then differentiating (14) with respect to y, we have

$$f_y(x, y, z) = 2xy + g_y(y, z)$$

and comparison with (12) gives

$$g_y(y, z) = e^{3z}$$

Thus, $g(y, z) = ye^{3z} + h(z)$ and we rewrite (14) as

$$f(x, y, z) = xy^2 + ye^{3z} + h(z)$$

Finally, differentiating with respect to z and comparing with (13), we obtain $h'(z) = 0$ and, therefore, $h(z) = K$, a constant. The desired function is

$$f(x, y, z) = xy^2 + ye^{3z} + K$$

It is easily verified that $\nabla f = \mathbf{F}$. ■ ■

■■ Conservation of Energy

Let's apply the ideas of this chapter to a continuous force field $\mathbf{F}$ that moves an object along a path C given by $\mathbf{r}(t)$, $a \leqslant t \leqslant b$, where $\mathbf{r}(a) = A$ is the initial point and $\mathbf{r}(b) = B$ is the terminal point of C. According to Newton's Second Law of Motion (see Section 10.4), the force $\mathbf{F}(\mathbf{r}(t))$ at a point on C is related to the acceleration $\mathbf{a}(t) = \mathbf{r}''(t)$ by the equation

$$\mathbf{F}(\mathbf{r}(t)) = m\mathbf{r}''(t)$$

So the work done by the force on the object is

$$W = \int_C \mathbf{F} \cdot d\mathbf{r} = \int_a^b \mathbf{F}(\mathbf{r}(t)) \cdot \mathbf{r}'(t)\, dt$$

$$= \int_a^b m\mathbf{r}''(t) \cdot \mathbf{r}'(t)\, dt$$

$$= \frac{m}{2} \int_a^b \frac{d}{dt} [\mathbf{r}'(t) \cdot \mathbf{r}'(t)]\, dt \qquad \text{(Theorem 10.2.3, Formula 4)}$$

$$= \frac{m}{2} \int_a^b \frac{d}{dt} |\mathbf{r}'(t)|^2\, dt$$

$$= \frac{m}{2} \Big[|\mathbf{r}'(t)|^2 \Big]_a^b \qquad \text{(Fundamental Theorem of Calculus)}$$

$$= \frac{m}{2} \left(|\mathbf{r}'(b)|^2 - |\mathbf{r}'(a)|^2 \right)$$

Therefore

$$\boxed{15} \qquad W = \tfrac{1}{2}m\,|\,\mathbf{v}(b)\,|^2 - \tfrac{1}{2}m\,|\,\mathbf{v}(a)\,|^2$$

where $\mathbf{v} = \mathbf{r}'$ is the velocity.

The quantity $\tfrac{1}{2}m\,|\,\mathbf{v}(t)\,|^2$, that is, half the mass times the square of the speed, is called the **kinetic energy** of the object. Therefore, we can rewrite Equation 15 as

$$\boxed{16} \qquad W = K(B) - K(A)$$

which says that the work done by the force field along C is equal to the change in kinetic energy at the endpoints of C.

Now let's further assume that $\mathbf{F}$ is a conservative force field; that is, we can write $\mathbf{F} = \nabla f$. In physics, the **potential energy** of an object at the point (x, y, z) is defined as $P(x, y, z) = -f(x, y, z)$, so we have $\mathbf{F} = -\nabla P$. Then by Theorem 2 we have

$$W = \int_C \mathbf{F} \cdot d\mathbf{r} = -\int_C \nabla P \cdot d\mathbf{r}$$
$$= -[P(\mathbf{r}(b)) - P(\mathbf{r}(a))]$$
$$= P(A) - P(B)$$

Comparing this equation with Equation 16, we see that

$$P(A) + K(A) = P(B) + K(B)$$

which says that if an object moves from one point A to another point B under the influence of a conservative force field, then the sum of its potential energy and its kinetic energy remains constant. This is called the **Law of Conservation of Energy** and it is the reason the vector field is called *conservative*.

13.3 Exercises

1. The figure shows a curve C and a contour map of a function f whose gradient is continuous. Find $\int_C \nabla f \cdot d\mathbf{r}$.

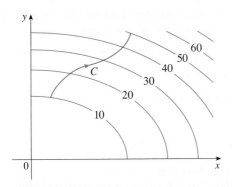

x \ y	0	1	2
0	1	6	4
1	3	5	7
2	8	2	9

2. A table of values of a function f with continuous gradient is given. Find $\int_C \nabla f \cdot d\mathbf{r}$, where C has parametric equations

$$x = t^2 + 1 \qquad y = t^3 + t \qquad 0 \le t \le 1$$

3–10 ■ Determine whether or not $\mathbf{F}$ is a conservative vector field. If it is, find a function f such that $\mathbf{F} = \nabla f$.

3. $\mathbf{F}(x, y) = (6x + 5y)\,\mathbf{i} + (5x + 4y)\,\mathbf{j}$

4. $\mathbf{F}(x, y) = (x^3 + 4xy)\,\mathbf{i} + (4xy - y^3)\,\mathbf{j}$

5. $\mathbf{F}(x, y) = xe^y\,\mathbf{i} + ye^x\,\mathbf{j}$

6. $\mathbf{F}(x, y) = e^y\,\mathbf{i} + xe^y\,\mathbf{j}$

7. $\mathbf{F}(x, y) = (2x\cos y - y\cos x)\,\mathbf{i} + (-x^2\sin y - \sin x)\,\mathbf{j}$

8. $\mathbf{F}(x, y) = (1 + 2xy + \ln x)\,\mathbf{i} + x^2\,\mathbf{j}$

9. $\mathbf{F}(x, y) = (ye^x + \sin y)\,\mathbf{i} + (e^x + x\cos y)\,\mathbf{j}$

10. $\mathbf{F}(x, y) = (xy \cos xy + \sin xy)\,\mathbf{i} + (x^2 \cos xy)\,\mathbf{j}$

11. The figure shows the vector field $\mathbf{F}(x, y) = \langle 2xy, x^2 \rangle$ and three curves that start at $(1, 2)$ and end at $(3, 2)$.
(a) Explain why $\int_C \mathbf{F} \cdot d\mathbf{r}$ has the same value for all three curves.
(b) What is this common value?

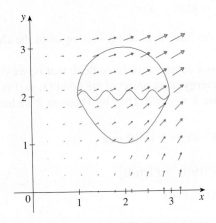

12–18 ■ (a) Find a function f such that $\mathbf{F} = \nabla f$ and (b) use part (a) to evaluate $\int_C \mathbf{F} \cdot d\mathbf{r}$ along the given curve C.

12. $\mathbf{F}(x, y) = y\,\mathbf{i} + (x + 2y)\,\mathbf{j}$,
C is the upper semicircle that starts at $(0, 1)$ and ends at $(2, 1)$

13. $\mathbf{F}(x, y) = x^3 y^4\,\mathbf{i} + x^4 y^3\,\mathbf{j}$,
C: $\mathbf{r}(t) = \sqrt{t}\,\mathbf{i} + (1 + t^3)\,\mathbf{j}$, $0 \le t \le 1$

14. $\mathbf{F}(x, y) = \dfrac{y^2}{1 + x^2}\,\mathbf{i} + 2y \arctan x\,\mathbf{j}$,
C: $\mathbf{r}(t) = t^2\,\mathbf{i} + 2t\,\mathbf{j}$, $0 \le t \le 1$

15. $\mathbf{F}(x, y, z) = yz\,\mathbf{i} + xz\,\mathbf{j} + (xy + 2z)\,\mathbf{k}$,
C is the line segment from $(1, 0, -2)$ to $(4, 6, 3)$

16. $\mathbf{F}(x, y, z) = (2xz + y^2)\,\mathbf{i} + 2xy\,\mathbf{j} + (x^2 + 3z^2)\,\mathbf{k}$,
C: $x = t^2$, $y = t + 1$, $z = 2t - 1$, $0 \le t \le 1$

17. $\mathbf{F}(x, y, z) = y^2 \cos z\,\mathbf{i} + 2xy \cos z\,\mathbf{j} - xy^2 \sin z\,\mathbf{k}$,
C: $\mathbf{r}(t) = t^2\,\mathbf{i} + \sin t\,\mathbf{j} + t\,\mathbf{k}$, $0 \le t \le \pi$

18. $\mathbf{F}(x, y, z) = e^y\,\mathbf{i} + xe^y\,\mathbf{j} + (z + 1)e^z\,\mathbf{k}$,
C: $\mathbf{r}(t) = t\,\mathbf{i} + t^2\,\mathbf{j} + t^3\,\mathbf{k}$, $0 \le t \le 1$

19–20 ■ Show that the line integral is independent of path and evaluate the integral.

19. $\int_C \tan y\,dx + x \sec^2 y\,dy$,
C is any path from $(1, 0)$ to $(2, \pi/4)$

20. $\int_C (1 - ye^{-x})\,dx + e^{-x}\,dy$,
C is any path from $(0, 1)$ to $(1, 2)$

21–22 ■ Find the work done by the force field $\mathbf{F}$ in moving an object from P to Q.

21. $\mathbf{F}(x, y) = 2y^{3/2}\,\mathbf{i} + 3x\sqrt{y}\,\mathbf{j}$; $P(1, 1)$, $Q(2, 4)$

22. $\mathbf{F}(x, y) = e^{-y}\,\mathbf{i} - xe^{-y}\,\mathbf{j}$; $P(0, 1)$, $Q(2, 0)$

23–24 ■ Is the vector field shown in the figure conservative? Explain.

23.

24.

25. If $\mathbf{F}(x, y) = \sin y\,\mathbf{i} + (1 + x \cos y)\,\mathbf{j}$, use a plot to guess whether $\mathbf{F}$ is conservative. Then determine whether your guess is correct.

26. Let $\mathbf{F} = \nabla f$, where $f(x, y) = \sin(x - 2y)$. Find curves C_1 and C_2 that are not closed and satisfy the equation.
(a) $\displaystyle \int_{C_1} \mathbf{F} \cdot d\mathbf{r} = 0$ (b) $\displaystyle \int_{C_2} \mathbf{F} \cdot d\mathbf{r} = 1$

27. Show that if the vector field $\mathbf{F} = P\,\mathbf{i} + Q\,\mathbf{j} + R\,\mathbf{k}$ is conservative and P, Q, R have continuous first-order partial derivatives, then
$$\frac{\partial P}{\partial y} = \frac{\partial Q}{\partial x} \qquad \frac{\partial P}{\partial z} = \frac{\partial R}{\partial x} \qquad \frac{\partial Q}{\partial z} = \frac{\partial R}{\partial y}$$

28. Use Exercise 27 to show that the line integral $\int_C y\,dx + x\,dy + xyz\,dz$ is not independent of path.

29–32 ■ Determine whether or not the given set is (a) open, (b) connected, and (c) simply-connected.

29. $\{(x, y) \mid x > 0, \ y > 0\}$ **30.** $\{(x, y) \mid x \ne 0\}$

31. $\{(x, y) \mid 1 < x^2 + y^2 < 4\}$

32. $\{(x, y) \mid x^2 + y^2 \le 1 \text{ or } 4 \le x^2 + y^2 \le 9\}$

33. Let $\mathbf{F}(x, y) = \dfrac{-y\,\mathbf{i} + x\,\mathbf{j}}{x^2 + y^2}$.
(a) Show that $\partial P/\partial y = \partial Q/\partial x$.
(b) Show that $\int_C \mathbf{F} \cdot d\mathbf{r}$ is not independent of path.
[*Hint:* Compute $\int_{C_1} \mathbf{F} \cdot d\mathbf{r}$ and $\int_{C_2} \mathbf{F} \cdot d\mathbf{r}$, where C_1 and C_2 are the upper and lower halves of the circle $x^2 + y^2 = 1$ from $(1, 0)$ to $(-1, 0)$.] Does this contradict Theorem 6?

34. (a) Suppose that $\mathbf{F}$ is an inverse square force field, that is,

$$\mathbf{F}(\mathbf{r}) = \frac{c\mathbf{r}}{|\mathbf{r}|^3}$$

for some constant c, where $\mathbf{r} = x\,\mathbf{i} + y\,\mathbf{j} + z\,\mathbf{k}$. Find the work done by $\mathbf{F}$ in moving an object from a point P_1 along a path to a point P_2 in terms of the distances d_1 and d_2 from these points to the origin.

(b) An example of an inverse square field is the gravitational field $\mathbf{F} = -(mMG)\mathbf{r}/|\mathbf{r}|^3$ discussed in Example 4 in Section 13.1. Use part (a) to find the work done by the gravitational field when the Earth moves from

aphelion (at a maximum distance of 1.52×10^8 km from the Sun) to perihelion (at a minimum distance of 1.47×10^8 km). (Use the values $m = 5.97 \times 10^{24}$ kg, $M = 1.99 \times 10^{30}$ kg, and $G = 6.67 \times 10^{-11}$ N·m²/kg².)

(c) Another example of an inverse square field is the electric force field $\mathbf{F} = \varepsilon q Q\mathbf{r}/|\mathbf{r}|^3$ discussed in Example 5 in Section 13.1. Suppose that an electron with a charge of -1.6×10^{-19} C is located at the origin. A positive unit charge is positioned a distance 10^{-12} m from the electron and moves to a position half that distance from the electron. Use part (a) to find the work done by the electric field. (Use the value $\varepsilon = 8.985 \times 10^9$.)

13.4 Green's Theorem

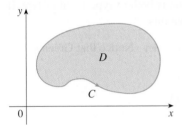

y

D

C

0   x

FIGURE 1

Green's Theorem gives the relationship between a line integral around a simple closed curve C and a double integral over the plane region D bounded by C. (See Figure 1. We assume that D consists of all points inside C as well as all points on C.) In stating Green's Theorem we use the convention that the **positive orientation** of a simple closed curve C refers to a single *counterclockwise* traversal of C. Thus, if C is given by the vector function $\mathbf{r}(t)$, $a \le t \le b$, then the region D is always on the left as the point $\mathbf{r}(t)$ traverses C. (See Figure 2.)

(a) Positive orientation

(b) Negative orientation

FIGURE 2

> **Green's Theorem** Let C be a positively oriented, piecewise-smooth, simple closed curve in the plane and let D be the region bounded by C. If P and Q have continuous partial derivatives on an open region that contains D, then
>
> $$\int_C P\,dx + Q\,dy = \iint_D \left(\frac{\partial Q}{\partial x} - \frac{\partial P}{\partial y} \right) dA$$

■ ■ Recall that the left side of this equation is another way of writing $\int_C \mathbf{F} \cdot d\mathbf{r}$, where $\mathbf{F} = P\,\mathbf{i} + Q\,\mathbf{j}$.

NOTE □ The notation

$$\oint_C P\,dx + Q\,dy \quad \text{or} \quad \oint_C P\,dx + Q\,dy$$

is sometimes used to indicate that the line integral is calculated using the positive orientation of the closed curve C. Another notation for the positively oriented boundary curve of D is ∂D, so the equation in Green's Theorem can be written as

$$\boxed{1} \qquad \iint_D \left(\frac{\partial Q}{\partial x} - \frac{\partial P}{\partial y} \right) dA = \int_{\partial D} P \, dx + Q \, dy$$

Green's Theorem should be regarded as the counterpart of the Fundamental Theorem of Calculus for double integrals. Compare Equation 1 with the statement of the Fundamental Theorem of Calculus, Part 2, in the following equation:

$$\int_a^b F'(x) \, dx = F(b) - F(a)$$

In both cases there is an integral involving derivatives (F', $\partial Q/\partial x$, and $\partial P/\partial y$) on the left side of the equation. And in both cases the right side involves the values of the original functions (F, Q, and P) only on the *boundary* of the domain. (In the one-dimensional case, the domain is an interval $[a, b]$ whose boundary consists of just two points, a and b.)

Green's Theorem is not easy to prove in the generality stated in Theorem 1, but we can give a proof for the special case where the region is both of type I and of type II (see Section 12.3). Let's call such regions **simple regions**.

Proof of Green's Theorem for the Case in Which D Is a Simple Region   Notice that Green's Theorem will be proved if we can show that

$$\boxed{2} \qquad \int_C P \, dx = -\iint_D \frac{\partial P}{\partial y} \, dA$$

and

$$\boxed{3} \qquad \int_C Q \, dy = \iint_D \frac{\partial Q}{\partial x} \, dA$$

We prove Equation 2 by expressing D as a type I region:

$$D = \{(x, y) \mid a \leqslant x \leqslant b, \ g_1(x) \leqslant y \leqslant g_2(x)\}$$

where g_1 and g_2 are continuous functions. This enables us to compute the double integral on the right side of Equation 2 as follows:

$$\boxed{4} \qquad \iint_D \frac{\partial P}{\partial y} \, dA = \int_a^b \int_{g_1(x)}^{g_2(x)} \frac{\partial P}{\partial y}(x, y) \, dy \, dx = \int_a^b [P(x, g_2(x)) - P(x, g_1(x))] \, dx$$

where the last step follows from the Fundamental Theorem of Calculus.

Now we compute the left side of Equation 2 by breaking up C as the union of the four curves C_1, C_2, C_3, and C_4 shown in Figure 3. On C_1 we take x as the parameter and write the parametric equations as $x = x$, $y = g_1(x)$, $a \leqslant x \leqslant b$. Thus

$$\int_{C_1} P(x, y) \, dx = \int_a^b P(x, g_1(x)) \, dx$$

Observe that C_3 goes from right to left but $-C_3$ goes from left to right, so we can write the parametric equations of $-C_3$ as $x = x$, $y = g_2(x)$, $a \leqslant x \leqslant b$. Therefore

$$\int_{C_3} P(x, y) \, dx = -\int_{-C_3} P(x, y) \, dx = -\int_a^b P(x, g_2(x)) \, dx$$

■ ■ Green's Theorem is named after the self-taught English scientist George Green (1793–1841). He worked full-time in his father's bakery from the age of nine and taught himself mathematics from library books. In 1828 he published privately *An Essay on the Application of Mathematical Analysis to the Theories of Electricity and Magnetism*, but only 100 copies were printed and most of those went to his friends. This pamphlet contained a theorem that is equivalent to what we know as Green's Theorem, but it didn't become widely known at that time. Finally, at age 40, Green entered Cambridge University as an undergraduate but died four years after graduation. In 1846 William Thomson (Lord Kelvin) located a copy of Green's essay, realized its significance, and had it reprinted. Green was the first person to try to formulate a mathematical theory of electricity and magnetism. His work was the basis for the subsequent electromagnetic theories of Thomson, Stokes, Rayleigh, and Maxwell.

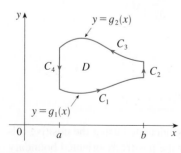

FIGURE 3

On C_2 or C_4 (either of which might reduce to just a single point), x is constant, so $dx = 0$ and

$$\int_{C_2} P(x, y)\, dx = 0 = \int_{C_4} P(x, y)\, dx$$

Hence

$$\int_C P(x, y)\, dx = \int_{C_1} P(x, y)\, dx + \int_{C_2} P(x, y)\, dx + \int_{C_3} P(x, y)\, dx + \int_{C_4} P(x, y)\, dx$$

$$= \int_a^b P(x, g_1(x))\, dx - \int_a^b P(x, g_2(x))\, dx$$

Comparing this expression with the one in Equation 4, we see that

$$\int_C P(x, y)\, dx = -\iint_D \frac{\partial P}{\partial y}\, dA$$

Equation 3 can be proved in much the same way by expressing D as a type II region (see Exercise 28). Then, by adding Equations 2 and 3, we obtain Green's Theorem. ■ ■

EXAMPLE 1 Evaluate $\int_C x^4\, dx + xy\, dy$, where C is the triangular curve consisting of the line segments from $(0, 0)$ to $(1, 0)$, from $(1, 0)$ to $(0, 1)$, and from $(0, 1)$ to $(0, 0)$.

SOLUTION Although the given line integral could be evaluated as usual by the methods of Section 13.2, that would involve setting up three separate integrals along the three sides of the triangle, so let's use Green's Theorem instead. Notice that the region D enclosed by C is simple and C has positive orientation (see Figure 4). If we let $P(x, y) = x^4$ and $Q(x, y) = xy$, then we have

$$\int_C x^4\, dx + xy\, dy = \iint_D \left(\frac{\partial Q}{\partial x} - \frac{\partial P}{\partial y} \right) dA = \int_0^1 \int_0^{1-x} (y - 0)\, dy\, dx$$

$$= \int_0^1 \left[\tfrac{1}{2} y^2 \right]_{y=0}^{y=1-x} dx = \tfrac{1}{2} \int_0^1 (1 - x)^2\, dx$$

$$= -\tfrac{1}{6} (1 - x)^3 \Big]_0^1 = \tfrac{1}{6} \qquad ■ ■$$

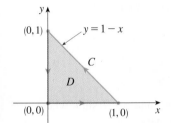

$(0, 1)$ $y = 1 - x$

C

D

$(0, 0)$ $(1, 0)$ x

FIGURE 4

Ⅴ **EXAMPLE 2** Evaluate $\oint_C (3y - e^{\sin x})\, dx + \left(7x + \sqrt{y^4 + 1} \right) dy$, where C is the circle $x^2 + y^2 = 9$.

SOLUTION The region D bounded by C is the disk $x^2 + y^2 \le 9$, so let's change to polar coordinates after applying Green's Theorem:

$$\oint_C (3y - e^{\sin x})\, dx + \left(7x + \sqrt{y^4 + 1} \right) dy$$

$$= \iint_D \left[\frac{\partial}{\partial x} \left(7x + \sqrt{y^4 + 1} \right) - \frac{\partial}{\partial y} (3y - e^{\sin x}) \right] dA$$

$$= \int_0^{2\pi} \int_0^3 (7 - 3)\, r\, dr\, d\theta$$

■ ■ Instead of using polar coordinates, we could simply use the fact that D is a disk of radius 3 and write

$$\iint_D 4\, dA = 4 \cdot \pi(3)^2 = 36\pi$$

$$= 4 \int_0^{2\pi} d\theta \int_0^3 r\, dr = 36\pi \qquad ■ ■$$

In Examples 1 and 2 we found that the double integral was easier to evaluate than the line integral. (Try setting up the line integral in Example 2 and you'll soon be convinced!) But sometimes it's easier to evaluate the line integral, and Green's Theorem is used in the reverse direction. For instance, if it is known that $P(x, y) = Q(x, y) = 0$ on the curve C, then Green's Theorem gives

$$\iint_D \left(\frac{\partial Q}{\partial x} - \frac{\partial P}{\partial y} \right) dA = \int_C P \, dx + Q \, dy = 0$$

no matter what values P and Q assume in the region D.

Another application of the reverse direction of Green's Theorem is in computing areas. Since the area of D is $\iint_D 1 \, dA$, we wish to choose P and Q so that

$$\frac{\partial Q}{\partial x} - \frac{\partial P}{\partial y} = 1$$

There are several possibilities:

$$P(x, y) = 0 \qquad P(x, y) = -y \qquad P(x, y) = -\tfrac{1}{2}y$$
$$Q(x, y) = x \qquad Q(x, y) = 0 \qquad Q(x, y) = \tfrac{1}{2}x$$

Then Green's Theorem gives the following formulas for the area of D:

> **5**
>
> $$A = \oint_C x \, dy = -\oint_C y \, dx = \tfrac{1}{2} \oint_C x \, dy - y \, dx$$

EXAMPLE 3 Find the area enclosed by the ellipse $\dfrac{x^2}{a^2} + \dfrac{y^2}{b^2} = 1$.

SOLUTION The ellipse has parametric equations $x = a \cos t$ and $y = b \sin t$, where $0 \leq t \leq 2\pi$. Using the third formula in Equation 5, we have

$$A = \tfrac{1}{2} \int_C x \, dy - y \, dx$$

$$= \tfrac{1}{2} \int_0^{2\pi} (a \cos t)(b \cos t) \, dt - (b \sin t)(-a \sin t) \, dt$$

$$= \frac{ab}{2} \int_0^{2\pi} dt = \pi ab$$ ■ ■

Although we have proved Green's Theorem only for the case where D is simple, we can now extend it to the case where D is a finite union of simple regions. For example, if D is the region shown in Figure 5, then we can write $D = D_1 \cup D_2$, where D_1 and D_2 are both simple. The boundary of D_1 is $C_1 \cup C_3$ and the boundary of D_2 is $C_2 \cup (-C_3)$ so, applying Green's Theorem to D_1 and D_2 separately, we get

$$\int_{C_1 \cup C_3} P \, dx + Q \, dy = \iint_{D_1} \left(\frac{\partial Q}{\partial x} - \frac{\partial P}{\partial y} \right) dA$$

$$\int_{C_2 \cup (-C_3)} P \, dx + Q \, dy = \iint_{D_2} \left(\frac{\partial Q}{\partial x} - \frac{\partial P}{\partial y} \right) dA$$

FIGURE 5

FIGURE 6

FIGURE 7

FIGURE 8

FIGURE 9

If we add these two equations, the line integrals along C_3 and $-C_3$ cancel, so we get

$$\int_{C_1 \cup C_2} P \, dx + Q \, dy = \iint_D \left(\frac{\partial Q}{\partial x} - \frac{\partial P}{\partial y} \right) dA$$

which is Green's Theorem for $D = D_1 \cup D_2$, since its boundary is $C = C_1 \cup C_2$.

The same sort of argument allows us to establish Green's Theorem for any finite union of nonoverlapping simple regions (see Figure 6).

V **EXAMPLE 4** Evaluate $\oint_C y^2 \, dx + 3xy \, dy$, where C is the boundary of the semiannular region D in the upper half-plane between the circles $x^2 + y^2 = 1$ and $x^2 + y^2 = 4$.

SOLUTION Notice that although D is not simple, the y-axis divides it into two simple regions (see Figure 7). In polar coordinates we can write

$$D = \{ (r, \theta) \mid 1 \le r \le 2, \ 0 \le \theta \le \pi \}$$

Therefore, Green's Theorem gives

$$\oint_C y^2 \, dx + 3xy \, dy = \iint_D \left[\frac{\partial}{\partial x} (3xy) - \frac{\partial}{\partial y} (y^2) \right] dA$$

$$= \iint_D y \, dA = \int_0^\pi \int_1^2 (r \sin \theta) \, r \, dr \, d\theta$$

$$= \int_0^\pi \sin \theta \, d\theta \int_1^2 r^2 \, dr = \left[-\cos \theta \right]_0^\pi \left[\tfrac{1}{3} r^3 \right]_1^2 = \frac{14}{3} \quad ■ ■$$

Green's Theorem can be extended to apply to regions with holes, that is, regions that are not simply-connected. Observe that the boundary C of the region D in Figure 8 consists of two simple closed curves C_1 and C_2. We assume that these boundary curves are oriented so that the region D is always on the left as the curve C is traversed. Thus, the positive direction is counterclockwise for the outer curve C_1 but clockwise for the inner curve C_2. If we divide D into two regions D' and D'' by means of the lines shown in Figure 9 and then apply Green's Theorem to each of D' and D'', we get

$$\iint_D \left(\frac{\partial Q}{\partial x} - \frac{\partial P}{\partial y} \right) dA = \iint_{D'} \left(\frac{\partial Q}{\partial x} - \frac{\partial P}{\partial y} \right) dA + \iint_{D''} \left(\frac{\partial Q}{\partial x} - \frac{\partial P}{\partial y} \right) dA$$

$$= \int_{\partial D'} P \, dx + Q \, dy + \int_{\partial D''} P \, dx + Q \, dy$$

Since the line integrals along the common boundary lines are in opposite directions, they cancel and we get

$$\iint_D \left(\frac{\partial Q}{\partial x} - \frac{\partial P}{\partial y} \right) dA = \int_{C_1} P \, dx + Q \, dy + \int_{C_2} P \, dx + Q \, dy = \int_C P \, dx + Q \, dy$$

which is Green's Theorem for the region D.

V **EXAMPLE 5** If $\mathbf{F}(x, y) = (-y \, \mathbf{i} + x \, \mathbf{j})/(x^2 + y^2)$, show that $\int_C \mathbf{F} \cdot d\mathbf{r} = 2\pi$ for every positively oriented simple closed path that encloses the origin.

SOLUTION Since C is an *arbitrary* closed path that encloses the origin, it's difficult to compute the given integral directly. So let's consider a counterclockwise-oriented circle

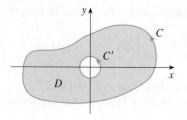

FIGURE 10

C' with center the origin and radius a, where a is chosen to be small enough that C' lies inside C. (See Figure 10.) Let D be the region bounded by C and C'. Then its positively oriented boundary is $C \cup (-C')$ and so the general version of Green's Theorem gives

$$\int_C P\,dx + Q\,dy + \int_{-C'} P\,dx + Q\,dy = \iint_D \left(\frac{\partial Q}{\partial x} - \frac{\partial P}{\partial y} \right) dA$$

$$= \iint_D \left[\frac{y^2 - x^2}{(x^2 + y^2)^2} - \frac{y^2 - x^2}{(x^2 + y^2)^2} \right] dA$$

$$= 0$$

Therefore $$\int_C P\,dx + Q\,dy = \int_{C'} P\,dx + Q\,dy$$

that is, $$\int_C \mathbf{F} \cdot d\mathbf{r} = \int_{C'} \mathbf{F} \cdot d\mathbf{r}$$

We now easily compute this last integral using the parametrization given by $\mathbf{r}(t) = a \cos t\,\mathbf{i} + a \sin t\,\mathbf{j}, 0 \le t \le 2\pi$. Thus

$$\int_C \mathbf{F} \cdot d\mathbf{r} = \int_{C'} \mathbf{F} \cdot d\mathbf{r} = \int_0^{2\pi} \mathbf{F}(\mathbf{r}(t)) \cdot \mathbf{r}'(t)\,dt$$

$$= \int_0^{2\pi} \frac{(-a \sin t)(-a \sin t) + (a \cos t)(a \cos t)}{a^2 \cos^2 t + a^2 \sin^2 t}\,dt$$

$$= \int_0^{2\pi} dt = 2\pi$$

■ ■

We end this section by using Green's Theorem to discuss a result that was stated in the preceding section.

Sketch of Proof of Theorem 13.3.6 We're assuming that $\mathbf{F} = P\,\mathbf{i} + Q\,\mathbf{j}$ is a vector field on an open simply-connected region D, that P and Q have continuous first-order partial derivatives, and that

$$\frac{\partial P}{\partial y} = \frac{\partial Q}{\partial x} \qquad \text{throughout } D$$

If C is any simple closed path in D and R is the region that C encloses, then Green's Theorem gives

$$\oint_C \mathbf{F} \cdot d\mathbf{r} = \oint_C P\,dx + Q\,dy = \iint_R \left(\frac{\partial Q}{\partial x} - \frac{\partial P}{\partial y} \right) dA = \iint_R 0\,dA = 0$$

A curve that is not simple crosses itself at one or more points and can be broken up into a number of simple curves. We have shown that the line integrals of $\mathbf{F}$ around these simple curves are all 0 and, adding these integrals, we see that $\int_C \mathbf{F} \cdot d\mathbf{r} = 0$ for any closed curve C. Therefore, $\int_C \mathbf{F} \cdot d\mathbf{r}$ is independent of path in D by Theorem 13.3.3. It follows that $\mathbf{F}$ is a conservative vector field. ■ ■

13.4 | Exercises

1–4 ■ Evaluate the line integral by two methods: (a) directly and (b) using Green's Theorem.

1. $\oint_C xy^2\,dx + x^3\,dy$,
C is the rectangle with vertices $(0, 0)$, $(2, 0)$, $(2, 3)$, and $(0, 3)$

2. $\oint_C y\,dx - x\,dy$,
C is the circle with center the origin and radius 1

3. $\oint_C xy\,dx + x^2y^3\,dy$,
C is the triangle with vertices $(0, 0)$, $(1, 0)$, and $(1, 2)$

4. $\oint_C x\,dx + y\,dy$, C consists of the line segments from $(0, 1)$ to $(0, 0)$ and from $(0, 0)$ to $(1, 0)$ and the parabola $y = 1 - x^2$ from $(1, 0)$ to $(0, 1)$

CAS 5–6 ■ Verify Green's Theorem by using a computer algebra system to evaluate both the line integral and the double integral.

5. $P(x, y) = x^4 y^5$, $Q(x, y) = -x^7 y^6$,
C is the circle $x^2 + y^2 = 1$

6. $P(x, y) = y^2 \sin x$, $Q(x, y) = x^2 \sin y$,
C consists of the arc of the parabola $y = x^2$ from $(0, 0)$ to $(1, 1)$ followed by the line segment from $(1, 1)$ to $(0, 0)$

7–12 ■ Use Green's Theorem to evaluate the line integral along the given positively oriented curve.

7. $\int_C e^y\,dx + 2xe^y\,dy$,
C is the square with sides $x = 0$, $x = 1$, $y = 0$, and $y = 1$

8. $\int_C x^2y^2\,dx + 4xy^3\,dy$,
C is the triangle with vertices $(0, 0)$, $(1, 3)$, and $(0, 3)$

9. $\int_C \left(y + e^{\sqrt{x}}\right)dx + (2x + \cos y^2)\,dy$,
C is the boundary of the region enclosed by the parabolas $y = x^2$ and $x = y^2$

10. $\int_C xe^{-2x}\,dx + (x^4 + 2x^2y^2)\,dy$,
C is the boundary of the region between the circles $x^2 + y^2 = 1$ and $x^2 + y^2 = 4$

11. $\int_C y^3\,dx - x^3\,dy$, C is the circle $x^2 + y^2 = 4$

12. $\int_C \sin y\,dx + x \cos y\,dy$, C is the ellipse $x^2 + xy + y^2 = 1$

13–16 ■ Use Green's Theorem to evaluate $\int_C \mathbf{F} \cdot d\mathbf{r}$. (Check the orientation of the curve before applying the theorem.)

13. $\mathbf{F}(x, y) = \langle \sqrt{x} + y^3, x^2 + \sqrt{y}\rangle$,
C consists of the arc of the curve $y = \sin x$ from $(0, 0)$ to $(\pi, 0)$ and the line segment from $(\pi, 0)$ to $(0, 0)$

14. $\mathbf{F}(x, y) = \langle y^2 \cos x, x^2 + 2y \sin x\rangle$,
C is the triangle from $(0, 0)$ to $(2, 6)$ to $(2, 0)$ to $(0, 0)$

15. $\mathbf{F}(x, y) = \langle e^x + x^2y, e^y - xy^2\rangle$,
C is the circle $x^2 + y^2 = 25$ oriented clockwise

16. $\mathbf{F}(x, y) = \langle y - \ln(x^2 + y^2), 2 \tan^{-1}(y/x)\rangle$, C is the circle $(x - 2)^2 + (y - 3)^2 = 1$ oriented counterclockwise

17. Use Green's Theorem to find the work done by the force $\mathbf{F}(x, y) = x(x + y)\,\mathbf{i} + xy^2\,\mathbf{j}$ in moving a particle from the origin along the x-axis to $(1, 0)$, then along the line segment to $(0, 1)$, and then back to the origin along the y-axis.

18. A particle starts at the point $(-2, 0)$, moves along the x-axis to $(2, 0)$, and then along the semicircle $y = \sqrt{4 - x^2}$ to the starting point. Use Green's Theorem to find the work done on this particle by the force field $\mathbf{F}(x, y) = \langle x, x^3 + 3xy^2\rangle$.

19. Use one of the formulas in (5) to find the area under one arch of the cycloid $x = t - \sin t$, $y = 1 - \cos t$.

20. If a circle C with radius 1 rolls along the outside of the circle $x^2 + y^2 = 16$, a fixed point P on C traces out a curve called an *epicycloid*, with parametric equations $x = 5 \cos t - \cos 5t$, $y = 5 \sin t - \sin 5t$. Graph the epicycloid and use (5) to find the area it encloses.

21. (a) If C is the line segment connecting the point (x_1, y_1) to the point (x_2, y_2), show that
$$\int_C x\,dy - y\,dx = x_1y_2 - x_2y_1$$
(b) If the vertices of a polygon, in counterclockwise order, are (x_1, y_1), (x_2, y_2), . . . , (x_n, y_n), show that the area of the polygon is
$$A = \tfrac{1}{2}[(x_1y_2 - x_2y_1) + (x_2y_3 - x_3y_2) + \cdots$$
$$+ (x_{n-1}y_n - x_ny_{n-1}) + (x_ny_1 - x_1y_n)]$$
(c) Find the area of the pentagon with vertices $(0, 0)$, $(2, 1)$, $(1, 3)$, $(0, 2)$, and $(-1, 1)$.

22. Let D be a region bounded by a simple closed path C in the xy-plane. Use Green's Theorem to prove that the coordinates of the centroid $(\bar{x}, \bar{y})$ of D are
$$\bar{x} = \frac{1}{2A}\oint_C x^2\,dy \qquad \bar{y} = -\frac{1}{2A}\oint_C y^2\,dx$$
where A is the area of D.

23. Use Exercise 22 to find the centroid of the triangle with vertices $(0, 0)$, $(1, 0)$, and $(0, 1)$.

24. Use Exercise 22 to find the centroid of a semicircular region of radius a.

25. A plane lamina with constant density $\rho(x, y) = \rho$ occupies a region in the xy-plane bounded by a simple closed path C. Show that its moments of inertia about the axes are
$$I_x = -\frac{\rho}{3}\oint_C y^3\,dx \qquad I_y = \frac{\rho}{3}\oint_C x^3\,dy$$

26. Use Exercise 25 to find the moment of inertia of a circular disk of radius a with constant density ρ about a diameter. (Compare with Example 4 in Section 12.5.)

27. If $\mathbf{F}$ is the vector field of Example 5, show that $\int_C \mathbf{F} \cdot d\mathbf{r} = 0$ for every simple closed path that does not pass through or enclose the origin.

28. Complete the proof of the special case of Green's Theorem by proving Equation 3.

29. Use Green's Theorem to prove the change of variables formula for a double integral (Formula 12.9.9) for the case

where $f(x, y) = 1$:

$$\iint_R dx\, dy = \iint_S \left| \frac{\partial(x, y)}{\partial(u, v)} \right| du\, dv$$

Here R is the region in the xy-plane that corresponds to the region S in the uv-plane under the transformation given by $x = g(u, v)$, $y = h(u, v)$.

[*Hint:* Note that the left side is $A(R)$ and apply the first part of Equation 5. Convert the line integral over ∂R to a line integral over ∂S and apply Green's Theorem in the uv-plane.]

13.5 Curl and Divergence

In this section we define two operations that can be performed on vector fields and that play a basic role in the applications of vector calculus to fluid flow and electricity and magnetism. Each operation resembles differentiation, but one produces a vector field whereas the other produces a scalar field.

■ Curl

If $\mathbf{F} = P\,\mathbf{i} + Q\,\mathbf{j} + R\,\mathbf{k}$ is a vector field on $\mathbb{R}^3$ and the partial derivatives of P, Q, and R all exist, then the **curl** of $\mathbf{F}$ is the vector field on $\mathbb{R}^3$ defined by

$$\boxed{1} \qquad \operatorname{curl} \mathbf{F} = \left(\frac{\partial R}{\partial y} - \frac{\partial Q}{\partial z} \right) \mathbf{i} + \left(\frac{\partial P}{\partial z} - \frac{\partial R}{\partial x} \right) \mathbf{j} + \left(\frac{\partial Q}{\partial x} - \frac{\partial P}{\partial y} \right) \mathbf{k}$$

As an aid to our memory, let's rewrite Equation 1 using operator notation. We introduce the vector differential operator ∇ ("del") as

$$\nabla = \mathbf{i}\, \frac{\partial}{\partial x} + \mathbf{j}\, \frac{\partial}{\partial y} + \mathbf{k}\, \frac{\partial}{\partial z}$$

It has meaning when it operates on a scalar function to produce the gradient of f:

$$\nabla f = \mathbf{i}\, \frac{\partial f}{\partial x} + \mathbf{j}\, \frac{\partial f}{\partial y} + \mathbf{k}\, \frac{\partial f}{\partial z} = \frac{\partial f}{\partial x}\,\mathbf{i} + \frac{\partial f}{\partial y}\,\mathbf{j} + \frac{\partial f}{\partial z}\,\mathbf{k}$$

If we think of ∇ as a vector with components $\partial/\partial x$, $\partial/\partial y$, and $\partial/\partial z$, we can also consider the formal cross product of ∇ with the vector field $\mathbf{F}$ as follows:

$$\nabla \times \mathbf{F} = \begin{vmatrix} \mathbf{i} & \mathbf{j} & \mathbf{k} \\ \dfrac{\partial}{\partial x} & \dfrac{\partial}{\partial y} & \dfrac{\partial}{\partial z} \\ P & Q & R \end{vmatrix}$$

$$= \left(\frac{\partial R}{\partial y} - \frac{\partial Q}{\partial z} \right) \mathbf{i} + \left(\frac{\partial P}{\partial z} - \frac{\partial R}{\partial x} \right) \mathbf{j} + \left(\frac{\partial Q}{\partial x} - \frac{\partial P}{\partial y} \right) \mathbf{k}$$

$$= \operatorname{curl} \mathbf{F}$$

Thus, the easiest way to remember Definition 1 is by means of the symbolic expression

$$\boxed{2} \qquad \boxed{\operatorname{curl} \mathbf{F} = \nabla \times \mathbf{F}}$$

EXAMPLE 1 If $\mathbf{F}(x, y, z) = xz\,\mathbf{i} + xyz\,\mathbf{j} - y^2\,\mathbf{k}$, find curl $\mathbf{F}$.

SOLUTION Using Equation 2, we have

$$
\operatorname{curl} \mathbf{F} = \nabla \times \mathbf{F} = \begin{vmatrix} \mathbf{i} & \mathbf{j} & \mathbf{k} \\ \dfrac{\partial}{\partial x} & \dfrac{\partial}{\partial y} & \dfrac{\partial}{\partial z} \\ xz & xyz & -y^2 \end{vmatrix}
$$

$$
= \left[\frac{\partial}{\partial y}(-y^2) - \frac{\partial}{\partial z}(xyz) \right] \mathbf{i} - \left[\frac{\partial}{\partial x}(-y^2) - \frac{\partial}{\partial z}(xz) \right] \mathbf{j}
$$

$$
+ \left[\frac{\partial}{\partial x}(xyz) - \frac{\partial}{\partial y}(xz) \right] \mathbf{k}
$$

$$
= (-2y - xy)\,\mathbf{i} - (0 - x)\,\mathbf{j} + (yz - 0)\,\mathbf{k}
$$

$$
= -y(2 + x)\,\mathbf{i} + x\,\mathbf{j} + yz\,\mathbf{k} \qquad\blacksquare\blacksquare
$$

■ ■ Most computer algebra systems have commands that compute the curl and divergence of vector fields. If you have access to a CAS, use these commands to check the answers to the examples and exercises in this section.

Recall that the gradient of a function f of three variables is a vector field on $\mathbb{R}^3$ and so we can compute its curl. The following theorem says that the curl of a gradient vector field is $\mathbf{0}$.

> $\boxed{3}$ **Theorem** If f is a function of three variables that has continuous second-order partial derivatives, then
>
> $$\operatorname{curl}(\nabla f) = \mathbf{0}$$

Proof We have

$$
\operatorname{curl}(\nabla f) = \nabla \times (\nabla f) = \begin{vmatrix} \mathbf{i} & \mathbf{j} & \mathbf{k} \\ \dfrac{\partial}{\partial x} & \dfrac{\partial}{\partial y} & \dfrac{\partial}{\partial z} \\ \dfrac{\partial f}{\partial x} & \dfrac{\partial f}{\partial y} & \dfrac{\partial f}{\partial z} \end{vmatrix}
$$

■ ■ Notice the similarity to what we know from Section 9.4: $\mathbf{a} \times \mathbf{a} = \mathbf{0}$ for every three-dimensional vector $\mathbf{a}$.

$$
= \left(\frac{\partial^2 f}{\partial y\, \partial z} - \frac{\partial^2 f}{\partial z\, \partial y} \right) \mathbf{i} + \left(\frac{\partial^2 f}{\partial z\, \partial x} - \frac{\partial^2 f}{\partial x\, \partial z} \right) \mathbf{j} + \left(\frac{\partial^2 f}{\partial x\, \partial y} - \frac{\partial^2 f}{\partial y\, \partial x} \right) \mathbf{k}
$$

$$
= 0\,\mathbf{i} + 0\,\mathbf{j} + 0\,\mathbf{k} = \mathbf{0}
$$

by Clairaut's Theorem. ■ ■

Since a conservative vector field is one for which $\mathbf{F} = \nabla f$, Theorem 3 can be rephrased as follows:

■ ■ Compare this with Exercise 27 in Section 13.3.

If $\mathbf{F}$ is conservative, then curl $\mathbf{F} = \mathbf{0}$.

This gives us a way of verifying that a vector field is not conservative.

☑ **EXAMPLE 2** Show that the vector field $\mathbf{F}(x, y, z) = xz\,\mathbf{i} + xyz\,\mathbf{j} - y^2\,\mathbf{k}$ is not conservative.

SOLUTION In Example 1 we showed that

$$\text{curl } \mathbf{F} = -y(2 + x)\,\mathbf{i} + x\,\mathbf{j} + yz\,\mathbf{k}$$

This shows that curl $\mathbf{F} \neq \mathbf{0}$ and so, by Theorem 3, $\mathbf{F}$ is not conservative. ■ ■

The converse of Theorem 3 is not true in general, but the following theorem says the converse is true if $\mathbf{F}$ is defined everywhere. (More generally it is true if the domain is simply-connected, that is, "has no hole.") Theorem 4 is the three-dimensional version of Theorem 13.3.6. Its proof requires Stokes' Theorem and is sketched at the end of Section 13.7.

> **4 Theorem** If $\mathbf{F}$ is a vector field defined on all of $\mathbb{R}^3$ whose component functions have continuous partial derivatives and curl $\mathbf{F} = \mathbf{0}$, then $\mathbf{F}$ is a conservative vector field.

☑ **EXAMPLE 3**
(a) Show that $\mathbf{F}(x, y, z) = y^2 z^3\,\mathbf{i} + 2xyz^3\,\mathbf{j} + 3xy^2 z^2\,\mathbf{k}$ is a conservative vector field.
(b) Find a function f such that $\mathbf{F} = \nabla f$.

SOLUTION
(a) We compute the curl of $\mathbf{F}$:

$$\text{curl } \mathbf{F} = \nabla \times \mathbf{F} = \begin{vmatrix} \mathbf{i} & \mathbf{j} & \mathbf{k} \\ \dfrac{\partial}{\partial x} & \dfrac{\partial}{\partial y} & \dfrac{\partial}{\partial z} \\ y^2 z^3 & 2xyz^3 & 3xy^2 z^2 \end{vmatrix}$$

$$= (6xyz^2 - 6xyz^2)\,\mathbf{i} - (3y^2 z^2 - 3y^2 z^2)\,\mathbf{j} + (2yz^3 - 2yz^3)\,\mathbf{k}$$

$$= \mathbf{0}$$

Since curl $\mathbf{F} = \mathbf{0}$ and the domain of $\mathbf{F}$ is $\mathbb{R}^3$, $\mathbf{F}$ is a conservative vector field by Theorem 4.

(b) The technique for finding f was given in Section 13.3. We have

$$\boxed{5} \qquad f_x(x, y, z) = y^2 z^3$$

$$\boxed{6} \qquad f_y(x, y, z) = 2xyz^3$$

$$\boxed{7} \qquad f_z(x, y, z) = 3xy^2 z^2$$

Integrating (5) with respect to x, we obtain

$$\boxed{8} \qquad f(x, y, z) = xy^2 z^3 + g(y, z)$$

Differentiating (8) with respect to y, we get $f_y(x, y, z) = 2xyz^3 + g_y(y, z)$, so comparison with (6) gives $g_y(y, z) = 0$. Thus, $g(y, z) = h(z)$ and

$$f_z(x, y, z) = 3xy^2 z^2 + h'(z)$$

Then (7) gives $h'(z) = 0$. Therefore

$$f(x, y, z) = xy^2z^3 + K$$ ■ ■

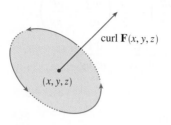

curl $\mathbf{F}(x, y, z)$

(x, y, z)

FIGURE 1

The reason for the name *curl* is that the curl vector is associated with rotations. One connection is explained in Exercise 35. Another occurs when $\mathbf{F}$ represents the velocity field in fluid flow (see Example 3 in Section 13.1). Particles near (x, y, z) in the fluid tend to rotate about the axis that points in the direction of curl $\mathbf{F}(x, y, z)$ and the length of this curl vector is a measure of how quickly the particles move around the axis (see Figure 1). If curl $\mathbf{F} = \mathbf{0}$ at a point P, then the fluid is free from rotations at P and $\mathbf{F}$ is called **irrotational** at P. In other words, there is no whirlpool or eddy at P. If curl $\mathbf{F} = \mathbf{0}$, then a tiny paddle wheel moves with the fluid but doesn't rotate about its axis. If curl $\mathbf{F} \neq \mathbf{0}$, the paddle wheel rotates about its axis. We give a more detailed explanation in Section 13.7 as a consequence of Stokes' Theorem.

■■
■■ **Divergence**

If $\mathbf{F} = P\,\mathbf{i} + Q\,\mathbf{j} + R\,\mathbf{k}$ is a vector field on $\mathbb{R}^3$ and $\partial P/\partial x$, $\partial Q/\partial y$, and $\partial R/\partial z$ exist, then the **divergence of F** is the function of three variables defined by

9
$$\text{div } \mathbf{F} = \frac{\partial P}{\partial x} + \frac{\partial Q}{\partial y} + \frac{\partial R}{\partial z}$$

Observe that curl $\mathbf{F}$ is a vector field but div $\mathbf{F}$ is a scalar field. In terms of the gradient operator $\nabla = (\partial/\partial x)\,\mathbf{i} + (\partial/\partial y)\,\mathbf{j} + (\partial/\partial z)\,\mathbf{k}$, the divergence of $\mathbf{F}$ can be written symbolically as the dot product of ∇ and $\mathbf{F}$:

10
$$\text{div } \mathbf{F} = \nabla \cdot \mathbf{F}$$

EXAMPLE 4 If $\mathbf{F}(x, y, z) = xz\,\mathbf{i} + xyz\,\mathbf{j} - y^2\,\mathbf{k}$, find div $\mathbf{F}$.

SOLUTION By the definition of divergence (Equation 9 or 10) we have

$$\text{div } \mathbf{F} = \nabla \cdot \mathbf{F} = \frac{\partial}{\partial x}\,(xz) + \frac{\partial}{\partial y}\,(xyz) + \frac{\partial}{\partial z}\,(-y^2)$$

$$= z + xz$$ ■ ■

If $\mathbf{F}$ is a vector field on $\mathbb{R}^3$, then curl $\mathbf{F}$ is also a vector field on $\mathbb{R}^3$. As such, we can compute its divergence. The next theorem shows that the result is 0.

11 **Theorem** If $\mathbf{F} = P\,\mathbf{i} + Q\,\mathbf{j} + R\,\mathbf{k}$ is a vector field on $\mathbb{R}^3$ and P, Q, and R have continuous second-order partial derivatives, then

$$\text{div curl } \mathbf{F} = 0$$

■ ■ Note the analogy with the scalar triple product: $\mathbf{a} \cdot (\mathbf{a} \times \mathbf{b}) = 0$.

Proof Using the definitions of divergence and curl, we have

$$\text{div curl } \mathbf{F} = \nabla \cdot (\nabla \times \mathbf{F})$$

$$= \frac{\partial}{\partial x}\left(\frac{\partial R}{\partial y} - \frac{\partial Q}{\partial z}\right) + \frac{\partial}{\partial y}\left(\frac{\partial P}{\partial z} - \frac{\partial R}{\partial x}\right) + \frac{\partial}{\partial z}\left(\frac{\partial Q}{\partial x} - \frac{\partial P}{\partial y}\right)$$

$$= \frac{\partial^2 R}{\partial x\, \partial y} - \frac{\partial^2 Q}{\partial x\, \partial z} + \frac{\partial^2 P}{\partial y\, \partial z} - \frac{\partial^2 R}{\partial y\, \partial x} + \frac{\partial^2 Q}{\partial z\, \partial x} - \frac{\partial^2 P}{\partial z\, \partial y}$$

$$= 0$$

because the terms cancel in pairs by Clairaut's Theorem.   ■ ■

|V| **EXAMPLE 5** Show that the vector field $\mathbf{F}(x, y, z) = xz\,\mathbf{i} + xyz\,\mathbf{j} - y^2\,\mathbf{k}$ can't be written as the curl of another vector field, that is, $\mathbf{F} \neq \text{curl } \mathbf{G}$.

SOLUTION In Example 4 we showed that

$$\text{div } \mathbf{F} = z + xz$$

and therefore div $\mathbf{F} \neq 0$. If it were true that $\mathbf{F} = \text{curl } \mathbf{G}$, then Theorem 11 would give

$$\text{div } \mathbf{F} = \text{div curl } \mathbf{G} = 0$$

which contradicts div $\mathbf{F} \neq 0$. Therefore, $\mathbf{F}$ is not the curl of another vector field.   ■ ■

■ ■ The reason for this interpretation of div $\mathbf{F}$ will be explained at the end of Section 13.8 as a consequence of the Divergence Theorem.

Again, the reason for the name *divergence* can be understood in the context of fluid flow. If $\mathbf{F}(x, y, z)$ is the velocity of a fluid (or gas), then div $\mathbf{F}(x, y, z)$ represents the net rate of change (with respect to time) of the mass of fluid (or gas) flowing from the point (x, y, z) per unit volume. In other words, div $\mathbf{F}(x, y, z)$ measures the tendency of the fluid to diverge from the point (x, y, z). If div $\mathbf{F} = 0$, then $\mathbf{F}$ is said to be **incompressible**.

Another differential operator occurs when we compute the divergence of a gradient vector field ∇f. If f is a function of three variables, we have

$$\text{div}(\nabla f) = \nabla \cdot (\nabla f) = \frac{\partial^2 f}{\partial x^2} + \frac{\partial^2 f}{\partial y^2} + \frac{\partial^2 f}{\partial z^2}$$

and this expression occurs so often that we abbreviate it as $\nabla^2 f$. The operator

$$\nabla^2 = \nabla \cdot \nabla$$

is called the **Laplace operator** because of its relation to **Laplace's equation**

$$\nabla^2 f = \frac{\partial^2 f}{\partial x^2} + \frac{\partial^2 f}{\partial y^2} + \frac{\partial^2 f}{\partial z^2} = 0$$

We can also apply the Laplace operator ∇^2 to a vector field

$$\mathbf{F} = P\,\mathbf{i} + Q\,\mathbf{j} + R\,\mathbf{k}$$

in terms of its components:

$$\nabla^2 \mathbf{F} = \nabla^2 P\,\mathbf{i} + \nabla^2 Q\,\mathbf{j} + \nabla^2 R\,\mathbf{k}$$

▪▪ Vector Forms of Green's Theorem

The curl and divergence operators allow us to rewrite Green's Theorem in versions that will be useful in our later work. We suppose that the plane region D, its boundary curve C, and the functions P and Q satisfy the hypotheses of Green's Theorem. Then we consider the vector field $\mathbf{F} = P\,\mathbf{i} + Q\,\mathbf{j}$. Its line integral is

$$\oint_C \mathbf{F} \cdot d\mathbf{r} = \oint_C P\,dx + Q\,dy$$

and, regarding $\mathbf{F}$ as a vector field on $\mathbb{R}^3$ with third component 0, we have

$$\text{curl } \mathbf{F} = \begin{vmatrix} \mathbf{i} & \mathbf{j} & \mathbf{k} \\ \dfrac{\partial}{\partial x} & \dfrac{\partial}{\partial y} & \dfrac{\partial}{\partial z} \\ P(x, y) & Q(x, y) & 0 \end{vmatrix} = \left(\dfrac{\partial Q}{\partial x} - \dfrac{\partial P}{\partial y} \right) \mathbf{k}$$

Therefore

$$(\text{curl } \mathbf{F}) \cdot \mathbf{k} = \left(\dfrac{\partial Q}{\partial x} - \dfrac{\partial P}{\partial y} \right) \mathbf{k} \cdot \mathbf{k} = \dfrac{\partial Q}{\partial x} - \dfrac{\partial P}{\partial y}$$

and we can now rewrite the equation in Green's Theorem in the vector form

12

$$\oint_C \mathbf{F} \cdot d\mathbf{r} = \iint_D (\text{curl } \mathbf{F}) \cdot \mathbf{k}\, dA$$

Equation 12 expresses the line integral of the tangential component of $\mathbf{F}$ along C as the double integral of the vertical component of curl $\mathbf{F}$ over the region D enclosed by C. We now derive a similar formula involving the *normal* component of $\mathbf{F}$.

If C is given by the vector equation

$$\mathbf{r}(t) = x(t)\,\mathbf{i} + y(t)\,\mathbf{j} \qquad a \le t \le b$$

then the unit tangent vector (see Section 10.2) is

$$\mathbf{T}(t) = \dfrac{x'(t)}{|\mathbf{r}'(t)|}\,\mathbf{i} + \dfrac{y'(t)}{|\mathbf{r}'(t)|}\,\mathbf{j}$$

You can verify that the outward unit normal vector to C is given by

$$\mathbf{n}(t) = \dfrac{y'(t)}{|\mathbf{r}'(t)|}\,\mathbf{i} - \dfrac{x'(t)}{|\mathbf{r}'(t)|}\,\mathbf{j}$$

(See Figure 2.) Then, from Equation 13.2.3, we have

$$\oint_C \mathbf{F} \cdot \mathbf{n}\, ds = \int_a^b (\mathbf{F} \cdot \mathbf{n})(t)\,|\mathbf{r}'(t)|\, dt$$

$$= \int_a^b \left[\dfrac{P(x(t), y(t))\, y'(t)}{|\mathbf{r}'(t)|} - \dfrac{Q(x(t), y(t))\, x'(t)}{|\mathbf{r}'(t)|} \right] |\mathbf{r}'(t)|\, dt$$

$$= \int_a^b P(x(t), y(t))\, y'(t)\, dt - Q(x(t), y(t))\, x'(t)\, dt$$

$$= \int_C P\, dy - Q\, dx = \iint_D \left(\dfrac{\partial P}{\partial x} + \dfrac{\partial Q}{\partial y} \right) dA$$

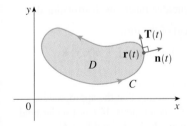

FIGURE 2

by Green's Theorem. But the integrand in this double integral is just the divergence of **F**. So we have a second vector form of Green's Theorem.

$$\boxed{13} \qquad \oint_C \mathbf{F} \cdot \mathbf{n} \, ds = \iint_D \text{div } \mathbf{F}(x, y) \, dA$$

This version says that the line integral of the normal component of **F** along C is equal to the double integral of the divergence of **F** over the region D enclosed by C.

13.5 Exercises

1–6 ■ Find (a) the curl and (b) the divergence of the vector field.

1. $\mathbf{F}(x, y, z) = xyz \, \mathbf{i} - x^2 y \, \mathbf{k}$

2. $\mathbf{F}(x, y, z) = x^2 yz \, \mathbf{i} + xy^2 z \, \mathbf{j} + xyz^2 \, \mathbf{k}$

3. $\mathbf{F}(x, y, z) = \mathbf{i} + (x + yz) \, \mathbf{j} + \left(xy - \sqrt{z}\right) \mathbf{k}$

4. $\mathbf{F}(x, y, z) = \cos xz \, \mathbf{j} - \sin xy \, \mathbf{k}$

5. $\mathbf{F}(x, y, z) = e^x \sin y \, \mathbf{i} + e^x \cos y \, \mathbf{j} + z \, \mathbf{k}$

6. $\mathbf{F}(x, y, z) = \dfrac{x}{x^2 + y^2 + z^2} \, \mathbf{i} + \dfrac{y}{x^2 + y^2 + z^2} \, \mathbf{j} + \dfrac{z}{x^2 + y^2 + z^2} \, \mathbf{k}$

- - - - - - - - - - -

7–9 ■ The vector field **F** is shown in the xy-plane and looks the same in all other horizontal planes. (In other words, **F** is independent of z and its z-component is 0.)
(a) Is div **F** positive, negative, or zero? Explain.
(b) Determine whether curl **F** $= \mathbf{0}$. If not, in which direction does curl **F** point?

7.

8.

9.

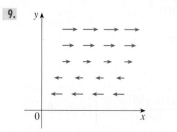

- - - - - - - - - - - - - - - - -

10. Let f be a scalar field and **F** a vector field. State whether each expression is meaningful. If not, explain why. If so, state whether it is a scalar field or a vector field.
(a) curl f (b) grad f

(c) div **F** (d) curl(grad f)
(e) grad **F** (f) grad(div **F**)
(g) div(grad f) (h) grad(div f)
(i) curl(curl **F**) (j) div(div **F**)
(k) (grad f) $\times$ (div **F**) (l) div(curl(grad f))

11–16 ■ Determine whether or not the vector field is conservative. If it is conservative, find a function f such that $\mathbf{F} = \nabla f$.

11. $\mathbf{F}(x, y, z) = yz \, \mathbf{i} + xz \, \mathbf{j} + xy \, \mathbf{k}$

12. $\mathbf{F}(x, y, z) = 3z^2 \, \mathbf{i} + \cos y \, \mathbf{j} + 2xz \, \mathbf{k}$

13. $\mathbf{F}(x, y, z) = 2xy \, \mathbf{i} + (x^2 + 2yz) \, \mathbf{j} + y^2 \, \mathbf{k}$

14. $\mathbf{F}(x, y, z) = e^z \, \mathbf{i} + \mathbf{j} + xe^z \, \mathbf{k}$

15. $\mathbf{F}(x, y, z) = ye^{-x} \, \mathbf{i} + e^{-x} \, \mathbf{j} + 2z \, \mathbf{k}$

16. $\mathbf{F}(x, y, z) = y \cos xy \, \mathbf{i} + x \cos xy \, \mathbf{j} - \sin z \, \mathbf{k}$

- - - - - - - - - - - - - - - - -

17. Is there a vector field **G** on $\mathbb{R}^3$ such that curl **G** $= xy^2 \, \mathbf{i} + yz^2 \, \mathbf{j} + zx^2 \, \mathbf{k}$? Explain.

18. Is there a vector field **G** on $\mathbb{R}^3$ such that curl **G** $= yz \, \mathbf{i} + xyz \, \mathbf{j} + xy \, \mathbf{k}$? Explain.

19. Show that any vector field of the form

$$\mathbf{F}(x, y, z) = f(x) \, \mathbf{i} + g(y) \, \mathbf{j} + h(z) \, \mathbf{k}$$

where f, g, h are differentiable functions, is irrotational.

20. Show that any vector field of the form

$$\mathbf{F}(x, y, z) = f(y, z) \, \mathbf{i} + g(x, z) \, \mathbf{j} + h(x, y) \, \mathbf{k}$$

is incompressible.

21–27 ■ Prove the identity, assuming that the appropriate partial derivatives exist and are continuous. If f is a scalar field and **F**, **G** are vector fields, then $f\mathbf{F}$, $\mathbf{F} \cdot \mathbf{G}$, and $\mathbf{F} \times \mathbf{G}$ are defined by

$$(f\mathbf{F})(x, y, z) = f(x, y, z)\mathbf{F}(x, y, z)$$

$$(\mathbf{F} \cdot \mathbf{G})(x, y, z) = \mathbf{F}(x, y, z) \cdot \mathbf{G}(x, y, z)$$

$$(\mathbf{F} \times \mathbf{G})(x, y, z) = \mathbf{F}(x, y, z) \times \mathbf{G}(x, y, z)$$

21. div(**F** + **G**) = div **F** + div **G**

22. curl(**F** + **G**) = curl **F** + curl **G**

23. $\operatorname{div}(f\mathbf{F}) = f \operatorname{div} \mathbf{F} + \mathbf{F} \cdot \nabla f$

24. $\operatorname{curl}(f\mathbf{F}) = f \operatorname{curl} \mathbf{F} + (\nabla f) \times \mathbf{F}$

25. $\operatorname{div}(\mathbf{F} \times \mathbf{G}) = \mathbf{G} \cdot \operatorname{curl} \mathbf{F} - \mathbf{F} \cdot \operatorname{curl} \mathbf{G}$

26. $\operatorname{div}(\nabla f \times \nabla g) = 0$

27. $\operatorname{curl}(\operatorname{curl} \mathbf{F}) = \operatorname{grad}(\operatorname{div} \mathbf{F}) - \nabla^2 \mathbf{F}$

28–30 ■ Let $\mathbf{r} = x\mathbf{i} + y\mathbf{j} + z\mathbf{k}$ and $r = |\mathbf{r}|$.

28. Verify each identity.
   (a) $\nabla \cdot \mathbf{r} = 3$ (b) $\nabla \cdot (r\mathbf{r}) = 4r$
   (c) $\nabla^2 r^3 = 12r$

29. Verify each identity.
   (a) $\nabla r = \mathbf{r}/r$ (b) $\nabla \times \mathbf{r} = \mathbf{0}$
   (c) $\nabla(1/r) = -\mathbf{r}/r^3$ (d) $\nabla \ln r = \mathbf{r}/r^2$

30. If $\mathbf{F} = \mathbf{r}/r^p$, find div $\mathbf{F}$. Is there a value of p for which div $\mathbf{F} = 0$?

31. Use Green's Theorem in the form of Equation 13 to prove **Green's first identity**:

$$\iint_D f\nabla^2 g \, dA = \oint_C f(\nabla g) \cdot \mathbf{n} \, ds - \iint_D \nabla f \cdot \nabla g \, dA$$

where D and C satisfy the hypotheses of Green's Theorem and the appropriate partial derivatives of f and g exist and are continuous. (The quantity $\nabla g \cdot \mathbf{n} = D_\mathbf{n} g$ occurs in the line integral. This is the directional derivative in the direction of the normal vector $\mathbf{n}$ and is called the **normal derivative** of g.)

32. Use Green's first identity (Exercise 31) to prove **Green's second identity**:

$$\iint_D (f\nabla^2 g - g\nabla^2 f) \, dA = \oint_C (f\nabla g - g\nabla f) \cdot \mathbf{n} \, ds$$

where D and C satisfy the hypotheses of Green's Theorem and the appropriate partial derivatives of f and g exist and are continuous.

33. Recall from Section 11.3 that a function g is called *harmonic* on D if it satisfies Laplace's equation, that is, $\nabla^2 g = 0$ on D. Use Green's first identity (with the same hypotheses as in Exercise 31) to show that if g is harmonic on D, then $\oint_C D_\mathbf{n} g \, ds = 0$. Here $D_\mathbf{n} g$ is the normal derivative of g defined in Exercise 31.

34. Use Green's first identity to show that if f is harmonic on D, and if $f(x, y) = 0$ on the boundary curve C, then $\iint_D |\nabla f|^2 \, dA = 0$. (Assume the same hypotheses as in Exercise 31.)

35. This exercise demonstrates a connection between the curl vector and rotations. Let B be a rigid body rotating about the z-axis. The rotation can be described by the vector $\mathbf{w} = \omega\mathbf{k}$, where ω is the angular speed of B, that is, the tangential speed of any point P in B divided by the distance d

from the axis of rotation. Let $\mathbf{r} = \langle x, y, z \rangle$ be the position vector of P.
   (a) By considering the angle θ in the figure, show that the velocity field of B is given by $\mathbf{v} = \mathbf{w} \times \mathbf{r}$.
   (b) Show that $\mathbf{v} = -\omega y\,\mathbf{i} + \omega x\,\mathbf{j}$.
   (c) Show that curl $\mathbf{v} = 2\mathbf{w}$.

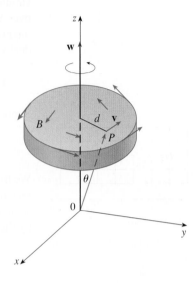

36. Maxwell's equations relating the electric field $\mathbf{E}$ and magnetic field $\mathbf{H}$ as they vary with time in a region containing no charge and no current can be stated as follows:

$$\operatorname{div} \mathbf{E} = 0 \qquad\qquad \operatorname{div} \mathbf{H} = 0$$

$$\operatorname{curl} \mathbf{E} = -\frac{1}{c}\frac{\partial \mathbf{H}}{\partial t} \qquad \operatorname{curl} \mathbf{H} = \frac{1}{c}\frac{\partial \mathbf{E}}{\partial t}$$

where c is the speed of light. Use these equations to prove the following:

   (a) $\nabla \times (\nabla \times \mathbf{E}) = -\dfrac{1}{c^2}\dfrac{\partial^2 \mathbf{E}}{\partial t^2}$

   (b) $\nabla \times (\nabla \times \mathbf{H}) = -\dfrac{1}{c^2}\dfrac{\partial^2 \mathbf{H}}{\partial t^2}$

   (c) $\nabla^2 \mathbf{E} = \dfrac{1}{c^2}\dfrac{\partial^2 \mathbf{E}}{\partial t^2}$ [*Hint:* Use Exercise 27.]

   (d) $\nabla^2 \mathbf{H} = \dfrac{1}{c^2}\dfrac{\partial^2 \mathbf{H}}{\partial t^2}$

37. We have seen that all vector fields of the form $\mathbf{F} = \nabla g$ satisfy the equation curl $\mathbf{F} = \mathbf{0}$ and that all vector fields of the form $\mathbf{F} = \operatorname{curl} \mathbf{G}$ satisfy the equation div $\mathbf{F} = 0$ (assuming continuity of the appropriate partial derivatives). This suggests the question: Are there any equations that all functions of the form $f = \operatorname{div} \mathbf{G}$ must satisfy? Show that the answer to this question is "No" by proving that *every* continuous function f on $\mathbb{R}^3$ is the divergence of some vector field. [*Hint:* Let $\mathbf{G}(x, y, z) = \langle g(x, y, z), 0, 0 \rangle$, where $g(x, y, z) = \int_0^x f(t, y, z) \, dt$.]

13.6 Surface Integrals

The relationship between surface integrals and surface area is much the same as the relationship between line integrals and arc length. Suppose f is a function of three variables whose domain includes a surface S. We will define the surface integral of f over S in such a way that, in the case where $f(x, y, z) = 1$, the value of the surface integral is equal to the surface area of S. We start with parametric surfaces and then deal with the special case where S is the graph of a function of two variables.

▪▪ Parametric Surfaces

Suppose that a surface S has a vector equation

$$\mathbf{r}(u, v) = x(u, v)\,\mathbf{i} + y(u, v)\,\mathbf{j} + z(u, v)\,\mathbf{k} \qquad (u, v) \in D$$

We first assume that the parameter domain D is a rectangle and we divide it into sub-rectangles R_{ij} with dimensions Δu and Δv. Then the surface S is divided into corresponding patches S_{ij} as in Figure 1. We evaluate f at a point P_{ij}^* in each patch, multiply by the area ΔS_{ij} of the patch, and form the Riemann sum

$$\sum_{i=1}^{m} \sum_{j=1}^{n} f(P_{ij}^*)\,\Delta S_{ij}$$

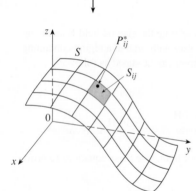

Then we take the limit as the number of patches increases and define the **surface integral of f over the surface S** as

$$\boxed{1} \qquad \iint\limits_{S} f(x, y, z)\, dS = \lim_{m, n \to \infty} \sum_{i=1}^{m} \sum_{j=1}^{n} f(P_{ij}^*)\,\Delta S_{ij}$$

Notice the analogy with the definition of a line integral (13.2.2) and also the analogy with the definition of a double integral (12.1.5).

To evaluate the surface integral in Equation 1 we approximate the patch area ΔS_{ij} by the area of an approximating parallelogram in the tangent plane. In our discussion of surface area in Section 12.6 we made the approximation

$$\Delta S_{ij} \approx |\mathbf{r}_u \times \mathbf{r}_v|\,\Delta u\,\Delta v$$

FIGURE 1

where $\qquad \mathbf{r}_u = \dfrac{\partial x}{\partial u}\,\mathbf{i} + \dfrac{\partial y}{\partial u}\,\mathbf{j} + \dfrac{\partial z}{\partial u}\,\mathbf{k} \qquad \mathbf{r}_v = \dfrac{\partial x}{\partial v}\,\mathbf{i} + \dfrac{\partial y}{\partial v}\,\mathbf{j} + \dfrac{\partial z}{\partial v}\,\mathbf{k}$

are the tangent vectors at a corner of S_{ij}. If the components are continuous and $\mathbf{r}_u$ and $\mathbf{r}_v$ are nonzero and nonparallel in the interior of D, it can be shown from Definition 1, even when D is not a rectangle, that

■■ We assume that the surface is covered only once as (u, v) ranges throughout D. The value of the surface integral does not depend on the parametrization that is used.

$$\boxed{2} \qquad \iint\limits_{S} f(x, y, z)\, dS = \iint\limits_{D} f(\mathbf{r}(u, v))\,|\mathbf{r}_u \times \mathbf{r}_v|\, dA$$

This should be compared with the formula for a line integral:

$$\int_C f(x, y, z)\, ds = \int_a^b f(\mathbf{r}(t))\, |\,\mathbf{r}'(t)\,|\, dt$$

Observe also that

$$\iint_S 1\, dS = \iint_D |\,\mathbf{r}_u \times \mathbf{r}_v\,|\, dA = A(S)$$

Formula 2 allows us to compute a surface integral by converting it into a double integral over the parameter domain D. When using this formula, remember that $f(\mathbf{r}(u, v))$ is evaluated by writing $x = x(u, v)$, $y = y(u, v)$, and $z = z(u, v)$ in the formula for $f(x, y, z)$.

EXAMPLE 1 Compute the surface integral $\iint_S x^2\, dS$, where S is the unit sphere $x^2 + y^2 + z^2 = 1$.

SOLUTION As in Example 4 in Section 10.5, we use the parametric representation

$$x = \sin\phi \cos\theta \quad y = \sin\phi \sin\theta \quad z = \cos\phi \quad 0 \le \phi \le \pi \quad 0 \le \theta \le 2\pi$$

that is,

$$\mathbf{r}(\phi, \theta) = \sin\phi \cos\theta\, \mathbf{i} + \sin\phi \sin\theta\, \mathbf{j} + \cos\phi\, \mathbf{k}$$

As in Example 1 in Section 12.6, we can compute that

$$|\,\mathbf{r}_\phi \times \mathbf{r}_\theta\,| = \sin\phi$$

Therefore, by Formula 2,

$$\iint_S x^2\, dS = \iint_D (\sin\phi \cos\theta)^2\, |\,\mathbf{r}_\phi \times \mathbf{r}_\theta\,|\, dA$$

■ ■ Here we use the identities

$$\cos^2\theta = \tfrac{1}{2}(1 + \cos 2\theta)$$

$$\sin^2\phi = 1 - \cos^2\phi$$

Instead, we could use Formulas 64 and 67 in the Table of Integrals.

$$= \int_0^{2\pi} \int_0^\pi \sin^2\phi \cos^2\theta \sin\phi\, d\phi\, d\theta = \int_0^{2\pi} \cos^2\theta\, d\theta \int_0^\pi \sin^3\phi\, d\phi$$

$$= \int_0^{2\pi} \tfrac{1}{2}(1 + \cos 2\theta)\, d\theta \int_0^\pi (\sin\phi - \sin\phi \cos^2\phi)\, d\phi$$

$$= \tfrac{1}{2}\Big[\theta + \tfrac{1}{2}\sin 2\theta\Big]_0^{2\pi} \Big[-\cos\phi + \tfrac{1}{3}\cos^3\phi\Big]_0^\pi = \frac{4\pi}{3}$$

■ ■

Surface integrals have applications similar to those for the integrals we have previously considered. For example, if a thin sheet (say, of aluminum foil) has the shape of a surface S and the density (mass per unit area) at the point (x, y, z) is $\rho(x, y, z)$, then the total **mass** of the sheet is

$$m = \iint_S \rho(x, y, z)\, dS$$

and the **center of mass** is $(\bar{x}, \bar{y}, \bar{z})$, where

$$\bar{x} = \frac{1}{m} \iint_S x\rho(x, y, z)\, dS \qquad \bar{y} = \frac{1}{m} \iint_S y\rho(x, y, z)\, dS \qquad \bar{z} = \frac{1}{m} \iint_S z\rho(x, y, z)\, dS$$

Moments of inertia can also be defined as before (see Exercise 35).

⠿ Graphs

Any surface S with equation $z = g(x, y)$ can be regarded as a parametric surface with parametric equations

$$x = x \qquad y = y \qquad z = g(x, y)$$

and so we have

$$\mathbf{r}_x = \mathbf{i} + \left(\frac{\partial g}{\partial x}\right) \mathbf{k} \qquad \mathbf{r}_y = \mathbf{j} + \left(\frac{\partial g}{\partial y}\right) \mathbf{k}$$

Thus

$$\boxed{3} \qquad \mathbf{r}_x \times \mathbf{r}_y = -\frac{\partial g}{\partial x}\mathbf{i} - \frac{\partial g}{\partial y}\mathbf{j} + \mathbf{k}$$

and

$$|\mathbf{r}_x \times \mathbf{r}_y| = \sqrt{\left(\frac{\partial z}{\partial x}\right)^2 + \left(\frac{\partial z}{\partial y}\right)^2 + 1}$$

Therefore, in this case, Formula 2 becomes

$$\boxed{4} \qquad \iint_S f(x, y, z)\, dS = \iint_D f(x, y, g(x, y)) \sqrt{\left(\frac{\partial z}{\partial x}\right)^2 + \left(\frac{\partial z}{\partial y}\right)^2 + 1}\; dA$$

Similar formulas apply when it is more convenient to project S onto the yz-plane or xz-plane. For instance, if S is a surface with equation $y = h(x, z)$ and D is its projection on the xz-plane, then

$$\iint_S f(x, y, z)\, dS = \iint_D f(x, h(x, z), z) \sqrt{\left(\frac{\partial y}{\partial x}\right)^2 + \left(\frac{\partial y}{\partial z}\right)^2 + 1}\; dA$$

EXAMPLE 2 Evaluate $\iint_S y\, dS$, where S is the surface $z = x + y^2, 0 \leqslant x \leqslant 1$, $0 \leqslant y \leqslant 2$. (See Figure 2.)

SOLUTION Since $\qquad \dfrac{\partial z}{\partial x} = 1 \qquad$ and $\qquad \dfrac{\partial z}{\partial y} = 2y$

Formula 4 gives

$$\iint_S y\, dS = \iint_D y \sqrt{1 + \left(\frac{\partial z}{\partial x}\right)^2 + \left(\frac{\partial z}{\partial y}\right)^2}\; dA$$

$$= \int_0^1 \int_0^2 y\sqrt{1 + 1 + 4y^2}\; dy\, dx$$

$$= \int_0^1 dx\, \sqrt{2} \int_0^2 y\sqrt{1 + 2y^2}\; dy$$

$$= \sqrt{2}\left(\tfrac{1}{4}\right)\tfrac{2}{3}(1 + 2y^2)^{3/2}\Big]_0^2 = \frac{13\sqrt{2}}{3}$$

FIGURE 2

If S is a piecewise-smooth surface, that is, a finite union of smooth surfaces S_1, $S_2, \ldots, S_n$ that intersect only along their boundaries, then the surface integral of f over S is defined by

$$\iint_S f(x, y, z) \, dS = \iint_{S_1} f(x, y, z) \, dS + \cdots + \iint_{S_n} f(x, y, z) \, dS$$

$\boxed{V}$ **EXAMPLE 3** Evaluate $\iint_S z \, dS$, where S is the surface whose sides S_1 are given by the cylinder $x^2 + y^2 = 1$, whose bottom S_2 is the disk $x^2 + y^2 \leqslant 1$ in the plane $z = 0$, and whose top S_3 is the part of the plane $z = 1 + x$ that lies above S_2.

SOLUTION The surface S is shown in Figure 3. (We have changed the usual position of the axes to get a better look at S.) For S_1 we use θ and z as parameters (see Example 5 in Section 10.5) and write its parametric equations as

$$x = \cos \theta \qquad y = \sin \theta \qquad z = z$$

where

$$0 \leqslant \theta \leqslant 2\pi \qquad \text{and} \qquad 0 \leqslant z \leqslant 1 + x = 1 + \cos \theta$$

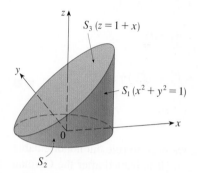

$S_3 \, (z = 1 + x)$

$S_1 \, (x^2 + y^2 = 1)$

S_2

FIGURE 3

Therefore

$$\mathbf{r}_\theta \times \mathbf{r}_z = \begin{vmatrix} \mathbf{i} & \mathbf{j} & \mathbf{k} \\ -\sin\theta & \cos\theta & 0 \\ 0 & 0 & 1 \end{vmatrix} = \cos\theta \, \mathbf{i} + \sin\theta \, \mathbf{j}$$

and

$$|\mathbf{r}_\theta \times \mathbf{r}_z| = \sqrt{\cos^2\theta + \sin^2\theta} = 1$$

Thus, the surface integral over S_1 is

$$\iint_{S_1} z \, dS = \iint_D z \, |\mathbf{r}_\theta \times \mathbf{r}_z| \, dA$$

$$= \int_0^{2\pi} \int_0^{1+\cos\theta} z \, dz \, d\theta = \int_0^{2\pi} \tfrac{1}{2}(1 + \cos\theta)^2 \, d\theta$$

$$= \tfrac{1}{2} \int_0^{2\pi} [1 + 2\cos\theta + \tfrac{1}{2}(1 + \cos 2\theta)] \, d\theta$$

$$= \tfrac{1}{2} \left[\tfrac{3}{2}\theta + 2\sin\theta + \tfrac{1}{4}\sin 2\theta \right]_0^{2\pi} = \frac{3\pi}{2}$$

Since S_2 lies in the plane $z = 0$, we have

$$\iint_{S_2} z \, dS = \iint_{S_2} 0 \, dS = 0$$

The top surface S_3 lies above the unit disk D and is part of the plane $z = 1 + x$. So, taking $g(x, y) = 1 + x$ in Formula 4 and converting to polar coordinates, we have

$$
\iint\limits_{S_3} z \, dS = \iint\limits_{D} (1 + x) \sqrt{1 + \left(\frac{\partial z}{\partial x}\right)^2 + \left(\frac{\partial z}{\partial y}\right)^2} \, dA
$$

$$
= \int_0^{2\pi} \int_0^1 (1 + r \cos \theta) \sqrt{1 + 1 + 0} \; r \, dr \, d\theta
$$

$$
= \sqrt{2} \int_0^{2\pi} \int_0^1 (r + r^2 \cos \theta) \, dr \, d\theta
$$

$$
= \sqrt{2} \int_0^{2\pi} \left(\tfrac{1}{2} + \tfrac{1}{3} \cos \theta\right) d\theta = \sqrt{2} \left[\frac{\theta}{2} + \frac{\sin \theta}{3}\right]_0^{2\pi} = \sqrt{2}\,\pi
$$

Therefore

$$
\iint\limits_{S} z \, dS = \iint\limits_{S_1} z \, dS + \iint\limits_{S_2} z \, dS + \iint\limits_{S_3} z \, dS
$$

$$
= \frac{3\pi}{2} + 0 + \sqrt{2}\,\pi = \left(\tfrac{3}{2} + \sqrt{2}\right)\pi
$$

■ ■

■■ Oriented Surfaces

In order to define surface integrals of vector fields, we need to rule out nonorientable surfaces such as the Möbius strip shown in Figure 4. [It is named after the German geometer August Möbius (1790–1868).] You can construct one for yourself by taking a long rectangular strip of paper, giving it a half-twist, and taping the short edges together as in Figure 5. If an ant were to crawl along the Möbius strip starting at a point P, it would end up on the "other side" of the strip (that is, with its upper side pointing in the opposite direction). Then, if the ant continued to crawl in the same direction, it would end up back at the same point P without ever having crossed an edge. (If you have constructed a Möbius strip, try drawing a pencil line down the middle.) Therefore, a Möbius strip really has only one side. You can graph the Möbius strip using the parametric equations in Exercise 30 in Section 10.5.

FIGURE 4

A Möbius strip

Visual 13.6 shows a Möbius strip with a normal vector that can be moved along the surface.

FIGURE 5

Constructing a Möbius strip

From now on we consider only orientable (two-sided) surfaces. We start with a surface S that has a tangent plane at every point (x, y, z) on S (except at any boundary point). There are two unit normal vectors $\mathbf{n}_1$ and $\mathbf{n}_2 = -\mathbf{n}_1$ at (x, y, z). (See Figure 6.)

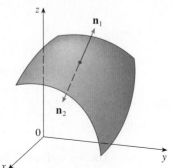

FIGURE 6

If it is possible to choose a unit normal vector **n** at every such point (x, y, z) so that **n** varies continuously over S, then S is called an **oriented surface** and the given choice of **n** provides S with an **orientation**. There are two possible orientations for any orientable surface (see Figure 7).

FIGURE 7
The two orientations of
an orientable surface

For a surface $z = g(x, y)$ given as the graph of g, we use Equation 3 to associate with the surface a natural orientation given by the unit normal vector

$$\boxed{5} \qquad \mathbf{n} = \frac{-\dfrac{\partial g}{\partial x}\,\mathbf{i} - \dfrac{\partial g}{\partial y}\,\mathbf{j} + \mathbf{k}}{\sqrt{1 + \left(\dfrac{\partial g}{\partial x}\right)^2 + \left(\dfrac{\partial g}{\partial y}\right)^2}}$$

Since the **k**-component is positive, this gives the *upward* orientation of the surface.

If S is a smooth orientable surface given in parametric form by a vector function $\mathbf{r}(u, v)$, then it is automatically supplied with the orientation of the unit normal vector

$$\boxed{6} \qquad \mathbf{n} = \frac{\mathbf{r}_u \times \mathbf{r}_v}{|\mathbf{r}_u \times \mathbf{r}_v|}$$

and the opposite orientation is given by $-\mathbf{n}$. For instance, in Example 4 in Section 10.5 we found the parametric representation

$$\mathbf{r}(\phi, \theta) = a \sin \phi \cos \theta \,\mathbf{i} + a \sin \phi \sin \theta \,\mathbf{j} + a \cos \phi \,\mathbf{k}$$

for the sphere $x^2 + y^2 + z^2 = a^2$. Then in Example 1 in Section 12.6 we found that

$$\mathbf{r}_\phi \times \mathbf{r}_\theta = a^2 \sin^2\phi \cos \theta \,\mathbf{i} + a^2 \sin^2\phi \sin \theta \,\mathbf{j} + a^2 \sin \phi \cos \phi \,\mathbf{k}$$

and

$$|\mathbf{r}_\phi \times \mathbf{r}_\theta| = a^2 \sin \phi$$

So the orientation induced by $\mathbf{r}(\phi, \theta)$ is defined by the unit normal vector

$$\mathbf{n} = \frac{\mathbf{r}_\phi \times \mathbf{r}_\theta}{|\mathbf{r}_\phi \times \mathbf{r}_\theta|} = \sin \phi \cos \theta \,\mathbf{i} + \sin \phi \sin \theta \,\mathbf{j} + \cos \phi \,\mathbf{k} = \frac{1}{a}\mathbf{r}(\phi, \theta)$$

Observe that **n** points in the same direction as the position vector, that is, outward from the sphere (see Figure 8). The opposite (inward) orientation would have been obtained (see Figure 9) if we had reversed the order of the parameters because $\mathbf{r}_\theta \times \mathbf{r}_\phi = -\mathbf{r}_\phi \times \mathbf{r}_\theta$.

For a **closed surface**, that is, a surface that is the boundary of a solid region E, the convention is that the **positive orientation** is the one for which the normal vectors point *outward* from E, and inward-pointing normals give the negative orientation (see Figures 8 and 9).

FIGURE 8
Positive orientation

FIGURE 9
Negative orientation

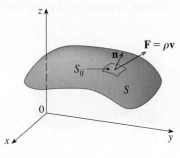

FIGURE 10

▪▪ Surface Integrals of Vector Fields

Suppose that S is an oriented surface with unit normal vector $\mathbf{n}$, and imagine a fluid with density $\rho(x, y, z)$ and velocity field $\mathbf{v}(x, y, z)$ flowing through S. (Think of S as an imaginary surface that doesn't impede the fluid flow, like a fishing net across a stream.) Then the rate of flow (mass per unit time) per unit area is $\rho\mathbf{v}$. If we divide S into small patches S_{ij}, as in Figure 10 (compare with Figure 1), then S_{ij} is nearly planar and so we can approximate the mass of fluid crossing S_{ij} in the direction of the normal $\mathbf{n}$ per unit time by the quantity

$$(\rho\mathbf{v} \cdot \mathbf{n})A(S_{ij})$$

where ρ, $\mathbf{v}$, and $\mathbf{n}$ are evaluated at some point on S_{ij}. (Recall that the component of the vector $\rho\mathbf{v}$ in the direction of the unit vector $\mathbf{n}$ is $\rho\mathbf{v} \cdot \mathbf{n}$.) By summing these quantities and taking the limit we get, according to Definition 1, the surface integral of the function $\rho\mathbf{v} \cdot \mathbf{n}$ over S:

$$\boxed{7} \qquad \iint_S \rho\mathbf{v} \cdot \mathbf{n}\, dS = \iint_S \rho(x, y, z)\mathbf{v}(x, y, z) \cdot \mathbf{n}(x, y, z)\, dS$$

and this is interpreted physically as the rate of flow through S.

If we write $\mathbf{F} = \rho\mathbf{v}$, then $\mathbf{F}$ is also a vector field on $\mathbb{R}^3$ and the integral in Equation 7 becomes

$$\iint_S \mathbf{F} \cdot \mathbf{n}\, dS$$

A surface integral of this form occurs frequently in physics, even when $\mathbf{F}$ is not $\rho\mathbf{v}$, and is called the *surface integral* (or *flux integral*) of $\mathbf{F}$ over S.

> $\boxed{8}$ **Definition** If $\mathbf{F}$ is a continuous vector field defined on an oriented surface S with unit normal vector $\mathbf{n}$, then the **surface integral of F over S** is
>
> $$\iint_S \mathbf{F} \cdot d\mathbf{S} = \iint_S \mathbf{F} \cdot \mathbf{n}\, dS$$
>
> This integral is also called the **flux** of $\mathbf{F}$ across S.

In words, Definition 8 says that the surface integral of a vector field over S is equal to the surface integral of its normal component over S (as previously defined).

If S is given by a vector function $\mathbf{r}(u, v)$, then $\mathbf{n}$ is given by Equation 6, and from Definition 8 and Equation 2 we have

$$\iint_S \mathbf{F} \cdot d\mathbf{S} = \iint_S \mathbf{F} \cdot \frac{\mathbf{r}_u \times \mathbf{r}_v}{|\mathbf{r}_u \times \mathbf{r}_v|}\, dS$$

$$= \iint_D \left[\mathbf{F}(\mathbf{r}(u, v)) \cdot \frac{\mathbf{r}_u \times \mathbf{r}_v}{|\mathbf{r}_u \times \mathbf{r}_v|} \right] |\mathbf{r}_u \times \mathbf{r}_v|\, dA$$

where D is the parameter domain. Thus, we have

■■ Compare Equation 9 to the similar expression for evaluating line integrals of vector fields in Definition 13.2.13:

$$\int_C \mathbf{F} \cdot d\mathbf{r} = \int_a^b \mathbf{F}(\mathbf{r}(t)) \cdot \mathbf{r}'(t)\, dt$$

■■ Figure 11 shows the vector field $\mathbf{F}$ in Example 4 at points on the unit sphere.

FIGURE 11

9

$$\iint_S \mathbf{F} \cdot d\mathbf{S} = \iint_D \mathbf{F} \cdot (\mathbf{r}_u \times \mathbf{r}_v)\, dA$$

EXAMPLE 4 Find the flux of the vector field $\mathbf{F}(x, y, z) = z\,\mathbf{i} + y\,\mathbf{j} + x\,\mathbf{k}$ across the unit sphere $x^2 + y^2 + z^2 = 1$.

SOLUTION Using the parametric representation

$$\mathbf{r}(\phi, \theta) = \sin\phi \cos\theta\,\mathbf{i} + \sin\phi \sin\theta\,\mathbf{j} + \cos\phi\,\mathbf{k} \qquad 0 \leqslant \phi \leqslant \pi \qquad 0 \leqslant \theta \leqslant 2\pi$$

we have

$$\mathbf{F}(\mathbf{r}(\phi, \theta)) = \cos\phi\,\mathbf{i} + \sin\phi \sin\theta\,\mathbf{j} + \sin\phi \cos\theta\,\mathbf{k}$$

and, from Example 1 in Section 12.6,

$$\mathbf{r}_\phi \times \mathbf{r}_\theta = \sin^2\phi \cos\theta\,\mathbf{i} + \sin^2\phi \sin\theta\,\mathbf{j} + \sin\phi \cos\phi\,\mathbf{k}$$

Therefore

$$\mathbf{F}(\mathbf{r}(\phi, \theta)) \cdot (\mathbf{r}_\phi \times \mathbf{r}_\theta) = \cos\phi \sin^2\phi \cos\theta + \sin^3\phi \sin^2\theta + \sin^2\phi \cos\phi \cos\theta$$

and, by Formula 9, the flux is

$$\iint_S \mathbf{F} \cdot d\mathbf{S} = \iint_D \mathbf{F} \cdot (\mathbf{r}_\phi \times \mathbf{r}_\theta)\, dA$$

$$= \int_0^{2\pi} \int_0^\pi (2\sin^2\phi \cos\phi \cos\theta + \sin^3\phi \sin^2\theta)\, d\phi\, d\theta$$

$$= 2\int_0^\pi \sin^2\phi \cos\phi\, d\phi \int_0^{2\pi} \cos\theta\, d\theta + \int_0^\pi \sin^3\phi\, d\phi \int_0^{2\pi} \sin^2\theta\, d\theta$$

$$= 0 + \int_0^\pi \sin^3\phi\, d\phi \int_0^{2\pi} \sin^2\theta\, d\theta \qquad \left(\text{since } \int_0^{2\pi} \cos\theta\, d\theta = 0 \right)$$

$$= \frac{4\pi}{3}$$

by the same calculation as in Example 1. ■■

If, for instance, the vector field in Example 4 is a velocity field describing the flow of a fluid with density 1, then the answer, $4\pi/3$, represents the rate of flow through the unit sphere in units of mass per unit time.

In the case of a surface S given by a graph $z = g(x, y)$, we can think of x and y as parameters and use Equation 3 to write

$$\mathbf{F} \cdot (\mathbf{r}_x \times \mathbf{r}_y) = (P\,\mathbf{i} + Q\,\mathbf{j} + R\,\mathbf{k}) \cdot \left(-\frac{\partial g}{\partial x}\,\mathbf{i} - \frac{\partial g}{\partial y}\,\mathbf{j} + \mathbf{k} \right)$$

Thus, Formula 9 becomes

10

$$\iint_S \mathbf{F} \cdot d\mathbf{S} = \iint_D \left(-P\frac{\partial g}{\partial x} - Q\frac{\partial g}{\partial y} + R \right) dA$$

This formula assumes the upward orientation of S; for a downward orientation we multiply by -1. Similar formulas can be worked out if S is given by $y = h(x, z)$ or $x = k(y, z)$. (See Exercises 31 and 32.)

 EXAMPLE 5 Evaluate $\iint_S \mathbf{F} \cdot d\mathbf{S}$, where $\mathbf{F}(x, y, z) = y\,\mathbf{i} + x\,\mathbf{j} + z\,\mathbf{k}$ and S is the boundary of the solid region E enclosed by the paraboloid $z = 1 - x^2 - y^2$ and the plane $z = 0$.

SOLUTION S consists of a parabolic top surface S_1 and a circular bottom surface S_2. (See Figure 12.) Since S is a closed surface, we use the convention of positive (outward) orientation. This means that S_1 is oriented upward and we can use Equation 10 with D being the projection of S_1 on the xy-plane, namely, the disk $x^2 + y^2 \leq 1$. Since

$$P(x, y, z) = y \qquad Q(x, y, z) = x \qquad R(x, y, z) = z = 1 - x^2 - y^2$$

on S_1 and

$$\frac{\partial g}{\partial x} = -2x \qquad \frac{\partial g}{\partial y} = -2y$$

we have

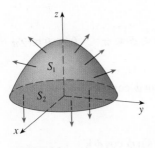

$$\iint_{S_1} \mathbf{F} \cdot d\mathbf{S} = \iint_D \left(-P\frac{\partial g}{\partial x} - Q\frac{\partial g}{\partial y} + R \right) dA$$

$$= \iint_D \left[-y(-2x) - x(-2y) + 1 - x^2 - y^2 \right] dA$$

$$= \iint_D (1 + 4xy - x^2 - y^2)\, dA$$

$$= \int_0^{2\pi} \int_0^1 (1 + 4r^2\cos\theta \sin\theta - r^2)\, r\, dr\, d\theta$$

$$= \int_0^{2\pi} \int_0^1 (r - r^3 + 4r^3\cos\theta \sin\theta)\, dr\, d\theta$$

$$= \int_0^{2\pi} \left(\tfrac{1}{4} + \cos\theta \sin\theta \right) d\theta = \tfrac{1}{4}(2\pi) + 0 = \frac{\pi}{2}$$

FIGURE 12

The disk S_2 is oriented downward, so its unit normal vector is $\mathbf{n} = -\mathbf{k}$ and we have

$$\iint_{S_2} \mathbf{F} \cdot d\mathbf{S} = \iint_{S_2} \mathbf{F} \cdot (-\mathbf{k})\, dS = \iint_D (-z)\, dA = \iint_D 0\, dA = 0$$

since $z = 0$ on S_2. Finally, we compute, by definition, $\iint_S \mathbf{F} \cdot d\mathbf{S}$ as the sum of the surface integrals of $\mathbf{F}$ over the pieces S_1 and S_2:

$$\iint_S \mathbf{F} \cdot d\mathbf{S} = \iint_{S_1} \mathbf{F} \cdot d\mathbf{S} + \iint_{S_2} \mathbf{F} \cdot d\mathbf{S} = \frac{\pi}{2} + 0 = \frac{\pi}{2} \qquad \blacksquare\blacksquare$$

Although we motivated the surface integral of a vector field using the example of fluid flow, this concept also arises in other physical situations. For instance, if $\mathbf{E}$ is an electric field (see Example 5 in Section 13.1), then the surface integral

$$\iint_S \mathbf{E} \cdot d\mathbf{S}$$

is called the **electric flux** of **E** through the surface S. One of the important laws of electrostatics is **Gauss's Law**, which says that the net charge enclosed by a closed surface S is

$$\boxed{11} \qquad Q = \varepsilon_0 \iint_S \mathbf{E} \cdot d\mathbf{S}$$

where ε_0 is a constant (called the permittivity of free space) that depends on the units used. (In the SI system, $\varepsilon_0 \approx 8.8542 \times 10^{-12} \, \text{C}^2/\text{N} \cdot \text{m}^2$.) Therefore, if the vector field **F** in Example 4 represents an electric field, we can conclude that the charge enclosed by S is $Q = 4\pi\varepsilon_0/3$.

Another application of surface integrals occurs in the study of heat flow. Suppose the temperature at a point (x, y, z) in a body is $u(x, y, z)$. Then the **heat flow** is defined as the vector field

$$\mathbf{F} = -K \, \nabla u$$

where K is an experimentally determined constant called the **conductivity** of the substance. The rate of heat flow across the surface S in the body is then given by the surface integral

$$\iint_S \mathbf{F} \cdot d\mathbf{S} = -K \iint_S \nabla u \cdot d\mathbf{S}$$

EXAMPLE 6 The temperature u in a metal ball is proportional to the square of the distance from the center of the ball. Find the rate of heat flow across a sphere S of radius a with center at the center of the ball.

SOLUTION Taking the center of the ball to be at the origin, we have

$$u(x, y, z) = C(x^2 + y^2 + z^2)$$

where C is the proportionality constant. Then the heat flow is

$$\mathbf{F}(x, y, z) = -K \, \nabla u = -KC(2x \, \mathbf{i} + 2y \, \mathbf{j} + 2z \, \mathbf{k})$$

where K is the conductivity of the metal. Instead of using the usual parametrization of the sphere as in Example 4, we observe that the outward unit normal to the sphere $x^2 + y^2 + z^2 = a^2$ at the point (x, y, z) is

$$\mathbf{n} = \frac{1}{a}(x \, \mathbf{i} + y \, \mathbf{j} + z \, \mathbf{k})$$

and so

$$\mathbf{F} \cdot \mathbf{n} = -\frac{2KC}{a}(x^2 + y^2 + z^2)$$

But on S we have $x^2 + y^2 + z^2 = a^2$, so $\mathbf{F} \cdot \mathbf{n} = -2aKC$. Therefore, the rate of heat flow across S is

$$\iint_S \mathbf{F} \cdot d\mathbf{S} = \iint_S \mathbf{F} \cdot \mathbf{n} \, dS = -2aKC \iint_S dS$$

$$= -2aKCA(S) = -2aKC(4\pi a^2) = -8KC\pi a^3 \qquad \blacksquare \ \blacksquare$$

13.6 Exercises

1. Let S be the cube with vertices $(\pm 1, \pm 1, \pm 1)$. Approximate $\iint_S \sqrt{x^2 + 2y^2 + 3z^2} \, dS$ by using a Riemann sum as in Definition 1, taking the patches S_{ij} to be the squares that are the faces of the cube and the points P_{ij}^* to be the centers of the squares.

2. A surface S consists of the cylinder $x^2 + y^2 = 1$, $-1 \leq z \leq 1$, together with its top and bottom disks. Suppose you know that f is a continuous function with $f(\pm 1, 0, 0) = 2$, $f(0, \pm 1, 0) = 3$, and $f(0, 0, \pm 1) = 4$. Estimate the value of $\iint_S f(x, y, z) \, dS$ by using a Riemann sum, taking the patches S_{ij} to be four quarter-cylinders and the top and bottom disks.

3. Let H be the hemisphere $x^2 + y^2 + z^2 = 50$, $z \geq 0$, and suppose f is a continuous function with $f(3, 4, 5) = 7$, $f(3, -4, 5) = 8$, $f(-3, 4, 5) = 9$, and $f(-3, -4, 5) = 12$. By dividing H into four patches, estimate the value of $\iint_H f(x, y, z) \, dS$.

4. Suppose that $f(x, y, z) = g(\sqrt{x^2 + y^2 + z^2})$, where g is a function of one variable such that $g(2) = -5$. Evaluate $\iint_S f(x, y, z) \, dS$, where S is the sphere $x^2 + y^2 + z^2 = 4$.

5–18 ■ Evaluate the surface integral.

5. $\iint_S yz \, dS$,
   S is the surface with parametric equations $x = u^2$, $y = u \sin v$, $z = u \cos v$, $0 \leq u \leq 1$, $0 \leq v \leq \pi/2$

6. $\iint_S \sqrt{1 + x^2 + y^2} \, dS$,
   S is the helicoid with vector equation $\mathbf{r}(u, v) = u \cos v \, \mathbf{i} + u \sin v \, \mathbf{j} + v \, \mathbf{k}$, $0 \leq u \leq 1$, $0 \leq v \leq \pi$

7. $\iint_S x^2 yz \, dS$,
   S is the part of the plane $z = 1 + 2x + 3y$ that lies above the rectangle $[0, 3] \times [0, 2]$

8. $\iint_S xy \, dS$,
   S is the triangular region with vertices $(1, 0, 0)$, $(0, 2, 0)$, and $(0, 0, 2)$

9. $\iint_S yz \, dS$,
   S is the part of the plane $x + y + z = 1$ that lies in the first octant

10. $\iint_S y \, dS$,
    S is the surface $z = \frac{2}{3}(x^{3/2} + y^{3/2})$, $0 \leq x \leq 1$, $0 \leq y \leq 1$

11. $\iint_S x^2 z^2 \, dS$,
    S is the part of the cone $z^2 = x^2 + y^2$ that lies between the planes $z = 1$ and $z = 3$

12. $\iint_S z \, dS$,
    S is the surface $x = y + 2z^2$, $0 \leq y \leq 1$, $0 \leq z \leq 1$

13. $\iint_S y \, dS$,
    S is the part of the paraboloid $y = x^2 + z^2$ that lies inside the cylinder $x^2 + z^2 = 4$

14. $\iint_S xy \, dS$,
    S is the boundary of the region enclosed by the cylinder $x^2 + z^2 = 1$ and the planes $y = 0$ and $x + y = 2$

15. $\iint_S (x^2 z + y^2 z) \, dS$,
    S is the hemisphere $x^2 + y^2 + z^2 = 4$, $z \geq 0$

16. $\iint_S xyz \, dS$,
    S is the part of the sphere $x^2 + y^2 + z^2 = 1$ that lies above the cone $z = \sqrt{x^2 + y^2}$

17. $\iint_S (x^2 y + z^2) \, dS$,
    S is the part of the cylinder $x^2 + y^2 = 9$ between the planes $z = 0$ and $z = 2$

18. $\iint_S (x^2 + y^2 + z^2) \, dS$,
    S consists of the cylinder in Exercise 17 together with its top and bottom disks

19–27 ■ Evaluate the surface integral $\iint_S \mathbf{F} \cdot d\mathbf{S}$ for the given vector field $\mathbf{F}$ and the oriented surface S. In other words, find the flux of $\mathbf{F}$ across S. For closed surfaces, use the positive (outward) orientation.

19. $\mathbf{F}(x, y, z) = xy \, \mathbf{i} + yz \, \mathbf{j} + zx \, \mathbf{k}$, S is the part of the paraboloid $z = 4 - x^2 - y^2$ that lies above the square $0 \leq x \leq 1$, $0 \leq y \leq 1$, and has upward orientation

20. $\mathbf{F}(x, y, z) = y \, \mathbf{i} + x \, \mathbf{j} + z^2 \, \mathbf{k}$,
    S is the helicoid of Exercise 6 with upward orientation

21. $\mathbf{F}(x, y, z) = xze^y \, \mathbf{i} - xze^y \, \mathbf{j} + z \, \mathbf{k}$,
    S is the part of the plane $x + y + z = 1$ in the first octant and has downward orientation

22. $\mathbf{F}(x, y, z) = x \, \mathbf{i} + y \, \mathbf{j} + z^4 \, \mathbf{k}$,
    S is the part of the cone $z = \sqrt{x^2 + y^2}$ beneath the plane $z = 1$ with downward orientation

23. $\mathbf{F}(x, y, z) = x \, \mathbf{i} - z \, \mathbf{j} + y \, \mathbf{k}$,
    S is the part of the sphere $x^2 + y^2 + z^2 = 4$ in the first octant, with orientation toward the origin

24. $\mathbf{F}(x, y, z) = xz \, \mathbf{i} + x \, \mathbf{j} + y \, \mathbf{k}$,
    S is the hemisphere $x^2 + y^2 + z^2 = 25$, $y \geq 0$, oriented in the direction of the positive y-axis

25. $\mathbf{F}(x, y, z) = y \, \mathbf{j} - z \, \mathbf{k}$,
    S consists of the paraboloid $y = x^2 + z^2$, $0 \leq y \leq 1$, and the disk $x^2 + z^2 \leq 1$, $y = 1$

26. $\mathbf{F}(x, y, z) = y \, \mathbf{i} + (z - y) \, \mathbf{j} + x \, \mathbf{k}$,
    S is the surface of the tetrahedron with vertices $(0, 0, 0)$, $(1, 0, 0)$, $(0, 1, 0)$, and $(0, 0, 1)$

27. $\mathbf{F}(x, y, z) = x^2 \, \mathbf{i} + y^2 \, \mathbf{j} + z^2 \, \mathbf{k}$, S is the boundary of the solid half-cylinder $0 \leq z \leq \sqrt{1 - y^2}$, $0 \leq x \leq 2$

CAS 28. Let S be the surface $z = xy$, $0 \leq x \leq 1$, $0 \leq y \leq 1$.
    (a) Evaluate $\iint_S xyz \, dS$ correct to four decimal places.
    (b) Find the exact value of $\iint_S x^2 yz \, dS$.

29. Find the value of $\iint_S x^2 y^2 z^2 \, dS$ correct to four decimal places, where S is the part of the paraboloid $z = 3 - 2x^2 - y^2$ that lies above the xy-plane.

30. Find the flux of $\mathbf{F}(x, y, z) = \sin(xyz) \, \mathbf{i} + x^2 y \, \mathbf{j} + z^2 e^{x/5} \, \mathbf{k}$ across the part of the cylinder $4y^2 + z^2 = 4$ that lies above the xy-plane and between the planes $x = -2$ and $x = 2$ with upward orientation. Illustrate by using a computer algebra system to draw the cylinder and the vector field on the same screen.

31. Find a formula for $\iint_S \mathbf{F} \cdot d\mathbf{S}$ similar to Formula 10 for the case where S is given by $y = h(x, z)$ and $\mathbf{n}$ is the unit normal that points toward the left.

32. Find a formula for $\iint_S \mathbf{F} \cdot d\mathbf{S}$ similar to Formula 10 for the case where S is given by $x = k(y, z)$ and $\mathbf{n}$ is the unit normal that points forward (that is, toward the viewer when the axes are drawn in the usual way).

33. Find the center of mass of the hemisphere $x^2 + y^2 + z^2 = a^2$, $z \geq 0$, if it has constant density.

34. Find the mass of a thin funnel in the shape of a cone $z = \sqrt{x^2 + y^2}$, $1 \leq z \leq 4$, if its density function is $\rho(x, y, z) = 10 - z$.

35. (a) Give an integral expression for the moment of inertia I_z about the z-axis of a thin sheet in the shape of a surface S if the density function is ρ.
(b) Find the moment of inertia about the z-axis of the funnel in Exercise 34.

36. The conical surface $z^2 = x^2 + y^2$, $0 \leq z \leq a$, has constant density k. Find (a) the center of mass and (b) the moment of inertia about the z-axis.

37. A fluid has density 870 kg/m³ and flows with velocity $\mathbf{v} = z \, \mathbf{i} + y^2 \, \mathbf{j} + x^2 \, \mathbf{k}$, where x, y, and z are measured in meters and the components of $\mathbf{v}$ in meters per second. Find the rate of flow outward through the cylinder $x^2 + y^2 = 4$, $0 \leq z \leq 1$.

38. Seawater has density 1025 kg/m³ and flows in a velocity field $\mathbf{v} = y \, \mathbf{i} + x \, \mathbf{j}$, where x, y, and z are measured in meters and the components of $\mathbf{v}$ in meters per second. Find the rate of flow outward through the hemisphere $x^2 + y^2 + z^2 = 9$, $z \geq 0$.

39. Use Gauss's Law to find the charge contained in the solid hemisphere $x^2 + y^2 + z^2 \leq a^2$, $z \geq 0$, if the electric field is $\mathbf{E}(x, y, z) = x \, \mathbf{i} + y \, \mathbf{j} + 2z \, \mathbf{k}$.

40. Use Gauss's Law to find the charge enclosed by the cube with vertices $(\pm 1, \pm 1, \pm 1)$ if the electric field is $\mathbf{E}(x, y, z) = x \, \mathbf{i} + y \, \mathbf{j} + z \, \mathbf{k}$.

41. The temperature at the point (x, y, z) in a substance with conductivity $K = 6.5$ is $u(x, y, z) = 2y^2 + 2z^2$. Find the rate of heat flow inward across the cylindrical surface $y^2 + z^2 = 6$, $0 \leq x \leq 4$.

42. The temperature at a point in a ball with conductivity K is inversely proportional to the distance from the center of the ball. Find the rate of heat flow across a sphere S of radius a with center at the center of the ball.

43. Let $\mathbf{F}$ be an inverse square field, that is, $\mathbf{F}(r) = c\mathbf{r}/|\mathbf{r}|^3$ for some constant c, where $r = x \, \mathbf{i} + y \, \mathbf{j} + z \, \mathbf{k}$. show that the flux of $\mathbf{F}$ across a sphere S with center the origin is independent of the radius S.

13.7 Stokes' Theorem

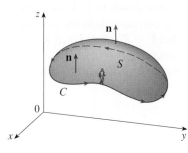

FIGURE 1

Stokes' Theorem can be regarded as a higher-dimensional version of Green's Theorem. Whereas Green's Theorem relates a double integral over a plane region D to a line integral around its plane boundary curve, Stokes' Theorem relates a surface integral over a surface S to a line integral around the boundary curve of S (which is a space curve). Figure 1 shows an oriented surface with unit normal vector $\mathbf{n}$. The orientation of S induces the **positive orientation of the boundary curve C** shown in the figure. This means that if you walk in the positive direction around C with your head pointing in the direction of $\mathbf{n}$, then the surface will always be on your left.

Stokes' Theorem Let S be an oriented piecewise-smooth surface that is bounded by a simple, closed, piecewise-smooth boundary curve C with positive orientation. Let $\mathbf{F}$ be a vector field whose components have continuous partial derivatives on an open region in $\mathbb{R}^3$ that contains S. Then

$$\int_C \mathbf{F} \cdot d\mathbf{r} = \iint_S \operatorname{curl} \mathbf{F} \cdot d\mathbf{S}$$

■ ■ Stokes' Theorem is named after the Irish mathematical physicist Sir George Stokes (1819–1903). Stokes was a professor at Cambridge University (in fact he held the same position as Newton, Lucasian Professor of Mathematics) and was especially noted for his studies of fluid flow and light. What we call Stokes' Theorem was actually discovered by the Scottish physicist Sir William Thomson (1824–1907, known as Lord Kelvin). Stokes learned of this theorem in a letter from Thomson in 1850 and asked students to prove it on an examination at Cambridge University in 1854. We don't know if any of those students was able to do so.

Since

$$\int_C \mathbf{F} \cdot d\mathbf{r} = \int_C \mathbf{F} \cdot \mathbf{T}\, ds \quad \text{and} \quad \iint_S \text{curl } \mathbf{F} \cdot d\mathbf{S} = \iint_S \text{curl } \mathbf{F} \cdot \mathbf{n}\, dS$$

Stokes' Theorem says that the line integral around the boundary curve of S of the tangential component of $\mathbf{F}$ is equal to the surface integral of the normal component of the curl of $\mathbf{F}$.

The positively oriented boundary curve of the oriented surface S is often written as ∂S, so Stokes' Theorem can be expressed as

$$\boxed{1} \qquad\qquad \iint_S \text{curl } \mathbf{F} \cdot d\mathbf{S} = \int_{\partial S} \mathbf{F} \cdot d\mathbf{r}$$

There is an analogy among Stokes' Theorem, Green's Theorem, and the Fundamental Theorem of Calculus. As before, there is an integral involving derivatives on the left side of Equation 1 (recall that curl $\mathbf{F}$ is a sort of derivative of $\mathbf{F}$) and the right side involves the values of $\mathbf{F}$ only on the *boundary* of S.

In fact, in the special case where the surface S is flat and lies in the xy-plane with upward orientation, the unit normal is $\mathbf{k}$, the surface integral becomes a double integral, and Stokes' Theorem becomes

$$\int_C \mathbf{F} \cdot d\mathbf{r} = \iint_S \text{curl } \mathbf{F} \cdot d\mathbf{S} = \iint_S (\text{curl } \mathbf{F}) \cdot \mathbf{k}\, dA$$

This is precisely the vector form of Green's Theorem given in Equation 13.5.12. Thus, we see that Green's Theorem is really a special case of Stokes' Theorem.

Although Stokes' Theorem is too difficult for us to prove in its full generality, we can give a proof when S is a graph and $\mathbf{F}$, S, and C are well behaved.

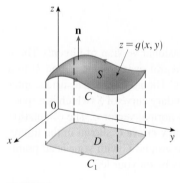

FIGURE 2

Proof of a Special Case of Stokes' Theorem We assume that the equation of S is $z = g(x, y)$, $(x, y) \in D$, where g has continuous second-order partial derivatives and D is a simple plane region whose boundary curve C_1 corresponds to C. If the orientation of S is upward, then the positive orientation of C corresponds to the positive orientation of C_1. (See Figure 2.) We are also given that $\mathbf{F} = P\,\mathbf{i} + Q\,\mathbf{j} + R\,\mathbf{k}$, where the partial derivatives of P, Q, and R are continuous.

Since S is a graph of a function, we can apply Formula 13.6.10 with $\mathbf{F}$ replaced by curl $\mathbf{F}$. The result is

$$\boxed{2} \quad \iint_S \text{curl } \mathbf{F} \cdot d\mathbf{S}$$

$$= \iint_D \left[-\left(\frac{\partial R}{\partial y} - \frac{\partial Q}{\partial z}\right)\frac{\partial z}{\partial x} - \left(\frac{\partial P}{\partial z} - \frac{\partial R}{\partial x}\right)\frac{\partial z}{\partial y} + \left(\frac{\partial Q}{\partial x} - \frac{\partial P}{\partial y}\right) \right] dA$$

where the partial derivatives of P, Q, and R are evaluated at $(x, y, g(x, y))$. If

$$x = x(t) \qquad y = y(t) \qquad a \leqslant t \leqslant b$$

is a parametric representation of C_1, then a parametric representation of C is

$$x = x(t) \qquad y = y(t) \qquad z = g(x(t), y(t)) \qquad a \le t \le b$$

This allows us, with the aid of the Chain Rule, to evaluate the line integral as follows:

$$\int_C \mathbf{F} \cdot d\mathbf{r} = \int_a^b \left(P \frac{dx}{dt} + Q \frac{dy}{dt} + R \frac{dz}{dt} \right) dt$$

$$= \int_a^b \left[P \frac{dx}{dt} + Q \frac{dy}{dt} + R \left(\frac{\partial z}{\partial x} \frac{dx}{dt} + \frac{\partial z}{\partial y} \frac{dy}{dt} \right) \right] dt$$

$$= \int_a^b \left[\left(P + R \frac{\partial z}{\partial x} \right) \frac{dx}{dt} + \left(Q + R \frac{\partial z}{\partial y} \right) \frac{dy}{dt} \right] dt$$

$$= \int_{C_1} \left(P + R \frac{\partial z}{\partial x} \right) dx + \left(Q + R \frac{\partial z}{\partial y} \right) dy$$

$$= \iint_D \left[\frac{\partial}{\partial x} \left(Q + R \frac{\partial z}{\partial y} \right) - \frac{\partial}{\partial y} \left(P + R \frac{\partial z}{\partial x} \right) \right] dA$$

where we have used Green's Theorem in the last step. Then, using the Chain Rule again and remembering that P, Q, and R are functions of x, y, and z and that z is itself a function of x and y, we get

$$\int_C \mathbf{F} \cdot d\mathbf{r} = \iint_D \left[\left(\frac{\partial Q}{\partial x} + \frac{\partial Q}{\partial z} \frac{\partial z}{\partial x} + \frac{\partial R}{\partial x} \frac{\partial z}{\partial y} + \frac{\partial R}{\partial z} \frac{\partial z}{\partial x} \frac{\partial z}{\partial y} + R \frac{\partial^2 z}{\partial x \, \partial y} \right) \right.$$

$$\left. - \left(\frac{\partial P}{\partial y} + \frac{\partial P}{\partial z} \frac{\partial z}{\partial y} + \frac{\partial R}{\partial y} \frac{\partial z}{\partial x} + \frac{\partial R}{\partial z} \frac{\partial z}{\partial y} \frac{\partial z}{\partial x} + R \frac{\partial^2 z}{\partial y \, \partial x} \right) \right] dA$$

Four of the terms in this double integral cancel and the remaining six terms can be arranged to coincide with the right side of Equation 2. Therefore

$$\int_C \mathbf{F} \cdot d\mathbf{r} = \iint_S \text{curl } \mathbf{F} \cdot d\mathbf{S}$$

■ ■

$\boxed{\text{V}}$ **EXAMPLE 1** Evaluate $\int_C \mathbf{F} \cdot d\mathbf{r}$, where $\mathbf{F}(x, y, z) = -y^2 \mathbf{i} + x \mathbf{j} + z^2 \mathbf{k}$ and C is the curve of intersection of the plane $y + z = 2$ and the cylinder $x^2 + y^2 = 1$. (Orient C to be counterclockwise when viewed from above.)

SOLUTION The curve C (an ellipse) is shown in Figure 3. Although $\int_C \mathbf{F} \cdot d\mathbf{r}$ could be evaluated directly, it's easier to use Stokes' Theorem. We first compute

$$\text{curl } \mathbf{F} = \begin{vmatrix} \mathbf{i} & \mathbf{j} & \mathbf{k} \\ \dfrac{\partial}{\partial x} & \dfrac{\partial}{\partial y} & \dfrac{\partial}{\partial z} \\ -y^2 & x & z^2 \end{vmatrix} = (1 + 2y) \mathbf{k}$$

Although there are many surfaces with boundary C, the most convenient choice is the elliptical region S in the plane $y + z = 2$ that is bounded by C. If we orient S upward, then C has the induced positive orientation. The projection D of S on the

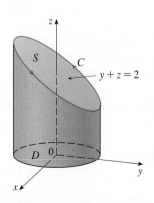

FIGURE 3

xy-plane is the disk $x^2 + y^2 \leqslant 1$ and so using Equation 13.6.10 with $z = g(x, y) = 2 - y$, we have

$$\int_C \mathbf{F} \cdot d\mathbf{r} = \iint_S \operatorname{curl} \mathbf{F} \cdot d\mathbf{S} = \iint_D (1 + 2y)\, dA$$

$$= \int_0^{2\pi} \int_0^1 (1 + 2r \sin \theta)\, r\, dr\, d\theta$$

$$= \int_0^{2\pi} \left[\frac{r^2}{2} + 2\,\frac{r^3}{3} \sin \theta \right]_0^1 d\theta = \int_0^{2\pi} \left(\tfrac{1}{2} + \tfrac{2}{3} \sin \theta \right) d\theta$$

$$= \tfrac{1}{2}(2\pi) + 0 = \pi$$

■ ■

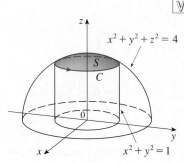

FIGURE 4

$x^2 + y^2 + z^2 = 4$

$x^2 + y^2 = 1$

V **EXAMPLE 2** Use Stokes' Theorem to compute the integral $\iint_S \operatorname{curl} \mathbf{F} \cdot d\mathbf{S}$, where $\mathbf{F}(x, y, z) = xz\, \mathbf{i} + yz\, \mathbf{j} + xy\, \mathbf{k}$ and S is the part of the sphere $x^2 + y^2 + z^2 = 4$ that lies inside the cylinder $x^2 + y^2 = 1$ and above the xy-plane. (See Figure 4.)

SOLUTION To find the boundary curve C we solve the equations $x^2 + y^2 + z^2 = 4$ and $x^2 + y^2 = 1$. Subtracting, we get $z^2 = 3$ and so $z = \sqrt{3}$ (since $z > 0$). Thus, C is the circle given by the equations $x^2 + y^2 = 1$, $z = \sqrt{3}$. A vector equation of C is

$$\mathbf{r}(t) = \cos t\, \mathbf{i} + \sin t\, \mathbf{j} + \sqrt{3}\, \mathbf{k} \qquad 0 \leqslant t \leqslant 2\pi$$

so

$$\mathbf{r}'(t) = -\sin t\, \mathbf{i} + \cos t\, \mathbf{j}$$

Also, we have

$$\mathbf{F}(\mathbf{r}(t)) = \sqrt{3} \cos t\, \mathbf{i} + \sqrt{3} \sin t\, \mathbf{j} + \cos t \sin t\, \mathbf{k}$$

Therefore, by Stokes' Theorem,

$$\iint_S \operatorname{curl} \mathbf{F} \cdot d\mathbf{S} = \int_C \mathbf{F} \cdot d\mathbf{r} = \int_0^{2\pi} \mathbf{F}(\mathbf{r}(t)) \cdot \mathbf{r}'(t)\, dt$$

$$= \int_0^{2\pi} \left(-\sqrt{3} \cos t \sin t + \sqrt{3} \sin t \cos t \right) dt$$

$$= \sqrt{3} \int_0^{2\pi} 0\, dt = 0$$

■ ■

Note that in Example 2 we computed a surface integral simply by knowing the values of $\mathbf{F}$ on the boundary curve C. This means that if we have another oriented surface with the same boundary curve C, then we get exactly the same value for the surface integral!

In general, if S_1 and S_2 are oriented surfaces with the same oriented boundary curve C and both satisfy the hypotheses of Stokes' Theorem, then

3

$$\iint_{S_1} \operatorname{curl} \mathbf{F} \cdot d\mathbf{S} = \int_C \mathbf{F} \cdot d\mathbf{r} = \iint_{S_2} \operatorname{curl} \mathbf{F} \cdot d\mathbf{S}$$

This fact is useful when it is difficult to integrate over one surface but easy to integrate over the other.

We now use Stokes' Theorem to throw some light on the meaning of the curl vector. Suppose that C is an oriented closed curve and $\mathbf{v}$ represents the velocity field in fluid flow. Consider the line integral

$$\int_C \mathbf{v} \cdot d\mathbf{r} = \int_C \mathbf{v} \cdot \mathbf{T}\, ds$$

and recall that $\mathbf{v} \cdot \mathbf{T}$ is the component of $\mathbf{v}$ in the direction of the unit tangent vector $\mathbf{T}$. This means that the closer the direction of $\mathbf{v}$ is to the direction of $\mathbf{T}$, the larger the value of $\mathbf{v} \cdot \mathbf{T}$. Thus, $\int_C \mathbf{v} \cdot d\mathbf{r}$ is a measure of the tendency of the fluid to move around C and is called the **circulation** of $\mathbf{v}$ around C. (See Figure 5.)

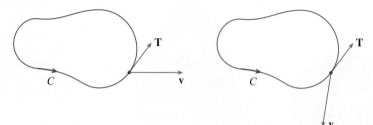

FIGURE 5   (a) $\int_C \mathbf{v} \cdot d\mathbf{r} > 0$, positive circulation   (b) $\int_C \mathbf{v} \cdot d\mathbf{r} < 0$, negative circulation

Now let $P_0(x_0, y_0, z_0)$ be a point in the fluid and let S_a be a small disk with radius a and center P_0. Then $(\text{curl } \mathbf{F})(P) \approx (\text{curl } \mathbf{F})(P_0)$ for all points P on S_a because curl $\mathbf{F}$ is continuous. Thus, by Stokes' Theorem, we get the following approximation to the circulation around the boundary circle C_a:

$$\int_{C_a} \mathbf{v} \cdot d\mathbf{r} = \iint_{S_a} \text{curl } \mathbf{v} \cdot d\mathbf{S} = \iint_{S_a} \text{curl } \mathbf{v} \cdot \mathbf{n}\, dS$$

$$\approx \iint_{S_a} \text{curl } \mathbf{v}(P_0) \cdot \mathbf{n}(P_0)\, dS = \text{curl } \mathbf{v}(P_0) \cdot \mathbf{n}(P_0)\pi a^2$$

This approximation becomes better as $a \to 0$ and we have

$$\boxed{4} \qquad \text{curl } \mathbf{v}(P_0) \cdot \mathbf{n}(P_0) = \lim_{a \to 0} \frac{1}{\pi a^2} \int_{C_a} \mathbf{v} \cdot d\mathbf{r}$$

■ ■ Imagine a tiny paddle wheel placed in the fluid at a point P, as in Figure 6; the paddle wheel rotates fastest when its axis is parallel to curl $\mathbf{v}$.

Equation 4 gives the relationship between the curl and the circulation. It shows that curl $\mathbf{v} \cdot \mathbf{n}$ is a measure of the rotating effect of the fluid about the axis $\mathbf{n}$. The curling effect is greatest about the axis parallel to curl $\mathbf{v}$.

Finally, we mention that Stokes' Theorem can be used to prove Theorem 13.5.4 (which states that if curl $\mathbf{F} = \mathbf{0}$ on all of $\mathbb{R}^3$, then $\mathbf{F}$ is conservative). From our previous work (Theorems 13.3.3 and 13.3.4), we know that $\mathbf{F}$ is conservative if $\int_C \mathbf{F} \cdot d\mathbf{r} = 0$ for every closed path C. Given C, suppose we can find an orientable surface S whose boundary is C. (This can be done, but the proof requires advanced techniques.) Then Stokes' Theorem gives

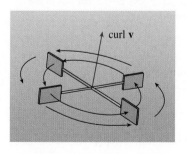

FIGURE 6

$$\int_C \mathbf{F} \cdot d\mathbf{r} = \iint_S \text{curl } \mathbf{F} \cdot d\mathbf{S} = \iint_S \mathbf{0} \cdot d\mathbf{S} = 0$$

A curve that is not simple can be broken into a number of simple curves, and the integrals around these simple curves are all 0. Adding these integrals, we obtain $\int_C \mathbf{F} \cdot d\mathbf{r} = 0$ for any closed curve C.

13.7 Exercises

1. A hemisphere H and a portion P of a paraboloid are shown. Suppose $\mathbf{F}$ is a vector field on $\mathbb{R}^3$ whose components have continuous partial derivatives. Explain why

$$\iint_H \operatorname{curl} \mathbf{F} \cdot d\mathbf{S} = \iint_P \operatorname{curl} \mathbf{F} \cdot d\mathbf{S}$$

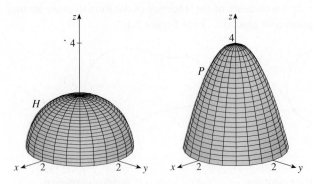

2–6 ■ Use Stokes' Theorem to evaluate $\iint_S \operatorname{curl} \mathbf{F} \cdot d\mathbf{S}$.

2. $\mathbf{F}(x, y, z) = yz\,\mathbf{i} + xz\,\mathbf{j} + xy\,\mathbf{k}$,
S is the part of the paraboloid $z = 9 - x^2 - y^2$ that lies above the plane $z = 5$, oriented upward

3. $\mathbf{F}(x, y, z) = x^2 e^{yz}\,\mathbf{i} + y^2 e^{xz}\,\mathbf{j} + z^2 e^{xy}\,\mathbf{k}$,
S is the hemisphere $x^2 + y^2 + z^2 = 4$, $z \geqslant 0$, oriented upward

4. $\mathbf{F}(x, y, z) = x^2 y^3 z\,\mathbf{i} + \sin(xyz)\,\mathbf{j} + xyz\,\mathbf{k}$,
S is the part of the cone $y^2 = x^2 + z^2$ that lies between the planes $y = 0$ and $y = 3$, oriented in the direction of the positive y-axis

5. $\mathbf{F}(x, y, z) = xyz\,\mathbf{i} + xy\,\mathbf{j} + x^2 yz\,\mathbf{k}$,
S consists of the top and the four sides (but not the bottom) of the cube with vertices $(\pm 1, \pm 1, \pm 1)$, oriented outward
[*Hint:* Use Equation 3.]

6. $\mathbf{F}(x, y, z) = e^{xy} \cos z\,\mathbf{i} + x^2 z\,\mathbf{j} + xy\,\mathbf{k}$,
S is the hemisphere $x = \sqrt{1 - y^2 - z^2}$, oriented in the direction of the positive x-axis [*Hint:* Use Equation 3.]

7–10 ■ Use Stokes' Theorem to evaluate $\int_C \mathbf{F} \cdot d\mathbf{r}$. In each case C is oriented counterclockwise as viewed from above.

7. $\mathbf{F}(x, y, z) = (x + y^2)\,\mathbf{i} + (y + z^2)\,\mathbf{j} + (z + x^2)\,\mathbf{k}$,
C is the triangle with vertices $(1, 0, 0)$, $(0, 1, 0)$, and $(0, 0, 1)$

8. $\mathbf{F}(x, y, z) = e^{-x}\,\mathbf{i} + e^x\,\mathbf{j} + e^z\,\mathbf{k}$,
C is the boundary of the part of the plane $2x + y + 2z = 2$ in the first octant

9. $\mathbf{F}(x, y, z) = yz\,\mathbf{i} + 2xz\,\mathbf{j} + e^{xy}\,\mathbf{k}$,
C is the circle $x^2 + y^2 = 16$, $z = 5$

10. $\mathbf{F}(x, y, z) = xy\,\mathbf{i} + 2z\,\mathbf{j} + 3y\,\mathbf{k}$,
C is the curve of intersection of the plane $x + z = 5$ and the cylinder $x^2 + y^2 = 9$

11. (a) Use Stokes' Theorem to evaluate $\int_C \mathbf{F} \cdot d\mathbf{r}$, where

$$\mathbf{F}(x, y, z) = x^2 z\,\mathbf{i} + xy^2\,\mathbf{j} + z^2\,\mathbf{k}$$

and C is the curve of intersection of the plane $x + y + z = 1$ and the cylinder $x^2 + y^2 = 9$ oriented counterclockwise as viewed from above.

(b) Graph both the plane and the cylinder with domains chosen so that you can see the curve C and the surface that you used in part (a).

(c) Find parametric equations for C and use them to graph C.

12. (a) Use Stokes' Theorem to evaluate $\int_C \mathbf{F} \cdot d\mathbf{r}$, where $\mathbf{F}(x, y, z) = x^2 y\,\mathbf{i} + \frac{1}{3} x^3\,\mathbf{j} + xy\,\mathbf{k}$ and C is the curve of intersection of the hyperbolic paraboloid $z = y^2 - x^2$ and the cylinder $x^2 + y^2 = 1$ oriented counterclockwise as viewed from above.

(b) Graph both the hyperbolic paraboloid and the cylinder with domains chosen so that you can see the curve C and the surface that you used in part (a).

(c) Find parametric equations for C and use them to graph C.

13–15 ■ Verify that Stokes' Theorem is true for the given vector field $\mathbf{F}$ and surface S.

13. $\mathbf{F}(x, y, z) = y^2\,\mathbf{i} + x\,\mathbf{j} + z^2\,\mathbf{k}$,
S is the part of the paraboloid $z = x^2 + y^2$ that lies below the plane $z = 1$, oriented upward

14. $\mathbf{F}(x, y, z) = x\,\mathbf{i} + y\,\mathbf{j} + xyz\,\mathbf{k}$,
S is the part of the plane $2x + y + z = 2$ that lies in the first octant, oriented upward

15. $\mathbf{F}(x, y, z) = y\,\mathbf{i} + z\,\mathbf{j} + x\,\mathbf{k}$,
S is the hemisphere $x^2 + y^2 + z^2 = 1$, $y \geqslant 0$, oriented in the direction of the positive y-axis

16. Let

$$\mathbf{F}(x, y, z) = \langle ax^3 - 3xz^2, x^2 y + by^3, cz^3 \rangle$$

Let C be the curve in Exercise 12 and consider all possible smooth surfaces S whose boundary curve is C. Find the values of a, b, and c for which $\iint_S \mathbf{F} \cdot d\mathbf{S}$ is independent of the choice of S.

17. A particle moves along line segments from the origin to the points $(1, 0, 0)$, $(1, 2, 1)$, $(0, 2, 1)$, and back to the

origin under the influence of the force field
$\mathbf{F}(x, y, z) = z^2\,\mathbf{i} + 2xy\,\mathbf{j} + 4y^2\,\mathbf{k}$. Find the work done.

18. Evaluate $\int_C (y + \sin x)\,dx + (z^2 + \cos y)\,dy + x^3\,dz$,
where C is the curve $\mathbf{r}(t) = \langle \sin t, \cos t, \sin 2t \rangle$,
$0 \le t \le 2\pi$. [*Hint:* Observe that C lies on the surface
$z = 2xy$.]

19. If S is a sphere and $\mathbf{F}$ satisfies the hypotheses of Stokes'
Theorem, show that $\iint_S \text{curl } \mathbf{F} \cdot d\mathbf{S} = 0$.

20. Suppose S and C satisfy the hypotheses of Stokes' Theorem
and f, g have continuous second-order partial derivatives.
Use Exercises 22 and 24 in Section 13.5 to show the
following.

(a) $\int_C (f\nabla g) \cdot d\mathbf{r} = \iint_S (\nabla f \times \nabla g) \cdot d\mathbf{S}$

(b) $\int_C (f\nabla f) \cdot d\mathbf{r} = 0$

(c) $\int_C (f\nabla g + g\nabla f) \cdot d\mathbf{r} = 0$

WRITING PROJECT

Three Men and Two Theorems

Although two of the most important theorems in vector calculus are named after George
Green and George Stokes, a third man, William Thomson (also known as Lord Kelvin),
played a large role in the formulation, dissemination, and application of both of these
results. All three men were interested in how the two theorems could help to explain and
predict physical phenomena in electricity and magnetism and fluid flow. The basic facts of
the story are given in the margin notes on pages 934 and 960.

Write a report on the historical origins of Green's Theorem and Stokes' Theorem. Explain
the similarities and relationship between the theorems. Discuss the roles that Green, Thomson, and Stokes played in discovering these theorems and making them widely known.
Show how both theorems arose from the investigation of electricity and magnetism and were
later used to study a variety of physical problems.

The dictionary edited by Gillispie [2] is a good source for both biographical and scientific
information. The book by Hutchinson [5] gives an account of Stokes' life and the book by
Thompson [8] is a biography of Lord Kelvin. The articles by Grattan-Guinness [3] and
Gray [4] and the book by Cannell [1] give background on the extraordinary life and works
of Green. Additional historical and mathematical information is found in the books by
Katz [6] and Kline [7].

■ ■ The photograph shows a stained-glass
window at Cambridge University in honor of
George Green.

Courtesy of the Masters and Fellows of Gonville and
Caius College, University of Cambridge, England

■ ■ Another source of information for this
project is the Internet. On the web site

www.stewartcalculus.com

click on **History of Mathematics.** Follow the
links to the St. Andrew's site and that of the
British Society for the History of Mathematics.

1. D. M. Cannell, *George Green, Mathematician and Physicist 1793–1841: The Background to His Life and Work* (Philadelphia: Society for Industrial and Applied Mathematics, 2001).

2. C. C. Gillispie, ed., *Dictionary of Scientific Biography* (New York: Scribner's, 1974).
See the article on Green by P. J. Wallis in Volume XV and the articles on Thomson by
Jed Buchwald and on Stokes by E. M. Parkinson in Volume XIII.

3. I. Grattan-Guinness, "Why did George Green write his essay of 1828 on electricity and
magnetism?" *Amer. Math. Monthly*, Vol. 102 (1995), pp. 387–396.

4. J. Gray, "There was a jolly miller." *The New Scientist*, Vol. 139 (1993), pp. 24–27.

5. G. E. Hutchinson, *The Enchanted Voyage and Other Studies* (Westport, CT: Green-
wood Press, 1978).

6. Victor Katz, *A History of Mathematics: An Introduction* (New York: HarperCollins,
1993), pp. 678–680.

7. Morris Kline, *Mathematical Thought from Ancient to Modern Times* (New York: Oxford
University Press, 1972), pp. 683–685.

8. Sylvanus P. Thompson, *The Life of Lord Kelvin* (New York: Chelsea, 1976).

13.8 The Divergence Theorem

In Section 13.5 we rewrote Green's Theorem in a vector version as

$$\int_C \mathbf{F} \cdot \mathbf{n}\, ds = \iint_D \operatorname{div} \mathbf{F}(x, y)\, dA$$

where C is the positively oriented boundary curve of the plane region D. If we were seeking to extend this theorem to vector fields on $\mathbb{R}^3$, we might make the guess that

$$\boxed{1} \qquad \iint_S \mathbf{F} \cdot \mathbf{n}\, dS = \iiint_E \operatorname{div} \mathbf{F}(x, y, z)\, dV$$

where S is the boundary surface of the solid region E. It turns out that Equation 1 is true, under appropriate hypotheses, and is called the Divergence Theorem. Notice its similarity to Green's Theorem and Stokes' Theorem in that it relates the integral of a derivative of a function (div $\mathbf{F}$ in this case) over a region to the integral of the original function $\mathbf{F}$ over the boundary of the region.

At this stage you may wish to review the various types of regions over which we were able to evaluate triple integrals in Section 12.7. We state and prove the Divergence Theorem for regions E that are simultaneously of types 1, 2, and 3 and we call such regions **simple solid regions**. (For instance, regions bounded by ellipsoids or rectangular boxes are simple solid regions.) The boundary of E is a closed surface, and we use the convention, introduced in Section 13.6, that the positive orientation is outward; that is, the unit normal vector $\mathbf{n}$ is directed outward from E.

■ ■ The Divergence Theorem is sometimes called Gauss's Theorem after the great German mathematician Karl Friedrich Gauss (1777–1855), who discovered this theorem during his investigation of electrostatics. In Eastern Europe the Divergence Theorem is known as Ostrogradsky's Theorem after the Russian mathematician Mikhail Ostrogradsky (1801–1862), who published this result in 1826.

> **The Divergence Theorem** Let E be a simple solid region and let S be the boundary surface of E, given with positive (outward) orientation. Let $\mathbf{F}$ be a vector field whose component functions have continuous partial derivatives on an open region that contains E. Then
>
> $$\iint_S \mathbf{F} \cdot d\mathbf{S} = \iiint_E \operatorname{div} \mathbf{F}\, dV$$

Thus, the Divergence Theorem states that, under the given conditions, the flux of $\mathbf{F}$ across the boundary surface of E is equal to the triple integral of the divergence of $\mathbf{F}$ over E.

Proof Let $\mathbf{F} = P\,\mathbf{i} + Q\,\mathbf{j} + R\,\mathbf{k}$. Then

$$\operatorname{div} \mathbf{F} = \frac{\partial P}{\partial x} + \frac{\partial Q}{\partial y} + \frac{\partial R}{\partial z}$$

so

$$\iiint_E \operatorname{div} \mathbf{F}\, dV = \iiint_E \frac{\partial P}{\partial x}\, dV + \iiint_E \frac{\partial Q}{\partial y}\, dV + \iiint_E \frac{\partial R}{\partial z}\, dV$$

If **n** is the unit outward normal of S, then the surface integral on the left side of the Divergence Theorem is

$$\iint_S \mathbf{F} \cdot d\mathbf{S} = \iint_S \mathbf{F} \cdot \mathbf{n}\, dS = \iint_S (P\,\mathbf{i} + Q\,\mathbf{j} + R\,\mathbf{k}) \cdot \mathbf{n}\, dS$$

$$= \iint_S P\,\mathbf{i} \cdot \mathbf{n}\, dS + \iint_S Q\,\mathbf{j} \cdot \mathbf{n}\, dS + \iint_S R\,\mathbf{k} \cdot \mathbf{n}\, dS$$

Therefore, to prove the Divergence Theorem, it suffices to prove the following three equations:

2
$$\iint_S P\,\mathbf{i} \cdot \mathbf{n}\, dS = \iiint_E \frac{\partial P}{\partial x}\, dV$$

3
$$\iint_S Q\,\mathbf{j} \cdot \mathbf{n}\, dS = \iiint_E \frac{\partial Q}{\partial y}\, dV$$

4
$$\iint_S R\,\mathbf{k} \cdot \mathbf{n}\, dS = \iiint_E \frac{\partial R}{\partial z}\, dV$$

To prove Equation 4 we use the fact that E is a type 1 region:

$$E = \big\{(x, y, z) \mid (x, y) \in D,\, u_1(x, y) \leqslant z \leqslant u_2(x, y)\big\}$$

where D is the projection of E onto the xy-plane. By Equation 12.7.6, we have

$$\iiint_E \frac{\partial R}{\partial z}\, dV = \iint_D \left[\int_{u_1(x, y)}^{u_2(x, y)} \frac{\partial R}{\partial z}(x, y, z)\, dz \right] dA$$

and, therefore, by the Fundamental Theorem of Calculus,

5
$$\iiint_E \frac{\partial R}{\partial z}\, dV = \iint_D [R(x, y, u_2(x, y)) - R(x, y, u_1(x, y))]\, dA$$

The boundary surface S consists of three pieces: the bottom surface S_1, the top surface S_2, and possibly a vertical surface S_3, which lies above the boundary curve of D. (See Figure 1. It might happen that S_3 doesn't appear, as in the case of a sphere.) Notice that on S_3 we have $\mathbf{k} \cdot \mathbf{n} = 0$, because $\mathbf{k}$ is vertical and $\mathbf{n}$ is horizontal, and so

$$\iint_{S_3} R\,\mathbf{k} \cdot \mathbf{n}\, dS = \iint_{S_3} 0\, dS = 0$$

Thus, regardless of whether there is a vertical surface, we can write

6
$$\iint_S R\,\mathbf{k} \cdot \mathbf{n}\, dS = \iint_{S_1} R\,\mathbf{k} \cdot \mathbf{n}\, dS + \iint_{S_2} R\,\mathbf{k} \cdot \mathbf{n}\, dS$$

The equation of S_2 is $z = u_2(x, y)$, $(x, y) \in D$, and the outward normal **n** points upward, so from Equation 13.6.10 (with **F** replaced by $R\,\mathbf{k}$) we have

$$\iint_{S_2} R\,\mathbf{k} \cdot \mathbf{n}\, dS = \iint_D R(x, y, u_2(x, y))\, dA$$

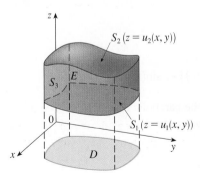

FIGURE 1

On S_1 we have $z = u_1(x, y)$, but here the outward normal $\mathbf{n}$ points downward, so we multiply by -1:

$$\iint_{S_1} R\,\mathbf{k} \cdot \mathbf{n}\,dS = -\iint_D R(x, y, u_1(x, y))\,dA$$

Therefore, Equation 6 gives

$$\iint_S R\,\mathbf{k} \cdot \mathbf{n}\,dS = \iint_D [R(x, y, u_2(x, y)) - R(x, y, u_1(x, y))]\,dA$$

Comparison with Equation 5 shows that

$$\iint_S R\,\mathbf{k} \cdot \mathbf{n}\,dS = \iiint_E \frac{\partial R}{\partial z}\,dV$$

■ ■ Notice that the method of proof of the Divergence Theorem is very similar to that of Green's Theorem.

Equations 2 and 3 are proved in a similar manner using the expressions for E as a type 2 or type 3 region, respectively. ■ ■

$\boxed{V}$ **EXAMPLE 1** Find the flux of the vector field $\mathbf{F}(x, y, z) = z\,\mathbf{i} + y\,\mathbf{j} + x\,\mathbf{k}$ over the unit sphere $x^2 + y^2 + z^2 = 1$.

SOLUTION First we compute the divergence of $\mathbf{F}$:

$$\operatorname{div}\mathbf{F} = \frac{\partial}{\partial x}(z) + \frac{\partial}{\partial y}(y) + \frac{\partial}{\partial z}(x) = 1$$

The unit sphere S is the boundary of the unit ball B given by $x^2 + y^2 + z^2 \leqslant 1$. Thus, the Divergence Theorem gives the flux as

■ ■ The solution in Example 1 should be compared with the solution in Example 4 in Section 13.6.

$$\iint_S \mathbf{F} \cdot d\mathbf{S} = \iiint_B \operatorname{div}\mathbf{F}\,dV = \iiint_B 1\,dV$$

$$= V(B) = \tfrac{4}{3}\pi(1)^3 = \frac{4\pi}{3}$$ ■ ■

$\boxed{V}$ **EXAMPLE 2** Evaluate $\displaystyle\iint_S \mathbf{F} \cdot d\mathbf{S}$, where

$$\mathbf{F}(x, y, z) = xy\,\mathbf{i} + \left(y^2 + e^{xz^2}\right)\mathbf{j} + \sin(xy)\,\mathbf{k}$$

and S is the surface of the region E bounded by the parabolic cylinder $z = 1 - x^2$ and the planes $z = 0$, $y = 0$, and $y + z = 2$. (See Figure 2.)

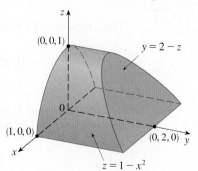

FIGURE 2

SOLUTION It would be extremely difficult to evaluate the given surface integral directly. (We would have to evaluate four surface integrals corresponding to the four pieces of S.) Furthermore, the divergence of $\mathbf{F}$ is much less complicated than $\mathbf{F}$ itself:

$$\operatorname{div} \mathbf{F} = \frac{\partial}{\partial x}(xy) + \frac{\partial}{\partial y}\left(y^2 + e^{xz^2}\right) + \frac{\partial}{\partial z}(\sin xy)$$

$$= y + 2y = 3y$$

Therefore, we use the Divergence Theorem to transform the given surface integral into a triple integral. The easiest way to evaluate the triple integral is to express E as a type 3 region:

$$E = \left\{(x, y, z) \mid -1 \le x \le 1, \ 0 \le z \le 1 - x^2, \ 0 \le y \le 2 - z\right\}$$

Then we have

$$\iint_S \mathbf{F} \cdot d\mathbf{S} = \iiint_E \operatorname{div} \mathbf{F} \, dV = \iiint_E 3y \, dV$$

$$= 3 \int_{-1}^{1} \int_{0}^{1-x^2} \int_{0}^{2-z} y \, dy \, dz \, dx$$

$$= 3 \int_{-1}^{1} \int_{0}^{1-x^2} \frac{(2-z)^2}{2} \, dz \, dx$$

$$= \frac{3}{2} \int_{-1}^{1} \left[-\frac{(2-z)^3}{3} \right]_{0}^{1-x^2} dx$$

$$= -\tfrac{1}{2} \int_{-1}^{1} \left[(x^2 + 1)^3 - 8 \right] dx$$

$$= -\int_{0}^{1} (x^6 + 3x^4 + 3x^2 - 7) \, dx = \tfrac{184}{35}$$

Although we have proved the Divergence Theorem only for simple solid regions, it can be proved for regions that are finite unions of simple solid regions. (The procedure is similar to the one we used in Section 13.4 to extend Green's Theorem.)

For example, let's consider the region E that lies between the closed surfaces S_1 and S_2, where S_1 lies inside S_2. Let $\mathbf{n}_1$ and $\mathbf{n}_2$ be outward normals of S_1 and S_2. Then the boundary surface of E is $S = S_1 \cup S_2$ and its normal $\mathbf{n}$ is given by $\mathbf{n} = -\mathbf{n}_1$ on S_1 and $\mathbf{n} = \mathbf{n}_2$ on S_2. (See Figure 3.) Applying the Divergence Theorem to S, we get

$$\boxed{7} \qquad \iiint_E \operatorname{div} \mathbf{F} \, dV = \iint_S \mathbf{F} \cdot d\mathbf{S} = \iint_S \mathbf{F} \cdot \mathbf{n} \, dS$$

$$= \iint_{S_1} \mathbf{F} \cdot (-\mathbf{n}_1) \, dS + \iint_{S_2} \mathbf{F} \cdot \mathbf{n}_2 \, dS$$

$$= -\iint_{S_1} \mathbf{F} \cdot d\mathbf{S} + \iint_{S_2} \mathbf{F} \cdot d\mathbf{S}$$

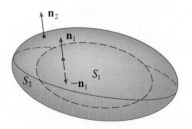

FIGURE 3

Let's apply this to the electric field (see Example 5 in Section 13.1):

$$\mathbf{E}(\mathbf{x}) = \frac{\varepsilon Q}{|\mathbf{x}|^3} \mathbf{x}$$

where S_1 is a small sphere with radius a and center the origin. You can verify that div $\mathbf{E} = 0$. (See Exercise 23.) Therefore, Equation 7 gives

$$\iint_{S_2} \mathbf{E} \cdot d\mathbf{S} = \iint_{S_1} \mathbf{E} \cdot d\mathbf{S} + \iiint_E \text{div } \mathbf{E} \, dV$$

$$= \iint_{S_1} \mathbf{E} \cdot d\mathbf{S} = \iint_{S_1} \mathbf{E} \cdot \mathbf{n} \, dS$$

The point of this calculation is that we can compute the surface integral over S_1 because S_1 is a sphere. The normal vector at $\mathbf{x}$ is $\mathbf{x}/|\mathbf{x}|$. Therefore

$$\mathbf{E} \cdot \mathbf{n} = \frac{\varepsilon Q}{|\mathbf{x}|^3} \mathbf{x} \cdot \left(\frac{\mathbf{x}}{|\mathbf{x}|} \right) = \frac{\varepsilon Q}{|\mathbf{x}|^4} \mathbf{x} \cdot \mathbf{x}$$

$$= \frac{\varepsilon Q}{|\mathbf{x}|^2} = \frac{\varepsilon Q}{a^2}$$

since the equation of S_1 is $|\mathbf{x}| = a$. Thus, we have

$$\iint_{S_2} \mathbf{E} \cdot d\mathbf{S} = \iint_{S_1} \mathbf{E} \cdot \mathbf{n} \, dS$$

$$= \frac{\varepsilon Q}{a^2} \iint_{S_1} dS = \frac{\varepsilon Q}{a^2} A(S_1)$$

$$= \frac{\varepsilon Q}{a^2} 4\pi a^2 = 4\pi \varepsilon Q$$

This shows that the electric flux of $\mathbf{E}$ is $4\pi\varepsilon Q$ through *any* closed surface S_2 that contains the origin. [This is a special case of Gauss's Law (Equation 13.6.11) for a single charge. The relationship between ε and ε_0 is $\varepsilon = 1/(4\pi\varepsilon_0)$.]

Another application of the Divergence Theorem occurs in fluid flow. Let $\mathbf{v}(x, y, z)$ be the velocity field of a fluid with constant density ρ. Then $\mathbf{F} = \rho\mathbf{v}$ is the rate of flow per unit area. If $P_0(x_0, y_0, z_0)$ is a point in the fluid and B_a is a ball with center P_0 and very small radius a, then div $\mathbf{F}(P) \approx$ div $\mathbf{F}(P_0)$ for all points in B_a since div $\mathbf{F}$ is continuous. We approximate the flux over the boundary sphere S_a as follows:

$$\iint_{S_a} \mathbf{F} \cdot d\mathbf{S} = \iiint_{B_a} \text{div } \mathbf{F} \, dV$$

$$\approx \iiint_{B_a} \text{div } \mathbf{F}(P_0) \, dV$$

$$= \text{div } \mathbf{F}(P_0) V(B_a)$$

This approximation becomes better as $a \to 0$ and suggests that

$$\boxed{8} \qquad \text{div } \mathbf{F}(P_0) = \lim_{a \to 0} \frac{1}{V(B_a)} \iint_{S_a} \mathbf{F} \cdot d\mathbf{S}$$

Equation 8 says that div $\mathbf{F}(P_0)$ is the net rate of outward flux per unit volume at P_0. (This is the reason for the name *divergence*.) If div $\mathbf{F}(P) > 0$, the net flow is outward near P and P is called a **source**. If div $\mathbf{F}(P) < 0$, the net flow is inward near P and P is called a **sink**.

For the vector field in Figure 4, it appears that the vectors that end near P_1 are shorter than the vectors that start near P_1. Thus, the net flow is outward near P_1, so div $\mathbf{F}(P_1) > 0$ and P_1 is a source. Near P_2, on the other hand, the incoming arrows are longer than the outgoing arrows. Here the net flow is inward, so div $\mathbf{F}(P_2) < 0$ and P_2 is a sink. We can use the formula for $\mathbf{F}$ to confirm this impression. Since $\mathbf{F} = x^2\,\mathbf{i} + y^2\,\mathbf{j}$, we have div $\mathbf{F} = 2x + 2y$, which is positive when $y > -x$. So the points above the line $y = -x$ are sources and those below are sinks.

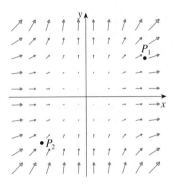

FIGURE 4
The vector field $\mathbf{F} = x^2\,\mathbf{i} + y^2\,\mathbf{j}$

13.8 Exercises

1–4 ■ Verify that the Divergence Theorem is true for the vector field $\mathbf{F}$ on the region E.

1. $\mathbf{F}(x, y, z) = 3x\,\mathbf{i} + xy\,\mathbf{j} + 2xz\,\mathbf{k}$,
E is the cube bounded by the planes $x = 0$, $x = 1$, $y = 0$, $y = 1$, $z = 0$, and $z = 1$

2. $\mathbf{F}(x, y, z) = x^2\,\mathbf{i} + xy\,\mathbf{j} + z\,\mathbf{k}$,
E is the solid bounded by the paraboloid $z = 4 - x^2 - y^2$ and the xy-plane

3. $\mathbf{F}(x, y, z) = xy\,\mathbf{i} + yz\,\mathbf{j} + zx\,\mathbf{k}$,
E is the solid cylinder $x^2 + y^2 \le 1$, $0 \le z \le 1$

4. $\mathbf{F}(x, y, z) = x\,\mathbf{i} + y\,\mathbf{j} + z\,\mathbf{k}$,
E is the unit ball $x^2 + y^2 + z^2 \le 1$

5–15 ■ Use the Divergence Theorem to calculate the surface integral $\iint_S \mathbf{F} \cdot d\mathbf{S}$; that is, calculate the flux of $\mathbf{F}$ across S.

5. $\mathbf{F}(x, y, z) = e^x \sin y\,\mathbf{i} + e^x \cos y\,\mathbf{j} + yz^2\,\mathbf{k}$,
S is the surface of the box bounded by the planes $x = 0$, $x = 1$, $y = 0$, $y = 1$, $z = 0$, and $z = 2$

6. $\mathbf{F}(x, y, z) = x^2z^3\,\mathbf{i} + 2xyz^3\,\mathbf{j} + xz^4\,\mathbf{k}$,
S is the surface of the box with vertices $(\pm1, \pm2, \pm3)$

7. $\mathbf{F}(x, y, z) = 3xy^2\,\mathbf{i} + xe^z\,\mathbf{j} + z^3\,\mathbf{k}$,
S is the surface of the solid bounded by the cylinder $y^2 + z^2 = 1$ and the planes $x = -1$ and $x = 2$

8. $\mathbf{F}(x, y, z) = x^3y\,\mathbf{i} - x^2y^2\,\mathbf{j} - x^2yz\,\mathbf{k}$,
S is the surface of the solid bounded by the hyperboloid $x^2 + y^2 - z^2 = 1$ and the planes $z = -2$ and $z = 2$

9. $\mathbf{F}(x, y, z) = xy \sin z\,\mathbf{i} + \cos(xz)\,\mathbf{j} + y \cos z\,\mathbf{k}$,
S is the ellipsoid $x^2/a^2 + y^2/b^2 + z^2/c^2 = 1$

10. $\mathbf{F}(x, y, z) = x^2y\,\mathbf{i} + xy^2\,\mathbf{j} + 2xyz\,\mathbf{k}$,
S is the surface of the tetrahedron bounded by the planes $x = 0$, $y = 0$, $z = 0$, and $x + 2y + z = 2$

11. $\mathbf{F}(x, y, z) = (\cos z + xy^2)\,\mathbf{i} + xe^{-z}\,\mathbf{j} + (\sin y + x^2z)\,\mathbf{k}$,
S is the surface of the solid bounded by the paraboloid $z = x^2 + y^2$ and the plane $z = 4$

12. $\mathbf{F}(x, y, z) = x^4\,\mathbf{i} - x^3z^2\,\mathbf{j} + 4xy^2z\,\mathbf{k}$,
S is the surface of the solid bounded by the cylinder $x^2 + y^2 = 1$ and the planes $z = x + 2$ and $z = 0$

13. $\mathbf{F}(x, y, z) = 4x^3z\,\mathbf{i} + 4y^3z\,\mathbf{j} + 3z^4\,\mathbf{k}$,
S is the sphere with radius R and center the origin

14. $\mathbf{F}(x, y, z) = (x^3 + y \sin z)\,\mathbf{i} + (y^3 + z \sin x)\,\mathbf{j} + 3z\,\mathbf{k}$,
S is the surface of the solid bounded by the hemispheres $z = \sqrt{4 - x^2 - y^2}$, $z = \sqrt{1 - x^2 - y^2}$ and the plane $z = 0$

CAS **15.** $\mathbf{F}(x, y, z) = e^y \tan z\,\mathbf{i} + y\sqrt{3 - x^2}\,\mathbf{j} + x \sin y\,\mathbf{k}$,
S is the surface of the solid that lies above the xy-plane and below the surface $z = 2 - x^4 - y^4$, $-1 \le x \le 1$, $-1 \le y \le 1$

CAS **16.** Use a computer algebra system to plot the vector field $\mathbf{F}(x, y, z) = \sin x \cos^2 y\,\mathbf{i} + \sin^3 y \cos^4 z\,\mathbf{j} + \sin^5 z \cos^6 x\,\mathbf{k}$ in the cube cut from the first octant by the planes $x = \pi/2$, $y = \pi/2$, and $z = \pi/2$. Then compute the flux across the surface of the cube.

17. Use the Divergence Theorem to evaluate $\iint_S \mathbf{F} \cdot d\mathbf{S}$, where

$$\mathbf{F}(x, y, z) = z^2 x\, \mathbf{i} + \left(\tfrac{1}{3}y^3 + \tan z\right)\mathbf{j} + (x^2 z + y^2)\,\mathbf{k}$$

and S is the top half of the sphere $x^2 + y^2 + z^2 = 1$. [*Hint:* Note that S is not a closed surface. First compute integrals over S_1 and S_2, where S_1 is the disk $x^2 + y^2 \le 1$, oriented downward, and $S_2 = S \cup S_1$.]

18. Let $\mathbf{F}(x, y, z) = z \tan^{-1}(y^2)\mathbf{i} + z^3 \ln(x^2 + 1)\mathbf{j} + z\,\mathbf{k}$. Find the flux of $\mathbf{F}$ across the part of the paraboloid $x^2 + y^2 + z = 2$ that lies above the plane $z = 1$ and is oriented upward.

19. A vector field $\mathbf{F}$ is shown. Use the interpretation of divergence derived in this section to determine whether div $\mathbf{F}$ is positive or negative at P_1 and at P_2.

20. (a) Are the points P_1 and P_2 sources or sinks for the vector field $\mathbf{F}$ shown in the figure? Give an explanation based solely on the picture.
(b) Given that $\mathbf{F}(x, y) = \langle x, y^2 \rangle$, use the definition of divergence to verify your answer to part (a).

CAS 21–22 ■ Plot the vector field and guess where div $\mathbf{F} > 0$ and where div $\mathbf{F} < 0$. Then calculate div $\mathbf{F}$ to check your guess.

21. $\mathbf{F}(x, y) = \langle xy, x + y^2 \rangle$ **22.** $F(x, y) = \langle x^2, y^2 \rangle$

23. Verify that div $\mathbf{E} = 0$ for the electric field

$$\mathbf{E}(\mathbf{x}) = \frac{\varepsilon Q}{|\mathbf{x}|^3}\,\mathbf{x}$$

24. Use the Divergence Theorem to evaluate

$$\iint_S (2x + 2y + z^2)\,dS$$

where S is the sphere $x^2 + y^2 + z^2 = 1$.

25–30 ■ Prove each identity, assuming that S and E satisfy the conditions of the Divergence Theorem and the scalar functions and components of the vector fields have continuous second-order partial derivatives.

25. $\iint_S \mathbf{a} \cdot \mathbf{n}\, dS = 0$, where $\mathbf{a}$ is a constant vector

26. $V(E) = \tfrac{1}{3}\iint_S \mathbf{F} \cdot d\mathbf{S}$, where $\mathbf{F}(x, y, z) = x\,\mathbf{i} + y\,\mathbf{j} + z\,\mathbf{k}$

27. $\iint_S \text{curl } \mathbf{F} \cdot d\mathbf{S} = 0$

28. $\iint_S D_{\mathbf{n}} f\, dS = \iiint_E \nabla^2 f\, dV$

29. $\iint_S (f \nabla g) \cdot \mathbf{n}\, dS = \iiint_E (f \nabla^2 g + \nabla f \cdot \nabla g)\, dV$

30. $\iint_S (f \nabla g - g \nabla f) \cdot \mathbf{n}\, dS = \iiint_E (f \nabla^2 g - g \nabla^2 f)\, dV$

31. Suppose S and E satisfy the conditions of the Divergence Theorem and f is a scalar function with continuous partial derivatives. Prove that

$$\iint_S f\mathbf{n}\, dS = \iiint_E \nabla f\, dV$$

These surface and triple integrals of vector functions are vectors defined by integrating each component function. [*Hint:* Start by applying the Divergence Theorem to $\mathbf{F} = f\mathbf{c}$, where $\mathbf{c}$ is an arbitrary constant vector.]

32. A solid occupies a region E with surface S and is immersed in a liquid with constant density ρ. We set up a coordinate system so that the xy-plane coincides with the surface of the liquid and positive values of z are measured downward into the liquid. Then the pressure at depth z is $p = \rho g z$, where g is the acceleration due to gravity (see Section 6.5). The total buoyant force on the solid due to the pressure distribution is given by the surface integral

$$\mathbf{F} = -\iint_S p\mathbf{n}\, dS$$

where $\mathbf{n}$ is the outer unit normal. Use the result of Exercise 31 to show that $\mathbf{F} = -W\mathbf{k}$, where W is the weight of the liquid displaced by the solid. (Note that $\mathbf{F}$ is directed upward because z is directed downward.) The result is *Archimedes' principle*: The buoyant force on an object equals the weight of the displaced liquid.

13.9 Summary

The main results of this chapter are all higher-dimensional versions of the Fundamental Theorem of Calculus. To help you remember them, we collect them together here (without hypotheses) so that you can see more easily their essential similarity. Notice that in each case we have an integral of a "derivative" over a region on the left side, and the right side involves the values of the original function only on the *boundary* of the region.

Fundamental Theorem of Calculus
$$\int_a^b F'(x)\, dx = F(b) - F(a)$$

Fundamental Theorem for Line Integrals
$$\int_C \nabla f \cdot d\mathbf{r} = f(\mathbf{r}(b)) - f(\mathbf{r}(a))$$

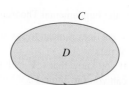

Green's Theorem
$$\iint_D \left(\frac{\partial Q}{\partial x} - \frac{\partial P}{\partial y} \right) dA = \int_C P\, dx + Q\, dy$$

Stokes' Theorem
$$\iint_S \operatorname{curl} \mathbf{F} \cdot d\mathbf{S} = \int_C \mathbf{F} \cdot d\mathbf{r}$$

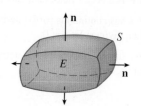

Divergence Theorem
$$\iiint_E \operatorname{div} \mathbf{F}\, dV = \iint_S \mathbf{F} \cdot d\mathbf{S}$$

13 Review

1. What is a vector field? Give three examples that have physical meaning.

2. (a) What is a conservative vector field?
   (b) What is a potential function?

3. (a) Write the definition of the line integral of a scalar function f along a smooth curve C with respect to arc length.
   (b) How do you evaluate such a line integral?
   (c) Write expressions for the mass and center of mass of a thin wire shaped like a curve C if the wire has linear density function $\rho(x, y)$.
   (d) Write the definitions of the line integrals along C of a scalar function f with respect to x, y, and z.
   (e) How do you evaluate these line integrals?

4. (a) Define the line integral of a vector field $\mathbf{F}$ along a smooth curve C given by a vector function $\mathbf{r}(t)$.
   (b) If $\mathbf{F}$ is a force field, what does this line integral represent?
   (c) If $\mathbf{F} = \langle P, Q, R \rangle$, what is the connection between the line integral of $\mathbf{F}$ and the line integrals of the component functions P, Q, and R?

5. State the Fundamental Theorem for Line Integrals.

6. (a) What does it mean to say that $\int_C \mathbf{F} \cdot d\mathbf{r}$ is independent of path?
   (b) If you know that $\int_C \mathbf{F} \cdot d\mathbf{r}$ is independent of path, what can you say about $\mathbf{F}$?

7. State Green's Theorem.

8. Write expressions for the area enclosed by a curve C in terms of line integrals around C.

9. Suppose $\mathbf{F}$ is a vector field on $\mathbb{R}^3$.
   (a) Define curl $\mathbf{F}$.
   (b) Define div $\mathbf{F}$.
   (c) If $\mathbf{F}$ is a velocity field in fluid flow, what are the physical interpretations of curl $\mathbf{F}$ and div $\mathbf{F}$?

10. If $\mathbf{F} = P\,\mathbf{i} + Q\,\mathbf{j}$, how do you test to determine whether $\mathbf{F}$ is conservative? What if $\mathbf{F}$ is a vector field on $\mathbb{R}^3$?

11. (a) Write the definition of the surface integral of a scalar function f over a surface S.
    (b) How do you evaluate such an integral if S is a parametric surface given by a vector function $\mathbf{r}(u, v)$?
    (c) What if S is given by an equation $z = g(x, y)$?
    (d) If a thin sheet has the shape of a surface S, and the density at (x, y, z) is $\rho(x, y, z)$, write expressions for the mass and center of mass of the sheet.

12. (a) What is an oriented surface? Give an example of a nonorientable surface.
    (b) Define the surface integral (or flux) of a vector field $\mathbf{F}$ over an oriented surface S with unit normal vector $\mathbf{n}$.
    (c) How do you evaluate such an integral if S is a parametric surface given by a vector function $\mathbf{r}(u, v)$?
    (d) What if S is given by an equation $z = g(x, y)$?

13. State Stokes' Theorem.

14. State the Divergence Theorem.

15. In what ways are the Fundamental Theorem for Line Integrals, Green's Theorem, Stokes' Theorem, and the Divergence Theorem similar?

Determine whether the statement is true or false. If it is true, explain why. If it is false, explain why or give an example that disproves the statement.

1. If $\mathbf{F}$ is a vector field, then div $\mathbf{F}$ is a vector field.

2. If $\mathbf{F}$ is a vector field, then curl $\mathbf{F}$ is a vector field.

3. If f has continuous partial derivatives of all orders on $\mathbb{R}^3$, then $\text{div}(\text{curl } \nabla f) = 0$.

4. If f has continuous partial derivatives on $\mathbb{R}^3$ and C is any circle, then $\int_C \nabla f \cdot d\mathbf{r} = 0$.

5. If $\mathbf{F} = P\,\mathbf{i} + Q\,\mathbf{j}$ and $P_y = Q_x$ in an open region D, then $\mathbf{F}$ is conservative.

6. $\int_{-C} f(x, y)\, ds = -\int_C f(x, y)\, ds$

7. If S is a sphere and $\mathbf{F}$ is a constant vector field, then $\iint_S \mathbf{F} \cdot d\mathbf{S} = 0$.

8. There is a vector field $\mathbf{F}$ such that
$$\text{curl } \mathbf{F} = x\,\mathbf{i} + y\,\mathbf{j} + z\,\mathbf{k}$$

—————————————————| EXERCISES |—————————————————

1. A vector field **F**, a curve C, and a point P are shown.
  (a) Is $\int_C \mathbf{F} \cdot d\mathbf{r}$ positive, negative, or zero? Explain.
  (b) Is div **F**(P) positive, negative, or zero? Explain.

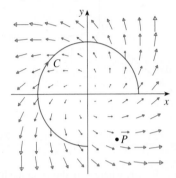

2–9 ■ Evaluate the line integral.

2. $\int_C x \, ds$,
  C is the arc of the parabola $y = x^2$ from $(0, 0)$ to $(1, 1)$

3. $\int_C yz \cos x \, ds$,
  $C : x = t,\ y = 3 \cos t,\ z = 3 \sin t,\ 0 \leqslant t \leqslant \pi$

4. $\int_C y \, dx + (x + y^2) \, dy$, C is the ellipse $4x^2 + 9y^2 = 36$
  with counterclockwise orientation

5. $\int_C y^3 \, dx + x^2 \, dy$, C is the arc of the parabola $x = 1 - y^2$
  from $(0, -1)$ to $(0, 1)$

6. $\int_C \sqrt{xy} \, dx + e^y \, dy + xz \, dz$,
  C is given by $\mathbf{r}(t) = t^4 \mathbf{i} + t^2 \mathbf{j} + t^3 \mathbf{k}, 0 \leqslant t \leqslant 1$

7. $\int_C xy \, dx + y^2 \, dy + yz \, dz$,
  C is the line segment from $(1, 0, -1)$, to $(3, 4, 2)$

8. $\int_C \mathbf{F} \cdot d\mathbf{r}$, where $\mathbf{F}(x, y) = xy \, \mathbf{i} + x^2 \mathbf{j}$ and C is given by
  $\mathbf{r}(t) = \sin t \, \mathbf{i} + (1 + t) \mathbf{j}, 0 \leqslant t \leqslant \pi$

9. $\int_C \mathbf{F} \cdot d\mathbf{r}$, where $\mathbf{F}(x, y, z) = e^z \mathbf{i} + xz \, \mathbf{j} + (x + y) \mathbf{k}$ and
  C is given by $\mathbf{r}(t) = t^2 \mathbf{i} + t^3 \mathbf{j} - t \, \mathbf{k}, 0 \leqslant t \leqslant 1$

■ ■ ■ ■ ■ ■ ■ ■ ■ ■ ■ ■ ■

10. Find the work done by the force field

$$\mathbf{F}(x, y, z) = z \, \mathbf{i} + x \, \mathbf{j} + y \, \mathbf{k}$$

  in moving a particle from the point $(3, 0, 0)$ to the point
  $(0, \pi/2, 3)$ along
  (a) A straight line
  (b) The helix $x = 3 \cos t,\ y = t,\ z = 3 \sin t$

11–12 ■ Show that **F** is a conservative vector field. Then find a
function f such that $\mathbf{F} = \nabla f$.

11. $\mathbf{F}(x, y) = (1 + xy)e^{xy} \, \mathbf{i} + (e^y + x^2 e^{xy}) \mathbf{j}$

12. $\mathbf{F}(x, y, z) = \sin y \, \mathbf{i} + x \cos y \, \mathbf{j} - \sin z \, \mathbf{k}$

■ ■ ■ ■ ■ ■ ■ ■ ■ ■ ■ ■ ■ ■

13–14 ■ Show that **F** is conservative and use this fact to evaluate $\int_C \mathbf{F} \cdot d\mathbf{r}$ along the given curve.

13. $\mathbf{F}(x, y) = (4x^3 y^2 - 2xy^3) \, \mathbf{i} + (2x^4 y - 3x^2 y^2 + 4y^3) \mathbf{j}$,
  $C : \mathbf{r}(t) = (t + \sin \pi t) \, \mathbf{i} + (2t + \cos \pi t) \mathbf{j},\ 0 \leqslant t \leqslant 1$

14. $\mathbf{F}(x, y, z) = e^y \mathbf{i} + (xe^y + e^z) \mathbf{j} + ye^z \mathbf{k}$,
  C is the line segment from $(0, 2, 0)$ to $(4, 0, 3)$

■ ■ ■ ■ ■ ■ ■ ■ ■ ■ ■ ■ ■ ■

15. Verify that Green's Theorem is true for the line integral
  $\int_C xy^2 \, dx - x^2 y \, dy$, where C consists of the parabola
  $y = x^2$ from $(-1, 1)$ to $(1, 1)$ and the line segment from
  $(1, 1)$ to $(-1, 1)$.

16. Use Green's Theorem to evaluate

$$\int_C \sqrt{1 + x^3} \, dx + 2xy \, dy$$

  where C is the triangle with vertices $(0, 0)$, $(1, 0)$, and $(1, 3)$.

17. Use Green's Theorem to evaluate $\int_C x^2 y \, dx - xy^2 \, dy$,
  where C is the circle $x^2 + y^2 = 4$ with counterclockwise
  orientation.

18. Find curl **F** and div **F** if

$$\mathbf{F}(x, y, z) = e^{-x} \sin y \, \mathbf{i} + e^{-y} \sin z \, \mathbf{j} + e^{-z} \sin x \, \mathbf{k}$$

19. Show that there is no vector field **G** such that

$$\operatorname{curl} \mathbf{G} = 2x \, \mathbf{i} + 3yz \, \mathbf{j} - xz^2 \, \mathbf{k}$$

20. Show that, under conditions to be stated on the vector fields
  F and **G**,

$$\operatorname{curl}(\mathbf{F} \times \mathbf{G}) = \mathbf{F} \operatorname{div} \mathbf{G} - \mathbf{G} \operatorname{div} \mathbf{F} + (\mathbf{G} \cdot \nabla)\mathbf{F} - (\mathbf{F} \cdot \nabla)\mathbf{G}$$

21. If C is any piecewise-smooth simple closed plane curve and
  f and g are differentiable functions, show that

$$\int_C f(x) \, dx + g(y) \, dy = 0$$

22. If f and g are twice differentiable functions, show that

$$\nabla^2(fg) = f\nabla^2 g + g\nabla^2 f + 2\nabla f \cdot \nabla g$$

23. If f is a harmonic function, that is, $\nabla^2 f = 0$, show that the
  line integral $\int f_y \, dx - f_x \, dy$ is independent of path in any
  simple region D.

24. (a) Sketch the curve C with parametric equations

$$x = \cos t \qquad y = \sin t \qquad z = \sin t \qquad 0 \leqslant t \leqslant 2\pi$$

  (b) Find $\int_C 2xe^{2y} \, dx + (2x^2 e^{2y} + 2y \cot z) \, dy - y^2 \csc^2 z \, dz$.

25–28 ■ Evaluate the surface integral.

25. $\iint_S z \, dS$, where S is the part of the paraboloid $z = x^2 + y^2$ that lies under the plane $z = 4$

26. $\iint_S (x^2z + y^2z) \, dS$, where S is the part of the plane $z = 4 + x + y$ that lies inside the cylinder $x^2 + y^2 = 4$

27. $\iint_S \mathbf{F} \cdot d\mathbf{S}$, where $\mathbf{F}(x, y, z) = xz \, \mathbf{i} - 2y \, \mathbf{j} + 3x \, \mathbf{k}$ and S is the sphere $x^2 + y^2 + z^2 = 4$ with outward orientation

28. $\iint_S \mathbf{F} \cdot d\mathbf{S}$, where $\mathbf{F}(x, y, z) = x^2 \, \mathbf{i} + xy \, \mathbf{j} + z \, \mathbf{k}$ and S is the part of the paraboloid $z = x^2 + y^2$ below the plane $z = 1$ with upward orientation

. . . .

29. Verify that Stokes' Theorem is true for the vector field

$$\mathbf{F}(x, y, z) = x^2 \, \mathbf{i} + y^2 \, \mathbf{j} + z^2 \, \mathbf{k}$$

where S is the part of the paraboloid $z = 1 - x^2 - y^2$ that lies above the xy-plane and S has upward orientation.

30. Use Stokes' Theorem to evaluate $\iint_S \text{curl } \mathbf{F} \cdot d\mathbf{S}$, where $\mathbf{F}(x, y, z) = x^2yz \, \mathbf{i} + yz^2 \, \mathbf{j} + z^3e^{xy} \, \mathbf{k}$, S is the part of the sphere $x^2 + y^2 + z^2 = 5$ that lies above the plane $z = 1$, and S is oriented upward.

31. Use Stokes' Theorem to evaluate $\int_C \mathbf{F} \cdot d\mathbf{r}$, where $\mathbf{F}(x, y, z) = xy \, \mathbf{i} + yz \, \mathbf{j} + zx \, \mathbf{k}$ and C is the triangle with vertices $(1, 0, 0)$, $(0, 1, 0)$, and $(0, 0, 1)$, oriented counterclockwise as viewed from above.

32. Use the Divergence Theorem to calculate the surface integral $\iint_S \mathbf{F} \cdot d\mathbf{S}$, where $\mathbf{F}(x, y, z) = x^3 \, \mathbf{i} + y^3 \, \mathbf{j} + z^3 \, \mathbf{k}$ and S is the surface of the solid bounded by the cylinder $x^2 + y^2 = 1$ and the planes $z = 0$ and $z = 2$.

33. Verify that the Divergence Theorem is true for the vector field

$$\mathbf{F}(x, y, z) = x \, \mathbf{i} + y \, \mathbf{j} + z \, \mathbf{k}$$

where E is the unit ball $x^2 + y^2 + z^2 \le 1$.

34. Compute the outward flux of

$$\mathbf{F}(x, y, z) = \frac{x \, \mathbf{i} + y \, \mathbf{j} + z \, \mathbf{k}}{(x^2 + y^2 + z^2)^{3/2}}$$

through the ellipsoid $4x^2 + 9y^2 + 6z^2 = 36$.

35. Let

$$\mathbf{F}(x, y, z) = (3x^2yz - 3y) \, \mathbf{i} + (x^3z - 3x) \, \mathbf{j} + (x^3y + 2z) \, \mathbf{k}$$

Evaluate $\int_C \mathbf{F} \cdot d\mathbf{r}$, where C is the curve with initial point $(0, 0, 2)$ and terminal point $(0, 3, 0)$ shown in the figure.

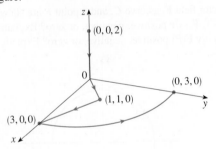

36. Let

$$\mathbf{F}(x, y) = \frac{(2x^3 + 2xy^2 - 2y) \, \mathbf{i} + (2y^3 + 2x^2y + 2x) \, \mathbf{j}}{x^2 + y^2}$$

Evaluate $\oint_C \mathbf{F} \cdot d\mathbf{r}$, where C is shown in the figure.

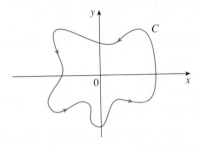

37. Find $\iint_S \mathbf{F} \cdot \mathbf{n} \, dS$, where $\mathbf{F}(x, y, z) = x \, \mathbf{i} + y \, \mathbf{j} + z \, \mathbf{k}$ and S is the outwardly oriented surface shown in the figure (the boundary surface of a cube with a unit corner cube removed).

38. If the components of $\mathbf{F}$ have continuous second partial derivatives and S is the boundary surface of a simple solid region, show that $\iint_S \text{curl } \mathbf{F} \cdot d\mathbf{S} = 0$.

1. Let S be a smooth parametric surface and let P be a point such that each line that starts at P intersects S at most once. The **solid angle** $\Omega(S)$ subtended by S at P is the set of lines starting at P and passing through S. Let $S(a)$ be the intersection of $\Omega(S)$ with the surface of the sphere with center P and radius a. Then the measure of the solid angle (in *steradians*) is defined to be

$$|\Omega(S)| = \frac{\text{area of } S(a)}{a^2}$$

Apply the Divergence Theorem to the part of $\Omega(S)$ between $S(a)$ and S to show that

$$|\Omega(S)| = \iint_S \frac{\mathbf{r} \cdot \mathbf{n}}{r^3}\, dS$$

where $\mathbf{r}$ is the radius vector from P to any point on S, $r = |\mathbf{r}|$, and the unit normal vector $\mathbf{n}$ is directed away from P.

This shows that the definition of the measure of a solid angle is independent of the radius a of the sphere. Thus, the measure of the solid angle is equal to the area subtended on a *unit* sphere. (Note the analogy with the definition of radian measure.) The total solid angle subtended by a sphere at its center is thus 4π steradians.

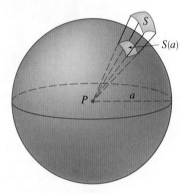

2. Prove the following identity:

$$\nabla(\mathbf{F} \cdot \mathbf{G}) = (\mathbf{F} \cdot \nabla)\mathbf{G} + (\mathbf{G} \cdot \nabla)\mathbf{F} + \mathbf{F} \times \text{curl } \mathbf{G} + \mathbf{G} \times \text{curl } \mathbf{F}$$

3. If $\mathbf{a}$ is a constant vector, $\mathbf{r} = x\,\mathbf{i} + y\,\mathbf{j} + z\,\mathbf{k}$, and S is an oriented, smooth surface with a simple, closed, smooth, positively oriented boundary curve C, show that

$$\iint_S 2\mathbf{a} \cdot d\mathbf{S} = \int_C (\mathbf{a} \times \mathbf{r}) \cdot d\mathbf{r}$$

4. Find the positively oriented simple closed curve C for which the value of the line integral

$$\int_C (y^3 - y)\, dx - 2x^3\, dy$$

is a maximum.

5. Let C be a simple closed piecewise-smooth space curve that lies in a plane with unit normal vector $\mathbf{n} = \langle a, b, c \rangle$ and has positive orientation with respect to $\mathbf{n}$. Show that the plane area enclosed by C is

$$\tfrac{1}{2} \int_C (bz - cy)\, dx + (cx - az)\, dy + (ay - bx)\, dz$$

6. The figure depicts the sequence of events in each cylinder of a four-cylinder internal combustion engine. Each piston moves up and down and is connected by a pivoted arm to a rotating crankshaft. Let $P(t)$ and $V(t)$ be the pressure and volume within a cylinder at time t, where $a \leqslant t \leqslant b$ gives the time required for a complete cycle. The graph shows how P and V vary through one cycle of a four-stroke engine.

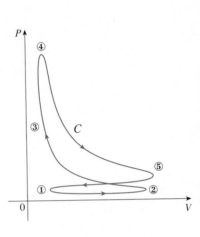

During the intake stroke (from ① to ②) a mixture of air and gasoline at atmospheric pressure is drawn into a cylinder through the intake valve as the piston moves downward. Then the piston rapidly compresses the mix with the valves closed in the compression stroke (from ② to ③) during which the pressure rises and the volume decreases. At ③ the sparkplug ignites the fuel, raising the temperature and pressure at almost constant volume to ④. Then, with valves closed, the rapid expansion forces the piston downward during the power stroke (from ④ to ⑤). The exhaust valve opens, temperature and pressure drop, and mechanical energy stored in a rotating flywheel pushes the piston upward, forcing the waste products out of the exhaust valve in the exhaust stroke. The exhaust valve closes and the intake valve opens. We're now back at ① and the cycle starts again.

(a) Show that the work done on the piston during one cycle of a four-stroke engine is $W = \int_C P\, dV$, where C is the curve in the PV-plane shown in the figure.

[*Hint:* Let $x(t)$ be the distance from the piston to the top of the cylinder and note that the force on the piston is $\mathbf{F} = AP(t)\, \mathbf{i}$, where A is the area of the top of the piston. Then $W = \int_{C_1} \mathbf{F} \cdot d\mathbf{r}$, where C_1 is given by $\mathbf{r}(t) = x(t)\, \mathbf{i}$, $a \leqslant t \leqslant b$. An alternative approach is to work directly with Riemann sums.]

(b) Use Formula 13.4.5 to show that the work is the difference of the areas enclosed by the two loops of C.

Appendixes

D | Precise Definitions of Limits

Here is a precise version of Definition 1 in Section 11.2:

1 **Definition** Let f be a function of two variables whose domain D includes points arbitrarily close to (a, b). Then we say that the **limit of $f(x, y)$ as (x, y) approaches (a, b)** is L and we write

$$\lim_{(x, y) \to (a, b)} f(x, y) = L$$

if for every number $\varepsilon > 0$ there is a corresponding number $\delta > 0$ such that

$$|f(x, y) - L| < \varepsilon \text{ whenever } (x, y) \in D \text{ and } 0 < \sqrt{(x - a)^2 + (y - b)^2} < \delta$$

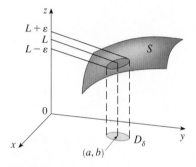

FIGURE 1

Because $|f(x, y) - L|$ is the distance between the numbers $f(x, y)$ and L, and $\sqrt{(x - a)^2 + (y - b)^2}$ is the distance between the point (x, y) and the point (a, b), Definition 1 says that the distance between $f(x, y)$ and L can be made arbitrarily small by making the distance from (x, y) to (a, b) sufficiently small (but not 0). An illustration of Definition 1 is given in Figure 1 where the surface S is the graph of f. If $\varepsilon > 0$ is given, we can find $\delta > 0$ such that if (x, y) is restricted to lie in the disk D_δ with center (a, b) and radius δ, and if $(x, y) \neq (a, b)$, then the corresponding part of S lies between the horizontal planes $z = L - \varepsilon$ and $z = L + \varepsilon$.

EXAMPLE 1 Prove that $\displaystyle\lim_{(x,y)\to(0,0)} \frac{3x^2 y}{x^2 + y^2} = 0$.

SOLUTION Let $\varepsilon > 0$. We want to find $\delta > 0$ such that

$$\left| \frac{3x^2 y}{x^2 + y^2} - 0 \right| < \varepsilon \qquad \text{whenever} \qquad 0 < \sqrt{x^2 + y^2} < \delta$$

that is,

$$\frac{3x^2 |y|}{x^2 + y^2} < \varepsilon \qquad \text{whenever} \qquad 0 < \sqrt{x^2 + y^2} < \delta$$

But $x^2 \leq x^2 + y^2$ since $y^2 \geq 0$, so $x^2/(x^2 + y^2) \leq 1$ and therefore

$$\frac{3x^2 |y|}{x^2 + y^2} \leq 3|y| = 3\sqrt{y^2} \leq 3\sqrt{x^2 + y^2}$$

Thus, if we choose $\delta = \varepsilon/3$ and let $0 < \sqrt{x^2 + y^2} < \delta$, then

$$\left| \frac{3x^2 y}{x^2 + y^2} - 0 \right| \leq 3\sqrt{x^2 + y^2} \leq 3\delta = 3\left(\frac{\varepsilon}{3}\right) = \varepsilon$$

Hence, by Definition 1,

$$\lim_{(x,y)\to(0,0)} \frac{3x^2 y}{x^2 + y^2} = 0$$

■ ■

D | Exercises

1. Use Definition 1 to prove that $\displaystyle\lim_{(x,y)\to(0,0)} \frac{xy}{\sqrt{x^2+y^2}} = 0$.

E | A Few Proofs

In this appendix we present proofs of some theorems that were stated in the main body of the text.

■ ■ Clairaut's Theorem was discussed in Section 11.3.

Clairaut's Theorem Suppose f is defined on a disk D that contains the point (a, b). If the functions f_{xy} and f_{yx} are both continuous on D, then $f_{xy}(a, b) = f_{yx}(a, b)$.

Proof For small values of h, $h \neq 0$, consider the difference

$$\Delta(h) = [f(a+h, b+h) - f(a+h, b)] - [f(a, b+h) - f(a, b)]$$

Notice that if we let $g(x) = f(x, b+h) - f(x, b)$, then

$$\Delta(h) = g(a+h) - g(a)$$

By the Mean Value Theorem, there is a number c between a and $a+h$ such that

$$g(a+h) - g(a) = g'(c)h = h[f_x(c, b+h) - f_x(c, b)]$$

Applying the Mean Value Theorem again, this time to f_x, we get a number d between b and $b+h$ such that

$$f_x(c, b+h) - f_x(c, b) = f_{xy}(c, d)h$$

Combining these equations, we obtain

$$\Delta(h) = h^2 f_{xy}(c, d)$$

If $h \to 0$, then $(c, d) \to (a, b)$, so the continuity of f_{xy} at (a, b) gives

$$\lim_{h\to 0} \frac{\Delta(h)}{h^2} = \lim_{(c,d)\to(a,b)} f_{xy}(c, d) = f_{xy}(a, b)$$

Similarly, by writing

$$\Delta(h) = [f(a+h, b+h) - f(a, b+h)] - [f(a+h, b) - f(a, b)]$$

and using the Mean Value Theorem twice and the continuity of f_{yx} at (a, b), we obtain

$$\lim_{h\to 0} \frac{\Delta(h)}{h^2} = f_{yx}(a, b)$$

It follows that $f_{xy}(a, b) = f_{yx}(a, b)$.

■ ■ This was stated as Theorem 8 in
Section 11.4.

Theorem If the partial derivatives f_x and f_y exist near (a, b) and are continuous at (a, b), then f is differentiable at (a, b).

Proof Let

$$\Delta z = f(a + \Delta x, b + \Delta y) - f(a, b)$$

According to Definition 11.4.7, to prove that f is differentiable at (a, b) we have to show that we can write Δz in the form

$$\Delta z = f_x(a, b)\,\Delta x + f_y(a, b)\,\Delta y + \varepsilon_1\,\Delta x + \varepsilon_2\,\Delta y$$

where ε_1 and $\varepsilon_2 \to 0$ as $(\Delta x, \Delta y) \to (0, 0)$.
Referring to Figure 1, we write

$$\boxed{1} \quad \Delta z = [f(a + \Delta x, b + \Delta y) - f(a, b + \Delta y)] + [f(a, b + \Delta y) - f(a, b)]$$

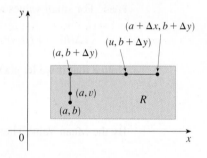

FIGURE 1

Observe that the function of a single variable

$$g(x) = f(x, b + \Delta y)$$

is defined on the interval $[a, a + \Delta x]$ and $g'(x) = f_x(x, b + \Delta y)$. If we apply the Mean Value Theorem to g, we get

$$g(a + \Delta x) - g(a) = g'(u)\,\Delta x$$

where u is some number between a and $a + \Delta x$. In terms of f, this equation becomes

$$f(a + \Delta x, b + \Delta y) - f(a, b + \Delta y) = f_x(u, b + \Delta y)\,\Delta x$$

This gives us an expression for the first part of the right side of Equation 1. For the second part we let $h(y) = f(a, y)$. Then h is a function of a single variable defined on the interval $[b, b + \Delta y]$ and $h'(y) = f_y(a, y)$. A second application of the Mean Value Theorem then gives

$$h(b + \Delta y) - h(b) = h'(v)\,\Delta y$$

where v is some number between b and $b + \Delta y$. In terms of f, this becomes

$$f(a, b + \Delta y) - f(a, b) = f_y(a, v)\,\Delta y$$

We now substitute these expressions into Equation 3 and obtain

$$\Delta z = f_x(u, b + \Delta y) \, \Delta x + f_y(a, v) \, \Delta y$$

$$= f_x(a, b) \, \Delta x + [f_x(u, b + \Delta y) - f_x(a, b)] \, \Delta x + f_y(a, b) \, \Delta y$$
$$+ [f_y(a, v) - f_y(a, b)] \, \Delta y$$

$$= f_x(a, b) \, \Delta x + f_y(a, b) \, \Delta y + \varepsilon_1 \, \Delta x + \varepsilon_2 \, \Delta y$$

where
$$\varepsilon_1 = f_x(u, b + \Delta y) - f_x(a, b)$$

$$\varepsilon_2 = f_y(a, v) - f_y(a, b)$$

Since $(u, b + \Delta y) \to (a, b)$ and $(a, v) \to (a, b)$ as $(\Delta x, \Delta y) \to (0, 0)$ and since f_x and f_y are continuous at (a, b), we see that $\varepsilon_1 \to 0$ and $\varepsilon_2 \to 0$ as $(\Delta x, \Delta y) \to (0, 0)$. Therefore, f is differentiable at (a, b). ■ ■

■ ■ The Second Derivatives Test was discussed in Section 11.7. Parts (b) and (c) have similar proofs.

> **Second Derivatives Test** Suppose the second partial derivatives of f are continuous on a disk with center (a, b), and suppose that $f_x(a, b) = 0$ and $f_y(a, b) = 0$ [that is, (a, b) is a critical point of f]. Let
>
> $$D = D(a, b) = f_{xx}(a, b)f_{yy}(a, b) - [f_{xy}(a, b)]^2$$
>
> (a) If $D > 0$ and $f_{xx}(a, b) > 0$, then $f(a, b)$ is a local minimum.
>
> (b) If $D > 0$ and $f_{xx}(a, b) < 0$, then $f(a, b)$ is a local maximum.
>
> (c) If $D < 0$, then $f(a, b)$ is not a local maximum or minimum.

Proof of part (a) We compute the second-order directional derivative of f in the direction of $\mathbf{u} = \langle h, k \rangle$. The first-order derivative is given by Theorem 11.6.3:

$$D_{\mathbf{u}} f = f_x h + f_y k$$

Applying this theorem a second time, we have

$$D_{\mathbf{u}}^2 f = D_{\mathbf{u}}(D_{\mathbf{u}} f) = \frac{\partial}{\partial x} (D_{\mathbf{u}} f)h + \frac{\partial}{\partial y} (D_{\mathbf{u}} f)k$$

$$= (f_{xx} h + f_{yx} k)h + (f_{xy} h + f_{yy} k)k$$

$$= f_{xx} h^2 + 2f_{xy} hk + f_{yy} k^2 \qquad \text{(by Clairaut's Theorem)}$$

If we complete the square in this expression, we obtain

$$\boxed{2} \qquad D_{\mathbf{u}}^2 f = f_{xx}\left(h + \frac{f_{xy}}{f_{xx}} k\right)^2 + \frac{k^2}{f_{xx}} (f_{xx} f_{yy} - f_{xy}^2)$$

We are given that $f_{xx}(a, b) > 0$ and $D(a, b) > 0$. But f_{xx} and $D = f_{xx}f_{yy} - f_{xy}^2$ are continuous functions, so there is a disk B with center (a, b) and radius $\delta > 0$ such that $f_{xx}(x, y) > 0$ and $D(x, y) > 0$ whenever (x, y) is in B. Therefore, by looking at Equation 2, we see that $D_\mathbf{u}^2 f(x, y) > 0$ whenever (x, y) is in B. This means that if C is the curve obtained by intersecting the graph of f with the vertical plane through $P(a, b, f(a, b))$ in the direction of $\mathbf{u}$, then C is concave upward on an interval of length 2δ. This is true in the direction of every vector $\mathbf{u}$, so if we restrict (x, y) to lie in B, the graph of f lies above its horizontal tangent plane at P. Thus, $f(x, y) \geq f(a, b)$ whenever (x, y) is in B. This shows that $f(a, b)$ is a local minimum. ■ ■

H   Polar Coordinates

Polar coordinates offer an alternative way of locating points in a plane. They are useful because, for certain types of regions and curves, polar coordinates provide very simple descriptions and equations. The principal applications of this idea occur in multivariable calculus: the evaluation of double integrals and the derivation of Kepler's laws of planetary motion.

H.1   Curves in Polar Coordinates

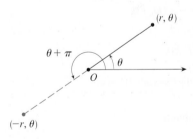

FIGURE 1

A coordinate system represents a point in the plane by an ordered pair of numbers called coordinates. Usually we use Cartesian coordinates, which are directed distances from two perpendicular axes. Here we describe a coordinate system introduced by Newton, called the **polar coordinate system**, which is more convenient for many purposes.

We choose a point in the plane that is called the **pole** (or origin) and is labeled O. Then we draw a ray (half-line) starting at O called the **polar axis**. This axis is usually drawn horizontally to the right and corresponds to the positive x-axis in Cartesian coordinates.

If P is any other point in the plane, let r be the distance from O to P and let θ be the angle (usually measured in radians) between the polar axis and the line OP as in Figure 1. Then the point P is represented by the ordered pair (r, θ) and r, θ are called **polar coordinates** of P. We use the convention that an angle is positive if measured in the counterclockwise direction from the polar axis and negative in the clockwise direction. If $P = O$, then $r = 0$ and we agree that $(0, \theta)$ represents the pole for any value of θ.

We extend the meaning of polar coordinates (r, θ) to the case in which r is negative by agreeing that, as in Figure 2, the points $(-r, \theta)$ and (r, θ) lie on the same line through O and at the same distance $|r|$ from O, but on opposite sides of O. If $r > 0$, the point (r, θ) lies in the same quadrant as θ; if $r < 0$, it lies in the quadrant on the opposite side of the pole. Notice that $(-r, \theta)$ represents the same point as $(r, \theta + \pi)$.

FIGURE 2

EXAMPLE 1 Plot the points whose polar coordinates are given.

(a) $(1, 5\pi/4)$ (b) $(2, 3\pi)$ (c) $(2, -2\pi/3)$ (d) $(-3, 3\pi/4)$

SOLUTION The points are plotted in Figure 3. In part (d) the point $(-3, 3\pi/4)$ is located three units from the pole in the fourth quadrant because the angle $3\pi/4$ is in the second quadrant and $r = -3$ is negative.

FIGURE 3

In the Cartesian coordinate system every point has only one representation, but in the polar coordinate system each point has many representations. For instance, the point $(1, 5\pi/4)$ in Example 1(a) could be written as $(1, -3\pi/4)$ or $(1, 13\pi/4)$ or $(-1, \pi/4)$. (See Figure 4.)

FIGURE 4

In fact, since a complete counterclockwise rotation is given by an angle 2π, the point represented by polar coordinates (r, θ) is also represented by

$$(r, \theta + 2n\pi) \qquad \text{and} \qquad (-r, \theta + (2n + 1)\pi)$$

where n is any integer.

The connection between polar and Cartesian coordinates can be seen from Figure 5, in which the pole corresponds to the origin and the polar axis coincides with the positive x-axis. If the point P has Cartesian coordinates (x, y) and polar coordinates (r, θ), then, from the figure, we have

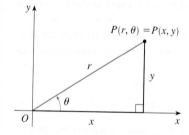

FIGURE 5

$$\cos \theta = \frac{x}{r} \qquad \sin \theta = \frac{y}{r}$$

and so

1
$$x = r \cos \theta \qquad y = r \sin \theta$$

Although Equations 1 were deduced from Figure 5, which illustrates the case where $r > 0$ and $0 < \theta < \pi/2$, these equations are valid for all values of r and θ. (See the general definition of $\sin \theta$ and $\cos \theta$ in Appendix C.)

Equations 1 allow us to find the Cartesian coordinates of a point when the polar coordinates are known. To find r and θ when x and y are known, we use the equations

2
$$r^2 = x^2 + y^2 \qquad \tan \theta = \frac{y}{x}$$

which can be deduced from Equations 1 or simply read from Figure 5.

EXAMPLE 2 Convert the point $(2, \pi/3)$ from polar to Cartesian coordinates.

SOLUTION Since $r = 2$ and $\theta = \pi/3$, Equations 1 give

$$x = r \cos \theta = 2 \cos \frac{\pi}{3} = 2 \cdot \frac{1}{2} = 1$$

$$y = r \sin \theta = 2 \sin \frac{\pi}{3} = 2 \cdot \frac{\sqrt{3}}{2} = \sqrt{3}$$

Therefore, the point is $\left(1, \sqrt{3}\right)$ in Cartesian coordinates.   ■ ■

EXAMPLE 3 Represent the point with Cartesian coordinates $(1, -1)$ in terms of polar coordinates.

SOLUTION If we choose r to be positive, then Equations 2 give

$$r = \sqrt{x^2 + y^2} = \sqrt{1^2 + (-1)^2} = \sqrt{2}$$

$$\tan \theta = \frac{y}{x} = -1$$

Since the point $(1, -1)$ lies in the fourth quadrant, we can choose $\theta = -\pi/4$ or $\theta = 7\pi/4$. Thus, one possible answer is $\left(\sqrt{2}, -\pi/4\right)$; another is $\left(\sqrt{2}, 7\pi/4\right)$.   ■ ■

NOTE ∘ Equations 2 do not uniquely determine θ when x and y are given because, as θ increases through the interval $0 \le \theta < 2\pi$, each value of $\tan \theta$ occurs twice. Therefore, in converting from Cartesian to polar coordinates, it's not good enough just to find r and θ that satisfy Equations 2. As in Example 3, we must choose θ so that the point (r, θ) lies in the correct quadrant.

The **graph of a polar equation** $r = f(\theta)$, or more generally $F(r, \theta) = 0$, consists of all points P that have at least one polar representation (r, θ) whose coordinates satisfy the equation.

$\boxed{V}$ **EXAMPLE 4** What curve is represented by the polar equation $r = 2$?

SOLUTION The curve consists of all points (r, θ) with $r = 2$. Since r represents the distance from the point to the pole, the curve $r = 2$ represents the circle with center O and radius 2. In general, the equation $r = a$ represents a circle with center O and radius $|a|$. (See Figure 6.)

FIGURE 6

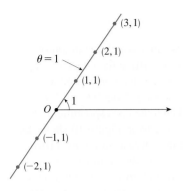

$\theta = 1$

FIGURE 7

EXAMPLE 5 Sketch the polar curve $\theta = 1$.

SOLUTION This curve consists of all points (r, θ) such that the polar angle θ is 1 radian. It is the straight line that passes through O and makes an angle of 1 radian with the polar axis (see Figure 7). Notice that the points $(r, 1)$ on the line with $r > 0$ are in the first quadrant, whereas those with $r < 0$ are in the third quadrant.   ■ ■

EXAMPLE 6
(a) Sketch the curve with polar equation $r = 2 \cos \theta$.
(b) Find a Cartesian equation for this curve.

SOLUTION

(a) In Figure 8 we find the values of r for some convenient values of θ and plot the corresponding points (r, θ). Then we join these points to sketch the curve, which appears to be a circle. We have used only values of θ between 0 and π, since if we let θ increase beyond π, we obtain the same points again.

FIGURE 8
Table of values and
graph of $r = 2 \cos \theta$

θ	$r = 2 \cos \theta$
0	2
$\pi/6$	$\sqrt{3}$
$\pi/4$	$\sqrt{2}$
$\pi/3$	1
$\pi/2$	0
$2\pi/3$	-1
$3\pi/4$	$-\sqrt{2}$
$5\pi/6$	$-\sqrt{3}$
π	-2

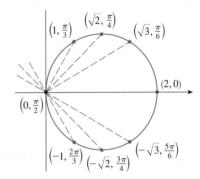

(b) To convert the given equation into a Cartesian equation we use Equations 1 and 2. From $x = r \cos \theta$ we have $\cos \theta = x/r$, so the equation $r = 2 \cos \theta$ becomes $r = 2x/r$, which gives

$$2x = r^2 = x^2 + y^2 \quad \text{or} \quad x^2 + y^2 - 2x = 0$$

Completing the square, we obtain

$$(x - 1)^2 + y^2 = 1$$

which is an equation of a circle with center $(1, 0)$ and radius 1.   ■ ■

■ ■ Figure 9 shows a geometrical illustration that the circle in Example 6 has the equation $r = 2 \cos \theta$. The angle OPQ is a right angle (Why?) and so $r/2 = \cos \theta$.

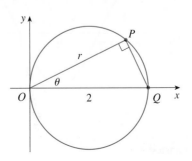

FIGURE 9

⟨V⟩ **EXAMPLE 7** Sketch the curve $r = 1 + \sin \theta$.

SOLUTION Instead of plotting points as in Example 6, we first sketch the graph of $r = 1 + \sin \theta$ in *Cartesian* coordinates in Figure 10 by shifting the sine curve up one unit. This enables us to read at a glance the values of r that correspond to increasing values of θ. For instance, we see that as θ increases from 0 to $\pi/2$, r (the distance from O) increases from 1 to 2, so we sketch the corresponding part of the polar curve in Figure 11(a). As θ increases from $\pi/2$ to π, Figure 10 shows that r decreases from 2 to 1, so we sketch the next part of the curve as in Figure 11(b). As θ increases from π to $3\pi/2$, r decreases from 1 to 0 as shown in part (c). Finally, as θ increases from $3\pi/2$ to 2π, r increases from 0 to 1 as shown in part (d). If we let θ increase beyond 2π or decrease beyond 0, we would simply retrace our path. Putting together the parts of the curve from Figure 11(a)–(d), we sketch the complete curve in part (e). It is called a **cardioid** because it's shaped like a heart.

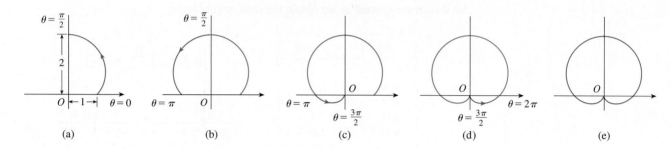

   (a)   (b)   (c)   (d)   (e)

EXAMPLE 8 Sketch the curve $r = \cos 2\theta$.

Module H helps you see how polar curves are traced out by showing animations similar to Figures 10–13. Tangents to these polar curves can also be visualized as in Figure 15 (see page A65).

SOLUTION As in Example 7, we first sketch $r = \cos 2\theta$, $0 \leq \theta \leq 2\pi$, in Cartesian coordinates in Figure 12. As θ increases from 0 to $\pi/4$, Figure 12 shows that r decreases from 1 to 0 and so we draw the corresponding portion of the polar curve in Figure 13 (indicated by ①). As θ increases from $\pi/4$ to $\pi/2$, r goes from 0 to -1. This means that the distance from O increases from 0 to 1, but instead of being in the first quadrant this portion of the polar curve (indicated by ②) lies on the opposite side of the pole in the third quadrant. The remainder of the curve is drawn in a similar fashion, with the arrows and numbers indicating the order in which the portions are traced out. The resulting curve has four loops and is called a **four-leaved rose**.

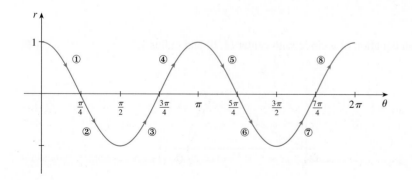

FIGURE 12
$r = \cos 2\theta$ in Cartesian coordinates

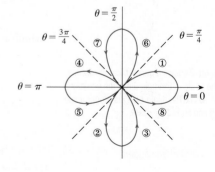

FIGURE 13
Four-leaved rose $r = \cos 2\theta$

When we sketch polar curves it is sometimes helpful to take advantage of symmetry. The following three rules are explained by Figure 14.

(a) If a polar equation is unchanged when θ is replaced by $-\theta$, the curve is symmetric about the polar axis.

(b) If the equation is unchanged when r is replaced by $-r$, or when θ is replaced by $\theta + \pi$, the curve is symmetric about the pole. (This means that the curve remains unchanged if we rotate it through 180° about the origin.)

(c) If the equation is unchanged when θ is replaced by $\pi - \theta$, the curve is symmetric about the vertical line $\theta = \pi/2$.

(a)

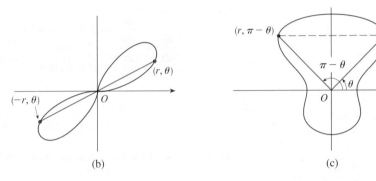

(b)   (c)

FIGURE 14

The curves sketched in Examples 6 and 8 are symmetric about the polar axis, since $\cos(-\theta) = \cos\theta$. The curves in Examples 7 and 8 are symmetric about $\theta = \pi/2$ because $\sin(\pi - \theta) = \sin\theta$ and $\cos 2(\pi - \theta) = \cos 2\theta$. The four-leaved rose is also symmetric about the pole. These symmetry properties could have been used in sketching the curves. For instance, in Example 6 we need only have plotted points for $0 \leqslant \theta \leqslant \pi/2$ and then reflected about the polar axis to obtain the complete circle.

▪▪ Tangents to Polar Curves

To find a tangent line to a polar curve $r = f(\theta)$ we regard θ as a parameter and write its parametric equations as

$$x = r\cos\theta = f(\theta)\cos\theta \qquad y = r\sin\theta = f(\theta)\sin\theta$$

Then, using the method for finding slopes of parametric curves (Equation 3.5.7) and the Product Rule, we have

$$\boxed{3} \qquad \frac{dy}{dx} = \frac{\dfrac{dy}{d\theta}}{\dfrac{dx}{d\theta}} = \frac{\dfrac{dr}{d\theta}\sin\theta + r\cos\theta}{\dfrac{dr}{d\theta}\cos\theta - r\sin\theta}$$

We locate horizontal tangents by finding the points where $dy/d\theta = 0$ (provided that $dx/d\theta \neq 0$). Likewise, we locate vertical tangents at the points where $dx/d\theta = 0$ (provided that $dy/d\theta \neq 0$).

Notice that if we are looking for tangent lines at the pole, then $r = 0$ and Equation 3 simplifies to

$$\frac{dy}{dx} = \tan\theta \qquad \text{if} \quad \frac{dr}{d\theta} \neq 0$$

For instance, in Example 8 we found that $r = \cos 2\theta = 0$ when $\theta = \pi/4$ or $3\pi/4$. This means that the lines $\theta = \pi/4$ and $\theta = 3\pi/4$ (or $y = x$ and $y = -x$) are tangent lines to $r = \cos 2\theta$ at the origin.

EXAMPLE 9
(a) For the cardioid $r = 1 + \sin \theta$ of Example 7, find the slope of the tangent line when $\theta = \pi/3$.
(b) Find the points on the cardioid where the tangent line is horizontal or vertical.

SOLUTION Using Equation 3 with $r = 1 + \sin \theta$, we have

$$\frac{dy}{dx} = \frac{\dfrac{dr}{d\theta} \sin \theta + r \cos \theta}{\dfrac{dr}{d\theta} \cos \theta - r \sin \theta} = \frac{\cos \theta \sin \theta + (1 + \sin \theta) \cos \theta}{\cos \theta \cos \theta - (1 + \sin \theta) \sin \theta}$$

$$= \frac{\cos \theta (1 + 2 \sin \theta)}{1 - 2 \sin^2\theta - \sin \theta} = \frac{\cos \theta (1 + 2 \sin \theta)}{(1 + \sin \theta)(1 - 2 \sin \theta)}$$

(a) The slope of the tangent at the point where $\theta = \pi/3$ is

$$\frac{dy}{dx}\bigg|_{\theta = \pi/3} = \frac{\cos(\pi/3)(1 + 2 \sin(\pi/3))}{(1 + \sin(\pi/3))(1 - 2 \sin(\pi/3))}$$

$$= \frac{\frac{1}{2}(1 + \sqrt{3})}{(1 + \sqrt{3}/2)(1 - \sqrt{3})} = \frac{1 + \sqrt{3}}{(2 + \sqrt{3})(1 - \sqrt{3})}$$

$$= \frac{1 + \sqrt{3}}{-1 - \sqrt{3}} = -1$$

(b) Observe that

$$\frac{dy}{d\theta} = \cos \theta (1 + 2 \sin \theta) = 0 \qquad \text{when } \theta = \frac{\pi}{2}, \frac{3\pi}{2}, \frac{7\pi}{6}, \frac{11\pi}{6}$$

$$\frac{dx}{d\theta} = (1 + \sin \theta)(1 - 2 \sin \theta) = 0 \qquad \text{when } \theta = \frac{3\pi}{2}, \frac{\pi}{6}, \frac{5\pi}{6}$$

Therefore, there are horizontal tangents at the points $(2, \pi/2)$, $(\frac{1}{2}, 7\pi/6)$, $(\frac{1}{2}, 11\pi/6)$ and vertical tangents at $(\frac{3}{2}, \pi/6)$ and $(\frac{3}{2}, 5\pi/6)$. When $\theta = 3\pi/2$, both $dy/d\theta$ and $dx/d\theta$ are 0, so we must be careful. Using l'Hospital's Rule, we have

$$\lim_{\theta \to (3\pi/2)^-} \frac{dy}{dx} = \left(\lim_{\theta \to (3\pi/2)^-} \frac{1 + 2 \sin \theta}{1 - 2 \sin \theta} \right)\left(\lim_{\theta \to (3\pi/2)^-} \frac{\cos \theta}{1 + \sin \theta} \right)$$

$$= -\frac{1}{3} \lim_{\theta \to (3\pi/2)^-} \frac{\cos \theta}{1 + \sin \theta}$$

$$= -\frac{1}{3} \lim_{\theta \to (3\pi/2)^-} \frac{-\sin \theta}{\cos \theta} = \infty$$

FIGURE 15

Tangent lines for $r = 1 + \sin\theta$

By symmetry,

$$\lim_{\theta \to (3\pi/2)^+} \frac{dy}{dx} = -\infty$$

Thus, there is a vertical tangent line at the pole (see Figure 15). ■ ■

NOTE ∘ Instead of having to remember Equation 3, we could employ the method used to derive it. For instance, in Example 9 we could have written

$$x = r\cos\theta = (1 + \sin\theta)\cos\theta = \cos\theta + \tfrac{1}{2}\sin 2\theta$$

$$y = r\sin\theta = (1 + \sin\theta)\sin\theta = \sin\theta + \sin^2\theta$$

Then we have

$$\frac{dy}{dx} = \frac{dy/d\theta}{dx/d\theta} = \frac{\cos\theta + 2\sin\theta\cos\theta}{-\sin\theta + \cos 2\theta} = \frac{\cos\theta + \sin 2\theta}{-\sin\theta + \cos 2\theta}$$

which is equivalent to our previous expression.

■ ■ Graphing Polar Curves with Graphing Devices

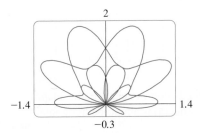

FIGURE 16

$r = \sin\theta + \sin^3(5\theta/2)$

Although it's useful to be able to sketch simple polar curves by hand, we need to use a graphing calculator or computer when we are faced with a curve as complicated as the one shown in Figure 16.

Some graphing devices have commands that enable us to graph polar curves directly. With other machines we need to convert to parametric equations first. In this case we take the polar equation $r = f(\theta)$ and write its parametric equations as

$$x = r\cos\theta = f(\theta)\cos\theta \qquad y = r\sin\theta = f(\theta)\sin\theta$$

Some machines require that the parameter be called t rather than θ.

EXAMPLE 10 Graph the curve $r = \sin(8\theta/5)$.

SOLUTION Let's assume that our graphing device doesn't have a built-in polar graphing command. In this case we need to work with the corresponding parametric equations, which are

$$x = r\cos\theta = \sin(8\theta/5)\cos\theta \qquad y = r\sin\theta = \sin(8\theta/5)\sin\theta$$

In any case we need to determine the domain for θ. So we ask ourselves: How many complete rotations are required until the curve starts to repeat itself? If the answer is n, then

$$\sin\frac{8(\theta + 2n\pi)}{5} = \sin\left(\frac{8\theta}{5} + \frac{16n\pi}{5}\right) = \sin\frac{8\theta}{5}$$

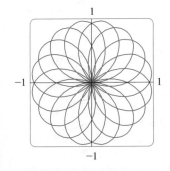

FIGURE 17

$r = \sin(8\theta/5)$

and so we require that $16n\pi/5$ be an even multiple of π. This will first occur when $n = 5$. Therefore, we will graph the entire curve if we specify that $0 \le \theta \le 10\pi$. Switching from θ to t, we have the equations

$$x = \sin(8t/5)\cos t \qquad y = \sin(8t/5)\sin t \qquad 0 \le t \le 10\pi$$

and Figure 17 shows the resulting curve. Notice that this rose has 16 loops. ■ ■

$\boxed{V}$ **EXAMPLE 11** Investigate the family of polar curves given by $r = 1 + c \sin \theta$. How does the shape change as c changes? (These curves are called **limaçons**, after a French word for snail, because of the shape of the curves for certain values of c.)

SOLUTION Figure 18 shows computer-drawn graphs for various values of c. For $c > 1$ there is a loop that decreases in size as c decreases. When $c = 1$ the loop disappears and the curve becomes the cardioid that we sketched in Example 7. For c between 1 and $\frac{1}{2}$ the cardioid's cusp is smoothed out and becomes a "dimple." When c decreases from $\frac{1}{2}$ to 0, the limaçon is shaped like an oval. This oval becomes more circular as $c \to 0$, and when $c = 0$ the curve is just the circle $r = 1$.

■ ■ In Exercise 45 you are asked to prove analytically what we have discovered from the graphs in Figure 18.

FIGURE 18

Members of the family of limaçons $r = 1 + c \sin \theta$

The remaining parts of Figure 18 show that as c becomes negative, the shapes change in reverse order. In fact, these curves are reflections about the horizontal axis of the corresponding curves with positive c. ■ ■

H.1 Exercises

1–2 ■ Plot the point whose polar coordinates are given. Then find two other pairs of polar coordinates of this point, one with $r > 0$ and one with $r < 0$.

1. (a) $(1, \pi/2)$ (b) $(-2, \pi/4)$ (c) $(3, 2)$

2. (a) $(3, 0)$ (b) $(2, -\pi/7)$ (c) $(-1, -\pi/2)$

3–4 ■ Plot the point whose polar coordinates are given. Then find the Cartesian coordinates of the point.

3. (a) $(3, \pi/2)$ (b) $(2\sqrt{2}, 3\pi/4)$ (c) $(-1, \pi/3)$

4. (a) $(2, 2\pi/3)$ (b) $(4, 3\pi)$ (c) $(-2, -5\pi/6)$

5–6 ■ The Cartesian coordinates of a point are given.
(i) Find polar coordinates (r, θ) of the point, where $r > 0$ and $0 \le \theta < 2\pi$.
(ii) Find polar coordinates (r, θ) of the point, where $r < 0$ and $0 \le \theta < 2\pi$.

5. (a) $(1, 1)$ (b) $(2\sqrt{3}, -2)$

6. (a) $\left(-1, -\sqrt{3}\right)$ (b) $(-2, 3)$

7–12 ■ Sketch the region in the plane consisting of points whose polar coordinates satisfy the given conditions.

7. $1 \le r \le 2$

8. $r \ge 0, \quad \pi/3 \le \theta \le 2\pi/3$

9. $0 \le r < 4, \quad -\pi/2 \le \theta < \pi/6$

10. $2 < r \le 5, \quad 3\pi/4 < \theta < 5\pi/4$

11. $2 < r < 3, \quad 5\pi/3 \le \theta \le 7\pi/3$

12. $-1 \le r \le 1, \quad \pi/4 \le \theta \le 3\pi/4$

13–16 ■ Identify the curve by finding a Cartesian equation for the curve.

13. $r = 3 \sin \theta$ **14.** $r = 2 \sin \theta + 2 \cos \theta$

15. $r = \csc\theta$ **16.** $r = \tan\theta \sec\theta$

17–20 ■ Find a polar equation for the curve represented by the given Cartesian equation.

17. $x = -y^2$ **18.** $x + y = 9$

19. $x^2 + y^2 = 2cx$ **20.** $x^2 - y^2 = 1$

21–22 ■ For each of the described curves, decide if the curve would be more easily given by a polar equation or a Cartesian equation. Then write an equation for the curve.

21. (a) A line through the origin that makes an angle of $\pi/6$ with the positive x-axis
    (b) A vertical line through the point $(3, 3)$

22. (a) A circle with radius 5 and center $(2, 3)$
    (b) A circle centered at the origin with radius 4

23–40 ■ Sketch the curve with the given polar equation.

23. $\theta = -\pi/6$ **24.** $r^2 - 3r + 2 = 0$

25. $r = \sin\theta$ **26.** $r = -3\cos\theta$

27. $r = 2(1 - \sin\theta)$, $\theta \geq 0$ **28.** $r = 1 - 3\cos\theta$

29. $r = \theta$, $\theta \geq 0$ **30.** $r = \ln\theta$, $\theta \geq 1$

31. $r = \sin 2\theta$ **32.** $r = 2\cos 3\theta$

33. $r = 2\cos 4\theta$ **34.** $r = \sin 5\theta$

35. $r^2 = 4\cos 2\theta$ **36.** $r^2 = \sin 2\theta$

37. $r = 2\cos(3\theta/2)$ **38.** $r^2\theta = 1$

39. $r = 1 + 2\cos 2\theta$ **40.** $r = 1 + 2\cos(\theta/2)$

41–42 ■ The figure shows the graph of r as a function of θ in Cartesian coordinates. Use it to sketch the corresponding polar curve.

41.   **42.**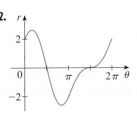

43. Show that the polar curve $r = 4 + 2\sec\theta$ (called a **conchoid**) has the line $x = 2$ as a vertical asymptote by showing that $\lim_{r \to \pm\infty} x = 2$. Use this fact to help sketch the conchoid.

44. Show that the curve $r = \sin\theta \tan\theta$ (called a **cissoid of Diocles**) has the line $x = 1$ as a vertical asymptote. Show also that the curve lies entirely within the vertical strip $0 \leq x < 1$. Use these facts to help sketch the cissoid.

45. (a) In Example 11 the graphs suggest that the limaçon $r = 1 + c\sin\theta$ has an inner loop when $|c| > 1$. Prove that this is true, and find the values of θ that correspond to the inner loop.
    (b) From Figure 18 it appears that the limaçon loses its dimple when $c = \frac{1}{2}$. Prove this.

46. Match the polar equations with the graphs labeled I–VI. Give reasons for your choices. (Don't use a graphing device.)
    (a) $r = \sin(\theta/2)$ (b) $r = \sin(\theta/4)$
    (c) $r = \sec(3\theta)$ (d) $r = \theta\sin\theta$
    (e) $r = 1 + 4\cos 5\theta$ (f) $r = 1/\sqrt{\theta}$

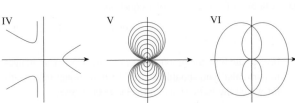

47–50 ■ Find the slope of the tangent line to the given polar curve at the point specified by the value of θ.

47. $r = 2\sin\theta$, $\theta = \pi/6$ **48.** $r = 2 - \sin\theta$, $\theta = \pi/3$

49. $r = 1/\theta$, $\theta = \pi$ **50.** $r = \sin 3\theta$, $\theta = \pi/6$

51–54 ■ Find the points on the given curve where the tangent line is horizontal or vertical.

51. $r = 3\cos\theta$ **52.** $r = e^\theta$

53. $r = 1 + \cos\theta$ **54.** $r^2 = \sin 2\theta$

55. Show that the polar equation $r = a\sin\theta + b\cos\theta$, where $ab \neq 0$, represents a circle, and find its center and radius.

56. Show that the curves $r = a\sin\theta$ and $r = a\cos\theta$ intersect at right angles.

57–60 ■ Use a graphing device to graph the polar curve. Choose the parameter interval to make sure that you produce the entire curve.

57. $r = e^{\sin\theta} - 2\cos(4\theta)$ (butterfly curve)

58. $r = \sin^2(4\theta) + \cos(4\theta)$

59. $r = 2 - 5\sin(\theta/6)$

60. $r = \cos(\theta/2) + \cos(\theta/3)$

61. How are the graphs of $r = 1 + \sin(\theta - \pi/6)$ and $r = 1 + \sin(\theta - \pi/3)$ related to the graph of $r = 1 + \sin\theta$? In general, how is the graph of $r = f(\theta - \alpha)$ related to the graph of $r = f(\theta)$?

62. Use a graph to estimate the y-coordinate of the highest points on the curve $r = \sin 2\theta$. Then use calculus to find the exact value.

63. (a) Investigate the family of curves defined by the polar equations $r = \sin n\theta$, where n is a positive integer. How is the number of loops related to n?
(b) What happens if the equation in part (a) is replaced by $r = |\sin n\theta|$?

64. A family of curves is given by the equations $r = 1 + c\sin n\theta$, where c is a real number and n is a positive integer. How does the graph change as n increases? How does it change as c changes? Illustrate by graphing enough members of the family to support your conclusions.

65. A family of curves has polar equations
$$r = \frac{1 - a\cos\theta}{1 + a\cos\theta}$$
Investigate how the graph changes as the number a changes. In particular, you should identify the transitional values of a for which the basic shape of the curve changes.

66. The astronomer Giovanni Cassini (1625–1712) studied the family of curves with polar equations
$$r^4 - 2c^2 r^2 \cos 2\theta + c^4 - a^4 = 0$$
where a and c are positive real numbers. These curves are called the **ovals of Cassini** even though they are oval shaped only for certain values of a and c. (Cassini thought that these curves might represent planetary orbits better than Kepler's ellipses.) Investigate the variety of shapes that these curves may have. In particular, how are a and c related to each other when the curve splits into two parts?

67. Let P be any point (except the origin) on the curve $r = f(\theta)$. If ψ is the angle between the tangent line at P and the radial line OP, show that
$$\tan\psi = \frac{r}{dr/d\theta}$$
[*Hint:* Observe that $\psi = \phi - \theta$ in the figure.]

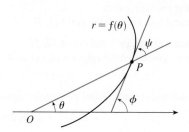

68. (a) Use Exercise 67 to show that the angle between the tangent line and the radial line is $\psi = \pi/4$ at every point on the curve $r = e^\theta$.
(b) Illustrate part (a) by graphing the curve and the tangent lines at the points where $\theta = 0$ and $\pi/2$.
(c) Prove that any polar curve $r = f(\theta)$ with the property that the angle ψ between the radial line and the tangent line is a constant must be of the form $r = Ce^{k\theta}$, where C and k are constants.

H.2 Areas and Lengths in Polar Coordinates

FIGURE 1

In this section we develop the formula for the area of a region whose boundary is given by a polar equation. We need to use the formula for the area of a sector of a circle

$$\boxed{1} \qquad A = \tfrac{1}{2}r^2\theta$$

where, as in Figure 1, r is the radius and θ is the radian measure of the central angle. Formula 1 follows from the fact that the area of a sector is proportional to its central angle: $A = (\theta/2\pi)\pi r^2 = \tfrac{1}{2}r^2\theta$.

Let $\mathcal{R}$ be the region, illustrated in Figure 2, bounded by the polar curve $r = f(\theta)$ and by the rays $\theta = a$ and $\theta = b$, where f is a positive continuous function and where $0 < b - a \leq 2\pi$. We divide the interval $[a, b]$ into subintervals with endpoints θ_0,

FIGURE 2

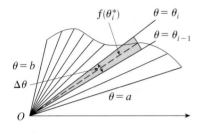

FIGURE 3

$\theta_1, \theta_2, \ldots, \theta_n$ and equal width $\Delta\theta$. The rays $\theta = \theta_i$ then divide $\mathcal{R}$ into n smaller regions with central angle $\Delta\theta = \theta_i - \theta_{i-1}$. If we choose θ_i^* in the ith subinterval $[\theta_{i-1}, \theta_i]$, then the area ΔA_i of the ith region is approximated by the area of the sector of a circle with central angle $\Delta\theta$ and radius $f(\theta_i^*)$. (See Figure 3.)

Thus, from Formula 1 we have

$$\Delta A_i \approx \tfrac{1}{2}[f(\theta_i^*)]^2 \Delta\theta$$

and so an approximation to the total area A of $\mathcal{R}$ is

2
$$A \approx \sum_{i=1}^{n} \tfrac{1}{2}[f(\theta_i^*)]^2 \Delta\theta$$

It appears from Figure 3 that the approximation in (2) improves as $n \to \infty$. But the sums in (2) are Riemann sums for the function $g(\theta) = \tfrac{1}{2}[f(\theta)]^2$, so

$$\lim_{n \to \infty} \sum_{i=1}^{n} \tfrac{1}{2}[f(\theta_i^*)]^2 \Delta\theta = \int_a^b \tfrac{1}{2}[f(\theta)]^2 \, d\theta$$

It therefore appears plausible (and can in fact be proved) that the formula for the area A of the polar region $\mathcal{R}$ is

3
$$A = \int_a^b \tfrac{1}{2}[f(\theta)]^2 \, d\theta$$

Formula 3 is often written as

4
$$A = \int_a^b \tfrac{1}{2}r^2 \, d\theta$$

with the understanding that $r = f(\theta)$. Note the similarity between Formulas 1 and 4.

When we apply Formula 3 or 4 it is helpful to think of the area as being swept out by a rotating ray through O that starts with angle a and ends with angle b.

 EXAMPLE 1 Find the area enclosed by one loop of the four-leaved rose $r = \cos 2\theta$.

SOLUTION The curve $r = \cos 2\theta$ was sketched in Example 8 in Section H.1. Notice from Figure 4 that the region enclosed by the right loop is swept out by a ray that rotates from $\theta = -\pi/4$ to $\theta = \pi/4$. Therefore, Formula 4 gives

$$A = \int_{-\pi/4}^{\pi/4} \tfrac{1}{2}r^2 \, d\theta = \tfrac{1}{2} \int_{-\pi/4}^{\pi/4} \cos^2 2\theta \, d\theta = \int_0^{\pi/4} \cos^2 2\theta \, d\theta$$

We could evaluate the integral using Formula 64 in the Table of Integrals. Or, as in Section 5.7, we could use the identity $\cos^2 x = \tfrac{1}{2}(1 + \cos 2x)$ to write

$$A = \int_0^{\pi/4} \tfrac{1}{2}(1 + \cos 4\theta) \, d\theta = \tfrac{1}{2}\left[\theta + \tfrac{1}{4}\sin 4\theta\right]_0^{\pi/4} = \frac{\pi}{8}$$

FIGURE 4

■ ■

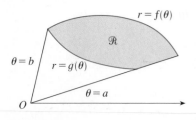

FIGURE 5

$\boxed{V}$ **EXAMPLE 2** Find the area of the region that lies inside the circle $r = 3 \sin \theta$ and outside the cardioid $r = 1 + \sin \theta$.

SOLUTION The cardioid (see Example 7 in Section H.1) and the circle are sketched in Figure 5 and the desired region is shaded. The values of a and b in Formula 4 are determined by finding the points of intersection of the two curves. They intersect when $3 \sin \theta = 1 + \sin \theta$, which gives $\sin \theta = \frac{1}{2}$, so $\theta = \pi/6, 5\pi/6$. The desired area can be found by subtracting the area inside the cardioid between $\theta = \pi/6$ and $\theta = 5\pi/6$ from the area inside the circle from $\pi/6$ to $5\pi/6$. Thus

$$A = \frac{1}{2} \int_{\pi/6}^{5\pi/6} (3 \sin \theta)^2 \, d\theta - \frac{1}{2} \int_{\pi/6}^{5\pi/6} (1 + \sin \theta)^2 \, d\theta$$

Since the region is symmetric about the vertical axis $\theta = \pi/2$, we can write

$$A = 2 \left[\frac{1}{2} \int_{\pi/6}^{\pi/2} 9 \sin^2 \theta \, d\theta - \frac{1}{2} \int_{\pi/6}^{\pi/2} (1 + 2 \sin \theta + \sin^2 \theta) \, d\theta \right]$$

$$= \int_{\pi/6}^{\pi/2} (8 \sin^2 \theta - 1 - 2 \sin \theta) \, d\theta$$

$$= \int_{\pi/6}^{\pi/2} (3 - 4 \cos 2\theta - 2 \sin \theta) \, d\theta \qquad \left[\text{because } \sin^2 \theta = \frac{1}{2}(1 - \cos 2\theta) \right]$$

$$= 3\theta - 2 \sin 2\theta + 2 \cos \theta \Big]_{\pi/6}^{\pi/2} = \pi \qquad \blacksquare \ \blacksquare$$

Example 2 illustrates the procedure for finding the area of the region bounded by two polar curves. In general, let $\mathcal{R}$ be a region, as illustrated in Figure 6, that is bounded by curves with polar equations $r = f(\theta)$, $r = g(\theta)$, $\theta = a$, and $\theta = b$, where $f(\theta) \geq g(\theta) \geq 0$ and $0 < b - a \leq 2\pi$. The area A of $\mathcal{R}$ is found by subtracting the area inside $r = g(\theta)$ from the area inside $r = f(\theta)$, so using Formula 3 we have

$$A = \int_a^b \frac{1}{2}[f(\theta)]^2 \, d\theta - \int_a^b \frac{1}{2}[g(\theta)]^2 \, d\theta$$

$$= \frac{1}{2} \int_a^b \left([f(\theta)]^2 - [g(\theta)]^2 \right) d\theta$$

FIGURE 6

$\boxed{\oslash}$ **CAUTION** ▫ The fact that a single point has many representations in polar coordinates sometimes makes it difficult to find all the points of intersection of two polar curves. For instance, it is obvious from Figure 5 that the circle and the cardioid have three points of intersection; however, in Example 2 we solved the equations $r = 3 \sin \theta$ and $r = 1 + \sin \theta$ and found only two such points, $\left(\frac{3}{2}, \pi/6 \right)$ and $\left(\frac{3}{2}, 5\pi/6 \right)$. The origin is also a point of intersection, but we can't find it by solving the equations of the curves because the origin has no single representation in polar coordinates that satisfies both equations. Notice that, when represented as $(0, 0)$ or $(0, \pi)$, the origin satisfies $r = 3 \sin \theta$ and so it lies on the circle; when represented as $(0, 3\pi/2)$, it satisfies $r = 1 + \sin \theta$ and so it lies on the cardioid. Think of two points moving along the curves as the parameter value θ increases from 0 to 2π. On one curve the origin is reached at $\theta = 0$ and $\theta = \pi$; on the other curve it is reached at $\theta = 3\pi/2$. The points don't collide at the origin because they reach the origin at different times, but the curves intersect there nonetheless.

Thus, to find *all* points of intersection of two polar curves, it is recommended that you draw the graphs of both curves. It is especially convenient to use a graphing calculator or computer to help with this task.

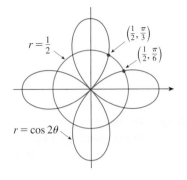

$r = \frac{1}{2}$

$\left(\frac{1}{2}, \frac{\pi}{3}\right)$

$\left(\frac{1}{2}, \frac{\pi}{6}\right)$

$r = \cos 2\theta$

FIGURE 7

EXAMPLE 3 Find all points of intersection of the curves $r = \cos 2\theta$ and $r = \frac{1}{2}$.

SOLUTION If we solve the equations $r = \cos 2\theta$ and $r = \frac{1}{2}$, we get $\cos 2\theta = \frac{1}{2}$ and, therefore, $2\theta = \pi/3, 5\pi/3, 7\pi/3, 11\pi/3$. Thus, the values of θ between 0 and 2π that satisfy both equations are $\theta = \pi/6, 5\pi/6, 7\pi/6, 11\pi/6$. We have found four points of intersection: $\left(\frac{1}{2}, \pi/6\right)$, $\left(\frac{1}{2}, 5\pi/6\right)$, $\left(\frac{1}{2}, 7\pi/6\right)$, and $\left(\frac{1}{2}, 11\pi/6\right)$.

However, you can see from Figure 7 that the curves have four other points of intersection—namely, $\left(\frac{1}{2}, \pi/3\right)$, $\left(\frac{1}{2}, 2\pi/3\right)$, $\left(\frac{1}{2}, 4\pi/3\right)$, and $\left(\frac{1}{2}, 5\pi/3\right)$. These can be found using symmetry or by noticing that another equation of the circle is $r = -\frac{1}{2}$ and then solving the equations $r = \cos 2\theta$ and $r = -\frac{1}{2}$. ■ ■

▪▪ Arc Length

To find the length of a polar curve $r = f(\theta)$, $a \le \theta \le b$, we regard θ as a parameter and write the parametric equations of the curve as

$$x = r \cos \theta = f(\theta) \cos \theta \qquad y = r \sin \theta = f(\theta) \sin \theta$$

Using the Product Rule and differentiating with respect to θ, we obtain

$$\frac{dx}{d\theta} = \frac{dr}{d\theta} \cos \theta - r \sin \theta \qquad \frac{dy}{d\theta} = \frac{dr}{d\theta} \sin \theta + r \cos \theta$$

so, using $\cos^2\theta + \sin^2\theta = 1$, we have

$$\left(\frac{dx}{d\theta}\right)^2 + \left(\frac{dy}{d\theta}\right)^2 = \left(\frac{dr}{d\theta}\right)^2 \cos^2\theta - 2r \frac{dr}{d\theta} \cos \theta \, \sin \theta + r^2 \sin^2\theta$$

$$+ \left(\frac{dr}{d\theta}\right)^2 \sin^2\theta + 2r \frac{dr}{d\theta} \sin \theta \, \cos \theta + r^2 \cos^2\theta$$

$$= \left(\frac{dr}{d\theta}\right)^2 + r^2$$

Assuming that f' is continuous, we can use Formula 6.3.1 to write the arc length as

$$L = \int_a^b \sqrt{\left(\frac{dx}{d\theta}\right)^2 + \left(\frac{dy}{d\theta}\right)^2} \, d\theta$$

Therefore, the length of a curve with polar equation $r = f(\theta)$, $a \le \theta \le b$, is

5
$$L = \int_a^b \sqrt{r^2 + \left(\frac{dr}{d\theta}\right)^2} \, d\theta$$

O

FIGURE 8
$r = 1 + \sin \theta$

$\boxed{V}$ **EXAMPLE 4** Find the length of the cardioid $r = 1 + \sin \theta$.

SOLUTION The cardioid is shown in Figure 8. (We sketched it in Example 7 in Section H.1.) Its full length is given by the parameter interval $0 \le \theta \le 2\pi$, so

Formula 5 gives

$$L = \int_0^{2\pi} \sqrt{r^2 + \left(\frac{dr}{d\theta}\right)^2}\, d\theta = \int_0^{2\pi} \sqrt{(1 + \sin\theta)^2 + \cos^2\theta}\, d\theta$$

$$= \int_0^{2\pi} \sqrt{2 + 2\sin\theta}\, d\theta$$

We could evaluate this integral by multiplying and dividing the integrand by $\sqrt{2 - 2\sin\theta}$, or we could use a computer algebra system. In any event, we find that the length of the cardioid is $L = 8$. ■ ■

H.2 Exercises

1–4 ■ Find the area of the region that is bounded by the given curve and lies in the specified sector.

1. $r = \sqrt{\theta}$, $0 \leqslant \theta \leqslant \pi/4$

2. $r = e^{\theta/2}$, $\pi \leqslant \theta \leqslant 2\pi$

3. $r = \sin\theta$, $\pi/3 \leqslant \theta \leqslant 2\pi/3$

4. $r = \sqrt{\sin\theta}$, $0 \leqslant \theta \leqslant \pi$

. . . .

5–8 ■ Find the area of the shaded region.

5.

$r = \theta$

6.

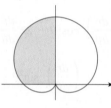

$r = 1 + \sin\theta$

7.

$r = 4 + 3\sin\theta$

8.

$r = \sin 4\theta$

. . . .

9–12 ■ Sketch the curve and find the area that it encloses.

9. $r^2 = 4\cos 2\theta$

10. $r = 3(1 + \cos\theta)$

11. $r = 2\cos 3\theta$

12. $r = 2 + \cos 2\theta$

. . . .

⊞ **13–14** ■ Graph the curve and find the area that it encloses.

13. $r = 1 + 2\sin 6\theta$

14. $r = 2\sin\theta + 3\sin 9\theta$

. . . .

15–18 ■ Find the area of the region enclosed by one loop of the curve.

15. $r = \sin 2\theta$

16. $r = 4\sin 3\theta$

17. $r = 1 + 2\sin\theta$ (inner loop)

18. $r = 2\cos\theta - \sec\theta$

. . . .

19–22 ■ Find the area of the region that lies inside the first curve and outside the second curve.

19. $r = 4\sin\theta$, $r = 2$

20. $r = 1 - \sin\theta$, $r = 1$

21. $r = 3\cos\theta$, $r = 1 + \cos\theta$

22. $r = 2 + \sin\theta$, $r = 3\sin\theta$

. . . .

23–26 ■ Find the area of the region that lies inside both curves.

23. $r = \sin\theta$, $r = \cos\theta$

24. $r = \sin 2\theta$, $r = \sin\theta$

25. $r = \sin 2\theta$, $r = \cos 2\theta$

26. $r^2 = 2\sin 2\theta$, $r = 1$

. . . .

27. Find the area inside the larger loop and outside the smaller loop of the limaçon $r = \frac{1}{2} + \cos\theta$.

28. When recording live performances, sound engineers often use a microphone with a cardioid pickup pattern because it suppresses noise from the audience. Suppose the microphone is placed 4 m from the front of the stage (as in the figure) and the boundary of the optimal pickup region is given by the cardioid $r = 8 + 8\sin\theta$, where r is measured in meters and the microphone is at the pole. The musicians want to know the area they will have on stage within the

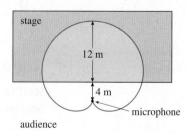

optimal pickup range of the microphone. Answer their question.

29–32 ■ Find all points of intersection of the given curves.

29. $r = \cos\theta, \quad r = 1 - \cos\theta$

30. $r = \cos 3\theta, \quad r = \sin 3\theta$

31. $r = \sin\theta, \quad r = \sin 2\theta$

32. $r^2 = \sin 2\theta, \quad r^2 = \cos 2\theta$

33. The points of intersection of the cardioid $r = 1 + \sin\theta$ and the spiral loop $r = 2\theta$, $-\pi/2 \leq \theta \leq \pi/2$, can't be found exactly. Use a graphing device to find the approximate values of θ at which they intersect. Then use these values to estimate the area that lies inside both curves.

34. Use a graph to estimate the values of θ for which the curves $r = 3 + \sin 5\theta$ and $r = 6\sin\theta$ intersect. Then estimate the area that lies inside both curves.

35–38 ■ Find the exact length of the polar curve.

35. $r = 3\sin\theta, \quad 0 \leq \theta \leq \pi/3$

36. $r = e^{2\theta}, \quad 0 \leq \theta \leq 2\pi$

37. $r = \theta^2, \quad 0 \leq \theta \leq 2\pi$

38. $r = \theta, \quad 0 \leq \theta \leq 2\pi$

39–40 ■ Use a calculator to find the length of the curve correct to four decimal places.

39. $r = 3\sin 2\theta$

40. $r = 4\sin 3\theta$

DISCOVERY PROJECT

Conic Sections in Polar Coordinates

In this project we give a unified treatment of all three types of conic sections in terms of a focus and directrix. We will see that if we place the focus at the origin, then a conic section has a simple polar equation. In Chapter 10 we will use the polar equation of an ellipse to derive Kepler's laws of planetary motion.

Let F be a fixed point (called the **focus**) and l be a fixed line (called the **directrix**) in a plane. Let e be a fixed positive number (called the **eccentricity**). Let C be the set of all points P in the plane such that

$$\frac{|PF|}{|Pl|} = e$$

(that is, the ratio of the distance from F to the distance from l is the constant e). Notice that if the eccentricity is $e = 1$, then $|PF| = |Pl|$ and so the given condition simply becomes the definition of a parabola.

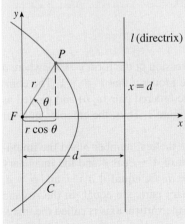

FIGURE 1

1. If we place the focus F at the origin and the directrix parallel to the y-axis and d units to the right, then the directrix has equation $x = d$ and is perpendicular to the polar axis. If the point P has polar coordinates (r, θ), use Figure 1 to show that

$$r = e(d - r\cos\theta)$$

2. By converting the polar equation in Problem 1 to rectangular coordinates, show that the curve C is an ellipse if $e < 1$.

3. Show that C is a hyperbola if $e > 1$.

4. Show that the polar equation

$$r = \frac{ed}{1 + e\cos\theta}$$

represents an ellipse if $e < 1$, a parabola if $e = 1$, or a hyperbola if $e > 1$.

5. For each of the following conics, find the eccentricity and directrix. Then identify and sketch the conic.

(a) $r = \dfrac{4}{1 + 3\cos\theta}$ (b) $r = \dfrac{8}{3 + 3\cos\theta}$ (c) $r = \dfrac{2}{2 + \cos\theta}$

6. Graph the conics $r = e/(1 - e\cos\theta)$ with $e = 0.4, 0.6, 0.8,$ and 1.0 on a common screen. How does the value of e affect the shape of the curve?

7. (a) Show that the polar equation of an ellipse with directrix $x = d$ can be written in the form

$$r = \frac{a(1 - e^2)}{1 - e \cos \theta}$$

(b) Find an approximate polar equation for the elliptical orbit of the planet Earth around the Sun (at one focus) given that the eccentricity is about 0.017 and the length of the major axis is about 2.99×10^8 km .

8. (a) The planets move around the Sun in elliptical orbits with the Sun at one focus. The positions of a planet that are closest to and farthest from the Sun are called its *perihelion* and *aphelion*, respectively. (See Figure 2.) Use Problem 7(a) to show that the perihelion distance from a planet to the Sun is $a(1 - e)$ and the aphelion distance is $a(1 + e)$.

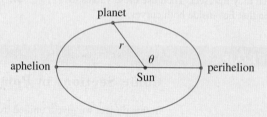

FIGURE 2

(b) Use the data of Problem 7(b) to find the distances from Earth to the Sun at perihelion and at aphelion.

9. (a) The planet Mercury travels in an elliptical orbit with eccentricity 0.206. Its minimum distance from the Sun is 4.6×10^7 km. Use the results of Problem 8(a) to find its maximum distance from the Sun.

(b) Find the distance traveled by the planet Mercury during one complete orbit around the Sun. (Use your calculator or computer algebra system to evaluate the definite integral.)

---

1 | Complex Numbers

FIGURE 1

Complex numbers as points in the Argand plane

A **complex number** can be represented by an expression of the form $a + bi$, where a and b are real numbers and i is a symbol with the property that $i^2 = -1$. The complex number $a + bi$ can also be represented by the ordered pair (a, b) and plotted as a point in a plane (called the Argand plane) as in Figure 1. Thus, the complex number $i = 0 + 1 \cdot i$ is identified with the point $(0, 1)$.

The **real part** of the complex number $a + bi$ is the real number a and the **imaginary part** is the real number b. Thus, the real part of $4 - 3i$ is 4 and the imaginary part is -3. Two complex numbers $a + bi$ and $c + di$ are **equal** if $a = c$ and $b = d$, that is, their real parts are equal and their imaginary parts are equal. In the Argand plane the horizontal axis is called the real axis and the vertical axis is called the imaginary axis.

The sum and difference of two complex numbers are defined by adding or subtracting their real parts and their imaginary parts:

$$(a + bi) + (c + di) = (a + c) + (b + d)i$$

$$(a + bi) - (c + di) = (a - c) + (b - d)i$$

For instance,

$$(1 - i) + (4 + 7i) = (1 + 4) + (-1 + 7)i = 5 + 6i$$

The product of complex numbers is defined so that the usual commutative and distributive laws hold:

$$(a + bi)(c + di) = a(c + di) + (bi)(c + di)$$
$$= ac + adi + bci + bdi^2$$

Since $i^2 = -1$, this becomes

$$(a + bi)(c + di) = (ac - bd) + (ad + bc)i$$

EXAMPLE 1

$$(-1 + 3i)(2 - 5i) = (-1)(2 - 5i) + 3i(2 - 5i)$$
$$= -2 + 5i + 6i - 15(-1) = 13 + 11i \qquad ■ ■$$

Division of complex numbers is much like rationalizing the denominator of a rational expression. For the complex number $z = a + bi$, we define its **complex conjugate** to be $\bar{z} = a - bi$. To find the quotient of two complex numbers we multiply numerator and denominator by the complex conjugate of the denominator.

EXAMPLE 2 Express the number $\dfrac{-1 + 3i}{2 + 5i}$ in the form $a + bi$.

SOLUTION We multiply numerator and denominator by the complex conjugate of $2 + 5i$, namely $2 - 5i$, and we take advantage of the result of Example 1:

$$\frac{-1 + 3i}{2 + 5i} = \frac{-1 + 3i}{2 + 5i} \cdot \frac{2 - 5i}{2 - 5i} = \frac{13 + 11i}{2^2 + 5^2} = \frac{13}{29} + \frac{11}{29}i \qquad ■ ■$$

The geometric interpretation of the complex conjugate is shown in Figure 2: $\bar{z}$ is the reflection of z in the real axis. We list some of the properties of the complex conjugate in the following box. The proofs follow from the definition and are requested in Exercise 18.

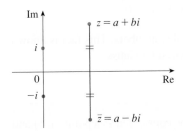

FIGURE 2

Properties of Conjugates

$$\overline{z + w} = \bar{z} + \bar{w} \qquad \overline{zw} = \bar{z}\,\bar{w} \qquad \overline{z^n} = \bar{z}^n$$

The **modulus**, or **absolute value**, $|z|$ of a complex number $z = a + bi$ is its distance from the origin. From Figure 3 we see that if $z = a + bi$, then

$$\boxed{|z| = \sqrt{a^2 + b^2}}$$

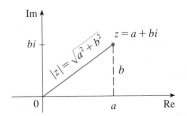

FIGURE 3

Notice that

$$z\bar{z} = (a + bi)(a - bi) = a^2 + abi - abi - b^2i^2 = a^2 + b^2$$

and so

$$\boxed{z\bar{z} = |z|^2}$$

This explains why the division procedure in Example 2 works in general:

$$\frac{z}{w} = \frac{z\overline{w}}{w\overline{w}} = \frac{z\overline{w}}{|w|^2}$$

Since $i^2 = -1$, we can think of i as a square root of -1. But notice that we also have $(-i)^2 = i^2 = -1$ and so $-i$ is also a square root of -1. We say that i is the **principal square root** of -1 and write $\sqrt{-1} = i$. In general, if c is any positive number, we write

$$\sqrt{-c} = \sqrt{c}\, i$$

With this convention, the usual derivation and formula for the roots of the quadratic equation $ax^2 + bx + c = 0$ are valid even when $b^2 - 4ac < 0$:

$$x = \frac{-b \pm \sqrt{b^2 - 4ac}}{2a}$$

EXAMPLE 3 Find the roots of the equation $x^2 + x + 1 = 0$.

SOLUTION Using the quadratic formula, we have

$$x = \frac{-1 \pm \sqrt{1^2 - 4 \cdot 1}}{2} = \frac{-1 \pm \sqrt{-3}}{2} = \frac{-1 \pm \sqrt{3}\, i}{2}$$

　■　■

We observe that the solutions of the equation in Example 3 are complex conjugates of each other. In general, the solutions of any quadratic equation $ax^2 + bx + c = 0$ with real coefficients a, b, and c are always complex conjugates. (If z is real, $\overline{z} = z$, so z is its own conjugate.)

We have seen that if we allow complex numbers as solutions, then every quadratic equation has a solution. More generally, it is true that every polynomial equation

$$a_n x^n + a_{n-1} x^{n-1} + \cdots + a_1 x + a_0 = 0$$

of degree at least one has a solution among the complex numbers. This fact is known as the Fundamental Theorem of Algebra and was proved by Gauss.

■■ Polar Form

We know that any complex number $z = a + bi$ can be considered as a point (a, b) and that any such point can be represented by polar coordinates (r, θ) with $r \geqslant 0$. In fact,

$$a = r \cos \theta \qquad b = r \sin \theta$$

as in Figure 4. Therefore, we have

$$z = a + bi = (r \cos \theta) + (r \sin \theta)i$$

Thus, we can write any complex number z in the form

$$\boxed{z = r(\cos \theta + i \sin \theta)}$$

FIGURE 4

where $\qquad r = |z| = \sqrt{a^2 + b^2} \qquad$ and $\qquad \tan \theta = \dfrac{b}{a}$

The angle θ is called the **argument** of z and we write $\theta = \arg(z)$. Note that $\arg(z)$ is not unique; any two arguments of z differ by an integer multiple of 2π.

EXAMPLE 4 Write the following numbers in polar form.

(a) $z = 1 + i$ (b) $w = \sqrt{3} - i$

SOLUTION

(a) We have $r = |z| = \sqrt{1^2 + 1^2} = \sqrt{2}$ and $\tan \theta = 1$, so we can take $\theta = \pi/4$. Therefore, the polar form is

$$z = \sqrt{2}\left(\cos\frac{\pi}{4} + i\sin\frac{\pi}{4}\right)$$

(b) Here we have $r = |w| = \sqrt{3 + 1} = 2$ and $\tan \theta = -1/\sqrt{3}$. Since w lies in the fourth quadrant, we take $\theta = -\pi/6$ and

$$w = 2\left[\cos\left(-\frac{\pi}{6}\right) + i\sin\left(-\frac{\pi}{6}\right)\right]$$

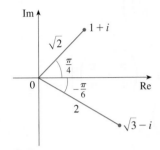

FIGURE 5

The numbers z and w are shown in Figure 5. ■ ■

The polar form of complex numbers gives insight into multiplication and division. Let

$$z_1 = r_1(\cos\theta_1 + i\sin\theta_1) \qquad z_2 = r_2(\cos\theta_2 + i\sin\theta_2)$$

be two complex numbers written in polar form. Then

$$z_1 z_2 = r_1 r_2(\cos\theta_1 + i\sin\theta_1)(\cos\theta_2 + i\sin\theta_2)$$

$$= r_1 r_2[(\cos\theta_1\cos\theta_2 - \sin\theta_1\sin\theta_2) + i(\sin\theta_1\cos\theta_2 + \cos\theta_1\sin\theta_2)]$$

Therefore, using the addition formulas for cosine and sine, we have

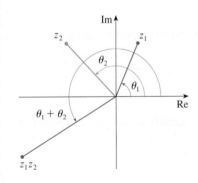

FIGURE 6

1
$$z_1 z_2 = r_1 r_2[\cos(\theta_1 + \theta_2) + i\sin(\theta_1 + \theta_2)]$$

This formula says that *to multiply two complex numbers we multiply the moduli and add the arguments.* (See Figure 6.)

A similar argument using the subtraction formulas for sine and cosine shows that *to divide two complex numbers we divide the moduli and subtract the arguments.*

$$\frac{z_1}{z_2} = \frac{r_1}{r_2}[\cos(\theta_1 - \theta_2) + i\sin(\theta_1 - \theta_2)] \qquad z_2 \neq 0$$

In particular, taking $z_1 = 1$ and $z_2 = z$, (and therefore $\theta_1 = 0$ and $\theta_2 = \theta$), we have the following, which is illustrated in Figure 7.

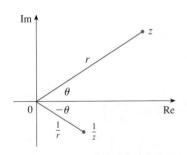

FIGURE 7

If $z = r(\cos\theta + i\sin\theta)$, then $\dfrac{1}{z} = \dfrac{1}{r}(\cos\theta - i\sin\theta)$.

EXAMPLE 5 Find the product of the complex numbers $1 + i$ and $\sqrt{3} - i$ in polar form.

SOLUTION From Example 4 we have

$$1 + i = \sqrt{2}\left(\cos\frac{\pi}{4} + i\sin\frac{\pi}{4}\right)$$

and

$$\sqrt{3} - i = 2\left[\cos\left(-\frac{\pi}{6}\right) + i\sin\left(-\frac{\pi}{6}\right)\right]$$

So, by Equation 1,

$$(1 + i)(\sqrt{3} - i) = 2\sqrt{2}\left[\cos\left(\frac{\pi}{4} - \frac{\pi}{6}\right) + i\sin\left(\frac{\pi}{4} - \frac{\pi}{6}\right)\right]$$

$$= 2\sqrt{2}\left(\cos\frac{\pi}{12} + i\sin\frac{\pi}{12}\right)$$

FIGURE 8

This is illustrated in Figure 8. ■ ■

Repeated use of Formula 1 shows how to compute powers of a complex number. If

$$z = r(\cos\theta + i\sin\theta)$$

then

$$z^2 = r^2(\cos 2\theta + i\sin 2\theta)$$

and

$$z^3 = zz^2 = r^3(\cos 3\theta + i\sin 3\theta)$$

In general, we obtain the following result, which is named after the French mathematician Abraham De Moivre (1667–1754).

> **2** **De Moivre's Theorem** If $z = r(\cos\theta + i\sin\theta)$ and n is a positive integer, then
>
> $$z^n = [r(\cos\theta + i\sin\theta)]^n = r^n(\cos n\theta + i\sin n\theta)$$

This says that *to take the nth power of a complex number we take the nth power of the modulus and multiply the argument by n.*

EXAMPLE 6 Find $\left(\frac{1}{2} + \frac{1}{2}i\right)^{10}$.

SOLUTION Since $\frac{1}{2} + \frac{1}{2}i = \frac{1}{2}(1 + i)$, it follows from Example 4(a) that $\frac{1}{2} + \frac{1}{2}i$ has the polar form

$$\frac{1}{2} + \frac{1}{2}i = \frac{\sqrt{2}}{2}\left(\cos\frac{\pi}{4} + i\sin\frac{\pi}{4}\right)$$

So by De Moivre's Theorem,

$$\left(\frac{1}{2} + \frac{1}{2}i\right)^{10} = \left(\frac{\sqrt{2}}{2}\right)^{10}\left(\cos\frac{10\pi}{4} + i\sin\frac{10\pi}{4}\right)$$

$$= \frac{2^5}{2^{10}}\left(\cos\frac{5\pi}{2} + i\sin\frac{5\pi}{2}\right) = \frac{1}{32}i$$

■ ■

De Moivre's Theorem can also be used to find the nth roots of complex numbers. An nth root of the complex number z is a complex number w such that

$$w^n = z$$

Writing these two numbers in trigonometric form as

$$w = s(\cos\phi + i\sin\phi) \qquad \text{and} \qquad z = r(\cos\theta + i\sin\theta)$$

and using De Moivre's Theorem, we get

$$s^n(\cos n\phi + i\sin n\phi) = r(\cos\theta + i\sin\theta)$$

The equality of these two complex numbers shows that

$$s^n = r \qquad \text{or} \qquad s = r^{1/n}$$

and

$$\cos n\phi = \cos\theta \qquad \text{and} \qquad \sin n\phi = \sin\theta$$

From the fact that sine and cosine have period 2π it follows that

$$n\phi = \theta + 2k\pi \qquad \text{or} \qquad \phi = \frac{\theta + 2k\pi}{n}$$

Thus

$$w = r^{1/n}\left[\cos\left(\frac{\theta + 2k\pi}{n}\right) + i\sin\left(\frac{\theta + 2k\pi}{n}\right)\right]$$

Since this expression gives a different value of w for $k = 0, 1, 2, \ldots, n - 1$, we have the following.

---

3 **Roots of a Complex Number** Let $z = r(\cos\theta + i\sin\theta)$ and let n be a positive integer. Then z has the n distinct nth roots

$$w_k = r^{1/n}\left[\cos\left(\frac{\theta + 2k\pi}{n}\right) + i\sin\left(\frac{\theta + 2k\pi}{n}\right)\right]$$

where $k = 0, 1, 2, \ldots, n - 1$.

Notice that each of the nth roots of z has modulus $|w_k| = r^{1/n}$. Thus, all the nth roots of z lie on the circle of radius $r^{1/n}$ in the complex plane. Also, since the argument of each successive nth root exceeds the argument of the previous root by $2\pi/n$, we see that the nth roots of z are equally spaced on this circle.

EXAMPLE 7 Find the six sixth roots of $z = -8$ and graph these roots in the complex plane.

SOLUTION In trigonometric form, $z = 8(\cos\pi + i\sin\pi)$. Applying Equation 3 with $n = 6$, we get

$$w_k = 8^{1/6}\left(\cos\frac{\pi + 2k\pi}{6} + i\sin\frac{\pi + 2k\pi}{6}\right)$$

We get the six sixth roots of -8 by taking $k = 0, 1, 2, 3, 4, 5$ in this formula:

$$w_0 = 8^{1/6}\left(\cos\frac{\pi}{6} + i\sin\frac{\pi}{6}\right) = \sqrt{2}\left(\frac{\sqrt{3}}{2} + \frac{1}{2}i\right)$$

$$w_1 = 8^{1/6}\left(\cos\frac{\pi}{2} + i\sin\frac{\pi}{2}\right) = \sqrt{2}\,i$$

$$w_2 = 8^{1/6}\left(\cos\frac{5\pi}{6} + i\sin\frac{5\pi}{6}\right) = \sqrt{2}\left(-\frac{\sqrt{3}}{2} + \frac{1}{2}i\right)$$

$$w_3 = 8^{1/6}\left(\cos\frac{7\pi}{6} + i\sin\frac{7\pi}{6}\right) = \sqrt{2}\left(-\frac{\sqrt{3}}{2} - \frac{1}{2}i\right)$$

$$w_4 = 8^{1/6}\left(\cos\frac{3\pi}{2} + i\sin\frac{3\pi}{2}\right) = -\sqrt{2}\,i$$

$$w_5 = 8^{1/6}\left(\cos\frac{11\pi}{6} + i\sin\frac{11\pi}{6}\right) = \sqrt{2}\left(\frac{\sqrt{3}}{2} - \frac{1}{2}i\right)$$

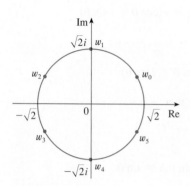

FIGURE 9

The six sixth roots of $z = -8$

All these points lie on the circle of radius $\sqrt{2}$ as shown in Figure 9. ■ ■

■■ Complex Exponentials

We also need to give a meaning to the expression e^z when $z = x + iy$ is a complex number. The theory of infinite series as developed in Chapter 8 can be extended to the case where the terms are complex numbers. Using the Taylor series for e^x (8.7.11) as our guide, we define

$$\boxed{4} \qquad e^z = \sum_{n=0}^{\infty} \frac{z^n}{n!} = 1 + z + \frac{z^2}{2!} + \frac{z^3}{3!} + \cdots$$

and it turns out that this complex exponential function has the same properties as the real exponential function. In particular, it is true that

$$\boxed{5} \qquad e^{z_1+z_2} = e^{z_1}e^{z_2}$$

If we put $z = iy$, where y is a real number, in Equation 4, and use the facts that

$$i^2 = -1, \quad i^3 = i^2 i = -i, \quad i^4 = 1, \quad i^5 = i, \quad \ldots$$

we get

$$e^{iy} = 1 + iy + \frac{(iy)^2}{2!} + \frac{(iy)^3}{3!} + \frac{(iy)^4}{4!} + \frac{(iy)^5}{5!} + \cdots$$

$$= 1 + iy - \frac{y^2}{2!} - i\frac{y^3}{3!} + \frac{y^4}{4!} + i\frac{y^5}{5!} + \cdots$$

$$= \left(1 - \frac{y^2}{2!} + \frac{y^4}{4!} - \frac{y^6}{6!} + \cdots\right) + i\left(y - \frac{y^3}{3!} + \frac{y^5}{5!} - \cdots\right)$$

$$= \cos y + i\sin y$$

Here we have used the Taylor series for cos y and sin y (Equations 8.7.16 and 8.7.15). The result is a famous formula called **Euler's formula**:

6
$$e^{iy} = \cos y + i \sin y$$

Combining Euler's formula with Equation 5, we get

7
$$e^{x+iy} = e^x e^{iy} = e^x(\cos y + i \sin y)$$

EXAMPLE 8 Evaluate: (a) $e^{i\pi}$ (b) $e^{-1+i\pi/2}$

■ ■ We could write the result of Example 8(a) as
$$e^{i\pi} + 1 = 0$$
This equation relates the five most famous numbers in all of mathematics: 0, 1, e, i, and π.

SOLUTION

(a) From Euler's equation (6) we have
$$e^{i\pi} = \cos \pi + i \sin \pi = -1 + i(0) = -1$$

(b) Using Equation 7 we get
$$e^{-1+i\pi/2} = e^{-1}\left(\cos \frac{\pi}{2} + i \sin \frac{\pi}{2}\right) = \frac{1}{e}[0 + i(1)] = \frac{i}{e}$$

■ ■

Finally, we note that Euler's equation provides us with an easier method of proving De Moivre's Theorem:
$$[r(\cos \theta + i \sin \theta)]^n = (re^{i\theta})^n = r^n e^{in\theta} = r^n(\cos n\theta + i \sin n\theta)$$

I Exercises

1–14 ■ Evaluate the expression and write your answer in the form $a + bi$.

1. $(5 - 6i) + (3 + 2i)$

2. $\left(4 - \frac{1}{2}i\right) - \left(9 + \frac{5}{2}i\right)$

3. $(2 + 5i)(4 - i)$

4. $(1 - 2i)(8 - 3i)$

5. $\overline{12 + 7i}$

6. $\overline{2i\left(\frac{1}{2} - i\right)}$

7. $\dfrac{1 + 4i}{3 + 2i}$

8. $\dfrac{3 + 2i}{1 - 4i}$

9. $\dfrac{1}{1 + i}$

10. $\dfrac{3}{4 - 3i}$

11. i^3

12. i^{100}

13. $\sqrt{-25}$

14. $\sqrt{-3}\sqrt{-12}$

■   ■   ■   ■   ■   ■   ■   ■   ■   ■   ■   ■   ■

15–17 ■ Find the complex conjugate and the modulus of the number.

15. $12 - 5i$

16. $-1 + 2\sqrt{2}\,i$

17. $-4i$

■   ■   ■   ■   ■   ■   ■   ■   ■   ■   ■   ■   ■

18. Prove the following properties of complex numbers.
   (a) $\overline{z + w} = \bar{z} + \bar{w}$ (b) $\overline{zw} = \bar{z}\,\bar{w}$
   (c) $\overline{z^n} = \bar{z}^n$, where n is a positive integer
       [*Hint:* Write $z = a + bi$, $w = c + di$.]

19–24 ■ Find all solutions of the equation.

19. $4x^2 + 9 = 0$

20. $x^4 = 1$

21. $x^2 + 2x + 5 = 0$

22. $2x^2 - 2x + 1 = 0$

23. $z^2 + z + 2 = 0$

24. $z^2 + \frac{1}{2}z + \frac{1}{4} = 0$

■   ■   ■   ■   ■   ■   ■   ■   ■   ■   ■   ■   ■

25–28 ■ Write the number in polar form with argument between 0 and 2π.

25. $-3 + 3i$

26. $1 - \sqrt{3}\,i$

27. $3 + 4i$

28. $8i$

■   ■   ■   ■   ■   ■   ■   ■   ■   ■   ■   ■   ■

29–32 ■ Find polar forms for zw, z/w, and $1/z$ by first putting z and w into polar form.

29. $z = \sqrt{3} + i$, $\quad w = 1 + \sqrt{3}\,i$

30. $z = 4\sqrt{3} - 4i$, $\quad w = 8i$

31. $z = 2\sqrt{3} - 2i, \quad w = -1 + i$

32. $z = 4(\sqrt{3} + i), \quad w = -3 - 3i$

· · · · · · · · · · · · · · · · · ·

33–36 ■ Find the indicated power using De Moivre's Theorem.

33. $(1 + i)^{20}$

34. $(1 - \sqrt{3}i)^5$

35. $(2\sqrt{3} + 2i)^5$

36. $(1 - i)^8$

· · · · · · · · · · · · · · · · · ·

37–40 ■ Find the indicated roots. Sketch the roots in the complex plane.

37. The eighth roots of 1

38. The fifth roots of 32

39. The cube roots of i

40. The cube roots of $1 + i$

· · · · · · · · · · · · · · · · · ·

41–46 ■ Write the number in the form $a + bi$.

41. $e^{i\pi/2}$

42. $e^{2\pi i}$

43. $e^{i\pi/3}$

44. $e^{-i\pi}$

45. $e^{2+i\pi}$

46. $e^{\pi+i}$

· · · · · · · · · · · · · · · · · ·

47. Use De Moivre's Theorem with $n = 3$ to express $\cos 3\theta$ and $\sin 3\theta$ in terms of $\cos \theta$ and $\sin \theta$.

48. Use Euler's formula to prove the following formulas for $\cos x$ and $\sin x$:

$$\cos x = \frac{e^{ix} + e^{-ix}}{2} \qquad \sin x = \frac{e^{ix} - e^{-ix}}{2i}$$

49. If $u(x) = f(x) + ig(x)$ is a complex-valued function of a real variable x and the real and imaginary parts $f(x)$ and $g(x)$ are differentiable functions of x, then the derivative of u is defined to be $u'(x) = f'(x) + ig'(x)$. Use this together with Equation 7 to prove that if $F(x) = e^{rx}$, then $F'(x) = re^{rx}$ when $r = a + bi$ is a complex number.

50. (a) If u is a complex-valued function of a real variable, its indefinite integral $\int u(x)\, dx$ is an antiderivative of u. Evaluate

$$\int e^{(1+i)x}\, dx$$

(b) By considering the real and imaginary parts of the integral in part (a), evaluate the real integrals

$$\int e^x \cos x\, dx \qquad \text{and} \qquad \int e^x \sin x\, dx$$

(c) Compare with the method used in Example 4 in Section 5.6.

J | Answers to Odd-Numbered Exercises

CHAPTER 8

Exercises 8.1 □ page 565

Abbreviations: C, convergent; D, divergent

1. (a) A sequence is an ordered list of numbers. It can also be defined as a function whose domain is the set of positive integers.
(b) The terms a_n approach 8 as n becomes large.
(c) The terms a_n become large as n becomes large.
3. $\frac{1}{3}, \frac{2}{5}, \frac{3}{7}, \frac{4}{9}, \frac{5}{11}, \frac{6}{13}$; yes; $\frac{1}{2}$ **5.** $a_n = \left(-\frac{2}{3}\right)^{n-1}$
7. $a_n = 5n - 3$ **9.** 5   **11.** 0
13. D   **15.** 0   **17.** 0   **19.** 0   **21.** 0
23. e^2 **25.** D   **27.** ln 2   **29.** D   **31.** $\pi/4$ **33.** 0
35. (a) 1060, 1123.60, 1191.02, 1262.48, 1338.23   (b) D
37. (a) D   (b) C   **39.** (b) $\left(1 + \sqrt{5}\right)/2$
41. Decreasing; yes   **43.** Not monotonic; yes
45. Convergent by the Monotonic Sequence Theorem; $5 \leqslant L < 8$
47. $\left(3 + \sqrt{5}\right)/2$ **49.** 62

Exercises 8.2 □ page 574

1. (a) A sequence is an ordered list of numbers whereas a series is the *sum* of a list of numbers.
(b) A series is convergent if the sequence of partial sums is a convergent sequence. A series is divergent if it is not convergent.
3. $-2.40000, -1.92000$
$-2.01600, -1.99680,$
$-2.00064, -1.99987,$
$-2.00003, -1.99999,$
$-2.00000, -2.00000;$
convergent, sum $= -2$

5. $1.55741, -0.62763,$
$-0.77018, 0.38764,$
$-2.99287, -3.28388,$
$-2.41243, -9.21214,$
$-9.66446, -9.01610;$
divergent

7. $0.64645, 0.80755,$
$0.87500, 0.91056,$
$0.93196, 0.94601,$
$0.95581, 0.96296,$
$0.96838, 0.97259;$
convergent, sum $= 1$

9. (a) C   (b) D   **11.** 3   **13.** 15   **15.** D   **17.** D
19. D   **21.** $\frac{5}{2}$ **23.** D   **25.** D   **27.** $\frac{3}{2}$
29. $\frac{11}{6}$ **31.** $\frac{2}{9}$ **33.** $\frac{1138}{333}$ **35.** $-3 < x < 3; x/(3 - x)$
37. All x; $\dfrac{2}{2 - \cos x}$ **39.** $\frac{1}{4}$
41. $a_1 = 0, a_n = 2/[n(n + 1)]$ for $n > 1$, sum $= 1$
43. (a) $S_n = D(1 - c^n)/(1 - c)$ (b) 5
45. $\left(\sqrt{3} - 1\right)/2$ **47.** $1/[n(n + 1)]$
49. The series is divergent.
53. $\{s_n\}$ is bounded and increasing.
55. (a) $0, \frac{1}{9}, \frac{2}{9}, \frac{1}{3}, \frac{2}{3}, \frac{7}{9}, \frac{8}{9}, 1$
57. (a) $\frac{1}{2}, \frac{5}{6}, \frac{23}{24}, \frac{119}{120}; [(n + 1)! - 1]/(n + 1)!$ (c) 1

Exercises 8.3 □ page 585

1. C

3. (a) Nothing   (b) C
5. p-series; geometric series; $b < -1; -1 < b < 1$
7. C   **9.** C   **11.** C   **13.** C   **15.** D   **17.** C
19. C   **21.** D   **23.** C   **25.** D   **27.** $p > 1$
29. (a) 1.54977, error $\leqslant 0.1$ (b) 1.64522, error $\leqslant 0.005$
(c) $n > 1000$
31. 2.61   **33.** 0.567975, error $\leqslant 0.000\overline{3}$ **39.** Yes

Exercises 8.4 □ page 592

1. (a) A series whose terms are alternately positive and negative
(b) $0 < b_{n+1} \leqslant b_n$ and $\lim_{n \to \infty} b_n = 0$, where $b_n = |a_n|$
(c) $|R_n| \leqslant b_{n+1}$
3. C   **5.** C   **7.** D   **9.** An underestimate
11. $p > 0$ **13.** 5
15. 0.8415

17. 0.0676   **19.** No   **21.** Yes   **23.** No   **25.** Yes
27. Yes   **29.** Yes   **31.** D   **33.** (a) and (d)

Exercises 8.5 □ **page 598**

1. A series of the form $\sum_{n=0}^{\infty} c_n(x - a)^n$, where x is a variable and a and the c_n's are constants

3. $1, [-1, 1)$　　**5.** $1, [-1, 1]$　　**7.** $\infty, (-\infty, \infty)$

9. $\frac{1}{2}, \left(-\frac{1}{2}, \frac{1}{2}\right]$　　**11.** $4, (-4, 4]$　　**13.** $2, (-4, 0]$

15. $b, (a - b, a + b)$　　**17.** $0, \left\{\frac{1}{2}\right\}$

19. (a) Yes　　(b) No　　**21.** k^k

23. (a) $(-\infty, \infty)$　　(b), (c)

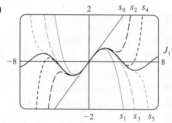

25. $(-1, 1), f(x) = (1 + 2x)/(1 - x^2)$　　**27.** 2

Exercises 8.6 □ **page 604**

1. 10　　**3.** $\sum_{n=0}^{\infty} (-1)^n x^n, (-1, 1)$　　**5.** $\sum_{n=0}^{\infty} x^{3n}, (-1, 1)$

7. $-\sum_{n=0}^{\infty} \frac{1}{5^{n+1}} x^n, (-5, 5)$　　**9.** $\sum_{n=0}^{\infty} (-1)^n \frac{1}{9^{n+1}} x^{2n+1}, (-3, 3)$

11. (a) $\sum_{n=0}^{\infty} (-1)^n (n + 1) x^n, R = 1$

(b) $\frac{1}{2} \sum_{n=0}^{\infty} (-1)^n (n + 2)(n + 1) x^n, R = 1$

(c) $\frac{1}{2} \sum_{n=2}^{\infty} (-1)^n n(n - 1) x^n, R = 1$

13. $\ln 5 - \sum_{n=1}^{\infty} \frac{x^n}{n5^n}, R = 5$　　**15.** $\sum_{n=3}^{\infty} \frac{n - 2}{2^{n-1}} x^n, R = 2$

17. $\ln 3 + \sum_{n=1}^{\infty} \frac{(-1)^{n-1}}{n3^n} x^n, R = 3$

19. $\sum_{n=0}^{\infty} \frac{2x^{2n+1}}{2n + 1}, R = 1$

21. $C + \sum_{n=0}^{\infty} \frac{t^{8n+2}}{8n + 2}, R = 1$

23. $C + \sum_{n=1}^{\infty} (-1)^{n+1} \frac{x^{2n-1}}{4n^2 - 1}, R = 1$

25. 0.199989　　**27.** 0.000983　　**29.** 0.09531

31. (b) 0.920　　**35.** $[-1, 1], [-1, 1), (-1, 1)$

Exercises 8.7 □ **page 615**

1. $b_8 = f^{(8)}(5)/8!$　　**3.** $\sum_{n=0}^{\infty} (n + 1)x^n, R = 1$

5. $\sum_{n=0}^{\infty} (-1)^n \frac{x^{2n}}{(2n)!}, R = \infty$　　**7.** $\sum_{n=0}^{\infty} \frac{5^n}{n!} x^n, R = \infty$

9. $7 + 5(x - 2) + (x - 2)^2, R = \infty$

11. $\sum_{n=0}^{\infty} \frac{e^3}{n!} (x - 3)^n, R = \infty$

13. $\sum_{n=0}^{\infty} (-1)^{n+1} \frac{1}{(2n)!} (x - \pi)^{2n}, R = \infty$

15. $\sum_{n=0}^{\infty} (-1)^n \frac{1 \cdot 3 \cdot 5 \cdot \cdots \cdot (2n - 1)}{2^n \cdot 3^{2n+1} \cdot n!} (x - 9)^n, R = 9$

19. $\sum_{n=0}^{\infty} (-1)^n \frac{\pi^{2n}}{(2n)!} x^{2n}, R = \infty$

21. $\sum_{n=0}^{\infty} (-1)^n \frac{1}{2n + 1} x^{2n+2}, R = 1$

23. $\sum_{n=0}^{\infty} (-1)^n \frac{1}{n!} x^{n+2}, R = \infty$

25. $\sum_{n=1}^{\infty} \frac{(-1)^{n+1} 2^{2n-1} x^{2n}}{(2n)!}, R = \infty$

27. $1 + \frac{x}{2} + \sum_{n=2}^{\infty} (-1)^{n-1} \frac{1 \cdot 3 \cdot 5 \cdot \cdots \cdot (2n - 3)}{2^n n!} x^n, R = 1$

29. $\sum_{n=0}^{\infty} (-1)^n \frac{1}{(2n)!} x^{4n}, R = \infty$

31. 0.81873

33. $C + \sum_{n=0}^{\infty} (-1)^n \frac{x^{6n+2}}{(6n + 2)(2n)!}$

35. $C + x + \frac{x^4}{8} + \sum_{n=2}^{\infty} (-1)^{n-1} \frac{1 \cdot 3 \cdot 5 \cdot \cdots \cdot (2n - 3)}{2^n n!(3n + 1)} x^{3n+1}$

37. 0.440　　**39.** 0.09998750　　**41.** $\frac{1}{3}$　　**43.** $\frac{1}{120}$

45. $1 - \frac{3}{2}x^2 + \frac{25}{24}x^4$ **47.** $1 + \frac{1}{6}x^2 + \frac{7}{360}x^4$
49. e^{-x^4} **51.** $1/\sqrt{2}$ **53.** $e^3 - 1$

Exercises 8.8 □ **page 620**

1. $1 + \frac{x}{2} + \sum_{n=2}^{\infty} (-1)^{n-1} \frac{1 \cdot 3 \cdot 5 \cdot \cdots \cdot (2n-3)}{2^n n!} x^n, R = 1$

3. $\sum_{n=0}^{\infty} (-1)^n \frac{(n+1)(n+2)}{2^{n+4}} x^n, R = 2$

5. $\frac{1}{2}x + \sum_{n=1}^{\infty} (-1)^n \frac{1 \cdot 3 \cdot 5 \cdot \cdots \cdot (2n-1)}{n! \, 2^{3n+1}} x^{2n+1}, R = 2$

7. $1 + \frac{3}{2}x, \; 1 + \frac{3}{2}x - \frac{3}{8}x^2, \; 1 + \frac{3}{2}x - \frac{3}{8}x^2 + \frac{5}{16}x^3$

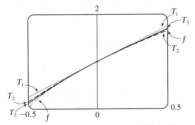

9. (a) $1 + \sum_{n=1}^{\infty} \frac{1 \cdot 3 \cdot 5 \cdot \cdots \cdot (2n-1)}{2^n n!} x^{2n}$

(b) $x + \sum_{n=1}^{\infty} \frac{1 \cdot 3 \cdot 5 \cdot \cdots \cdot (2n-1)}{2^n n!} \frac{x^{2n+1}}{2n+1}$

11. (a) $\sum_{n=1}^{\infty} nx^n$ (b) 2

13. (a) $1 + \frac{x^2}{2} + \sum_{n=2}^{\infty} (-1)^{n-1} \frac{1 \cdot 3 \cdot 5 \cdot \cdots \cdot (2n-3)}{2^n n!} x^{2n}$

(b) 99,225

Exercises 8.9 □ **page 628**

1. (a) $T_0(x) = 1 = T_1(x), \; T_2(x) = 1 - \frac{1}{2}x^2 = T_3(x),$
$T_4(x) = 1 - \frac{1}{2}x^2 + \frac{1}{24}x^4 = T_5(x),$
$T_6(x) = 1 - \frac{1}{2}x^2 + \frac{1}{24}x^4 - \frac{1}{720}x^6$

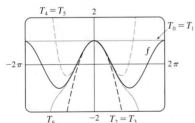

(b)

x	f	$T_0 = T_1$	$T_2 = T_3$	$T_4 = T_5$	T_6
$\frac{\pi}{4}$	0.7071	1	0.6916	0.7074	0.7071
$\frac{\pi}{2}$	0	1	−0.2337	0.0200	−0.0009
π	−1	1	−3.9348	0.1239	−1.2114

(c) As n increases, $T_n(x)$ is a good approximation to $f(x)$ on a larger and larger interval.

3. $\frac{1}{2} + \frac{\sqrt{3}}{2}\left(x - \frac{\pi}{6}\right) - \frac{1}{4}\left(x - \frac{\pi}{6}\right)^2 - \frac{\sqrt{3}}{12}\left(x - \frac{\pi}{6}\right)^3$

5. $x + \frac{1}{6}x^3$

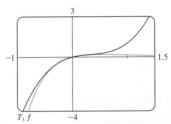

7. $x - 2x^2 + 2x^3$

9. $T_8(x) = 1 + \frac{1}{2}x^2 + \frac{5}{24}x^4 + \frac{61}{720}x^6 + \frac{277}{8064}x^8$

11. (a) $2 + \frac{1}{4}(x - 4) - \frac{1}{64}(x - 4)^2$ (b) 1.5625×10^{-5}
13. (a) $1 + \frac{2}{3}(x - 1) - \frac{1}{9}(x - 1)^2 + \frac{4}{81}(x - 1)^3$
(b) 0.000097
15. (a) $1 + x^2$ (b) 0.00006
17. (a) $x^2 - \frac{1}{6}x^4$ (b) 0.042
19. 0.57358 **21.** Four
23. $-1.037 < x < 1.037$ **25.** 21 m, no
29. (c) They differ by about 8×10^{-9} km.

Chapter 8 Review □ **page 632**

True-False Quiz

1. False **3.** True **5.** False **7.** False **9.** False
11. True **13.** True **15.** False **17.** True

Exercises

1. $\frac{1}{2}$ **3.** D   **5.** 0   **7.** e^{12} **9.** C   **11.** C
13. D   **15.** C   **17.** C   **19.** 8   **21.** $\pi/4$ **23.** $\frac{4111}{3330}$
25. 0.9721   **27.** 0.18976224, $|\text{error}| < 6.4 \times 10^{-7}$
31. 4, $[-6, 2)$ **33.** 0.5, $[2.5, 3.5)$

35. $\frac{1}{2} \sum_{n=0}^{\infty} (-1)^n \left[\frac{1}{(2n)!} \left(x - \frac{\pi}{6} \right)^{2n} + \frac{\sqrt{3}}{(2n+1)!} \left(x - \frac{\pi}{6} \right)^{2n+1} \right]$

37. $\sum_{n=0}^{\infty} (-1)^n x^{n+2}, R = 1$ **39.** $-\sum_{n=1}^{\infty} \frac{x^n}{n}, R = 1$

41. $\sum_{n=0}^{\infty} (-1)^n \frac{x^{8n+4}}{(2n+1)!}, R = \infty$

43. $\frac{1}{2} + \sum_{n=1}^{\infty} \frac{1 \cdot 5 \cdot 9 \cdots (4n-3)}{n! 2^{6n+1}} x^n, R = 16$

45. $C + \ln|x| + \sum_{n=1}^{\infty} \frac{x^n}{n \cdot n!}$

47. (a) $1 + \frac{1}{2}(x-1) - \frac{1}{8}(x-1)^2 + \frac{1}{16}(x-1)^3$
(b)   (c) 0.000006

49. $-\frac{1}{6}$
51. (b) 0 if $x = 0$, $(1/x) - \cot x$ if $x \neq k\pi$, k an integer

Focus on Problem Solving □ page 634

1. $15!/5! = 10,897,286,400$
3. (a) $s_n = 3 \cdot 4^n, l_n = 1/3^n, p_n = 4^n/3^{n-1}$ (c) $2\sqrt{3}/5$
5. $\ln \frac{1}{2}$

CHAPTER 9

Exercises 9.1 □ page 641

1. $(4, 0, -3)$ **3.** $Q; R$
5. A vertical plane that
intersects the xy-plane in
the line $y = 2 - x, z = 0$
(see graph at right)

7. (a) $|PQ| = 6, |QR| = 2\sqrt{10}, |RP| = 6$;
isosceles triangle
(b) $|PQ| = 3, |QR| = 3\sqrt{5}, |RP| = 6$; right triangle
9. (a) No   (b) Yes
11. $(x-3)^2 + (y-8)^2 + (z-1)^2 = 30$
13. $(3, -2, 1), 5$ **15.** (b) $\frac{5}{2}, \frac{1}{2}\sqrt{94}, \frac{1}{2}\sqrt{85}$
17. (a) $(x-2)^2 + (y+3)^2 + (z-6)^2 = 36$
(b) $(x-2)^2 + (y+3)^2 + (z-6)^2 = 4$
(c) $(x-2)^2 + (y+3)^2 + (z-6)^2 = 9$

19. A plane parallel to the xz-plane and 4 units to the left of it
21. A half-space consisting of all points in front of the plane
$x = 3$
23. All points on or between the horizontal planes $z = 0$ and
$z = 6$
25. All points on or inside a sphere with radius $\sqrt{3}$ and center O
27. All points on or inside a circular cylinder of radius 3 with
axis the y-axis
29. $y < 0$ **31.** $r^2 < x^2 + y^2 + z^2 < R^2$
33. (a) $(2, 1, 4)$ (b)

35. $14x - 6y - 10z = 9$, a plane perpendicular to AB

Exercises 9.2 □ page 649

1. (a) Scalar   (b) Vector   (c) Vector   (d) Scalar
3. $\vec{AB} = \vec{DC}, \vec{DA} = \vec{CB}, \vec{DE} = \vec{EB}, \vec{EA} = \vec{CE}$
5. (a)   (b)

(c)   (d)

7. $\mathbf{a} = \langle -4, -2 \rangle$ **9.** $\mathbf{a} = \langle 2, 0, -2 \rangle$

11. $\langle 1, 3 \rangle$ **13.** $\langle 0, 1, -1 \rangle$

15. $\langle 2, -18 \rangle, \langle 1, -42 \rangle, 13, 10$
17. $-\mathbf{i} + \mathbf{j} + 2\mathbf{k}, -4\mathbf{i} + \mathbf{j} + 9\mathbf{k}, \sqrt{14}, \sqrt{82}$
19. $\frac{8}{9}\mathbf{i} - \frac{1}{9}\mathbf{j} + \frac{4}{9}\mathbf{k}$ **21.** $\langle 2, 2\sqrt{3} \rangle$
23. $\mathbf{F} = (6\sqrt{3} - 5\sqrt{2})\mathbf{i} + (6 + 5\sqrt{2})\mathbf{j} \approx 3.32\mathbf{i} + 13.07\mathbf{j}$,
$|\mathbf{F}| \approx 13.5$ lb, $\theta \approx 76°$
25. $\sqrt{493} \approx 22.2$ mi/h, N8°W

27. $T_1 \approx -196\,i + 3.92\,j$, $T_2 \approx 196\,i + 3.92\,j$

29. (a), (b)   (d) $s = \frac{9}{7}$, $t = \frac{11}{7}$

31. $a \approx \langle 0.50, 0.31, 0.81 \rangle$

33. A sphere with radius 1, centered at (x_0, y_0, z_0)

Exercises 9.3 □ page 655

1. (b), (c), (d) are meaningful

3. -15 **5.** 19   **7.** 32   **9.** $u \cdot v = \frac{1}{2}$, $u \cdot w = -\frac{1}{2}$

13. $\cos^{-1}\left(\dfrac{9 - 4\sqrt{7}}{20}\right) \approx 95°$ **15.** $\cos^{-1}\left(\dfrac{-1}{2\sqrt{7}}\right) \approx 101°$

17. (a) Neither   (b) Orthogonal
(c) Orthogonal   (d) Parallel

19. Yes   **21.** $(i - j - k)/\sqrt{3}$ $\left[\text{or } (-i + j + k)/\sqrt{3}\right]$

23. 3, $\left\langle \frac{9}{5}, -\frac{12}{5} \right\rangle$ **25.** $\frac{9}{7}$, $\left\langle \frac{27}{49}, \frac{54}{49}, -\frac{18}{49} \right\rangle$

29. $\langle 0, 0, -2\sqrt{10} \rangle$ or any vector of the form $\langle s, t, 3s - 2\sqrt{10} \rangle$, $s, t \in \mathbb{R}$

31. 38 J   **33.** $250 \cos 20° \approx 235$ ft-lb   **35.** $\frac{13}{5}$

37. $\cos^{-1}\left(1/\sqrt{3}\right) \approx 55°$

Exercises 9.4 □ page 664

1. (a) Scalar   (b) Meaningless   (c) Vector
(d) Meaningless   (e) Meaningless   (f) Scalar

3. 24; into the page   **5.** $10.8 \sin 80° \approx 10.6$ N·m

7. $2i - j + 3k$ **9.** $t^4 i - 2t^3 j + t^2 k$

11. $2i + 13j - 8k$

13. $\langle 12/\sqrt{209}, -1/\sqrt{209}, 8/\sqrt{209} \rangle$, $\langle -12/\sqrt{209}, 1/\sqrt{209}, -8/\sqrt{209} \rangle$

15. 16   **17.** (a) $\langle 13, -14, 5 \rangle$ (b) $\frac{1}{2}\sqrt{390}$

19. ≈ 417 N   **21.** 82   **23.** 3

33. (a) No   (b) No   (c) Yes

Exercises 9.5 □ page 673

1. (a) True   (b) False   (c) True   (d) False   (e) False
(f) True   (g) False   (h) True   (i) True   (j) False
(k) True

3. $r = (-2i + 4j + 10k) + t(3i + j - 8k)$;
$x = -2 + 3t$, $y = 4 + t$, $z = 10 - 8t$

5. $r = (i + 6k) + t(i + 3j + k)$;
$x = 1 + t$, $y = 3t$, $z = 6 + t$

7. $x = 2 + 2t$, $y = 1 + \frac{1}{2}t$, $z = -3 - 4t$;
$(x - 2)/2 = 2y - 2 = (z + 3)/(-4)$

9. $x = 1 + t$, $y = -1 + 2t$, $z = 1 + t$;
$x - 1 = (y + 1)/2 = z - 1$

11. Yes

13. (a) $x/2 = (y - 2)/3 = (z + 1)/(-7)$
(b) $\left(-\frac{2}{7}, \frac{11}{7}, 0\right)$, $\left(-\frac{4}{3}, 0, \frac{11}{3}\right)$, $(0, 2, -1)$

15. $r(t) = (2i - j + 4k) + t(2i + 7j - 3k)$, $0 \leq t \leq 1$

17. Parallel   **19.** Skew

21. $-2x + y + 5z = 1$ **23.** $2x - y + 3z = 0$

25. $x + y + z = 2$ **27.** $33x + 10y + 4z = 190$

29. $x - 2y + 4z = -1$ **31.** $(2, 3, 5)$

33. Neither, $\approx 70.5°$ **35.** Parallel

37. (a) $x - 2 = y/(-8) = z/(-7)$
(b) $\cos^{-1}(-\sqrt{6/5}) \approx 119°$ (or 61°)

39. $(x/a) + (y/b) + (z/c) = 1$

41. $x = 3t$, $y = 1 - t$, $z = 2 - 2t$

43. P_1 and P_3 are parallel, P_2 and P_4 are identical

45. $\sqrt{22/5}$ **47.** $\frac{25}{3}$ **49.** $7\sqrt{6}/18$ **53.** $1/\sqrt{6}$

Exercises 9.6 □ page 683

1. (a) 25; a 40-knot wind blowing in the open sea for 15 h will create waves about 25 ft high.
(b) $f(30, t)$ is a function of t giving the wave heights produced by 30-knot winds blowing for t hours.
(c) $f(v, 30)$ is a function of v giving the wave heights produced by winds of speed v blowing for 30 hours.

3. (a) 4   (b) $\mathbb{R}^2$ (c) $[0, \infty)$

5. $\{(x, y) \mid y \geq x^2, x \neq \pm 1\}$

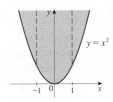

7. $\{(x, y) \mid -1 \leq x \leq 1, -1 \leq y \leq 1\}$

9. $z = 3$, horizontal plane   **11.** $3x + 2y + z = 6$, plane

13. $z = y^2 + 1$, parabolic cylinder

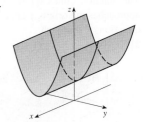

15. (a) VI   (b) V   (c) I   (d) IV   (e) II   (f) III

17. $z = \sqrt{4x^2 + y^2}$

19.

21. $x^2 + \dfrac{(y-2)^2}{4} + (z-3)^2 = 1$

Ellipsoid with center (0, 2, 3)

23. (a) A circle of radius 1 centered at the origin
(b) A circular cylinder of radius 1 with axis the z-axis
(c) A circular cylinder of radius 1 with axis the y-axis

25. (a) $x = k$, $y^2 - z^2 = 1 - k^2$, hyperbola ($k \neq \pm 1$);
$y = k$, $x^2 - z^2 = 1 - k^2$, hyperbola ($k \neq \pm 1$);
$z = k$, $x^2 + z^2 = 1 + k^2$, circle
(b) The hyperboloid is rotated so that it has axis the y-axis
(c) The hyperboloid is shifted one unit in the negative y-direction

27.

f appears to have a maximum value of about 15. There are two local maximum points but no local minimum point.

29.

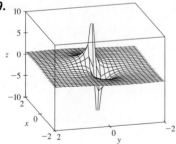

The function values approach 0 as x, y become large; as (x, y) approaches the origin, f approaches $\pm\infty$ or 0, depending on the direction of approach.

31.

Exercises 9.7 □ **page 689**

1. See pages 685–86.

3.

$\left(\sqrt{2}, \sqrt{2}, 1\right)$

5. (a) $\left(\sqrt{2}, 7\pi/4, 4\right)$ (b) $\left(2, 4\pi/3, 2\right)$

7. (a)

(0, 0, 1)

(b)

$\left(\tfrac{1}{2}\sqrt{2}, \tfrac{1}{2}\sqrt{6}, \sqrt{2}\right)$

9. (a) $(4, \pi/3, \pi/6)$ (b) $\left(\sqrt{2}, 3\pi/2, 3\pi/4\right)$
11. Circular cylinder, radius 3, axis the z-axis
13. Half-cone   **15.** Circular paraboloid
17. Circular cylinder, radius 1, axis parallel to the z-axis
19. Sphere, radius 5, center the origin
21. (a) $z = r^2$ (b) $\rho \sin^2\phi = \cos\phi$
23. (a) $r = 2\sin\theta$ (b) $\rho\sin\phi = 2\sin\theta$
25.

27.

29.

31. Cylindrical coordinates:
$6 \leq r \leq 7$, $0 \leq \theta \leq 2\pi$,
$0 \leq z \leq 20$

33. $0 \leq \phi \leq \pi/4$, $0 \leq \rho \leq \cos\phi$ **35.**

Chapter 9 Review □ page 691

True-False Quiz

1. True **3.** True **5.** True **7.** True **9.** True
11. False **13.** False **15.** False

Exercises

1. (a) $(x + 1)^2 + (y - 2)^2 + (z - 1)^2 = 69$
(b) $(y - 2)^2 + (z - 1)^2 = 68, x = 0$
(c) center $(4, -1, -3)$, radius 5
3. $\mathbf{u} \cdot \mathbf{v} = 3\sqrt{2}$; $|\mathbf{u} \times \mathbf{v}| = 3\sqrt{2}$; out of the page
5. $-2, -4$ **7.** (a) 2 (b) -2 (c) -2 (d) 0
9. $\cos^{-1}\left(\frac{1}{3}\right) \approx 71°$ **11.** (a) $\langle 4, -3, 4 \rangle$ (b) $\sqrt{41}/2$
13. 166 N, 114 N
15. $x = 4 - 3t, y = -1 + 2t, z = 2 + 3t$
17. $x = -2 + 2t, y = 2 - t, z = 4 + 5t$
19. $-4x + 3y + z = -14$ **21.** $x + y + z = 4$
23. Skew **25.** $22/\sqrt{26}$
27. $\{(x, y) \mid x > y^2\}$

29.

31.

33. Ellipsoid

35. Circular cylinder

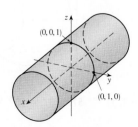

37. $\left(\sqrt{3}, 3, 2\right), (4, \pi/3, \pi/3)$
39. $\left(2\sqrt{2}, 2\sqrt{2}, 4\sqrt{3}\right), \left(4, \pi/4, 4\sqrt{3}\right)$
41. $r^2 + z^2 = 4, \rho = 2$ **43.** $z = 4r^2$

Focus on Problem Solving □ page 693

1. $\left(\sqrt{3} - 1.5\right)$ m
3. (a) $(x + 1)/(-2c) = (y - c)/(c^2 - 1) = (z - c)/(c^2 + 1)$
(b) $x^2 + y^2 = t^2 + 1, z = t$ (c) $4\pi/3$

CHAPTER 10

Exercises 10.1 □ page 700

1. $[1, 5]$ **3.** $\langle 1, 0, 0 \rangle$
5.

7.

9.

11.

13. $\mathbf{r}(t) = \langle t, 2t, 3t \rangle, 0 \leq t \leq 1$;
$x = t, y = 2t, z = 3t, 0 \leq t \leq 1$
15. $\mathbf{r}(t) = \langle 3t + 1, 2t - 1, 5t + 2 \rangle, 0 \leq t \leq 1$;
$x = 3t + 1, y = 2t - 1, z = 5t + 2, 0 \leq t \leq 1$
17. VI **19.** IV **21.** V
23.

25. $(0, 0, 0), (1, 0, 1)$

27.

29.

33. $\mathbf{r}(t) = t\,\mathbf{i} + \frac{1}{2}(t^2 - 1)\,\mathbf{j} + \frac{1}{2}(t^2 + 1)\,\mathbf{k}$
35. $x = 2\cos t,\ y = 2\sin t,\ z = 4\cos^2 t$ **37.** Yes

Exercises 10.2 □ **page 707**

1. (a)

(b), (d)

(c) $\mathbf{r}'(4) = \lim\limits_{h \to 0} \dfrac{\mathbf{r}(4 + h) - \mathbf{r}(4)}{h}$; $\mathbf{T}(4) = \dfrac{\mathbf{r}'(4)}{|\mathbf{r}'(4)|}$

3. (a), (c)

(b) $\mathbf{r}'(t) = \langle 1, 2t \rangle$

5. (a), (c)

7. (a), (c)

(b) $\mathbf{r}'(t) = \cos t\,\mathbf{i} - 2\sin t\,\mathbf{j}$ (b) $\mathbf{r}'(t) = e^t\,\mathbf{i} + 3e^{3t}\,\mathbf{j}$
9. $\mathbf{r}'(t) = \langle 2t, -1, 1/(2\sqrt{t}) \rangle$
11. $\mathbf{r}'(t) = 2te^{t^2}\,\mathbf{i} + [3/(1 + 3t)]\,\mathbf{k}$ **13.** $\mathbf{r}'(t) = \mathbf{b} + 2t\mathbf{c}$
15. $\frac{3}{5}\mathbf{j} + \frac{4}{5}\mathbf{k}$ **17.** $\langle 1, 2t, 3t^2 \rangle,\ \langle 1/\sqrt{14}, 2/\sqrt{14}, 3/\sqrt{14} \rangle,$
$\langle 0, 2, 6t \rangle,\ \langle 6t^2, -6t, 2 \rangle$
19. $x = 1 + 5t,\ y = 1 + 4t,\ z = 1 + 3t$
21. $x = 1 - t,\ y = t,\ z = 1 - t$
23. $x = t,\ y = 1 - t,\ z = 2t$

25. (a) Not smooth   (b) Smooth   (c) Not smooth
27. $66°$ **29.** $4\mathbf{i} - 3\mathbf{j} + 5\mathbf{k}$ **31.** $\mathbf{i} + \mathbf{j} + \mathbf{k}$
33. $e^t\,\mathbf{i} + t^2\,\mathbf{j} + (t\ln t - t)\,\mathbf{k} + \mathbf{C}$
35. $t^2\,\mathbf{i} + t^3\,\mathbf{j} + \left(\frac{2}{3}t^{3/2} - \frac{2}{3}\right)\mathbf{k}$
41. $1 - 4t\cos t + 11t^2\sin t + 3t^3\cos t$

Exercises 10.3 □ **page 714**

1. $20\sqrt{29}$ **3.** $\frac{1}{27}(13^{3/2} - 8)$ **5.** 9.5706

7. $\mathbf{r}(t(s)) = \dfrac{2}{\sqrt{29}}s\,\mathbf{i} + \left(1 - \dfrac{3}{\sqrt{29}}s\right)\mathbf{j} + \left(5 + \dfrac{4}{\sqrt{29}}s\right)\mathbf{k}$

9. $(3\sin 1, 4, 3\cos 1)$
11. (a) $\langle (2/\sqrt{29})\cos t, 5/\sqrt{29}, (-2/\sqrt{29})\sin t \rangle,$
$\langle -\sin t, 0, -\cos t \rangle$ (b) $\frac{2}{29}$

13. (a) $\dfrac{1}{e^{2t} + 1}\langle 1 - e^{2t}, \sqrt{2}e^t, \sqrt{2}e^t \rangle$ (b) $\sqrt{2}e^{2t}/(e^{2t} + 1)^2$

15. $2/(4t^2 + 1)^{3/2}$ **17.** $\frac{4}{25}$ **19.** $\frac{1}{7}\sqrt{\frac{19}{14}}$
21. $e^x|x + 2|/[1 + (x + 1)^2e^{2x}]^{3/2}$
23. $15\sqrt{x}/(1 + 100x^3)^{3/2}$
25. $(-\frac{1}{2}\ln 2, 1/\sqrt{2})$; approaches 0
27. (a) P (b) $1.3, 0.7$
29.

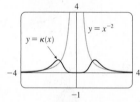

31. a is $y = f(x)$, b is $y = \kappa(x)$

33. $\kappa(t) = \dfrac{6\sqrt{4\cos^2 t - 12\cos t + 13}}{(17 - 12\cos t)^{3/2}}$

integer multiples of 2π

35. $1/(\sqrt{2}e^t)$ **37.** $\langle \frac{2}{3}, \frac{2}{3}, \frac{1}{3} \rangle, \langle -\frac{1}{3}, \frac{2}{3}, -\frac{2}{3} \rangle, \langle -\frac{2}{3}, \frac{1}{3}, \frac{2}{3} \rangle$
39. $y = 6x + \pi,\ x + 6y = 6\pi$
41. $\left(x + \frac{5}{2}\right)^2 + y^2 = \frac{81}{4},\ x^2 + \left(y - \frac{5}{3}\right)^2 = \frac{16}{9}$

43. $(-1, -3, 1)$ **51.** 2.07×10^{10} Å ≈ 2 m

Exercises 10.4 □ **page 725**

1. (a) $1.8\mathbf{i} - 3.8\mathbf{j} - 0.7\mathbf{k}$, $2.0\mathbf{i} - 2.4\mathbf{j} - 0.6\mathbf{k}$,
$2.8\mathbf{i} + 1.8\mathbf{j} - 0.3\mathbf{k}$, $2.8\mathbf{i} + 0.8\mathbf{j} - 0.4\mathbf{k}$
(b) $2.4\mathbf{i} - 0.8\mathbf{j} - 0.5\mathbf{k}$, 2.58

3. $\mathbf{v}(t) = \langle -t, 1 \rangle$
$\mathbf{a}(t) = \langle -1, 0 \rangle$
$|\mathbf{v}(t)| = \sqrt{t^2 + 1}$

5. $\mathbf{v}(t) = -3 \sin t\,\mathbf{i} + 2 \cos t\,\mathbf{j}$
$\mathbf{a}(t) = -3 \cos t\,\mathbf{i} - 2 \sin t\,\mathbf{j}$
$|\mathbf{v}(t)| = \sqrt{5 \sin^2 t + 4}$

7. $\mathbf{v}(t) = \mathbf{i} + 2t\,\mathbf{j}$
$\mathbf{a}(t) = 2\,\mathbf{j}$
$|\mathbf{v}(t)| = \sqrt{1 + 4t^2}$

9. $\langle 2t, 3t^2, 2t \rangle$, $\langle 2, 6t, 2 \rangle$, $|t|\sqrt{9t^2 + 8}$
11. $\sqrt{2}\,\mathbf{i} + e^t\,\mathbf{j} - e^{-t}\,\mathbf{k}$, $e^t\,\mathbf{j} + e^{-t}\,\mathbf{k}$, $e^t + e^{-t}$
13. $\mathbf{v}(t) = t\,\mathbf{i} + 2t\,\mathbf{j} + \mathbf{k}$, $\mathbf{r}(t) = \left(\tfrac{1}{2}t^2 + 1\right)\mathbf{i} + t^2\,\mathbf{j} + t\,\mathbf{k}$
15. (a) $\mathbf{r}(t) = \left(\tfrac{1}{3}t^3 + t\right)\mathbf{i} + (t - \sin t + 1)\,\mathbf{j} + \left(\tfrac{1}{4} - \tfrac{1}{4}\cos 2t\right)\mathbf{k}$
(b)

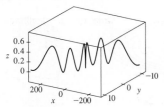

17. $t = 4$ **19.** $\mathbf{r}(t) = t\,\mathbf{i} - t\,\mathbf{j} + \tfrac{5}{2}t^2\,\mathbf{k}$, $|\mathbf{v}(t)| = \sqrt{25t^2 + 2}$
21. (a) ≈ 22 km (b) ≈ 3.2 km (c) 500 m/s
23. 30 m/s **25.** $\approx 10.2°$, $\approx 79.8°$
27. $13.0° < \theta < 36.0°$, $55.4° < \theta < 85.5°$
29. (a) 16 m (b) $\approx 23.6°$ upstream

31. $6t, 6$ **33.** $0, 1$ **35.** 4.5 cm/s^2, 9.0 cm/s^2 **37.** $t = 1$

Exercises 10.5 □ **page 732**

1. Plane through $(0, 3, 1)$ containing vectors $\langle 1, 0, 4 \rangle$,
$\langle 1, -1, 5 \rangle$
3. Hyperbolic paraboloid
5.

7.

9.

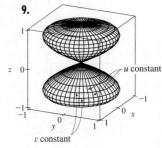

11. IV **13.** I **15.** II
17. $x = 1 + u + v$, $y = 2 + u - v$, $z = -3 - u + v$
19. $x = x$, $z = z$, $y = \sqrt{1 - x^2 + z^2}$
21. $x = 2 \sin \phi \cos \theta$, $y = 2 \sin \phi \sin \theta$,
$z = 2 \cos \phi$, $0 \le \phi \le \pi/4$, $0 \le \theta \le 2\pi$
$\left[\text{or } x = x, y = y, z = \sqrt{4 - x^2 - y^2}, x^2 + y^2 \le 2\right]$
23. $x = x$, $y = 4 \cos \theta$, $z = 4 \sin \theta$, $0 \le x \le 5$, $0 \le \theta \le 2\pi$
27. $x = x$, $y = e^{-x} \cos \theta$,
$z = e^{-x} \sin \theta$, $0 \le x \le 3$,
$0 \le \theta \le 2\pi$

29. (b)

31. (a) Direction reverses (b) Number of coils doubles

Chapter 10 Review □ **page 734**

True-False Quiz

1. True **3.** False **5.** False **7.** False **9.** True

Exercises

1. (a)

(b) $\mathbf{r}'(t) = \mathbf{i} - \pi \sin \pi t\,\mathbf{j} + \pi \cos \pi t\,\mathbf{k}$,
$\mathbf{r}''(t) = -\pi^2 \cos \pi t\,\mathbf{j} - \pi^2 \sin \pi t\,\mathbf{k}$
3. $\mathbf{r}(t) = 4 \cos t\,\mathbf{i} + 4 \sin t\,\mathbf{j} + (5 - 4\cos t)\mathbf{k}$, $0 \le t \le 2\pi$
5. $\frac{1}{3}\mathbf{i} - (2/\pi^2)\mathbf{j} + (2/\pi)\mathbf{k}$ **7.** 86.631 **9.** $\pi/2$
11. (a) $\langle t^2, t, 1 \rangle / \sqrt{t^4 + t^2 + 1}$
(b) $\langle 2t, 1 - t^4, -2t^3 - t \rangle / \sqrt{t^8 + 4t^6 + 2t^4 + 5t^2}$
(c) $\sqrt{t^8 + 4t^6 + 2t^4 + 5t^2} / (t^4 + t^2 + 1)^2$
13. $12/17^{3/2}$ **15.** $x - 2y + 2\pi = 0$
17. $\mathbf{v}(t) = (1 + \ln t)\mathbf{i} + \mathbf{j} - e^{-t}\mathbf{k}$,
$|\mathbf{v}(t)| = \sqrt{2 + 2\ln t + (\ln t)^2 + e^{-2t}}$, $\mathbf{a}(t) = (1/t)\mathbf{i} + e^{-t}\mathbf{k}$
19. (a) About 3.8 ft above the ground, 60.8 ft from the athlete
(b) ≈ 21.4 ft (c) ≈ 64.2 ft from the athlete
21. $x = 2 \sin \phi \cos \theta$, $y = 2 \sin \phi \sin \theta$, $z = 2 \cos \phi$,
$0 \le \theta \le 2\pi$, $\pi/3 \le \phi \le 2\pi/3$
23. $\pi |t|$

Focus on Problem Solving □ **page 736**

1. (a) $\mathbf{v} = \omega R(-\sin \omega t\,\mathbf{i} + \cos \omega t\,\mathbf{j})$ (c) $\mathbf{a} = -\omega^2 \mathbf{r}$
3. (a) $90°$, $v_0^2/(2g)$
5. (a) ≈ 0.94 ft to the right of the table's edge, ≈ 15 ft/s
(b) $\approx 7.6°$ (c) ≈ 2.13 ft to the right of the table's edge
7. $56°$

CHAPTER 11

Exercises 11.1 □ page 746

1. (a) -27; a temperature of $-15°C$ with wind blowing at
40 km/h feels equivalent to about $-27°C$ without wind.
(b) When the temperature is $-20°C$, what wind speed gives a
wind chill of $-30°C$? 20 km/h
(c) With a wind speed of 20 km/h, what temperature gives a
wind chill of $-49°C$? $-35°C$
(d) A function of wind speed that gives wind-chill values when
the temperature is $-5°C$
(e) A function of temperature that gives wind-chill values when
the wind speed is 50 km/h
3. Yes
5. $\{(x, y) \mid \frac{1}{9}x^2 + y^2 < 1\}$

7. (a) e (b) $\{(x, y, z) \mid z \ge x^2 + y^2\}$ (c) $[1, \infty)$
9. ≈ 56, ≈ 35 **11.** Steep; nearly flat
13.

15. $(y - 2x)^2 = k$

17. $y = \ln x + k$

19. $y = ke^{-x}$

21. $y^2 - x^2 = k^2$

23. $x^2 + 9y^2 = k$

25.

27.

29.

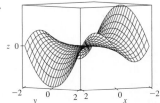

31. (a) C   (b) II   **33.** (a) F   (b) I

35. (a) B   (b) VI

37. Family of parallel planes

39. Family of hyperboloids of one or two sheets with axis the y-axis

41. (a) Shift the graph of f upward 2 units

(b) Stretch the graph of f vertically by a factor of 2

(c) Reflect the graph of f about the xy-plane

(d) Reflect the graph of f about the xy-plane and then shift it upward 2 units

43. If $c = 0$, the graph is a cylindrical surface. For $c > 0$, the level curves are ellipses. The graph curves upward as we leave the origin, and the steepness increases as c increases. For $c < 0$, the level curves are hyperbolas. The graph curves upward in the y-direction and downward, approaching the xy-plane, in the x-direction giving a saddle-shaped appearance near $(0, 0, 1)$.

45. (b) $y = 0.75x + 0.01$

Exercises 11.2 □ page 755

1. Nothing; if f is continuous, $f(3, 1) = 6$ **3.** $-\frac{5}{2}$

5. 2025   **7.** Does not exist   **9.** Does not exist   **11.** 0

13. Does not exist   **15.** 2   **17.** Does not exist

19. The graph shows that the function approaches different numbers along different lines.

21. $h(x, y) = (2x + 3y - 6)^2 + \sqrt{2x + 3y - 6}$;
$\{(x, y) \mid 2x + 3y \geqslant 6\}$

23. Along the line $y = x$ **25.** $\{(x, y) \mid y \neq \pm e^{x/2}\}$

27. $\{(x, y) \mid x^2 + y^2 > 4\}$

29. $\{(x, y, z) \mid y \geqslant 0, y \neq \sqrt{x^2 + z^2}\}$

31. $\{(x, y) \mid (x, y) \neq (0, 0)\}$ **33.** 0   **35.** 0

Exercises 11.3 □ page 766

1. (a) The rate of change of temperature as longitude varies, with latitude and time fixed; the rate of change as only latitude varies; the rate of change as only time varies.

(b) Positive, negative, positive

3. (a) $f_T(-15, 30) \approx 1.3$; for a temperature of $-15°C$ and wind speed of 30 km/h, the wind-chill index rises by $1.3°C$ for each degree the temperature increases. $f_v(-15, 30) \approx -0.15$; for a temperature of $-15°C$ and wind speed of 30 km/h, the wind-chill index decreases by $0.15°C$ for each km/h the wind speed increases.

(b) Positive, negative   (c) 0

5. (a) Positive   (b) Negative   **7.** $c = f, b = f_x, a = f_y$

9. $f_x(1, 2) = -8 = $ slope of C_1, $f_y(1, 2) = -4 = $ slope of C_2

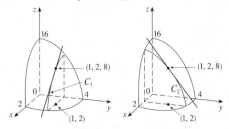

11. $f_x = 2x + 2xy, f_y = 2y + x^2$

13. $f_x(x, y) = 3, f_y(x, y) = -8y^3$

15. $\partial z/\partial x = e^{3y}, \partial z/\partial y = 3xe^{3y}$

17. $f_x(x, y) = 2y/(x + y)^2, f_y(x, y) = -2x/(x + y)^2$

19. $\partial w/\partial \alpha = \cos \alpha \cos \beta, \partial w/\partial \beta = -\sin \alpha \sin \beta$

21. $f_r(r, s) = \dfrac{2r^2}{r^2 + s^2} + \ln(r^2 + s^2), f_s(r, s) = \dfrac{2rs}{r^2 + s^2}$

23. $\partial u/\partial t = e^{w/t}(1 - w/t), \partial u/\partial w = e^{w/t}$

25. $f_x = y^2z^3, f_y = 2xyz^3 + 3z, f_z = 3xy^2z^2 + 3y$

27. $\partial w/\partial x = 1/(x + 2y + 3z), \partial w/\partial y = 2/(x + 2y + 3z),$
$\partial w/\partial z = 3/(x + 2y + 3z)$

29. $\partial u/\partial x = e^{-t} \sin \theta, \partial u/\partial t = -xe^{-t} \sin \theta, \partial u/\partial \theta = xe^{-t} \cos \theta$

31. $f_x = yz^2 \tan(yt)$, $f_y = xyz^2t \sec^2(yt) + xz^2 \tan(yt)$,
$f_z = 2xyz \tan(yt)$, $f_t = xy^2z^2 \sec^2(yt)$
33. $\partial u/\partial x_i = x_i/\sqrt{x_1^2 + x_2^2 + \cdots + x_n^2}$ **35.** $\frac{3}{5}$ **37.** $-\frac{1}{3}$
39. $f_x(x, y) = y^2 - 3x^2y$, $f_y(x, y) = 2xy - x^3$
41. $\dfrac{\partial z}{\partial x} = \dfrac{3yz - 2x}{2z - 3xy}$, $\dfrac{\partial z}{\partial y} = \dfrac{3xz - 2y}{2z - 3xy}$
43. $\dfrac{\partial z}{\partial x} = \dfrac{1 + y^2z^2}{1 + y + y^2z^2}$, $\dfrac{\partial z}{\partial y} = \dfrac{-z}{1 + y + y^2z^2}$
45. (a) $f'(x), g'(y)$ (b) $f'(x + y), f'(x + y)$
47. $f_{xx} = 12x^2 - 6y^3$, $f_{xy} = -18xy^2 = f_{yx}$, $f_{yy} = -18x^2y$
49. $z_{xx} = -2y/(x + y)^3$, $z_{xy} = (x - y)/(x + y)^3 = z_{yx}$,
$z_{yy} = 2x/(x + y)^3$
51. $u_{ss} = e^{-s} \sin t$, $u_{st} = -e^{-s} \cos t = u_{ts}$, $u_{tt} = -e^{-s} \sin t$
55. $12xy$, $72xy$
57. $24 \sin(4x + 3y + 2z)$, $12 \sin(4x + 3y + 2z)$
59. $\theta e^{r\theta}(2 \sin \theta + \theta \cos \theta + r\theta \sin \theta)$
61. ≈ 12.2, ≈ 16.8, ≈ 23.25 **71.** R^2/R_1^2
77. No   **79.** $x = 1 + t, y = 2, z = 2 - 2t$ **81.** -2
83. (a)

(b) $f_x(x, y) = \dfrac{x^4y + 4x^2y^3 - y^5}{(x^2 + y^2)^2}$, $f_y(x, y) = \dfrac{x^5 - 4x^3y^2 - xy^4}{(x^2 + y^2)^2}$

(c) $0, 0$ (e) No, since f_{xy} and f_{yx} are not continuous.

Exercises 11.4 □ **page 778**

1. $z = -8x - 2y$ **3.** $z = y$
5.

7.

9. $2x + \frac{1}{4}y - 1$ **11.** $\frac{1}{2}x + y + \frac{1}{4}\pi - \frac{1}{2}$
13. $-\frac{2}{3}x - \frac{7}{3}y + \frac{20}{3}$; 2.846   **15.** $\frac{3}{7}x + \frac{2}{7}y + \frac{6}{7}z$; 6.9914
17. $4T + H - 329$; 129°F
19. $dz = 3x^2 \ln(y^2)\, dx + (2x^3/y)\, dy$
21. $dR = \beta^2 \cos \gamma\, d\alpha + 2\alpha\beta \cos \gamma\, d\beta - \alpha\beta^2 \sin \gamma\, d\gamma$
23. $\Delta z = 0.9225, dz = 0.9$ **25.** 5.4 cm²   **27.** 16 cm³
29. 2.3%   **31.** $\frac{1}{17} \approx 0.059\ \Omega$ **33.** $3x - y + 3z = 3$
35. $-x + 2z = 1$ **37.** $x - y + z = 2$
39. $\varepsilon_1 = \Delta x, \varepsilon_2 = \Delta y$

Exercises 11.5 □ **page 786**

1. $\pi \cos x \cos y - (\sin x \sin y)/(2\sqrt{t})$
3. $e^{y/z}[2t - (x/z) - (2xy/z^2)]$
5. $\partial z/\partial s = 2x + y + xt + 2yt$, $\partial z/\partial t = 2x + y + xs + 2ys$

7. $\dfrac{\partial z}{\partial s} = e^r\left(t \cos \theta - \dfrac{s}{\sqrt{s^2 + t^2}} \sin \theta\right)$,
$\dfrac{\partial z}{\partial t} = e^r\left(s \cos \theta - \dfrac{t}{\sqrt{s^2 + t^2}} \sin \theta\right)$
9. 62   **11.** 7, 2
13. $\dfrac{\partial u}{\partial r} = \dfrac{\partial u}{\partial x} \dfrac{\partial x}{\partial r} + \dfrac{\partial u}{\partial y} \dfrac{\partial y}{\partial r}$, $\dfrac{\partial u}{\partial s} = \dfrac{\partial u}{\partial x} \dfrac{\partial x}{\partial s} + \dfrac{\partial u}{\partial y} \dfrac{\partial y}{\partial s}$,
$\dfrac{\partial u}{\partial t} = \dfrac{\partial u}{\partial x} \dfrac{\partial x}{\partial t} + \dfrac{\partial u}{\partial y} \dfrac{\partial y}{\partial t}$
15. $\dfrac{\partial v}{\partial x} = \dfrac{\partial v}{\partial p} \dfrac{\partial p}{\partial x} + \dfrac{\partial v}{\partial q} \dfrac{\partial q}{\partial x} + \dfrac{\partial v}{\partial r} \dfrac{\partial r}{\partial x}$,
$\dfrac{\partial v}{\partial y} = \dfrac{\partial v}{\partial p} \dfrac{\partial p}{\partial y} + \dfrac{\partial v}{\partial q} \dfrac{\partial q}{\partial y} + \dfrac{\partial v}{\partial r} \dfrac{\partial r}{\partial y}$,
$\dfrac{\partial v}{\partial z} = \dfrac{\partial v}{\partial p} \dfrac{\partial p}{\partial z} + \dfrac{\partial v}{\partial q} \dfrac{\partial q}{\partial z} + \dfrac{\partial v}{\partial r} \dfrac{\partial r}{\partial z}$
17. 85, 178, 54   **19.** $\frac{9}{7}, \frac{9}{7}$ **21.** 36, 24, 30
23. $\dfrac{4(xy)^{3/2} - y}{x - 2x^2\sqrt{xy}}$ **25.** $\dfrac{3yz - 2x}{2z - 3xy}, \dfrac{3xz - 2y}{2z - 3xy}$
27. $\dfrac{1 + y^2z^2}{1 + y + y^2z^2}, -\dfrac{z}{1 + y + y^2z^2}$
29. 2°C/s   **31.** ≈ -0.33 m/s per minute
33. (a) 6 m³/s   (b) 10 m²/s   (c) 0 m/s   **35.** ≈ -0.27 L/s
37. (a) $\partial z/\partial r = (\partial z/\partial x) \cos \theta + (\partial z/\partial y) \sin \theta$,
$\partial z/\partial \theta = -(\partial z/\partial x)r \sin \theta + (\partial z/\partial y)r \cos \theta$
43. $4rs\, \partial^2z/\partial x^2 + (4r^2 + 4s^2)\partial^2z/\partial x\, \partial y + 4rs\, \partial^2z/\partial y^2 + 2\, \partial z/\partial y$

Exercises 11.6 □ **page 799**

1. ≈ -0.1 millibar/mi   **3.** ≈ 0.778 **5.** $\frac{5}{16}\sqrt{3} + \frac{1}{4}$
7. (a) $\nabla f(x, y) = \langle 5y^2 - 12x^2y, 10xy - 4x^3 \rangle$
(b) $\langle -4, 16 \rangle$ (c) $172/13$
9. (a) $\langle e^{2yz}, 2xze^{2yz}, 2xye^{2yz} \rangle$ (b) $\langle 1, 12, 0 \rangle$ (c) $-\frac{22}{3}$
11. $23/10$ **13.** $4\sqrt{2}$ **15.** $9/(2\sqrt{5})$ **17.** $2/5$
19. $4\sqrt{2}, \langle -1, 1 \rangle$ **21.** $\sqrt{3}, \langle 1, -1, -1 \rangle$
23. (b) $\langle -12, 92 \rangle$ **25.** All points on the line $y = x + 1$
27. (a) $-40/(3\sqrt{3})$
29. (a) $32/\sqrt{3}$ (b) $\langle 38, 6, 12 \rangle$ (c) $2\sqrt{406}$ **31.** $\frac{327}{13}$
35. (a) $4x - 5y - z = 4$ (b) $\dfrac{x - 2}{4} = \dfrac{y - 1}{-5} = \dfrac{z + 1}{-1}$
37. (a) $x + y - z = 1$ (b) $x - 1 = y = -z$
39.

41. $\langle 4, 8 \rangle, x + 2y = 4$

45. $(\pm\sqrt{6}/3, \mp 2\sqrt{6}/3, \pm\sqrt{6}/2)$
49. $x = -1 - 10t, y = 1 - 16t, z = 2 - 12t$

53. If $\mathbf{u} = \langle a, b \rangle$ and $\mathbf{v} = \langle c, d \rangle$, then $af_x + bf_y$ and $cf_x + df_y$ are known, so we solve linear equations for f_x and f_y.

Exercises 11.7 □ page 809

1. (a) f has a local minimum at $(1, 1)$.
(b) f has a saddle point at $(1, 1)$.
3. Local minimum at $(1, 1)$, saddle point at $(0, 0)$
5. Maximum $f\left(-1, \frac{1}{2}\right) = 11$
7. Minima $f(1, 1) = 0$, $f(-1, -1) = 0$, saddle point at $(0, 0)$
9. Saddle points at $(1, -1)$, $(-1, 1)$
11. None
13. Saddle points $(0, n\pi)$, n an integer
15. Minimum $f(0, 0) = 0$, saddle points at $(\pm 1, 0)$
17. Maximum $f(0, 0) = 2$, minimum $f(0, 2) = -2$, saddle points $(\pm 1, 1)$
19. Maximum $f(\pi/3, \pi/3) = 3\sqrt{3}/2$, minimum $f(5\pi/3, 5\pi/3) = -3\sqrt{3}/2$
21. Minima $f(-1.714, 0) \approx -9.200$, $f(1.402, 0) \approx 0.242$, saddle point $(0.312, 0)$, lowest point $(-1.714, 0, -9.200)$
23. Maxima $f(-1.267, 0) \approx 1.310$, $f(1.629, \pm 1.063) \approx 8.105$, saddle points $(-0.259, 0)$, $(1.526, 0)$, highest points $(1.629, \pm 1.063, 8.105)$
25. Maximum $f(2, 0) = 9$, minimum $f(0, 3) = -14$
27. Maximum $f(\pm 1, 1) = 7$, minimum $f(0, 0) = 4$
29. Maximum $f(3, 0) = 83$, minimum $f(1, 1) = 0$
31.

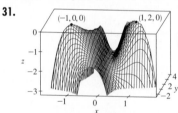

33. $\sqrt{3}$ **35.** $(2, 1, \sqrt{5}), (2, 1, -\sqrt{5})$ **37.** $\frac{100}{3}, \frac{100}{3}, \frac{100}{3}$
39. $16/\sqrt{3}$ **41.** $\frac{4}{3}$ **43.** Cube, edge length $c/12$
45. Square base of side 40 cm, height 20 cm **47.** $L^3/(3\sqrt{3})$

Exercises 11.8 □ page 818

1. $\approx 59, 30$
3. No maximum, minima $f(1, 1) = f(-1, -1) = 2$
5. Maxima $f(\pm 2, 1) = 4$, minima $f(\pm 2, -1) = -4$
7. Maximum $f(1, 3, 5) = 70$, minimum $f(-1, -3, -5) = -70$
9. Maximum $2/\sqrt{3}$, minimum $-2/\sqrt{3}$
11. Maximum $\sqrt{3}$, minimum 1
13. Maximum $f\left(\frac{1}{2}, \frac{1}{2}, \frac{1}{2}, \frac{1}{2}\right) = 2$, minimum $f\left(-\frac{1}{2}, -\frac{1}{2}, -\frac{1}{2}, -\frac{1}{2}\right) = -2$
15. Maximum $f(1, \sqrt{2}, -\sqrt{2}) = 1 + 2\sqrt{2}$, minimum $f(1, -\sqrt{2}, \sqrt{2}) = 1 - 2\sqrt{2}$
17. Maximum $\frac{3}{2}$, minimum $\frac{1}{2}$
19. Maxima $f(\pm 1/\sqrt{2}, \mp 1/(2\sqrt{2})) = e^{1/4}$, minima $f(\pm 1/\sqrt{2}, \pm 1/(2\sqrt{2})) = e^{-1/4}$
25–37. See Exercises 33–47 in Section 11.7.

39. Nearest $\left(\frac{1}{2}, \frac{1}{2}, \frac{1}{2}\right)$, farthest $(-1, -1, 2)$
41. Maximum ≈ 9.7938, minimum ≈ -5.3506
43. (a) c/n (b) When $x_1 = x_2 = \cdots = x_n$

Chapter 11 Review □ page 823

True-False Quiz

1. True **3.** False **5.** False **7.** True **9.** False
11. True

Exercises

1. $\{(x, y) \mid y > -x - 1\}$ **3.**

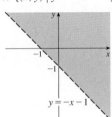

5. **7.**

9. $\frac{2}{3}$
11. (a) $\approx 3.5°C/m$, $-3.0°C/m$ (b) $\approx 0.35°C/m$ by Equation 11.6.9 (Definition 11.6.2 gives $\approx 1.1°C/m$.)
(c) -0.25
13. $f_x = 1/\sqrt{2x + y^2}$, $f_y = y/\sqrt{2x + y^2}$
15. $g_u = \tan^{-1} v$, $g_v = u/(1 + v^2)$
17. $T_p = \ln(q + e^r)$, $T_q = p/(q + e^r)$, $T_r = pe^r/(q + e^r)$
19. $f_{xx} = 24x$, $f_{xy} = -2y = f_{yx}$, $f_{yy} = -2x$
21. $f_{xx} = k(k - 1)x^{k-2}y^l z^m$, $f_{xy} = klx^{k-1}y^{l-1}z^m = f_{yx}$,
$f_{xz} = kmx^{k-1}y^l z^{m-1} = f_{zx}$, $f_{yy} = l(l - 1)x^k y^{l-2}z^m$,
$f_{yz} = lmx^k y^{l-1}z^{m-1} = f_{zy}$, $f_{zz} = m(m - 1)x^k y^l z^{m-2}$
25. (a) $z = 8x + 4y + 1$ (b) $\dfrac{x - 1}{8} = \dfrac{y + 2}{4} = 1 - z$
27. (a) $2x - 2y - 3z = 3$ (b) $\dfrac{x - 2}{4} = \dfrac{y + 1}{-4} = \dfrac{z - 1}{-6}$
29. (a) $4x - y - 2z = 6$
(b) $x = 3 + 8t, y = 4 - 2t, z = 1 - 4t$
31. $\left(2, \frac{1}{2}, -1\right), \left(-2, -\frac{1}{2}, 1\right)$
33. $60x + \frac{24}{5}y + \frac{32}{5}z - 120$; 38.656
35. $2xy^3(1 + 6p) + 3x^2y^2(pe^p + e^p) + 4z^3(p \cos p + \sin p)$
37. -47, 108
43. $ze^{x\sqrt{y}}\langle z\sqrt{y}, xz/(2\sqrt{y}), 2\rangle$ **45.** $\frac{43}{5}$ **47.** $\sqrt{145}/2, \langle 4, \frac{9}{2}\rangle$
49. $\approx \frac{5}{8}$ knot/mi **51.** Minimum $f(-4, 1) = -11$

53. Maximum $f(1, 1) = 1$; saddle points $(0, 0)$, $(0, 3)$, $(3, 0)$
55. Maximum $f(1, 2) = 4$, minimum $f(2, 4) = -64$
57. Maximum $f(-1, 0) = 2$, minima $f(1, \pm 1) = -3$,
saddle points $(-1, \pm 1)$, $(1, 0)$
59. Maximum $f(\pm\sqrt{2/3}, 1/\sqrt{3}) = 2/(3\sqrt{3})$,
minimum $f(\pm\sqrt{2/3}, -1/\sqrt{3}) = -2/(3\sqrt{3})$
61. Maximum 1, minimum -1
63. $(\pm 3^{-1/4}, 3^{-1/4}\sqrt{2}, \pm 3^{1/4})$, $(\pm 3^{-1/4}, -3^{-1/4}\sqrt{2}, \pm 3^{1/4})$
65. $P(2 - \sqrt{3})$, $P(3 - \sqrt{3})/6$, $P(2\sqrt{3} - 3)/3$

Focus on Problem Solving □ **page 826**

1. $L^2W^2, \frac{1}{4}L^2W^2$ **3.** (a) $x = w/3$, base $= w/3$ (b) Yes
9. $\sqrt{6}/2, 3\sqrt{2}/2$

CHAPTER 12

Exercises 12.1 □ page 836

1. (a) 288   (b) 144   **3.** (a) $\pi^2/2 \approx 4.935$ (b) 0
5. (a) -6 (b) -3.5 **7.** $U < V < L$
9. (a) ≈ 248 (b) 15.5   **11.** 60   **13.** 3
15. 1.141606, 1.143191, 1.143535, 1.143617, 1.143637, 1.143642

Exercises 12.2 □ page 842

1. $9 + 27y, 8x + 24x^2$ **3.** 10   **5.** 2
7. $261{,}632/45$ **9.** $\frac{21}{2}\ln 2$ **11.** 6   **13.** $9 \ln 2$
15. $[(\sqrt{3} - 1)/2] - (\pi/12)$ **17.** $\frac{1}{2}(e^2 - 3)$
19.

21. 47.5   **23.** $\frac{166}{27}$ **25.** $\frac{4}{15}(2\sqrt{2} - 1)$ **27.** 36
29. $21e - 57$

31. $\frac{5}{6}$
33. Fubini's Theorem does not apply. The integrand has an infinite discontinuity at the origin.

Exercises 12.3 □ page 850

1. $\frac{9}{20}$ **3.** $\frac{4}{9}e^{3/2} - \frac{32}{45}$ **5.** $e - 1$ **7.** $\frac{256}{21}$
9. $\frac{1}{2}\ln 2$ **11.** $(1 - \cos 1)/2$ **13.** $\frac{147}{20}$
15. 0   **17.** $\frac{7}{18}$ **19.** $\frac{31}{8}$ **21.** $\frac{1}{6}$ **23.** $\frac{128}{15}$

25. $\frac{1}{3}$ **27.** 0, 1.213, 0.713
29. $\frac{64}{3}$ **31.** $\pi/2$
33. $\int_0^2 \int_{y^2}^4 f(x, y)\, dx\, dy$ **35.** $\int_{-3}^3 \int_0^{\sqrt{9-x^2}} f(x, y)\, dy\, dx$

37. $\int_0^{\ln 2} \int_{e^y}^2 f(x, y)\, dx\, dy$

39. $(e^9 - 1)/6$ **41.** $\frac{1}{4}\sin 81$ **43.** $(2\sqrt{2} - 1)/3$ **45.** 1
47. $0 \le \iint_D \sqrt{x^3 + y^3}\, dA \le \sqrt{2}$ **51.** 8π **53.** $2\pi/3$

Exercises 12.4 □ page 856

1. $\int_0^{2\pi} \int_0^2 f(r\cos\theta, r\sin\theta)\, r\, dr\, d\theta$ **3.** $\int_{-2}^2 \int_x^2 f(x, y)\, dy\, dx$
5. $\int_0^{2\pi} \int_2^5 f(r\cos\theta, r\sin\theta)\, r\, dr\, d\theta$
7. $33\pi/2$

9. 0   **11.** $\frac{1}{2}\pi\sin 9$ **13.** $\frac{3}{64}\pi^2$ **15.** $\frac{16}{3}\pi$
17. $\frac{4}{3}\pi a^3$ **19.** $(2\pi/3)[1 - (1/\sqrt{2})]$
21. $(8\pi/3)(64 - 24\sqrt{3})$ **23.** $\pi/12$
25. $\frac{1}{2}\pi(1 - \cos 9)$ **27.** $2\sqrt{2}/3$
29. 1800π ft^3 **31.** $\frac{15}{16}$ **33.** (a) $\sqrt{\pi}/4$ (b) $\sqrt{\pi}/2$

Exercises 12.5 □ page 866

1. $\frac{64}{3}$ C   **3.** $\frac{4}{3}, (\frac{4}{3}, 0)$ **5.** $6, (\frac{3}{4}, \frac{3}{2})$
7. $\frac{1}{4}(e^2 - 1), \left(\dfrac{e^2 + 1}{2(e^2 - 1)}, \dfrac{4(e^3 - 1)}{9(e^2 - 1)}\right)$
9. $\frac{27}{2}, (\frac{8}{5}, \frac{1}{2})$ **11.** $(\frac{3}{8}, 3\pi/16)$
13. $(2a/5, 2a/5)$ if vertex is $(0, 0)$ and sides are along positive axes
15. $\frac{1}{16}(e^4 - 1), \frac{1}{8}(e^2 - 1), \frac{1}{16}(e^4 + 2e^2 - 3)$
17. $7ka^6/180, 7ka^6/180, 7ka^6/90$ if vertex is $(0, 0)$ and sides are along positive axes
19. $m = \pi^2/8, (\bar{x}, \bar{y}) = \left(\dfrac{2\pi}{3} - \dfrac{1}{\pi}, \dfrac{16}{9\pi}\right), I_x = 3\pi^2/64,$
$I_y = (\pi^4 - 3\pi^2)/16, I_0 = \pi^4/16 - 9\pi^2/64$
21. (a) $\frac{1}{2}$ (b) 0.375   (c) $\frac{5}{48} \approx 0.1042$

23. (b) (i) $e^{-0.2} \approx 0.8187$
(ii) $1 + e^{-1.8} - e^{-0.8} - e^{-1} \approx 0.3481$ (c) 2, 5
25. (a) ≈ 0.500 (b) ≈ 0.632
27. (a) $\iint_D (k/20)[20 - \sqrt{(x - x_0)^2 + (y - y_0)^2}] \, dA$, where D is the disk with radius 10 mi centered at the center of the city
(b) $200\pi k/3 \approx 209k, 200(\pi/2 - \frac{8}{9})k \approx 136k$, on the edge

Exercises 12.6 □ page 870

1. $15\sqrt{26}$ **3.** $3\sqrt{14}$ **5.** $(\pi/6)(17\sqrt{17} - 5\sqrt{5})$
7. 4   **9.** $(\sqrt{21}/2) + \frac{17}{4}[\ln(2 + \sqrt{21}) - \ln\sqrt{17}]$
11. $(2\pi/3)(2\sqrt{2} - 1)$ **13.** 13.9783
15. (a) 24.2055   (b) 24.2476   **17.** 4.4506
19. $\frac{45}{8}\sqrt{14} + \frac{15}{16}\ln[(11\sqrt{5} + 3\sqrt{70})/(3\sqrt{5} + \sqrt{70})]$
21. (b)

(c) $\int_0^{2\pi} \int_0^{\pi} \sqrt{36 \sin^4 u \cos^2 v + 9 \sin^4 u \sin^2 v + 4 \cos^2 u \sin^2 u} \, du \, dv$
23. 4π **27.** $(\pi/6)(37\sqrt{37} - 17\sqrt{17})$

Exercises 12.7 □ page 879

3. 1   **5.** $\frac{1}{3}(e^3 - 1)$ **7.** 4   **9.** $\frac{65}{28}$ **11.** $\frac{1}{10}$
13. $8/(3e)$ **15.** $16\pi/3$ **17.** $\frac{16}{3}$ **19.** 36π
21. (a) $\int_0^1 \int_0^x \int_0^{\sqrt{1-y^2}} dz \, dy \, dx$ (b) $\frac{1}{4}\pi - \frac{1}{3}$
23. 60.533
25.

27. $\int_{-2}^2 \int_0^6 \int_{-\sqrt{4-x^2}}^{\sqrt{4-x^2}} f(x, y, z) \, dz \, dy \, dx$
$= \int_0^6 \int_{-2}^2 \int_{-\sqrt{4-x^2}}^{\sqrt{4-x^2}} f(x, y, z) \, dz \, dx \, dy$
$= \int_{-2}^2 \int_0^6 \int_{-\sqrt{4-z^2}}^{\sqrt{4-z^2}} f(x, y, z) \, dx \, dy \, dz$
$= \int_0^6 \int_{-2}^2 \int_{-\sqrt{4-z^2}}^{\sqrt{4-z^2}} f(x, y, z) \, dx \, dz \, dy$
$= \int_{-2}^2 \int_{-\sqrt{4-x^2}}^{\sqrt{4-x^2}} \int_0^6 f(x, y, z) \, dy \, dz \, dx$
$= \int_{-2}^2 \int_{-\sqrt{4-z^2}}^{\sqrt{4-z^2}} \int_0^6 f(x, y, z) \, dy \, dx \, dz$
29. $\int_{-1}^1 \int_0^{1-x^2} \int_0^y f(x, y, z) \, dz \, dy \, dx$
$= \int_0^1 \int_{-\sqrt{1-y}}^{\sqrt{1-y}} \int_0^y f(x, y, z) \, dz \, dx \, dy$
$= \int_0^1 \int_z^1 \int_{-\sqrt{1-y}}^{\sqrt{1-y}} f(x, y, z) \, dx \, dy \, dz$
$= \int_0^1 \int_0^y \int_{-\sqrt{1-y}}^{\sqrt{1-y}} f(x, y, z) \, dx \, dz \, dy$
$= \int_{-1}^1 \int_0^{1-x^2} \int_z^{1-x^2} f(x, y, z) \, dy \, dz \, dx$
$= \int_0^1 \int_{-\sqrt{1-z}}^{\sqrt{1-z}} \int_z^{1-x^2} f(x, y, z) \, dy \, dx \, dz$

31. $\int_0^1 \int_{\sqrt{x}}^1 \int_0^{1-y} f(x, y, z) \, dz \, dy \, dx$
$= \int_0^1 \int_0^{y^2} \int_0^{1-y} f(x, y, z) \, dz \, dx \, dy$
$= \int_0^1 \int_0^{1-z} \int_0^{y^2} f(x, y, z) \, dx \, dy \, dz$
$= \int_0^1 \int_0^{1-y} \int_0^{y^2} f(x, y, z) \, dx \, dz \, dy$
$= \int_0^1 \int_0^{1-\sqrt{x}} \int_{\sqrt{x}}^{1-z} f(x, y, z) \, dy \, dz \, dx$
$= \int_0^1 \int_0^{(1-z)^2} \int_{\sqrt{x}}^{1-z} f(x, y, z) \, dy \, dx \, dz$

33. $\int_0^1 \int_z^1 \int_0^y f(x, y, z) \, dz \, dx \, dy = \int_0^1 \int_0^x \int_0^y f(x, y, z) \, dz \, dy \, dx$
$= \int_0^1 \int_z^1 \int_y^1 f(x, y, z) \, dx \, dy \, dz = \int_0^1 \int_0^y \int_y^1 f(x, y, z) \, dx \, dz \, dy$
$= \int_0^1 \int_0^x \int_z^x f(x, y, z) \, dy \, dz \, dx = \int_0^1 \int_z^1 \int_z^x f(x, y, z) \, dy \, dx \, dz$

35. $\frac{79}{30}, (\frac{358}{553}, \frac{33}{79}, \frac{571}{553})$ **37.** $a^5, (7a/12, 7a/12, 7a/12)$

39. (a) $m = \int_{-3}^3 \int_{-\sqrt{9-x^2}}^{\sqrt{9-x^2}} \int_1^{5-y} \sqrt{x^2 + y^2} \, dz \, dy \, dx$
(b) $(\bar{x}, \bar{y}, \bar{z})$, where
$\bar{x} = (1/m)\int_{-3}^3 \int_{-\sqrt{9-x^2}}^{\sqrt{9-x^2}} \int_1^{5-y} x\sqrt{x^2 + y^2} \, dz \, dy \, dx$
$\bar{y} = (1/m)\int_{-3}^3 \int_{-\sqrt{9-x^2}}^{\sqrt{9-x^2}} \int_1^{5-y} y\sqrt{x^2 + y^2} \, dz \, dy \, dx$
$\bar{z} = (1/m)\int_{-3}^3 \int_{-\sqrt{9-x^2}}^{\sqrt{9-x^2}} \int_1^{5-y} z\sqrt{x^2 + y^2} \, dz \, dy \, dx$
(c) $\int_{-3}^3 \int_{-\sqrt{9-x^2}}^{\sqrt{9-x^2}} \int_1^{5-y} (x^2 + y^2)^{3/2} \, dz \, dy \, dx$
41. (a) $\frac{3}{32}\pi + \frac{11}{24}$

(b) $(\bar{x}, \bar{y}, \bar{z}) = \left(\dfrac{28}{9\pi + 44}, \dfrac{30\pi + 128}{45\pi + 220}, \dfrac{45\pi + 208}{135\pi + 660} \right)$

(c) $\frac{1}{240}(68 + 15\pi)$
43. $I_x = I_y = I_z = \frac{2}{3}kL^5$
45. (a) $\frac{1}{8}$ (b) $\frac{1}{64}$ (c) $\frac{1}{5760}$ **47.** $L^3/8$
49. The region bounded by the ellipsoid $x^2 + 2y^2 + 3z^2 = 1$

Exercises 12.8 □ page 886

1.

$64\pi/3$

3.

$(9\pi/4)(2 - \sqrt{3})$

5. $\int_0^{\pi/2} \int_0^3 \int_0^2 f(r \cos\theta, r \sin\theta, z) \, r \, dz \, dr \, d\theta$ **7.** 384π
9. $\pi(e^6 - e - 5)$ **11.** $2\pi/5$
13. (a) 162π (b) $(0, 0, 15)$
15. $\pi Ka^2/8, (0, 0, 2a/3)$ **17.** $4\pi/5$
19. $15\pi/16$ **21.** $1562\pi/15$
23. (a) 10π (b) $(0, 0, 2.1)$
25. (a) $(0, 0, \frac{3}{8}a)$ (b) $4K\pi a^5/15$
27. $(2\pi/3)[1 - (1/\sqrt{2})], (0, 0, 3/[8(2 - \sqrt{2})])$
29. $5\pi/6$ **31.** 0   **33.** $(4\sqrt{2} - 5)/15$ **35.** $136\pi/99$
37. (a) $\iiint_C h(P)g(P) \, dV$, where C is the cone
(b) $\approx 3.1 \times 10^{19}$ ft-lb

Exercises 12.9 □ page 897

1. -14 **3.** 0 **5.** $2uvw$

7. The parallelogram with vertices $(0, 0)$, $(6, 3)$, $(12, 1)$, $(6, -2)$

9. The region bounded by the line $y = 1$, the y-axis, and $y = \sqrt{x}$

11. -3 **13.** 6π **15.** $2 \ln 3$

17. (a) $\frac{4}{3}\pi abc$ (b) 1.083×10^{12} km^3

19. $\frac{8}{5} \ln 8$ **21.** $\frac{3}{2} \sin 1$ **23.** $e - e^{-1}$

Chapter 12 Review □ page 899

True-False Quiz

1. True   **3.** True   **5.** True   **7.** False

Exercises

1. ≈ 64.0 **3.** $4e^2 - 4e + 3$ **5.** $\frac{1}{2} \sin 1$ **7.** $\frac{2}{3}$

9. $\int_0^\pi \int_2^4 f(r \cos \theta, r \sin \theta) \, r \, dr \, d\theta$

11. The region inside the loop of the four-leaved rose $r = \sin 2\theta$ in the first quadrant

13. $\frac{1}{2} \sin 1$ **15.** $\frac{1}{2}e^6 - \frac{7}{2}$ **17.** $\frac{1}{4} \ln 2$

19. 8 **21.** $81\pi/5$ **23.** 40.5 **25.** $\pi/96$

27. $\frac{64}{15}$ **29.** 176 **31.** $\frac{2}{3}$ **33.** $2ma^3/9$

35. (a) $\frac{1}{4}$ (b) $\left(\frac{1}{3}, \frac{8}{15}\right)$

(c) $I_x = \frac{1}{12}, I_y = \frac{1}{24}; \overline{\overline{y}} = 1/\sqrt{3}, \overline{\overline{x}} = 1/\sqrt{6}$

37. (a) $(0, 0, h/4)$ (b) $\pi a^4 h/10$

39. $\ln(\sqrt{2} + \sqrt{3}) + \sqrt{2}/3$

41. 97.2 **43.** 0.0512 **45.** (a) $\frac{1}{15}$ (b) $\frac{1}{3}$ (c) $\frac{1}{45}$

47. $\int_0^1 \int_0^{1-z} \int_{-\sqrt{y}}^{\sqrt{y}} f(x, y, z) \, dx \, dy \, dz$ **49.** $-\ln 2$ **51.** 0

Focus on Problem Solving □ page 902

1. 30 **3.** $\frac{1}{2} \sin 1$ **7.** (b) 0.90

CHAPTER 13

Exercises 13.1 □ page 910

1.

3.

5.

7.

9.

11. II   **13.** I   **15.** IV   **17.** III

19.

The line $y = 2x$

21. $\nabla f(x, y) = \dfrac{1}{x + 2y}\mathbf{i} + \dfrac{2}{x + 2y}\mathbf{j}$

23. $\nabla f(x, y, z) = \dfrac{x}{\sqrt{x^2 + y^2 + z^2}}\mathbf{i}$

$\quad + \dfrac{y}{\sqrt{x^2 + y^2 + z^2}}\mathbf{j} + \dfrac{z}{\sqrt{x^2 + y^2 + z^2}}\mathbf{k}$

25. $\nabla f(x, y) = (y - 2)\mathbf{i} + x\mathbf{j}$

27.

29. IV   **31.** II   **33.** $(2.04, 1.03)$

35. (a)   (b) $y = 1/x, x > 0$

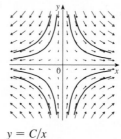

$y = C/x$

Exercises 13.2 □ page 921

1. $\frac{1}{12}(17\sqrt{17} - 1)$ **3.** 1638.4   **5.** $\frac{17}{3}$
7. 320   **9.** $\frac{1}{12}\sqrt{14}(e^6 - 1)$ **11.** $\frac{1}{5}$ **13.** $\frac{97}{3}$
15. (a) Positive   (b) Negative   **17.** $-\frac{59}{105}$
19. $\frac{6}{5} - \cos 1 - \sin 1$
21. $3\pi + \frac{2}{3}$

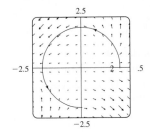

23. (a) $\frac{11}{8} - 1/e$ (b)

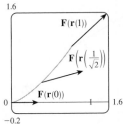

25. $\frac{945}{16,777,216}\pi$ **27.** $2\pi k, (4/\pi, 0)$
29. (a) $\bar{x} = (1/m)\int_C x\rho(x, y, z)\,ds,$
$\bar{y} = (1/m)\int_C y\rho(x, y, z)\,ds,$
$\bar{z} = (1/m)\int_C z\rho(x, y, z)\,ds,$ where $m = \int_C \rho(x, y, z)\,ds$
(b) $(0, 0, 3\pi)$
31. $I_x = k((\pi/2) - \frac{4}{3}), I_y = k((\pi/2) - \frac{2}{3})$
33. $2\pi^2$ **35.** 26   **37.** 1.67×10^4 ft-lb   **39.** (b) Yes
41. ≈ 22 J

Exercises 13.3 □ page 931

1. 40   **3.** $f(x, y) = 3x^2 + 5xy + 2y^2 + K$
5. Not conservative   **7.** $f(x, y) = x^2 \cos y - y \sin x + K$
9. $f(x, y) = ye^x + x \sin y + K$ **11.** (b) 16
13. (a) $f(x, y) = \frac{1}{4}x^4y^4$ (b) 4
15. (a) $f(x, y, z) = xyz + z^2$ (b) 77
17. (a) $f(x, y, z) = xy^2 \cos z$ (b) 0
19. 2   **21.** 30   **23.** No   **25.** Conservative
29. (a) Yes   (b) Yes   (c) Yes
31. (a) Yes   (b) Yes   (c) No

Exercises 13.4 □ page 939

1. 6   **3.** $\frac{2}{3}$ **7.** $e - 1$ **9.** $\frac{1}{3}$ **11.** -24π
13. $\frac{4}{3} - 2\pi$ **15.** $\frac{625}{2}\pi$ **17.** $-\frac{1}{12}$
19. 3π **21.** (c) $\frac{9}{2}$ **23.** $\left(\frac{1}{3}, \frac{1}{3}\right)$

Exercises 13.5 □ page 946

1. (a) $-x^2\,\mathbf{i} + 3xy\,\mathbf{j} - xz\,\mathbf{k}$ (b) yz
3. (a) $(x - y)\,\mathbf{i} - y\,\mathbf{j} + \mathbf{k}$ (b) $z - 1/(2\sqrt{z})$
5. (a) $\mathbf{0}$ (b) 1

7. (a) Negative   (b) curl $\mathbf{F} = \mathbf{0}$
9. (a) Zero   (b) curl $\mathbf{F}$ points in the negative z-direction
11. $f(x, y, z) = xyz + K$ **13.** $f(x, y, z) = x^2y + y^2z + K$
15. Not conservative   **17.** No

Exercises 13.6 □ page 958

1. $8(1 + \sqrt{2} + \sqrt{3}) \approx 33.17$ **3.** 900π
5. $5\sqrt{5}/48 + 1/240$ **7.** $171\sqrt{14}$ **9.** $\sqrt{3}/24$
11. $364\sqrt{2}\pi/3$ **13.** $(\pi/60)(391\sqrt{17} + 1)$
15. 16π **17.** 16π **19.** $\frac{713}{180}$ **21.** $-\frac{1}{6}$ **23.** $-\frac{4}{3}\pi$
25. 0   **27.** $2\pi + \frac{8}{3}$ **29.** 3.4895
31. $\iint_S \mathbf{F} \cdot d\mathbf{S} = \iint_D [P(\partial h/\partial x) - Q + R(\partial h/\partial z)]\,dA,$
where D = projection on xz-plane
33. $(0, 0, a/2)$
35. (a) $I_z = \iint_S (x^2 + y^2)\rho(x, y, z)\,dS$ (b) $4329\sqrt{2}\pi/5$
37. 0 kg/s   **39.** $8\pi a^3\varepsilon_0/3$ **41.** 1248π

Exercises 13.7 □ page 964

3. 0   **5.** 0   **7.** -1 **9.** 80π
11. (a) $81\pi/2$ (b)

(c) $x = 3\cos t, y = 3\sin t,$
$z = 1 - 3(\cos t + \sin t),$
$0 \le t \le 2\pi$

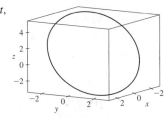

17. 3

Exercises 13.8 □ page 971

5. 2   **7.** $9\pi/2$
9. 0   **11.** $32\pi/3$ **13.** 0
15. $341\sqrt{2}/60 + \frac{81}{20}\arcsin(\sqrt{3}/3)$ **17.** $13\pi/20$
19. Negative at P_1, positive at P_2
21. div $\mathbf{F} > 0$ in quadrants I, II; div $\mathbf{F} < 0$ in quadrants III, IV

Chapter 13 Review □ page 974

True-False Quiz

1. False   **3.** True   **5.** False   **7.** True

Exercises

1. (a) Negative   (b) Positive   **3.** $6\sqrt{10}$ **5.** $\frac{4}{15}$
7. $\frac{110}{3}$ **9.** $\frac{11}{12} - 4/e$ **11.** $f(x, y) = e^y + xe^{xy}$ **13.** 0

17. -8π **25.** $\pi\left(391\sqrt{17}+1\right)/60$
27. $-64\pi/3$ **31.** $-\frac{1}{2}$ **35.** -4 **37.** 21

APPENDIXES

Exercises H.1 □ **page A14**

1. (a)

(b)

$(1,5\pi/2),\ (-1,3\pi/2)$ $(2,5\pi/4),\ (-2,9\pi/4)$

(c)

$(3,2+2\pi),\ (-3,2+\pi)$

3. (a)

(b)

$(0,3)$ $(-2,2)$

(c)

$\left(-\frac{1}{2},\ -\sqrt{3}/2\right)$

5. (a) (i) $\left(\sqrt{2},\ \pi/4\right)$ (ii) $\left(-\sqrt{2},\ 5\pi/4\right)$
(b) (i) $(4,\ 11\pi/6)$ (ii) $(-4,\ 5\pi/6)$

7.

9.

11.

13. Circle, center $\left(0,\frac{3}{2}\right)$, radius $\frac{3}{2}$
15. Horizontal line, 1 unit above the x-axis
17. $r=-\cot\theta\ \csc\theta$ **19.** $r=2c\cos\theta$
21. (a) $\theta=\pi/6$ (b) $x=3$
23.

25.

27.

29.

31.

33.

35.

37.

39.

41.

43.

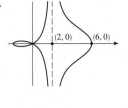

45. (a) For $c < -1$, the loop begins at $\theta = \sin^{-1}(-1/c)$ and ends at $\theta = \pi - \sin^{-1}(-1/c)$; for $c > 1$, it begins at $\theta = \pi + \sin^{-1}(1/c)$ and ends at $\theta = 2\pi - \sin^{-1}(1/c)$.

47. $\sqrt{3}$ **49.** $-\pi$

51. Horizontal at $\left(3/\sqrt{2}, \pi/4\right)$, $\left(-3/\sqrt{2}, 3\pi/4\right)$; vertical at $(3, 0)$, $(0, \pi/2)$

53. Horizontal at $\left(\frac{3}{2}, \pi/3\right)$, $\left(\frac{3}{2}, 5\pi/3\right)$, and the pole; vertical at $(2, 0)$, $\left(\frac{1}{2}, 2\pi/3\right)$, $\left(\frac{1}{2}, 4\pi/3\right)$

55. Center $(b/2, a/2)$, radius $\sqrt{a^2 + b^2}/2$

57.

59.

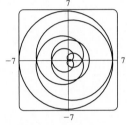

61. By counterclockwise rotation through angle $\pi/6$, $\pi/3$, or α about the origin

63. (a) A rose with n loops if n is odd and $2n$ loops if n is even
(b) Number of loops is always $2n$

65. For $0 < a < 1$, the curve is an oval, which develops a dimple as $a \to 1^-$. When $a > 1$, the curve splits into two parts, one of which has a loop.

Exercises H.2 □ **page A20**

1. $\pi^2/64$ **3.** $\pi/12 + \sqrt{3}/8$ **5.** $\pi^3/6$ **7.** $41\pi/4$
9. 4

11. π

13. 3π

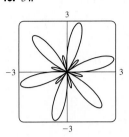

15. $\pi/8$ **17.** $\pi - \left(3\sqrt{3}/2\right)$
19. $(4\pi/3) + 2\sqrt{3}$ **21.** π
23. $(\pi - 2)/8$ **25.** $(\pi/2) - 1$
27. $(\pi + 3\sqrt{3})/4$ **29.** $\left(\frac{1}{2}, \pi/3\right)$, $\left(\frac{1}{2}, 5\pi/3\right)$, and the pole
31. $\left(\sqrt{3}/2, \pi/3\right)$, $\left(\sqrt{3}/2, 2\pi/3\right)$, and the pole
33. Intersection at $\theta \approx 0.89, 2.25$; area ≈ 3.46
35. π **37.** $\frac{8}{3}[(\pi^2 + 1)^{3/2} - 1]$ **39.** 29.0653

Exercises I □ **page A29**

1. $8 - 4i$ **3.** $13 + 18i$ **5.** $12 - 7i$ **7.** $\frac{11}{13} + \frac{10}{13}i$
9. $\frac{1}{2} - \frac{1}{2}i$ **11.** $-i$ **13.** $5i$ **15.** $12 + 5i$; 13
17. $4i, 4$ **19.** $\pm\frac{3}{2}i$ **21.** $-1 \pm 2i$
23. $-\frac{1}{2} \pm \left(\sqrt{7}/2\right)i$ **25.** $3\sqrt{2}\left[\cos(3\pi/4) + i\sin(3\pi/4)\right]$
27. $5\left\{\cos\left[\tan^{-1}\left(\frac{4}{3}\right)\right] + i\sin\left[\tan^{-1}\left(\frac{4}{3}\right)\right]\right\}$
29. $4[\cos(\pi/2) + i\sin(\pi/2)], \cos(-\pi/6) + i\sin(-\pi/6),$
$\frac{1}{2}[\cos(-\pi/6) + i\sin(-\pi/6)]$
31. $4\sqrt{2}\left[\cos(7\pi/12) + i\sin(7\pi/12)\right],$
$(2\sqrt{2})[\cos(13\pi/12) + i\sin(13\pi/12)], \frac{1}{4}[\cos(\pi/6) + i\sin(\pi/6)]$
33. -1024 **35.** $-512\sqrt{3} + 512i$
37. $\pm1, \pm i, (1/\sqrt{2})(\pm1 \pm i)$ **39.** $\pm(\sqrt{3}/2) + \frac{1}{2}i, -i$

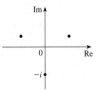

41. i **43.** $\frac{1}{2} + \left(\sqrt{3}/2\right)i$ **45.** $-e^2$
47. $\cos 3\theta = \cos^3\theta - 3\cos\theta\sin^2\theta$,
$\sin 3\theta = 3\cos^2\theta\sin\theta - \sin^3\theta$

Credits

cover, iv, v, x □ © 2003 Tom Bonner

ii □ Walt Disney Concert Hall, CA / Frank O. Gehry & Associates

iii □ *(top, middle)* Walt Disney Concert Hall, CA / Frank O. Gehry & Associates; © 2003 Tom Bonner

vii □ *(top right, middle)* © 2003 Tom Bonner; Photography by Federico Zignani (www.Artdrive.org) Courtesy of the Los Angeles Philharmonic

viii □ Walt Disney Concert Hall, CA / Frank O. Gehry & Associates

ix □ *(top to bottom)* © 2003 Tom Bonner; Walt Disney Concert Hall, CA / Frank O. Gehry & Associates

556 □ © 2003 Tom Bonner

595 □ Membrane courtesy of National Film Board of Canada; Mathematica model by Brian C. Betsill

627 □ Courtesy of Eugene Hecht

630 □ © Luke Dodd, Photo Researchers, Inc.

636 □ *(top left)* © 2003 Tom Bonner; Photography by Federico Zignani (www.Artdrive.org) Courtesy of the Los Angeles Philharmonic

694 □ *(top left)* © 2003 Tom Bonner; Walt Disney Concert Hall, CA / Frank O. Gehry & Associates

738 □ *(top left)* © 2003 Tom Bonner; Walt Disney Concert Hall, CA / Frank O. Gehry & Associates

742 □ Tarbuck, *Atmosphere: An Introduction to Meteorology,* 4th Edition, © 1989. Reprinted by permission of Pearson Education, Inc., Upper Saddle River, NJ

788, 790 □ Copyright 1997, USA Today. Reprinted with permission

799 □ Tarbuck, *Atmosphere: An Introduction to Meteorology,* 4th Edition, © 1989. Reprinted by permission of Pearson Education, Inc., Upper Saddle River, NJ

820 □ Courtesy of Orbital Sciences Corporation

825 □ NOAA / AOML / Hurricane Research Division

828 □ © 2003 Tom Bonner

889 □ Courtesy of dan clegg

904 □ *(top left)* © 2003 Tom Bonner; Walt Disney Concert Hall, CA / Frank O. Gehry & Associates

905 □ *ONERA* photograph, Werlé 1974

965 □ Courtesy of the Masters and Fellows of Gonville and Caius College, University of Cambridge, England

A51, A53 □ © 2003 Tom Bonner

Index

DIFFERENTIATION RULES

GENERAL FORMULAS

1. $\dfrac{d}{dx}(c) = 0$

2. $\dfrac{d}{dx}[cf(x)] = cf'(x)$

3. $\dfrac{d}{dx}[f(x) + g(x)] = f'(x) + g'(x)$

4. $\dfrac{d}{dx}[f(x) - g(x)] = f'(x) - g'(x)$

5. $\dfrac{d}{dx}[f(x)g(x)] = f(x)g'(x) + g(x)f'(x)$ (Product Rule)

6. $\dfrac{d}{dx}\left[\dfrac{f(x)}{g(x)}\right] = \dfrac{g(x)f'(x) - f(x)g'(x)}{[g(x)]^2}$ (Quotient Rule)

7. $\dfrac{d}{dx}f(g(x)) = f'(g(x))g'(x)$ (Chain Rule)

8. $\dfrac{d}{dx}(x^n) = nx^{n-1}$ (Power Rule)

EXPONENTIAL AND LOGARITHMIC FUNCTIONS

9. $\dfrac{d}{dx}(e^x) = e^x$

10. $\dfrac{d}{dx}(a^x) = a^x \ln a$

11. $\dfrac{d}{dx}\ln|x| = \dfrac{1}{x}$

12. $\dfrac{d}{dx}(\log_a x) = \dfrac{1}{x \ln a}$

TRIGONOMETRIC FUNCTIONS

13. $\dfrac{d}{dx}(\sin x) = \cos x$

14. $\dfrac{d}{dx}(\cos x) = -\sin x$

15. $\dfrac{d}{dx}(\tan x) = \sec^2 x$

16. $\dfrac{d}{dx}(\csc x) = -\csc x \cot x$

17. $\dfrac{d}{dx}(\sec x) = \sec x \tan x$

18. $\dfrac{d}{dx}(\cot x) = -\csc^2 x$

INVERSE TRIGONOMETRIC FUNCTIONS

19. $\dfrac{d}{dx}(\sin^{-1}x) = \dfrac{1}{\sqrt{1 - x^2}}$

20. $\dfrac{d}{dx}(\cos^{-1}x) = -\dfrac{1}{\sqrt{1 - x^2}}$

21. $\dfrac{d}{dx}(\tan^{-1}x) = \dfrac{1}{1 + x^2}$

22. $\dfrac{d}{dx}(\csc^{-1}x) = -\dfrac{1}{x\sqrt{x^2 - 1}}$

23. $\dfrac{d}{dx}(\sec^{-1}x) = \dfrac{1}{x\sqrt{x^2 - 1}}$

24. $\dfrac{d}{dx}(\cot^{-1}x) = -\dfrac{1}{1 + x^2}$

HYPERBOLIC FUNCTIONS

25. $\dfrac{d}{dx}(\sinh x) = \cosh x$

26. $\dfrac{d}{dx}(\cosh x) = \sinh x$

27. $\dfrac{d}{dx}(\tanh x) = \operatorname{sech}^2 x$

28. $\dfrac{d}{dx}(\operatorname{csch} x) = -\operatorname{csch} x \coth x$

29. $\dfrac{d}{dx}(\operatorname{sech} x) = -\operatorname{sech} x \tanh x$

30. $\dfrac{d}{dx}(\coth x) = -\operatorname{csch}^2 x$

INVERSE HYPERBOLIC FUNCTIONS

31. $\dfrac{d}{dx}(\sinh^{-1}x) = \dfrac{1}{\sqrt{1 + x^2}}$

32. $\dfrac{d}{dx}(\cosh^{-1}x) = \dfrac{1}{\sqrt{x^2 - 1}}$

33. $\dfrac{d}{dx}(\tanh^{-1}x) = \dfrac{1}{1 - x^2}$

34. $\dfrac{d}{dx}(\operatorname{csch}^{-1}x) = -\dfrac{1}{|x|\sqrt{x^2 + 1}}$

35. $\dfrac{d}{dx}(\operatorname{sech}^{-1}x) = -\dfrac{1}{x\sqrt{1 - x^2}}$

36. $\dfrac{d}{dx}(\coth^{-1}x) = \dfrac{1}{1 - x^2}$

TABLE OF INTEGRALS

BASIC FORMS

1. $\int u\, dv = uv - \int v\, du$ *Integration by Parts*

2. $\int u^n\, du = \dfrac{u^{n+1}}{n+1} + C, \quad n \neq -1$

3. $\int \dfrac{du}{u} = \ln|u| + C$

4. $\int e^u\, du = e^u + C$

5. $\int a^u\, du = \dfrac{a^u}{\ln a} + C$

6. $\int \sin u\, du = -\cos u + C$

7. $\int \cos u\, du = \sin u + C$

8. $\int \sec^2 u\, du = \tan u + C$

9. $\int \csc^2 u\, du = -\cot u + C$

10. $\int \sec u \tan u\, du = \sec u + C$

11. $\int \csc u \cot u\, du = -\csc u + C$

12. $\int \tan u\, du = \ln|\sec u| + C$

13. $\int \cot u\, du = \ln|\sin u| + C$

14. $\int \sec u\, du = \ln|\sec u + \tan u| + C$

15. $\int \csc u\, du = \ln|\csc u - \cot u| + C$

16. $\int \dfrac{du}{\sqrt{a^2 - u^2}} = \sin^{-1}\dfrac{u}{a} + C$

17. $\int \dfrac{du}{a^2 + u^2} = \dfrac{1}{a}\tan^{-1}\dfrac{u}{a} + C$

18. $\int \dfrac{du}{u\sqrt{u^2 - a^2}} = \dfrac{1}{a}\sec^{-1}\dfrac{u}{a} + C$

19. $\int \dfrac{du}{a^2 - u^2} = \dfrac{1}{2a}\ln\left|\dfrac{u+a}{u-a}\right| + C$

20. $\int \dfrac{du}{u^2 - a^2} = \dfrac{1}{2a}\ln\left|\dfrac{u-a}{u+a}\right| + C$

FORMS INVOLVING $\sqrt{a^2 + u^2}$, $a > 0$

21. $\int \sqrt{a^2 + u^2}\, du = \dfrac{u}{2}\sqrt{a^2 + u^2} + \dfrac{a^2}{2}\ln\left(u + \sqrt{a^2 + u^2}\right) + C$

22. $\int u^2 \sqrt{a^2 + u^2}\, du = \dfrac{u}{8}(a^2 + 2u^2)\sqrt{a^2 + u^2} - \dfrac{a^4}{8}\ln\left(u + \sqrt{a^2 + u^2}\right) + C$

23. $\int \dfrac{\sqrt{a^2 + u^2}}{u}\, du = \sqrt{a^2 + u^2} - a\ln\left|\dfrac{a + \sqrt{a^2 + u^2}}{u}\right| + C$

24. $\int \dfrac{\sqrt{a^2 + u^2}}{u^2}\, du = -\dfrac{\sqrt{a^2 + u^2}}{u} + \ln\left(u + \sqrt{a^2 + u^2}\right) + C$

25. $\int \dfrac{du}{\sqrt{a^2 + u^2}} = \ln\left(u + \sqrt{a^2 + u^2}\right) + C$

26. $\int \dfrac{u^2\, du}{\sqrt{a^2 + u^2}} = \dfrac{u}{2}\sqrt{a^2 + u^2} - \dfrac{a^2}{2}\ln\left(u + \sqrt{a^2 + u^2}\right) + C$

27. $\int \dfrac{du}{u\sqrt{a^2 + u^2}} = -\dfrac{1}{a}\ln\left|\dfrac{\sqrt{a^2 + u^2} + a}{u}\right| + C$

28. $\int \dfrac{du}{u^2\sqrt{a^2 + u^2}} = -\dfrac{\sqrt{a^2 + u^2}}{a^2 u} + C$

29. $\int \dfrac{du}{(a^2 + u^2)^{3/2}} = \dfrac{u}{a^2\sqrt{a^2 + u^2}} + C$

6

TABLE OF INTEGRALS

FORMS INVOLVING $\sqrt{a^2 - u^2}$, $a > 0$

30. $\displaystyle \int \sqrt{a^2 - u^2}\, du = \frac{u}{2}\sqrt{a^2 - u^2} + \frac{a^2}{2}\sin^{-1}\frac{u}{a} + C$

31. $\displaystyle \int u^2\sqrt{a^2 - u^2}\, du = \frac{u}{8}(2u^2 - a^2)\sqrt{a^2 - u^2} + \frac{a^4}{8}\sin^{-1}\frac{u}{a} + C$

32. $\displaystyle \int \frac{\sqrt{a^2 - u^2}}{u}\, du = \sqrt{a^2 - u^2} - a\ln\left|\frac{a + \sqrt{a^2 - u^2}}{u}\right| + C$

33. $\displaystyle \int \frac{\sqrt{a^2 - u^2}}{u^2}\, du = -\frac{1}{u}\sqrt{a^2 - u^2} - \sin^{-1}\frac{u}{a} + C$

34. $\displaystyle \int \frac{u^2\, du}{\sqrt{a^2 - u^2}} = -\frac{u}{2}\sqrt{a^2 - u^2} + \frac{a^2}{2}\sin^{-1}\frac{u}{a} + C$

35. $\displaystyle \int \frac{du}{u\sqrt{a^2 - u^2}} = -\frac{1}{a}\ln\left|\frac{a + \sqrt{a^2 - u^2}}{u}\right| + C$

36. $\displaystyle \int \frac{du}{u^2\sqrt{a^2 - u^2}} = -\frac{1}{a^2 u}\sqrt{a^2 - u^2} + C$

37. $\displaystyle \int (a^2 - u^2)^{3/2}\, du = -\frac{u}{8}(2u^2 - 5a^2)\sqrt{a^2 - u^2} + \frac{3a^4}{8}\sin^{-1}\frac{u}{a} + C$

38. $\displaystyle \int \frac{du}{(a^2 - u^2)^{3/2}} = \frac{u}{a^2\sqrt{a^2 - u^2}} + C$

FORMS INVOLVING $\sqrt{u^2 - a^2}$, $a > 0$

39. $\displaystyle \int \sqrt{u^2 - a^2}\, du = \frac{u}{2}\sqrt{u^2 - a^2} - \frac{a^2}{2}\ln\left|u + \sqrt{u^2 - a^2}\right| + C$

40. $\displaystyle \int u^2\sqrt{u^2 - a^2}\, du = \frac{u}{8}(2u^2 - a^2)\sqrt{u^2 - a^2} - \frac{a^4}{8}\ln\left|u + \sqrt{u^2 - a^2}\right| + C$

41. $\displaystyle \int \frac{\sqrt{u^2 - a^2}}{u}\, du = \sqrt{u^2 - a^2} - a\cos^{-1}\frac{a}{|u|} + C$

42. $\displaystyle \int \frac{\sqrt{u^2 - a^2}}{u^2}\, du = -\frac{\sqrt{u^2 - a^2}}{u} + \ln\left|u + \sqrt{u^2 - a^2}\right| + C$

43. $\displaystyle \int \frac{du}{\sqrt{u^2 - a^2}} = \ln\left|u + \sqrt{u^2 - a^2}\right| + C$

44. $\displaystyle \int \frac{u^2\, du}{\sqrt{u^2 - a^2}} = \frac{u}{2}\sqrt{u^2 - a^2} + \frac{a^2}{2}\ln\left|u + \sqrt{u^2 - a^2}\right| + C$

45. $\displaystyle \int \frac{du}{u^2\sqrt{u^2 - a^2}} = \frac{\sqrt{u^2 - a^2}}{a^2 u} + C$

46. $\displaystyle \int \frac{du}{(u^2 - a^2)^{3/2}} = -\frac{u}{a^2\sqrt{u^2 - a^2}} + C$

TABLE OF INTEGRALS

FORMS INVOLVING $a + bu$

47. $\displaystyle\int \frac{u\,du}{a + bu} = \frac{1}{b^2}\left(a + bu - a\ln|a + bu|\right) + C$

48. $\displaystyle\int \frac{u^2\,du}{a + bu} = \frac{1}{2b^3}\left[(a + bu)^2 - 4a(a + bu) + 2a^2\ln|a + bu|\right] + C$

49. $\displaystyle\int \frac{du}{u(a + bu)} = \frac{1}{a}\ln\left|\frac{u}{a + bu}\right| + C$

50. $\displaystyle\int \frac{du}{u^2(a + bu)} = -\frac{1}{au} + \frac{b}{a^2}\ln\left|\frac{a + bu}{u}\right| + C$

51. $\displaystyle\int \frac{u\,du}{(a + bu)^2} = \frac{a}{b^2(a + bu)} + \frac{1}{b^2}\ln|a + bu| + C$

52. $\displaystyle\int \frac{du}{u(a + bu)^2} = \frac{1}{a(a + bu)} - \frac{1}{a^2}\ln\left|\frac{a + bu}{u}\right| + C$

53. $\displaystyle\int \frac{u^2\,du}{(a + bu)^2} = \frac{1}{b^3}\left(a + bu - \frac{a^2}{a + bu} - 2a\ln|a + bu|\right) + C$

54. $\displaystyle\int u\sqrt{a + bu}\,du = \frac{2}{15b^2}(3bu - 2a)(a + bu)^{3/2} + C$

55. $\displaystyle\int \frac{u\,du}{\sqrt{a + bu}} = \frac{2}{3b^2}(bu - 2a)\sqrt{a + bu} + C$

56. $\displaystyle\int \frac{u^2\,du}{\sqrt{a + bu}} = \frac{2}{15b^3}(8a^2 + 3b^2u^2 - 4abu)\sqrt{a + bu} + C$

57. $\displaystyle\int \frac{du}{u\sqrt{a + bu}} = \frac{1}{\sqrt{a}}\ln\left|\frac{\sqrt{a + bu} - \sqrt{a}}{\sqrt{a + bu} + \sqrt{a}}\right| + C, \quad\text{if } a > 0$

$\displaystyle\qquad\qquad\quad = \frac{2}{\sqrt{-a}}\tan^{-1}\sqrt{\frac{a + bu}{-a}} + C, \qquad\text{if } a < 0$

58. $\displaystyle\int \frac{\sqrt{a + bu}}{u}\,du = 2\sqrt{a + bu} + a\int\frac{du}{u\sqrt{a + bu}}$

59. $\displaystyle\int \frac{\sqrt{a + bu}}{u^2}\,du = -\frac{\sqrt{a + bu}}{u} + \frac{b}{2}\int\frac{du}{u\sqrt{a + bu}}$

60. $\displaystyle\int u^n\sqrt{a + bu}\,du = \frac{2}{b(2n + 3)}\left[u^n(a + bu)^{3/2} - na\int u^{n-1}\sqrt{a + bu}\,du\right]$

61. $\displaystyle\int \frac{u^n\,du}{\sqrt{a + bu}} = \frac{2u^n\sqrt{a + bu}}{b(2n + 1)} - \frac{2na}{b(2n + 1)}\int\frac{u^{n-1}\,du}{\sqrt{a + bu}}$

62. $\displaystyle\int \frac{du}{u^n\sqrt{a + bu}} = -\frac{\sqrt{a + bu}}{a(n - 1)u^{n-1}} - \frac{b(2n - 3)}{2a(n - 1)}\int\frac{du}{u^{n-1}\sqrt{a + bu}}$

TABLE OF INTEGRALS

TRIGONOMETRIC FORMS

63. $\displaystyle\int \sin^2 u \, du = \tfrac{1}{2}u - \tfrac{1}{4}\sin 2u + C$

64. $\displaystyle\int \cos^2 u \, du = \tfrac{1}{2}u + \tfrac{1}{4}\sin 2u + C$

65. $\displaystyle\int \tan^2 u \, du = \tan u - u + C$

66. $\displaystyle\int \cot^2 u \, du = -\cot u - u + C$

67. $\displaystyle\int \sin^3 u \, du = -\tfrac{1}{3}(2 + \sin^2 u)\cos u + C$

68. $\displaystyle\int \cos^3 u \, du = \tfrac{1}{3}(2 + \cos^2 u)\sin u + C$

69. $\displaystyle\int \tan^3 u \, du = \tfrac{1}{2}\tan^2 u + \ln|\cos u| + C$

70. $\displaystyle\int \cot^3 u \, du = -\tfrac{1}{2}\cot^2 u - \ln|\sin u| + C$

71. $\displaystyle\int \sec^3 u \, du = \tfrac{1}{2}\sec u \tan u + \tfrac{1}{2}\ln|\sec u + \tan u| + C$

72. $\displaystyle\int \csc^3 u \, du = -\tfrac{1}{2}\csc u \cot u + \tfrac{1}{2}\ln|\csc u - \cot u| + C$

73. $\displaystyle\int \sin^n u \, du = -\frac{1}{n}\sin^{n-1} u \cos u + \frac{n-1}{n}\int \sin^{n-2} u \, du$

74. $\displaystyle\int \cos^n u \, du = \frac{1}{n}\cos^{n-1} u \sin u + \frac{n-1}{n}\int \cos^{n-2} u \, du$

75. $\displaystyle\int \tan^n u \, du = \frac{1}{n-1}\tan^{n-1} u - \int \tan^{n-2} u \, du$

76. $\displaystyle\int \cot^n u \, du = \frac{-1}{n-1}\cot^{n-1} u - \int \cot^{n-2} u \, du$

77. $\displaystyle\int \sec^n u \, du = \frac{1}{n-1}\tan u \sec^{n-2} u + \frac{n-2}{n-1}\int \sec^{n-2} u \, du$

78. $\displaystyle\int \csc^n u \, du = \frac{-1}{n-1}\cot u \csc^{n-2} u + \frac{n-2}{n-1}\int \csc^{n-2} u \, du$

79. $\displaystyle\int \sin au \sin bu \, du = \frac{\sin(a-b)u}{2(a-b)} - \frac{\sin(a+b)u}{2(a+b)} + C$

80. $\displaystyle\int \cos au \cos bu \, du = \frac{\sin(a-b)u}{2(a-b)} + \frac{\sin(a+b)u}{2(a+b)} + C$

81. $\displaystyle\int \sin au \cos bu \, du = -\frac{\cos(a-b)u}{2(a-b)} - \frac{\cos(a+b)u}{2(a+b)} + C$

82. $\displaystyle\int u \sin u \, du = \sin u - u \cos u + C$

83. $\displaystyle\int u \cos u \, du = \cos u + u \sin u + C$

84. $\displaystyle\int u^n \sin u \, du = -u^n \cos u + n \int u^{n-1} \cos u \, du$

85. $\displaystyle\int u^n \cos u \, du = u^n \sin u - n \int u^{n-1} \sin u \, du$

86. $\displaystyle\int \sin^n u \cos^m u \, du = -\frac{\sin^{n-1} u \cos^{m+1} u}{n+m} + \frac{n-1}{n+m}\int \sin^{n-2} u \cos^m u \, du$

$\displaystyle\qquad = \frac{\sin^{n+1} u \cos^{m-1} u}{n+m} + \frac{m-1}{n+m}\int \sin^n u \cos^{m-2} u \, du$

INVERSE TRIGONOMETRIC FORMS

87. $\displaystyle\int \sin^{-1} u \, du = u \sin^{-1} u + \sqrt{1 - u^2} + C$

88. $\displaystyle\int \cos^{-1} u \, du = u \cos^{-1} u - \sqrt{1 - u^2} + C$

89. $\displaystyle\int \tan^{-1} u \, du = u \tan^{-1} u - \tfrac{1}{2}\ln(1 + u^2) + C$

90. $\displaystyle\int u \sin^{-1} u \, du = \frac{2u^2 - 1}{4}\sin^{-1} u + \frac{u\sqrt{1-u^2}}{4} + C$

91. $\displaystyle\int u \cos^{-1} u \, du = \frac{2u^2 - 1}{4}\cos^{-1} u - \frac{u\sqrt{1-u^2}}{4} + C$

92. $\displaystyle\int u \tan^{-1} u \, du = \frac{u^2 + 1}{2}\tan^{-1} u - \frac{u}{2} + C$

93. $\displaystyle\int u^n \sin^{-1} u \, du = \frac{1}{n+1}\left[u^{n+1}\sin^{-1} u - \int \frac{u^{n+1}\,du}{\sqrt{1-u^2}}\right], \quad n \neq -1$

94. $\displaystyle\int u^n \cos^{-1} u \, du = \frac{1}{n+1}\left[u^{n+1}\cos^{-1} u + \int \frac{u^{n+1}\,du}{\sqrt{1-u^2}}\right], \quad n \neq -1$

95. $\displaystyle\int u^n \tan^{-1} u \, du = \frac{1}{n+1}\left[u^{n+1}\tan^{-1} u - \int \frac{u^{n+1}\,du}{1+u^2}\right], \quad n \neq -1$

TABLE OF INTEGRALS

EXPONENTIAL AND LOGARITHMIC FORMS

96. $\displaystyle\int u e^{au}\, du = \frac{1}{a^2}(au - 1)e^{au} + C$

97. $\displaystyle\int u^n e^{au}\, du = \frac{1}{a}u^n e^{au} - \frac{n}{a}\int u^{n-1}e^{au}\, du$

98. $\displaystyle\int e^{au}\sin bu\, du = \frac{e^{au}}{a^2 + b^2}(a\sin bu - b\cos bu) + C$

99. $\displaystyle\int e^{au}\cos bu\, du = \frac{e^{au}}{a^2 + b^2}(a\cos bu + b\sin bu) + C$

100. $\displaystyle\int \ln u\, du = u\ln u - u + C$

101. $\displaystyle\int u^n \ln u\, du = \frac{u^{n+1}}{(n+1)^2}[(n+1)\ln u - 1] + C$

102. $\displaystyle\int \frac{1}{u\ln u}\, du = \ln|\ln u| + C$

HYPERBOLIC FORMS

103. $\displaystyle\int \sinh u\, du = \cosh u + C$

104. $\displaystyle\int \cosh u\, du = \sinh u + C$

105. $\displaystyle\int \tanh u\, du = \ln\cosh u + C$

106. $\displaystyle\int \coth u\, du = \ln|\sinh u| + C$

107. $\displaystyle\int \operatorname{sech} u\, du = \tan^{-1}|\sinh u| + C$

108. $\displaystyle\int \operatorname{csch} u\, du = \ln\left|\tanh\tfrac{1}{2}u\right| + C$

109. $\displaystyle\int \operatorname{sech}^2 u\, du = \tanh u + C$

110. $\displaystyle\int \operatorname{csch}^2 u\, du = -\coth u + C$

111. $\displaystyle\int \operatorname{sech} u \tanh u\, du = -\operatorname{sech} u + C$

112. $\displaystyle\int \operatorname{csch} u \coth u\, du = -\operatorname{csch} u + C$

FORMS INVOLVING $\sqrt{2au - u^2}$, $a > 0$

113. $\displaystyle\int \sqrt{2au - u^2}\, du = \frac{u - a}{2}\sqrt{2au - u^2} + \frac{a^2}{2}\cos^{-1}\left(\frac{a - u}{a}\right) + C$

114. $\displaystyle\int u\sqrt{2au - u^2}\, du = \frac{2u^2 - au - 3a^2}{6}\sqrt{2au - u^2} + \frac{a^3}{2}\cos^{-1}\left(\frac{a - u}{a}\right) + C$

115. $\displaystyle\int \frac{\sqrt{2au - u^2}}{u}\, du = \sqrt{2au - u^2} + a\cos^{-1}\left(\frac{a - u}{a}\right) + C$

116. $\displaystyle\int \frac{\sqrt{2au - u^2}}{u^2}\, du = -\frac{2\sqrt{2au - u^2}}{u} - \cos^{-1}\left(\frac{a - u}{a}\right) + C$

117. $\displaystyle\int \frac{du}{\sqrt{2au - u^2}} = \cos^{-1}\left(\frac{a - u}{a}\right) + C$

118. $\displaystyle\int \frac{u\, du}{\sqrt{2au - u^2}} = -\sqrt{2au - u^2} + a\cos^{-1}\left(\frac{a - u}{a}\right) + C$

119. $\displaystyle\int \frac{u^2\, du}{\sqrt{2au - u^2}} = -\frac{(u + 3a)}{2}\sqrt{2au - u^2} + \frac{3a^2}{2}\cos^{-1}\left(\frac{a - u}{a}\right) + C$

120. $\displaystyle\int \frac{du}{u\sqrt{2au - u^2}} = -\frac{\sqrt{2au - u^2}}{au} + C$